INTRODUCTION TO
NUMERICAL PROGRAMMING

SERIES IN COMPUTATIONAL PHYSICS

Steven A. Gottlieb and Rubin H. Landau
Series Editors

SERIES IN COMPUTATIONAL PHYSICS

Steven A. Gottlieb and Rubin H. Landau, Series Editors

INTRODUCTION TO
NUMERICAL PROGRAMMING

A Practical Guide for Scientists and Engineers
Using Python and C/C++

Titus Adrien Beu

Babeș-Bolyai University
Faculty of Physics
Cluj-Napoca, Romania

 CRC Press
Taylor & Francis Group
Boca Raton   London   New York

CRC Press is an imprint of the
Taylor & Francis Group, an **informa** business

CRC Press
Taylor & Francis Group
6000 Broken Sound Parkway NW, Suite 300
Boca Raton, FL 33487-2742

© 2015 by Taylor & Francis Group, LLC
CRC Press is an imprint of Taylor & Francis Group, an Informa business

Printed on acid-free paper
Version Date: 20140716

International Standard Book Number-13: 978-1-4665-6967-6 (Paperback)

Library of Congress Cataloging-in-Publication Data

Beu, Titus A., author.
  Introduction to numerical programming : a practical guide for scientists and engineers using Python and C/C++ / Titus A. Beu.
      pages cm. -- (Series in computational physics)
  Includes bibliographical references and index.
  ISBN 978-1-4665-6967-6 (paperback : acid-free paper) 1. Physics--Data processing. 2. Engineering--Data processing. 3. Computer programming. 4. Python (Computer program language) 5. C (Computer program language) 6. C++ (Computer program language) I. Title.

QC52.B48 2015
005.13'3--dc23

2014022779

Visit the Taylor & Francis Web site at
http://www.taylorandfrancis.com

and the CRC Press Web site at
http://www.crcpress.com

To Mihaela and Victor

Contents

Series Preface

There can be little argument that computation has become an essential element in all areas of physics, be it via simulation, symbolic manipulations, data manipulations, equipment interfacing, or something with which we are not yet familiar. Nevertheless, even though the style of teaching and organization of subjects being taught by physics departments has changed in recent times, the actual content of the courses has been slow to incorporate the new-found importance of computation. Yes, there are now speciality courses and many textbooks in *Computational Physics*, but that is not the same thing as incorporating computation into the very heart of a modern physics curriculum so that the physics being taught today more closely resembles the physics being done today. Not only will such integration provide valuable professional skills to students, but it will also help keep physics alive by permitting new areas to be studied and old problems to be solved.

This series is intended to provide undergraduate and graduate level textbooks for a modern physics curriculum in which computation is incorporated within the traditional subjects of physics, or in which there are new, multidisciplinary subjects in which physics and computation are combined as a "computational science." The level of presentation will allow for their use as primary or secondary textbooks for courses that wish to emphasize the importance of numerical methods and computational tools in science. They will offer essential foundational materials for students and instructors in the physical sciences as well as academic and industry professionals in physics, engineering, computer science, applied math, and biology.

Titles in the series are targeted to specific disciplines that currently lack a textbook with a computational physics approach. Among these subject areas are condensed matter physics, materials science, particle physics, astrophysics, mathematical methods of computational physics, quantum mechanics, plasma physics, fluid dynamics, statistical physics, optics, biophysics, electricity and magnetism, gravity, cosmology, and high-performance computing in physics. We aim for a presentation that is concise and practical, often including solved problems and examples. The books are meant for teaching, although researchers may find them useful as well. In select cases, we have allowed more advanced, edited works to be included when they share the spirit of the series—to contribute to wider application of computational tools in the classroom as well as research settings.

Although the series editors had been all-too-willing to express the need for change in the physics curriculum, the actual idea for this series came from the series manager, Luna Han of Taylor & Francis Publishers. We wish to thank her sincerely for that, as well as for encouragement and direction throughout the project.

Steve Gottlieb
Bloomington
Rubin H. Landau
Corvallis
Series Editors

Preface

This book is devoted to the general field of numerical programming, with emphasis on methods specific to computational physics and engineering.

While tremendous advances of computer performances have been achieved in recent years, numerical methods still remain virtually inexhaustible resources for extending the range of challenging real-life problems made tractable. Along these lines, the book sets as its primordial goal to contribute to fostering interest in numerical programming by making computations accessible and appealing to broader categories of scientists and engineers.

I have written this book for advanced undergraduate and graduate students in natural sciences and engineering, with the aim of being suited as curriculum material for a one- or two-semester course in numerical programming based on Python or C/C++. The book may also be used for independent study or as a reference material beyond academic studies. It may be useful, for instance, as an introductory text for researchers preparing to engage in scientific computing, or engineers needing effective numerical tools for applicative calculations.

I have placed emphasis on the rigorous, yet accessible presentation of the fundamental numerical methods, which are complemented with implementations and applications, highlighting specific numerical behavior and often featuring graphical output. Although the material requires basic knowledge of calculus, linear algebra, and differential equations, it does not assume previous acquaintance with numerical techniques, leading the reader all the way from elementary algorithms to elaborate methods of relevance for modern numerical programming. Notably, there is a steady increase of complexity of the methods which are dealt with, both within each chapter and along the sequence of chapters.

In essence, the book aims at assisting the reader in developing their capacity of algorithmic reasoning, a specific coding dexterity, and an efficient scientific programming style, which are essential prerequisites for subsequent involvement in comprehensive programming projects.

Relation to reference works. The literature abounds in excellent books on numerical programming. To mention just two works that have set milestones in the field, the book makes frequent reference to *Numerical Analysis* by R.L. Burden and J.D. Faires, and to *Numerical Recipes* by W. H. Press, S. A. Teukolsky, W. T. Vetterling, and B. P. Flannery. Some reference texts excel in the rigorous presentation of the mathematical foundations and algorithms, yet with generic implementations and limited justifications of the concretely employed coding solutions. Other works highlight the subtleties and performance of concrete implementations, being mainly addressed to experienced users.

The goal of this book is to bridge the two mentioned options, providing the reader with an adequate level of proficiency by following a route which develops a well-balanced combination of basic theoretical knowledge, as well as code design and efficiency analysis skills. Along the way, constant emphasis is laid on the fact that good programming is neither barely a mechanistic translation of mathematical formulas into a programming language, nor should effective implementations be complicated by principle.

In all respects, the book follows as a guiding principle the "beauty of simplicity" with a view to achieving effectiveness and long-lasting usability of numerical applications. As a distinctive feature, the book also attaches particular importance to graphical representations, as a means of synthetic assessment of numerical behavior, the material deliberately being highly visual and brimming with graphs.

Programming languages. In terms of programming languages, the book relies on *standard* Python (version 3.3) and ANSI C/C++, not making use of any additional packages. On the one hand, the choice is motivated by their modern "minimalism," which enables the systematic presentation of programming concepts with relatively simple means, and on the other, by their ubiquity, which ensures the cross-platform portability of the applications, as well as the potential mobility of the users to other fields of applied programming.

As for Python, it proves its virtues ever more increasingly, not only as an easy-to-use teaching language, but also in high-performance computing for creating complementary scripts and graphical user interfaces in conjunction with compiled languages, or for building stand-alone applications in various fields of present-day science and technology. With this in mind, the book addresses both the effectiveness of the compiled C/C++ code and the ease-of-use and portable graphical features of the interpreter-based Python applications.

Starting from the fact that code indentation is compulsory in Python, but optional in C/C++ (being merely a means of enhancing code readability), throughout the book we employ a unified code formatting style for Python and C/C++, implying an *indentation size of three spaces* for marking consecutive subordination levels.

Numerical methods. The numerical methods are presented in detail, although avoiding an excess of purely mathematical aspects. The implementations are designed so as to highlight the specificity of scientific programming, and, for the sake of creating a consistent programming style, they assume a compromise between clarity and performance. In certain cases, without sacrificing the execution speed, less efficient memory usage but more elegant and illustrative implementations are preferred.

The routines are richly commented in order to clarify the role of the various parameters, variables, and code sequences. The comments on individual code lines are generally right aligned in order not to disperse the executable sequences and to save on source file size.

Addition of delicate algorithmic elements is carried out in a gradual manner, through successive versions of the routines of increasing complexity. The routines are complemented with numerical examples, designed to elucidate the data input/output and to highlight specific numerical behavior of the solutions.

The material also emphasizes the *cross-disciplinary usability* of numerical methods originally developed for specific fields of computational physics or engineering. As a concrete example, the Verlet propagator, which is a well-established method for solving equations of motion in molecular dynamics simulations, can additionally be applied successfully to study the motion of celestial systems. The text also points repeatedly to the fact that mastering numerical methods implies to a large extent the in-depth understanding and control of errors.

Python versions of the developed routines are listed separately in each chapter, alongside the relevant theoretical and algorithmic aspects, while C/C++ versions are listed jointly as header files in separate sections, generically titled "Implementations in C/C++." In perfect symmetry, the Python routines are also grouped in modules corresponding to each of the treated chapters.

In fact, the book brings forward two similar Python and C/C++ *numerical* libraries, numxlib.py and numxlib.h, which include the partial libraries developed in the separate chapters (see Appendix D for a description). Even though the applications illustrating the various numerical methods exposed in the book employ the partial numerical libraries (as, for instance, ode.py or ode.h is included for solving ordinary differential equations), the overall library numxlib can well be used in their place, and this is actually the recommended option, once the numerical methods presented in the book have been mastered.

Graphics. Bearing in mind that in modern research the amount of output data quite often exceeds the user's capacity to analyze individual figures, as already mentioned, the book places importance on graphical representation methods, which are able to reveal global trends, otherwise difficult to extract from numeric sequences. The graphics is based on the remarkable capabilities of Python's Tkinter package, not in the least taking advantage of its wide-ranging portability. In Chapter 3, the book puts forward a versatile Python graphics library, `graphlib.py` (see also Appendix E), which is used throughout the material to produce runtime graphics both from Python and C/C++. In particular, the C/C++ applications gain access to `graphlib.py` via a dedicated interface file, `graphlib.h` (see Appendix F).

The persistent use of graphical presentation methods is meant to stimulate the development of abilities for building large-scale scientific applications with graphical user interfaces.

Exercise problems. The applications in each chapter complementing theoretical aspects and implemented routines mainly address illustrative topics from undergraduate-level courses, such as Mechanics, Electricity, Optics, and Quantum Physics. More complex topics, involving concepts beyond the undergraduate curriculum, are typically presented in the final sections of the chapters, being mainly intended to highlight the full potential of the basic numerical methods and to provide instructors and researchers with additional material.

The input to the problems is deliberately kept simple, often merely as bare assignments to the problem parameters, in order to allow the reader to focus on the essential aspects. The hands-on style of presenting the material, with trial-and-error routes, with emphasis on advantages and pitfalls, encourages the reader to develop skills by experimenting.

As a rule, the C/C++ versions of the programs are not included in the "Problems" section of the chapters. Instead, along with their Python counterparts, they are to be found as supplementary material on the book's website. Also as a general rule, all the routines, libraries, and programs come in pairs of Python and C/C++ files having identical names and appropriate extensions.

Book structure. The book is divided into 13 chapters. Following this introduction, Chapter 1 deals with the main aspects of approximate numbers and propagation of associated errors. Without giving a systematic presentation of the Python and C languages, Chapter 2 reviews the language elements of interest to coding the numerical methods developed in later chapters. Chapter 3 is devoted to introducing elements of graphic functionality in Python and C/C++ applications. In Chapter 4 basic sorting and indexing methods are presented.

Chapter 5 describes prototypical techniques of function evaluation (in particular, for special functions). Chapter 6 describes methods for solving algebraic and transcendental equations, while Chapter 7 is devoted to describing the most important methods for solving systems of linear equations. Chapter 8 deals with eigenvalue problems, in particular, for symmetric matrices. Chapter 9 is concerned with approximating tabulated functions, specifically by interpolation (including by spline functions) and regression (including nonlinear regression).

Chapter 10 describes methods of integrating one- and multi-dimensional functions by classical and Gaussian quadratures, while Chapter 11 is devoted to Monte Carlo integration techniques and generation of random variables with various distributions. In Chapter 12 methods for solving ordinary differential equations are presented, ranging from Runge–Kutta to Numerov's method, and Chapter 13 discusses the most important discretization methods for elliptic, parabolic, and hyperbolic partial differential equations, familiarizing the reader with the stability analysis.

Supplementary material and the book's website. The book is complemented with material provided on the book's website,

http://www.crcpress.com/product/isbn/9781466569676/

The material is organized by chapters in a simple folder structure, each chapter folder containing routine libraries, example programs, and sample graphical output. The root folder is /INP/ (from "Introduction to Numerical Programming") and the chapter folders are named Chnn, with nn representing the two-digit chapter number.

The file names associated with the application programs contain the complete information on their location. The general format is Pnn-name.py or Pnn-name.cpp, whereby nn indicates, as before, the chapter folder /INP/Chnn/, and the file's extension indicates the respective subfolder /INP/Chnn/Python/ or /INP/Chnn/C/. To locate the file corresponding to a particular program listing from the text, the reader is advised to use the cross-reference list provided in Appendix G, "List of Programs by Chapter."

All the general Python libraries (*.py) developed in the book are to be found in the folder /INP/modules/, while the corresponding C/C++ header files (*.h) reside in the folder /INP/include/.

Access to the libraries numxlib and graphlib. Accessing the libraries on a particular computer is a technical problem. It basically boils down to adding the folders /INP/modules/ and /INP/include/ to the search paths of the respective programming environments.

For Python in particular, this operation requires certain knowledge of operating systems and there are several methods in use. For reasons of simplicity, two cross-platform methods are recommended. The first method of gaining access to the developed modules is to simply copy them from the distribution folder /INP/modules/ into the standard folder of the Python 3.3 installation,

/Python33/Lib/site-packages/

which is the environment implied by the applications developed in the book.

The second method relies on Python's ability to dynamically redefine search paths by invoking the sys.path.append method. Assuming the folder structure presented above and the script to be executed located in /INP/Ch14/Python/, gaining access to numxlib.py might be achieved by adding to an executable Python script:

```
import sys
sys.path.append("/INP/modules/")
from numxlib import *
```

It is important, however, to append the path up to and including the root folder for both the libraries and the executable script (/INP/, in this case).

An essential aspect to be noted is that the use of the graphics library graphlib.py also requires access to the utilities file utils.py. Failure to find utils.py in Python's search path results in graphlib.py aborting. However, performing graphics with graphlib from C/C++ requires the interface file graphlib.h to have access to the basic Python library graphlib.py.

Author

Titus Adrian Beu, professor of theoretical and computational physics at the University "Babeş-Bolyai" from Cluj-Napoca, Romania, has been active in the broader field of computational physics for more than 30 years, both in research and academia. His research has been theoretical and computational in nature, covering a multitude of subjects. His research topics have evolved from Tokamak plasma and nuclear reactor calculations in the 1980s, collision theory and molecular cluster spectroscopy in the 1990s, to fullerenes and nanofluidics simulations in recent years. Development of ample computer codes has permanently been at the core of all research projects the author has conducted. In parallel, he has delivered lectures on general programming techniques and advanced numerical methods, general simulation methods, and advanced molecular dynamics.

The material covered by the book stems jointly from the author's research experience and the computational physics courses taught at the university. The basic material has undergone a continuous process of distillation and reformulation over the years, following the evolution of modern programming concepts and technologies.

Acknowledgments

Special thanks are due to Professor Rubin Landau, for having had the idea of the project in the first place, and for providing feedback on subtle aspects during the elaboration of the material. The author wishes to express his gratitude to Luna Han, for the encouragement and forthcoming support throughout the realization of the book. The author wishes to acknowledge Professor Manuel José Páez of Universidad de Antioquia for his contribution to the preparation of some problems in Chapters 5, 6, 7, 9, and 10 during the early stages of manuscript development. Special thanks are also due to Dr. Octavian More for proofreading the final manuscript.

1

Approximate Numbers

1.1 Sources of Errors in Numerical Calculations

Unlike the mathematical formulations of scientific and engineering problems, which virtually lead to exact results, concrete numerical calculations are affected by inherent inaccuracies. The errors occurring in computations can be both genuinely numerical and nonspecific in nature. The *nonspecific errors* are rather diverse and may comprise, among others, errors related to the formulation of the problem, the employed methodology, or the input data. In their turn, the *specific numerical errors*, which are of primary concern to numerical analysis, include roundoff and truncation errors.

Formulation errors are typically involved when either the complexity of the studied phenomenon cannot be grasped by a closed theoretical formulation, or, the exact mathematical formulation that faithfully describes the modeled process leads to an intractable numerical implementation. For the sake of reducing the complexity of the problem, one is compelled in such cases to accept simplifying assumptions, or even a simplified and less realistic theoretical formulation. By way of consequence, no matter how accurate the employed numerical methodology is, the solution is likely to deviate to some extent from the exact description of the real phenomenon.

Method errors arise in situations (and the practice abounds with examples) in which the exact mathematical formulation cannot be directly transposed into an existing numerical methodology. In such situations, the original problem is replaced by an approximate problem, for which there exists an adequate numerical methodology yielding a close-enough solution. From this perspective, numerical methods are in their vast majority approximation methods. For instance, a boundary value problem for a differential equation is typically solved numerically by approximating the derivatives with finite-difference schemes based on discrete solution values and solving the resulting system of linear equations by specific methods.

Data errors refer to uncertainties impairing the quality of the primary data, or their inherently finite (and therefore approximate) machine representation.

Roundoff errors are a direct consequence of the internal representation of numerical data, with a finite number of significant digits. For example, the result of the operation $1/3$ is 0.3333333333333333 (both in Python and C/C++) and implies a roundoff error of approximately 3×10^{-17}. Considered individually, rounding errors may seem harmless and manageable. Nevertheless, with increasing number of operations, particularly in repetitive calculations implying differences of close operands or divisions with relatively small divisors, they tend to accumulate and may eventually compromise the results.

Truncation (residual) errors are characteristic of theoretically infinite, repetitive numerical processes, which have to be interrupted after a large, yet, unavoidably, finite number of calculation steps. The incomplete calculations induce truncation errors, which can be virtually minimized by continuing the process.

Provided they are adequately implemented, numerical methods enable, in principle, the control of the roundoff and truncation errors. In particular, to reduce roundoff errors, one has to consider from among equivalent mathematical formulations the ones that avoid operations prone to induce and propagate such errors, such as differences of close values (see Section 1.5). As for the truncation errors, these are, in principle, under the full control of the programmer, whose only concern remains the construction of finite and efficient algorithms. While finite algorithms are compulsory for reaching converged results, efficiency directly reflects the programmer's experience and coding style. It has been stated, under different forms, that the minimization of truncation errors lies at the core of the whole numerical analysis. In any case, mastering numerical methods implies to a large extent an in-depth knowledge of estimating and reducing errors.

Numerous examples can be given in which the same mathematical formalism, coded in different ways, yields qualitatively different results due to the different propagation of the errors. In fact, in developing the Jacobi method for solving eigenvalue problems for symmetric matrices (see Section 8.3), one reaches the equation

$$\cot^2 \varphi + \frac{a_{jj} - a_{ii}}{a_{ji}} \cot \varphi - 1 = 0,$$

whose *minimum* solution φ is sought for. Solving this equation relative to $\cot \varphi$, we obtain by subsequent inversion:

$$\tan \varphi = \text{sign} \left(\frac{a_{ii} - a_{jj}}{2a_{ji}} \right) \left[\left| \frac{a_{ii} - a_{jj}}{2a_{ji}} \right| + \sqrt{\left(\frac{a_{ii} - a_{jj}}{2a_{ji}} \right)^2 + 1} \right]^{-1}. \tag{1.1}$$

On the other hand, solving the equation directly in terms of $\tan \varphi$, we obtain the alternative, mathematically equivalent expression:

$$\tan \varphi = \text{sign} \left(\frac{a_{jj} - a_{ii}}{2a_{ji}} \right) \left[\left| \frac{a_{jj} - a_{ii}}{2a_{ji}} \right| - \sqrt{\left(\frac{a_{jj} - a_{ii}}{2a_{ji}} \right)^2 + 1} \right]. \tag{1.2}$$

While Equation 1.1 leads as part of the iterative Jacobi procedure to a stable algorithm, the use of Formula 1.2 causes severe instabilities, arising from the aberrant propagation of roundoff errors, favored, in turn, by the implied subtraction of possibly close operands.

1.2 Absolute and Relative Errors

Let x^* be the exact value of a certain quantity and x a known approximation thereof. By definition, the *absolute error* associated with x is

$$\Delta^* = |x^* - x|. \tag{1.3}$$

In general, numerical calculations operate with approximations to both the quantities and their absolute errors. In the ideal case, in which, besides the approximation x, the exact absolute error would also be known, the exact number would be exactly expressible as

$$x^* = x \pm \Delta^*. \tag{1.4}$$

However, in general, only an *estimate* of the absolute error Δ is available and, correspondingly, only an estimate of the exact value can be determined:

$$x - \Delta \leq x^* \leq x + \Delta. \tag{1.5}$$

For the estimate x to be reliable, it is important that Δ does not underestimate the true error, that is, $\Delta \geq \Delta^*$, case in which it is called *limiting absolute error*.

By definition, the *relative error* δ^* of the approximation x to x^* is equal to the ratio of the absolute error to the modulus of the exact number, provided the latter is nonzero:

$$\delta^* = \frac{\Delta^*}{|x^*|}, \quad (x^* \neq 0). \tag{1.6}$$

Replacing Δ^* in Equation 1.4 from Equation 1.6, the following expression for the exact number results:

$$x^* = x(1 \pm \delta^*), \tag{1.7}$$

and, based on the approximate relative error $\delta \geq \delta^*$, one can express the exact value as

$$x^* \approx x(1 \pm \delta). \tag{1.8}$$

Since, by virtue of the division by the reference value, the relative error does no longer depend on the magnitude of the number, it is a useful measure for establishing general precision or convergence criteria, which are applicable irrespective of the reference values. Moreover, as shown in Section 1.4, the relative error can be rigorously linked to the number of significant digits in a result, whereas, for the absolute error, such a connection is not possible.

Let us consider as an example the approximation $x = 2.71$ to the exact number $x^* \equiv e = 2.71828\ldots$. Given the limiting values $2.71 < e < 2.72$, it results that $|e - x| < 0.01$. We can thus use as an upper bound for the absolute error the width of the bounding interval, $\Delta = 0.01$, and we write $e \approx 2.71 \pm 0.01$. Correspondingly, the estimate of the relative error is $\delta = 0.01/2.71 \approx 0.004$ and we have $e \approx 2.71 \times (1 \pm 0.004)$.

Illustrative estimates of truncation errors are provided by the evaluation of functions from their Taylor series expansions (see Section 5.2). Specifically, let us evaluate the exponential function

$$e^x = 1 + x + \frac{x^2}{2!} + \frac{x^3}{3!} \cdots$$

for the particular argument $x = 1$:

$$e = 1 + 1 + \frac{1}{2} + \frac{1}{6} + \cdots = 2.7182\ldots$$

This is an absolutely convergent series and the successive terms add decreasing positive contributions to the result. Considering only the terms up to the third order, the approximate value of the exponential is

$$\bar{e} = 1 + 1 + \frac{1}{2} + \frac{1}{6} = 2.6666\ldots,$$

and a rough estimate of the corresponding truncation error is just the value of the last term added, namely $\Delta = 1/6 = 0.1666\ldots$. This is an overestimate, since the true truncation error is lower, namely $\Delta^* = e - (1 + 1 + 1/2 + 1/6) = 0.0516\ldots$. Anyway, a more "pessimistic" error estimate is rather preferable, since it is more likely to enforce within an iterative procedure the achievement of the aimed level of precision. However, it should be obvious that the true function value is typically not known, and considering as an educated guess for the truncation error, just the last term added is the pragmatic approach adopted in practice.

For the sake of clarity, let us consider in the following analysis errors with respect to the true value e. The exact, relative truncation error amounts in this case to $\delta^* = |(e - \bar{e})/e| = 1.90 \times 10^{-2}$.

Now considering the exponential for $x = -1$,

$$e^{-1} = 1 - 1 + \frac{1}{2} - \frac{1}{6} + \cdots = 0.3678\ldots,$$

the terms can be seen to alternate in sign, and the approximated function turns out to behave in a qualitatively different way. In fact, the $O(x^3)$ approximation yields in this case:

$$\bar{e}^{-1} = 1 - 1 + \frac{1}{2} - \frac{1}{6} = 0.3333\ldots,$$

and the relative truncation error amounts to $\delta^* = |(e^{-1} - \bar{e}^{-1})/e^{-1}| = 9.39 \times 10^{-2}$. Thus, while for $x = 1$, the approximation is in error roughly by 2%, for $x = -1$, the same number of terms brings about an error exceeding 9%. A simple change, that is, considering for e^{-1} the inverse of the approximation $\bar{e}$,

$$\bar{e}^{-1} = \frac{1}{\bar{e}} = \left(1 + 1 + \frac{1}{2} + \frac{1}{6}\right)^{-1} = 0.375,$$

again lowers the relative truncation error to $\delta^* = |(e^{-1} - \bar{e}^{-1})/e^{-1}| = 1.94 \times 10^{-2}$. The important conclusion to be drawn from this simple example is that equivalent mathematical formulations may lead to results featuring different degrees of accuracy. The less involved the subtraction operations of close operands, the more likely a good level of precision (for quantitative arguments, see Section 1.5).

1.3 Representation of Numbers

In computers, numbers are represented in binary formats. The most common machine representation of *signed integers* (type `int` in C/C++) requires 32 bits, with the left-most bit (31) representing the sign (0 stands for "+" and 1, for "−") and the bits 0–30 representing the coefficients (0 or 1) of the corresponding powers of 2. For example, the binary representation of the base-10 integer 100 is

$$100_{10} = 00000000000000000000000001100100.$$

Indeed, considering only the significant seven right-most digits, we have:

$$1 \cdot 2^6 + 1 \cdot 2^5 + 0 \cdot 2^4 + 0 \cdot 2^3 + 1 \cdot 2^2 + 0 \cdot 2^1 + 0 \cdot 2^0 = 100.$$

The integer values representable on 32 bits range from $-2^{31} = -2\,147\,483\,648$ to $2^{31} - 1 = 2\,147\,483\,647$.

Another data type that is most useful in numerical calculations is the *double-precision floating-point* type defined by the IEEE Standard for Floating-Point Arithmetic (IEEE 754). This type, denoted as `double` in C/C++, uses 64 bits and has the following layout:

- Sign s: bit 63 (left most).
- Exponent e: bits 52–62 representing the *biased exponent* of 2.
- Fraction f: bits 0–51, representing the *fractional part* of the mantissa.

The actual binary encoding of double-precision data is no longer straightforward, such as in the case of integers, but, for reasons of optimizing the representable range of values, the general format is

$$x = (-1)^s (1 + f) \cdot 2^{e-1023},$$

$$f = \sum_{i=1}^{52} \frac{b_{52-i}}{2^i},$$

where the "1023" in the exponent is the so-called *bias*, which guarantees that the stored exponent e is always positive, and b_i are the binary digits of the fractional part of the mantissa f. The "1" in the mantissa is not actually stored, but it only provides an extra bit of precision.

The double-precision floating-point format gives 15–17 decimal digits precision in the mantissa and supports exponents of 10 in the range -324 to $+308$. Correspondingly, the representable values range between $\pm 1.80 \cdot 10^{308}$, excluding the range $\pm 4.94 \cdot 10^{-324}$ (with approximate mantissas).

The same sample number as above, that is, $100 = 1 \cdot 10^2$, considered this time as a double-precision value, has the machine representation:

$$
\begin{array}{ccc}
0 & 10000000101 & 1001000000000000000000000000000000000000000000000000 \\
\text{sign} & \text{exponent} & \text{mantissa}
\end{array}
$$

which is, obviously, less intelligible than the integer format. In fact, from the above IEEE 745 representation, we determine for the exponent and the fractional part of the mantissa:

$$e = 2^{10} + 2^2 + 2^0 = 1029,$$

$$f = \frac{1}{2^1} + \frac{1}{2^4} = \frac{9}{16}.$$

Therewith we recover, indeed, the encoded number:

$$x = \left(1 + \frac{9}{16}\right) \cdot 2^{1029-1023} = \frac{25}{16} \cdot 2^6 = 100.$$

In spite of the efficiency of the binary machine representation, the results of computations, in general, and their error analysis, in particular, are, obviously, more easily comprehensible in base 10. In this context, any positive number may be represented in base 10 as

$$x = d_m \cdot 10^m + d_{m-1} \cdot 10^{m-1} + \cdots + d_{m-n+1} \cdot 10^{m-n+1} + \cdots, \tag{1.9}$$

where d_i represent the decimal digits, with $d_m \neq 0$. For example,

$$273.15 = 2 \cdot 10^2 + 7 \cdot 10^1 + 3 \cdot 10^0 + 1 \cdot 10^{-1} + 5 \cdot 10^{-2}.$$

1.4 Significant Digits

Concrete computations imply using approximate numbers with a finite number of digits. All the n decimal digits $d_m, d_{m-1}, \ldots, d_{m-n+1}$, uniquely defining the number, are called *significant digits (figures)*. Some of the significant digits may be 0, but not, however, the leading digit, d_m. For example, in the number $0.01030 = 1 \cdot 10^{-2} + 0 \cdot 10^{-3} + 3 \cdot 10^{-4}$, the leading and the trailing zeros are not significant, since the first two zeros just serve to fix the position of the decimal point in a particular fixed-point representation, while the last zero carries no information at all.

Definition 1.1

An approximate number x is said to have n *exact significant digits*, $d_m, d_{m-1}, \ldots, d_{m-n+1}$, if its absolute error Δ does not exceed one-half-unit in the nth position from left to right, that is,

$$\Delta \leq \frac{1}{2} \cdot 10^{m-n+1}. \tag{1.10}$$

For example, the number $x = 272.75$ approximates the exact value $x^* = 273.15$ to three significant digits. Indeed, having in view that $\Delta = |x^* - x| = 0.4 \leq 1/2 \cdot 10^0$ and $m = 2$, it follows that $m - n + 1 = 0$ and, from here, the number of exact significant digits is $n = m + 1 = 3$.

The formulation "n exact significant digits" should not be taken literally, since there can be cases in which, according to the above definition, the approximation x has n exact significant digits, without any of them actually coinciding with corresponding digits of the exact number x^*. For example, the number $x = 299.5$ approximates to three digits the exact value $x^* = 300.0$, and yet, its digits are all different ($\Delta = |300 - 299.5| = 0.5 \leq 1/2 \cdot 10^0$ and it follows that $n = m + 1 = 3$).

Theorem 1.1

If an approximate number x has n exact significant digits, the associated relative error δ satisfies the inequality

$$\delta \leq \frac{1}{2d_m} \cdot 10^{-(n-1)}, \tag{1.11}$$

where $d_m > 0$ is the first significant digit of the number.

Indeed, using the upper bound to the absolute error Δ provided by inequality (1.10), as well as the decimal representation (1.9) of the approximate number x, we obtain, successively:

$$\delta = \frac{\Delta}{|x|} \leq \frac{(1/2) \cdot 10^{m-n+1}}{d_m \cdot 10^m + d_{m-1} \cdot 10^{m-1} + \cdots} \leq \frac{1}{2d_m} \cdot 10^{-(n-1)},$$

where the denominator was minimized (to maximize the fraction) by retaining only the leading term $d_m \cdot 10^m$.

The importance of Relation 1.11 lies in the fact that, by taking its logarithm, it provides an operational estimate for the number of exact significant digits in terms of the relative error:

$$n \leq 1 - \log_{10}(2\delta d_m). \tag{1.12}$$

For example, used instead of the exact value $e = 2.71828\ldots$, the approximate number $x = 2.71$ has the associated relative error $\delta = |e - 2.71|/e = 3.05 \times 10^{-3}$, which corresponds to $n \leq 1 - \log_{10}(2 \times 3.05 \times 10^{-3} \times 2) = 2.91$, namely, two exact significant digits. The approximation $x = 2.72$, with the associated relative error $\delta = 6.32 \times 10^{-4}$, has $n \leq 3.60$, that is, three exact significant digits.

From Inequality 1.12, we can obtain a less rigorous, but very serviceable result. Considering for the relative error the simplified form $\delta = 10^{-p}$ and for the leading decimal digit the "average" value $d_m = 5$, we are left with

$$n \approx p. \tag{1.13}$$

This rule of thumb, useful for quick estimates, simply identifies the number of significant digits of an approximate number with the absolute value of the power of 10 from its relative error.

1.5 Errors of Elementary Operations

Let us establish, in the following, the way in which various elementary operations propagate the relative errors of their operands. In relation to numerical calculations, the considerations developed below specifically refer to roundoff and truncation errors.

1.5.1 Error of a Sum

We consider first the sum of two approximate numbers, x_1 and x_2, having the *same sign*. The sum evidently preserves the sign and cumulates the magnitudes of the operands:

$$x = x_1 + x_2. \tag{1.14}$$

For the total error, we have

$$\Delta x = \Delta x_1 + \Delta x_2,$$

but, since the individual errors can, depending on their signs, both compensate each other and accumulate, the sum of the absolute errors of the operands actually provides an *upper bound* for the total absolute error of the sum:

$$\Delta \leq \Delta_1 + \Delta_2. \tag{1.15}$$

Let us demonstrate the following theorem.

Theorem 1.2

The relative error of the sum of two approximate numbers of the same sign does not exceed the largest relative error of the operands:

$$\delta \leq \max(\delta_1, \delta_2). \tag{1.16}$$

Indeed, using Relation 1.15, the relative error $\delta = \Delta / |x|$ results to be bounded from above:

$$\delta \leq \frac{\Delta_1 + \Delta_2}{|x_1 + x_2|} = \frac{|x_1|\delta_1 + |x_2|\delta_2}{|x_1 + x_2|}, \tag{1.17}$$

or, by replacing both δ_1 and δ_2 with their maximum:

$$\delta \leq \frac{|x_1| + |x_2|}{|x_1 + x_2|} \max(\delta_1, \delta_2),$$

which proves the theorem. The importance of Relation 1.16 lies in the fact that it ensures that the relative error of the sum of two or more terms of the same sign does not exceed the largest of the individual relative errors. In other words, the least-accurate term dictates the accuracy of the sum. This turns summation into an operation that does not build up errors.

1.5.2 Error of a Difference

Let us now consider the difference of two approximate numbers of the *same sign*:

$$x = x_1 - x_2. \tag{1.18}$$

It should be self-evident that Equations 1.14 and 1.18 cover all the possible algebraic situations. Just like in the case of the summation, the most unfavorable case is the one in which the absolute errors enhance each other, and, therefore, the sum of the absolute errors is, again, the upper bound for the absolute error of the difference:

$$\Delta \leq \Delta_1 + \Delta_2.$$

Maximizing the total absolute error in the definition of the relative error $\delta = \Delta/x$, there results the limiting relation:

$$\delta \leq \frac{\Delta_1 + \Delta_2}{|x_1 - x_2|} = \frac{|x_1|\delta_1 + |x_2|\delta_2}{|x_1 - x_2|}. \tag{1.19}$$

If the operands are close in value, the absolute difference $|x_1 - x_2|$ is small and, even for small individual relative errors, δ_1 and δ_2, the relative error of the difference, δ, can become significant.

To exemplify, let us consider the numbers $x_1 = 2.725$ and $x_2 = 2.715$, which we consider correct to four significant digits. According to Equation 1.11, the individual relative errors satisfy:

$$\delta_{1,2} \leq \frac{1}{2 \times 2} \cdot 10^{-(4-1)} = 2.5 \times 10^{-4}.$$

Nevertheless, based on Equation 1.19, the difference $x = x_1 - x_2 = 0.01$ features a significantly larger relative error:

$$\delta \leq \frac{(2.725 + 2.715) \times 2.5 \times 10^{-4}}{0.01} = 1.4 \times 10^{-1},$$

and the corresponding number of exact significant digits decreases, accordingly, to

$$n \leq 1 - \log_{10}(2 \times 1.4 \times 10^{-1} \times 1) \approx 1.5.$$

To sum up, the relative error of the difference, δ, increases by more than three orders of magnitude as compared to those of the individual terms, while the number of exact significant digits drops from 4 to 1.

Thus, when subtracting nearly equal numbers, a significant *loss of accuracy* arises, and, therefore, it is of utmost importance to reformulate prior to the implementation, whenever possible, both the mathematical formalism and the algorithm so as to avoid such critical operations.

1.5.3 Error of a Product

Theorem 1.3

The relative error of a product of approximate numbers, $x = x_1 x_2$, does not exceed the sum of the relative errors of the factors, δ_1 and δ_2:

$$\delta \leq \delta_1 + \delta_2. \tag{1.20}$$

To demonstrate the statement, we assume that the *factors are positive*. In general, the overall sign can be equated on both sides and, thus, x, x_1, and x_2 can be identified with their absolute values. Applying the logarithm to the product, we get:

$$\log x = \log x_1 + \log x_2. \tag{1.21}$$

Using the basic differentiation formula

$$\frac{d}{dx} \log x = \frac{1}{x},$$

and passing from differentials to finite differences, we can establish the approximation formula:

$$\Delta \log x \approx \frac{\Delta}{x}. \tag{1.22}$$

Application of Approximation 1.22 on both sides of Equation 1.21 leads to

$$\frac{\Delta x}{x} = \frac{\Delta x_1}{x_1} + \frac{\Delta x_2}{x_2}. \tag{1.23}$$

Taking the absolute values of the terms obviously maximizes the right-hand side and we obtain the inequality:

$$\frac{|\Delta x|}{x} \leq \frac{|\Delta x_1|}{x_1} + \frac{|\Delta x_2|}{x_2}.$$

If x_1 and x_2 are affected by small absolute errors $\Delta_1 = |\Delta x_1|$ and $\Delta_2 = |\Delta x_2|$, the above inequality also remains valid when replacing the approximate numbers by the corresponding exact values:

$$\frac{\Delta}{x^*} \leq \frac{\Delta_1}{x_1^*} + \frac{\Delta_2}{x_2^*}. \tag{1.24}$$

Finally, identifying the terms with the corresponding relative errors demonstrates the theorem.

Result 1.20 can be readily extended to products of several factors of arbitrary signs. If all but one factor is exact, the relative error of the product equates to the relative error of the approximate factor.

1.5.4 Error of a Quotient

Let us consider the division $x = x_1/x_2$, of two nonzero approximate numbers. For the sake of simplicity, as in the case of the product, we derive the bounding relation for the relative error of the quotient for *positive operands* x_1 and x_2. Applying the logarithm to both sides of the division,

$$\log x = \log x_1 - \log x_2, \tag{1.25}$$

and using Approximation 1.22, we arrive to

$$\frac{\Delta x}{x} = \frac{\Delta x_1}{x_1} - \frac{\Delta x_2}{x_2}. \tag{1.26}$$

Replacement of the absolute values of the terms maximizes the right-hand side and leads to an inequality, which is similar to the one obtained for the product:

$$\frac{|\Delta x|}{x} \leq \frac{|\Delta x_1|}{x_1} + \frac{|\Delta x_2|}{x_2}.$$

Considering, again, that x_1 and x_2 are affected by small absolute errors $\Delta_1 = |\Delta x_1|$ and $\Delta_2 = |\Delta x_2|$, we finally obtain the bounding relation:

$$\delta \leq \delta_1 + \delta_2, \tag{1.27}$$

which states that the relative error of the quotient does not exceed the cumulated relative errors of the dividend and divisor.

References and Suggested Further Reading

Burden, R. and J. Faires. 2010. *Numerical Analysis* (9th ed.). Boston: Brooks/Cole, Cengage Learning.

Demidovich, B. and I. Maron. 1987. *Computational Mathematics* (4th ed.). Moscow: MIR Publishers.

Hildebrand, F. B. 1987. *Introduction to Numerical Analysis* (9th ed.). New York: Dover Publications.

Zarowski, C. J. 2004. *An Introduction to Numerical Analysis for Electrical and Computer Engineers.* Hoboken, NJ: Wiley-Interscience.

2

Basic Programming Techniques

2.1 Programming Concepts

In a broad sense, computer programs are sequences of instructions meant to be executed in a certain order, so as to perform specific tasks or solve the given problems. The ways in which the same task can be properly coded are literally innumerable, being largely a matter of programming style and experience. On a higher level of abstraction, a program can be conceived as a collection of *functions* performing well-defined tasks. One of the functions, called *main program*, is compelling and controls the overall execution of the program, while the other functions carry out subordinated tasks. The number, functionality, and implementation of the program units are not imposed by any strict rules and may depend on various factors, not in the least on the programmer's proficiency.

Even though there are no infallible, general recipes for writing programs in science and engineering, since the early days of computing, certain general principles have crystallized. Two of the basic concepts for writing good codes are *modularization* and *encapsulation* (Knuth 1998; Press et al. 2002). *Modularization* implies, essentially, decomposing the program into functionally relevant program units (i.e., functions, subroutines, procedures, etc.), which should interact solely through well-defined interfaces (headers with parameter lists). *Encapsulation*, on the other hand, requires that the scopes of all hierarchically subordinated structures must be completely included in one another, ensuring a traceable domain of "visibility" for all the program definitions.

Rational modularization is particularly beneficial in the case of complex computer applications developed by teams of programmers. Quite on the contrary, excessive detailing into units lacking relevance is prone to producing inefficient programs. Among the benefits of equilibrated modularization, worth mentioning are improved readability (even at later times and by different programmers), simplified troubleshooting, debugging, maintenance and modification, and, not in the least, reusability of the general-purpose routines in other applications. The latter is, in fact, the concept lying on the basis of standard libraries, which enable the incorporation of preprogrammed algorithms in comprehensive applications without knowledge of their implementation details.

While indentation is not a requirement in C/C++, it is compulsory in Python, since it determines the program structure itself. Aiming, insofar as possible, to provide an equivalent code in Python and C/C++, we consistently apply for both languages an *indent of three spaces* for each successive subordination level. The chosen indent ensures good readability of the code, avoiding, at the same time, a too large shift of the "deeper" code blocks to the right.

In the following section, we discuss two issues that are fundamental to scientific programming, namely, mechanisms of passing arguments to functions and allocation of arrays. In Appendix C, we touch upon the more complex problem of mixed programming, specifically addressing the embedding of Python routines in C/C++ codes. The techniques developed are then used in the subsequent chapters for creating portable graphics from C/C++ programs using a Python library.

2.2 Functions and Parameters

As an illustration of the programming concepts mentioned above, next, we consider the elementary task of calculating factorials, $n! = 1 \cdot 2 \cdots n$, which can be conveniently implemented based on the recurrence relation

$$n! = n \cdot (n-1)!$$

with the limiting case $0! = 1$.

Despite its tempting plainness, the straightforward implementation of the factorial in the main program would result in poor programming. This is mainly because it would imply redundant code sequences for every instance in which the factorial would be necessary. Instead, the reasonable idea is to devise a specialized function to calculate the factorial. This is materialized in Listings 2.1 (in Python) and 2.2 (in C/C++).

As a first attempt, for reasons of simplicity, the function Factorial is designed to communicate with the main program through the *global variables* f and n. The main program assigns to n the integer value for which the factorial is to be calculated. Upon its call, Factorial gains access to n, calculates the factorial, and stores it in f. Once the control is returned, the main program "finds" the result in the variable f.

Beyond its simplicity, the communication through global variables has several obvious downsides. First, the fixed variables it operates with prevent the function Factorial from being called as part of an expression (like, for instance, for calculating combinations). Second, the routine remains functional within different programs only if it is imported along with the global variables it depends on, which, on the other hand, may, incidentally, conflict with variables having the same name, but with different significance. Third, being accessible from any program unit between calls to Factorial, the global variables n and f are prone to unaccounted modification. In sum, despite being in broad use due to its simplicity, communication based on global variables does not comply with the concept of modular programming and should be limited to well-justified cases, in which the benefits outweigh the potentially unsafe and difficult-to-control side effects.

The recommendable implementation of the factorial is provided in Listings 2.3 and 2.4, in the context of calculating combinations. The function Fact receives the integer n through its parameter list and returns the factorial by way of its name. In this manner, the communication between Fact and the calling program unit is restricted to the function's header, avoiding undesirable interferences. The final return statement is essential for effectively conveying the result back to the caller. A more elaborate

Listing 2.1   Factorial Function Communicating through Global Variables (Python Coding)

```python
# Factorial with global variables

def Factorial():
    global f, n                                    # global variables
    f = 1e0
    for i in range(2,n+1): f *= i

# main

n = int(input("n = "))

Factorial()

print(n,"! = ",f)
```

Listing 2.2   Factorial Function Communicating through Global Variables (C/C++ Coding)

```c
// Factorial with global variables
#include <stdio.h>

float f;                                                    // global variables
int n;

void Factorial()
{
   int i;

   f = 1e0;
   for (i=2; i<=n; i++)  f *= i;
}

int main()
{
   printf("n = ");  scanf("%d",&n);

   Factorial();

   printf("%d! = %f\n",n,f);
}
```

version of Fact would first validate the input ($n \geq 0$). For simplicity, however, this feature was not implemented.

Taking advantage of the built-in recursivity of the Python and C/C++ functions, that is to say, of their ability to call themselves directly or indirectly, algorithms involving mathematical recursivity can be implemented in a more natural and compact way. Basically, each recursive call to a function creates

Listing 2.3   Use of the Recommended Implementation of the Factorial (Python Coding)

```python
# Calculates combinations using a factorial function

#=============================================================================
def Fact(n):
#-----------------------------------------------------------------------------
#  Returns the factorial of n
#-----------------------------------------------------------------------------
   f = 1e0
   for i in range(2,n+1): f *= i
   return f

# main

n = int(input("n = "))
k = int(input("k = "))

C = Fact(n)/(Fact(k)*Fact(n-k))

print("C({0:d},{1:d}) = {2:f}".format(n,k,C))
```

Listing 2.4   Use of the Recommended Implementation of the Factorial (C/C++ Coding)

```c
// Calculates combinations using a factorial function
#include <stdio.h>

//===========================================================================
double Fact(int n)
//---------------------------------------------------------------------------
// Returns the factorial of n
//---------------------------------------------------------------------------
{
   double f;
   int i;

   f = 1e0;
   for (i=2; i<=n; i++) f *= i;
   return f;
}

int main()
{
   double C;
   int k, n;

   printf("n = "); scanf("%d",&n);
   printf("k = "); scanf("%d",&k);

   C = Fact(n)/(Fact(k)*Fact(n-k));

   printf("C(%d,%d) = %f\n",n,k,C);
}
```

on the *stack* (specialized memory region) a full set of copies of the locals and parameters associated with the particular level of recursion. To avoid infinite processes, and, thereby, the danger of overrunning the stack's capacity, any recursive routine needs to implement a control mechanism for the depth of the self-calls and the eventual exit from the process in reversed order of the recursion levels.

   The recursive implementation of the factorial as function FactRec is provided in Listings 2.5 and 2.6. It should be pointed out that, although more compact, due to the overhead brought about by the repeated self-calls and replication of the local variables on the stack, the recursive implementation is less efficient than the iterative implementation used in function Fact. Obviously, the speed difference is relevant not for single executions, but for intensive use of the functions, when the recursive approach proves to be significantly slower.

Listing 2.5   Factorial Function Using Recursivity (Python Coding)

```python
#==========================================================================
def FactRec(n):
#--------------------------------------------------------------------------
#  Returns the factorial of n using recursivity
#--------------------------------------------------------------------------
   return (n * FactRec(n-1) if (n > 1) else 1e0)
```

Listing 2.6   Factorial Function Using Recursivity (C/C++ Coding)

```
//=============================================================================
double FactRec(int n)
//-----------------------------------------------------------------------------
// Returns the factorial of n using recursivity
//-----------------------------------------------------------------------------
{
   return ((n > 1) ? n * FactRec(n-1) : 1e0);
}
```

2.3 Passing Arguments to Python Functions

For the purpose of the following discussion, we make a distinction between *parameters* and *arguments* of functions. In fact, by *parameters*, we refer to variables declared in the parameter list of a function and operated upon in its body, whereas by *arguments*, we denote concrete values or variables that are set in correspondence with the parameters in a calling statement.

Modern programming languages provide two basic *mechanisms for passing arguments* to functions: *by value* and *by reference*. The *pass-by-value* mechanism essentially implies *copying the values* of the arguments into the corresponding parameters upon function call, without subsequent reflection outside the function of the changes suffered by the parameters. The *pass-by-reference* mechanism assumes, on the other hand, *copying the references* of the arguments (the memory addresses, not the values!) into the corresponding parameters, with any changes suffered by the parameters directly reflected by the corresponding arguments. In fact, through the pass-by-reference mechanism, the parameters become references to the arguments, or, in other words, local aliases, which enable the function to directly operate upon the arguments.

The choice of the appropriate mechanism to be used for passing arguments should rely on a thorough understanding of both their advantages and the possible side effects.

For *scalar constants*, the only usable mechanism is pass-by-value. For *scalar variables*, both mechanisms are applicable and the choice depends on whether the arguments are to be purposely "protected" from the changes of the linked parameters—case, in which the pass-by-value mechanism is appropriate, or, on the contrary, should reflect the changes, actually retrieving results from the function—case, in which the pass-by-reference method should be applied.

Compound data types (arrays, lists, tuples, structures, classes, etc.) are always passed *by reference*, no local copies of the arguments are being created, but rather references to the data structures are being transmitted. Therefore, any operations performed within the function on a compound parameter are automatically reflected by changes of the corresponding argument in the calling program.

Let us now provide a few examples to illustrate the mechanisms of passing arguments.

Listing 2.7   Returning Interchanged Argument Values from a Function (Python Coding)

```
# Returning swapped arguments from a function

def Swap(x, y):                           # scalar arguments passed by value
   temp = x; x = y; y = temp
   return (x, y)                          # return interchanged values

# main

a = 1e0; b = 2e0                          # values to be swapped
(a,b) = Swap(a,b)      # left side - return values; right side - input values
print(a,b)
```

Retrieving single results via the function's name was shown in Section 2.2 to be readily accomplished by a `return` statement specifying the value to be returned. Python takes this further by generalizing the mechanism and enabling the retrieval of several pieces of data through the function's name. For illustration, we provide a classical example program in Listing 2.7, in which the function `Swap` is supposed to interchange the values of two variables. The parameters `x` and `y` acquire their initial values as copies of the arguments a and b, which are passed *by value*. At the same time, by virtue of the `return` statement, `x` and `y` become references to the elements in the list on the left-hand side of the call to `Swap` (in this particular case, still a and b), which receive the changed values *by reference*.

Thus, the process involves both the pass-by-value and pass-by-reference mechanisms and actually separates the arguments into two lists: the *input list* (on the right side), whose arguments are passed *by value*, and the *output list* (on the left side), whose arguments are passed *by reference*. This becomes particularly obvious if the interchanged values are returned to a different pair of arguments:

```
(c,d) = Swap(a,b)
```

Indeed, while, upon return from `Swap`, c and d receive by reference the interchanged input values, a and b remain unchanged, since they are passed by value.

The fact that the input and output lists are functionally disjoint is also clarified by the simplified version of the function `Swap` shown below, in which the parameters create an inverted correspondence between the input and output arguments.

```
def Swap(x, y):                            # scalar arguments passed by value
    return (y, x)                               # return interchanged values
```

An important point to make is that in Python, due to the characteristic dynamic typing, the same function can operate with arguments of different types, which actually determine the types of the corresponding parameters. In particular, the function `Swap` can equally well interchange numeric values or strings, so that the arguments in Program 2.7 might, as well, have been strings:

```
a = "1st"; b = "2nd"
```

Let us now replace the pair of arguments to be interchanged with the elements of a list, as in Program 2.8. Being a compound data type, the list `alist` is passed by default by reference. The function actually operates by means of its local alias `xlist` directly upon `alist` itself. Correspondingly, the mechanism of returning data via the function's name is no longer applied and, therefore, a return statement becomes unnecessary.

Listing 2.8   Returning Interchanged List Components from a Function (Python Coding)

```
# Returning swapped list elements from a function

def Swap(xlist):                                   # list passed by reference
    temp = xlist[0]; xlist[0] = xlist[1]; xlist[1] = temp

# main

alist = [1e0, 2e0]                              # list of values to be swapped
Swap(alist)
print(alist)
```

2.4 Passing Arguments to C/C++ Functions

Retrieving a single result from a function via the function's name was illustrated in Section 2.2. Unlike Python, in C/C++, this mechanism cannot be directly generalized for arbitrary data sets. The only way of returning data sets in compliance with modular programming, that is, through the parameter list, involves explicit or disguised use of *pointers*.

Pointers are particular types of variables, which are specialized in storing *memory addresses* (Kernighan and Ritchie 1988). As an immediate application, pointers can be used to reference variables by their addresses, rather than by their names, thus providing external aliases to variables and enabling the retrieval of data from functions via their parameters.

The address of an object in the memory (e.g., a variable or an array element) can be retrieved by applying the *address operator* &. Accordingly, the statement

```
p = &x;
```

assigns the address of the variable x to the pointer p, and p is said to *point to* x (see Figure 2.1). The address operator cannot be applied to constants or expressions.

To access the object a pointer points to, one applies the *indirection operator* *. Concretely, if p contains the address of the variable x, then *p precisely represents the *value* of x.

The declaration of pointers is similar to that of usual variables, *type* *name*, specifying, notably, not the type of the pointer itself, but rather the type of the indirected pointer, or, in other words, of the variables the pointer is defined to point to. Hence, the declaration

```
double *p;
```

defines a pointer specialized in pointing to variables (locations) of type double. The type of variables pointed to is essential both in indirecting the pointer and in performing arithmetic operations therewith.

Incrementing a pointer, for example p++, yields the address of the next memory location of the type the pointer points to. Similarly, decrementing a pointer, p--, results in the address of the previous location. Incrementing, instead, the *variable* pointed to by pointer p is obtained by ++(*p).

Two of the major uses of pointers in scientific programming concern the mechanism of returning values from functions via their parameter lists (discussed in the following paragraph) and the dynamic memory allocation for arrays (discussed in the next section).

Since the only mechanism implemented in ANSI C for passing arguments to functions is the pass-by-value mechanism, called functions cannot directly modify variables defined in the calling function, thereby returning results. In other words, the transmission of data invariably occurs from the caller to the called function. Nevertheless, the use of pointers enables an indirect implementation of the pass-by-reference mechanism. The basic idea is to pass (again by value!) the *addresses* of the locations where the calling function expects the called function to save the results, so as to "find" and use them subsequently. Obviously, to hold addresses of memory locations, the output parameters of the called function have to be pointers.

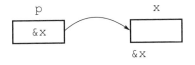

FIGURE 2.1   Pointers contain the addresses of the memory locations they point to.

Listing 2.9   Function Returning Modified Argument Values (C Coding)

```
// Returning swapped arguments from a function (C syntax)
#include <stdio.h>

void Swap(double *x, double *y)            // formal arguments are pointers
{
   double temp;                                    // temporary variable

   temp = *x; *x = *y; *y = temp;   // variable values accessed by indirection
}

int main()
{
   double a, b;

   a = 1e0; b = 2e0;                              // values to be swapped
   Swap(&a,&b);                          // actual arguments are addresses
   printf("%f %f\n",a,b);
}
```

For illustration of the described pass-by-reference mechanism, we refer again to the example of swapping variables, implemented in Python in Section 2.3. The ANSI C version shown in Listing 2.9 defines a function Swap for interchanging values of type double (similar to the swap function for integers defined in Kernighan and Ritchie 1988). As explained above, the main program does not pass to Swap the *values* of the arguments a and b, but rather their *addresses*, &a and &b, so that Swap can directly operate on the original locations. Correspondingly, to be able to receive the addresses of the operands, the parameters of Swap have to be declared pointers. Swap operates with the actual *values* of the arguments a and b from the main program by indirecting the pointers x and y, which contain their addresses.

Aside from returning results by the function's name, the method based on pointer-type parameters represents the main technique for retrieving results from functions in the spirit of modular programming.

Not changing the essence of the standard mechanism based on pointers, C++ significantly simplifies the syntax related to the pass-by-reference mechanism, improving the readability of the code (see

Listing 2.10   Function Returning Modified Argument Values (C++ Coding)

```
// Returning swapped arguments from a function (C++ syntax)
#include <stdio.h>

void Swap(double &x, double &y)            // arguments passed by reference
{
   double temp;                                    // temporary variable

   temp = x; x = y; y = temp;
}

void main()
{
   double a, b;

   a = 1e0; b = 2e0;                              // values to be swapped
   Swap(a,b);
   printf("%f %f\n",a,b);
}
```

Listing 2.10). The C++ syntax requires the "address" symbol & to be placed in front of the parameters that are to receive arguments by reference and any other explicit reference to pointers to be removed.

2.5 Arrays in Python

Numerical applications in science and engineering often involve tensor quantities, which are closely represented by *arrays*. More than other languages, Python offers a wealth of compound data types that are useful for grouping basic data types. The most versatile thereof is certainly the *list* type, which even enables grouping objects of different types. The list type implements a good number of methods for creating, accessing, modifying, and sorting lists. Owing to their intuitive nature and ease of use, throughout the book, we employ *lists* to represent arrays in Python programs.

As a general rule, relying on Python's feature of dynamical typing, we initialize and dimension the lists at run time to their full extent as soon as their size becomes available. We avoid to gradually expand lists by using, for example, the append method. For example, defining a numerical vector a with n components amounts in our programs to constructing it as an n-membered list by multiplication of a single-membered list:

```
a = [0]*n
```

It is implied that the valid components are a[0] through a[n-1], which are all set to 0. The components can be obviously initialized with any convenient constant of compatible type.

A rather controversial issue concerns the range of valid indexes in relation to the default *offset* implemented by the language, which is 0 both for Python and C/C++. Since the vast majority of the numerical methods presented in the book involve natural index ranges starting at 1, with just a few clearly marked exceptions, all the developed implementations consistently use as valid components a[1] through a[n], deliberately leaving unused the component a[0]. Therefore, the typical declaration of such a one-dimensional array takes the form:

```
a = [0]*(n+1)
```

The straightforward generalization of the declaration for a two-dimensional (2D) array reads:

```
a = [[0]*(n+1) for i in range(n+1)]
```

and evidences the topological structure of the array, as a collection of row vectors, the last usable element being a[n][n]. A detailed example of using arrays for basic matrix operations is given in Section 2.7.

2.6 Dynamic Array Allocation in C/C++

In C/C++, arrays are intrinsically linked to pointers. Indeed, the name of an array is by definition a pointer pointing to the beginning of the allocated memory block, that is to say, the pointer contains the address of the first array element. For example, for allocating an array of 10 components of type double, one might use the declaration:

```
double a[10];
```

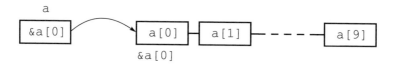

FIGURE 2.2   The name of an array is a pointer containing the address of the first array element.

Technically, the above statement allocates a memory block of 10 consecutive locations of commensurate size to accommodate values of type `double`, defines a pointer a to values of type `double`, and stores therein the starting address of the memory block. Owing to the null offset featured by default by the C/C++ arrays, the valid components are in this case a[0] through a[9]. As indicated above, the pointer a associated with the array's name contains the address of the first element, that is, &a[0] (see also Figure 2.2) and the allocation of the corresponding memory block takes place *statically*, during the compilation phase of the program.

Any operation with array elements may also be carried out by means of the associated pointer. According to pointer arithmetic, adding 1 to a pointer results in the memory address following the one contained in the pointer. From this perspective, since a contains the address of a[0], the operation a+1 yields the address of a[1], that is, &a[1], and, correspondingly, *(a+1) represents the value a[1]. In general, the element a[i] may also be referenced by the indirection *(a+i), and this is the actual way in which the compiler operates with references to array elements.

Although simple and easy to operate with, the static allocation of arrays is not always suitable. The storage requirements of large present-day computer applications can be very demanding and the efficient use of the primary memory often implies reusing it dynamically to store different objects, according to the instant needs. Therefore, a methodology for dynamic allocation of arrays, enabling, in addition, arbitrary offsets and index ranges, adapted to the natural mathematical notations, is highly useful.

The technical solution to the dynamic memory allocation in C/C++ is provided by the function `malloc`, defined in `stdlib.h`:

```
void *malloc((size_t) n)
```

`malloc` returns a pointer to an uninitialized contiguous block of n bytes in the *heap* (memory region used for dynamic objects), or NULL if the request cannot be satisfied ((size_t) converts n into the type used in C for sizing memory objects. In particular, to obtain a pointer to a memory block of sufficient size to hold n components of type `double`, one may issue the call

```
(double*) malloc((size_t) (n*sizeof(double)));
```

The function `sizeof(type)` returns the length in bytes of the internal representation of the type *type* and, hence, n*sizeof(double) represents the number of bytes necessary to store n components of type `double`. Finally, (double*) converts the unspecialized pointer returned by `malloc` into a pointer to locations of type `double`.

With these elements at hand, we build the function `Vector0` (see Listing 2.11), which dynamically allocates a block of n `double`-type locations, and returns the starting address of the block via its name. If the allocation fails, possibly due to the unavailability of a suitable contiguous block in the heap, the execution is terminated by a call to the standard function `exit()`, and an error code is returned to the operation system for further action.

In the calling program, the pointer returned by `Vector0` may be assigned to an array, whose components then correspond to successive locations in the allocated memory block. The array's name must be first declared as a pointer and, once the size n becomes available, the array is effectively allocated by the

Listing 2.11   Example of Dynamic Allocation of an Array (C/C++ Coding)

```c
double *Vector0(int n)
{
   double *p;
                                         // assign block start to array pointer
   p = (double*) malloc((size_t) (n*sizeof(double)));
   if (!p) {
      printf("Vector0: allocation error !\n");
      exit(1);
   }
   return p;
}

int main()
{
   int i, n;
   double *a;                            // array name is a pointer

   n = ...;                              // array size results during execution
   a = Vector0(n);                       // a points to the allocated block

   for (i=0; i<=n-1; i++) a[i] = ...;    // a is used as array
}
```

statement a = Vector0(n). Further on, a can be handled as a usual null-offset array, with the valid components a[0] through a[n-1].

In practice, it is of particular interest to dynamically allocate arrays indexed not between 0 and n-1, but rather between some arbitrary imin and imax. This is precisely the task solved by the function Vector given in Listing 2.12. The number of components to be allocated in this case is imax-imin+1.

Listing 2.12   Functions for the Dynamic Allocation/Deallocation of a Vector (C/C++ Coding)

```c
//===============================================================================
double *Vector(int imin, int imax)
//-------------------------------------------------------------------------------
// Allocates a double vector with indices in the range [imin,imax]
//-------------------------------------------------------------------------------
{
   double *p;
                                         // assign block start to array pointer
   p = (double*) malloc((size_t) ((imax-imin+1)*sizeof(double)));
   if (!p) {
      printf("Vector: allocation error !\n");
      exit(1);
   }
   return p - imin;                      // adjust for offset
}

//===============================================================================
void FreeVector(double *p, int imin)
//-------------------------------------------------------------------------------
// Deallocates a double vector allocated with Vector, with offset imin
//-------------------------------------------------------------------------------
{
   free((void*) (p+imin));               // compensate for offset
}
```

A more subtle aspect concerns the fact that, by the assignment a = Vector(...), the beginning of the block is automatically associated with the component a[0] and by no means with a[imin], as desired. Since the elements a[0] through a[imin-1] are anyway irrelevant, a simple solution to this problem is to instruct the function Vector to artificially return an address with imin locations ahead of the true block's origin. This address being associated by default with a[0] by way of the array's name, the beginning of the allocated block is actually set in correspondence with a[imin]. Technically, the artificial shift of the block's origin is achieved in Vector by simply subtracting imin from the pointer to be returned.

Like any other routine creating dynamic objects in the heap, the function Vector is supposed to be complemented with a pair function designed to dispose the created objects (once they become unnecessary) and to free the heap block for further use. The function FreeVector, provided in Listing 2.12, exactly plays this role, namely, it releases the heap zone allocated by a previous call to Vector by calling the standard function free. As a first thing, however, FreeVector recovers the true origin of the allocated block by adding imin to the received pointer, to compensate for its artificial shift applied in function Vector. The pointer is then converted into the generic type void* before being disposed.

A typical use of the presented allocation/deallocation routines is given in Listing 2.13. Any attempt to employ the components a[0] through a[imin-1], located outside the allocated block, is hazardous and may produce unexpected results.

Let us now generalize the one-dimensional technique developed so far, for the case of 2D arrays. Adopting a storage model similar to the one implemented by Press et al. (2002), the name of a 2D array is a *pointer to an array of row pointers*, each pointing to the first element on one of the rows of the 2D array (see Figure 2.3). Hence, the correct declaration of a 2D array with double components is double **a and reflects the "pointer to pointers" philosophy. The array as a whole is referenced by its name (or pointer) a, the entire row i is referred to by the row pointer a[i], and a particular element is accessed as a[i][j].

The function Matrix presented in Listing 2.14 is designed to allocate a matrix with double-type components, with the row index ranging from imin to imax and the column index ranging from jmin to jmax. The routine starts by defining the numbers of rows and columns ni = imax-imin+1 and nj = jmax-jmin+1, respectively. Then it defines a pointer p to an array of ni row pointers, shifting its starting address by imin to account for the row offset. Next, the routine allocates a block of ni*nj double-type locations for the array components and assigns the block's origin (shifted to account for the row offset) to the first row pointer. To point correctly to the beginning of each successive row, the addresses assigned to the rest of the row pointers are evenly incremented by the row length nj. The pair of Matrix is FreeMatrix and it deallocates in reversed order the allocated objects: first the main block and then the array of row pointers.

Listing 2.13   Example of Dynamic Allocation of a Vector (C/C++ Coding)

```
int i, imin, imax;
double *a;                                    // array name is a pointer

imin = ...;                      // min. and max. indexes (during execution)
imax = ...;
                    // allocate heap block and assign starting address to a
a = Vector(imin,imax);

for (i=imin; i<=imax; i++) a[i] = ...;               // use a as array

FreeVector(a,imin);                     // free heap block occupied by a
```

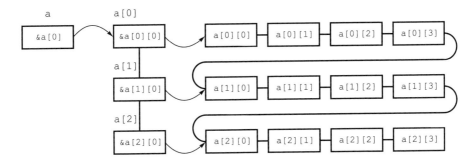

FIGURE 2.3 The storage scheme of a 3 × 4 matrix. The name of the array is a pointer to an array of row pointers, each of which contains the address of the first element in the corresponding row.

Listing 2.14 Functions for the Dynamic Allocation/Deallocation of a Matrix (C/C++ Coding)

```
//==============================================================================
double **Matrix(int imin, int imax, int jmin, int jmax)
//------------------------------------------------------------------------------
// Allocates a double matrix, with row and column indices in the range
// [imin,imax] x [jmin,jmax]
//------------------------------------------------------------------------------
{
   int i, ni = imax-imin+1, nj = jmax-jmin+1;   // numbers of rows and columns
   double **p;
                                                 // allocate array of row pointers
   p = (double**) malloc((size_t)(ni*sizeof(double*)));
   if (!p) {
      printf("Matrix: level 1 allocation error !\n");
      exit(1);
   }
   p -= imin;                                    // adjust for row offset
                                        // assign block start to 1st row pointer
   p[imin] = (double*) malloc((size_t)(ni*nj*sizeof(double)));
   if (!p[imin]) {
      printf("Matrix: level 2 allocation error !\n");
      exit(2);
   }
   p[imin] -= jmin;                    // adjust 1st row pointer for column offset
                                       // define row pointers spaced by row length
   for (i=imin+1; i<=imax; i++) p[i] = p[i-1] + nj;

   return p;
}

//==============================================================================
void FreeMatrix(double **p, int imin, int jmin)
//------------------------------------------------------------------------------
// Deallocates a double matrix allocated with Matrix, with row and column
// offsets imin and jmin
//------------------------------------------------------------------------------
{
   free((void*) (p[imin]+jmin));                 // deallocate block
   free((void*) (p+imin));                       // deallocate array of row pointers
}
```

A typical context of using the functions `Matrix` and `FreeMatrix` is the following:

```
int i, imin, imax, j, jmin, jmax;
double **a;                         // array name is a pointer to row pointers

imin = ...; imax = ...;             // min. and max. indexes (during execution)
jmin = ...; jmax = ...;
                            // allocate heap block and assign starting address to a
a = Matrix(imin,imax,jmin,jmax);

for (i=imin; i<=imax; i++)
    for (j=jmin; j<=jmax; j++)  a[i][j] = ...;          // use a as array

FreeMatrix(a,imin,jmin);                        // free heap block occupied by a
```

Here, again, the actual limiting indexes are assumed to become available only at run time. Following the actual allocation, the array's name can be used as a regular array. If necessary, the allocated heap block can be reused after deallocation.

The functions `Vector`, `FreeVector`, `Matrix`, and `FreeMatrix` can also serve as templates for building allocation/deallocation functions for arrays of other types, which practically amount to merely replacing the type `double` with the appropriate type. In fact, the book also makes use of integer and complex vectors and matrices. The functions dealing with `int`-type arrays are named `IVector`, `FreeIVector`, `IMatrix`, and `FreeIMatrix`, while those operating with complex arrays are called `CVector`, `FreeCVector`, `CMatrix`, and `FreeCMatrix`. Owing to their general utility, they are collected in the header file `memalloc.h`, listed in full length in Appendix A, and which is included in all our further C/C++ programs requiring dynamic allocation of arrays.

The overall structure of the file `memalloc.h` is indicated in Listing 2.15. For the sake of compatibility with ANSI C, we define a `complex` type named `dcmplx`, with real and imaginary components of type `double`. With a view to avoiding multiple definitions of the contained functions, possibly generated by several include directives, `memalloc.h` implements the standard C/C++ technique of conditional inclusion based on directives addressed to the preprocessor. The preprocessor, working on the code before compilation pretty much like a text editor, checks upon each inclusion attempt that the distinctive variable _MEMALLOC_ is not already defined (`#ifndef _MEMALLOC_`) and, if so, continues with defining it and including the content of the header file up to the directive `#endif` in the edited code. On the contrary, on subsequent inclusion attempts, the preprocessor finds the variable _MEMALLOC_ already defined and abandons the inclusion of `memalloc.h`, thereby its content being effectively included only once.

The same conditional inclusion technique is consistently used throughout the book for collecting all the developed routines in libraries pertaining to each chapter. Obviously, the inclusion of each such header file is controlled by a distinct variable.

2.7 Basic Matrix Operations

Elementary matrix operations, such as summation, multiplication, and transposition, offer simple and illustrative examples for programs importing procedures from modules or header files. Even though the concrete implementation of such routines is not of primary practical interest, we present two libraries based on them (`matutil.py` and `matutil.h`), which are included below in Python and C/C++ main programs for checking the simple matrix identity:

$$(\mathbf{A} \cdot \mathbf{B})^T - \mathbf{B}^T \cdot \mathbf{A}^T = \mathbf{0}, \tag{2.1}$$

where $\mathbf{A}$ and $\mathbf{B}$ are arbitrary (random) matrices, and $\mathbf{0}$ is the null matrix.

Listing 2.15 Functions for Dynamic Allocation of Vectors and Matrices (`memalloc.h` Partial)

```
//------------------------------ memalloc.h ------------------------------
// Functions for dynamic memory allocation of vectors and matrices
//------------------------------------------------------------------------
#ifndef _MEMALLOC_
#define _MEMALLOC_

#include <stdlib.h>
#include <stdio.h>
#include <complex>

using namespace std;
typedef complex<double> dcmplx;

double *Vector(int imin, int imax)
// Allocates a double vector with indices in the range [imin,imax]
...
void FreeVector(double *p, int imin)
// Deallocates a double vector allocated with Vector, with offset imin
...
double **Matrix(int imin, int imax, int jmin, int jmax)
// Allocates a double matrix, with row and column indices in the range
// [imin,imax] x [jmin,jmax]
...
void FreeMatrix(double **p, int imin, int jmin)
// Deallocates a double matrix allocated with Matrix, with row and column
// offsets imin and jmin
...
int *IVector(int imin, int imax)
// Allocates an int vector with indices in the range [imin,imax]
...
void FreeIVector(int *p, int imin)
// Deallocates an int vector allocated with IVector, with offset imin
...
int **IMatrix(int imin, int imax, int jmin, int jmax)
// Allocates an int matrix, with row and column indices in the range
// [imin,imax] x [jmin,jmax]
...
void FreeIMatrix(int **p, int imin, int jmin)
// Deallocates an int matrix allocated with IMatrix, with row and column
// offsets imin and jmin
...
dcmplx *CVector(int imin, int imax)
// Allocates a complex vector with indices in the range [imin,imax]
...
void FreeCVector(dcmplx *p, int imin)
// Deallocates a complex vector allocated with CVector, with offset imin
...
dcmplx **CMatrix(int imin, int imax, int jmin, int jmax)
// Allocates a complex matrix, with row and column indices in the range
// [imin,imax] x [jmin,jmax]
...
void FreeCMatrix(dcmplx **p, int imin, int jmin)
// Deallocates a complex matrix allocated with CMatrix, with row and column
// offsets imin and jmin
...

#endif
```

The Python module containing the auxiliary functions is given in Listing 2.16, while the corresponding C/C++ header file is given in Listing 2.17. Both files are actually partial versions of more comprehensive libraries, listed in full length in Appendix B.

Listing 2.16   Routines for Basic Matrix Operations (`matutil.py` Partial)

```python
#--------------------------------- matutil.h ---------------------------------
#  Contains utility routines for basic operations with vectors and matrices
#-----------------------------------------------------------------------------
from math import *

#=============================================================================
def MatPrint(a, n, m):
#-----------------------------------------------------------------------------
#  Prints the elements of matrix a[1:n][1:m] on the display
#-----------------------------------------------------------------------------
   for i in range(1,n+1):
      for j in range (1,m+1): print('{0:11.2e}'.format(a[i][j]),end="")
      print()

#=============================================================================
def MatTrans(a, n):
#-----------------------------------------------------------------------------
#  Replaces the square matrix a[1:n][1:n] by its transpose
#-----------------------------------------------------------------------------
   for i in range(2,n+1):
      for j in range(1,i):
         t = a[i][j]; a[i][j] = a[j][i]; a[j][i] = t

#=============================================================================
def MatDiff(a, b, c, n, m):
#-----------------------------------------------------------------------------
#  Returns the difference of matrices a and b in c[1:n][1:m]
#-----------------------------------------------------------------------------
   for i in range(1,n+1):
      for j in range (1,m+1): c[i][j] = a[i][j] - b[i][j]

#=============================================================================
def MatProd(a, b, c, n, l, m):
#-----------------------------------------------------------------------------
#  Returns the product of matrices a[1:n][1:l] and b[1:l][1:m] in c[1:n][1:m]
#-----------------------------------------------------------------------------
   for i in range(1,n+1):
      for j in range (1,m+1):
         t = 0e0
         for k in range(1,l+1): t += a[i][k] * b[k][j]
         c[i][j] = t

#=============================================================================
def MatNorm(a, n, m):
#-----------------------------------------------------------------------------
#  Returns the max norm of matrix a[1:n][1:m], i.e. max|a[i][j]|
#-----------------------------------------------------------------------------
   norm = 0e0
   for i in range(1,n+1):
      for j in range(1,m+1):
         if (norm < fabs(a[i][j])): norm - fabs(a[i][j])
   return norm
```

Listing 2.17 Routines for Basic Matrix Operations (`matutil.h` Partial)

```
//-------------------------- matutil.h --------------------------
// Contains utility routines for basic operations with vectors and matrices
//-----------------------------------------------------------------
#ifndef _MATUTIL_
#define _MATUTIL_

#include <stdio.h>


//================================================================
void MatPrint(double **a, int n, int m)
//-----------------------------------------------------------------
// Prints the elements of matrix a[1:n][1:m] on the display
//-----------------------------------------------------------------
{
   int i, j;

   for (i=1; i<=n; i++) {
      for (j=1; j<=m; j++) printf("%12.2e",a[i][j]);
      printf("\n");
   }
}


//================================================================
void MatTrans(double **a, int n)
//-----------------------------------------------------------------
// Replaces the square matrix a[1:n][1:n] by its transpose
//-----------------------------------------------------------------
{
   double t;
   int i, j;

   for (i=2; i<=n; i++)
      for (j=1; j<=(i-1); j++) {
         t = a[i][j]; a[i][j] = a[j][i]; a[j][i] = t;
      }
}


//================================================================
void MatDiff(double **a, double **b, double **c, int n, int m)
//-----------------------------------------------------------------
// Returns the difference of matrices a and b in c[1:n][1:m]
//-----------------------------------------------------------------
{
   int i, j;

   for (i=1; i<=n; i++)
      for (j=1; j<=m; j++) c[i][j] = a[i][j] - b[i][j];
}


//================================================================
void MatProd(double **a, double **b, double **c, int n, int l, int m)
//-----------------------------------------------------------------
// Returns the product of matrices a[1:n][1:l] and b[1:l][1:m] in c[1:n][1:m]
//-----------------------------------------------------------------
{
   double t;
   int i, j, k;
```

```
      for (i=1; i<=n; i++)
         for (j=1; j<=m; j++) {
            t = 0e0;
            for (k=1; k<=1; k++) t += a[i][k] * b[k][j];
            c[i][j] = t;
         }
   }


//==============================================================
double MatNorm(double **a, int n, int m)
//--------------------------------------------------------------
// Returns the max norm of matrix a[1:n][1:m], i.e. max|a[i][j]|
//--------------------------------------------------------------
{
   double norm;
   int i, j;

   norm = 0e0;
   for (i=1; i<=n; i++)
      for (j=1; j<=m; j++)
         if (norm < fabs(a[i][j])) norm = fabs(a[i][j]);
   return norm;
}

#endif
```

The function `MatPrint` displays matrices in formatted tabular form. While in Python, iterative printing of several items on the same line is obtained by explicitly inhibiting the default new line after each item, in C/C++, one specifically sends a final newline character to complete the line.

`MatTrans` transposes its square input matrix of order n by swapping the pairs of symmetric elements. The function `MatDiff` returns the difference between two n×m matrices, a and b. The output matrix c can safely coincide with either of the input matrices. The function `MatProd` calculates the product of the n×1 matrix a with the 1×m matrix b. The n×m result is returned in array c.

Last but not the least, the function `MatNorm` returns the max-norm of the input matrix:

$$\|A\|_{\max} = \max\{|a_{ij}|\},$$

that is, the maximum absolute value of its elements.

The Python and C/C++ versions of the program for checking the matrix identity (2.1), given in Listings 2.18 and 2.19, request as sole input the size of the square matrices **A** and **B**. The program then

Listing 2.18   Check of Matrix Identity (Python Coding)

```
# Checks identity (A B)_trans = B_trans A_trans for random arrays A and B
from random import *
from matutil import *

# main

n = int(input("n = "))

A = [[0]*(n+1) for i in range(n+1)]                    # define arrays
B = [[0]*(n+1) for i in range(n+1)]
C = [[0]*(n+1) for i in range(n+1)]
D - [[0]*(n+1) for i in range(n+1)]
```

```
for i in range(1,n+1):                    # array A: random sub-unitary elements
    for j in range(1,n+1): A[i][j] = random()
print("Array A:")
MatPrint(A,n,n)

for i in range(1,n+1):                    # array B: random sub-unitary elements
    for j in range(1,n+1): B[i][j] = random()
print("Array B:")
MatPrint(B,n,n)

MatProd(A,B,C,n,n,n)                                                        # A*B
MatTrans(C,n)                                                        # (A*B)_trans

MatTrans(A,n)                                                             # A_trans
MatTrans(B,n)                                                             # B_trans
MatProd(B,A,D,n,n,n)                                             # B_trans* A_trans

MatDiff(C,D,D,n,n)                            # (A*B)_trans - B_trans * A_trans
print("Norm ((A*B)_trans - B_trans * A_trans) = ",MatNorm(D,n,n))
```

Listing 2.19   Check of Matrix Identity (C/C++ Coding)

```
// Checks identity (A B)_trans = B_trans A_trans for random matrices A and B
#include <stdio.h>
#include "memalloc.h"
#include "matutil.h"

int main()
{
   double **A, **B, **C, **D;
   int i, j, n;

   printf("n = "); scanf("%i",&n);

   A = Matrix(1,n,1,n); B = Matrix(1,n,1,n);            // allocates arrays
   C = Matrix(1,n,1,n); D = Matrix(1,n,1,n);

   for (i=1; i<=n; i++)                  // array A: random sub-unitary elements
      for (j=1; j<=n; j++) A[i][j] = rand()/(RAND_MAX+1e0);
   printf("Array A:\n");
   MatPrint(A,n,n);

   for (i=1; i<=n; i++)                  // array B: random sub-unitary elements
      for (j=1; j<=n; j++) B[i][j] = rand()/(RAND_MAX+1e0);
   printf("Array B:\n");
   MatPrint(B,n,n);

   MatProd(A,B,C,n,n,n);                                                  // A*B
   MatTrans(C,n);                                                 // (A*B)_trans

   MatTrans(A,n);                                                      // A_trans
   MatTrans(B,n);                                                      // B_trans
   MatProd(B,A,D,n,n,n);                                      // B_trans* A_trans

   MatDiff(C,D,D,n,n);                          // (A*B)_trans - B_trans * A_trans
   printf("Norm ((A*B)_trans - B_trans * A_trans) = %e\n",MatNorm(D,n,n));
}
```

allocates four arrays, A and B, which are initialized with random subunitary values, and C and D, to hold the two terms of the identity, that is, $\mathbf{C} = (\mathbf{A} \cdot \mathbf{B})^T$ and $\mathbf{D} = \mathbf{B}^T \cdot \mathbf{A}^T$, which are calculated separately. Instead of checking in element-wise manner the cancellation of the difference $\mathbf{C} - \mathbf{D}$, the program prints out a more informative and easy-to-check quantity, namely, the *norm of the difference*, that is, the maximum absolute value of the elements of the difference. Obviously, independently of the size of the matrices, the expected norm is numerically zero, that is, of the order of 10^{-15}.

Besides the routines employed in the above example program, the actual library modules `matutil.py` and `matutil.h` contain several other functions for operations with matrices: `MatRead`–reads in the elements of a matrix from the keyboard, `MatZero`—zeros the elements of a matrix, `MatCopy`—copies a matrix into another matrix, and `MatPow`—raises a matrix to a power. Yet another set of functions contained in the modules operate with vectors: `VecPrint`—prints the elements of a vector on the display, `VecZero`—zeros the elements of a vector, `VecCopy`—copies a vector into another vector, `VecDiff`—returns the difference between two vectors, `VecNorm`—calculates the norm of a vector, and, finally, `MatVecProd`—returns the matrix product between a matrix and a vector.

References and Suggested Further Reading

Beu, T. A. 2004. *Numerical Calculus in C* (3rd ed., in Romanian). Cluj-Napoca: MicroInformatica.

Kernighan, B. and D. M. Ritchie. 1988. *The C Programming Language* (2nd ed.). Englewood Cliffs, NJ: Prentice-Hall.

Knuth, D. E. 1998. *The Art of Computer Programming* (3rd ed.), vol. 1: Fundamental Algorithms. Reading, MA: Addison-Wesley Professional.

Press, W. H., S. A. Teukolsky, W. T. Vetterling, and B. P. Flannery. 2002. *Numerical Recipes: The Art of Scientific Computing* (2nd ed.). Cambridge: Cambridge University Press.

3

Elements of Scientific Graphics

3.1 The Tkinter Package

Tk is an open-source *cross-platform toolkit* for building graphical user interfaces (GUIs) for various programming languages. Developed in 1991 by John Ousterhout, Tk provides a library of basic *widgets* (such as frames, buttons, menus, etc.), which are commonly used in modern application software. *Widget* is the generic term used for the building blocks of GUI applications, or, in general, for stand-alone applications that can be embedded in third-party applications.

Tkinter (from "Tk interface") is the object-oriented *interface* of Python to the Tk toolkit. Tkinter was written by Fredrik Lundh in 1999 (Lundh, 2013a) and has become Python's standard GUI package, being available, along with Tk, on most Unix-like, Windows, and Macintosh systems. Tkinter consists of several Python modules, the most important being the `tkinter` module. It contains all the classes and functions required to work with the Tk toolkit.

The simplest Tkinter application is the one creating an empty Tk widget (Listing 3.1). Any program resting upon Tkinter first needs to import the `tkinter` module. Tkinter is initialized by creating a Tk *root widget*, and this is accomplished in our sample program by the statement `root = Tk()`. The root widget (see Figure 3.1) is, basically, an ordinary window with the "look and feel" specific to the particular operating system and window manager that are being used. Notably, any program should operate with a single root widget, and this should be created prior to any other widgets, which would relate to the root as subordinated "child" objects.

With the code composed of just the first two statements and the root widget merely created, the window would not be effectively displayed. For the application window to appear, the execution must enter the Tkinter *event loop* and this is implemented as a call to the method `root.mainloop()`. The program will stay idle in the event loop until the window is closed. In the broader picture, the event loop handles the operations queued by Tkinter along with events triggered by the user (e.g., mouse clicks) or the windowing system (e.g., window redraws). Typical operations queued by Tkinter concern widget resizing (as initiated by the `pack` method) and widget redrawing (as caused by the `update` method).

The root widget can control other subordinated widgets. A Label widget, for instance, can be used to display text or an image on the screen. In Listing 3.2, we create a Label widget as a child to the root

Listing 3.1 Creating a Tk Root Widget (Python Coding)

```
from tkinter import *        # import Tkinter module

root = Tk()                  # create Tk root widget

root.mainloop()              # enter Tkinter event loop
```

FIGURE 3.1   Empty Tk widget created with Tkinter.

Listing 3.2   Creating a `tkinter` Label Widget (Python Coding)

```
from tkinter import *                              # import Tkinter module

root = Tk()                                        # create Tk root widget
                                    # create label widget: child to root
w = Label(root, text="This is a Label widget")
w.pack()                             # resize label to fit text; make it visible

root.mainloop()                                   # enter Tkinter event loop
```

FIGURE 3.2   `Label` widget created with Tkinter.

window. By the first argument passed to the constructor, we specify the root as parent and use the `text` option to specify the text to be displayed.

As in the case of the root widget, the mere creation of a child widget does not make it automatically visible. The program needs to call the `pack()` method on each new widget to resize it and to make it visible. In particular, the `pack()` method resizes the `Label` widget to fit the text and renders it visible. The packer is one of Tk's geometry management mechanisms that queue operations in the event loop. The `Label` widget created by Program 3.2 can be seen in Figure 3.2.

3.2 The Canvas Widget

The Canvas is a general-purpose widget, providing versatile graphics facilities. The Canvas widget is most appropriate for drawing pictures, graphs, plots, and even more complex layouts, and also for creating custom widgets.

The image displayed on the Canvas results by the successive creation of *canvas items*. The items are placed in a *stack*, with the new items drawn, by default, on top of those already existing. The standard items supported by the Canvas widget are arc, bitmap, image, line, oval, polygon, rectangle, text, and window. Several items defined by closed boundaries (such as chords, pie slices, ovals, polygons, and rectangles) feature an *outline* and an *interior area*, which can be colored or made transparent separately. Window items can be used to display other Tkinter widgets on the canvas.

For positioning items within its rectangular area, the Canvas widget provides two types of coordinates: the *canvas coordinates* and the *window coordinates*. The existence of the two coordinate systems accounts for the fact that the canvas may sometimes be larger than the actual window displaying it. In such cases, the scrollbars help indeed moving the canvas around in the window, but positioning objects may be more meaningful with respect to a reference point related to the window itself. Specifically, *canvas coordinates* are defined relative to the top left corner of the *entire canvas*, which is assigned the coordinates $(0,0)$, whereas *window coordinates* are defined relative to the top left corner of the *displayed area* of the canvas. Since in the graphical applications developed in the book the canvas is displayed entirely, we will actually make no distinction between the two coordinate systems.

A Canvas object is created according to the syntax:

```
w = Canvas(parent, option=value,...)
```

The first argument specifies the parent object, which is typically the Tk root widget, but can equally well be another subordinated widget. Common options usable with the Canvas widget are

- `bd` = *border width*—width of the canvas border in pixels. The default is 2.
- `bg` = *color*—background color of the canvas in #rgb (4 bit), #rrggbb (8 bit), or #rrrgggbbb (12 bit) format, or as standard color name.
- `height` = *height*—vertical size of the canvas. An integer value is assumed to be in pixels.
- `width` = *width*—horizontal size of the canvas.

3.2.1 Commonly Used Canvas Items

Canvas line objects are used for drawing lines on the canvas. A line can consist of any arbitrary number of segments connected end to end. The syntax for creating a line object connecting a series of points $(x1,y1)$, $(x2,y2)$,..., (xn,yn) on the Canvas widget w is

```
id = w.create_line(x1, y1, x2, y2,..., xn, yn, option=value,...)
```

and returns the `id` of the created item.

Options (selected):

- `arrow` = FIRST, LAST, or BOTH—places an arrowhead on the first, second, or on both end points. The default is a line without arrowheads.
- `dash` = *pattern*—a list of segment lengths (to be repeated if necessary), whereby the odd segments are drawn and the even ones are skipped. The default is a solid line.
- `fill` = *color*—line color. The default is `"black"`.
- `width` = *width*—line width in pixels. The default is 1.

Canvas oval objects are used for drawing ellipses or, in particular, circles on the canvas. The shape and size of the ellipse is defined by specifying the top left corner $(x1,y1)$ and the bottom right corner $(x2,y2)$ of its bounding rectangle. The syntax for creating an oval object on the canvas w is

```
id = w.create_oval(x1, y1, x2, y2, option=value,...)
```

and returns the `id` of the created item.

Options (selected):

- `fill` = *color*—fill color of the ellipse's interior. An empty string " " means "transparent" and this is the default.
- `outline` = *color*—color of the border around the ellipse. The default outline color is `"black"`.
- `width` = *width*—width in pixels of the border around the ellipse. The default is 1.

Canvas rectangle objects are used for drawing rectangles on the canvas, specified by the top left corner (`x1,y1`) and the bottom right corner (`x2,y2`). The syntax for creating a rectangle object on the canvas w is

```
id = w.create_rectangle(x1, y1, x2, y2, option=value,...)
```

and returns the `id` of the created item.

Options (selected):

- `fill` = *color*—fill color of the rectangle's interior. An empty string " " means "transparent" and this is the default.
- `outline` = *color*—color of the border around the rectangle. The default outline color is `"black"`.
- `width` = *width*—width in pixels of the border around the rectangle. The default is 1.

Canvas text objects are used for displaying one or more lines of text on the canvas. The syntax for producing the text on the canvas w at position (`x,y`) is

```
id = w.create_text(x, y, option=value,...)
```

and returns the `id` of the created item.

Options (selected):

- `anchor` = `CENTER`—by default, the text is centered horizontally and vertically about (`x,y`). For example, for `anchor` = `NW`, the upper left corner of the text-bounding rectangle is placed at the reference point (`x,y`). Other legal arguments are `N`, `NE`, `E`, `SE`, `S`, `SW`, and `W`.
- `fill` = *color*—text color. The default is `"black"`.
- `font` = *font*—tuple of font attributes. For example, to create a 14-point Helvetica regular font, one would use `font` = (`"Helvetica",14`).
- `justify` = `LEFT` (default), `CENTER`, or `RIGHT`. For multiline texts, the option controls how the lines are justified.
- `text` = *string*—text to be displayed. The newline character `"\n"` can be used to force line breaks.
- `width` = *width*—maximum line width. Lines longer than this value are wrapped. By default, the text width is set by the longest line.

3.2.2 Methods for Updating Canvas Widgets

In its most rudimentary form, updating or animating the display implies deleting the content of widgets and redrawing it.

The Canvas widget, in particular, supports the method `delete`, which enables deleting selectively or entirely the displayed items. The general syntax is

```
w.delete(item)
```

where *item* is a tag or ID, and it causes the matching items to be deleted. If *all* items drawn on the canvas need to be removed, the *item* should be set to ALL.

Updating the content of widgets is slightly more delicate, because it may possibly affect intertwined processes enqueued in the event loop. Tkinter provides two general methods that enable updating the display. The most straightforward is update, having the syntax

```
w.update()
```

that processes all pending events and redraws the widgets. Nevertheless, it should be used with care, since it may lead to unpredictable effects if called from an inappropriate place (e.g., from an event callback). A typical infinite loop of deleting and redrawing a Canvas widget w is the following:

```
while (1):                              # infinite loop
    w.delete(ALL)                       # delete canvas
    .....                               # drawing operations
    w.update()                          # update canvas
```

The after method, having the syntax

```
w.after(delay_ms, callback=None, *args)
```

registers a *callback function* that will be called with arguments args with a delay of at least delay_ms milliseconds. The actual delay may exceed the requested one, depending on how busy the system is and the callback is only called once. An infinite loop of deleting and redrawing a Canvas widget w could be implemented as a callback function calling itself via the after method:

```
def Clock(w):                           # callback function
    w.delete(ALL)                       # delete canvas
    .....                               # drawing operations
    w.after(10,Clock,w)                 # call callback after 10 ms
```

Once the callback function (Clock, in the above example) is called, it starts calling itself iteratively, however not recursively. The mechanism of exiting the infinite loop should be explicitly coded, or else, the process is terminated once the close button of the widget is pressed. The safer and, therefore, recommendable implementation of repetitive sequences of redrawing Canvas widgets is by means of the after method.

3.3 Simple Tkinter Applications

Program 3.3 creates an empty Canvas widget, without any functionality. The Tk root widget needs to be created before any other object. The Canvas widget may then be created as a child to the root.

For the canvas to become visible, it has to be resized by the packer, pretty much like stretching painting canvas on a frame (Figure 3.3). The overall appearance of the window is finally ensured by the program's entrance into the event loop.

In the next example, presented in Listing 3.4, we use the canvas to display several graphic items: two arrows, a rectangle, an oval, and a piece of text. Once the canvas is packed, these items are created by invoking the corresponding methods. The result is shown in Figure 3.4.

Let us now build a Tkinter widget for displaying an analog clock, like the one in Figure 3.5. Obviously, more than just drawing the graphic elements (clock face and hands), we have to animate the widget, by periodically deleting the moving objects (the clock hands), redrawing them, and updating the widget. The simple idea behind the implementation shown in Listing 3.5 is to use a routine, Clock0, for drawing the face and hands of the clock in their instant positions. An infinite loop, coded in the main program,

Listing 3.3   Creating a Tkinter Canvas Widget (Python Coding)

```
from tkinter import *                                      # import Tkinter module

root = Tk()                                                # create Tk root widget
root.title("Canvas")
                                                           # create canvas: child to root
w = Canvas(root, width=300, height=200, bg = "white")
w.pack()                                                   # resize and make canvas visible

root.mainloop()                                            # enter Tkinter event loop
```

FIGURE 3.3   Canvas widget created with Tkinter.

controls the updating of the canvas, by deleting its content altogether (for simplicity), redrawing the clock by calling the routine Clock0, and applying the update method on the Canvas widget to make the changes visible.

The arguments passed to routine Clock0 are the Canvas object, w, and the widget sizes (in pixels), nx and ny. Clock0 determines from the latter the coordinates x0 and y0 of the center and the half-widths

Listing 3.4   Tkinter Canvas Widget with Simple Graphic Elements (Python Coding)

```
from tkinter import *                                      # import Tkinter module

root = Tk()                                                # create Tk root widget
root.title("Canvas")

nx = 300; ny = 300                                         # canvas size
w = Canvas(root, width=nx, height=ny, bg = "white")        # create canvas
w.pack()                                                   # resize and make canvas visible

w.create_line(150, 150, 100, 100, fill="red" , arrow=LAST, width=2)   # arrows
w.create_line(150, 150, 150, 250, fill="blue", arrow=LAST)

w.create_rectangle(50, 50, 250, 250, width=2)              # square
w.create_oval(50, 50, 250, 250)                            # circle

w.create_text(150, 75, text="(10:30)")                     # text

root.mainloop()                                            # enter Tkinter event loop
```

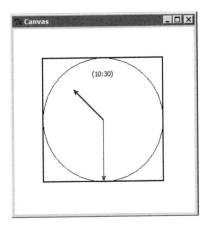

FIGURE 3.4 Canvas widget created with Tkinter, containing simple graphic elements.

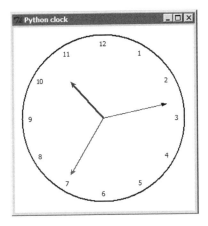

FIGURE 3.5 Python clock widget built with Tkinter.

`lx` and `ly` of the clock's face. These dimensions are then used to position the rest of the graphic elements. The hour labels are placed in equal angular distances of $\pi/6$ on a circle of radius `r0`.

The current time is retrieved using `localtime()`, one of the several time functions provided by the module `time.py`. The routine returns a tuple of nine integers, with the third, fourth, and fifth components representing, respectively, the current hour (range $[0, 23]$), minute, and second. The integer minutes and hours are completed with their fractional parts. The resulting fractional hours `t_h`, minutes `t_m`, and seconds `t_s` are then converted, in sequence, into angular and, finally, Cartesian positions of the arrowheads of the corresponding clock hands.

In a more elaborate implementation of the routine `Clock0`, only the clock hands would be deleted and redrawn at every update, while the clock's face would be drawn only once, at the beginning of the run.

As mentioned in Section 3.2.1, the `update` method is susceptible of producing unexpected results if called from unsuitable places (e.g., from an event callback). In fact, moving the infinite loop (for some reason) to a deeper hierarchical level may result in malfunction of the clock widget. A more robust implementation of the repetitive updating process is based on a callback function, as in the variant presented in Listing 3.6. The routine `Clock0` is no longer called directly from the main program and the infinite loop is replaced by *a single call* to an additional *callback function*, `Clock`, which acts as an interface to `Clock0`. Once `Clock` is called, it starts calling itself iteratively until the window is closed. The delay

Listing 3.5   Python Clock Application Built with Tkinter (Python Coding)

```python
# Analog clock
from tkinter import *                               # import Tkinter module
from math import *
from time import *

def Clock0(w, nx, ny):                              # clock draw function
    x0 = nx/2; lx = 9*nx/20             # center and half-width of clock face
    y0 = ny/2; ly = 9*ny/20
    r0 = 0.9 * min(lx,ly)               # distance of hour labels from center
    r1 = 0.6 * min(lx,ly)                      # length of hour hand
    r2 = 0.8 * min(lx,ly)                      # length of minute hand

    w.create_oval(x0-lx, y0-ly, x0+lx, y0+ly, width=2)          # clock face
    for i in range(1,13):                              # label the clock face
        phi = pi/6 * i                         # angular position of label
        x = x0 + r0 * sin(phi)                # Cartesian position of label
        y = y0 - r0 * cos(phi)
        w.create_text(x, y, text=str(i))                       # hour label

    t = localtime()                                            # current time
    t_s = t[5]                                                    # seconds
    t_m = t[4] + t_s/60                                           # minutes
    t_h = t[3] % 12 + t_m/60                                 # hours [0,12]

    phi = pi/6 * t_h                                       # hour hand angle
    x = x0 + r1 * sin(phi)                           # position of arrowhead
    y = y0 - r1 * cos(phi)                                # draw hour hand
    w.create_line(x0, y0, x, y, arrow=LAST, fill="red", width=3)

    phi = pi/30 * t_m                                    # minute hand angle
    x = x0 + r2 * sin(phi)                           # position of arrowhead
    y = y0 - r2 * cos(phi)                               # draw minute hand
    w.create_line(x0, y0, x, y, arrow=LAST, fill="blue", width=2)

    phi = pi/30 * t_s                                    # second hand angle
    x = x0 + r2 * sin(phi)                           # position of arrowhead
    y = y0 - r2 * cos(phi)
    w.create_line(x0, y0 , x, y, arrow=LAST)               # draw second hand

# main

root = Tk()                                          # create Tk root widget
root.title("Python clock")

nx = 300; ny = 300                                         # canvas size
w = Canvas(root, width=nx, height=ny, bg = "white")      # create canvas
w.pack()                                             # make canvas visible

while (1):                                              # infinite loop
    w.delete(ALL)                                      # delete canvas
    Clock0(w, nx, ny)                                    # draw clock
    w.update()                                         # update canvas

root.mainloop()                                     # enter Tkinter event loop
```

Listing 3.6 Python Clock Using Callback Function (Python Coding)

```
# Analog clock with callback function
from tkinter import *                    # import Tkinter module
from math import *
from time import *

def Clock0(w, nx, ny):                   # clock draw function
    .....                                # drawing operations

def Clock(w, nx, ny):                    # clock callback function
    w.delete(ALL)                        # delete canvas
    Clock0(w, nx, ny)                    # draw clock
    w.after(10, Clock, w, nx, ny)        # call callback after 10 ms

# main

root = Tk()                              # create Tk root widget
root.title("Python clock")

nx = 300; ny = 300                       # canvas size
w = Canvas(root, width=nx, height=ny, bg = "white")   # create canvas w
w.pack()                                 # make canvas visible

Clock(w, nx, ny)                         # call clock function

root.mainloop()                          # enter Tkinter event loop
```

for the callback call (10 ms in our example) should be chosen as sufficiently small to ensure at least one updating of the clock widget per second.

3.4 Plotting Functions of One Variable

Let us consider the problem of plotting a real function of one variable,

$$f : [x_{\min}, x_{\max}] \to [y_{\min}, y_{\max}],$$

tabulated for a set of n arguments x_i:

$$f(x_i) = y_i, \quad i = 1, 2, \ldots, n.$$

In general, the plot of the function onto the graphic display cannot make direct use of the problem-specific coordinates (x_i, y_i), called *user coordinates* in the following diagrams. To fit into the admissible ranges of arguments used by the Canvas methods of Tkinter, the user coordinates have to be first transformed into *canvas coordinates* (see Figure 3.6).

With a view to having a proportionate graphical representation on the screen, there must exist a linear relationship between the user coordinates and the canvas coordinates:

$$\bar{x} = A_x x + B_x$$
$$\bar{y} = A_y y + B_y.$$

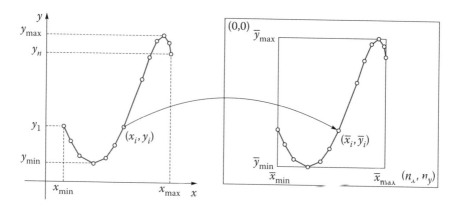

FIGURE 3.6 There exists a *linear correspondence* between the *user coordinates* (x_i, y_i) (on the left) and the *canvas coordinates* $(\bar{x}_i, \bar{y}_i)$ (on the right).

Considering, in particular for the x-axis, that the limits x_{min} and x_{max} of the user domain can be determined from the set of arguments x_i, and that the limits $\bar{x}_{min}$ and $\bar{x}_{max}$ of the canvas domain are set independently, in accordance with the desired positioning of the plot, the scaling coefficients A_x and B_x can be determined from the linear conditions:

$$\bar{x}_{min} = A_x x_{min} + B_x$$
$$\bar{x}_{max} = A_x x_{max} + B_x,$$

wherefrom, the scaling coefficients for the x-axis are

$$A_x = (\bar{x}_{max} - \bar{x}_{min})/(x_{max} - x_{min})$$
$$B_x = \bar{x}_{min} - A_x x_{max}. \tag{3.1}$$

Analogously, the scaling coefficients A_y and B_y for the y-axis are given by

$$A_y = (\bar{y}_{max} - \bar{y}_{min})/(y_{max} - y_{min})$$
$$B_y = \bar{y}_{min} - A_y y_{min}. \tag{3.2}$$

There exists, though, a lack of symmetry between the x- and y-axes. This difference stems, on the one hand, from the fact that, while the set of arguments x_i is supposed to be ascendingly ordered, with x_{min} and x_{max} coinciding, respectively, with x_1 and x_n, the set of ordinates y_i is not necessarily ordered (except for the case of monotonic functions), and hence, the limiting values y_{min} and y_{max} have to be explicitly determined. On the other hand, since the vertical axes in user and canvas coordinates have opposite directions, ascending canvas $\bar{y}$-coordinates actually decrease (down to 0) for increasing y-coordinates.

With the scaling coefficients given by Equations 3.1 and 3.2, the canvas coordinates corresponding to the n-tabulated function values are determined from the relations:

$$\bar{x}_i = A_x x_i + B_x, \ i = 1, 2, \ldots, n$$
$$\bar{y}_i = A_y y_i + B_y.$$

The minimal Python graphics library `graphlib0.py` presented in Listing 3.7 is meant to illustrate the basic concepts underlying scientific graphics, by providing elementary functionalities for plotting

Listing 3.7 Minimal Python Library for Plotting Functions (graphlib0.py)

```python
#----------------------------- graphlib0.py -----------------------------
# Graphics library based on Tkinter
#------------------------------------------------------------------------
from math import *
from tkinter import *

root = Tk()                                         # create Tk root widget
root.title("graphlib0 v1.0")


#========================================================================
def MainLoop():                                     # creates Tk event loop
   root.mainloop()


#========================================================================
def GraphInit(nxwin, nywin):         # creates Canvas widget and returns object
   global w, nxw, nyw                         # make canvas object available

   nxw = nxwin; nyw = nywin                                    # canvas size
   w = Canvas(root, width=nxw, height=nyw, bg = "white")     # create canvas
   w.pack()                                          # make canvas visible
   return w

def Nint(x): return int(floor(x + 0.5))                    # nearest integer


#========================================================================
def Plot0(x, y, n, col, fxmin, fxmax, fymin, fymax, xtext, ytext, title):
#------------------------------------------------------------------------
#  Plots a real function of one variable specified by a set of (x,y) points.
#
#  x[]    - abscissas of tabulation points (x[1] through x[n])
#  y[]    - ordinates of tabulation points (y[1] through y[n])
#  n      - number of tabulation points
#  col    - plot color ("red", "green", "blue" etc.)
#  fxmin  - min fractional x-limit of viewport (0 < fxmin < fxmax < 1)
#  fxmax  - max fractional x-limit of viewport
#  fymin  - min fractional y-limit of viewport (0 < fymin < fymax < 1)
#  fymax  - max fractional y-limit of viewport
#  xtext  - x-axis title
#  ytext  - y-axis title
#  title  - plot title
#------------------------------------------------------------------------
   global w, nxw, nyw                             # canvas object and size

   xmin = min(x[1:n+1]); xmax = max(x[1:n+1])         # user domain limits
   ymin = min(y[1:n+1]); ymax = max(y[1:n+1])
                                            # corrections for horizontal plots
   if (ymin == 0.0 and ymax == 0.0): ymin = -1e0; ymax = 1e0
   if (ymin == ymax): ymin *= 0.9; ymax *= 1.1

   ixmin = Nint(fxmin*nxw); iymin = Nint((1e0-fymin)*nyw)    # canvas domain
   ixmax = Nint(fxmax*nxw); iymax = Nint((1e0-fymax)*nyw)
   w.create_rectangle(ixmin,iymax,ixmax,iymin)            # draw plot frame

   w.create_text((ixmin+ixmax)/2,iymin+10,text=xtext,anchor=N) # x-axis title
   w.create_text(ixmin-10,(iymin+iymax)/2,text=ytext,anchor=E) # y-axis title
   w.create_text((ixmin+ixmax)/2,iymax-10,text=title,anchor=S)  # plot title
                                                    # labels axes ends
   w.create_text(ixmin,iymin+10,text="{0:5.2f}".format(xmin),anchor=NW)
```

```
      w.create_text(ixmax,iymin+10,text="{0:5.2f}".format(xmax),anchor=NE)
      w.create_text(ixmin-10,iymin,text="{0:5.2f}".format(ymin),anchor=E)
      w.create_text(ixmin-10,iymax,text="{0:5.2f}".format(ymax),anchor=E)

      ax = (ixmax-ixmin)/(xmax-xmin)                    # x-axis scaling coefficients
      bx = ixmin - ax*xmin
      ay = (iymax-iymin)/(ymax-ymin)                    # y-axis scaling coefficients
      by = iymin - ay*ymin
                                                                        # draw axes
      if (xmin*xmax < 0): w.create_line(Nint(bx),iymin,Nint(bx),iymax)  # y-axis
      if (ymin*ymax < 0): w.create_line(ixmin,Nint(by),ixmax,Nint(by))  # x-axis

      ix0 = Nint(ax*x[1]+bx); iy0 = Nint(ay*y[1]+by)                     # 1st point
      for i in range(2,n+1):
         ix = Nint(ax*x[i]+bx); iy = Nint(ay*y[i]+by)                    # new point
         w.create_line(ix0,iy0,ix,iy,fill=col)                          # draw line
         ix0 = ix; iy0 = iy                                             # save point
```

real functions of one variable. One of the basic ideas is to "hide" the technicalities involved by invoking the Tkinter methods underneath of simple wrapper functions, which should be called instead. Second, to simplify the calling and the argument lists of the graphic functions, context-related variables (such as Tk root object, canvas object, canvas size, etc.) are conveyed between the library's functions behind the scenes, by `global` variables. Third, taking advantage of the remarkable portability of Python, the library's functions are built so as to be readily embeddable in programs written in other languages, such as C++. The embedding technique is applied in Section 3.6.

The module `graphlib0.py` creates, first of all, the Tk root widget and wraps its `mainloop` method into the function `MainLoop`. The whole procedure of creating and packing the Canvas widget is wrapped into the function `GraphInit`, having as sole arguments the sizes `nxwin` and `nywin` of the canvas. These values are copied into the variables `nxw` and `nyw`, which are made accessible to all the functions contained in `graphlib0.py`, by way of `global` statements. The module also provides a general-purpose function, `Nint(x)`, which returns the nearest integer to its real argument `x`.

The actual plotting function, `Plot0`, is based on the simple scaling procedure described above. It receives as a main input the n pairs of coordinates, (`x[1]`,`y[1]`) through (`x[n]`,`y[n]`), of the points that sample the function to be plotted. The color to be used for the plot is specified through the parameter `col`, and the corresponding arguments can be colors specified both in RGB format or by standard names (`"red"`, `"green"`, `"blue"`, etc.).

Instead of passing actual *canvas coordinates* to `Plot0` for positioning the plot onto the display, it is considerably more practical to convey *fractional coordinates* (comprised between 0 and 1). Indeed, while absolute coordinates produce different layouts on canvasses of different sizes, fractional coordinates are size independent. In addition, fractional coordinates also enable an intuitive positioning of several plots onto the same canvas. In fact, the parameters `fxmin` and `fxmax` specify, respectively, the minimum and maximum *fractional limits* of the $\bar{x}$-interval of canvas coordinates, that is $\bar{x}_{min}/n_x$ and $\bar{x}_{max}/n_x$. Similarly, `fymin` and `fymax` indicate, respectively, the minimum and maximum fractional canvas coordinates, $\bar{y}_{min}$ and $\bar{y}_{max}$. Finally, the list of parameters of `Plot0` contains titles for the x-axis (`xtext`), y-axis (`ytext`), and the whole plot (`title`).

The routine `Plot0` starts by determining the minima and maxima of the x and y user coordinates (`xmin`, `xmax`, `ymin`, and `ymax`). All operations concern, as mentioned, only the components of `x[]` and `y[]` indexed from 1 to n. On the basis of the bounding user coordinates, the corresponding absolute canvas coordinates are determined (`ixmin`, `ixmax`, `iymin`, and `iymax`). This is done by simply multiplying the fractional canvas coordinates (`fxmin`, `fxmax`, `fymin`, and `fymax`) with the absolute size of the canvas for each separate direction. In the case of the *y*-axis, to account for the opposite directions of the user and canvas axes, the *complement* with respect to 1 of the input fractional coordinates is actually used. To transform with minimum distortions the real user coordinates into the integer canvas coordinates, instead of simply applying the built-in function int, we use the Nint function defined within the module. The limiting canvas coordinates are used for drawing the bounding rectangle of the plot using the create_rectangle method.

The particular case of horizontal plots requires special treatment, namely, an artificial supplementary rescaling of the bounding values, to avoid the otherwise zero vertical size of the bounding box. Specifically, in the case of a plot coinciding with the *x*-axis, the vertical limits are arbitrarily changed to $y_{min} = -1$ and $y_{max} = 1$. In the case of a horizontal plot with some nonzero *y*-intercept, the limiting values are scaled at haphazard to 0.9 and 1.1, respectively, of their initial common value.

The next section of Plot0 labels the plot, centering the corresponding strings (xtext, ytext, and title) about the bottom, left, and top boundaries, respectively. While, the coordinates of the boundary centers are readily obtainable from the limiting canvas coordinates, the accurate justifying of the text about these points is achieved by using the anchor option of the create_text method with the arguments N, E, and S, respectively. The ends of the coordinate axes are then labeled with the limiting user values.

The scaling coefficients are calculated separately for the two directions and, if they fit into the plotted domain, the coordinate axes are drawn.

The actual plotting of the function starts by calculating the canvas coordinates ix0 and iy0 of the first point. In a loop running over the rest of the points (from index 2 to n), a line segment is drawn from the previous point to the current point (ix,iy) using the method create_line. Before going over to a new point, the canvas coordinates of the current point are saved into ix0 and iy0.

Program 3.8 illustrates the use of the library module graphlib.py by the example of the function $f(x) = x^3 \exp(-x)$.

Listing 3.8   Plotting Functions of One Variable (Python Coding)

```
# Plot a function of one variable
from math import *
from graphlib0 import *

def Func(x):                                    # function to be plotted
   return pow(x,3) * exp(-x)

# main

xmin = -0.8; xmax = 7.8              # limits of the plotting interval
n = 50                                         # number of points

x = [0]*(n+1)                              # coordinates of points
y = [0]*(n+1)

h = (xmax-xmin)/(n-1)                             # argument spacing
for i in range(1,n+1):
   x[i] = xmin + (i-1)*h                               # arguments
   y[i] = Func(x[i])                            # function values

GraphInit(800,600)                                    # create canvas

Plot0(x,y,n,"blue",0.15,0.95,0.15,0.85,"x","y","x^3 * exp(-x)")  # create plot

MainLoop()                                  # enter Tkinter event loop
```

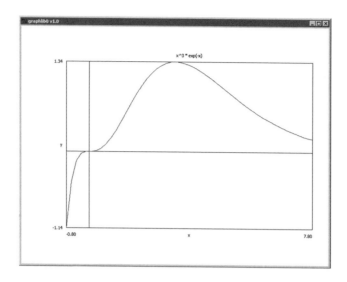

FIGURE 3.7   Plot of the function $x^3 e^{-x}$ created using routine Plot0 from module graphlib0.py.

The program defines the function Func to be plotted and, for simplicity, initializes by assignments the limits xmin and xmax of the plotting interval and the number n of the points at which the function is to be evaluated and plotted. Two arrays, x and y, with usable components with indexes 1 to n (index 0 is not used) are defined for holding the arguments and the corresponding function values. On the basis of the constant spacing h, equally spaced arguments are generated between xmin and xmax and the function is evaluated for them.

The Canvas widget is created by a call to function GraphInit, whose arguments specify the canvas size. The call to Plot0 specifies, besides the arrays x and y and the number of points n to be plotted, also the color of the plot and the fractional position of the plot on the canvas (0.15,0.95,0.15,0.85), which are respectively passed to the function parameters fxmin, fxmax, fymin, and fymax. The graph resulting by running program 3.8 with xmin=-0.8, xmax=7.8, and n-50 is shown in Figure 3.7.

3.5 Graphics Library graphlib.py

The function Plot0, developed in Section 3.4, fully demonstrates the capabilities of Tkinter to produce useful scientific graphics with relatively simple means (see Figure 3.7). Nevertheless, the plots produced with Plot0 are yet susceptible to enhancement, not in the least by a more appropriate labeling of the axes. Along this line, the plot shown in Figure 3.8 is created with a more elaborate version of Plot0, namely Plot, and is, indeed, more informative. Essentially, the axes are extended (the plot appears contracted), so as to include a whole number of equal intervals of simply expressible length, whose ends can be labeled with a reduced number of significant digits. Along with several other graphics functions, Plot forms the core of the module graphlib.py, which is consistently used throughout the book for creating runtime graphical representations.

Owing to their relative complexity, the routines of the module graphlib.py are not described in detail and only their functionality is briefly explained below. Anyway, the module is listed in full length in Appendix D (Listing D.1) and is available as supplementary material on the book's website. The significant implementation details are documented as comments in the source code. Not only that graphlib.py is used as runtime graphics library for the Python programs, but, adequately interfaced

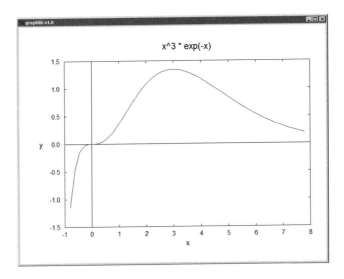

FIGURE 3.8   Plot of the function $x^3 e^{-x}$ created using routine `Plot` from module `graphlib.py`.

(as explained in Section 3.6), it is also used as standard graphics library for the C++ programs developed in the book.

Listing 3.9 provides an abbreviated form of the `graphlib.py` module, presenting, with a few exceptions, only the headers of the member functions.

Listing 3.9   The Python Graphics Library `graphlib.py` (Headers Only)

```
#------------------------------ graphlib.py ---------------------------------
#  Graphics library based on Tkinter.
#----------------------------------------------------------------------------
from math import *
from utils import *
from tkinter import *

root = Tk()                                            # create Tk root widget
root.title("graphlib v1.0")

#============================================================================
def MainLoop():                                        # creates Tk event loop
   root.mainloop()

#============================================================================
def GraphUpdate():                                     # updates Tk root widget
   root.update()

#============================================================================
def GraphInit(nxwin, nywin):        # creates Canvas widget and returns object
   global w, nxw, nyw                           # make canvas object available

   nxw = nxwin; nyw = nywin                                    # canvas size
   w = Canvas(root, width=nxw, height=nyw, bg = "white")     # create canvas
   w.pack()                                             # make canvas visible
   return w
```

```
#==============================================================================
def GraphClear():                                   # deletes content of Canvas widget
    global w, nxw, nyw                                       # canvas object and size
    w.delete(ALL)


#==============================================================================
def Limits(xmin, xmax, maxint):
#------------------------------------------------------------------------------
#  Replaces the limits xmin and xmax of a real interval with the limits of
#  the smallest extended inteval which includes a number <= maxint of
#  subintervals of length d * 10**p, with d = 1, 2, 5 and p integer.
#
#  scale - scale factor (10**p)
#  nsigd - relevant number of significant digits
#  nintv - number of subintervals
#
#  Returns: xmin, xmax, scale, nsigd, nintv


#==============================================================================
def FormStr(x, scale, nsigd):
#------------------------------------------------------------------------------
#  Formats the number x (with factor scale) to nsigd significant digits
#  returning the mantissa in mant[] and the exponent of 10 in expn[].
#
#  Returns: mant, expn


#==============================================================================
def Plot(x, y, n, col, sty, fxmin, fxmax, fymin, fymax, xtext, ytext, title):
#------------------------------------------------------------------------------
#  Plots a real function of one variable specified by a set of (x,y) points.
#  The x and y-domains are extended to fit at most 10 intervals expressible
#  as d * 10^p, with d = 1, 2, 5 and p integer.
#
#  x[]   - abscissas of tabulation points (x[1] through x[n])
#  y[]   - ordinates of tabulation points (y[1] through y[n])
#  n     - number of tabulation points
#  col   - plot color ("red", "green", "blue" etc.)
#  sty   - plot style: 0 - scatter plot, 1 - line plot, 2 - polar plot,
#                      3 - drop lines, 4 - histogram
#  fxmin - min fractional x-limit of viewport (0 < fxmin < fxmax < 1)
#  fxmax - max fractional x-limit of viewport
#  fymin - min fractional y-limit of viewport (0 < fymin < fymax < 1)
#  fymax - max fractional y-limit of viewport
#  xtext - x-axis title; for "None" - axis is not labeled
#  ytext - y-axis title; for "None" - axis is not labeled
#  title - plot title


#==============================================================================
def MultiPlot(x, y, sig, n, col, sty, nplot, maxint, \
              xminp, xmaxp, ioptx, yminp, ymaxp, iopty, \
              fxmin, fxmax, fymin, fymax, xtext, ytext, title):
#------------------------------------------------------------------------------
#  Plots nplot real functions of one variable given by sets of (x,y) points.
#  The coordinate sets are stored contiguously in the arrays x[] and y[].
#  The x and y-domains are extended to fit at most maxint intervals
#  expressible as d * 10^p, with d = 1, 2, 5 and p integer.
#
#  x[]   - abscissas of tabulation points for all functions
#  y[]   - ordinates of tabulation points
```

```
#   sig[]   - error bars of the tabulation points (useful for sty == 4)
#   n[]     - ending index for the individual plots (nmax = n[nplot])
#   col[]   - plot color ("red", "green", "blue" etc.)
#   sty[]   - plot style 0 - scatter plot with squares
#                         1 - line plot;  -1 - dashed line
#                         2 - polar plot; -2 - dashed line
#                         3 - drop lines
#                         4 - error bars; -4 - including line plot
#   nplot   - number of plots
#   maxint  - max. number of labeling intervals
#   ioptx   - 0 - resize x-axis automatically
#             1 - resize x-axis based on user interval [xminp,xmaxp]
#   iopty   - 0 - resize y-axis automatically
#             1 - resize y-axis based on user interval [yminp,ymaxp]
#   fxmin   - min fractional x-limit of viewport (0 < fxmin < fxmax < 1)
#   fxmax   - max fractional x-limit of viewport
#   fymin   - min fractional y-limit of viewport (0 < fymin < fymax < 1)
#   fymax   - max fractional y-limit of viewport
#   xtext   - x-axis title; for "None" - axis is not labeled
#   ytext   - y-axis title; for "None" - axis is not labeled
#   title   - plot title


#===========================================================================
def RGBcolors(ncolstep):
#---------------------------------------------------------------------------
#  Generates ncol = 1280/ncolsep RGB colors in icol[1] through icol[ncol]
#
#  Returns: icol, ncol


#===========================================================================
def ColorLegend(fmin, fmax, ixmin, ixmax, iymin, iymax):
#---------------------------------------------------------------------------
#  Draws and labels the color legend for the interval [fmin,fmax] in the
#  rectangle [ixmin,ixmax] x [iymin,iymax]. [fmin,fmax] is extended to fit at
#  most 10 intervals expressible as d * 10^p, with d = 1, 2, 5 and p integer.
#
#  Returns: fmin, fmax, icol, ncol, nintv


#===========================================================================
def Contour(z, nx, ny, xmin, xmax, ymin, ymax, zmin, zmax, \
            fxmin, fxmax, fymin, fymax, xtext, ytext, title):
#---------------------------------------------------------------------------
#  Plots a function z(x,y) defined in [xmin,xmax] x [ymin,ymax] and tabulated
#  on a regular Cartesian grid with (nx-1)x(ny-1) mesh cells as contour plot.
#  The level curves result by inverse linear interpolation inside the cells.
#
#  z     - tabulated function values
#  nx    - number of x-mesh points
#  ny    - number of y-mesh points
#  zmin  - minimum level considered
#  zmin  - maximum level considered
#  fxmin - minimum relative viewport abscissa (0 < fxmin < fxmax < 1)
#  fxmax - maximum relative viewport abscissa
#  fymin - minimum relative viewport ordinate (0 < fymin < fymax < 1)
#  fymax - maximum relative viewport ordinate
#  xtext - x-axis title; xtext = "" - axis is not labeled
#  ytext - y-axis title; ytext = "" - axis is not labeled
#  title - plot title
```

```
#============================================================================
def PlotParticles(x, y, z, r, col, n, dmax, \
                  xminp, xmaxp, ioptx, yminp, ymaxp, iopty, \
                  fxmin, fxmax, fymin, fymax, title):
#----------------------------------------------------------------------------
#   Plots a system of particles as connected colored spheres
#
#   x,y,z[] - coordinates of particles
#   r[]     - radii of particles
#   col[]   - colors of particles ("red", "green", "blue" etc.)
#   n       - number of particles
#   dmax    - max inter-distance for which particles are connected
#   ioptx   - 0 - resize x-axis automatically
#             1 - resize x-axis to provided user interval (xminp,xmaxp)
#   iopty   - 0 - resize y-axis automatically
#             1 - resize y-axis to provided user interval (yminp,ymaxp)
#   fxmin   - min fractional x-limit of viewport (0 < fxmin < fxmax < 1)
#   fxmax   - max fractional x-limit of viewport
#   fymin   - min fractional y-limit of viewport (0 < fymin < fymax < 1)
#   fymax   - max fractional y-limit of viewport
#   title   - plot title


#============================================================================
def PlotStruct(x, y, z, n, ind1, ind2, ind3, n3, \
               xminp, xmaxp, ioptx, yminp, ymaxp, iopty, \
               fxmin, fxmax, fymin, fymax, title):
#----------------------------------------------------------------------------
#   Renders a 3D structure defined by nodes and triangular surfaces
#
#   x,y,z[] - coordinates of nodes
#   n       - number of nodes
#   ind1[]  - index of 1st node of each triangle
#   ind2[]  - index of 2nd node of each triangle
#   ind3[]  - index of 3rd node of each triangle
#   n3      - number of triangles
#   ioptx   - 0 - resize x-axis automatically
#             1 - resize x-axis to provided user interval (xminp,xmaxp)
#   iopty   - 0 - resize y-axis automatically
#             1 - resize y-axis to provided user interval (yminp,ymaxp)
#   fxmin   - min fractional x-limit of viewport (0 < fxmin < fxmax < 1)
#   fxmax   - max fractional x-limit of viewport
#   fymin   - min fractional y-limit of viewport (0 < fymin < fymax < 1)
#   fymax   - max fractional y-limit of viewport
#   title   - plot title
```

3.5.1 Auxiliary Functions

The functions MainLoop and GraphInit are identical with those from module graphlib0.py (see Section 3.4). Two additional functions, GraphUpdate and GraphClear, are defined to help animating runtime plots. GraphUpdate calls the update method of the Tk root widget to process all pending events and redraws the widget, as explained in Section 3.2.1. GraphClear clears the drawing area by deleting all the items in the display list of the Canvas widget.

Limits is an auxiliary function, designed to *extend* an arbitrary real interval, specified by the limits xmin and xmax, so as to include an integer number, not exceeding maxint, of equal subintervals of length expressible as $d \cdot 10^p$, with $d = 2, 5$, or 10 and p integer. The output consists of the adjusted values of xmin and xmax, the scale factor scale (equal to 10^p), the number of significant digits required to optimally represent the limits, `nsigd`, and the resulting integer number of subintervals, `nintv`. The function `Limits` is typically called to adjust the x and y user intervals for optimal labeling of plots created with routines `Plot` and `MultiPlot`. It is also used to resize contour plots (see function `Contour`) and color legends (see function `ColorLegend`).

On the basis of the previous output from function `Limits`, namely, the scale factor `scale` and the number of significant digits `nsigd`, the function `FormStr` determines the optimal labeling format for a given argument `x`, separating the mantissa in `mant[]` and the exponent of 10 in `expn[]`.

The functions `ColorLegend` and `RGBcolors` are used to display a color legend alongside a contour plot (see Figure 3.11). Upon receiving the color step size `ncolstep`, `RGBcolors` returns in vector `icol[]` the RGB representation of `1280/ncolstep` color shades. `ColorLegend` adjusts the user interval `[fmin,fmax]` using the function `Limits` and maps the resized interval onto the interval of colors returned by `RGBcolors`. The colors are represented by adjacent line segments within the rectangle (in canvas coordinates) `[ixmin,ixmax]` × `[iymin,iymax]`, which, in turn, are labeled according to the output returned by `Limits`.

The routines composing `graphlib.py` also call the general-purpose auxiliary functions defined in the module `utils.py`, which is listed in Appendix D. These utility functions are `Nint`—rounds its real argument to the closest integer, `Sign`—transfers onto its first argument the sign of the second one, `Magn`—returns the order of magnitude of its argument as a power of 10, and `Index`—creates the index list for the ascending ordering of a vector.

3.5.2 Routines for Plotting Functions

The routine `Plot` is conceived for plotting a real function of one variable, being, as pointed out above, an improved version of `Plot0`. Apart from a single supplementary parameter, namely the plot style, the input of the two functions is identical. The arrays `x[]` and `y[]` are expected to contain in their components 1 to n the coordinates of the representative points of the function. The plot color is specified by the parameter `col`, and the corresponding arguments can be RGB colors or standard color names. The plot-style parameter `sty` can take one of the integer values: 0—scatter plot, 1—line plot, 2—polar plot, 3—drop lines, and 4—histogram. `fxmin`, `fxmax`, `fymin`, and `fymax` represent the fractional limits of the plotting domain on the canvas, and, finally, `xtext`, `ytext`, and `title` are the titles of the axes and of the whole representation, respectively.

In particular, the call producing the line plot shown in Figure 3.8 is (`sty=1`)

```
Plot(x,y,n,"blue",1,0.15,0.95,0.15,0.85,"x","y","x^3 * exp(-x)")
```

Comparative graphs of the same function, with different style parameters, are shown in Figure 3.9 and can be obtained using the code sequence.

```
GraphInit(1200,800)                              # create canvas

Plot(x,y,n,"blue",0,0.08,0.48,0.56,0.92,"None","y","Scatter plot sty = 0")
Plot(x,y,n,"red" ,2,0.56,0.96,0.56,0.92,"x","y","Polar plot sty = 2")
Plot(x,y,n,"red" ,3,0.08,0.48,0.08,0.44,"x","y","Drop lines sty = 3")
Plot(x,y,n,"blue",4,0.56,0.96,0.08,0.44,"x","None","Histogram sty = 4")
```

Using `"None"` as axis title inhibits the labeling of the respective axis altogether. A particular feature of polar plots (`sty=2`) is that, irrespective of the function values, the plotting domain is resized to a square, with the x- and y-axes made equal and centered about the origin. The independent coordinate `x[]` is not supposed to vary monotonically, and this also enables the representation of cyclic functions.

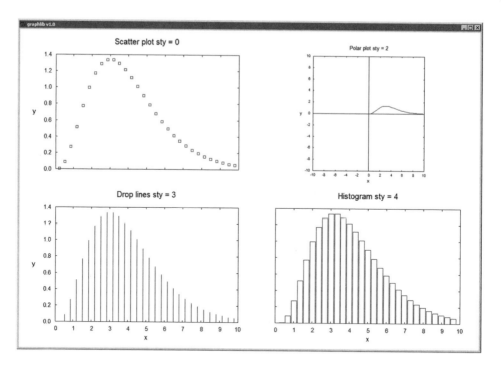

FIGURE 3.9   Plots of function x^3e^{-x} created using routine Plot with different style parameters.

The routine MultiPlot generalizes the functionalities of routine Plot, being designed to depict *several functions* of one variable on a single graph. The functions are specified by separate sets of (x, y) points, each defined on independent x-meshes. For an efficient use of the memory, it is not advisable to employ two-dimensional arrays to store the coordinates, with one dimension running over the individual plots and the other, running over the number of points in the largest set. Instead, the representative points of the nplot plots are contiguously packed in sequenced sets in the same vectors x[] and y[], with an additional array, n[], specifying the *ending index* for each set.

The array sig[] is only used when error bars (sty[i]=4) are employed for at least one of the plots. The array col[] contains the colors of the individual plots and sty[] specifies the corresponding style parameters: 0—scatter plot, 1—line plot, −1—line plot with dashed line, 2—polar plot, −2—polar plot with dashed line, 3—drop lines, 4—error bars, and −4—error bars and continuous line. For ioptx set to 0, the x-axis is resized automatically, while for nonzero ioptx, the axis is resized based on the provided user interval [xminp,xmaxp]. The y-axis is resized analogously, in accordance with the option iopty. The rest of the parameters have the same significance as for function Plot.

The use of routine MultiPlot can be grasped from the program presented in Listing 3.10, which is aimed to plot two functions on the same graph (nplot=2). The arrays n[], col[], and sty[] are defined with nplot+1 (i.e., 3) components to compensate for the 0-offset and to make the components of indexes 1 and 2 usable.

With the variables n1 and n2 specifying the number of representative points for each function, the arrays x[] and y[] are initialized with n1+n2 components, so as to be able to store the data for the concatenated sets. The indexes for the first set range between 1 and n1, and those for the second set range between n1+1 and n1+n2. Correspondingly, the ending indexes for the two plots are n[1]=n1 and n[2]=n1+n2. The x-mesh and the respective function values are generated separately for each set.

The color and style parameters for the plots are set in the corresponding components of the arrays col[] and sty[] and the graph produced by the routine MultiPlot is shown in Figure 3.10.

Listing 3.10   Plotting Two Functions with the Routine `MultiPlot` (Python Coding)

```
# Plots two functions of one variable on the same graph
from math import *
from graphlib import *

def Func(x):                                            # function to be plotted
    return pow(x,3) * exp(-x)

# main

nplot = 2                                                     # number of plots
n = [0]*(nplot+1)                                     # ending indexes of plots
col = [""]*(nplot+1)                                           # plot colors
sty = [0]*(nplot+1)                                            # plot styles

xmin = -0.82; xmax = 7.8                         # limits of plotting interval
n1 = 30                                         # number of points for set 1
n2 = 50                                         # number of points for set 2

x = [0]*(n1+n2+1)                                             # coordinates
y = [0]*(n1+n2+1)

h = (xmax-xmin)/(n1-1)                                    # spacing for set 1
for i in range(1,n1+1):
    x[i] = xmin + (i-1)*h                                       # arguments
    y[i] = Func(x[i])                                     # function values

h = (xmax-xmin)/(n2-1)                                    # spacing for set 2
for i in range(1,n2+1):
    x[n1+i] = xmin + (i-1)*h                                    # arguments
    y[n1+i] = Func(x[n1+i]) * 0.9                         # function values

GraphInit(800,600)                                         # create canvas

n[1] = n1    ; col[1] = "red" ; sty[1] = 0                 # scatter plot
n[2] = n1+n2; col[2] = "blue"; sty[2] = 1                     # line plot
MultiPlot(x,y,y,n,col,sty,nplot,10,0e0,0e0,0,0e0,0e0,0,    # create plots
          0.15,0.95,0.15,0.85,"x","y","Multiple plots")

MainLoop()                                          # enter Tkinter event loop
```

The routine `Contour` creates the contour plot of a real function of two variables, defined in the domain $[xmin, xmax] \times [ymin, ymax]$ and specified by its values $z[i][j]$ at the $nx \times ny$ nodes of a regular grid. The representation is confined to z-values between `zmin` and `zmax`. The rest of the parameters, `fxmin` through `title`, have the same significance as for `Plot` and `MultiPlot`.

In broad lines, `Contour` scans the z-levels in increasing order, between `zmin` and `zmax`. For each level, the code runs sequentially through the square cells formed by four neighboring grid nodes. In each cell, the occurrence of the current z-level is looked for along each of the four sides, whereby the coordinates of the occurrence points are determined by inverse linear interpolation. Each pair of such points located on different cell sides is connected by a colored line segment in the color corresponding to the current level. Finally, the overall contour plot results by the concatenation of the level segments from the individual square cells.

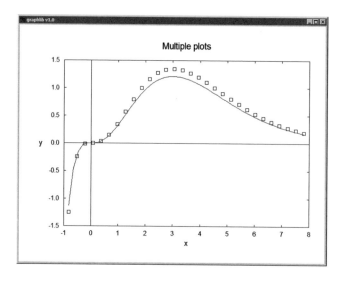

FIGURE 3.10   Plots of two functions on the same graph, created with routine MultiPlot.

Listing 3.11   Contour Plot of the Function $\cos x^2 + \cos y^2$ Using Routine Contour (Python Coding)

```
# Contour plot of a function of two variables
from math import *
from graphlib import *

def Func(x, y):                                          # function to be plotted
    return cos(x*x) + cos(y*y)

# main

xmin = -pi; xmax = pi; ymin = -pi; ymax = pi             # domain boundaries
nx = 41; ny = 41                                         # number of mesh points

u = [[0]*(ny+1) for i in range(nx+1)]                    # array of function values

hx = (xmax-xmin)/(nx-1)                                  # x-mesh spacing
hy = (ymax-ymin)/(ny-1)                                  # y-mesh spacing
for i in range(1,nx+1):
    x = xmin + (i-1)*hx                                  # x-mesh point
    for j in range(1,ny+1):
        y = ymin + (j-1)*hy                              # y-mesh point
        u[i][j] = Func(x,y)                              # function value

umin = umax = u[1][1]                                    # minimum and maximum of function
for i in range(1,nx+1):
    for j in range(1,ny+1):
        if (u[i][j] < umin): umin = u[i][j]
        if (u[i][j] > umax): umax = u[i][j]

GraphInit(800,800)                                       # create canvas

Contour(u,nx,ny,xmin,xmax,ymin,ymax,umin,umax,          # create contour plot
        0.15,0.85,0.15,0.85,"x","y","cos(x*x) + cos(y*y)")

MainLoop()                                               # enter Tkinter event loop
```

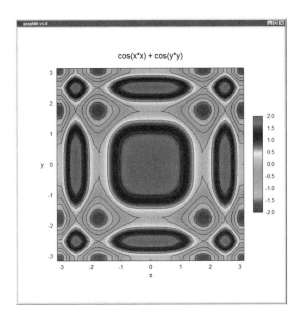

FIGURE 3.11 Contour plot of function $\cos x^2 + \cos y^2$ created using routine `Contour`.

An example program, creating the contour plot of the function $f(x, y) = \cos x^2 + \cos y^2$ in the domain $[-\pi, \pi] \times [-\pi, \pi]$, is given in Listing 3.11 and the resulting graphics is shown in Figure 3.11. The spacings `hx` and `hy` generate `nx × ny` pairs of regularly distributed `(x,y)` coordinates in the domain, for which the user function is evaluated and stored in the array `u[][]`.

The limiting plotting levels conveyed to the routine are, in general, just the minimum and maximum values of the user function, `umin` and `umax`. However, in the case of functions featuring singularities, the limiting plotting levels may be set to convenient finite values, so as to exclude the singular regions from representation.

3.5.3 Simple 3D Rendering Routines

Although not of primary interest to the book, the module `graphlib.py` provides two simple routines for rendering 3D (three-dimensional) structures as 2D images.

The first 3D rendering routine, called `PlotParticles`, depicts systems of particles as spheres, optionally connected by cylindrical bonds (see left panel of Figure 3.12). The main input consists of the coordinate vectors `x[]`, `y[]`, and `z[]` of the centers of the n particles, their radii `r[]`, and colors `col[]`. `dmax` is a cutoff distance beyond which the particles are no longer linked by bonds (with `dmax` set to 0, no bonds at all are depicted). The `ioptx` and `iopty` parameters control whether the x- and y-axis, respectively, are to be sized automatically to reveal the entire structure (`ioptx=0` and `iopty=0`), or, instead, the input user coordinates (`xminp,xmaxp`) and (`yminp,ymaxp`) are to be considered to crop the image. The rest of the parameters, `fxmin` through `title`, have the same meaning as for the routines `Plot` and `MultiPlot`.

The visibility of the different spheres is handled by the simple, yet not very efficient, *priority fill algorithm*, which implies drawing the spheres in decreasing order of their distance to the onlooker. Technically, this aspect is solved by indexing the particles in increasing order of their z-coordinates, with the z-axis pointing towards the observer, and drawing them farthest to closest, one on top of the previous ones. More distant spheres will be painted over and, thus, will be partially or entirely hidden by closer spheres.

Listing 3.12 3D rendering of a Molecule Using the Routine `PlotParticles` (Python Coding)

```
# 3D rendering of a molecule
from graphlib import *

n = 5                                        # number of particles
x = [0, 0.00, 0.52, 0.52,-1.04, 0.00]               # coordinates
y = [0, 0.00, 0.90,-0.90, 0.00, 0.00]               # [0] not used
z = [0, 0.00,-0.37,-0.37,-0.37, 1.10]
r = [0, 0.30, 0.20, 0.20, 0.20, 0.20]                     # radii
col = ["", "red", "blue", "blue", "blue", "blue"]        # colors

dmax = 1.5e0                           # cutoff distance for bonds

GraphInit(600,600)                              # create canvas

PlotParticles(x,y,z,r,col,n,dmax,0e0,0e0,0,0e0,0e0,0,   # plot particles
              0.15,0.85,0.15,0.85,"CH4 molecule")

MainLoop()                             # enter Tkinter event loop
```

The example program given in Listing 3.12 renders the CH$_4$ (methane) molecule and the resulting image is shown in the left panel of Figure 3.12.

`PlotStruct` is a simple 3D rendering routine for structures specified by nodes and triangular surfaces defined between them. The input to `PlotStruct` consists of the coordinate vectors x[], y[], and z[] of the n nodes and the indexes ind1[], ind2[], and ind3[] of the nodes defining each of the n3 triangles. The rest of the parameters have the same significance as for routine `PlotParticles`. The same hiding technique based on the priority fill algorithm is employed, considering, this time, the triangles' closest node to the observer for establishing the drawing order.

In the example provided in Listing 3.13, a simplified airplane is rendered and the resulting image is shown in the right panel of Figure 3.12.

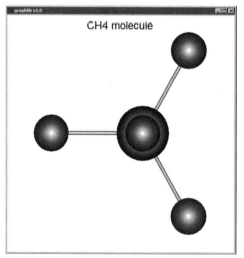 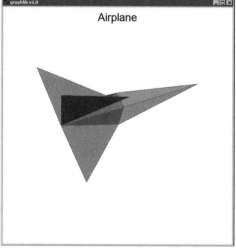

FIGURE 3.12 3D rendering of a molecule using the routine `PlotParticles` (left panel), and of a rigid body using the routine `PlotStruct` (right panel).

Listing 3.13   3D Rendering of a Rigid body Using the Routine `PlotStruct` (Python Coding)

```
# 3D rendering of an airplane
from graphlib import *

n = 10                                                    # number of nodes
x = [0, -2.7,-0.7,-4.6, 1.8, 0.8, 8.0, 2.2, 1.7,-2.7, 2.7]   # coordinates
y = [0, -0.8,-5.1, 3.6,-0.7, 1.5, 2.4, 0.6, 1.7, 1.5, 1.4]   # [0] not used
z = [0,  0.5, 3.6,-2.5, 0.5,-1.0,-1.7, 0.8, 0.0, 3.8, 0.3]

n3 = 9                                          # number of defining triangles
ind1 = [0, 1, 1, 1, 1, 1, 1, 1, 6, 1]           # indexes of triangle corners
ind2 = [0, 2, 3, 4, 5, 4, 5, 7, 7, 9]
ind3 = [0, 4, 5, 6, 6, 7, 8, 8, 8,10]

GraphInit(600,600)                                          # create canvas

PlotStruct(x,y,z,n,ind1,ind2,ind3,n3,0e0,0e0,0,0e0,0e0,0,   # plot structure
           0.15,0.85,0.15,0.85,"Airplane")

MainLoop()                                          # enter Tkinter event loop
```

3.5.4 Histograms

A histogram is a special type of plot, which is usable for *discrete distributions of data* and has the appearance of adjacent rectangles raised over finite data intervals, called *bins*. By definition, the area of a particular rectangle is equal to the frequency of observations falling in that interval, and the cumulated area of the histogram is, correspondingly, equal to the total number of observations. The height of a rectangle represents the frequency of observations divided by the interval width and has the significance of a frequency density. In situations in which it is more useful to display *relative frequencies* of data occurring in the individual bins, the histogram may be normalized by simply dividing all the bin values by the total number of observations.

Histograms can be plotted using the routine `Plot` from `graphlib.py` with the style parameter `sty=4`, and Figure 3.13 shows an example plot. However, before any discrete distribution can be actually plotted as a histogram, the data need to be binned (assigned to separate bins) and the histogram must be normalized. The idea underpinning the routine `HistoBin`, which is contained in the module `graphlib.py` (Listing 3.14), being designed to fulfill these tasks, is that, in many applications, the discrete observations are not available in their entirety right from the beginning. Therefore, the routine is normally operated in three successive modes:

- *Initializing* the histogram, that is, defining the boundaries of the bins and zeroing the corresponding frequencies.
- *Binning* new observations, that is, assigning them to the appropriate bins and incrementing the corresponding occurrence frequencies.
- *Normalizing* the histogram, that is, dividing the frequencies by the total number of observations.

The input parameters of the routine `HistoBin` are `xnew`—the new value to be binned, a and b—the limits of the total binning interval (data falling outside are ignored), n—the number of bin boundaries (n−1 is the number of bins), and `iopt`—the option specifying the mode in which the routine is to be used: 0—to initialize the histogram, 1—to bin the new value `xnew`, and 2—to normalize the histogram. The output consists of the vectors `x[]`, containing the bin boundaries, and `y[]`, representing the (normalized or unnormalized) frequencies. Specifically, `y[i]` is the frequency of observations occurring in the interval $[x[i], x[i+1])$.

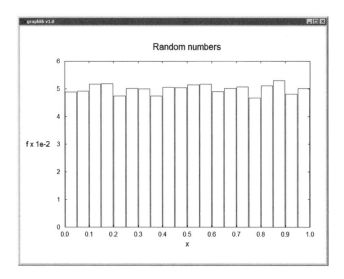

FIGURE 3.13 Histogram of 10,000 random numbers, binned with the function `HistoBin` and plotted with the function `Plot`.

Listing 3.14 Function for Binning Data for Histograms (Python Coding)

```
#===============================================================================
def HistoBin(xnew, a, b, x, y, n, iopt):
#-------------------------------------------------------------------------------
#   Bins data for a histogram to be plotted by function Plot (with sty = 4)
#
#   xnew - new value to be binned
#   a, b - limits of total binning interval
#   x[]  - bin boundaries (x[1] = a, x[n] = b)
#   y[]  - frequency of values in the bins: y[i] in [x[i],x[i+1]), i = 1,n-1
#   n    - number of bin boundaries
#   iopt - option: 0 - zeros bins
#                  1 - bins new value xnew
#                  2 - normalizes histogram
#-------------------------------------------------------------------------------
   h = (b-a)/(n-1)                                              # bin size

   if (iopt == 0):                                         # initialize bins
      for i in range(1,n+1):
         x[i] = a + (i-1)*h                                  # bin boundaries
         y[i] = 0e0

   elif (iopt == 1):                                       # bin new values
      i = (int) ((xnew-a)/h) + 1                               # bin index
      if ((i >= 1) and (i < n)): y[i] += 1           # increment bin value

   elif (iopt == 2):                                      # normalize histogram
      s = 0e0
      for i in range(1,n): s += y[i]           # sum of unnormalized bin values
      for i in range(1,n): y[i] /= s                    # normalize bin values
```

Listing 3.15   Building the Histogram of a Sequence of Random Numbers (Python Coding)

```
# Histogram of random numbers
from random import *
from graphlib import *

n = 21                                          # number of bin boundaries
nrnd = 10000                                    # number of random numbers
x = [0]*(n+1); y = [0]*(n+1)                        # plotting points

a = 0e0; b = 1e0                                     # domain limits

HistoBin(0e0,a,b,x,y,n,0)                       # initialize histogram

for irnd in range(1,nrnd+1): HistoBin(random(),a,b,x,y,n,1)  # bin new values

HistoBin(0e0,a,b,x,y,n,2)                       # normalize histogram

GraphInit(800,600)                                  # create canvas
                                                    # create histogram
Plot(x,y,n,"blue",4,0.15,0.95,0.15,0.85,"x","f x","Random numbers")

MainLoop()                                      # enter Tkinter event loop
```

The equidistant bin boundaries are defined as integer multiples of a spacing h relative to the left boundary a. The index of the bin to which a new observable xnew is to be assigned and whose frequency is to be incremented is calculated based on the relative distance (xnew-a)/h.

Let us consider for illustration the distribution of a sequence of random numbers in the interval [0, 1], as the histogram of their occurrence frequencies in 20 equal subintervals (Listing 3.15). The random numbers are generated by means of the function random(), defined in the standard module random.py. The typical use of HistoBin implies an initial call with iopt=0 to zero the histogram, repeated calls with iopt=1 for binning the observations as they become available, and a final call with iopt=2 for normalizing the histogram.

Figure 3.13 shows the histogram resulting by generating nrnd=10000 random values and using n=21 bin boundaries, or, in other words, n-1 sampling intervals (20 bins). Since the function random() returns uniformly distributed numbers, the larger the number of generated values, the more the envelope of the histogram approaches a constant function.

3.6 Creating Plots in C++ Using the Library **graphlib.py**

The Python graphics library graphlib.py described in Section 3.5 is conceived to work not only with Python programs, but, properly interfaced, with other languages, as well. Specifically, any C++ program including the interface file graphlib.h (listed in Appendix E) can seamlessly call with similar syntax the Python graphics functions defined in the module graphlib.py. The significant advantage lies in the fact that C++ programs thus benefit from the remarkable portability of Python graphics based on the standard Tkinter package.

The basic concepts for embedding Python code in C++ programs are explained in Appendix C and the procedures described below are a direct generalization. Technically, the proper functioning of graphlib.h requires access to graphlib.py (and to the therein imported module utils.py) either by search path settings or by simply placing the three files in the same folder.

To enable the communication with the module graphlib.py, graphlib.h defines two wrapper classes: Python and PyGraph. The class Python is used for invoking the Python interpreter, while the methods of PyGraph are wrappers (interfaces) to the graphics functions contained in graphlib.py. The counterpart functions from the two files have identical names and parameter lists: MainLoop, GraphUpdate, GraphInit, GraphClear, Plot, MultiPlot, Contour, PlotParticles, and PlotStruct. HistoBin is both in graphlib.h and graphlib.py a stand-alone routine.

For the purpose of clarifying the use of the graphics library graphlib.py in C++ programs, we consider again the simple example of plotting a function of one variable, similar to the implementation given in Listing 3.16. The arrays needed for storing the coordinates of the representation points are dynamically allocated using the function Vector from the header file memalloc.h, which needs to be included in the program. The Tkinter root widget w is explicitly created as an object of type PyGraph. The Canvas widget is created invoking the function GraphInit from graphlib.py via its interface function from graphlib.h, which is called as method of the root widget. Likewise, the routines Plot and MainLoop from graphlib.py are called as methods of the root widget via their wrapper functions from graphlib.h. The graphic representation produced by the C++ program 3.16 is, obviously, identical with the one shown in Figure 3.8.

Listing 3.16   Plotting a Function of One Variable (C/C++ Coding)

```
// Plot a function of one variable
#include <math.h>
#include "memalloc.h"
#include "graphlib.h"

double Func(double x)                          // function to be plotted
   { return pow(x,3) * exp(-x); }

int main(int argc, wchar_t** argv)
{
   double *x, *y;                   // declare dynamic arrays as pointers
   double h, xmin, xmax;
   int i, n;

   xmin = -0.8; xmax = 7.8;              // limits of the plotting interval
   n = 50;                                        // number of points

   x = Vector(1,n);             // allocate arrays for coordinates of points
   y = Vector(1,n);

   h = (xmax-xmin)/(n-1);                            // argument spacing
   for (i=1; i<=n; i++) {
      x[i] = xmin + (i-1)*h;                              // arguments
      y[i] = Func(x[i]);                            // function values
   }

   PyGraph w(argc, argv);                      // create Tkinter root widget
   w.GraphInit(800,600);                               // create canvas
                                                        // create plot
   w.Plot(x,y,n,"blue",1,0.15,0.95,0.15,0.85,"x","y","x^3 * exp(-x)");

   w.MainLoop();                              // enter Tkinter event loop
}
```

Listing 3.17   Plotting Two Functions with the Routine `MultiPlot` (C/C++ Coding)

```
// Plots two functions of one variable on the same graph
#include <math.h>
#include "memalloc.h"
#include "graphlib.h"

double Func(double x)                          // function to be plotted
   { return pow(x,3) * exp(-x); }

int main(int argc, wchar_t** argv)
{
   double *x, *y;                    // declare dynamic arrays as pointers
   double h, xmin, xmax;
   int i, n1, n2;
   const int nplot = 2;                             // number of plots
   int n[nplot+1];                              // ending indexes of plots
   int sty[nplot+1];                                // styles of plots
   const char* col[nplot+1];                        // colors of plots

   xmin = -0.82; xmax = 7.8;               // limits of plotting interval
   n1 = 30;                             // number of points for set 1
   n2 = 50;                             // number of points for set 2

   x = Vector(1,n1+n2);          // allocate arrays for coordinates of points
   y = Vector(1,n1+n2);

   h = (xmax-xmin)/(n1-1);                           // spacing for set 1
   for (i=1; i<=n1; i++) {
      x[i] = xmin + (i-1)*h;                            // arguments
      y[i] = Func(x[i]);                           // function values
   }

   h = (xmax-xmin)/(n2-1);                           // spacing for set 2
   for (i=1; i<=n2; i++) {
      x[n1+i] = xmin + (i-1)*h;                          // arguments
      y[n1+i] = Func(x[n1+i]) * 0.9;               // function values
   }

   PyGraph w(argc, argv);                   // create Tkinter root widget
   w.GraphInit(800,600);                            // create canvas

   n[1] = n1   ; col[1] = "red" ; sty[1] = 0;          // scatter plot
   n[2] = n1+n2; col[2] = "blue"; sty[2] = 1;             // line plot
   w.MultiPlot(x,y,y,n,col,sty,nplot,10,0e0,0e0,0,0e0,0e0,0,  // create plots
            0.15,0.95,0.15,0.85,"x","y","Multiple plots");

   w.MainLoop();                            // enter Tkinter event loop
}
```

As a further example, illustrating this time the use of the routine `MultiPlot`, we give in Listing 3.17 the one-to-one translation of Program 3.10. The program plots two functions on the same graph and we allocate two integer arrays, n and `sty`, for storing the ending indexes and style parameters for the two sets of values to be plotted. With n1 and n2 specifying the number of representation points for each function, the arrays `x[]` and `y[]` are declared as pointers and are then allocated by means of the function

Listing 3.18   Contour Plot of the Function $\cos x^2 + \cos y^2$ Using Routine Contour (C/C++ Coding)

```
// Contour plot of a function of two variables
#include <math.h>
#include "memalloc.h"
#include "graphlib.h"

#define pi 3.141592653589793

double Func(double x, double y)                        // function to be plotted
   { return cos(x*x) + cos(y*y); }

int main(int argc, wchar_t** argv)
{
   double **u,                              // declare dynamic array as pointer
   double hx, hy, umin, umax, x, xmin, xmax, y, ymin, ymax;
   int i, j, nx, ny;

   xmin = -pi; xmax = pi; ymin = -pi; ymax = pi;        // domain boundaries
   nx = 41; ny = 41;                                    // number of mesh points

   u = Matrix(1,nx,1,ny);                       // allocate array of function values

   hx = (xmax-xmin)/(nx-1);                                 // x-mesh spacing
   hy = (ymax-ymin)/(ny-1);                                 // y-mesh spacing
   for (i=1; i<=nx; i++) {
      x = xmin + (i-1)*hx;                                  // x-mesh point
      for (j=1; j<=ny; j++) {
         y = ymin + (j-1)*hy;                               // y-mesh point
         u[i][j] = Func(x,y);                            // function value
      }
   }

   umin = umax = u[1][1];                     // minimum and maximum of function
   for (i=1; i<=nx; i++)
      for (j=1; j<=ny; j++) {
         if (u[i][j] < umin) umin = u[i][j];
         if (u[i][j] > umax) umax = u[i][j];
      }

   PyGraph w(argc, argv);                         // create Tkinter root widget
   w.GraphInit(800,800);                                  // create canvas

   w.Contour(u,nx,ny,xmin,xmax,ymin,ymax,umin,umax,     // create contour plot
          0.15,0.85,0.15,0.85,"x","y","cos(x*x) + cos(y*y)");

   w.MainLoop();                                   // enter Tkinter event loop
}
```

Vector to have n1+n2 components, so as to hold the concatenated coordinate sets. The ending indexes for the two sets are n[1]=n1 and n[2]=n1+n2. The *x*-mesh and the corresponding function values are generated separately for each set. Further explanations to the arguments of the routine MultiPlot are given in Section 3.5.2. The graph produced by running Program 3.17 is identical with the one in Figure 3.10.

The last example concerns plotting the function of two variables $\cos x^2 + \cos y^2$ as a contour plot. The corresponding program, given in Listing 3.18 is the faithful translation of the Python program 3.11.

Here, we use the function `Matrix` from the include file `memalloc.h` to dynamically allocate the array `u[][]` of the function values. The function is sampled on a regular grid with nx × ny nodes. As limiting values for the contour plot, the routine `Contour` receives the minimum and maximum function values. The resulting plot is identical with the one shown in Figure 3.11.

References and Suggested Further Reading

Beu, T. A. 2004. *Numerical Calculus in C* (3rd ed., in Romanian). Cluj-Napoca: MicroInformatica.

Lundh, F. 2013a. An introduction to Tkinter. http://effbot.org/tkinterbook/tkinter-index.htm.

Lundh, F. 2013b. An Introduction to Tkinter: Basic widget methods. http://effbot.org/tkinterbook/widget.htm.

Lundh, F. 2013c. An introduction to Tkinter: The Tkinter canvas widget. http://effbot.org/tkinterbook/canvas.htm.

Python Wiki. 2013. TkInter. https://wiki.python.org/moin/TkInter.

Shipman, J. W. 2013. Tkinter 8.5 reference: A GUI for Python. http://infohost.nmt.edu/tcc/help/pubs/tkinter/web/index.html.

4

Sorting and Indexing

4.1 Introduction

Even though sorting and indexing are not numerical algorithms by nature, they are essential for manipulating large amounts of data, stored either in the main computer memory or as magnetic files. The data structures are generally organized as collections of *records* or *tuples*, each composed of several *fields* of various types. Perhaps, the most common example of a data structure is one composed of records of personal data, with the fields holding separate details such as given name, family name, age, E-mail address, and so on. The information from each record that is used for classifying the records is called *key* and it is typically identified with one of the fields, which can be numeric or nonnumeric. The order of the records in the collection is indicated by an *index*.

Sorting designates in general the process of ordering the items of a collection according to certain *order relations*. To name just a few typical reasons for sorting data, let us mention optimized sequential access to the data elements, speedy identification of particular numerical values (e.g., minimum, maximum, and median), separation of the data in relevant categories (e.g., quartiles), and so on.

Sorting can be carried out relative to a single key (e.g., family name or age), or according to multiple hierarchically subordinated keys (e.g., family name as primary key, given name as secondary key, and E-mail address as tertiary key). In the case of multiple-key sorting, the records sorted by the primary key are subsorted for equal primary keys by the secondary key, and so on.

Especially in the case of very large data collections, instead of the effective physical rearrangement of the records, it may prove more efficient to build an *index table*, holding the record *indexes* in the order in which the records would appear in the sorted collection. Specifically, the first entry of the index table represents the index of the record coming first in sorted order, the second entry represents the index of the record coming second, and so on. Using as indexes the successive elements of the index table, the records may be accessed in sorted order. Indexing algorithms can be directly adapted from sorting algorithms.

Indexing is particularly useful when the records of several linked collections need to be processed in parallel relative to the same sorting criterion. Further useful information that can be derived from the index table are the *ranks* of the records, which simply represent the *serial numbers* of the sorted records.

In the numerical programming specific to science and engineering, structured data have most commonly a tensor character, and storing it in the main memory is realized in the form of arrays. In such a context, the commonly employed key is just the magnitude of the elements. Below, we consider for illustration the ascending sorting of a numerical array and provide the index table, the rank table, and the sorted array:

$$\text{Original array:} \quad \begin{array}{|cccccc|} \hline 30 & 60 & 50 & 20 & 10 & 40 \\ \hline \end{array}$$

$$\text{Indexes:} \quad \begin{array}{|cccccc|} \hline 5 & 4 & 1 & 6 & 3 & 2 \\ \hline \end{array}$$

Ranks:	3	6	5	2	1	4

Sorted array:	10	20	30	40	50	60

The first entry in the index array is 5, indicating that the first element in the sorted array is the fifth element of the original array, that is, 10. The first element of the rank array is 3, indicating that the first element of the original array, that is, 30, would represent the third entry in the sorted array.

Since the early days of computing, there has been developed a wealth of sorting algorithms, each standing out by specific features: simplicity, low key comparison count, low record swap count, stability, and so on. In spite of the vivid interest for this topic, none of the currently available sorting algorithms manage to excel in all respects. In the following section, we illustrate this vast subject by three algorithms, covering a rather broad spectrum of performance: *bubble sort* (featuring simple coding, but increased operation count), *insertion sort* (featuring moderate complexity and intermediate operation count), and *Quicksort* (featuring increased complexity and low operation count).

The Python routines developed in this chapter are members of the module `sort.py`, and their C/C++ counterparts are included in the header file `sort.h` (listed in Section 4.6).

4.2 Bubble Sort

Bubble sort has the simplest and most intuitive underlying idea among all the sorting algorithms. The method's name suggests its modus operandi, which is similar to bubbling in a liquid. Considering standard (ascending) sorting, the larger values (bubbles) displace the smaller ones to get first to their final sorted position at the end of the list (at the surface). Despite its simple and stable implementation, which reduces to not more than a few code lines, bubble sort is quite rarely used in practice due to its considerable $O(n^2)$ worst-case operation count, where n is the number of items to be sorted. The only noteworthy advantage of bubble sort over other algorithms is the simple embedding of an additional mechanism for *early detection* of nearly-sorted lists, case, in which, the operation count drops to $O(n)$.

Concretely, bubble sort implies repeated passes through the list, whereby the pairs of adjacent elements are compared in sequence from left to right and swapped if the desired order relation is not met. By the successive swaps, the largest element already reaches its final position at the end of the first pass, as the last element of the sorted list. Correspondingly, the second pass needs to operate on a list reduced to its (presumably) not yet sorted first $n - 1$ elements. Carrying out the same operations on lists of lengths gradually reducing down to 2, the process yields the sorted list in at most $n - 1$ passes.

The worst-case scenario corresponds to a list of elements in reverse order. It is easy to check that, in this case, the algorithm will perform $n(n - 1)/2$ comparisons and $n(n - 1)/2$ swaps ($3n(n - 1)/2$ save operations).

The next sorting example illustrates the operation of bubble sort.

Original array:

30	60	50	20	10	40

Pass 1:

30	**50**	**60**	20	10	40

30	50	**20**	**60**	10	40

| 30 | 50 | 20 | **10** | **60** | 40 ‖ |

| 30 | 50 | 20 | 10 | **40** | **60** ‖ |

Pass 2:

| 30 | **20** | **50** | 10 | 40 ‖ | 60 |

| 30 | 20 | **10** | **50** | 40 ‖ | 60 |

| 30 | 20 | 10 | **40** | **50** ‖ | 60 |

Pass 3:

| **20** | **30** | 10 | 40 ‖ | 50 | 60 |

| 20 | **10** | **30** | 40 ‖ | 50 | 60 |

Pass 4:

| **10** | **20** | 30 ‖ | 40 | 50 | 60 |

The vertical double bar separates the unsorted lower values (on the left) from the sorted higher values (on the right). Swapped pairs of values are represented as boxed and bolded.

At the first pass, the algorithm sweeps through the whole list, comparing the adjacent values, swapping them if necessary, and placing finally the largest value (60) at the end of the list. The second pass operates along the same lines, but only with the first five elements, swapping the second-largest value (50) all the way to the end of the sublist. The last theoretically required sweep ("Pass 5") would operate on the two left-most elements, but is no longer necessary since the array is already sorted. An optimized actual implementation should recognize that the elements are already in order and should drop this last pass.

Listing 4.1 presents three implementations of bubble sort. BubbleSort0 represents the basic implementation, without embedded detection of already-sorted arrays. The outer loop of passes is controlled by the counter ipass, and the length of the inner loop, over the pairs of adjacent elements, is gradually reduced from $n - 1$ down to 1. Irrespective of the particular array, BubbleSort0 performs all the $n(n - 1)/2$ theoretical comparisons before returning the sorted array.

An improved version, known as *modified bubble sort*, is implemented in the routine BubbleSort1. It uses a flag, named swap, which is initialized with 0 and is set to 1 if at least a swap is made during an entire pass. If by the end of a pass no swap has been performed (swap remains 0), it is implied that the array is already in order and the process is terminated. The best case, namely an already-sorted array, obviously requires just $n - 1$ compares during the first pass, but no swaps, reducing the total operation count to $O(n)$.

The routine BubbleSort represents a yet improved variant of BubbleSort1, since it eliminates the redundancy caused by the double control of the outer loop, both by the counter ipass and the flag swap. The chosen coding solution is the replacement of the for-loop by a while-loop controlled by the flag swap solely.

Listing 4.1   Sorting Routines Based on Bubble Sort (Python Coding)

```
#==============================================================================
def BubbleSort0(x, n):
#------------------------------------------------------------------------------
#  Ascending sort of array x[1..n] by bubble sort
#------------------------------------------------------------------------------
    for ipass in range(1,n):                                    # loop of passes
        for i in range(1,n-ipass+1):                 # loop over unsorted sublists
            if (x[i] > x[i+1]):                              # compare neighbors
                xi = x[i]; x[i] = x[i+1]; x[i+1] = xi           # swap neighbors

#==============================================================================
def BubbleSort1(x, n):
#------------------------------------------------------------------------------
#  Ascending sort of array x[1..n] by modified bubble sort
#------------------------------------------------------------------------------
    for ipass in range(1,n):                                    # loop of passes
        swap = 0                                          # initialize swap flag
        for i in range(1,n-ipass+1):                 # loop over unsorted sublists
            if (x[i] > x[i+1]):                              # compare neighbors
                xi = x[i]; x[i] = x[i+1]; x[i+1] = xi           # swap neighbors
                swap = 1                                          # set swap flag

        if (not swap): break                        # exit loop if no swap occurs

#==============================================================================
def BubbleSort(x, n):
#------------------------------------------------------------------------------
#  Ascending sort of array x[1..n] by modified bubble sort
#------------------------------------------------------------------------------
    ipass = 0                                           # initialize pass counter
    swap = 1                                # initialize swap flag to enter loop
    while (swap):                              # perform passes while swaps occur
        ipass += 1                                        # in crease pass counter
        swap = 0                                          # initialize swap flag
        for i in range(1,n-ipass+1):                 # loop over unsorted sublists
            if (x[i] > x[i+1]):                              # compare neighbors
                xi = x[1]; x[1] = x[1+1]; x[i+1] = x1           # swap neighbors
                swap = 1                                          # set swap flag
```

Listing 4.2 presents a generic main program, designed to sort the input array `x[]` of n elements by calling the routine `BubbleSort`, which is assumed to be imported from module `sort.py`.

4.3 Insertion Sort

Despite being less efficient for large lists than more sophisticated methods, as, for instance, Quicksort and Heapsort, insertion sort nonetheless represents a reliable combination of simplicity and performance. Likewise as for bubble sort, the operation count amounts for worst case (list in reversed order) to $O(n^2)$; yet, insertion sort is generally more efficient than most quadratic algorithms, including bubble sort (see Figure 4.1). In addition, insertion sort is *stable* (it does not change the initial relative order of records with equal keys) and *adaptive* (it takes advantage of lists that are nearly sorted). Indeed, the best-case operation count (for an already-sorted list) drops to $O(n)$. Yet another advantage is that insertion sort can be easily adapted for indexing and multiple-key sorting.

Listing 4.2 Generic Use of Sorting Routines (Python Coding)

```
# Ascending sort of an array
from sort import *

# main

n = 6                                   # number of values to be sorted
x = [0., 30., 60., 50., 20., 10., 40.]          # array to be sorted

print("Original array:")
for i in range(1,n+1): print("{0:6.2f}".format(x[i]),end="")
print()

BubbleSort(x,n)

print("Sorted array:")
for i in range(1,n+1): print("{0:6.2f}".format(x[i]),end="")
print()
```

The basic idea behind insertion sort is to progressively extend an already-sorted sublist (starting with a sublist composed of just the first element), by inserting each added element directly into the proper location.

Technically, each insertion step starts by appending to the previously sorted sublist the next element from the entire list. This so-called *pivot* is *saved* in the first instance to a temporary location to free its original location in the list. Next, by *comparisons* with the already-sorted elements, in decreasing order of their values, the proper location of the pivot is determined. In parallel, the larger elements from the sorted sublist are *shifted* one by one to higher-rank positions, starting with the largest one, which moves to the original location of the pivot. Finally, the pivot is *inserted* into the location last vacated by the smallest of the larger values found.

The operation of insertion sort can also be grasped from the next example, which refers to the sequence that was previously used for illustrating bubble sort.

Original array:

| | 30 | 60 | 50 | 20 | 10 | 40 | |

Pivot 2:

| | 30 | **60** | | 50 | 20 | 10 | 40 | | save pivot | **60** |

Pivot 3:

	30	60	**50**		20	10	40		save pivot	**50**
	30	60	**60**		20	10	40		60 > 50; shift 60
	30	**50**	60		20	10	40		insert pivot

Pivot 4:

| 30 | 50 | 60 | **20** ‖ | 10 | 40 | save pivot | **20** |

| 30 | 50 | 60 | **60** ‖ | 10 | 40 | $60 > 20$; shift 60 |

| 30 | 50 | **50** | 60 ‖ | 10 | 40 | $50 > 20$; shift 50 |

| 30 | **30** | 50 | 60 ‖ | 10 | 40 | $30 > 20$; shift 30 |

| **20** | 30 | 50 | 60 ‖ | 10 | 40 | insert pivot |

Pivot 5:

| 20 | 30 | 50 | 60 | **10** ‖ | 40 | save pivot | **10** |

| 20 | 30 | 50 | 60 | **60** ‖ | 40 | $60 > 10$; shift 60 |

| 20 | 30 | 50 | **50** | 60 ‖ | 40 | $50 > 10$; shift 50 |

| 20 | 30 | **30** | 50 | 60 ‖ | 40 | $30 > 10$; shift 30 |

| 20 | **20** | 30 | 50 | 60 ‖ | 40 | $20 > 10$; shift 20 |

| **10** | 20 | 30 | 50 | 60 ‖ | 40 | insert pivot |

Pivot 6:

| 10 | 20 | 30 | 50 | 60 | **40** ‖ | save pivot | **40** |

| 10 | 20 | 30 | 50 | 60 | **60** ‖ | $60 > 40$; shift 60 |

| 10 | 20 | 30 | 50 | **50** | 60 ‖ | $50 > 40$; shift 50 |

| 10 | 20 | 30 | **40** | 50 | 60 ‖ | insert pivot |

In the above diagrams, the successive phases are denoted by the index of the element acting as the pivot. The vertical double bar separates on the left the sublist on which the algorithm operates (including the previously sorted sequence and the pivot), while the unsorted elements reside on the right side. The framed pairs of locations mark the promotion of elements larger than the pivot to higher-rank positions and imply overwriting the location on the right by the element on the left.

During the phase "Pivot 2", the sequence composed of just the first element (30) is expanded by the *second* element (60), which acts as a *pivot* and is saved to a temporary location to make its location available. Next, the pivot is compared with the elements already sorted (for now, just 30) and, if no larger element exists (which is the case), the pass is terminated with the pivot unmoved.

During the phase "Pivot 3", the sorted sequence (30, 60) is augmented by the *third* element (50), which acts as *pivot* and is saved to free its location. The pivot is then compared in reversed order with the sorted elements. Since $60 > 50$, 60 is shifted to the right, occupying the pivot's initial location. Since $30 < 50$, the search is finished and the pivot is inserted in the second location, just vacated by 60. The pass is completed with a new sorted sublist (30, 50, 60), and the execution continues with the next pivot.

Listing 4.3 Ascending Sort by Direct Insertion (Python Coding)

```
#==============================================================================
def InsertSort(x, n):
#------------------------------------------------------------------------------
#  Ascending sort of array x[1..n] by direct insertion
#------------------------------------------------------------------------------
   for ipiv in range(2,n+1):                      # loop over pivots
      xpiv = x[ipiv]                        # save pivot to free its location
      i = ipiv - 1                           # initialize sublist counter
      while ((i > 0) and (x[i] > xpiv)):        # scan to the left of pivot
         x[i+1] = x[i]                      # item > pivot: shift to the right
         i -= 1

      x[i+1] = xpiv                    # insert pivot into last freed location
```

The whole procedure is repeated until the last unsorted element acts as pivot and is inserted in its proper location.

The described operations can be distinctly identified in the implementation provided in Listing 4.3. The variable xpiv acts as temporary storage for the current pivot. Inside the while-loop, the pivot undergoes comparisons with the already-sorted values, which are accompanied by the ascending shift (from x[i] to x[i+1]) of the values larger than the pivot. Once a smaller element is encountered, the pivot is inserted into the last freed position, namely x[i+1] (the index i+1 compensates for the last, unnecessary decrement of the sublist counter i).

The average performance of sorting algorithms is descriptively revealed by the operation count for random sequences of varying lengths (see Problem 4.3). Figure 4.1 shows the log–log plots of the total number of comparisons, N_{comp}, and total operation count (comparisons plus save operations), N_{op}, resulting by using in turn bubble sort, insertion sort, and Quicksort (presented in the next section) for random sequences of lengths ranging between $n = 100$ and $10,000$. The larger count values reveal the lower efficiency of bubble sort. Moreover, the steeper slope also indicates a more rapid deterioration of the performance with increasing sequence length. As for insertion sort, it performs best in terms of comparisons, but ranks second overall, being outperformed by Quicksort.

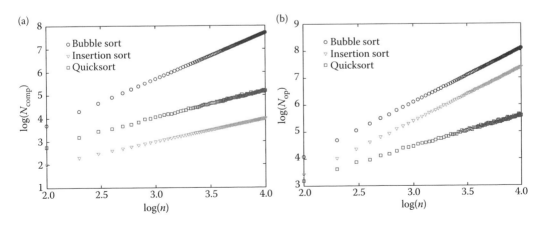

FIGURE 4.1 Operation counts as functions of the number of elements for random sequences ordered by bubble sort, insertion sort, and Quicksort: total number of compares (a) and cumulated number of comparisons and swaps (b).

4.4 Quicksort

Quicksort (Hoare 1961) is one of the most robust and efficient sorting algorithms available. On average, it performs $O(n \log n)$ comparisons to sort n items, being faster than most of the algorithms in its category.

Notwithstanding that it is *neither stable* (the initial relative order of records with equal keys is not guaranteed), *nor adaptive* (nearly sorted lists are not being taken advantage of), Quicksort represents one of the best illustrations of the "divide-and-conquer" algorithm design paradigm. Essentially, it works recursively, breaking down the sorting problem into two similar smaller-size problems, until sublists are created that do not need to be sorted.

Technically, at every level of operation, the algorithm picks from the entire list a distinct element, called *pivot*, and separates two *adjoining sublists* of elements smaller ("<"-list) and, respectively, larger or equal to the pivot ("≥"-list). Once the two lists are separated, the pivot is inserted in between, in its final sorted position. The partitioning process is recursively continued at lower levels of subordination, separately for each created sublist, until sublists of size zero or one are created (which, obviously, do not need sorting) and, thus, upon return to the top level, the whole list becomes sorted.

Various implementations of Quicksort essentially differ by their strategy of choosing the pivot (Sedgewick 1978). Among the common choices, noteworthy are the first, the median, the last, or a random element. Shuffling the array prior to sorting may improve the performance, especially for intensive uses of the algorithm, and may also circumvent worst-case situations, with the lists initially in reverse order. For the sake of simplicity, however, we adopt the choice of the *last element as pivot*. Relocating the pivot from its end position in between the "<"- and "≤"-lists amounts, in this case, simply to swapping the pivot with the first element of the "≥"-list.

Let us now illustrate the operation of Quicksort by the same example of ascending sorting that was also considered for bubble sort and insertion sort.

Level 0: Process the whole array:

$$\boxed{30 \quad 60 \quad 50 \quad 20 \quad 10 \quad 40}$$

Use as pivot the last element in the list:

$$30 \quad 60 \quad 50 \quad 20 \quad 10 \quad \boxed{\mathbf{40}}$$
$$\text{pv}$$

Separate by swaps the "<"- and "≤"-lists (visually separated by a double bar):

‖ **30** 60 50 20 10 | **40** |	$30 < 40$; move boundary
| 30 ‖ 60 50 20 10 | **40** |	add 30 to "<"-sublist
| 30 ‖ **60** 50 20 10 | **40** |	$60 > 40$; do nothing
| 30 ‖ 60 **50** 20 10 | **40** |	$50 > 40$; do nothing
| 30 ‖ 60 50 **20** 10 | **40** |	$20 < 40$; swap 60 with 20
| 30 | **20** ‖ 50 | **60** | 10 | **40** |	add 20 to "<"-sublist
| 30 20 ‖ 50 60 **10** | **40** |	$10 < 40$; swap 50 with 10
| 30 20 | **10** ‖ 60 | **50** | **40** |	add 10 to "<"-sublist
| 30 20 10 ‖ 60 50 **40** |	array partitioned

Insert the pivot between the "$<$"- and "$\leq$"-lists in its final sorted position by swapping it with the first element of the "$\geq$"-list:

$$\begin{array}{|ccc|c|cc|}\hline 30 & 20 & 10 & \mathbf{40} & 50 & \mathbf{60} \\\hline\end{array}$$
$$\quad\quad\quad < \quad\quad\quad \text{pv} \quad\quad \geq$$

Level 1: Sort recursively the "$<$"- and "$\leq$"-lists (in particular, the "$\geq$"-list is already sorted):

$$\begin{array}{|ccc|}\hline 30 & 20 & 10 \\\hline\end{array}\quad\quad\begin{array}{|cc|}\hline 50 & 60 \\\hline\end{array}$$

Use as pivot the last element in the "$<$"-list:

$$\begin{array}{|cc|c|}\hline 30 & 20 & \boxed{10} \\\hline\end{array}$$
$$\quad\quad\quad\quad \text{pv}$$

Separate by swaps the "$<$"- and "$\leq$"-sublists:

$$\begin{array}{|ccc|c|}\hline\hline \mathbf{30} & 20 & \boxed{10} \\\hline\end{array}\quad 30 > 10; \text{ do nothing}$$
$$\begin{array}{|ccc|c|}\hline\hline 30 & \mathbf{20} & \boxed{10} \\\hline\end{array}\quad 20 > 10; \text{ do nothing}$$

Array partitioned (in particular, the "$<$"-sublist is missing); swap the pivot with the first element of the "$\geq$"-sublist to place it in its sorted position:

$$\begin{array}{|c|cc|}\hline\hline \mathbf{10} & 20 & \mathbf{30} \\\hline\end{array}$$
$$\quad \text{pv} \quad\quad \geq$$

Level 2: Sort the "$<$"- and "$\leq$"-sublists created at Level 1 (in particular, the "$<$"-sublist is missing and the "$\geq$"-sublist is already sorted).

Level 0: Return in reversed order to the top level with the sorted list in place.

The routine QuickSort (Listing 4.4) is a one-to-one implementation of the described algorithm. The indexes l and n, passed as arguments, define the subarray of the array x[] that is to be sorted at a given call, and they are particularly meaningful in the context of recursivity. Considering as pivot the last element, x[n], Quicksort starts with a void "$<$"-array, identifying its upper index m with the lower array index, l. The algorithm then scans for elements smaller than the pivot. Each such element is swapped to the end of the "$<$"-list, which is thus expanded. Once no element smaller than the pivot is found, the partitioning into the "$<$"- and "$\geq$"-arrays is considered finalized and the pivot is swapped with the first element of the "$\geq$"-array, acquiring its final sorted position between the "$<$"- and "$\geq$"-arrays. The "$<$"- and "$\geq$"-arrays are next sorted recursively by self-calls of Quicksort. The exit from recursivity occurs as soon as the condition l>=n no longer holds.

4.5 Indexing and Ranking

Let us consider an unsorted sequence X_i, $i = 1, \ldots, n$, which needs to be processed in ascending order of its component values. Instead of the actual physical sorting, one can build an index table with components I_i, so that the indirectly referenced sequence

$$X_{I_i}, \quad i = 1, \ldots, n$$

Listing 4.4   Sorting Based on Quicksort (Python Coding)

```
#===========================================================================
def QuickSort(x, l, n):
#---------------------------------------------------------------------------
#  Ascending sort of array x[1..n] by Quicksort
#---------------------------------------------------------------------------
   if (l >= n): return
                                         # pivot = x[n]; create "<" and ">=" lists
   m = l                                 # upper index in "<"-list
   for i in range(l,n):                  # scan entire list, excepting pivot
      if (x[i] < x[n]):                  # compare current value with pivot
         t = x[i]; x[i] = x[m]; x[m] = t    # swap < value to end of "<"-list
         m += 1                          # extend "<"-list: increase upper index

   t = x[m]; x[m] = x[n]; x[n] = t       # swap pivot between "<" and ">=" lists

   QuickSort(x,l,m-1)                                        # sort "<"-list

   QuickSort(x,m+1,n)                                        # sort ">="-list
```

is ascendingly ordered relative to the index i. That is to say, I_1 represents the index of the smallest component, X_{I_1}, whereas I_n is the index of the largest component, X_{I_n}. The index table I thus contains the entire information on the ordering of the sequence X relative to the considered sorting criterion (in the discussed case, increasing component value).

Indexing is particularly useful when the same sequence needs to be processed relative to several different sorting criteria. To this end, one establishes an index table for each single criterion, instead of physically reordering the sequence relative to each of them.

Index tables can be built, in principle, based on any sorting algorithm. Owing to its straightforward concept, its versatility and, not in the least, its stability, we implement two indexing algorithms based on *insertion sort* (see Section 4.3). The routine Index given in Listing 4.5 represents the single-key implementation thereof. The first step in our implementation is the initialization of the components of the index array, ind[i], by the indexes i themselves, as for an already-sorted sequence. The routine then performs a generic insertion sort, however, with any operation implying the component x[i] replaced by an operation on the corresponding component of the index array ind[i]. Concrete references to the sorted components are done by indirect referencing by way of the index array, x[ind[i]]. Such

Listing 4.5   Indexing Based on Direct Insertion (Python Coding)

```
#===========================================================================
def Index(x, ind, n):
#---------------------------------------------------------------------------
#  Ascending indexing of array x[1..n] in ind[] by insertion sort
#---------------------------------------------------------------------------
   for i in range(1,n+1): ind[i] = i                  # initialize index array

   for ipiv in range(2,n+1):                            # loop over pivots
      xpiv = x[ipiv]                         # save pivot to free its location
      i = ipiv - 1                           # initialize sublist counter
      while ((i > 0) and (x[ind[i]] > xpiv)):    # scan to the left of pivot
         ind[i+1] = ind[i]                   # item > pivot: shift to the right
         i -= 1

      ind[i+1] = ipiv                    # insert pivot into last freed location
```

Listing 4.6   Ranking Based on the Array of Indexes (Python Coding)

```
#===============================================================================
def Rank(ind, rnk, n):
#-------------------------------------------------------------------------------
#  Returns the ranks rnk[1..n] for a list indexed in ind[1..n]
#-------------------------------------------------------------------------------
   for i in range(1,n+1): rnk[ind[i]] = i
```

indirect referencing is, however, unnecessary in the case of the pivot x[ipiv], since ind[ipiv] still contains the initial value ipiv and, at the preceding steps, only the components ind[i] with i<ipiv have been modified.

In many applications, it may prove useful to build the array of the *ranks* of the elements relative to a sorting criterion. The rank simply represents the serial number that an element would have in the sorted sequence. For a sequence of length n, the ranks are integers from 1 to n: rank 1 corresponds to the smallest element, while rank n corresponds to the largest. The array of ranks can be readily obtained from an existing index array. Concretely, the routine Rank (Listing 4.6) assigns the rank i to the element having the index ind[i].

Listing 4.7   Simple Application of Sorting and Indexing (Python Coding)

```
# Ascending sort and indexing of an array
from sort import *

# main

n = 6                                      # number of values to be sorted
x = [0., 30., 60., 50., 20., 10., 40.]              # array to be sorted
ind = [0]*(n+1); rnk = [0]*(n+1)            # arrays of indexes and ranks

print("Original array:")
for i in range(1,n+1): print("{0:6.2f}".format(x[i]),end="")
print()

Index(x,ind,n)
print("Indexes:")
for i in range(1,n+1): print("{0:6d}".format(ind[i]),end="")
print()

Rank(ind,rnk,n)
print("Ranks:")
for i in range(1,n+1): print("{0:6d}".format(rnk[i]),end="")
print()

print("Indexed array:")
for i in range(1,n+1): print("{0:6.2f}".format(x[ind[i]]),end="")
print()

InsertSort(x,n)
print("Sorted array:")
for i in range(1,n+1): print("{0:6.2f}".format(x[i]),end="")
print()
```

Listing 4.8   Indexing Routine with Two Keys Based on Direct Insertion (Python Coding)

```python
#===========================================================================
def Index2(x, y, ind, n):
#---------------------------------------------------------------------------
#  Ascending indexing of correlated arrays x[1..n] and y[1..n] in ind[] by
#  insertion sort using x[] as primary key and y[] as secondary key
#---------------------------------------------------------------------------
   for i in range(1,n+1): ind[i] = i                 # initialize index array

   for ipiv in range(2,n+1):                          # loop over pivots
      xpiv = x[ipiv]; ypiv = y[ipiv]
      i = ipiv - 1                               # initialize sublist counter
      while ((i > 0) and                           # scan to the left of pivot
            (x[ind[i]] > xpiv or (x[ind[i]] == xpiv and y[ind[i]] > ypiv))):
         ind[i+1] = ind[i]                    # item > pivot: shift to the right
         i -= 1

      ind[i+1] = ipiv                 # insert pivot into last freed location
```

Listing 4.9   Indexing of Two Correlated Arrays (Python Coding)

```python
# Ascending indexing of two correlated arrays
from sort import *

# main

n = 6                                       # number of values to be sorted
x = [0., 40., 60., 40., 20., 60., 20.]                       # primary array
y = [0.,  2.,  2.,  1.,  2.,  1.,  1.]                     # secondaryy array
ind = [0]*(n+1)                                          # array of indexes

print("Original arrays:")
print("x",end="")
for i in range(1,n+1): print("{0:6.2f}".format(x[i]),end="")
print()
print("y",end="")
for i in range(1,n+1): print("{0:6.2f}".format(y[i]),end="")
print()

Index2(x,y,ind,n)
print("Indexes:")
for i in range(1,n+1): print("{0:6d}".format(ind[i]),end="")
print()

print("Indexed arrays:")
print("x",end="")
for i in range(1,n+1): print("{0:6.2f}".format(x[ind[i]]),end="")
print()
print("y",end="")
for i in range(1,n+1): print("{0:6.2f}".format(y[ind[i]]),end="")
print()
```

On the basis of the techniques presented in this chapter, ascendingly processing the components of an array x[] may be accomplished either by actual sorting, using, for example, routine InsertSort, or by indexing the sequence by means of routine Index and referencing the components indirectly,

Listing 4.10 Output of Program 4.9.

```
Original arrays:
x 40.00 60.00 40.00 20.00 60.00 20.00
y  2.00  2.00  1.00  2.00  1.00  1.00
Indexes:
      6     4     3     1     5     2
Indexed arrays:
x 20.00 20.00 40.00 40.00 60.00 60.00
y  1.00  2.00  1.00  2.00  1.00  2.00
```

using the created index array ind[], namely as x[ind[i]]. Program 4.7 exemplifies both options. However, given that the routine InsertSort reorders the sequence x[] irrecoverably, the program actually starts by indexing and ranking the original sequence by calls to the routines Index and Rank.

Let us finally deal with the simultaneous indexing of two correlated arrays, x[] and y[], whereby y[] is hierarchically subordinated to x[]. In such a case, the values of the corresponding components of x[] and y[] play the roles of *primary key* and *secondary key*, respectively. The routine Index2 provided in Listing 4.8 represents the straightforward generalization of the single-key routine Index. The main difference regards the fact that shifting is performed not only for components of the primary array x[] larger than the current primary pivot xpiv, but also for equal primary key values and components of the secondary array y[] larger than the secondary pivot ypiv.

The use of the routine Index2 is illustrated in Program 4.9 and the resulting output is shown in Listing 4.10. Along with the index table ind[] returned by Index2 for the two arrays x[] and y[], the program also lists the actually sorted values x[ind[i]] and y[ind[i]].

4.6 Implementations in C/C++

Listing 4.11 shows the content of the file sort.h, which contains equivalent C/C++ implementations of the Python functions developed in the main text and included in the module sort.py. The corresponding routines have identical names, parameters, and functionalities.

Listing 4.11 Sorting, Indexing, and Ranking Routines (sort.h)

```c
//-------------------------------- sort.h --------------------------------
// Contains routines for sorting, indexing and ranking numeric sequences.
// Part of the numxlib numerics library. Author: Titus Beu, 2013
//------------------------------------------------------------------------
#ifndef _SORT_
#define _SORT_

//========================================================================
void BubbleSort0(double x[], int n)
//------------------------------------------------------------------------
// Ascending sort of array x[1..n] by bubble sort
//------------------------------------------------------------------------
{
   double xi;
   int i, ipass;

   for (ipass=1; ipass<=(n-1); ipass++)            // loop of passes
      for (i=1; i<=n-ipass; i++)              // loop over unsorted sublists
         if (x[i] > x[i+1])                         // compare neighbors
            { xi = x[i]; x[i] = x[i+1]; x[i+1] = xi; }   // swap neighbors
}
```

```
//=============================================================================
void BubbleSort1(double x[], int n)
//-----------------------------------------------------------------------------
// Ascending sort of array x[1..n] by modified bubble sort
//-----------------------------------------------------------------------------
{
   double xi;
   int i, ipass, swap;

   for (ipass=1; ipass<=(n-1); ipass++) {                        // loop of passes
      swap = 0;                                          // initialize swap flag
      for (i=1; i<=n-ipass; i++)                   // loop over unsorted sublists
         if (x[i] > x[i+1]) {                               // compare neighbors
            xi = x[i]; x[i] = x[i+1]; x[i+1] = xi;           // swap neighbors
            swap = 1;                                        // set swap flag
         }
      if (!swap) break;                          // exit loop if no swap occurs
   }
}


//=============================================================================
void BubbleSort(double x[], int n)
//-----------------------------------------------------------------------------
// Ascending sort of array x[1..n] by modified bubble sort
//-----------------------------------------------------------------------------
{
   double xi;
   int i, ipass, swap;

   ipass = 0;                                      // initialize pass counter
   swap = 1;                            // initialize swap flag to enter loop
   while (swap) {                          // perform passes while swaps occur
      ipass ++;                                       // increase pass counter
      swap = 0;                                        // initialize swap flag
      for (i=1; i<=n-ipass; i++)                   // loop over unsorted sublists
         if (x[i] > x[i+1]) {                               // compare neighbors
            xi = x[i]; x[i] = x[i+1]; x[i+1] = xi;           // swap neighbors
            swap = 1;                                        // set swap flag
         }
   }
}


//=============================================================================
void InsertSort(double x[], int n)
//-----------------------------------------------------------------------------
// Ascending sort of array x[1..n] by direct insertion
//-----------------------------------------------------------------------------
{
   double xpiv;
   int i, ipiv;

   for (ipiv=2; ipiv<=n; ipiv++) {                           // loop over pivots
      xpiv = x[ipiv];                          // save pivot to free its location
      i = ipiv - 1;                            // initialize sublist counter
      while ((i > 0) && (x[i] > xpiv)) {          // scan to the left of pivot
         x[i+1] = x[i];                      // item > pivot: shift to the right
         i--;
      }
```

```
         x[i+1] = xpiv;                        // insert pivot into last freed location
   }
}

//===============================================================================
void QuickSort(double x[], int l, int n)
//-------------------------------------------------------------------------------
// Ascending sort of array x[1..n] by Quicksort
//-------------------------------------------------------------------------------
{
   double t;
   int i, m;

   if (l >= n) return;
                                   // pivot = x[n]; create "<" and ">=" lists
   m = l;                                            // upper index in "<"-list
   for (i=l; i<=n-1; i++)                   // scan entire list, excepting pivot
      if (x[i] < x[n]) {                     // compare current value with pivot
         t = x[i]; x[i] = x[m]; x[m] = t;   // swap < value to end of "<"-list
         m++;                               // extend "<"-list: increase upper index
      }
   t = x[m]; x[m] = x[n]; x[n] = t;   // swap pivot between "<" and ">=" lists

   QuickSort(x,l,m-1);                                        // sort "<"-list

   QuickSort(x,m+1,n);                                        // sort ">="-list
}

//===============================================================================
void Index(double x[], int ind[], int n)
//-------------------------------------------------------------------------------
// Ascending indexing of array x[1..n] in ind[] by insertion sort
//-------------------------------------------------------------------------------
{
   double xpiv;
   int i, ipiv;

   for (i=1; i<=n; i++) ind[i] = i;                     // initialize index array

   for (ipiv=2; ipiv<=n; ipiv++) {                            // loop over pivots
      xpiv = x[ipiv];                          // save pivot to free its location
      i = ipiv - 1;                               // initialize sublist counter
      while ((i > 0) && (x[ind[i]] > xpiv)) {    // scan to the left of pivot
         ind[i+1] = ind[i];                       // item > pivot: shift to the right
         i--;
      }
      ind[i+1] = ipiv;                   // insert pivot into last freed location
   }
}

//===============================================================================
void Rank(int ind[], int rnk[], int n)
//-------------------------------------------------------------------------------
// Returns the ranks rnk[1..n] for a list indexed in ind[1..n]
//-------------------------------------------------------------------------------
{
   int i;

   for (i=1; i<=n; i++) rnk[ind[i]] = i;
```

```
}

//==============================================================================
void Index2(double x[], double y[], int ind[], int n)
//------------------------------------------------------------------------------
// Ascending indexing of correlated arrays x[1..n] and y[1..n] in ind[] by
// insertion sort using x[] as primary key and y[] as secondary key
//------------------------------------------------------------------------------
{
   double xpiv, ypiv;
   int i, ipiv;

   for (i=1; i<=n; i++) ind[i] = i;                        // initialize index array

   for (ipiv=2; ipiv<=n; ipiv++) {                             // loop over pivots
      xpiv = x[ipiv]; ypiv = y[ipiv];
      i = ipiv - 1;                                 // initialize sublist counter
      while ((i > 0) &&                             // scan to the left of pivot
              (x[ind[i]] > xpiv || (x[ind[i]] == xpiv && y[ind[i]] > ypiv))) {
         ind[i+1] = ind[i];                  // item > pivot: shift to the right
         i--;
      }
      ind[i+1] = ipiv;                 // insert pivot into last freed location
   }
}

#endif
```

4.7 Problems

The Python and C/C++ programs for the following problems may import the functions developed in this chapter from the modules sort.py and sort.h, respectively, which are available as supplementary material. For creating runtime plots, the graphical routines contained in the libraries graphlib.py and graphlib.h may be employed.

PROBLEM 4.1

The four quartiles divide a ranked data set into four groups, composed of equal number of values (25% of the values, each). For example, for a set of length n, the second quartile includes the values ranked between $[n/4] + 1$ and $[n/2]$ (the square brackets designate the integer part), or, in other words, having in the sorted sequence indexes between $[n/4] + 1$ and $[n/2]$.

Generate index and rank tables for the ascending ordering of a sequence of $n = 12$ random numbers using the routines Index and Rank. Group the values by quartile and list them along with their index and rank.

Solution

The implementations are given in Listings 4.12 and 4.13, and a typical output is shown in Listing 4.14.

Listing 4.12   Sorting the Elements of an Array into Quartiles (Python Coding)

```
# Sorting the elements of an array into quartiles
from sort import *
from random import *

# main
```

```
n = 12                                    # number of values to be sorted
x = [0]*(n+1)                                    # array to be sorted
ind = [0]*(n+1); rnk = [0]*(n+1)        # arrays of indexes and ranks

print("Original array:")
for i in range(1,n+1):
   x[i] = random()                              # random sub-unitary values
   print("{0:6.2f}".format(x[i]),end="")
print("\n")

Index(x,ind,n)                                   # create array of indexes
Rank(ind,rnk,n)                                  # create array of ranks

print("Quartile 1")
print("     i    ind    rnk          x")
for i in range(1,int(n/4)+1):
   indi = ind[i]
   print("{0:6d}{1:6d}{2:6d}{3:10.2f}".format(i,ind[i],rnk[indi],x[indi]))

print("Quartile 2")
print("     i    ind    rnk          x")
for i in range(int(n/4)+1,int(n/2)+1):
   indi = ind[i]
   print("{0:6d}{1:6d}{2:6d}{3:10.2f}".format(i,ind[i],rnk[indi],x[indi]))

print("Quartile 3")
print("     i    ind    rnk          x")
for i in range(int(n/2)+1,int(3*n/4)+1):
   indi = ind[i]
   print("{0:6d}{1:6d}{2:6d}{3:10.2f}".format(i,ind[i],rnk[indi],x[indi]))

print("Quartile 4")
print("     i    ind    rnk          x")
for i in range(int(3*n/4)+1,n+1):
   indi = ind[i]
   print("{0:6d}{1:6d}{2:6d}{3:10.2f}".format(i,ind[i],rnk[indi],x[indi]))
```

Listing 4.13   Sorting the Elements of a Sequence into Quartiles (C/C++ Coding)

```
// Sorting the elements of an array into quartiles
#include <stdio.h>
#include <stdlib.h>
#include "sort.h"

int main()
{
   const int n = 12;                        // number of values to be sorted
   double x[n+1];                                    // array to be sorted
   int ind[n+1], rnk[n+1];                  // arrays of indexes and ranks
   int i, indi;

   printf("Original array:\n");
   for (i=1; i<=n; i++) {
      x[i] = rand()/(RAND_MAX+1e0);               // random sub-unitary values
      printf("%6.2f",x[i]);
   }
   printf("\n");
```

```
    Index(x,ind,n);                          // create array of indexes
    Rank(ind,rnk,n);                         // create array of ranks

    printf("Quartile 1\n");
    printf("     i   ind   rnk          x\n");
    for (i=1; i<=int(n/4); i++) {
       indi = ind[i];
       printf("%6i%6i%6i%10.2f\n",i,ind[i],rnk[indi],x[indi]);
    }

    printf("Quartile 2\n");
    printf("     i   ind   rnk          x\n");
    for (i=int(n/4)+1; i<=int(n/2); i++) {
       indi = ind[i];
       printf("%6i%6i%6i%10.2f\n",i,ind[i],rnk[indi],x[indi]);
    }

    printf("Quartile 3\n");
    printf("     i   ind   rnk          x\n");
    for (i=int(n/2)+1; i<=int(3*n/4); i++) {
       indi = ind[i];
       printf("%6i%6i%6i%10.2f\n",i,ind[i],rnk[indi],x[indi]);
    }

    printf("Quartile 4\n");
    printf("     i   ind   rnk          x\n");
    for (i=int(3*n/4)+1; i<=n; i++) {
       indi = ind[i];
       printf("%6i%6i%6i%10.2f\n",i,ind[i],rnk[indi],x[indi]);
    }
}
```

Listing 4.14   Typical Output of the Sorting Programs 4.12 and 4.13

```
Original array:
  0.00  0.56  0.19  0.81  0.58  0.48  0.35  0.90  0.82  0.75  0.17  0.86
Quartile 1
     i   ind   rnk          x
     1     1     1       0.00
     2    11     2       0.17
     3     3     3       0.19
Quartile 2
     i   ind   rnk          x
     4     7     4       0.35
     5     6     5       0.48
     6     2     6       0.56
Quartile 3
     i   ind   rnk          x
     7     5     7       0.58
     8    10     8       0.75
     9     4     9       0.81
Quartile 4
     i   ind   rnk          x
    10     9    10       0.82
    11    12    11       0.86
    12     8    12       0.90
```

PROBLEM 4.2

Generate the index table for a sequence of $n = 50$ random numbers using the routine Index, and also sort the sequence using routine InsertSort. Plot the original sequence, along with the sorted sequence and the one ordered by means of the index table, as functions of the component index, using routine Plot from the library graphlib. Check visually, by the monotonous increase of the values, that the sequences are actually sorted.

Solution

The implementations are shown in Listings 4.15 and 4.16, and the original/sorted sequences are plotted in Figure 4.2.

Listing 4.15   Sorting and Indexing of a Random Sequence Using Direct Insertion (Python Coding)

```python
# Ascending sort of an array
from random import *
from sort import *
from graphlib import *

# main

n = 50;                                         # number of values to be sorted
x = [0]*(n+1); x0 = [0]*(n+1)                   # array to be sorted and copy
ix = [0]*(n+1)                                  # array of sequential indexes
ind = [0]*(n+1)                                 # array of indexes

for i in range(1,n+1):
    x[i] = x0[i] = random()                     # array to be sorted
    ix[i] = float(i)

GraphInit(1200,600)

Plot(ix,x,n,"blue",3,0.06,0.32,0.25,0.80,"i","x","Initial array")

InsertSort(x,n)

Plot(ix,x,n,"blue",3,0.39,0.65,0.25,0.80,"i","x","Sorted array")

Index(x0,ind,n)
for i in range(1,n+1): x[i] = x0[ind[i]]        # sort by indexing

Plot(ix,x,n,"blue",3,0.72,0.98,0.25,0.80,"i","x","Indexed array")

MainLoop()
```

Listing 4.16   Sorting and Indexing of a Random Sequence Using Direct Insertion (C/C++ Coding)

```c
// Ascending sort of an array
#include "memalloc.h"
#include "sort.h"
#include "graphlib.h"

int main(int argc, wchar_t** argv)
{
    double *ix, *x, *x0;
    int *ind;
    int i, n;
```

```
      n = 50;                                      // number of values to be sorted
      x = Vector(1,n); x0 = Vector(1,n);           // array to be sorted and copy
      ix = Vector(1,n);                            // array of sequential indexes
      ind = IVector(1,n);                                    // array of indexes

      for (i=1; i<=n; i++) {
         x[i] = x0[i] = rand();                          // array to be sorted
         ix[i] = double(i);
      }

      PyGraph c(argc, argv);
      c.GraphInit(1200,600);

      c.Plot(ix,x,n,"blue",3,0.06,0.32,0.25,0.80,"i","x","Initial array");

      InsertSort(x,n);

      c.Plot(ix,x,n,"blue",3,0.39,0.65,0.25,0.80,"i","x","Sorted array");

      Index(x0,ind,n);                                       // sort by indexing
      for (i=1; i<=n; i++) x[i] = x0[ind[i]];

      c.Plot(ix,x,n,"blue",3,0.72,0.98,0.25,0.80,"i","x","Indexed array");

      c.MainLoop();
   }
```

PROBLEM 4.3

Generate sequences of $n = 100$–$10,000$ random numbers and sort them using comparatively bubble sort, insert sort, and Quicksort.

    a. Modify the sorting functions BubbleSort, InsertSort, and QuickSort developed in the main text, in such a way as to separately return the total number of performed comparisons and save operations, respectively, and print these to a file.

    b. Add plotting capabilities to the implementations by using the routine MultiPlot from the library graphlib and plot the dependences for the three sorting methods on the same graph.

    c. Assess qualitatively the performance of the sorting algorithms for small and extensive sequences, respectively.

Solution

The Python sorting routines, modified so as to return the operation counts, are given in Listing 4.17. The global variables ncomp and nsave are used to retrieve, respectively, the total number of comparisons and save operations from the sorting routines into the calling program.

Listing 4.17   Python Sorting Routines Modified to Return Operation Counts (Python Coding)

```
#=================================================================================
def BubbleSort(x, n):
#---------------------------------------------------------------------------------
#  Ascending sort of array x[1..n] by modified bubble sort
#---------------------------------------------------------------------------------
   global ncomp, nsave                            # no. of compares and saves
   ipass = 0                                       # initialize pass counter
   swap = 1                                 # initialize swap flag to enter loop
   while (swap):                            # perform passes while swaps occur
```

```
        ipass += 1                                    # increase pass counter
        swap = 0                                      # initialize swap flag
        for i in range(1,n-ipass+1):            # loop over unsorted sublists
            ncomp += 1     #---------------------------------------------------
            if (x[i] > x[i+1]):                         # compare neighbors
                nsave += 3     #-----------------------------------------------
                xi = x[i]; x[i] = x[i+1]; x[i+1] = xi        # swap neighbors
                swap = 1                                     # set swap flag


#=============================================================================
def InsertSort(x, n):
#-----------------------------------------------------------------------------
#  Ascending sort of array x[1..n] by direct insertion
#-----------------------------------------------------------------------------
    global ncomp, nsave                        # no. of compares and saves
    for ipiv in range(2,n+1):                        # loop over pivots
        nsave += 1     #---------------------------------------------------
        xpiv = x[ipiv]                      # save pivot to free its location
        i = ipiv - 1                        # initialize sublist counter
        ncomp += 1     #-------------------------------------------------------
        while ((i > 0) and (x[i] > xpiv)):      # scan to the left of pivot
            nsave += 1     #---------------------------------------------------
            x[i+1] = x[i]               # item > pivot: shift to the right
            i -= 1

        nsave += 1     #---------------------------------------------------
        x[i+1] = xpiv              # insert pivot into last freed location


#=============================================================================
def QuickSort(x, l, n):
#-----------------------------------------------------------------------------
#  Ascending sort of array x[1..n] by Quicksort
#-----------------------------------------------------------------------------
    global ncomp, nsave                             # no. of compares and saves
    if (l >= n): return
                                   # pivot = x[n]; create "<" and ">=" lists
    m = l                                       # upper index in "<"-list
    for i in range(l,n):                        # scan entire list, excepting pivot
        ncomp += 1     #------------------------------------------------------
        if (x[i] < x[n]):                       # compare current value with pivot
            nsave += 3     #--------------------------------------------------
            t = x[i]; x[i] = x[m]; x[m] = t     # swap < value to end of "<"-list
            m += 1                              # extend "<"-list: increase upper index

    nsave += 3     #------------------------------------------------------
    t = x[m]; x[m] = x[n]; x[n] = t     # swap pivot between "<" and ">=" lists

    QuickSort(x,l,m-1)                                      # sort "<"-list

    QuickSort(x,m+1,n)                                      # sort ">="-list
```

The complete programs pertaining to the first task are to be found, respectively, in the files P04-SortComp0.py and P04-SortComp0.cpp and they print the output to file sort.txt. The program variants contained in the files P04-SortComp1.py and P04-SortComp1.cpp have been added graphics capabilities and produce the plots requested by the second task. Typical dependences of the operation counts on the sequence length resulting by running these programs are shown in Figure 4.3.

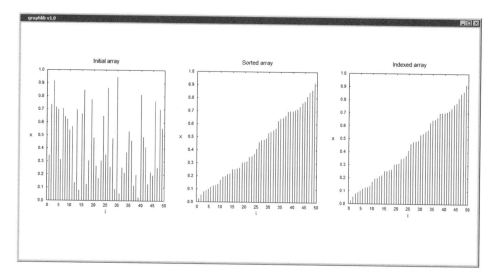

FIGURE 4.2   Original and sorted sequence of random numbers, as returned by routine `InsertSort` and indexed by routine `Index`.

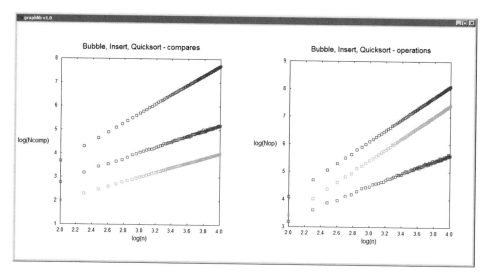

FIGURE 4.3   Comparison of operation counts for random sequences (compares and compares+swaps) for bubble sort, insert sort, and Quicksort.

References and Suggested Further Reading

Hoare, C. A. R. 1961. Algorithm 64: Quicksort. *Communications of the ACM* **4**(7), 321.

Knuth, D. E. 1998. *The Art of Computer Programming* (2nd ed.), vol. 3: *Sorting and Searching*. Reading, MA: Addison-Wesley Professional.

Press, W. H., S. A. Teukolsky, W. T. Vetterling, and B. P. Flannery. 2007. *Numerical Recipes* (3rd ed.): *The Art of Scientific Computing*. Cambridge: Cambridge University Press.

Sedgewick, R. 1978. Implementing Quicksort programs. *Communications of the ACM* **21**(10), 847–857.

5

Evaluation of Functions

Function evaluations are virtually ubiquitous in numerical programming and their *accuracy* and *efficiency* determine, to a large extent, the overall performance of the computer codes. While for individual evaluations a predefined accuracy is commonly a sufficient requirement, for massive computations, the efficiency may decide the very feasibility of the calculations. Even though the same techniques may sometimes beneficially impact on both aspects, accuracy and efficiency are typically achieved by distinct methods.

The specific form under which calculations are expressed and coded is essential, since, due to purely numerical phenomena, perfectly equivalent mathematical formulations may lead to disparate results, or, again, to similar results, but within rather different amounts of time. Indeed, due to the finite machine representation of numbers (as discussed in Chapter 1), given the unavoidably limited number of feasible algorithmic steps, numerical calculations are approximate by nature. From this perspective, the straightforward programming of mathematical formulas, without an in-depth analysis of the possible redundancies and propagation of errors, is rarely the method of choice. Quite on the contrary, well-founded mathematical formulations, which avoid error-prone and time-consuming operations, are desirable with a view to accurate and efficient programming. The next sections reveal pitfalls and provide remedies for common methods of evaluating functions.

5.1 Evaluation of Polynomials by Horner's Scheme

Polynomials are among the most frequently evaluated functions in numerical applications. Owing to their apparent conceptual simplicity, they are, however, prone to inefficient coding by the possible overuse of the built-in power function, which is known to be time consuming. However, representing the powers as repeated multiplications may marginally improve the efficiency, but lacks elegance and generality.

It is Horner's scheme that enables the transformation of the polynomial from the monomial form (linear combination of monomials) into a computationally efficient form, involving only elementary arithmetic operations.

An nth-degree polynomial with real coefficients,

$$P(x) = \sum_{i=0}^{n} a_i x^{n-i} = a_0 x^n + a_1 x^{n-1} + \cdots + a_{n-1} x + a_n, \tag{5.1}$$

may be rewritten by successively factoring out the argument x as

$$P(x) = (\cdots((a_0 x + a_1)x + a_2)x + \cdots + a_{n-1})x + a_n.$$

Based on this factorized form, the evaluation of the polynomial for a *particular argument x* reduces in Horner's scheme to the recursive computation of the *numbers p_i*:

$$\begin{cases} p_0 = a_0, \\ p_i = p_{i-1}x + a_i, \quad i = 1, 2, \ldots, n \end{cases} \tag{5.2}$$

with the last one representing the sought polynomial value:

$$P(x) = p_n. \tag{5.3}$$

Horner's scheme illustrates the beneficial combination of a simple algorithmic formulation with the efficiency gained by using only elementary operations. It not only reduces the number of necessary multiplications, but also results in less numerical instability, which might, otherwise, possibly be caused by subtractions involving large monomials of comparable magnitudes.

One can show that the numbers $p_0, p_1, \ldots, p_{n-1}$ for a particular $x = x_0$ are the coefficients of the quotient polynomial $Q(x)$ for the *synthetic division* of $P(x)$ by the binomial $(x - x_0)$, whereby p_n is the corresponding remainder and $P(x_0) = p_n$, that is,

$$P(x) = (x - x_0)Q(x) + p_n, \tag{5.4}$$

$$Q(x) = p_0 x^{n-1} + \cdots + p_i x^{n-i-1} + \cdots + p_{n-1}. \tag{5.5}$$

Indeed, by replacing Equation 5.5 into Equation 5.4, one obtains:

$$P(x) = p_0 x^n + \cdots + (p_i - p_{i-1}x_0)x^{n-i} + \cdots + (p_n - p_{n-1}x_0).$$

Identifying the coefficients with those of $P(x)$ from Equation 5.1 results in the recurrence

$$\begin{cases} a_0 = p_0, \\ a_i = p_i - p_{i-1}x, \quad i = 1, 2, \ldots, n, \end{cases} \tag{5.6}$$

which is obviously equivalent to Equation 5.2 and proves the statement. Hence, Formulas 5.2 provide for $x = x_0$ the coefficients of the quotient polynomial $Q(x)$ of the division of the polynomial $P(x)$ by the binomial $(x - x_0)$.

The derivative of a polynomial can be readily calculated from

$$P'(x) = \sum_{i=0}^{n-1} (n - i) a_i x^{n-i-1} \tag{5.7}$$

based equally on Horner's scheme, either separately, or in parallel with the evaluation of the polynomial $P(x)$ itself.

Any rational fraction $R(x)$ can be represented as a fraction of two polynomials,

$$R(x) = \frac{P(x)}{Q(x)}, \tag{5.8}$$

where $P(x)$ and $Q(x)$ can be calculated separately using Horner's scheme. Alternatively, polynomial fractions can be transformed into continued fractions, the evaluation methodology of which is described in Section 5.3.

Listing 5.1 implements Horner's scheme (5.2) through (5.3) in functions for evaluating polynomials (`Poly`) and their derivatives (`PolyDerive`), as well as for performing synthetic divisions (`PolyDivide`). All three routines are part of the module `elemfunc.py`.

The function `Poly` evaluates for the argument `x` the polynomial of degree n having the coefficients `a[0]` through `a[n]`. The quantities p_i are implemented as a single scalar `p`, since they are used only once during the iterative calculation. The last value of `p` (i.e., p_n) represents just the polynomial value, being returned.

The function `PolyDerive` returns through the components `b[0]` to `b[n]` the coefficients of the derivative of the input polynomial having the coefficients `a[]`.

The routine `PolyDivide` returns in `b[0]` to `b[n-1]` the coefficients of the quotient for the division of the polynomial having the coefficients `a[]` by the binomial `(x-x0)`. The function can be equally used to evaluate the dividend polynomial for `x0`, since the last component `b[n]` represents both the remainder and the polynomial value $P(x_0)$. The arrays `a` and `b` may coincide in the call, with the input array being simply overwritten.

Let us now employ for exemplification the polynomial

$$P(x) = 6x^3 - 11x^2 + 6x + 3,$$

with the coefficients $a_0 = 6$, $a_1 = -11$, $a_2 = 6$, and $a_3 = 3$, and having the derivative:

$$P'(x) = 18x^2 - 22x + 6.$$

Considering the synthetic division by the binomial $(x - x_0)$ with $x_0 = 1$, we have according to Equation 5.2 the following coefficients for the quotient: $b_0 = a_0 = 6$, $b_1 = b_0 x_0 + a_1 = 6 \cdot 1 - 11 = -5$, $b_2 = b_1 x_0 + a_2 = -5 \cdot 1 + 6 = 1$, and the remainder $b_3 = b_2 x_0 + a_3 = 1 \cdot 1 + 3 = 4$, which is, obviously,

Listing 5.1 Evaluation of Polynomials by Horner's Scheme (Python Coding)

```
#===========================================================================
def Poly(x, a, n):
#---------------------------------------------------------------------------
#  Evaluates the polynomial P(x) = a[0] x^n + a[1] x^(n-1) + ... + a[n] with
#  real coefficients in x using Horner's scheme
#---------------------------------------------------------------------------
   p = a[0]
   for i in range(1,n+1): p = p*x + a[i]
   return p


#===========================================================================
def PolyDerive(a, b, n):
#---------------------------------------------------------------------------
#  For the real polynomial P(x) = a[0] x^n + ... + a[n], the function returns
#  the coefficients of the derivative P'(x) = b[0] x^(n-1) + ... + b[n-1]
#---------------------------------------------------------------------------
   for i in range(0,n+1): b[i] = (n-i) * a[i]


#===========================================================================
def PolyDivide(x0, a, b, n):
#---------------------------------------------------------------------------
#  For the real polynomial P(x) = a[0] x^n + ... + a[n], the function returns
#  the coefficients of the division by the binomial (x-x0):
#  P(x) = (x-x0) (b[0] x^(n-1) + ... + b[n-1]) + b[n] (b[n] = P(x0))
#---------------------------------------------------------------------------
   b[0] = a[0]
   for i in range(1,n+1): b[i] = b[i-1]*x0 + a[i]
```

Listing 5.2   Elementary Operations with Polynomials (Python Coding)

```
# Operations with polynomials
from elemfunc import *

n = 3                                              # degree of polynomial
a = [6e0, -11e0, 6e0, 3e0]                             # coefficients
b = [0]*(n+1)                                           # work array
x0 = 1e0

print("Coefficients of polynomial:")
print(a)

print("\nPolynomial value:")                    # evaluation of polynomial
print("P(",x0,") =",Poly(x0, a, n))

PolyDerive(a, b, n)                               # derivative of polynomial
print("\nCoefficients of derivative:")
print(b)

PolyDivide(x0, a, b, n)                          # synthetic division by (x-x0)
print("\nCoefficients of quotient and remainder:")
print(b)
R = b[n]                                                # remainder
print("\nCheck of remainder:")
print("R - P(x0) =",R-Poly(x0,a,n))

x1 = 2e0                                    # check synthetic division for x1
P = Poly(x1,a,n)                                      # polynomial value
Q = Poly(x1,b,n-1)                                      # quotient value
print("\nCheck of division for x1 =",x1,":")
print("P(x1) - (x1-x0)Q(x1) - R =",P - (x1-x0)*Q - R)
```

equal to $P(x_0)$. The quotient polynomial and remainder are thus

$$Q(x) = 6x^2 - 5x + 1, \quad R = P(x_0) = 4.$$

A direct check of the synthetic division for a particular argument x_1 would imply verifying that $P(x_1) - (x_1 - x_0)Q(x_1) - R = 0$ holds.

The above polynomial operations are implemented in Program 5.2, which assumes that the called routines are contained in the module `elemfunc.py`.

5.2 Evaluation of Analytic Functions

An infinitely differentiable real function $f(x)$ is said to be *analytic* in a neighborhood $|x - x_0| < R$ of some point x_0 if it can be represented by a convergent Taylor series:

$$f(x) = f(x_0) + \frac{f'(x_0)}{1!}(x - x_0) + \frac{f''(x_0)}{2!}(x - x_0)^2 + \cdots = \sum_{i=0}^{\infty} \frac{f^{(i)}(x_0)}{i!}(x - x_0)^i. \tag{5.9}$$

In the particular case when the reference point is $x_0 = 0$, the expansion is called Maclaurin's series.

The nth-order partial sum of a Taylor-series expansion is called *Taylor polynomial* of degree n:

$$F_n(x) = \sum_{i=0}^{n} \frac{f^{(i)}(x_0)}{i!}(x - x_0)^i = \sum_{i=0}^{n} t_i(x). \tag{5.10}$$

The *n*th-order *remainder* is defined as the difference between the function $f(x)$ and the *n*th-degree Taylor polynomial,

$$R_n(x) = f(x) - F_n(x) = \sum_{i=n+1}^{\infty} t_i(x) \tag{5.11}$$

and it represents the error of replacing function $f(x)$ with the *n*th-degree Taylor polynomial, or, equivalently, the error of truncating the Taylor series after the *n*th-order term. Given that, for a convergent alternating series, the truncation error is rigorously bounded from above by the absolute value of the last term added (Demidovich and Maron 1987), that is, $|R_n(x)| \le |t_n(x)|$, the latter provides, in general, practical estimates for the absolute and relative errors:

$$\Delta^{(n)} = |t_n(x)|, \quad \delta^{(n)} = \left| \frac{t_n(x)}{F_n(x)} \right|. \tag{5.12}$$

Let us consider for illustration the power-series expansion of the exponential function,

$$e^x = \sum_{i=0}^{\infty} \frac{x^i}{i!} = 1 + x + \frac{x^2}{2!} + \cdots,$$

having as interval of convergence $(-\infty, +\infty)$ (the series converges for any argument). The *n*th-degree Taylor polynomial is in this case

$$F_n(x) = \sum_{i=0}^{n} t_i(x), \quad t_i(x) = \frac{x^i}{i!}. \tag{5.13}$$

When evaluating functions from series expansions, it is always desirable to take advantage of possible recurrence relations between the consecutive terms. In particular, for the exponential, the simple recurrence relation $t_i = (x/i)\, t_{i-1}$ brings about the significant benefit that it avoids evaluating factorials, which is slow and may lead to overflows. Correspondingly, the calculations can be efficiently performed based on the recurrent scheme:

$$\begin{cases} t_0 = 1, & F_0 = 1, \\ t_i = \dfrac{x}{i}\, t_{i-1}, & F_i = F_{i-1} + t_i, \quad i = 1, 2, \ldots, n, \ldots \end{cases} \tag{5.14}$$

The iterative process should be continued until the relative error estimate reduces below a predefined tolerance ε, that is,

$$|t_n/F_n| \le \varepsilon.$$

To treat in a unifying manner also the cases when $F_n = 0$, avoiding additional tests or possible divisions by 0, the convergence criterion may be reformulated as

$$|t_n| \le \varepsilon |F_n|. \tag{5.15}$$

The function Exp0 from Listing 5.3 exemplifies the straightforward implementation of the recurrent process (5.14) through (5.15). The routine receives the argument x and returns by its name the value of the exponential. The relative tolerance eps is set to take advantage of the double precision representation of the variables (essentially requesting 14 exact decimal digits), but a more flexible implementation,

Listing 5.3   Exponential from Power-Series Expansion (Python Coding)

```
#===============================================================================
def Exp0(x):
#-------------------------------------------------------------------------------
#  Evaluates exp(x) from its power-series expansion
#-------------------------------------------------------------------------------
   eps = 1e-14                                              # relative precision

   i = 0
   f = t = 1e0
   while (fabs(t) > eps*fabs(f)):
      i += 1
      t *= x/i
      f += t

   return f
```

possibly useful in more complex cases, would pass eps through the argument list. The summation loop is coded as a while block since the number of iterations is not known in advance and the convergence speed (as number of terms necessary for reaching a given precision) largely depends on the particular argument. The terms t_i are all stored sequentially in the same scalar variable t, since, once using t_{i-1} for calculating t_i, the former becomes redundant.

Even though the function Exp0 works faultlessly for positive arguments, reproducing the results of the built-in exponential function exp with all significant digits, it progressively exhibits its major weakness for negative arguments. Whereas, down to $x = -8$, Exp0 still provides accurately all significant figures, for $x = -15$ only 7 digits remain exact, while, for $x = -20$, the returned value hardly reproduces 3 digits of the exact result. Indeed, for negative arguments, the terms in the series are alternatively positive and negative. As discussed in Chapter 1, differences of close values are affected by larger relative errors than those of the operands. Hence, for increasing negative arguments, the series shows a progressive tendency of accumulating roundoff errors by the subtraction of the increasingly large neighboring high-order terms.

Listing 5.4   Exponential from Power-Series Expansion (Corrected) (Python Coding)

```
#===============================================================================
def Exp(x):
#-------------------------------------------------------------------------------
#  Evaluates exp(x) from its power-series expansion
#  For x < 0 avoids potential instabilities due to subtractions
#-------------------------------------------------------------------------------
   eps = 1e-14                                              # relative precision

   i = 0
   f = t = 1e0
   absx = fabs(x)
   while (fabs(t) > eps*fabs(f)):
      i += 1
      t *= absx/i
      f += t

   return f if (x >= 0e0) else 1e0/f
```

A simple solution to the poor behavior of the Algorithm 5.14 for negative arguments, still maintaining the flawless operation for positive ones, is to redefine the exponential for a negative argument as the reciprocal of the exponential for the absolute argument:

$$
e^x = \begin{cases} e^{|x|} & \text{for } x \geq 0, \\[2mm] 1/e^{|x|} & \text{for } x < 0. \end{cases}
$$

The resulting routine, called Exp, is given in Listing 5.4.

5.3 Continued Fractions

A mathematical expression implying recursive summations of fractions in the successive denominators, that is,

$$
a_0 + \cfrac{b_1}{a_1 + b_2/(a_2 + \cdots)} \equiv \left[a_0; \frac{b_1}{a_1}, \frac{b_2}{a_2}, \ldots, \frac{b_i}{a_i}, \ldots \right] \tag{5.16}
$$

is called a *continued fraction*. The elements a_i and b_i ($i = 0, 1, 2, \ldots$) can be either numbers or functions. A *finite* (or *terminating*) *continued fraction* is one having a finite number of elements and it is identical with the common fraction obtained by successive reductions to a common denominator, starting from the highest-index elements. In the case of an *infinite* (or *nonterminating*) *continued fraction*, the evaluation cannot be simplified to reductions to a common denominator, and certainly not in a finite number of steps.

For example, the number $e = 2.71828\ldots$ admits the continued-fraction representation:

$$
e = \left[1; \frac{1}{1}, \frac{-1}{2}, \frac{1}{3}, \frac{-1}{2}, \frac{1}{5}, \frac{-1}{2}, \ldots \right].
$$

We call *convergents* of a continued fraction the expressions:

$$
R_1 = a_0 + \frac{b_1}{a_1}, \quad R_2 = a_0 + \cfrac{b_1}{a_1 + b_2/a_2}, \ldots
$$

R_i being the ith-order convergent.

A continued fraction is said to be *convergent* if the series of convergents has a limit

$$
A = \lim_{i \to \infty} R_i,
$$

and this value is attributed to the fraction. If the limit does not exist, the fraction is termed *divergent*.

Theorem 5.1

For a continued fraction, the numerators P_i and denominators Q_i of the convergents

$$R_i = \frac{P_i}{Q_i}, \quad i = 1, 2, 3, \dots \tag{5.17}$$

satisfy the recurrence relations

$$P_i = a_i P_{i-1} + b_i P_{i-2}, \tag{5.18}$$

$$Q_i = a_i Q_{i-1} + b_i Q_{i-2}, \tag{5.19}$$

with

$$\begin{cases} P_0 = a_0, & P_{-1} = 1, \\ Q_0 = 1, & Q_{-1} = 0. \end{cases} \tag{5.20}$$

The proof can be readily carried out by mathematical induction. For $i = 1$ we have

$$R_1 = a_0 + \frac{b_1}{a_1} = \frac{a_1 a_0 + b_1}{a_1} = \frac{a_1 P_0 + b_1 P_{-1}}{a_1 Q_0 + b_1 Q_{-1}} = \frac{P_1}{Q_1}$$

and the first statement is true. Assuming that the theorem holds for all indexes up to i, we show that it equally holds for $i + 1$. By hypothesis:

$$R_i = \frac{P_i}{Q_i} = \frac{a_i P_{i-1} + b_i P_{i-2}}{a_i Q_{i-1} + b_i Q_{i-2}}.$$

R_{i+1} is then directly obtained by replacing here a_i with $(a_i + b_{i+1}/a_{i+1})$ and making use of Equations 5.18 and 5.19 one obtains:

$$R_{i+1} = \frac{(a_i + b_{i+1}/a_{i+1})P_{i-1} + b_i P_{i-2}}{(a_i + b_{i+1}/a_{i+1})Q_{i-1} + b_i Q_{i-2}} = \frac{P_i + (b_{i+1}/a_{i+1})P_{i-1}}{Q_i + (b_{i+1}/a_{i+1})Q_{i-1}}.$$

Finally, we get

$$R_{i+1} = \frac{a_{i+1}P_i + b_{i+1}P_{i-1}}{a_{i+1}Q_i + b_{i+1}Q_{i-1}} = \frac{P_{i+1}}{Q_{i+1}},$$

and the theorem is thus demonstrated.

Continued fractions have the significant advantage of being more rapidly converging than any other infinite representation of functions, such as the power-series expansions.

In the following, let us consider as an example the continued fraction representation of $\tan x$:

$$\tan x = \left[0; \frac{x}{1}, \frac{-x^2}{3}, \frac{-x^2}{5}, \dots, \frac{-x^2}{2i-1}, \dots \right]. \tag{5.21}$$

The elements of the fraction can be identified to be

$$\begin{cases} b_1 = x, & b_i = -x^2, \\ a_0 = 0, & a_1 = 1, \quad a_i = a_{i-1} + 2, \quad i = 2, 3, \dots. \end{cases}$$

Observing that $P_1 = b_1 = x$ and $Q_1 = a_1 = 1$, we have the following starting values:

$$\begin{cases} a_1 = 1, & b = -x^2, \\ P_0 = 0, & P_1 = x, \\ Q_0 = 1, & Q_1 = 1. \end{cases} \tag{5.22}$$

The recursive evaluation of the convergents R_i is accomplished using the relations:

$$\begin{cases} a_i = a_{i-1} + 2, & i = 2, 3, \dots, \\ P_i = a_i P_{i-1} + b P_{i-2}, \\ Q_i = a_i Q_{i-1} + b Q_{i-2}, \\ R_i = P_i/Q_i. \end{cases} \tag{5.23}$$

To achieve a given relative precision ε in the value of $\tan x$, the iterations have to be continued as long as the relative difference between two consecutive convergents exceeds ε, or, equivalently, until

$$|R_i - R_{i-1}| \leq \varepsilon |R_i|. \tag{5.24}$$

The routine Tan presented in Listing 5.5 evaluates the $\tan x$ function from its continued fraction representation, based on the Relations 5.22 through 5.24. For storing the numerators P_i, P_{i-1}, and P_{i-2} of the convergents, a stack of variables (p, pm1 and, respectively, pm2) is set up. A similar stack is employed also for the denominators Q_i. As for the convergents R_i, two values, the most recent and the preceding one, are sufficient for the evaluation of the relative change between successive iterations. The values in the stacks are shifted backward at the beginning of each iteration in order to free storage for the most recent values, P_i, Q_i, and R_i, respectively in the variables p, q, and r.

Listing 5.5   Tangent Function Using Continued Fraction Representation (Python Coding)

```
#==============================================================================
def Tan(x):
#------------------------------------------------------------------------------
#   Evaluates tan(x) from its continued fraction representation
#------------------------------------------------------------------------------
   eps = 1e-14                                             # relative precision

   a = 1e0; b = -x*x
   pm1 = 0e0; p = x
   qm1 = 1e0; q = 1e0
   rm1 = 0e0; r = x
   while (fabs(r-rm1) > eps*fabs(r)):
      pm2 = pm1; pm1 = p                                   # shift the stack
      qm2 = qm1; qm1 = q
      rm1 = r
      a += 2e0
      p = a*pm1 + b*pm2
      q = a*qm1 + b*qm2
      r = p/q                                              # new convergent value

   return r
```

5.4 Orthogonal Polynomials

A system of polynomials $\{f_n(x)\}$ is said to be *orthogonal* on the interval $[a, b]$ with respect to the *weight function* $w(x)$ if each of the polynomials is square-integrable and the scalar product of any two different polynomials vanishes:

$$\int_a^b w(x)f_n(x)f_m(x)\, dx = N_n\delta_{nm}, \quad n, m = 0, 1, 2, \ldots, \tag{5.25}$$

where δ_{nm} is Kronecker's delta. The positive-defined weight function $w(x)$ determines each polynomial $f_n(x)$ up to a constant normalization factor N_n, typically depending on the order n. The specification of the normalization factor is called *standardization*.

Table 5.1 summarizes the defining elements for some of the representative orthogonal polynomials—Chebyshev of the first kind, Legendre, Laguerre, and Hermite—used in many fields of science and engineering.

Notably, irrespective of the family they belong to, the orthogonal polynomials satisfy certain general relationships which have the same functional form (Arfken and Weber 2005; Abramowitz and Stegun 1972). Mathematically, all these relations can be derived from the so-called *generating function*, which completely characterizes a polynomial family. One of the defining relations is the second-order *differential equation* satisfied by the polynomials:

$$g_2(x)\frac{d^2 f_n}{dx^2} + g_1(x)\frac{df_n}{dx} + h_n f_n(x) = 0. \tag{5.26}$$

The generic functions $g_2(x)$ and $g_1(x)$ depend on the variable x, however, not on the degree n, whereas h_n is a constant depending *only* on n.

Two further general relationships satisfied by the orthogonal polynomials of a certain family, of foremost importance for their practical evaluation, are the *recurrence relation* with respect to the order n,

$$a_n f_n(x) = (b_n + c_n x)f_{n-1}(x) - d_n f_{n-2}(x), \tag{5.27}$$

and the expression of the *first derivative* of the nth-order polynomial in terms of the polynomials of orders n and $n - 1$:

$$g_2(x)\frac{df_n}{dx} = g_1(x)f_n(x) + g_0(x)f_{n-1}(x). \tag{5.28}$$

It is noteworthy that the derivatives df_n/dx of a system of orthogonal polynomials compose an orthogonal system themselves.

TABLE 5.1 Definition of Some Orthogonal Polynomials (Abramowitz and Stegun 1972)

$f_n(x)$	Name	a	b	$w(x)$	N_n
$T_n(x)$	Chebyshev of the 1st kind	-1	1	$(1-x^2)^{-1/2}$	$\begin{cases}\pi & \text{if } n = 0 \\ \pi/2 & \text{if } n \neq 0\end{cases}$
$P_n(x)$	Legendre	-1	1	1	$2/(2n+1)$
$L_n(x)$	Laguerre	0	∞	e^{-x}	1
$H_n(x)$	Hermite	$-\infty$	∞	e^{-x^2}	$\sqrt{\pi}2^n n!$

Note: a and b are the limits of the domain of definition, $w(x)$ is the associated weight function, and N_n represents the normalization factor.

TABLE 5.2　Coefficients for the Recurrence Relation (5.27) and
Expressions of the Polynomials of Degrees 0 and 1

$f_n(x)$	a_n	b_n	c_n	d_n	$f_0(x)$	$f_1(x)$
$T_n(x)$	1	0	2	1	1	x
$P_n(x)$	n	0	$2n-1$	$n-1$	1	x
$L_n(x)$	n	$2n-1$	-1	$n-1$	1	$1-x$
$H_n(x)$	1	0	2	$2(n-1)$	1	$2x$

Source: Abramowitz and Stegun (1972).

TABLE 5.3　Functions Defining Relation (5.28)
between the First Derivative and Orthogonal
Polynomials of Two Consecutive Orders

$f_n(x)$	$g_2(x)$	$g_1(x)$	$g_0(x)$
$T_n(x)$	$1-x^2$	$-nx$	n
$P_n(x)$	$1-x^2$	$-nx$	n
$L_n(x)$	x	n	$-n$
$H_n(x)$	1	0	$2n$

Source: Abramowitz and Stegun (1972).

The coefficients for the recurrence relation (5.27) and the expressions for the orders 0 and 1 of the mentioned polynomials, useful for starting the recurrence, are specified in Table 5.2. Table 5.3 gathers the functions $g_0(x)$, $g_1(x)$, and $g_2(x)$, necessary to evaluate the first derivative based on Formula 5.28.

We now turn to the problem of evaluating orthogonal polynomials by using their recurrence relations. In order to calculate the value of $f_n(x)$ for a particular argument x_0, one option is to successively propagate the lower-order polynomial expressions, starting with $f_0(x)$ and $f_1(x)$, and to evaluate in the end the resulting expression of $f_n(x)$. Such an approach becomes, however, impracticable already for not very large orders, due to the rapid increase in complexity of the manipulated expressions. A much more efficient way is, instead, to recurrently propagate the *numerical values* of the implied polynomials according to the steps:

1. Evaluate the starting values $f_0(x_0)$ and $f_1(x_0)$.
2. Apply repeatedly the recurrence relation

$$f_i(x_0) = \frac{1}{a_i}[(b_i + c_i x_0)f_{i-1}(x_0) - d_i f_{i-2}(x_0)], \quad i = 2, 3, \ldots, n. \tag{5.29}$$

In general, *recurrence relations are computationally more efficient* than power-series expansions, for which the convergence rate markedly depends on the particular argument value. Even though they provide the results within a predetermined number of operations, not being affected by truncation errors, recurrence relations may still exhibit in certain situations *instabilities* caused by the accumulation of roundoff errors. Nevertheless, the instabilities can be circumvented by taking advantage of the general property that any recurrence relation, which is *unstable under forward iteration* (determining f_i from f_{i-1} and f_{i-2}), can be shown to be certainly *stable for backward iteration* (determining f_i from f_{i+1} and f_{i+2}). For the backward recursion, a delicate problem, though, remains the choice of two convenient higher-order polynomials that efficiently provide the starting values.

As an illustration of evaluation based on forward recursion, we consider the Chebyshev polynomials of the first kind. This system of polynomials is useful as basis set for expanding functions defined on the interval $[-1, 1]$ due to the particularly rapid convergence of the resulting series. The concrete recurrence relation and the formula for the first derivative are in this case:

$$T_i(x) = 2xT_{i-1}(x) - T_{i-2}(x), \quad i = 2, 3, \ldots, n \tag{5.30}$$

$$\frac{dT_n}{dx} = \frac{n}{x^2 - 1}[xT_n(x) - T_{n-1}(x)]. \tag{5.31}$$

The lowest-order polynomials necessary to start the recurrence are

$$T_0(x) = 1, \quad T_1(x) = x. \tag{5.32}$$

Function Chebyshev included in Listing 5.6 implements the described procedure (5.30) through (5.32) and returns, alongside with the polynomial value, its first derivative via parameter d.

Since it cannot be unified algorithmically, the case $n = 0$ is treated separately and the corresponding derivative is set to zero for any argument x. In the case $n = 1$, the iteration loop is not executed at all and the same line of code assigning the value x to variable f ($P_1(x) = x$) represents the initialization of the recurrence for higher-order polynomials.

For $n > 1$, the variables f, fm1, and fm2 temporarily store the values $T_i(x)$, $T_{i-1}(x)$ and, respectively, $T_{i-2}(x)$. This stack of variables is updated at the beginning of every iteration by shifting the values downward, so as to free the variable f for the new value generated by the recurrence relation. To avoid the singularities in the cases $x = \pm 1$, the derivative is calculated from an alternative form obtained by applying l'Hôpital's rule (the limit of an indeterminate quotient of functions equals the limit of the quotient of their derivatives),

$$\frac{dT_n}{dx} = \frac{n^2}{x}T_n(x). \tag{5.33}$$

Figure 5.1 shows the plots of the Chebyshev polynomials of the first kind up to order $n = 4$ obtained by using function Chebyshev, and the number of zeros of each polynomial can be seen to match its degree.

The above considerations can be applied identically to all of the mentioned orthogonal polynomials and the implemented routines are collected in Listing 5.6.

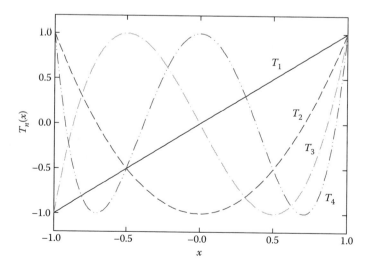

FIGURE 5.1   Chebyshev polynomials of the first kind $T_n(x)$ up to order $n = 4$, as calculated by function Chebyshev.

Listing 5.6 Orthogonal Polynomials from Recurrence Relations (Python Coding)

```
#=================================================================================
def Chebyshev(n, x):
#---------------------------------------------------------------------------------
#  Evaluates the n-th order Chebyshev polynomial and its derivative d in x
#  using the recurrence relation
#---------------------------------------------------------------------------------
   if (n == 0):
      f = 1e0; d = 0e0
   else:
      f = x; fm1 = 1e0; x2 = 2*x
      for i in range(2,n+1):
         fm2 = fm1; fm1 = f
         f = x2*fm1 - fm2

      d = n*(x*f-fm1)/(x*x-1e0) if (x*x-1e0) else n*n*f/x

   return (f, d)


#=================================================================================
def Legendre(n, x):
#---------------------------------------------------------------------------------
#  Evaluates the n-th order Legendre polynomial and its derivative d in x
#  using the recurrence relation
#---------------------------------------------------------------------------------
   if (n == 0):
      f = 1e0; d = 0e0
   else:
      f = x; fm1 = 1e0
      for i in range(2,n+1):
         fm2 = fm1; fm1 = f
         f = ((2*i-1)*x*fm1 - (i-1)*fm2)/i

      d = n*(x*f-fm1)/(x*x-1e0) if (x*x-1e0) else 0.5*n*(n+1)*f/x

   return (f, d)


#=================================================================================
def Laguerre(n, x):
#---------------------------------------------------------------------------------
#  Evaluates the n-th order Laguerre polynomial and its derivative d in x
#  using the recurrence relation
#---------------------------------------------------------------------------------
   if (n == 0):
      f = 1e0; d = 0e0
   else:
      f = 1e0 - x; fm1 = 1e0
      for i in range(2,n+1):
         fm2 = fm1; fm1 = f
         f = ((2*i-1-x)*fm1 - (i-1)*fm2)/i

      d = n*(f-fm1)/x if x else -n*f

   return (f, d)


#=================================================================================
def Hermite(n, x):
#---------------------------------------------------------------------------------
#  Evaluates the n-th order Hermite polynomial and its derivative d in x
```

```
#   using the recurrence relation
#---------------------------------------------------------------------------
    if (n == 0):
        f = 1e0; d = 0e0
    else:
        f = 2*x; fm1 = 1e0; x2 = 2*x
        for i in range(2,n+1):
            fm2 = fm1; fm1 = f
            f = x2*fm1 - 2*(i-1)*fm2
        d = 2*n*fm1

    return (f, d)
```

5.5 Spherical Harmonics—Associated Legendre Functions

The spherical harmonics and the associated Legendre functions play a central role in numerous fundamental problems of mathematical physics. One of the typical occurrences is linked to solving partial differential equations involving the Laplace operator in spherical coordinates by the method of separation of variables, such as Poisson's equation in electrostatics or Schrödinger's equation in quantum mechanics.

Essentially, the spherical harmonics $Y_{lm}(\theta, \varphi)$ are the solutions of the angular part of Laplace's equation in spherical coordinates,

$$\frac{1}{\sin\theta}\frac{\partial}{\partial\theta}\left(\sin\theta\frac{\partial Y_{lm}}{\partial\theta}\right) + \frac{1}{\sin^2\theta}\frac{\partial^2 Y_{lm}}{\partial\varphi^2} + l(l+1)Y_{lm} = 0, \tag{5.34}$$

with $l = 0, 1, 2, \ldots$ and $m = -l, -l+1, \ldots, l-1, l$. The spherical harmonics can be expressed by means of the associated Legendre functions $P_l^m(\cos\theta)$, as

$$Y_{lm}(\theta, \varphi) = \sqrt{\frac{2l+1}{4\pi}\frac{(l-m)!}{(l+m)!}}P_l^m(\cos\theta)e^{im\varphi}, \tag{5.35}$$

and the property $Y_{l,-m}(\theta, \varphi) = (-1)^m Y_{lm}^*(\theta, \varphi)$ enables them to be expressed solely by associated Legendre functions with $m \geq 0$.

The spherical harmonics form a complete orthonormal basis set in the space $\mathcal{L}_2(S_1)$ of the square-integrable functions defined on the unit sphere (functions depending only on angular variables, that is, $f(\theta, \varphi)$) and they satisfy the *orthonormalization condition*:

$$\int_0^{2\pi} d\varphi \int_0^\pi Y_{l'm'}^*(\theta, \varphi)Y_{lm}(\theta, \varphi)\sin\theta \, d\theta = \delta_{l'l}\delta_{m'm}, \tag{5.36}$$

and the *closure (completeness) relation*:

$$\sum_{l=0}^\infty \sum_{m=-l}^l Y_{lm}^*(\theta', \varphi')Y_{lm}(\theta, \varphi) = \frac{\delta(\theta - \theta')\delta(\varphi - \varphi')}{\sin\theta}. \tag{5.37}$$

The completeness and orthonormalization guarantee that any function $f \in \mathcal{L}_2(S_1)$ can be expanded uniquely in a converging Fourier series with respect to the spherical harmonics:

$$f(\theta, \varphi) = \sum_{l=0}^{\infty} \sum_{m=-l}^{l} a_{lm} Y_{lm}(\theta, \varphi). \tag{5.38}$$

The associated Legendre functions $P_l^m(x)$ satisfy the differential equation

$$\left(1 - x^2\right) \frac{d^2 P_l^m}{dx^2} - 2x \frac{dP_l^m}{dx} + \left[l(l+1) - \frac{m^2}{1 - x^2} \right] P_l^m = 0, \tag{5.39}$$

which can be shown to result by replacing the spherical harmonics (5.35) in Equation 5.34 and noting $x = \cos\theta$. For $m \geq 0$, the associated Legendre functions can be expressed in terms of the derivatives of the basic Legendre polynomials $P_l(x)$:

$$P_l^m(x) = (-1)^m \left(1 - x^2\right)^{m/2} \frac{d^m}{dx^m} P_l(x). \tag{5.40}$$

The corresponding orthogonality relation reads:

$$\int_{-1}^{1} P_{l'}^m(x) P_l^m(x) dx = \frac{2}{(2l+1)} \frac{(l+m)!}{(l-m)!} \delta_{l'l}. \tag{5.41}$$

For negative upper indexes, the convenient way to evaluate the associated Legendre functions is provided by the relation:

$$P_l^{-m}(x) = (-1)^m \frac{(l-m)!}{(l+m)!} P_l^m(x), \quad m = 0, 1, \ldots, l. \tag{5.42}$$

The associated Legendre functions form a complete orthonormal basis set in the space $\mathcal{L}_2[-1, 1]$ of the square integrable functions defined on interval $[-1, 1]$, having as a direct consequence the fact that any function $f \in \mathcal{L}_2[-1, 1]$ may be uniquely expanded in a converging Fourier series in terms of these functions:

$$f(x) = \sum_{l=0}^{\infty} \sum_{m=-l}^{l} a_{lm} P_l^m(x). \tag{5.43}$$

The efficient and stable method to evaluate the associated Legendre functions is based on the recurrence relation with respect to their lower order:

$$(i - m) P_i^m(x) = (2i - 1) x P_{i-1}^m(x) - (i + m - 1) P_{i-2}^m(x). \tag{5.44}$$

The complication arisen from the necessity of starting the recurrence from two values corresponding to functions of given superior index m can be overcome by considering the particularly simple expressions of $P_{m-1}^m(x)$ and $P_m^m(x)$ resulting from their definition (5.40). Indeed, by setting $l = m - 1$, we have

$$P_{m-1}^m = 0, \tag{5.45}$$

since it implies differentiating m times a polynomial of degree $m - 1$, and, by considering $l = m$, we get

$$P_m^m(x) = (-1)^m (2m - 1)!!(1 - x^2)^{m/2},\tag{5.46}$$

where $(2m - 1)!! = 1 \cdot 3 \cdot 5 \cdots (2m - 1)$ is the double factorial (in this case, the product of the odd positive integers up to $2m - 1$). The evaluation of $P_m^m(x)$ can be carried out most efficiently by using the equivalent product form:

$$P_m^m(x) = \prod_{i=1}^{m} \left\{ (2i - 1) \left[-\sqrt{1 - x^2} \right] \right\}.\tag{5.47}$$

Once the values of $P_{m-1}^m(x)$ and $P_m^m(x)$ are calculated, the recurrence relation (5.44) has to be applied for the lower index running over the values $i = m + 1, \ldots, l$.

The routine for evaluating associated Legendre functions based on relations (5.44) through (5.47) is given in Listing 5.7. For reasons of simplicity, the function aLegendre is designed only for $m \geq 0$, since

Listing 5.7   Associated Legendre Functions and Spherical Harmonics (Python Coding)

```
#========================================================================
def aLegendre(l, m, x):
#------------------------------------------------------------------------
#  Evaluates the associated Legendre function of orders l and m >= 0 in x
#------------------------------------------------------------------------
   if (l < m): return 0e0

   p = 1e0; pm1 = 0e0                        # seed values: P(m,m,x), P(m-1,m,x)
   if (m):
      sqx = -sqrt(1e0-x*x)
      for i in range(1,m+1): p *= (2*i-1) * sqx

   for i in range(m+1,l+1):                                    # recurrence
      pm2 = pm1; pm1 = p
      p = ((2*i-1)*x*pm1 - (i+m-1)*pm2)/(i-m)

   return p

#========================================================================
def SpherY(l, m, theta, phi):
#------------------------------------------------------------------------
#  Evaluates the real and imaginary parts (ReY and ImY) of the spherical
#  harmonic of orders l and m for arguments theta and phi.
#  Calls aLegendre to calculate associated Legendre polynomials.
#------------------------------------------------------------------------
   mabs = abs(m)

   fact = 1e0
   for i in range(l-mabs+1,l+mabs+1): fact *= i         # (l+|m|)!/(l-|m|)!

   fact = sqrt((2*l+1)/(4e0*pi*fact)) * aLegendre(l,mabs,cos(theta))
   if (m < 0 and m % 2): fact = -fact

   ReY = fact * cos(m*phi)
   ImY = fact * sin(m*phi)

   return (ReY,ImY)
```

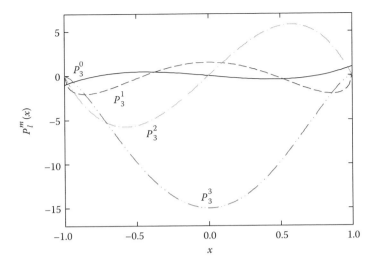

FIGURE 5.2 Plots of associated Legendre functions $P_3^0(x)$, $P_3^1(x)$, $P_3^2(x)$, and $P_3^3(x)$ as calculated by function aLegendre.

values for $m < 0$ can be simply obtained by applying property (5.42). For $l < m$, the function returns 0, as this amounts to differentiating the Legendre polynomial $P_l(x)$ a number m of times which exceeds its degree l. For $m > 0$, the starting value $P_m^m(x)$ is calculated according to Equation 5.47 and stored in variable p, while for $l > m$, the recurrence relation (5.44) is applied for $i = m + 1, \ldots, l$.

Figure 5.2 displays as examples the plots of the associated Legendre functions $P_3^0(x)$, $P_3^1(x)$, $P_3^2(x)$, and $P_3^3(x)$, obtained by using function aLegendre. Obviously, while increasing the upper index, the number of zeros reduces alongside with the reduction of the degree of the polynomial composing the associated Legendre function.

The real and imaginary parts of the spherical harmonics require the direct use of the associated Legendre functions and, for $m \geq 0$, we have

$$\text{Re } Y_{lm}(\theta, \varphi) = \sqrt{\frac{2l+1}{4\pi} \frac{(l-m)!}{(l+m)!}} P_l^m(\cos\theta) \cos m\varphi, \tag{5.48}$$

$$\text{Im } Y_{lm}(\theta, \varphi) = \sqrt{\frac{2l+1}{4\pi} \frac{(l-m)!}{(l+m)!}} P_l^m(\cos\theta) \sin m\varphi. \tag{5.49}$$

Function SpherY from Listing 5.7 implements these relations, using for output the parameters ReY and ImY. In calculating $(l + m)!/(l - m)!$, which occurs in the normalization factor, instead of evaluating factorials (which are time consuming if invoked often), only the distinct factors are multiplied:

$$\frac{(l+m)!}{(l-m)!} = (l-m+1)(l-m+2)\cdots(l+m) = \prod_{i=l-m+1}^{l+m} i.$$

For negative values of m, the spherical harmonics can be evaluated by making use of the property $Y_{l,-m}(\theta,\varphi) = (-1)^m Y_{lm}^*(\theta,\varphi)$. Again, for reasons of efficiency, in determining the sign factor $(-1)^m$ we avoid calling the built-in function pow and check the *parity* of m, instead.

5.6 Spherical Bessel Functions

Bessel functions occur in a wide range of mathematical physics problems, from sound vibrations to quantum mechanical scattering (Arfken and Weber 2005). In particular, in spherical coordinates, the *spherical Bessel functions* compose the radial solutions to the quantum mechanical problems of the spherically symmetric square potential well and of the elastic scattering in atomic and nuclear physics.

The special attention devoted in this section to the spherical Bessel functions also stems from the fact that they provide illustrative examples of numerical instabilities arising during the *forward* iteration of certain recurrence relations. Nevertheless, the instabilities can be circumvented by *backward* recursion (for decreasing orders), thereby resulting stable processes.

The differential equation of the spherical Bessel functions is:

$$x^2 \frac{d^2 f_n}{dx^2} + 2x f_n'(x) + \left[x^2 - n(n+1)\right] f_n(x) = 0, \quad n = 0, \pm 1, \pm 2, \ldots \tag{5.50}$$

and its real, linearly independent solutions are the *spherical Bessel function* $j_n(x)$ (regular solution) and the *spherical Neumann function* $y_n(x)$ (irregular solution, which diverges at the origin).

Both the spherical Bessel and Neumann functions satisfy the recurrence relation:

$$f_i(x) = \frac{2i-1}{x} f_{i-1}(x) - f_{i-2}(x), \tag{5.51}$$

which can be conveniently used to evaluate them for given arguments and orders, by the forward propagation of the function values from lower to higher orders. The simple expressions for the lowest orders 0 and 1,

$$j_0(x) = \frac{\sin x}{x}, \quad j_1(x) = \frac{1}{x}\left[j_0(x) - \cos x\right], \tag{5.52}$$

$$y_0(x) = -\frac{\cos x}{x}, \quad y_1(x) = \frac{1}{x}\left[y_0(x) - \sin x\right], \tag{5.53}$$

can be actually used to start the recursive evaluation.

For the spherical Neumann functions $y_n(x)$, the forward recurrence based on Equation 5.51 and started from initial values provided by Expressions 5.53 does not pose any stability problems and the direct implementation is given as function SBessy in Listing 5.8.

Quite in contrast to the spherical Neumann functions, for the spherical Bessel functions $j_n(x)$, the stability of the forward iteration in the recurrence Relation 5.51 is not automatically granted. Specifically, small arguments with $|x| < n$ can cause an appreciable accumulation of roundoff errors even for small orders, rendering the forward recursion inapplicable. An effective way to avoid these instabilities is provided by *Miller's method* (Miller 1952), which basically relies on the stability of the *backward recurrence* (for decreasing function orders):

$$j_i(x) = \frac{2i+3}{x} j_{i+1}(x) - j_{i+2}(x). \tag{5.54}$$

The unavailability of compact expressions for large-order Bessel functions, to be used for initiating this recurrence, can be overcome by the insight that, for a given x, the absolute values of the spherical Bessel

Listing 5.8 Spherical Neumann Functions $y_n(x)$ (Python Coding)

```
#================================================================================
def SBessy(n, x):
#--------------------------------------------------------------------------------
#   Evaluates iteratively the spherical Neumann function of order n in x
#--------------------------------------------------------------------------------
    y0 = -cos(x)/x
    if (n == 0): return y0
    y1 = (y0 - sin(x))/x
    if (n == 1): return y1

    for i in range(2,n+1):
        y = (2*i-1)/x*y1 - y0
        y0 = y1; y1 = y

    return y
```

functions of orders $n > |x|$ decrease as their order increases and, eventually, for some large enough order, $N \gg n$, the current function $j_N(x)$ becomes numerically negligible. Hence, considering the educated guess for the starting values

$$\tilde{j}_N = 0, \quad \tilde{j}_{N-1} = 1, \tag{5.55}$$

which is consistent with the *homogeneous* character of the differential equation (5.50), and iterating in Equation 5.54 backwards, the process ends up with a sequence of values, $\tilde{j}_N, \tilde{j}_{N-1}, \ldots, \tilde{j}_n, \ldots, \tilde{j}_1, \tilde{j}_0$, which are all affected by the same unknown scaling factor k. In particular,

$$\tilde{j}_0 = k j_0(x), \tag{5.56}$$

and, by using the known expression $j_0(x) = \sin x / x$, one can readily find the scaling factor and, thereby, the correct value of the nth-order spherical Bessel function:

$$j_n(x) = \frac{1}{k} \tilde{j}_n = \frac{j_0(x)}{\tilde{j}_0} \tilde{j}_n. \tag{5.57}$$

Another asset of Miller's algorithm consists in that it also prescribes the conditions under which the backward recursion should be applied and, in such a case, it provides a recipe for choosing the appropriate starting order N, so as to achieve the aimed precision in the evaluation of $j_n(x)$. As already mentioned, the forward iteration is always stable for $|x| > n$. A simple method to determine whether it remains stable also for $|x| < n$ is to retain on the right-hand side of the recurrence relation 5.51 only the first term, which is dominant for small x-values:

$$j_i(x) \simeq \frac{2i - 1}{x} j_{i-1}(x). \tag{5.58}$$

By starting with the arbitrary value $\tilde{j}_n = 1$ and iterating over increasing orders $i = n + 1, n + 2, \ldots,$ one can monitor the *net increase* of the successive spherical Bessel functions with respect to j_n:

$$\prod_{i=n}^{N} \frac{2i - 1}{x}. \tag{5.59}$$

Listing 5.9   Spherical Bessel Functions $j_n(x)$ (Python Coding)

```
#=============================================================================
def SBessj(n, x):
#-----------------------------------------------------------------------------
#  Evaluates iteratively the spherical Bessel function of order n in x
#-----------------------------------------------------------------------------
   if (x == 0e0): return 1e0 if (n == 0) else 0e0
   j0 = sin(x)/x
   if (n == 0): return j0
   j1 = (j0 - cos(x))/x
   if (n == 1): return j1

   nmax = 0                                 # finds direction of stable recurrence
   if (n >= fabs(x)):                       # nmax - 0 forward, nmax /- 0 backward
      jn = 1.
      for i in range(n,n+51):
         jn *= (2*i-1)/x                              # net factor of increase
         if (jn >= 1e8): nmax = i + 10; break         # for forward iteration

   if (nmax == 0):                                      # forward iteration
      for i in range(2,n+1):
         j = (2*i-1)/x*j1 - j0
         j0 = j1; j1 = j
      return j
   else:                                               # backward iteration
      j2 = 0.; j1 = 1e-20
      for i in range(nmax,-1,-1):
         j = (2*i+3)/x*j1 - j2
         j2 = j1; j1 = j
         if (i == n): jn = j                      # non-normalized jn

      return (j0/j)*jn                             # normalized jn
```

If, from a certain order N onward, this factor exceeds a reasonable value (such as 10^8), the forward recurrence proves unstable and, consequently, the backward recurrence needs to be applied starting from the found order N. In concrete implementations of the downward recursion, in order to dissipate even more the effect of the particular choice of the initial values (5.55), the starting order is yet increased above the found value, typically to $N + 10$.

The function SBessj from Listing 5.9 implements Miller's algorithm for calculating the nth-order spherical Bessel function. For the lowest orders, the function uses the analytical expressions (5.52). The variable nmax is used to flag the direction of stable iteration in the recurrence relation and, when used for descending iteration, it corresponds to the deliberately overestimated starting order $N > n$.

Figure 5.3 displays the spherical Neumann functions of orders up to $n = 3$, calculated using routine SBessy. Confirming their status of irregular solutions, they can be seen to diverge at the origin. Figure 5.4 shows the plots of the corresponding spherical Bessel functions, calculated using function SBessj. By contrast with the spherical Neumann functions, their regular character follows from their finite behavior at the origin.

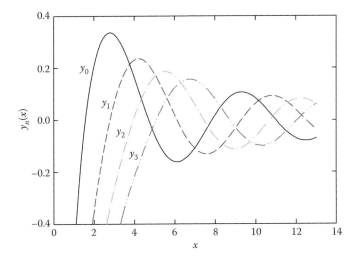

FIGURE 5.3 Spherical Neumann functions $y_n(x)$ ($n = 0 - 3$), as calculated by function SBessy.

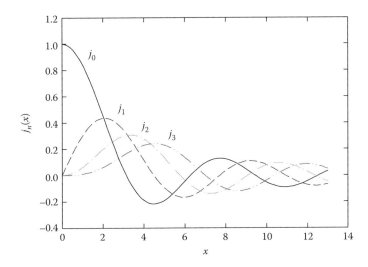

FIGURE 5.4 Spherical Bessel functions $j_n(x)$ ($n = 0 - 3$), as calculated by function SBessj.

5.7 Implementations in C/C++

Listings 5.10 and 5.11 show the content of the files elemfunc.h and specfunc.h, which contain equivalent C/C++ implementations of the Python functions developed in the main text and included in the modules elemfunc.py and specfunc.py. The corresponding routines have identical names, parameters, and functionalities.

Listing 5.10 Elementary Functions (elemfunc.h)

```
//--------------------------------- elemfunc.h ---------------------------------
// Contains routines for evaluating elementary functions.
// Author: Titus Beu, 2013
//------------------------------------------------------------------------------
#ifndef _ELEMFUNC_
```

```
#define _ELEMFUNC_

#include <math.h>

//===========================================================================
double Poly(double x, double a[], int n)
//---------------------------------------------------------------------------
// Evaluates the polynomial P(x) = a[0] x^n + a[1] x^(n-1) + ... + a[n] with
// real coefficients in x using Horner's scheme
//---------------------------------------------------------------------------
{
   double p;
   int i;

   p = a[0];
   for (i=1; i<=n; i++) p = p*x + a[i];
   return p;
}

//===========================================================================
void PolyDerive(double a[], double b[], int n)
//---------------------------------------------------------------------------
// For the real polynomial P(x) = a[0] x^n + ... + a[n], the function returns
// the coefficients of the derivative P'(x) = b[0] x^(n-1) + ... + b[n-1]
//---------------------------------------------------------------------------
{
   int i;

   for (i=0; i<=n; i++) b[i] = (n-i) * a[i];
}

//===========================================================================
void PolyDivide(double x0, double a[], double b[], int n)
//---------------------------------------------------------------------------
// For the real polynomial P(x) = a[0] x^n + ... + a[n], the function returns
// the coefficients of the division by the binomial (x-x0):
// P(x) = (x-x0) (b[0] x^(n-1) + ... + b[n-1]) + b[n] (b[n] = P(x0))
//---------------------------------------------------------------------------
{
   int i;

   b[0] = a[0];
   for (i=1; i<=n; i++) b[i] = b[i-1]*x0 + a[i];
}

//===========================================================================
double Exp0(double x)
//---------------------------------------------------------------------------
// Evaluates exp(x) from its power-series expansion
//---------------------------------------------------------------------------
{
   const double eps = 1e-14;                              // relative precision
   double f, t;
   int i;

   i = 0;
   f = t = 1e0;
   while (fabs(t) > eps*fabs(f)) {
      i++;
```

```
      t *= x/i;
      f += t;
   }
   return f;
}


//=============================================================================
double Exp(double x)
//-----------------------------------------------------------------------------
// Evaluates exp(x) from its power-series expansion
// For x < 0 avoids potential instabilities due to subtractions
//-----------------------------------------------------------------------------
{
   const double eps = 1e-14;                              // relative precision
   double absx, f, t;
   int i;

   i = 0;
   f = t = 1e0;
   absx = fabs(x);
   while (fabs(t) > eps*fabs(f)) {
      i++;
      t *= absx/i;
      f += t;
   }
   return (x >= 0.0 ? f : 1e0/f);
}


//=============================================================================
double Sin(double x)
//-----------------------------------------------------------------------------
// Evaluates sin(x) from its power-series expansion
//-----------------------------------------------------------------------------
{
   const double eps = 1e-14;                              // relative precision
   double f, t, x2;
   int i;

   i = 1;
   f = t = x;
   x2 = x*x;
   while (fabs(t) > eps*fabs(f)) {
      i += 2;
      t *= -x2/((i-1)*i);
      f += t;
   }
   return f;
}


//=============================================================================
double ArcSin(double x)
//-----------------------------------------------------------------------------
// Evaluates arcsin(x) from its power-series expansion (|x| < 1)
//-----------------------------------------------------------------------------
{
   const double eps = 1e-14;                              // relative precision
   double f, t, x2;
   int i, i2;
```

```
   i = 1;
   f = t = x;
   x2 = x*x;
   while (fabs(t) > eps*fabs(f)) {
      i2 = i*i;
      i += 2;
      t *= i2*x2/((i-1)*i);
      f += t;
   }
   return f;
}

//==========================================================================
double Tan(double x)
//--------------------------------------------------------------------------
// Evaluates tan(x) from its continued fraction representation
//--------------------------------------------------------------------------
{
   const double eps = 1e-14;                          // relative precision
   double a, b, p, pm1, pm2, q, qm1, qm2, r, rm1;

   a = 1e0; b = -x*x;
   pm1 = 0e0; p = x;
   qm1 = 1e0; q = 1e0;
   rm1 = 0e0; r = p/q;
   while (fabs(r-rm1) > eps*fabs(r)) {
      pm2 = pm1; pm1 = p;                             // shift the stack
      qm2 = qm1; qm1 = q;
      rm1 = r;
      a += 2e0;
      p = a*pm1 + b*pm2;
      q = a*qm1 + b*qm2;
      r = p/q;                                        // new convergent value
   }
   return r;
}

//==========================================================================
double Exp1(double x)
//--------------------------------------------------------------------------
// Evaluates exp(x) from its continued fraction representation
//--------------------------------------------------------------------------
{
   const double eps = 1e-14;                          // relative precision
   double a, b, p, pm1, pm2, q, qm1, qm2, r, rm1;
   int i;

   a = 1e0; b = x;
   pm1 = 1e0; p = 1e0 + x;
   qm1 = 1e0; q = 1e0;
   rm1 = 1e0; r = p/q;
   i = 1;
   while (fabs(r-rm1) > eps*fabs(r)) {
      i++;
      pm2 = pm1; pm1 = p;                             // shift the stack
      qm2 = qm1; qm1 = q;
      rm1 = r;
      a - (i%2 ? i : 2e0);
      b = -b;
```

```
      p = a*pm1 + b*pm2;
      q = a*qm1 + b*qm2;
      r = (q ? p/q : 9e99);                        // new convergent value
   }
   return r;
}

#endif
```

Listing 5.11 Special Functions (`specfunc.h`)

```
//-------------------------------- specfunc.h --------------------------------
// Contains routines for evaluating special functions.
// Part of the numxlib numerics library. Author: Titus Beu, 2013
//----------------------------------------------------------------------------
#ifndef _SPECFUNC_
#define _SPECFUNC_

#include <stdlib.h>
#include <math.h>


//============================================================================
double Chebyshev(int n, double x, double d)
//----------------------------------------------------------------------------
// Evaluates the n-th order Chebyshev polynomial and its derivative d in x
// using the recurrence relation
//----------------------------------------------------------------------------
{
   double f, fm1, fm2, x2;
   int i;

   if (n == 0) {
      f = 1e0; d = 0e0;
   } else {
      f = x; fm1 = 1e0; x2 = 2*x;
      for (i=2; i<=n; i++) {
         fm2 = fm1; fm1 = f;
         f = x2*fm1 - fm2;
      }
      d = (x*x-1e0) ? n*(x*f-fm1)/(x*x-1e0) : n*n*f/x;
   }
   return f;
}


//============================================================================
double Legendre(int n, double x, double &d)
//----------------------------------------------------------------------------
// Evaluates the n-th order Legendre polynomial and its derivative d in x
// using the recurrence relation
//----------------------------------------------------------------------------
{
   double f, fm1, fm2;
   int i;

   if (n == 0) {
      f = 1e0; d = 0e0;
   } else {
      f = x; fm1 = 1e0;
      for (i=2; i<=n; i++) {
```

```
            fm2 = fm1; fm1 = f;
            f = ((2*i-1)*x*fm1 - (i-1)*fm2)/i;
         }
         d = (x*x-1e0) ? n*(x*f-fm1)/(x*x-1e0) : 0.5*n*(n+1)*f/x;
      }
      return f;
}


//==========================================================================
double aLegendre(int l, int m, double x)
//--------------------------------------------------------------------------
// Evaluates the associated Legendre function of orders l and m >= 0 in x
//--------------------------------------------------------------------------
{
   double p, pm1, pm2, sqx;
   int i;

   if (l < m) return 0e0;

   p = 1e0; pm1 = 0e0;                          // seed values: P(m,m,x), P(m-1,m,x)
   if (m) {
      sqx = -sqrt(1e0-x*x);
      for (i=1; i<=m; i++) p *= (2*i-1) * sqx;
   }

   for (i=m+1; i<=l; i++) {                                          // recurrence
      pm2 = pm1; pm1 = p;
      p = ((2*i-1)*x*pm1 - (i+m-1)*pm2)/(i-m);
   }
   return p;
}


//==========================================================================
double Laguerre(int n, double x, double &d)
//--------------------------------------------------------------------------
// Evaluates the n-th order Laguerre polynomial and its derivative d in x
// using the recurrence relation
//--------------------------------------------------------------------------
{
   double f, fm1, fm2;
   int i;

   if (n == 0) {
      f = 1e0; d = 0e0;
   } else {
      f = 1e0 - x; fm1 = 1e0;
      for (i=2; i<=n; i++) {
         fm2 = fm1; fm1 = f;
         f = ((2*i-1-x)*fm1 - (i-1)*fm2)/i;
      }
      d = x ? n*(f-fm1)/x : -n*f;
   }
   return f;
}


//==========================================================================
double aLaguerre(int n, int k, double x)
//--------------------------------------------------------------------------
// Evaluates the associated Laguerre polynomial of orders n and k in x
```

```
// using the recurrence relation
//------------------------------------------------------------------------
{
   double f, fm1, fm2;
   int i;

   if (n == 0) {
      f = 1e0;
   } else {
      f = 1e0 + k - x; fm1 = 1e0;
      for (i=2; i<=n; i++) {
         fm2 = fm1; fm1 = f;
         f = ((2*i+k-1-x)*fm1 - (i+k-1)*fm2)/i;
      }
   }
   return f;
}

//========================================================================
double Hermite(int n, double x, double &d)
//------------------------------------------------------------------------
// Evaluates the n-th order Hermite polynomial and its derivative d in x
// using the recurrence relation
//------------------------------------------------------------------------
{
   double f, fm1, fm2, x2;
   int i;

   if (n == 0) {
      f = 1e0; d = 0e0;
   } else {
      f = 2*x; fm1 = 1e0; x2 = 2*x;
      for (i=2; i<=n; i++) {
         fm2 = fm1; fm1 = f;
         f = x2*fm1 - 2*(i-1)*fm2;
      }
      d = 2*n*fm1;
   }
   return f;
}

//========================================================================
void SpherY(int l, int m, double theta, double phi, double &ReY, double &ImY)
//------------------------------------------------------------------------
// Evaluates the real and imaginary parts (ReY and ImY) of the spherical
// harmonic of orders l and m for arguments theta and phi.
// Calls aLegendre to calculate associated Legendre polynomials.
//------------------------------------------------------------------------
{
#define pi 3.141592653589793
   double fact;
   int i, mabs;

   mabs = abs(m);

   fact = 1e0;
   for (i=l-mabs+1; i<=l+mabs; i++) fact *= i;              // (l+|m|)!/(l-|m|)!

   fact = sqrt((2*l+1)/(4e0*pi*fact)) * aLegendre(l,mabs,cos(theta));
```

```
      if (m < 0 && m % 2) fact = -fact;

   ReY = fact * cos(m*phi);
   ImY = fact * sin(m*phi);
}


//===========================================================================
double SBessj(int n, double x)
//---------------------------------------------------------------------------
// Evaluates iteratively the spherical Bessel function of order n in x
//---------------------------------------------------------------------------
{
   double j, j0, j1, j2, jn;
   int i, nmax;

   if (x == 0e0) return (n == 0 ? 1e0 : 0e0);
   j0 = sin(x)/x; if (n == 0) return j0;
   j1 = (j0 - cos(x))/x; if (n == 1) return j1;

   nmax = 0;                          // finds direction of stable recurrence
   if ((double)n >= fabs(x)) {        // nmax = 0 forward, nmax /= 0 backward
      jn = 1.;
      for (i=n; i<=(n+50); i++) {
         jn *= (2*i-1)/x;                        // net factor of increase
         if (jn >= 1e8) {nmax = i + 10; break;}  // for forward iteration
      }
   }

   if (nmax == 0) {                                       // forward iteration
      for (i=2; i<=n; i++) {
         j = (2*i-1)/x*j1 - j0;
         j0 = j1; j1 = j;
      }
      return j;
   } else {                                               // backward iteration
      j2 = 0.; j1 = 1e-20;
      for (i=nmax; i>=0; i--) {
         j = (2*i+3)/x*j1 - j2;
         j2 = j1; j1 = j;
         if (i == n) jn = j;                               // non-normalized jn
      }
      return (j0/j)*jn;                                    // normalized jn
   }
}


//===========================================================================
double SBessy(int n, double x)
//---------------------------------------------------------------------------
// Evaluates iteratively the spherical Neumann function of order n in x
//---------------------------------------------------------------------------
{
   double y, y0, y1;
   int i;

   y0 = -cos(x)/x; if (n == 0) return y0;
   y1 = (y0 - sin(x))/x; if (n == 1) return y1;

   for (i=2; i<=n; i++) {
      y = (2*i-1)/x*y1 - y0;
```

```
        y0 = y1; y1 = y;
    }
    return y;
}

#endif
```

5.8 Problems

The Python and C/C++ programs for the following problems may import the functions developed in this chapter from the modules elemfunc.py (.h) and, respectively, for the special functions, from specfunc.py (.h), which are available as supplementary material. For runtime plots, the graphical routines contained in the libraries graphlib.py and graphlib.h may be employed.

PROBLEM 5.1

Write a function to calculate the coefficients of the first derivative of polynomials and use it in a program to plot the fifth-degree Legendre polynomial,

$$P_5(x) = \left(63x^5 - 70x^3 + 15x\right)/8, \tag{5.60}$$

and its first derivative.

Listing 5.12   Plot of a Polynomial and Its Derivative (Python Coding)

```
# Plot polynomial and derivative
from elemfunc import *
from graphlib import *

# main

np = 5                                              # degree of polynomial
a = [63e0/8e0, 0e0, -70e0/8e0, 0e0, 15e0/8e0, 0e0]        # coefficients
b = [0]*(np+1)                                      # coeffs of derivative

xmin = -1e0; xmax = 1e0                                # plotting domain
h = 0.01e0                                           # argument spacing
n = int((xmax-xmin)/h) + 1                          # number of points

x = [0]*(n+1); y = [0]*(n+1); z = [0]*(n+1)            # arrays for plots

PolyDerive(a,b,np)                        # coefficients of derivative in b

for i in range(1,n+1):
    x[i] = xmin + (i-1)*h                                    # argument
    y[i] = Poly(x[i],a,np)                                 # polynomial
    z[i] = Poly(x[i],b,np-1)                               # derivative

GraphInit(1200,600)

Plot(x,y,n,"blue",1,0.10,0.45,0.15,0.85,"x","P(x)","Polynomial")
Plot(x,z,n,"red" ,1,0.60,0.95,0.15,0.85,"x","P'(x)","Derivative")

MainLoop()
```

Solution

Implementations are provided in Listings 5.12 and 5.13, and graphical output, in Figure 5.5.

Listing 5.13   Plot of a Polynomial and Its Derivative (C/C++ Coding)

```
// Plot polynomial and derivative
#include "memalloc.h"
#include "elemfunc.h"
#include "graphlib.h"

int main(int argc, wchar_t** argv)
{
   const int np = 5;                                      // degree of polynomial
   double a[] = {63e0/8e0, 0e0, -70e0/8e0, 0e0, 15e0/8e0, 0e0};   // coeffs
   double b[np+1];                                             // work array
   double h, xmin, xmax, *x, *y, *z;
   int i, n;

   xmin = -1e0; xmax = 1e0;                                 // plotting domain
   h = 0.01e0;                                              // argument spacing
   n = int((xmax-xmin)/h + 0.5) + 1;                        // number of points

   x = Vector(1,n); y = Vector(1,n); z = Vector(1,n);       // arrays for plots

   PolyDerive(a,b,np);                              // coefficients of derivative in b

   for (i=1; i<=n; i++) {
      x[i] = xmin + (i-1)*h;                                        // argument
      y[i] = Poly(x[i],a,np);                                       // polynomial
      z[i] = Poly(x[i],b,np-1);                                     // derivative
   }

   PyGraph w(argc, argv);
   w.GraphInit(1200,600);

   w.Plot(x,y,n,"blue",1,0.10,0.45,0.15,0.85,"x","P(x)","Polynomial");
   w.Plot(x,z,n,"red" ,1,0.60,0.95,0.15,0.85,"x","P'(x)","Derivative");

   w.MainLoop();
}
```

PROBLEM 5.2

Write routines for calculating the functions $\sin x$ and $\arcsin x$ from their power-series expansions:

$$\sin x = \sum_{i=0}^{\infty} \frac{(-1)^i}{(2i+1)!} x^{2i+1} = x - \frac{x^3}{3!} + \frac{x^5}{5!} - \cdots, \tag{5.61}$$

$$\arcsin x = \sum_{i=0}^{\infty} \frac{(2i)!}{2^{2i}(i!)^2(2i+1)} x^{2i+1} = x + \frac{1}{6}x^3 + \frac{3}{40}x^5 + \cdots. \tag{5.62}$$

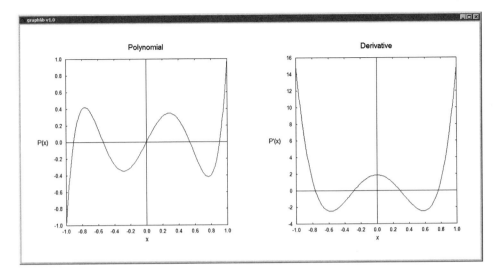

FIGURE 5.5 Fifth-degree Legendre polynomial and its first derivative, as calculated with functions `Poly` and `PolyDerive`.

Solution

Simplify the power series to have an index running over the exponent values:

$$\sin x = \sum_{i=1,3}^{\infty} \frac{(-1)^{(i-1)/2}}{i!} x^i,$$

$$\arcsin x = \sum_{i=0}^{\infty} \frac{(2i-1)!!}{(2i)!!(2i+1)} x^{2i+1} = \sum_{i=1,3}^{\infty} \frac{(i-2)!!}{(i-1)!!i} x^i.$$

Recursive process for $\sin x$:

$$\begin{cases} t_0 = x, & F_0 = x, \\ t_i = \dfrac{x^2}{(i-1)i} t_{i-1}, & F_i = F_{i-1} + t_i, \quad i = 1, 3, 5, \ldots. \end{cases} \tag{5.63}$$

Recursive process for $\arcsin x$:

$$\begin{cases} t_0 = x, & F_0 = x, \\ t_i = \dfrac{(i-2)^2 x^2}{(i-1)i} t_{i-1}, & F_i = F_{i-1} + t_i, \quad i = 1, 3, 5, \ldots. \end{cases} \tag{5.64}$$

The Python implementations of the routines are given in Listing 5.14, being included in the module `elemfunc.py`, and the corresponding C/C++ variants are part of the header file `elemfunc.h`.

Listing 5.14   Sine and Arcsine Functions from Their Power-Series Expansions (Python Coding)

```
#=========================================================================
def Sin(x):
#-------------------------------------------------------------------------
#  Evaluates sin(x) from its power-series expansion
#-------------------------------------------------------------------------
   eps = 1e-14                                          # relative precision

   i = 1
   f = t = x
   x2 = x*x
   while (fabs(t) > eps*fabs(f)):
      i += 2
      t *= -x2/((i-1)*i)
      f += t

   return f

#=========================================================================
def ArcSin(x):
#-------------------------------------------------------------------------
#  Evaluates arcsin(x) from its power-series expansion
#-------------------------------------------------------------------------
   eps = 1e-14                                          # relative precision

   i = 1
   f = t = x
   x2 = x*x
   while (fabs(t) > eps*fabs(f)):
      i2 = i*i
      i += 2
      t *= i2*x2/((i-1)*i)
      f += t

   return f
```

PROBLEM 5.3

Write a function to evaluate the function $\exp(x)$ from its continued fraction representation (Abramowitz and Stegun 1972, Equation 4.2.40):

$$e^x = \left[1; \frac{x}{1}, \frac{-x}{2}, \frac{x}{3}, \frac{-x}{2}, \frac{x}{5}, \frac{-x}{2}, \ldots\right]. \tag{5.65}$$

Write a main program to call this routine, as well as function Exp (based on the power-series expansion and given in Listing 5.4), and compare the number of iterations needed for convergence with the same relative precision $\varepsilon = 10^{-14}$, for arguments $x \in [-10, 10]$ spaced by 1.

Solution

The elements of the continued fraction are

$$\begin{cases} b_1 = x, \quad b_i = -b_{i-1}, \\ a_0 = 1, \quad a_1 = 1, \quad a_i = \begin{cases} 2, & i = 2, 4, \ldots, \\ i, & i = 3, 5, \ldots. \end{cases} \end{cases} \tag{5.66}$$

Listing 5.15 Exponential Function from Its Continued Fraction Representation (Python Coding)

```
#===============================================================================
def Exp1(x):
#-------------------------------------------------------------------------------
#  Evaluates exp(x) from its continued fraction representation
#-------------------------------------------------------------------------------
   eps = 1e-14                                            # relative precision

   a = 1e0; b = x
   pm1 = 1e0; p = 1e0 + x
   qm1 = 1e0; q = 1e0
   rm1 = 1e0; r = p/q
   i = 1
   while (fabs(r-rm1) > eps*fabs(r)):
      i += 1
      pm2 = pm1; pm1 = p                                  # shift the stack
      qm2 = qm1; qm1 = q
      rm1 = r
      a = (i if i%2 else 2e0)
      b = -b
      p = a*pm1 + b*pm2
      q = a*qm1 + b*qm2
      r = (p/q if q else 9e99)                            # new convergent value

   return r
```

The following starting values can be used:

$$\begin{cases} a_1 = 1, & b_1 = x, \\ P_0 = 1, & P_1 = 1 + x, \\ Q_0 = 1, & Q_1 = 1. \end{cases} \tag{5.67}$$

The recursive evaluation of the convergents R_i can be carried out based on the relations:

$$\begin{cases} a_i = \begin{cases} 2, & i = 2, 4, \ldots, \\ i, & i = 3, 5, \ldots, \end{cases} \\ b_i = -b_{i-1}, \end{cases} \tag{5.68}$$

and

$$\begin{cases} P_i = a_i P_{i-1} + b_i P_{i-2}, \\ Q_i = a_i Q_{i-1} + b_i Q_{i-2}, \\ R_i = P_i/Q_i. \end{cases} \tag{5.69}$$

The Python implementation of the routine is given in Listing 5.15, being included in the module elemfunc.py, and the corresponding C/C++ variant is part of the header file elemfunc.h.

PROBLEM 5.4

Write a program to calculate the squared norms $|Y_{lm}(\theta, \varphi)|^2$ of the spherical harmonics (defined by Equation 5.35), for $l = 0, 1, 2$ and $m = -l, -l+1, \ldots, l$, using the function SpherY from the modules specfunc.py and, respectively, specfunc.h. Display the norms as polar plots using the function Plot from the graphics library graphlib, with the input parameter polar set to a nonzero value.

Solution

The implementations are given in Listings 5.16 and 5.17, and the graphical output is shown in Figure 5.6.

Listing 5.16   Squared Spherical Harmonics (Python Coding)

```
# Plot squared spherical harmonics
from math import *
from specfunc import *
from graphlib import *

GraphInit(1200,1000)

lmax = 2                                              # maximum value of l
maxplotx = 3                                      # max. no. of plots along x

h = pi/180                                           # theta angle spacing
n = int(2*pi/h) + 1                                # no. of theta values

x = [0]*(n+1); y = [0]*(n+1)

nplot = 0
for l in range(0,lmax+1): nplot += (2*l+1)               # total no. of plots

nplotx = min(nplot,maxplotx)                        # no. of plots along x
nploty = int(nplot/nplotx)                          # no. of plots along y
if (nplot % nplotx): nploty += 1                 # incomplete row of plots

dplotx = 1e0/nplotx                    # fractional width of a plot along x
dploty = 1e0/nploty                    # fractional width of a plot along y

xplot = 0; yplot = 0                             # lower-left corner of plot
for l in range(0,lmax+1):                                          # l-loop
   for m in range (-l,l+1):                                        # m-loop
      for i in range(1,n+1):                                  # theta-loop
         theta = i * h
         (ReY,ImY) = SpherY(l,m,theta,0e0)              # spherical harmonic
         f = ReY * ReY + ImY * ImY                           # squared norm
         x[i] = f * sin(theta)                       # Cartesian projections
         y[i] = f * cos(theta)

      fxmin = xplot + 0.1*dplotx; fxmax = xplot + 0.9*dplotx         # viewport
      fymin = yplot + 0.1*dploty; fymax = yplot + 0.9*dploty
      title = "l = " + repr(l) + ",   m = " + repr(m)
      Plot(x,y,n,"blue",2,fxmin,fxmax,fymin,fymax,"","",title)

      xplot += dplotx
      if (xplot >= 1):                               # reached the right margin
         xplot = 0                                    # begin a new row of plots
         yplot += dploty

MainLoop()
```

Listing 5.17 Squared Spherical Harmonics (C/C++ Coding)

```
// Plot squared spherical harmonics
#include <math.h>
#include "memalloc.h"
#include "specfunc.h"
#include "graphlib.h"
#define pi 3.141592653589793

int main(int argc, wchar_t** argv)
{
   int i, l, lmax, m, maxplotx, n, nplot, nplotx, nploty;
   double dplotx, dploty, f, h, ReY, ImY, theta, xplot, yplot;
   double fxmin, fxmax, fymin, fymax;
   double *x, *y;
   char title[20], lchr[5], mchr[5];

   PyGraph w(argc, argv);
   w.GraphInit(1200,1000);

   lmax = 2;                                        // maximum value of l
   maxplotx = 3;                               // max. no. of plots along x
   h = pi/180;                                      // theta angle spacing
   n = int(2*pi/h) + 1;                             // no. of theta values
   x = Vector(1,n); y = Vector(1,n);

   nplot = 0;
   for (l=0; l<=lmax; l++) nplot += (2*l+1);        // total no. of plots
   nplotx = (nplot <= maxplotx ? nplot : maxplotx); // no. of plots along x
   nploty = int(nplot/nplotx);                      // no. of plots along y
   if (nplot % nplotx) nploty++;                  // incomplete row of plots

   dplotx = 1e0/nplotx;                    // fractional width of a plot along x
   dploty = 1e0/nploty;                    // fractional width of a plot along y

   xplot = 0; yplot = 0;                           // lower-left corner of plot
   for (l=0; l<=lmax; l++) {                                        // l-loop
      for (m=-l; m<=l; m++) {                                       // m-loop
         for (i=1; i<=n; i++) {                                 // theta-loop
            theta = i * h;
            SpherY(l,m,theta,0e0,ReY,ImY);                 // spherical harmonic
            f = ReY * ReY + ImY * ImY;                          // squared norm
            x[i] = f * sin(theta);                      // Cartesian projections
            y[i] = f * cos(theta);
         }
         fxmin = xplot + 0.1*dplotx; fxmax = xplot + 0.9*dplotx;   // viewport
         fymin = yplot + 0.1*dploty; fymax = yplot + 0.9*dploty;
         sprintf(lchr,"%i",l); sprintf(mchr,"%i",m);
         strcpy(title,"l = ");
         strcat(strcat(strcat(title,lchr),",   m = "),mchr);
         w.Plot(x,y,n,"blue",2,fxmin,fxmax,fymin,fymax,"","",title);

         xplot += dplotx;
         if (xplot >= 1) {                             // reached the right margin
            xplot = 0;                                 // begin a new row of plots
            yplot += dploty;
         }
      }
   }
   w.MainLoop();
}
```

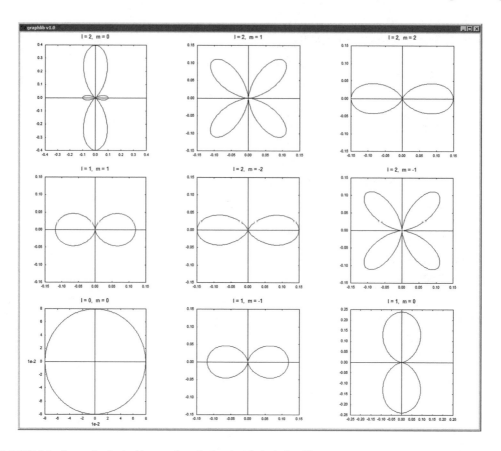

FIGURE 5.6   Squared spherical harmonics calculated with the help of function SpherY.

PROBLEM 5.5

The *addition theorem for the spherical harmonics* equates the summed products of all spherical harmonics for a particular order l and given angles (θ_1, φ_1) and (θ_2, φ_2) with the Legendre polynomial of order l,

$$P_l(\cos \gamma) = \frac{4\pi}{2l+1} \sum_{m=-l}^{m=l} Y_{lm}^*(\theta_2, \varphi_2) Y_{lm}(\theta_1, \varphi_1), \tag{5.70}$$

where γ is the angle between directions (θ_1, φ_1) and (θ_2, φ_2), which can be expressed as

$$\cos \gamma = \cos \theta_1 \cos \theta_2 + \sin \theta_1 \sin \theta_2 \cos (\varphi_1 - \varphi_2), \tag{5.71}$$

and the spherical harmonics $Y_{lm}(\theta, \varphi)$ are defined in Equation 5.35. Using routines Legendre and SpherY included in files specfunc.py and, respectively, specfunc.h, write Python and C/C++ programs to check the addition theorem for $l = 5$ and the angle combination $\theta_1 = \pi/9$, $\varphi_1 = \pi/3$, $\theta_2 = \pi/3$, $\varphi_2 = \pi/8$.

Solution

The real and imaginary parts in the addition theorem may be separated as

$$\frac{4\pi}{2l+1} \sum_{m=-l}^{m=l} [\operatorname{Re} Y_{lm}(\theta_2,\varphi_2)\operatorname{Re} Y_{lm}(\theta_1,\varphi_1) + \operatorname{Im} Y_{lm}(\theta_2,\varphi_2)\operatorname{Im} Y_{lm}(\theta_1,\varphi_1)] = P_l(\cos\gamma),$$

$$\frac{4\pi}{2l+1} \sum_{m=-l}^{m=l} [\operatorname{Re} Y_{lm}(\theta_2,\varphi_2)\operatorname{Im} Y_{lm}(\theta_1,\varphi_1) - \operatorname{Im} Y_{lm}(\theta_2,\varphi_2)\operatorname{Re} Y_{lm}(\theta_1,\varphi_1)] = 0.$$

The programs for checking of the above relations are given in Listings 5.18 and 5.19.

Listing 5.18 Addition Theorem for Spherical Harmonics (Python Coding)

```python
# Checks the addition theorem for spherical harmonics
from math import *
from specfunc import *

l = 5
theta1 = pi/5; phi1 = pi/9
theta2 = pi/3; phi2 = pi/8

cosgam = cos(theta1) * cos(theta2) \
       + sin(theta1) * sin(theta2) * cos(phi2 - phi1)
(P,d)  = Legendre(l,cosgam)

sumRe = 0e0
sumIm = 0e0
for m in range(-l,l+1):
    (ReY1,ImY1) = SpherY(l, m, theta1, phi1)
    (ReY2,ImY2) = SpherY(l, m, theta2, phi2)

    sumRe += ReY2 * ReY1 + ImY2 * ImY1
    sumIm += ReY2 * ImY1 - ImY2 * ReY1

sumRe *= 4*pi/(2*l+1)
sumIm *= 4*pi/(2*l+1)

print(P-sumRe,sumIm)                      # check: P - sumRe = 0, sumIm = 0
```

Listing 5.19 Addition Theorem for Spherical Harmonics (C/C++ Coding)

```c
// Checks the addition theorem for spherical harmonics
#include <stdio.h>
#include <math.h>
#include "specfunc.h"

#define pi 3.141592653589793

int main()
{
   double cosgam, phi1, phi2, theta1, theta2;
   double d, P, ImY1, ImY2, ReY1, ReY2, sumIm, sumRe;
   int l, m;
```

```
    l = 5;
    theta1 = pi/5; phi1 = pi/9;
    theta2 = pi/3; phi2 = pi/8;

    cosgam = cos(theta1) * cos(theta2) \
           + sin(theta1) * sin(theta2) * cos(phi2 - phi1);
    P = Legendre(l,cosgam,d);

    sumRe = 0e0;
    sumIm = 0e0;
    for (m=-l; m<=l; m++) {
        SpherY(l, m, theta1, phi1, ReY1, ImY1);
        SpherY(l, m, theta2, phi2, ReY2, ImY2);

        sumRe += ReY2 * ReY1 + ImY2 * ImY1;
        sumIm += ReY2 * ImY1 - ImY2 * ReY1;
    }
    sumRe *= 4*pi/(2*l+1);
    sumIm *= 4*pi/(2*l+1);

    printf("%f  %f\n",P-sumRe,sumIm);              // P - sumRe = 0, sumIm ~ 0
}
```

PROBLEM 5.6

The propagation of spherical electromagnetic waves is described by the scalar Helmholtz equation, which, upon separation of variables in spherical coordinates, leads to the radial equation (see Arfken and Weber 2005, Equation 11.140):

$$\left[r^2 \frac{d^2}{dr^2} + r\frac{d}{dr} + k^2 r^2 - \left(n + \frac{1}{2}\right)^2 \right] Z(r) = 0. \tag{5.72}$$

The linearly independent solutions of this equation are the so-called *cylindrical* Bessel and Neumann functions of fractional orders, $J_{n+1/2}(kr)$ and $N_{n+1/2}(kr)$, which are related to the spherical Bessel and

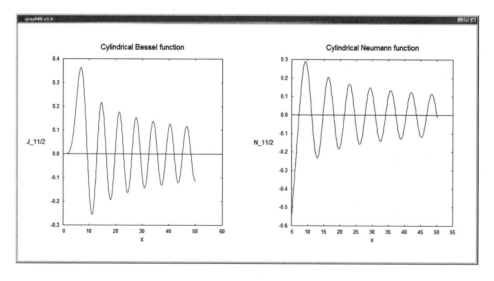

FIGURE 5.7 Cylindrical Bessel functions $J_{11/2}(x)$ and $N_{11/2}(x)$ calculated according to Equation 5.73 by using functions SBessj and SBessy.

Neumann functions by

$$j_n(x) = \sqrt{\frac{\pi}{2x}} J_{n+1/2}(x), \quad y_n(x) = \sqrt{\frac{\pi}{2x}} N_{n+1/2}(x). \tag{5.73}$$

Write routines for calculating cylindrical Bessel functions by calling functions SBessj and SBessy defined in specfunc.py and, respectively, specfunc.h. Plot the cylindrical Bessel functions $J_{11/2}(x)$ for $0 \le x \le 50$ and $N_{11/2}(x)$ for $x \in [0, 50]$.

Solution

The Python and C/C++ programs are available as supplementary material in the files P05-CBess Plot.py and P05-CBessPlot.cpp, and the graphical output is shown in Figure 5.7. Sample definitions for the cylindrical Bessel and Neumann functions, based on the corresponding spherical functions, can be seen in Listing 5.20.

Listing 5.20   Cylindrical Bessel and Neumann functions (Python Coding)

```
def CBessJ(n, x):                   # cylindrical Bessel function of order n+1/2
    return sqrt(2e0*x/pi) * SBessj(n,x)

def CBessN(n, x):                   # cylindrical Neumann function of order n+1/2
    return sqrt(2e0*x/pi) * SBessy(n,x)
```

PROBLEM 5.7

The associated Laguerre polynomials are defined in terms of Laguerre polynomials as (see Arfken and Weber 2005)

$$L_n^k(x) = (-1)^k \frac{d^k}{dx^k} L_{n+k}(x). \tag{5.74}$$

The associated Laguerre polynomials satisfy the recurrence relation:

$$n L_n^k(x) = (2n + k - 1 - x) L_{n-1}^k(x) - (n + k - 1) L_{n-1}^k(x) \tag{5.75}$$

with respect to the lower order, and the simple expressions for the orders $n = 0$ and 1,

$$L_0^k(x) = 1, \quad L_1^k(x) = -x + k + 1, \tag{5.76}$$

can be used to start the recurrence and calculate the polynomials for any given orders.

One of the most illustrative occurrences of the associated Laguerre functions is as part of the wave function of the hydrogen atom:

$$R_{nl}(r) = \left\{ \left(\frac{2}{na_0}\right)^3 \frac{(n - l - 1)!}{2n[(n + l)!]^3} \right\}^{1/2} \rho^l e^{-\rho/2} L_{n+l}^{2l+1}(\rho), \tag{5.77}$$

where $a_0 = 0.529177249$ Å is the Bohr radius and the dimensionless radial variable ρ is defined as

$$\rho = \frac{2}{na_0} r. \tag{5.78}$$

Write a routine to evaluate the associated Laguerre polynomials using Equations 5.75 and 5.76 and use it in a program to plot the radial probability density $\rho^2 |R_{nl}(\rho)|^2$ in dimensionless units, setting $a_0 = 1$.

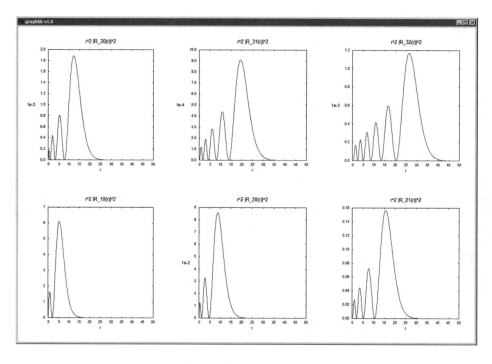

FIGURE 5.8 Radial density probabilities $\rho^2|R_{nl}(\rho)|^2$ for the electron in the hydrogen atom.

Listing 5.21 Radial Wave Functions for the Hydrogen Atom (Python Coding)

```
#===============================================================================
def aLaguerre(n, k, x):
#-------------------------------------------------------------------------------
#  Evaluates the associated Laguerre polynomial of orders n and k in x
#  using the recurrence relation
#-------------------------------------------------------------------------------
   if (n == 0):
      f = 1e0
   else:
      f = 1e0 + k - x; fm1 = 1e0
      for i in range(2,n+1):
         fm2 = fm1; fm1 = f
         f = ((2*i+k-1-x)*fm1 - (i+k-1)*fm2)/i

   return f


#===============================================================================
def RadPsiH(n, l, r):
#-------------------------------------------------------------------------------
#  Evaluates the radial wave function of the hydrogen atom for principal
#  quantum number n and orbital quantum number l at radius r (a0 is taken 1)
#-------------------------------------------------------------------------------
   fNorm = sqrt(8e0/(n*n*n)*Fact(n-l-1)/(2e0*n*pow(Fact(n+l),3)))

   return fNorm * pow(r,l) * exp(-0.5e0*r) * aLaguerre(n+l,2*l+1,r)
```

Solution

Python and C/C++ sample programs are provided as supplementary material in the files P05-RadProbH.py and P05-RadProbH.cpp, and the graphical output can be seen in Figure 5.8. The routines for calculating associated Laguerre polynomials are included under the name aLaguerre in the libraries specfunc.py and specfunc.h.

The Python version of aLaguerre is listed along with a routine for radial wave functions of the hydrogen atom, RadPsiH, in Listing 5.21. Fact is supposed to be a function returning the factorial of its argument (for an example, see Listing 2.3).

References and Suggested Further Reading

Abramowitz, M. and I. A. Stegun (Eds.). 1972. *Handbook of Mathematical Functions: With Formulas, Graphs, and Mathematical Tables*. New York: Dover Publications.

Acton, F. S. 1997. *Numerical Methods That Work*. Spectrum Series. Washington: Mathematical Association of America.

Arfken, G. B. and H. J. Weber. 2005. *Mathematical Methods for Physicists* (6th ed.). Burlington: Academic Press.

Beu, T. A. 2004. *Numerical Calculus in C* (3rd ed., in Romanian). Cluj-Napoca: MicroInformatica.

Demidovich, B. and I. Maron. 1987. *Computational Mathematics* (4th ed.). Moscow: MIR Publishers.

Knuth, D. E. 1998. *The Art of Computer Programming* (3rd ed.), Vol. 2: *Seminumerical Algorithms*. Reading, MA: Addison-Wesley Professional.

Miller, J. C. P. 1952. *British Association for the Advancement of Science, Mathematical Tables, Vol. X, Bessel Functions, Part II*. Cambridge: Cambridge University Press.

6

Algebraic and Transcendental Equations

6.1 Root Separation

An equation is said to be *algebraic* or *polynomial* if it has the form

$$f(x) = 0, \tag{6.1}$$

and the function $f(x)$ is a polynomial or is convertible into polynomial form. For example, the equation $x^2 - (x-1)^{1/2} - 3 = 0$ is, indeed, algebraic, since, by separating the square root and squaring both sides of the resulted equation, it becomes $x^4 - 6x^2 - x + 10 = 0$. If $f(x)$ cannot be transformed into a polynomial, the equation is called *transcendental*.

Finding roots of polynomials up to the fourth order is a straightforward task and it can be accomplished, in general, by factorization into radicals. However, starting with the quintic equations (with $f(x)$ being a fifth-order polynomial), there is no general method for an analytic solution and one has to resort to numerical methods.

Given a real-valued function

$$f \colon [x_{\min}, x_{\max}] \to \mathbb{R},$$

defined on some subinterval of the real axis, any value $\xi \in [x_{\min}, x_{\max}]$ satisfying $f(\xi) = 0$ is called a *root* of equation $f(x) = 0$ or a *zero* of function $f(x)$.

In general, by *approximate root* of equation $f(x) = 0$, one understands a value ξ' "close" to the exact root ξ. An approximate root can be actually defined numerically in two ways: the number ξ' contained in some small interval around the exact root, that is, so that $|\xi' - \xi| < \varepsilon$ for a small $\varepsilon > 0$, or, the number ξ' making the function numerically equal to zero, that is, $|f(\xi')| < \varepsilon$. Obviously, as suggested by Figures 6.1a and b, these two definitions are not equivalent. While Figure 6.1a illustrates the situation in which the approximate root satisfies $|\xi' - \xi| < \varepsilon$, yet not also $|f(\xi')| < \varepsilon$, Figure 6.1b depicts the case in which, on the contrary, the second criterion is satisfied, however, not also the first one.

The practical calculation of the real roots of an equation $f(x) = 0$ typically implies two distinct phases:

1. *Separation of the roots*, that is, establishment of a partition of the entire search interval, $x_{\min} = x_1, x_2, \ldots, x_M = x_{\max}$, so that any particular subinterval $[x_m, x_{m+1}]$ contains at most one root (or none);
2. *Numerical calculation of the roots* already separated in distinct subintervals by iterative refinement procedures.

Many practical approaches for separating the real roots of algebraic or transcendental equations rely on consequences of Rolle's theorem, which basically states that the first derivative $f'(x)$ of a real-valued

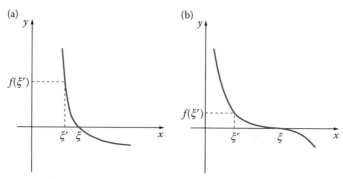

FIGURE 6.1 Approximate roots not simultaneously satisfying the criteria $|\xi' - \xi| < \varepsilon$ and $|f(\xi')| < \varepsilon$.

continuous and differentiable function $f(x)$ has an odd number of zeros between any two consecutive zeros of the function. As a consequence, $f(x)$ itself has *at most one zero* between any two *consecutive* zeros of its first derivative. If $x_1 < x_2 < \cdots < x_M$ are the zeros of $f'(x)$ and one forms Rolle's sequence, $f(-\infty), f(x_1), f(x_2), \ldots, f(x_M), f(+\infty)$, then, according to the mentioned consequence, each subinterval,

$$(-\infty, x_1), (x_1, x_2), \ldots, (x_m, x_{m+1}), \ldots, (x_M, +\infty),$$

contains one real root of the equation $f(x) = 0$, provided the function takes opposite-sign values at the end points of the subinterval, that is, if $f(x_m)f(x_{m+1}) < 0$. Consequently, the equation $f(x) = 0$ possesses as many zeros as the sign changes shown by Rolle's sequence. From a practical point of view, this approach has an important drawback, since it implies solving the equation $f'(x) = 0$, which can be sometimes as cumbersome as solving the original equation $f(x) = 0$.

The following theorem has direct practical utility for separating zeros of functions.

Theorem 6.1

If a continuous real-valued function $f(x)$ changes sign at the ends of a closed interval $[a, b]$, that is, $f(a)f(b) < 0$, then the interval contains *at least* a real zero of $f(x)$; in other words, there exists at least an interior point $\xi \in (a, b)$ so that $f(\xi) = 0$.

The root ξ is unique in $[a, b]$ if the derivative $f'(x)$ exists and *preserves its sign*. Otherwise, it is possible that $[a, b]$ contains multiple roots (as in Figure 6.2a) and, moreover, this is possible even if $f(a)f(b) > 0$ (as in Figure 6.2b).

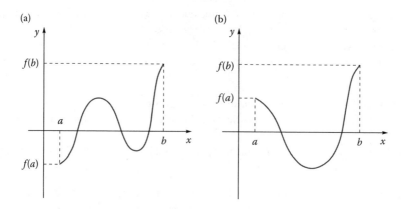

FIGURE 6.2 Examples of nonseparated roots.

Practically, the process of root separation implies determining the signs of the function $f(x)$ at the points of a sufficiently fine partition $\{x_m\}$, the choice of which largely depends on the particularities of the function. If the subintervals $[x_m, x_{m+1}]$ are sufficiently small, according to the above theorem, each subinterval contains *at most* a real root of the equation $f(x) = 0$. As such, the intervals for which $f(x_m)f(x_{m+1}) > 0$ do not contain any root, while those for which $f(x_m)f(x_{m+1}) \leq 0$ contain a single root.

6.2 Bisection Method

Let us consider the equation

$$f(x) = 0, \tag{6.2}$$

in which the real function $f(x)$ is continuous on the interval $[a, b]$. We assume that, as a result of a previous root-separation process, Equation 6.2 has at most one root within $[a, b]$. Denoting this root by ξ, the following cases are possible:

a. If $f(a) = 0$, then the root is $\xi = a$;
b. If $f(b) = 0$, then the root is $\xi = b$;
c. If $f(a)f(b) < 0$, then the root is one of the interior points, $\xi \in (a, b)$;
d. If $f(a)f(b) > 0$, then there is no root $\xi \in [a, b]$.

Considering the occurrence of case (c), which is the only nontrivial one requiring actual computations, with a view to finding the contained root, we divide the interval $[a, b]$ into two equal subintervals by the midpoint $x_0 = (a+b)/2$, as can be seen in Figure 6.3. In particular, if $f(x_0) = 0$, then $\xi = x_0$ is the sought root. In the opposite case, we choose the subinterval $[a_1, b_1]$ at whose ends the function has opposite signs and which, obviously, continues to confine the root, and we store its boundaries. One of the following cases can come about

$$a_1 = a_0, \quad b_1 = x_0, \quad \text{if } f(a)f(x_0) < 0,$$
$$a_1 = x_0, \quad b_1 = b_0, \quad \text{if } f(a)f(x_0) > 0.$$

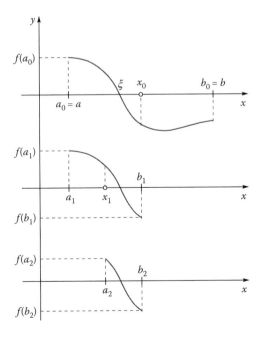

FIGURE 6.3   Halving process of the search interval for the bisection method.

The new interval $[a_1, b_1]$ is halved anew, performing similar sign tests as above. By continuing this procedure, one obtains after i iterations the bounding interval $[a_i, b_i]$ as a consequence of halving the interval $[a_{i-1}, b_{i-1}]$ by the point $x_{i-1} = (a_{i-1} + b_{i-1})/2$. As such, the new boundaries read:

$$a_i = a_{i-1}, \quad b_i = x_{i-1}, \quad \text{if } f(a_{i-1})f(x_{i-1}) < 0,$$
$$a_i = x_{i-1}, \quad b_i = b_{i-1}, \quad \text{if } f(a_{i-1})f(x_{i-1}) > 0. \tag{6.3}$$

As for all previous bounding intervals, it is obvious that the function $f(x)$ changes sign at the ends of the most recent one,

$$f(a_i)f(b_i) < 0, \tag{6.4}$$

whose length relates to that of the initial interval by

$$b_i - a_i = \frac{b - a}{2^i}. \tag{6.5}$$

As a result of halving the most recent interval by its midpoint,

$$x_i = \frac{1}{2}(a_i + b_i), \tag{6.6}$$

one obtains either the numerical root $\xi = x_i$, or a new bounding interval $[a_{i+1}, b_{i+1}]$.

Since the left boundaries of the successive intervals, $a_0, a_1, \ldots, a_i$, form a *bounded nondecreasing sequence*, and the right boundaries, $b_0, b_1, \ldots, b_i$, form a *bounded nonincreasing sequence*, it follows by taking the limit in Equation 6.5 that the sequences $\{a_i\}$ and $\{b_i\}$ have a common limit

$$\xi = \lim_{i \to \infty} a_i = \lim_{i \to \infty} b_i.$$

Furthermore, by virtue of the continuity of function $f(x)$, inequality (6.4) leads by taking the limit as $i \to \infty$ to $[f(\xi)]^2 \leq 0$, whence $f(\xi) = 0$, meaning that ξ is the root of Equation 6.2.

Obviously, one can perform in practice just a finite number of iterations and, thus, only an approximation to the root can be determined. The iterative procedure may be terminated, considering as root $\xi = x_i$, when one or both of the following conditions are met:

$$|f(x_i)| \leq \varepsilon, \qquad b_i - a_i \leq \varepsilon, \tag{6.7}$$

where $\varepsilon > 0$ is a predefined tolerance. For actual implementations, a more useful stopping criterion, which can be directly related to the expected number of exact significant digits in the calculated roots, is provided by the *estimate of the relative error* based on the width of the bounding interval:

$$\left| \frac{b_i - a_i}{x_i} \right| \leq \varepsilon.$$

An even more convenient form of this criterion, still relying on the relative error, but avoiding incidental divisions by zero, is

$$|b_i - a_i| \leq \varepsilon |x_i|. \tag{6.8}$$

The direct implementation of the exposed algorithm is represented by the function `Bisect` from Listing 6.1, which receives as input parameters the actual name of the user function, `Func`, whose zero is to be determined, and limits a and b of the search interval in which the separated root is supposed to exist. `Bisect` returns, in addition to the calculated root x, also an error index equal to 0—for normal

Listing 6.1 Bisection Method (Python Coding)

```
#===========================================================================
def Bisect(Func, a, b):
#---------------------------------------------------------------------------
#  Determines a real root x of function Func isolated in interval [a,b] by
#  the bisection method
#  Error code: 0 - normal execution
#              1 - [a,b] does not isolate one root or contains several roots
#              2 - max. number of iterations exceeded
#---------------------------------------------------------------------------
   eps = 1e-10                                    # relative precision of root
   itmax = 100                                       # max. no. of iterations

   x = a; fa = Func(x)                                       # is a the root?
   if (fabs(fa) == 0e0): return (x,0)
   x = b; fb = Func(x)                                       # is b the root?
   if (fabs(fb) == 0e0): return (x,0)

   if (fa*fb > 0): return (x,1)                  # [a,b] does not contain a root
                                                 # or contains several roots
   for it in range(1,itmax+1):
      x = 0.5e0 * (a + b)                                   # new approximation
      fx = Func(x)
      if (fa*fx > 0): a = x                    # choose new bounding interval
      else: b = x
      if (((b-a) <= eps*fabs(x)) or (fabs(fx) <= eps)): return (x,0)

   print("Bisect: max. no. of iterations exceeded !"); return (x,2)
```

execution, 1—if the search interval does not isolate a single root, or 2—if the convergence of the iterative halving procedure is poor, possibly due to an ill-conditioned function.

As a technical detail, in determining whether the sought root coincides with one of the boundaries of the search interval, it is useless to check the exact "mathematical" cancellation of the function, since, due to the possible propagation of roundoff errors in evaluating the function, such condition might not be met. Instead, one should compare the corresponding function values with some small "numerical zero," represented by the positive constant eps. For reasons of efficiency, there is no need to update the *function values* for the updated boundaries, $f(a_i)$, and $f(b_i)$, since the only aspect controlling the algorithm is their relative *sign*, and this remains by definition invariant throughout the procedure, with the root confined within the sequence of intervals $[a_i, b_i]$.

As a practical example, Program 6.2 uses the routine Bisect to determine the zeros of the user function func. The function main implements as its major part a *root separation loop*, which sequentially partitions the total search interval [xmin, xmax] into subintervals [a, b] of width h, which are supposed to isolate single roots. The root determined in a given subinterval is printed out only if the error index returned by Bisect is 0 (normal execution) and if the root does not coincide with the right limit of the subinterval. The letter condition avoids the double printing of roots, both as right and left boundaries of neighboring subintervals. Obviously, in a more elaborate implementation of the driver loop, instead of printing out the roots, they might be stored or further processed.

In the chosen example, the user function is $f(x) = (x^4 - 5x^2 + 4) \sin 2x$ and the program yields the simple zeros $\{-\pi, -2, -\pi/2, -1, 1, \pi/2, 2, \pi\}$. As an indication of the accuracy, the function $f(x)$ evaluates for all zeros to values lower than 10^{-8}.

Listing 6.2   Real Roots of Functions by the Bisection Method (Python Coding)

```python
# Real roots of a real function by the bisection method
from math import *
from roots import *

def func(x):
    return (x*x*x*x - 5e0*x*x + 4e0) * sin(2e0*x)

# main

xmin = -3.5e0                                   # limits of root search domain
xmax =  3.5e0
h = 0.1e0                                        # width of root separation intervals

a = xmin
while (a < xmax):                                # root separation loop
    b = a + h                                    # search interval [a,b]
    (x,ierr) = Bisect(func,a,b)
    if ((ierr == 0) and (x != b)):
        print('x = {0:8.5f} in ({1:6.2f},{2:6.2f})  f(x) = {3:7.0e}'.
            format(x,a,b,func(x)))
    a = b                                        # shift left boundary
```

6.3 Method of False Position

Similar to the bisection method, the method of false position (or, regula falsi method) is aimed at finding a real root ξ of an equation $f(x) = 0$ that is already isolated in an interval $[a, b]$, at whose ends the function takes values of opposite sign ($f(a)f(b) < 0$). The main difference consists in the way of choosing the successive bounding intervals, $[a_1, b_1], [a_2, b_2], \ldots, [a_i, b_i]$, which in the false position method do not result by bisection, but by partitioning the current interval $[a_i, b_i]$ in the ratio $f(a_i)/f(b_i)$ of the function values at the two ends. This procedure is repeated in a sequence of trial-and-error steps, substituting "false" test values for the unknown quantities to find successively improved approximations of the solution.

Geometrically, as can be seen from Figure 6.4, the false position method implies replacing the function $f(x)$ by the chord defined by the points $(a_i, f(a_i))$ and $(b_i, f(b_i))$. From the chord equation,

$$\frac{x_i - a_i}{b_i - a_i} = \frac{-f(a_i)}{f(b_i) - f(a_i)},$$

one can readily obtain the coordinate of the intersection point with the x-axis:

$$x_i = \frac{a_i f(b_i) - b_i f(a_i)}{f(b_i) - f(a_i)}. \tag{6.9}$$

After a certain number of iterations, one obtains either a numerically exact root $\xi = x_i$, so that $f(x_i) \leq \varepsilon$, or a sequence of intervals $[a_0, b_0], [a_1, b_1], \ldots, [a_i, b_i], \ldots$ defined by

$$\begin{aligned} a_{i+1} = a_i, \quad b_{i+1} = x_i, \quad \text{if } f(a_i)f(x_i) < 0, \\ a_{i+1} = x_i, \quad b_{i+1} = b_i, \quad \text{if } f(a_i)f(x_i) > 0, \end{aligned} \tag{6.10}$$

and satisfying

$$f(a_{i+1})f(b_{i+1}) < 0. \tag{6.11}$$

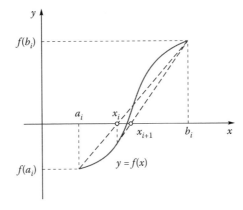

FIGURE 6.4 Division of the search interval by the chord connecting the points $(a_i, f(a_i))$, și $(b_i, f(b_i))$ in the false position method.

The false position method is coded as routine `FalsPos` in Listing 6.3. The general construction principles, input/output parameters, and functionalities of this routine are identical with those of the function `Bisect` described in the previous section and, as such, the two routines are interchangeable in applications.

Listing 6.3 Method of False Position (Python Coding)

```
#============================================================================
def FalsPos(Func, a, b):
#----------------------------------------------------------------------------
#  Determines a real root x of function Func isolated in interval [a,b] by
#  the false position method
#  Error code: 0 - normal execution
#              1 - [a,b] does not isolate one root or contains several roots
#              2 - max. number of iterations exceeded
#----------------------------------------------------------------------------
   eps = 1e-10                             # relative precision of root
   itmax = 100                             # max. no. of iterations

   x = a; fa = Func(x)                            # is a the root?
   if (fabs(fa) == 0e0): return (x,0)
   x = b; fb = Func(x)                            # is b the root?
   if (fabs(fb) == 0e0): return (x,0)

   if (fa*fb > 0): return (x,1)              # [a,b] does not contain a root
                                             # or contains several roots
   for it in range(1,itmax+1):
      x = (a*fb - b*fa)/(fb - fa)                  # new approximation
      fx = Func(x)
      if (fa*fx > 0):                        # choose new bounding interval
         dx = x - a; a = x; fa = fx
      else:
         dx = b - x; b = x; fb = fx
      if ((fabs(dx) <= eps*fabs(x)) or (fabs(fx) <= eps)): return (x,0)

   printf("FalsPos: max. no. of iterations exceeded !"); return (x,2)
```

One can notice, though, that, as a difference relative to the function `Bisect`, the routine `FalsPos` updates after each partitioning of the bounding interval not only the boundaries a_i and b_i, but also the corresponding function values, $f(a_i)$ and $f(b_i)$ (`fa` and `fb` in the code), since these are involved in calculating the next partitioning point x_{i+1}.

Even though the false position method is generally more efficient than the bisection method in terms of interval divisions necessary to achieve a given accuracy, it is likewise considered useful mainly for determining initial approximations for iterative methods, which converge more rapidly in the vicinities of the roots.

6.4 Method of Successive Approximations

One of the most representative numerical methods for finding zeros of real-valued functions is the *method of successive approximations*. Its importance, nevertheless, extends beyond the particular field of transcendental equations, since it actually represents a general strategy of solving mathematical problems by means of sequences of approximations that recursively converge to the solution. The classes of problems to which the method of successive approximations is applicable include among others differential, integral, and integro-differential equations.

Let us consider the transcendental equation

$$f(x) = 0, \tag{6.12}$$

where $f(x)$ is a continuous real-valued function defined in some interval $[a, b]$. The method of successive approximations requires transforming the equation into the equivalent form

$$x = \varphi(x), \tag{6.13}$$

so that Equations 6.12 and 6.13 have the same roots. Obviously, it is always possible to cast Equation 6.12 under form (6.13) and this is feasible even if the function $f(x)$ does not explicitly contain the term x, for example by adding it to both sides of the equation.

Let x_0 be an initial approximation of the exact root ξ of Equation 6.13. By replacing it in the right-hand side function $\varphi(x)$, one obtains a new approximation of the root:

$$x_1 = \varphi(x_0).$$

Iterating this procedure, there results a sequence of numbers:

$$x_{i+1} = \varphi(x_i), \quad i = 0, 1, 2, \ldots \tag{6.14}$$

If this sequence converges, then the limit $\xi = \lim x_i$ exists. Passing in Equation 6.14 to the limit and assuming $\varphi(x)$ to be continuous, we find

$$\lim_{i \to \infty} x_{i+1} = \varphi(\lim_{i \to \infty} x_i),$$

and herefrom

$$\xi = \varphi(\xi).$$

Hence, ξ is a root of Equation 6.13 and it can be calculated, in principle, to any accuracy by using the recurrence relation (6.14).

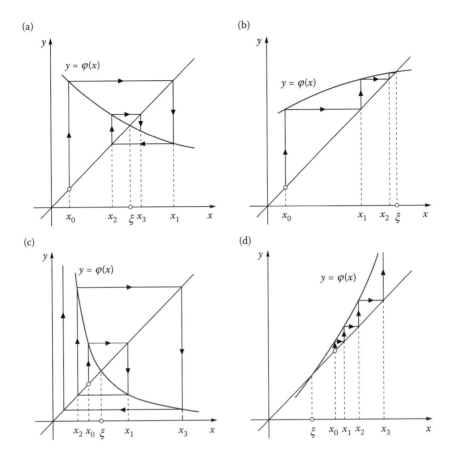

FIGURE 6.5   Iterative processes in the method of successive approximations applied to equation $x = \varphi(x)$ for (a) $-1 < \varphi'(x) < 0$; (b) $0 < \varphi'(x) < 1$; (c) $\varphi'(x) < -1$; (d) $\varphi'(x) > 1$. Cases (a) and (b) converge, while (c) and (d) diverge.

From a geometrical perspective, a real root ξ of equation $x = \varphi(x)$ is the abscissa of the intersection of the curve $y = \varphi(x)$ with the first bisector $y = x$. Notably, depending on the value of the first derivative $\varphi'(x)$ in the vicinity of the root, recurrence (6.14) can converge to the root or, on the contrary, can diverge irrefutably. Basically, four qualitatively different scenarios are possible, and the manner in which the sequence of approximations $x_1, x_2, x_3, \ldots$ results starting from the initial approximation x_0 is illustrated in Figure 6.5. Specifically, the iterative process converges in Figures 6.5a and b and diverges in Figsure 6.5c and d, and one can conclude that the process converges and, hence, the method of successive approximations is applicable successfully, for roots enclosed in intervals in which

$$|\varphi'(x)| < 1. \tag{6.15}$$

To practically apply the method of successive approximations, the conditions that have to be met are prescribed by the next theorem, which provides a sufficient condition for the convergence of the iteration sequence defined by Relation 6.14.

Theorem 6.2

Given the equation

$$x = \varphi(x), \tag{6.16}$$

with the function $\varphi(x)$ defined and differentiable in the interval $[a, b]$, if the inequality

$$|\varphi'(x)| \leq \lambda < 1 \tag{6.17}$$

holds true for any $x \in [a, b]$, then the iteration sequence defined by the recurrence relation

$$x_{i+1} = \varphi(x_i), \quad i = 0, 1, 2, \ldots \tag{6.18}$$

converges to the root $\xi \in [a, b]$ of the equation (unique, if it exists), regardless of the initial approximation x_0.

Proof. By subtracting the obvious equality $\xi = \varphi(\xi)$ from the recurrence relation (6.18), one obtains

$$x_{i+1} - \xi = \varphi(x_i) - \varphi(\xi).$$

By virtue of the mean value theorem, there exists a point ζ included between x_i and ξ, so that

$$\varphi(x_i) - \varphi(\xi) = \varphi'(\zeta)(x_i - \xi).$$

From the two relations above and having in view inequality (6.17), it results:

$$|x_{i+1} - \xi| \leq \lambda |x_i - \xi|.$$

Taking the values $0, 1, 2, \ldots$ in sequence for i, one obtains successively:

$$|x_1 - \xi| \leq \lambda |x_0 - \xi|,$$
$$|x_2 - \xi| \leq \lambda |x_1 - \xi| \leq \lambda^2 |x_0 - \xi|,$$
$$\cdots\cdots\cdots\cdots\cdots$$
$$|x_{i+1} - \xi| \leq \lambda |x_i - \xi| \leq \cdots \leq \lambda^{i+1} |x_0 - \xi|.$$

Passing to the limit in the last inequality and using the fact that the upper bound λ of $|\varphi'(x)|$ is subunitary and, therefore, $\lim_{i \to \infty} \lambda^{i+1} = 0$, it follows:

$$\lim_{i \to \infty} x_{i+1} = \xi.$$

Thus, the limit of the iteration sequence is precisely the root ξ of the considered equation. $\qquad\square$

To demonstrate that, under the provisions of the theorem, ξ is the only root within the interval $[a, b]$, we assume the contrary, that is, that there exists another root $\tilde{\xi} \in [a, b]$, so that $\tilde{\xi} = \varphi(\tilde{\xi})$. Hence, by using again the mean value theorem:

$$\xi - \tilde{\xi} = \varphi(\xi) - \varphi(\tilde{\xi}) = \varphi'(\zeta)(\xi - \tilde{\xi}),$$

where the point ζ is enclosed between ξ and $\tilde{\xi}$. One can further rewrite the above equation as

$$(\xi - \tilde{\xi})[1 - \varphi'(\zeta)] = 0.$$

Since by hypothesis $1 - \varphi'(\zeta) \neq 0$, it follows that $\xi = \tilde{\xi}$, implying that the root ξ is unique.

From a practical standpoint, the equation $f(x) = 0$ can be transformed into $x = \varphi(x)$ in several ways. Nonetheless, with a view to having the method of successive approximations applicable, Condition 6.17 has to be met. The smaller the number λ, the quicker the convergence of the iterative process to the root ξ. In general, the equation $f(x) = 0$ may be replaced by the equivalent equation

$$x = x - qf(x), \quad q > 0, \tag{6.19}$$

which implies that $\varphi(x) = x - qf(x)$. The parameter q has to be chosen in such a way that Condition 6.17 is satisfied, namely,

$$|\varphi'(x)| = |1 - qf'(x)| \le \lambda < 1.$$

In the trivial case in which $q = 1$ renders the process converging, Equation 6.19 takes the form $x = x - f(x)$ and the actual recurrence relation reads:

$$x_{i+1} = x_i - f(x_i), \quad i = 0, 1, 2, \ldots \tag{6.20}$$

Practically, one iterates this relation until the *relative difference* between two consecutive approximations of the root becomes smaller than an acceptable tolerance ε,

$$\delta_i \equiv |\Delta x_i / x_{i+1}| \le \varepsilon,$$

whereby, the absolute correction of the root can be determined as

$$\Delta x_i \equiv x_{i+1} - x_i = -f(x_i). \tag{6.21}$$

Reformulating the stopping condition as

$$|\Delta x_i| \le \varepsilon |x_{i+1}| \tag{6.22}$$

avoids possible divisions by zero (when $x_{i+1} = 0$). In the particular case of a root $\xi = 0$, both sides eventually vanish and the criterion is satisfied. Limiting the number of iterations enables detection of cases in which the method is slowly, if at all, converging.

The implementation of the method of successive approximations is illustrated by the function `Iter` listed in Listing 6.4. The procedural argument `Func` of the solver `Iter` receives the actual name of the user function (`func` in the program) that evaluates the function $f(x)$ whose zeros are to be determined. On entry to function `Iter`, the argument x should contain an initial approximation to the sought root. The successive corrections of the root are simply calculated as the current function value `f` with reversed sign. Once the relative correction drops under the tolerance `eps`, the iteration loop is left and an error code of 0 is returned, indicating normal execution. The complete execution of all `itmax` iterations actually implies failure to reach the prescribed precision due to the slow convergence and the error code returned in this case is 1.

To timely detect diverging processes and not to perform uselessly all `itmax`-allowed iterations, as soon as a new function value `f` and, implicitly, a new error estimate (`fabs(f)`) become available, it is compared with the error estimate from the previous iteration (`fabs(dx)`), since dx still contains the previous correction. If the absolute error estimate rises (`fabs(f) > fabs(dx)`), the recurrence diverges and the routine is left with error code 2.

We consider in the following the simple transcendental equation

$$x - e^{-x} = 0,$$

Listing 6.4   Method of Successive Approximations (Python Coding)

```
#=============================================================================
def Iter(Func, a, b, x):
#-----------------------------------------------------------------------------
#  Determines a root x of function Func isolated in [a,b] by the method of
#  successive approximations. x contains on entry an initial approximation.
#  Error code: 0 - normal execution
#              1 - interval does not contain a root
#              2 - max. number of iterations exceeded
#              3 - diverging process
#-----------------------------------------------------------------------------
   eps = 1e-10                                      # relative precision of root
   itmax = 100                                         # max. no. of iterations

   dx = -Func(x);                                      # initialize correction
   for it in range(1,itmax+1):
      f = Func(x)
      if (fabs(f) > fabs(dx)): break        # compare new with old correction
      dx = -f                                          # update correction
      x += dx                                          # new approximation
      if ((x < a) or (x > b)): return (x,1)    # [a,b] does not contain a root
      if (fabs(dx) <= eps*fabs(x)): return (x,0)       # check convergence

   if (it > itmax):
      print("Iter: max. no. of iterations exceeded !"); return (x,2)
   if (fabs(f) > fabs(dx)):
      print("Iter: diverging process !"); return (x,3)
```

illustrating the crucial manner in which the fulfillment of condition $|\varphi'(x)| \equiv |1 - f'(x)| < 1$ determines the convergence of the algorithm. Choosing, as in Listing 6.5,

$$f(x) = x - e^{-x},$$

which corresponds to $|\varphi'(x)| = e^{-x} < 1$, the successive approximations converge from $x_0 = 0$ toward the root $x = 0.56714$. On the contrary, considering $f(x) = e^{-x} - x$, the process diverges rapidly, since

Listing 6.5   Real Roots of Functions by the Method of Successive Approximations (Python Coding)

```
# Real root of a real function by the method of successive approximations
from math import *
from roots import *

def func(x): return x - exp(-x)

# main

a = -1e10; b = 1e10                                      # search interval
x = 0e0                                               # initial approximation

(x,ierr) = Iter(func,a,b,x)
if (ierr == 0):
   print("x = {0:8.5f}    f(x) = {1:7.0e}".format(x,func(x)))
else: print("No solution found !")
```

$|\varphi'(x)| = 2 + e^{-x} > 1$. Thus, mathematically equivalent forms of transcendental equations may lead to qualitatively different numerical behavior.

More than by practical efficiency, the method of successive approximations stands out by its distinctive methodological virtues. It evidences both general numerical phenomena, such as conditional convergence, and the commonly useful healing strategies. In applications, however, it can generally neither match the robustness of the bisection and false position methods, doubled by their ability to maintain the root isolated, nor the convergence speed of the methods based on the local derivatives, which are discussed in the following sections.

6.5 Newton's Method

Known also as the *Newton–Raphson method* (named after Isaac Newton and Joseph Raphson), Newton's method is probably the most famous algorithm for the iterative calculation of zeros of real-valued functions. Its wide use is largely due to the simplicity of the underlying idea, adopted under various forms by different variants of the method, but also due to the remarkable efficiency and stability of the corresponding implementations.

Let us assume that the equation

$$f(x) = 0 \tag{6.23}$$

for the real-valued differentiable function $f(x)$ possesses a real zero ξ in the interval $[a, b]$ and that $f'(x)$ and $f''(x)$ are continuous and conserve their sign for all $x \in [a, b]$.

Starting from an initial approximation x_0 of the root ξ, an improved approximation x_1 can be determined as the x-intercept of the tangent line at the curve $y = f(x)$ at the point $(x_0, f(x_0))$, as can be seen in Figure 6.6. By choosing, in particular, $x_0 = b$, the local derivative $f'(x_0)$ can be identified with the slope of the tangent line,

$$f'(x_0) = \frac{f(x_0)}{x_0 - x_1},$$

wherefrom we get the improved approximation

$$x_1 = x_0 - \frac{f(x_0)}{f'(x_0)}.$$

By using the tangent line at the point $(x_1, f(x_1))$, a new approximation x_2 results by a similar construction.

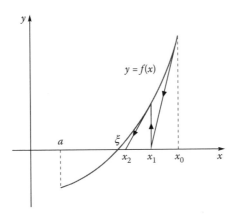

FIGURE 6.6 Sequence of approximate roots in the Newton–Raphson method.

Employing the outlined procedure iteratively, one can determine a sequence of approximations $x_1, x_2, x_3, \ldots$ of the zero ξ of function $f(x)$ based on the *recurrence relation*:

$$x_{i+1} = x_i - \frac{f(x_i)}{f'(x_i)}, \quad i = 0, 1, 2, \ldots \tag{6.24}$$

This formula can also be justified by considering the truncated Taylor expansion:

$$f(x_{i+1}) = f(x_i) + f'(x_i)(x_{i+1} - x_i) + \cdots$$

Admitting that x_{i+1} represents an improved approximation (or, even the solution ξ itself), one can consider that $f(x_{i+1}) \simeq 0$ and expresses x_{i+1} to give the recurrence relation (6.24).

Regarding Newton's method as a method of successive approximations and introducing the *iteration function*,

$$\varphi(x) = x - \frac{f(x)}{f'(x)}, \tag{6.25}$$

the convergence conditions of the sequence of approximations (6.24) to the solution ξ are specified by the same Theorem 6.2 as for the method of successive approximations.

As in the method of successive approximations, one typically iterates until the relative correction decreases under a prescribed tolerance ε, that is,

$$|\Delta x_i / x_{i+1}| \leq \varepsilon,$$

with the absolute correction Δx_i resulting from the recurrence relation (6.24):

$$\Delta x_i \equiv x_{i+1} - x_i = -\frac{f(x_i)}{f'(x_i)}. \tag{6.26}$$

Consistently handling the case $x_{i+1} = 0$ implies reformulating the stopping condition as

$$|\Delta x_i| \leq \varepsilon |x_{i+1}|. \tag{6.27}$$

With differences exclusively limited to the calculation of the consecutive root corrections, the implementation of Newton's method illustrated in Listing 6.6 is identical with the one for the method of successive approximations.

The user function called by the routine Newton is supposed to return besides the value of function $f(x)$, whose zero is to be determined, also the first derivative through the second parameter. Its actual name is received by routine Newton via the procedural parameter Func. The input value x, enclosed by the boundaries a and b of the search interval, is used as an initial approximation, being replaced on exit by the found zero.

The exit from the iteration loop occurs either normally, when the root correction drops below the specified tolerance eps, case in which the error code 0 is returned, or when the maximum allowed number of iterations is exhausted due to the slow convergence or even divergence of the process, case in which the returned error code is 1.

If the derivative of the function vanishes numerically in the course of the recursion (fabs(df)<eps), the routine Newton executes a "rescue" iteration that calculates the root correction as in the method of successive approximations, that is, without dividing the function by its derivative. This approach conveniently treats the cases of higher-order zeros, in which both the function and its derivative vanish.

Listing 6.6 Newton–Raphson Method (Python Coding)

```
#============================================================================
def Newton(Func, a, b, x):
#----------------------------------------------------------------------------
#  Determines a real root x of function Func isolated in interval [a,b] by
#  the Newton-Raphson method using the analytical derivative. x contains on
#  input an initial approximation.
#  Error code: 0 - normal execution
#              1 - interval does not contain a root
#              2 - max. number of iterations exceeded
#----------------------------------------------------------------------------
   eps = 1e-10                                       # relative precision of root
   itmax = 100                                       # max. no. of iterations

   for it in range(1,itmax+1):
      (f,df) = Func(x)                               # function and derivative
      dx = -f/df if fabs(df) > eps else -f           # root correction
      x += dx                                        # new approximation
      if ((x < a) or (x > b)): return (x,1)   # [a,b] does not contain a root
      if (fabs(dx) <= eps*fabs(x)): return (x,0)     # check convergence

   print("Newton: max. no. of iterations exceeded !"); return (x,2)
```

Listing 6.7 User Function with Derivative for Newton's Method (Python Coding)

```
def func(x):
   df = 1e0 + exp(-x)
   return (x - exp(-x), df)
```

Listing 6.7 exemplifies the construction of the user function for the routine Newton, in the particular case of the equation $x - e^{-x} = 0$.

In many practical cases, it is difficult or even impossible to express the derivative $f'(x)$ in a directly programmable form. In such situations, one can resort to evaluating it numerically. Routine NewtonNumDrv, presented in Listing 6.8, uses a finite forward difference representation of the derivative, based on the current x and on this value incremented by a small dx, respectively. Apart from the different evaluation of the derivative and the fact that the user function is no longer expected to return the derivative, the routines Newton and NewtonNumDrv are identical.

Unlike the method of successive approximations, the convergence of Newton's method is insensitive to the way of writing of the function $f(x)$, since the additional division by the derivative in the correction term, $-f(x_i)/f'(x_i)$, adjusts the iteration function in the vicinity of the roots so as to meet the requirements of Theorem 6.2.

Considering specifically the equation

$$x - e^{-x} = 0,$$

in contrast to the method of successive approximations, Newton's method converges to the root $x = 0.567143$ irrespective of the function definition as $f(x) = x - e^{-x}$ or $f(x) = e^{-x} - x$. Indeed, in both cases, the derivative of the iteration function has the same expression $\varphi'(x) = e^{-x}(x - e^{-x})/(1 + e^{-x})^2$ and, for any positive x, it is subunitary, as the theorem requires.

Listing 6.8   Newton's Method with Numerical Derivative (Python Coding)

```
#=============================================================================
def NewtonNumDrv(Func, a, b, x):
#-----------------------------------------------------------------------------
#  Determines a real root x of function Func isolated in interval [a,b] by
#  the Newton-Raphson method using the numerical derivative. x contains on
#  input an initial approximation.
#  Error code: 0 - normal execution
#              1 - interval does not contain a root
#              2 - max. number of iterations exceeded
#-----------------------------------------------------------------------------
   eps = 1e-10                                     # relative precision of root
   itmax = 100                                     # max. no. of iterations

   for it in range(1,itmax+1):
      f = Func(x)
      dx = eps*fabs(x) if x else eps                       # derivation step
      df = (Func(x+dx)-f)/dx                           # numerical derivative
      dx = -f/df if fabs(df) > eps else -f                # root correction
      x += dx                                           # new approximation
      if ((x < a) or (x > b)): return (x,1)    # [a,b] does not contain a root
      if (fabs(dx) <= eps*fabs(x)): return (x,0)        # check convergence

   print("NewtonNumDrv: max. no. of iterations exceeded !"); return (x,2)
```

Newton's method converges in general significantly more rapidly than the method of successive approximations. For example, in the case of the equation

$$x - \sin x - 0.25 = 0,$$

starting from the same initial approximation $x_0 = 0$, Newton's method requires 10 iterations to yield the solution $x = 1.171229653$ with 10 exact decimal digits (relative precision $\varepsilon = 10^{-10}$), while the method of successive approximations needs 28 iterations to complete the calculation with the same degree of accuracy. To complete the picture, see Problem 7.2 for a further comparison between the bisection and secant methods.

6.6 Secant Method

The secant method represents a trade-off between the efficiency of Newton's method and the inconveniences thereof caused by the necessity for the supplementary evaluation of the first derivative of the function whose zeros are being determined. A significant simplification is that the derivative no longer enters the successive corrections, which are, nevertheless, obtained from a recurrence relation involving *three* (instead of two) successive approximations of the root.

The function $f(x)$ is assumed to be nearly linear in the vicinity of the root and the chord determined by the points $(x_{i-1}, f(x_{i-1}))$ and $(x_i, f(x_i))$ corresponding to the most recent approximations (see Figure 6.7) is used instead of the tangent line to provide the new approximation x_{i+1} by its x-intercept. Alternatively, the recurrence relation for the secant method can be obtained by simply approximating the derivative in the Newton–Raphson formula,

$$x_{i+1} = x_i - \frac{f(x_i)}{f'(x_i)},$$

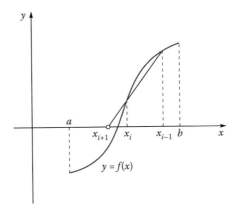

FIGURE 6.7 Sequence of approximate roots in the secant method.

by the finite (forward) difference expression involving the most recent approximations:

$$f'(x_i) \approx \frac{f(x_i) - f(x_{i-1})}{x_i - x_{i-1}}. \tag{6.28}$$

The recurrence relation thus obtained,

$$x_{i+1} = x_i - f(x_i)\frac{x_i - x_{i-1}}{f(x_i) - f(x_{i-1})}, \tag{6.29}$$

enables the calculation of the approximation x_{i+1} of the root on the basis of the previous approximations x_i and x_{i-1}. The iterative process is terminated, as for all other methods described in this chapter, when the relative correction of the root decreases below a predefined tolerance ε:

$$|x_{i+1} - x_i| \leq \varepsilon |x_{i+1}|. \tag{6.30}$$

For initiating the recurrence, one can take, in principle, any arbitrary initial approximations x_0 and x_1. However, starting from a single initial approximation x_0, one can obtain the second approximation by applying a single step of the method of successive approximations:

$$x_1 = x_0 - f(x_0). \tag{6.31}$$

The main difference between the secant and the false position methods regards the fact that, while the secant method always employs the two most recent approximations, which, however, do not necessarily bracket the root, the false position method retains two, not necessarily consecutive approximations, which, however, correspond to function values of opposite sign and thus implicitly bracket the root. Even though it generally converges faster, by not necessarily bracketing the root, the secant method does not guarantee the convergence for insufficiently smooth functions.

The algorithmic principles discussed above are coded in Listing 6.9 as the function Secant, which can be seen to perform, as discussed above, an initial iteration for producing the second root approximation according to the method of successive approximations.

Listing 6.9   Secant Method (Python Coding)

```
#===========================================================================
def Secant(Func, a, b, x):
#---------------------------------------------------------------------------
#   Determines a real root x of function Func isolated in interval [a,b] by
#   the secant method. x contains on entry an initial approximation.
#   Error code: 0 - normal execution
#               1 - interval does not contain a root
#               2 - max. number of iterations exceeded
#---------------------------------------------------------------------------
    eps = 1e-10                                  # relative precision of root
    itmax = 1000                                    # max. no. of iterations

    x0 = x; f0 = Func(x0)
    x = x0 - f0                                         # first approximation
    for it in range(1,itmax+1):
        f = Func(x)
        df = (f-f0)/(x-x0)                          # approximate derivative
        x0 = x; f0 = f                      # store abscissa and function value
        dx = -f/df if fabs(df) > eps else -f             # root correction
        x += dx                                        # new approximation
        if ((x < a) or (x > b)): return (x,1)   # [a,b] does not contain a root
        if (fabs(dx) <= eps*fabs(x)): return (x,0)      # check convergence

    print("Secant: max. no. of iterations exceeded !"); return (x,2)
```

In the case of the test equation

$$x - \sin x - 0.25 = 0, \tag{6.32}$$

the secant method performs almost as well as Newton's method, requiring 12 instead of 10 iterations starting from the initial approximation $x_0 = 0$ to yield the solution $x = 1.171229653$ with 10 exact decimal digits.

6.7 Birge–Vieta Method

The Birge–Vieta method represents the application of the Newton–Raphson method, in particular for determining the zeros of real polynomials. Specifically, we consider the algebraic equation

$$P_n(x) = 0, \tag{6.33}$$

where $P_n(x)$ is an n-th degree polynomial with real coefficients:

$$P_n(x) \equiv a_0 x^n + a_1 x^{n-1} + \cdots + a_{n-1} x + a_n. \tag{6.34}$$

The sequence of approximations of a specific root ξ is given in this case by the recurrence relation

$$x_{i+1} = x_i - \frac{P_n(x_i)}{P_n'(x_i)}, \quad i = 0, 1, 2, \ldots \tag{6.35}$$

and implies the evaluation of both the polynomial and its derivative for the sequence of arguments x_i. In principle, starting from various initial approximations, this relation could be used to determine all the roots. However, the algorithm can be rendered significantly more efficient by using Horner's scheme not only for the implied polynomial evaluations, but also for reducing the polynomial order by synthetic division (see Section 6.1) once a root is found.

Let x_i be the value for which $P_n(x_i)$ and $P'_n(x_i)$ have to be evaluated. According to the *polynomial remainder theorem*, by dividing $P_n(x)$ by the binomial $(x - x_i)$, one obtains a quotient polynomial of order $(n - 1)$, $P_{n-1}(x)$, and a remainder, p_n, which is nonzero if x_i is not a zero of $P_n(x)$. Accordingly, the *synthetic division* can be represented as

$$P_n(x) = (x - x_i)P_{n-1}(x) + p_n, \qquad (6.36)$$

with

$$P_{n-1}(x) = p_0 x^{n-1} + p_1 x^{n-2} + \cdots + p_{n-2} x + p_{n-1}. \qquad (6.37)$$

The coefficients $p_0, p_1, \ldots, p_n$ can be determined by equating the expanded expression resulting from the synthetic division (6.36) with the definition (6.34) of $P_n(x)$:

$$\begin{cases} p_0 & = & a_0, \\ p_j & = & p_{j-1} x_i + a_j, \quad j = 1, 2, \ldots, n. \end{cases} \qquad (6.38)$$

Indeed, from Equation 6.36, it follows that the polynomial value $P_n(x_i)$ is just the remainder of the division of $P_n(x)$ by $(x - x_i)$:

$$P_n(x_i) = p_n. \qquad (6.39)$$

Turning to the evaluation of the first derivative, $P'_n(x)$ results by differentiating the synthetic division rule (6.36):

$$P'_n(x) = (x - x_i)P'_{n-1}(x) + P_{n-1}(x), \qquad (6.40)$$

and, since $P'_n(x_i) = P_{n-1}(x_i)$, the evaluation of $P'_n(x_i)$ actually amounts to the evaluation of $P_{n-1}(x_i)$. By applying Horner's scheme, this calculation consists of iteratively evaluating the numbers

$$\begin{cases} d_0 = p_0, \\ d_j = d_{j-1} x_i + p_j, \quad j = 1, 2, \ldots, n - 1, \end{cases} \qquad (6.41)$$

which finally yield the desired result:

$$P'_n(x_i) = d_{n-1}. \qquad (6.42)$$

In summary, with a view to efficiently calculating a particular root ξ of the polynomial $P_n(x)$ by means of the recurrence relation (6.35), $P_n(x_i)$ and $P'_n(x_i)$ have to be evaluated using Formulas 6.38 through 6.39 and 6.41 through 6.42, respectively. Once a root ξ has been determined, the remaining roots can be economically calculated from the reduced polynomial $P_{n-1}(x)$ obtained by the synthetic division of $P_n(x)$ by $(x - \xi)$. The coefficients of $P_{n-1}(x)$ result from the same relations (6.38) as the ones used for evaluating $P_n(x_i)$, however, with x_i replaced by the root ξ.

The described procedure can be applied repeatedly, reducing in parallel the polynomial by synthetic division, until all real roots of the initial polynomial $P_n(x)$ have been found. The routine `BirgeVieta` presented in Listing 6.10 reflects the discussed technical details, while Listing 6.11 provides an example of a main program that makes use of the routine, which is assumed to be included in the module `roots.py`.

The main body of the routine `BirgeVieta` consists of a loop that runs over the decreasing orders m of the reduced polynomials, with the first one being the original polynomial $P_n(x)$. For each reduced polynomial, a single zero is determined by Newton–Raphson iterations. The successive numbers p_j and d_j generated as part of Horner's scheme to evaluate the polynomial and its derivative are recursively stored in scalars (p and d), with the last, unpaired p_m, calculated separately outside the loop.

Listing 6.10   Birge–Vieta Method (Python Coding)

```
#===========================================================================
def BirgeVieta(a, n, xx):
#---------------------------------------------------------------------------
#  Determines the real roots of a real polynomial by the Birge-Vieta method
#  a[]    - coefficients of the polynomial
#  n      - order of the polynomial
#  xx[]   - array of roots (output)
#  nx     - number of found roots (output)
#---------------------------------------------------------------------------
   eps = 1e-10                                      # relative precision of root
   itmax = 100                                         # max. no. of iterations

   nx = 0
   if (n <= 1): return

   x = 0                                    # initial approximation for 1st root
   for m in range(n,1,-1):                        # loop over reduced polynomials
      for it in range(1,itmax+1):                      # Newton-Raphson method
         p = a[0]; d = p                                    # Horner's scheme
         for j in range(1,m):
            p = p*x + a[j]                                       # polynomial
            d = d*x + p                                          # derivative
         p = p*x + a[m]
         d = -p/d if d else -p                               # root correction
         x += d
         if (fabs(d) <= eps*fabs(x)): break             # check convergence

      if (it == itmax):
         print("Birge: max. no. of iterations exceeded !"); return

      nx += 1
      xx[nx] = x                                                # store root
                                     # coefficients of new reduced polynomial
      for j in range(1,m): a[j] += a[j-1]*x

   nx += 1
   xx[nx] = -a[1]/a[0]                             # root of 1st order polynomial

   return nx
```

Listing 6.11   Real Roots of Polynomials by the Birge–Vieta Method (Python Coding)

```
# Real roots of a polynomial by the Birge-Vieta method
from math import *
from roots import *

n = 4                                                    # degree of polynomial
ax = [1e0, 0e0, -5e0, 0e0, 4e0]                               # coefficients
xx = [0]*(n+1)                                                     # zeros

nx = BirgeVieta(ax,n,xx)                                       # nx zeros found

for i in range(1,nx+1): print("x[",i,"] = ",xx[i])
if (nx <= 0): print(" Non-existing real zeros")
```

The Newton–Raphson iteration loop is left either normally, when the expected relative precision of the current root is achieved, or when all itmax-allowed iterations have been exhausted. The latter case signifies that there are no remaining real zeros of the polynomial $P_n(x)$, and the control is returned to the main program with the roots found so far. On the contrary, if the Newton–Raphson iterations converge, the counter nx of the found roots is increased, the last root is stored in the corresponding component of array xx, and the coefficients of the next reduced polynomial $P_{m-1}(x)$ are prepared. The root of the last first-order reduced polynomial is simply calculated from the ratio of its coefficients.

6.8 Newton's Method for Systems of Nonlinear Equations

The Newton–Raphson method can be generalized in a straightforward manner for systems of nonlinear equations (Press et al. 2007). The convergence conditions for the multidimensional algorithm are, nevertheless, considerably more demanding than for the one-dimensional case. In fact, the multidimensional method is generally very sensitive to the initial approximation, being efficient only if one starts sufficiently close to the exact solution. Moreover, it may well rapidly diverge instead for inadequate initial approximations, due to the rather poor global convergence.

Let us consider the system of nonlinear equations

$$f_i(x_1, x_2, \ldots, x_n) = 0, \quad i = 1, 2, \ldots, n, \tag{6.43}$$

with respect to the unknowns $x_1, x_2, \ldots, x_n$. In matrix notations, this system may be rewritten compactly as

$$\mathbf{f}(\mathbf{x}) = 0, \tag{6.44}$$

where $\mathbf{x}$ is the vector of the unknowns, whose components are x_i, while the components of the vector $\mathbf{f}$ are the functions $f_i(x_1, x_2, \ldots, x_n)$.

We assume that, in the vicinity of the exact solution ξ, each of the functions f_i can be expanded in a Taylor series:

$$f_i(\mathbf{x} + \Delta\mathbf{x}) = f_i(\mathbf{x}) + \sum_{j=1}^{n} \frac{\partial f_i}{\partial x_j} \Delta x_j + O((\Delta\mathbf{x})^2). \tag{6.45}$$

$\Delta\mathbf{x}$ denotes here the vector of corrections corresponding to the components of the vector $\mathbf{x}$ and the derivatives $\partial f_i/\partial x_j$ represent the matrix elements of the Jacobian of the system of functions $f_i(\mathbf{x})$:

$$\mathbf{J}(\mathbf{x}) = \begin{bmatrix} \dfrac{\partial f_1(\mathbf{x})}{\partial x_1} & \dfrac{\partial f_1(\mathbf{x})}{\partial x_2} & \cdots & \dfrac{\partial f_1(\mathbf{x})}{\partial x_n} \\[2ex] \dfrac{\partial f_2(\mathbf{x})}{\partial x_1} & \dfrac{\partial f_2(\mathbf{x})}{\partial x_2} & \cdots & \dfrac{\partial f_2(\mathbf{x})}{\partial x_n} \\[2ex] \vdots & \vdots & & \vdots \\[2ex] \dfrac{\partial f_n(\mathbf{x})}{\partial x_1} & \dfrac{\partial f_n(\mathbf{x})}{\partial x_2} & \cdots & \dfrac{\partial f_n(\mathbf{x})}{\partial x_n} \end{bmatrix}. \tag{6.46}$$

With the above notations, the Taylor-series expansion (6.45) reads:

$$\mathbf{f}(\mathbf{x} + \Delta\mathbf{x}) = \mathbf{f}(\mathbf{x}) + \mathbf{J} \cdot \Delta\mathbf{x} + O((\Delta\mathbf{x})^2). \tag{6.47}$$

Limiting the expansion to the linear approximation and imposing the condition $\mathbf{f}(\mathbf{x} + \Delta\mathbf{x}) \simeq 0$, that is, requiring that the corrections advance the solution toward the exact solution, $\mathbf{x} + \Delta\mathbf{x} \simeq \xi$, the new corrections result by solving the following system of linear equations having the Jacobian as matrix:

$$\mathbf{J(x)} \cdot \Delta \mathbf{x} = -\mathbf{f(x)}. \tag{6.48}$$

The corrected solution $\mathbf{x} + \Delta \mathbf{x}$ provides in general only an *improved* approximation, not directly the exact roots of the system of functions. This is due to the fact that Equation 6.48 was obtained by truncating Expansion 6.47. As such, rather than a single-step approach, an iterative process is conceivable, in which every new correction $\Delta \mathbf{x}^{(k)}$, resulted by solving the system

$$\mathbf{J(x}^{(k)}) \cdot \Delta \mathbf{x}^{(k)} = -\mathbf{f(x}^{(k)}), \tag{6.49}$$

advances the vector $\mathbf{x}^{(k)}$ closer to the exact solution of the system as a new approximation

$$\mathbf{x}^{(k+1)} = \mathbf{x}^{(k)} + \Delta \mathbf{x}^{(k)}. \tag{6.50}$$

By formally replacing the solution of system (6.49), one obtains in a compact form the characteristic recurrence relation for the multidimensional Newton–Raphson method:

$$\mathbf{x}^{(k+1)} = \mathbf{x}^{(k)} - \left[\mathbf{J(x}^{(k)}) \right]^{-1} \cdot \mathbf{f(x}^{(k)}), \tag{6.51}$$

which obviously generalizes the one-dimensional Newton's formula (6.24).

Analogously to the one-dimensional Newton–Raphson algorithm, in the multidimensional method, one typically iterates based on Relations 6.49 and 6.50 until the relative corrections of all solution components decrease below a chosen tolerance ε:

$$|\Delta x_i^{(k)}| \le \varepsilon |x_i^{(k+1)}|, \quad i = 1, 2, \ldots, n. \tag{6.52}$$

This formulation of the stopping criterion, which deliberately avoids divisions, also allows for the cases with vanishing solution components to be treated unitarily. Another useful criterion, complementary to the former, checks the degree in which the functions of the system have converged to zero:

$$|f_i(\mathbf{x}^{(k+1)}| \le \varepsilon, \quad i = 1, 2, \ldots, n. \tag{6.53}$$

While, in general, it is sufficient to ascertain that either of these two criteria is satisfied, in cases of problematic convergence, it may be advisable to check both. Nevertheless, once either the solution components or the functions reach the full floating point accuracy, the others do not change any more.

To approximate the partial derivatives of the functions f_i with respect to the variables x_j, which represent the matrix elements of the Jacobian, we employ *finite difference schemes*. Such methods are detailed in Chapter 13, in the context of differential equations with partial derivatives. Yet, we derive in what follows a useful approximation for the first derivative by considering the Taylor expansions of each function f_i for the symmetrically shifted arguments $x_j - h$ and $x_j + h$, respectively:

$$f_i(\ldots, x_j - h, \ldots) = f_i(\ldots, x_j, \ldots) - \frac{h}{1!} \frac{\partial f_i}{\partial x_j} + \frac{h^2}{2!} \frac{\partial^2 f_i}{\partial x_j^2} + O(h^3), \tag{6.54}$$

$$f_i(\ldots, x_j + h, \ldots) = f_i(\ldots, x_j, \ldots) + \frac{h}{1!} \frac{\partial f_i}{\partial x_j} + \frac{h^2}{2!} \frac{\partial^2 f_i}{\partial x_j^2} + O(h^3). \tag{6.55}$$

Here, all derivatives are taken at $\mathbf{x} = (x_1, \ldots, x_j, \ldots, x_n)$ and the neglected terms are of the order $O(h^3)$. One can deduce an explicit expression for the first derivative from any of these two expansions by *neglecting* also the second derivative. However, due to the division by h, the obtained approximations have $O(h)$ error:

$$\frac{\partial f_i}{\partial x_j} = \frac{f_i(\ldots, x_j, \ldots) - f_i(\ldots, x_j - h, \ldots)}{h} + O(h),$$

$$\frac{\partial f_i}{\partial x_j} = \frac{f_i(\ldots, x_j + h, \ldots) - f_i(\ldots, x_j, \ldots)}{h} + O(h).$$

The first is termed *backward difference scheme* and the second one is a *forward difference scheme*. In contrast, due to the exact cancellation of the second-order derivative, the *difference* of Expansions 6.54 and 6.55 remains of order $O(h^3)$, and hence, by expressing the first derivative, one obtains for the matrix elements of the Jacobian an improved approximation of order $O(h^2)$:

$$J_{ij} \equiv \frac{\partial f_i}{\partial x_j} = \frac{f_i(\ldots, x_j + h, \ldots) - f_i(\ldots, x_j - h, \ldots)}{2h} + O(h^2). \tag{6.56}$$

Such a discretized formula is called a *central difference scheme* because it involves symmetric arguments about the reference point $\mathbf{x}$.

Listing 6.12 illustrates the calculation of the matrix elements of the Jacobian by using Scheme 6.56. The reference point is received in the function `Jacobian` via array `x[]`, whose components are decremented and, respectively, incremented in sequence by a step-size `h`, which is dynamically adjusted in

Listing 6.12 Jacobian of a System of Multidimensional Functions (Python Coding)

```
#===========================================================================
def Jacobian(x, jac, n, Func):
#---------------------------------------------------------------------------
#  Calculates the Jacobian of a system of n real functions with n variables
#  using central finite differences
#  x[]      - point at which the Jacobian is evaluated
#  jac[][] - Jacobian
#  n        - space dimension
#  Func     - user function returning the function values at point x
#                Func(f, x, n)
#---------------------------------------------------------------------------
   eps = 1e-10

   fm = [0]*(n+1)
   fp = [0]*(n+1)

   for j in range(1,n+1):                          # loop over coordinates
      x0 = x[j]                                              # store x[j]
      h = eps*fabs(x0) if x0 else eps                       # step-size
      x[j] = x0 - h; Func(fm,x,n)              # decrement x[j]
      x[j] = x0 + h; Func(fp,x,n)              # increment x[j]
      h2 = 1e0/(2e0*h)
      for i in range(1,n+1): jac[i][j] = (fp[i] - fm[i]) * h2     # Jacobian
      x[j] = x0                                      # restore x[j]
```

accordance with the component value. Once all matrix elements of the Jacobian involving shifts along a particular coordinate are calculated, the initial value of that coordinate is restored to its initial value.

The routine NewtonSys from Listing 6.13 is conceived to find the multidimensional root of a system of nonlinear equations, and, structurally and functionally, it resembles the function Newton, described in Section 6.5. Vector x[] is supposed to contain on entry an initial approximation of the root.

The function Jacobi is called at each Newton–Raphson iteration to calculate the Jacobian matrix, which is then inverted by means of routine MatInv. The latter is discussed in detail in Chapter 7 and is

Listing 6.13   Newton's Method for Systems of Nonlinear Equations (Python Coding)

```
#=============================================================================
def NewtonSys(Func, x, n):
#-----------------------------------------------------------------------------
#  Determines a n-dimensional real zero of a system of n real functions by
#  Newton-Raphson method.
#  Func - user function returning the function values f[] for arguments x[]
#            Func(f, x, n)
#  x[]  - initial approximation (input), solution (output)
#  dx[] - error estimates of the solution components (output)
#  n    - order of system
#  ierr - error code: 0 - normal execution
#                     1 - max. number of iterations exceeded
#  Calls: Jacobian - computes Jacobian
#         MatInv   - inverts (n x n) matrix (in linsys.h)
#-----------------------------------------------------------------------------
   eps = 1e-14                              # precision for cumulated error
   itmax = 200                                    # max. no. of iterations

   f   = [0]*(n+1)
   dx  = [0]*(n+1)
   jac = [[0]*(n+1) for i in range(n+1)]

   for it in range(1,itmax+1):
      Func(f,x,n)                                              # functions
      Jacobian(x,jac,n,Func)                                   # Jacobian
      det = MatInv(jac,n)                               # inverse Jacobian

      if (det):                                            # corrections
         for i in range(1,n+1):                  # non-singular Jacobian
            sum = 0e0
            for j in range(1,n+1): sum -= jac[i][j] * f[j]
            dx[i] = sum
      else:
         for i in range(1,n+1): dx[i] = -f[i]           # singular Jacobian

      err = 0e0
      for i in range(1,n+1):
         x[i] += dx[i]                               # new approximation
         err += fabs(f[i])                               # cumulate error
      if (err <= eps): break                          # check convergence

   ierr = 0
   if (it >= itmax):
      ierr = 1; print("NewtonSys: max. no. of iterations exceeded !")
   return (dx,ierr)
```

Listing 6.14 Real Roots of Systems of Nonlinear Equations (Python Coding)

```
# Newton-Raphson method for systems of non-linear equations
from math import *
from roots import *

def Func(f, x, n):                                  # zeros: (1, 2), (2, 1)
    f[1] = pow(x[1],3) * x[2] + x[1] * pow(x[2],3) - 10e0
    f[2] = pow(x[1],2) * x[2] + x[1] * pow(x[2],2) - 6e0

# main

n = 2
f = [0]*(n+1); x = [0]*(n+1); dx = [0]*(n+1)

x[1] = 0e0; x[2] = 0e0                               # initial approximation
(dx,ierr) = NewtonSys(Func,x,n)
Func(f,x,n)

print("\nSolution:")
print("              x              dx           f")
for i in range(1,n+1):
    print("{0:d}   {1:15.7e}   {2:7.0e}   {3:7.0}".format(i,x[i],dx[i],f[i]))
```

assumed to be included in the module `linsys.py` and, respectively, in the header file `linsys.h`. The corrections of the individual root components are obtained by multiplying the inverted Jacobian matrix with the array of current function values. In the case of a singular Jacobian, the corrections are determined, as in the method of successive approximations, simply by the function values. The implemented stopping criterion checks the decrease of the cumulated absolute function values below the chosen tolerance `eps` and, when satisfied, the routine exits with error code 0. The completion of all `itmax`-allowed iterations occurs in cases of weak or absent convergence, in which an error message is issued and the error code 1 is returned.

The user function `Func` returns by array `f[]` the values of the n functions $f_i(\mathbf{x})$ corresponding to the arguments transferred via array `x[]`. As a concrete example, Listing 6.14 implements the left-hand side functions of the system

$$x^3y + xy^3 - 10 = 0,$$
$$x^2y + xy^2 - 6 = 0, \tag{6.57}$$

whose exact solutions are $(x = 1, y = 2)$ and $(x = 2, y = 1)$. The variables corresponding to the unknowns x and y are `x[1]` and `x[2]`, respectively.

The method is, indeed, sensitive to the initial approximation. For instance, starting from $(x = 0, y = 0)$, the program yields the first solution, whereas, from $(x = 4, y = 0)$, it converges to the second solution.

6.9 Implementations in C/C++

Listing 6.15 shows the content of the file `roots.h`, which contains equivalent C/C++ implementations of the Python functions developed in the main text and included in the module `roots.py`. The corresponding routines have identical names, parameters, and functionalities.

Listing 6.15   Solvers for Algebraic and Transcendental Equations (`roots.h`)

```
//-------------------------------- roots.h --------------------------------
// Contains routines for determining real roots of real functions.
// Part of the numxlib numerics library. Author: Titus Beu, 2013
//-------------------------------------------------------------------------
#ifndef _ROOTS_
#define _ROOTS_

#include <stdio.h>
#include <math.h>
#include "linsys.h"


//===========================================================================
int Bisect(double Func(double), double a, double b, double &x)
//---------------------------------------------------------------------------
// Determines a real root x of function Func isolated in interval [a,b] by
// the bisection method
// Error code: 0 - normal execution
//             1 - [a,b] does not isolate one root or contains several roots
//             2 - max. number of iterations exceeded
//---------------------------------------------------------------------------
{
   const double eps = 1e-10;                     // relative precision of root
   const int itmax = 100;                             // max. no. of iterations
   double fa, fb, fx;
   int it;

   x = a; fa = Func(x);                                       // is a the root?
   if (fabs(fa) == 0e0) return 0;
   x = b; fb = Func(x);                                       // is b the root?
   if (fabs(fb) == 0e0) return 0;

   if (fa*fb > 0) return 1;                     // [a,b] does not contain a root
                                                // or contains several roots
   for (it=1; it<=itmax; it++) {
      x = 0.5e0 * (a + b);                              // new approximation
      fx = Func(x);
      if (fa*fx > 0) a = x; else b = x;        // choose new bounding interval
      if (((b-a) <= eps*fabs(x)) || (fabs(fx) <= eps)) return 0;
   }
   printf("Bisect: max. no. of iterations exceeded !\n"); return 2;
}


//===========================================================================
int FalsPos(double Func(double), double a, double b, double &x)
//---------------------------------------------------------------------------
// Determines a real root x of function Func isolated in interval [a,b] by
// the false position method
// Error code: 0 - normal execution
//             1 - interval does not contain a root
//             2 - max. number of iterations exceeded
//---------------------------------------------------------------------------
{
   const double eps = 1e-10;                     // relative precision of root
   const int itmax = 100;                             // max. no. of iterations
   double dx, fa, fb, fx;
   int it;

   x = a; fa = Func(x);                                       // is a the root?
```

```
      if (fabs(fa) == 0e0) return 0;
      x = b; fb = Func(x);                                 // is b the root?
      if (fabs(fb) == 0e0) return 0;

      if (fa*fb > 0) return 1;                 // [a,b] does not contain a root
                                               // or contains several roots

      for (it=1; it<=itmax; it++) {
         x = (a*fb - b*fa)/(fb - fa);                      // new approximation
         fx = Func(x);
         if (fa*fx > 0) {                      // choose new bounding interval
            dx = x - a; a = x; fa = fx;
         } else {
            dx = b - x; b = x; fb = fx;
         }
         if ((fabs(dx) <= eps*fabs(x)) || (fabs(fx) <= eps)) return 0;
      }
      printf("FalsPos: max. no. of iterations exceeded !\n"); return 2;
   }


   //============================================================================
   int Iter(double Func(double), double a, double b, double &x)
   //----------------------------------------------------------------------------
   // Determines a root x of function Func isolated in [a,b] by the method of
   // successive approximations. x contains on entry an initial approximation.
   // Error code: 0 - normal execution
   //             1 - interval does not contain a root
   //             2 - max. number of iterations exceeded
   //             3 - diverging process
   //----------------------------------------------------------------------------
   {
      const double eps = 1e-10;                  // relative precision criterion
      const int itmax = 100;                              // max. no. of iterations
      double dx, f;
      int it;

      dx = -Func(x);                                       // initialize correction
      for (it=1; it<=itmax; it++) {
         f = Func(x);
         if (fabs(f) > fabs(dx)) break;       // compare new with old correction
         dx = -f;                                          // update correction
         x += dx;                                          // new approximation
         if ((x < a) || (x > b)) return 1;   // interval does not contain a root
         if (fabs(dx) <= eps*fabs(x)) return 0;            // check convergence
      }
      if (it > itmax)
         { printf("Iter: max. no. of iterations exceeded !\n"); return 2; }
      if (fabs(f) > fabs(dx))
         { printf("Iter: diverging process !\n"); return 3; }
   }


   //============================================================================
   int Newton(double Func(double, double &), double a, double b, double &x)
   //----------------------------------------------------------------------------
   // Determines a real root x of function Func isolated in interval [a,b] by
   // the Newton-Raphson method using the analytical derivative. x contains on
   // input an initial approximation.
   // Error code: 0 - normal execution
   //             1 - interval does not contain a root
   //             2 - max. number of iterations exceeded
```

```
//-------------------------------------------------------------------------------
{
   const double eps = 1e-10;                        // relative precision criterion
   const int itmax = 100;                           // max. no. of iterations
   double df, dx, f;
   int it;

   for (it=1; it<=itmax; it++) {
      f = Func(x,df);                               // function and derivative
      dx = (fabs(df) > eps) ? -f/df : -f;           // root correction
      x += dx;                                      // new approximation
      if ((x < a) || (x > b)) return 1;        // [a,b] does not contain a root
      if (fabs(dx) <= eps*fabs(x)) return 0;        // check convergence
   }
   printf("Newton: max. no. of iterations exceeded !\n"); return 2;
}


//===============================================================================
int NewtonNumDrv(double Func(double), double a, double b, double &x)
//-------------------------------------------------------------------------------
// Determines a real root x of function Func isolated in interval [a,b] by
// the Newton-Raphson method using the numerical derivative. x contains on
// input an initial approximation.
// Error code: 0 - normal execution
//             1 - interval does not contain a root
//             2 - max. number of iterations exceeded
//-------------------------------------------------------------------------------
{
   const double eps = 1e-10;                        // relative precision criterion
   const int itmax = 100;                           // max. no. of iterations
   double df, dx, f;
   int it;

   for (it=1; it<=itmax; it++) {
      f = Func(x);
      dx = x ? eps*fabs(x) : eps;                   // derivation step
      df = (Func(x+dx)-f)/dx;                       // numerical derivative
      dx = (fabs(df) > eps) ? -f/df : -f;           // root correction
      x += dx;                                      // new approximation
      if ((x < a) || (x > b)) return 1;        // [a,b] does not contain a root
      if (fabs(dx) <= eps*fabs(x)) return 0;        // check convergence
   }
   printf("NewtonNumDrv: max. no. of iterations exceeded !\n"); return 2;
}


//===============================================================================
int Secant(double Func(double), double a, double b, double &x)
//-------------------------------------------------------------------------------
// Determines a real root x of function Func isolated in interval [a,b] by
// the secant method. x contains on entry an initial approximation.
// Error code: 0 - normal execution
//             1 - interval does not contain a root
//             2 - max. number of iterations exceeded
//-------------------------------------------------------------------------------
{
   const double eps = 1e-10;                        // relative precision criterion
   const int itmax = 100;                           // max. no. of iterations
   double df, dx, f, f0, x0;
   int it;
```

```
      x0 = x; f0 = Func(x0);
      x = x0 - f0;
      for (it=1; it<=itmax; it++) {
         f = Func(x);
         df = (f-f0)/(x-x0);                              // approximate derivative
         x0 = x; f0 = f;                        // store abscissa and function value
         dx = (fabs(df) > eps) ? -f/df : -f;                     // root correction
         x += dx;                                           // new approximation
         if ((x < a) || (x > b)) return 1;      // [a,b] does not contain a root
         if (fabs(dx) <= eps*fabs(x)) return 0;           // check convergence
      }
      printf("Secant: max. no. of iterations exceeded !\n"); return 2;
   }

   //============================================================================
   void BirgeVieta(double a[], int n, double xx[], int &nx)
   //----------------------------------------------------------------------------
   // Determines the real roots of a real polynomial by the Birge-Vieta method
   // a[]   - coefficients of the polynomial
   // n     - order of the polynomial
   // xx[]  - array of roots (output)
   // nx    - number of found roots (output)
   //----------------------------------------------------------------------------
   {
      const double eps = 1e-10;                     // relative precision criterion
      const int itmax = 100;                             // max. no. of iterations
      double d, p, x;
      int it, j, m;

      nx = 0;
      if (n <= 1) return;

      x = 0;                                      // initial approximation for 1st root
      for (m=n; m>=2; m--) {                       // loop over reduced polynomials
         for (it=1; it<=itmax; it++) {                    // Newton-Raphson method
            p = a[0]; d = p;                                   // Horner's scheme
            for (j=1; j<=m-1; j++) {
               p = p*x + a[j];                                     // polynomial
               d = d*x + p;                                        // derivative
            }
            p = p*x + a[m];
            d = (d) ? -p/d : -p;                             // root correction
            x += d;
            if (fabs(d) <= eps*fabs(x)) break;           // check convergence
         }
         if (it == itmax) {
            printf("Birge: max. no. of iterations exceeded !\n"); return;
         }
         nx++;
         xx[nx] = x;                                               // store root
                                          // coefficients of new reduced polynomial
         for (j=1; j<=(m-1); j++) a[j] += a[j-1]*x;
      }
      nx++;
      xx[nx] = -a[1]/a[0];                           // root of 1st order polynomial
   }

   //============================================================================
```

```
void Jacobian(double x[], double **jac, int n,
              void Func(double[],double[],int))
//---------------------------------------------------------------------------
// Calculates the Jacobian of a system of n real functions with n variables
// using central finite differences
// x[]      - point at which the Jacobian is evaluated
// jac[][] - Jacobian
// n        - space dimension
// Func     - user function returning the function values at point x
//                void Func(double f[], double x[], int n);
//---------------------------------------------------------------------------
{
   const double eps = 1e-10;
   int i, j;
   double h, h2, *fm, *fp, x0;

   fm = Vector(1,n);
   fp = Vector(1,n);

   for (j=1; j<=n; j++) {                              // loop over coordinates
      x0 = x[j];                                              // store x[j]
      h = x0 ? eps*fabs(x0) : eps;                            // step-size
      x[j] = x0 - h; Func(fm,x,n);                      // decrement x[j]
      x[j] = x0 + h; Func(fp,x,n);                      // increment x[j]
      h2 = 1e0/(2e0*h);
      for (i=1; i<=n; i++) jac[i][j] = (fp[i] - fm[i]) * h2;      // Jacobian
      x[j] = x0;                                         // restore x[j]
   }

   FreeVector(fm,1);
   FreeVector(fp,1);
}


//===========================================================================
int NewtonSys(void Func(double [], double [], int),
              double x[], double dx[], int n)
//---------------------------------------------------------------------------
// Determines a n-dimensional real zero of a system of n real functions by
// Newton-Raphson method.
// Func - user function returning the function values f[] for arguments x[]
//           void Func(double f[], double x[], int n);
// x[]  - initial approximation (input), solution (output)
// dx[] - error estimates of the solution components (output)
// n    - order of system
// ierr - error code: 0 - normal execution
//                    1 - max. number of iterations exceeded
// Calls: Jacobian - computes Jacobian
//        MatInv   - inverts (n x n) matrix (in linsys.h)
//---------------------------------------------------------------------------
{
   const double eps = 1e-14;                  // precision for cumulated error
   const int itmax = 100;                          // max. no. of iterations
   double *f, **jac;
   double det, err, sum;
   int i, ierr, it, j;

   f   = Vector(1,n);
   jac - Matrix(1,n,1,n);
```

```
   for (it=1; it<=itmax; it++) {
      Func(f,x,n);                                              // functions
      Jacobian(x,jac,n,Func);                                   // Jacobian
      MatInv(jac,n,det);                                // inverse Jacobian

      if (det) {                                             // corrections
         for (i=1; i<=n; i++) {                     // non-singular Jacobian
            sum = 0e0;
            for (j=1; j<=n; j++) sum -= jac[i][j] * f[j];
            dx[i] = sum;
         }
      } else { for (i=1; i<=n; i++) dx[i] = -f[i]; }     // singular Jacobian

      err = 0e0;
      for (i=1; i<=n; i++) {
         x[i] += dx[i];                               // new approximation
         err += fabs(f[i]);                               // cumulate error
      }
      if (err <= eps) break;                            // check convergence
   }

   FreeVector(f,1);
   FreeMatrix(jac,1,1);

   ierr = 0;
   if (it >= itmax)
      { ierr = 1; printf("NewtonSys: max. no. of iterations exceeded !\n"); }
   return ierr;
}

#endif
```

6.10 Problems

The Python and C/C++ programs for the following problems may import the functions developed in this chapter from the modules `roots.py` and `roots.h`, respectively, which are available as supplementary material. For creating runtime plots, the graphical routines contained in the libraries `graphlib.py` and `graphlib.h` may be employed.

PROBLEM 6.1

Using the routine `Bisect`, write a C/C++ program (analogous to the Python program 6.2) for finding the zeros of the function

$$f(x) = \left(x^4 - 5x^2 + 4\right) \sin 2x.$$

Consider the search interval $[-3.5, 3.5]$ and, prior to applying the bisection method, separate the zeros in subintervals of width $h = 0.1$. For each found zero, check the accuracy by evaluating the corresponding function value.

Solution
The implementation is shown in Listing 6.16.

Listing 6.16   Real Roots of Functions (C/C++ Coding)

```
// Real roots of a real function by the bisection method
#include <stdio.h>
#include <math.h>
#include "roots.h"

double func(double x)
{
   return (x*x*x*x - 5e0*x*x + 4e0) * sin(2e0*x);
}

int main()
{
   double a, b, h, x, xmin, xmax;
   int ierr;

   xmin = -3.5e0;                              // limits of root search domain
   xmax =  3.5e0;
   h = 0.1e0;                       // width of root separation intervals

   a = xmin;
   while (a < xmax) {                                  // root separation loop
      b = a + h;                                       // search interval [a,b]
      ierr = Bisect(func,a,b,x);
      if ((ierr == 0) && (x != b))
         printf("x = %8.5f in (%6.2f,%6.2f)  f(x) = %7.0e\n",x,a,b,func(x));
      a = b;                                           // shift left boundary
   }
}
```

PROBLEM 6.2

Modify the routines `Bisect` and `Secant` in such a way as to return the number of iterations they require to achieve convergence, namely the final value of the iteration counter `it`. For instance, declare `it` global, return it via the parameter list, or, simply print out its values in the routines.

Use the modified routines comparatively to solve the equation

$$x - \sin x - 0.25 = 0,$$

and assess the convergence speed of the two methods, using the search interval $[-30, 30]$ for the bisection method, and the initial approximation $x_0 = -30$ for the secant method.

Solution

Python and C/C++ sample programs are provided as supplementary material in the files `P06-CompareZeros.py` and `P06-CompareZeros.cpp`.

PROBLEM 6.3

Kepler's equation (derived by Johannes Kepler in 1609) plays a central role in classical celestial mechanics, because it enables the calculation of angular positions for orbiting objects. For an elliptical orbit, it relates the *mean anomaly M*, the *eccentric anomaly E*, and the *eccentricity* $e = \sqrt{1 - b^2/a^2}$ ($a > b$ are semiaxes):

$$M = E - e \sin E. \tag{6.58}$$

The "anomalies" are defined by astronomers as angular positions: the mean anomaly M is the angular position of the object on a fictitious circular orbit, and, assuming constant angular velocity, it can be related to the lapse of time since the passage through the perihelion:

$$M = \frac{2\pi}{T}(t - t_0), \tag{6.59}$$

where T is the orbital period and t_0 is the time when the object was at the perihelion.

For the Halley comet (Mottmann 1986), the relevant parameters have the values: $e = 0.9672671$, $T = 75.9600$ years, and $t_0 = 1986.1113$ years (on February 9, 1986).

a. Find the eccentric anomaly E of the Halley comet on April 1, 1986, for $t = 1986.2491$. To this end, plot the function

$$f(E) = E - e \sin E - M, \tag{6.60}$$

for $E \in [0, 1]$ to make sure that the interval contains a single zero and determine it by Newton's method.

b. Plot $E = E(t)$ for a whole period of revolution by using Newton's method to solve

$$E - e \sin E - M(t) = 0, \tag{6.61}$$

with a time step $\Delta t = T/100$. Use as initial approximation for each new moment of time, $t_i = t_0 + (i - 1)\Delta t$, the previous value $E(t_{i-1})$, starting with $E(t_0) = 0$.

Solution

The Python implementation is given in Listing 6.17, the C/C++ version is available as supplementary material (P06-Kepler.cpp), and the graphical output is depicted in Figure 6.8.

Listing 6.17 Solution of Kepler's Equation (Python Coding)

```python
# Solution of Kepler's equation
from math import *
from roots import *
from graphlib import *

pi2 = 2e0 * pi

def Func(E):
   global e, M                              # eccentricity, mean anomaly
   f = E - e * sin(E) - M
   df = 1e0 - e * cos(E)
   return (f, df)

# main
n = 101
x = [0]*(n+1);  y = [0]*(n+1)

GraphInit(1200,600)
                                            # Halley comet (Mottmann 1986):
e  = 0.9672671e0                            # eccentricity
T = 75.96e0                                 # period
t0 = 1986.1113e0                            # time at perihelion

#----------------------------------------------------------------------
t = 1986.2491e0
M = pi2 / T * (t - t0)                      # mean anomaly
```

```
Emax = 1e0                                    # eccentric anomaly vs. Kepler's function
h = Emax / (n-1)
for i in range(1,n+1):
  E = (i-1)*h
  x[i] = E; (y[i],df) = Func(E)

Plot(x,y,n,"blue",1,0.10,0.45,0.15,0.85,
     "E (rad)","f(E)","Kepler's function")

(E,ierr) = Newton(Func,0e0,Emax,E)            # solve Kepler's equation
print("E = ",E," rad at t = ",t," years")

#--------------------------------------------------------------------------
h = T / (n-1)                                 # time dependence of eccentric anomaly
x[1] = t0; y[1] = E = 0e0
for i in range(2,n+1):
   t = t0 + (i-1)*h
   M = pi2 / T * (t - t0)
   (E,ierr) = Newton(Func,0e0,pi2,E)          # solve Kepler's equation
   x[i] = t; y[i] = E

Plot(x,y,n,"red",1,0.60,0.95,0.15,0.85,
     "t (years)","E (rad)","Eccentric anomaly")

MainLoop()
```

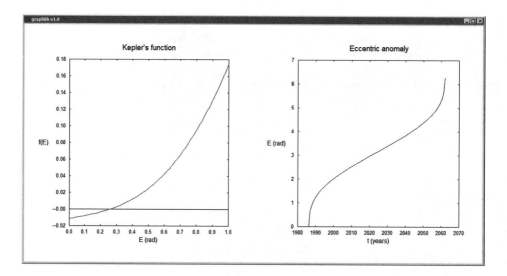

FIGURE 6.8   Solution of Kepler's equation for the Halley comet: Kepler's function $f(E)$ as a function of the eccentric anomaly E for $t = 1986.2491$ (left panel) and evolution of the eccentric anomaly over a period (right panel).

PROBLEM 6.4

The intensity distribution produced in the *Fraunhofer diffraction* by a slit of width W and infinite height, when illuminated with monochromatic light of wavelength λ, is (Hecht 2002):

$$I(\theta) = I_0 \left[\frac{\sin(\pi W \sin\theta/\lambda)}{\pi W \sin\theta/\lambda} \right]^2, \tag{6.62}$$

where θ is the angle with respect to the incident light at which the diffracted wave is observed. Introducing

$$x = \pi W \sin \theta / \lambda, \tag{6.63}$$

the intensity and its first derivative with respect to x may be written as

$$I(x) = I_0 \left(\frac{\sin x}{x} \right)^2, \quad I(0) = I_0, \tag{6.64}$$

$$I'(x) = \frac{2I_0}{x} \left(\cos x - \frac{\sin x}{x} \right) \frac{\sin x}{x}, \quad I'(0) = 0. \tag{6.65}$$

The *half-width* of the central intensity maximum represents the positive value $x_{1/2}$ for which the intensity drops to half its maximum value I_0 and it is, obviously, the solution of the equation

$$I(x) - I_0/2 = 0. \tag{6.66}$$

The positions of the *intensity maxima* x_i can be obtained as zeros of the first derivative (6.65), which amounts to solving:

$$I'(x) = 0. \tag{6.67}$$

a. Plot the distributions $I(x)$ and $I'(x)$ for $x \in [-10, 10]$ using the routine `MultiPlot` from the graphics library `graphlib`.
b. Find the half-width $x_{1/2}$ of the major diffraction maximum by solving Equation 6.66 using the secant method and the initial approximation $x_0 = \pi$ (the first zero of $I(x)$).
c. Find the intensity maxima $x_i \in [-10, 10]$ by solving Equation 6.67; separate the roots with a step-size $h = 0.5$ (which is lower than the distance between the zeros of $\sin x$) and calculate them with the method of false position to keep them bracketed.

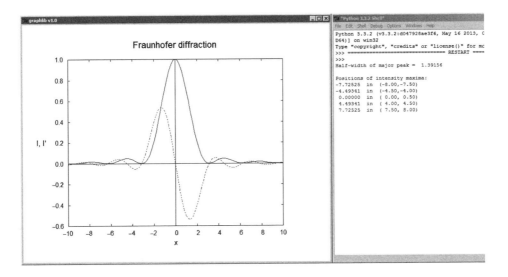

FIGURE 6.9 Intensity distribution $I(x)$ (continuous) and derivative $I'(x)$ (dashed) for the Fraunhofer diffraction by a rectangular slit of finite width and infinite height.

Solution

The Python implementation is given in Listing 6.18, the C/C++ version is available as supplementary material (P06-Fraunhofer.cpp), and the graphical output is depicted in Figure 6.9.

Listing 6.18   Fraunhofer Diffraction (Python Coding)

```python
# Intensity distribution for Fraunhofer diffraction
from math import *
from roots import *
from graphlib import *

def Intens(x):                                          # intensity
   global I0                                        # maximum intensity
   sinc = sin(x)/x if x else 1e0
   return I0 * sinc * sinc
def dIntens(x):                                    # derivative of Intens
   global I0                                        # maximum intensity
   if (x):
      sinc = sin(x)/x
      return 2e0 * I0 * (cos(x) - sinc) * sinc / x
   else: return 0e0
def func(x):                              # function for half-width calculation
   global I0                                        # maximum intensity
   return Intens(x) - 0.5e0 * I0

# main

GraphInit(800,600)

I0 = 1e0                                             # maximum intensity
xmin = -10e0; xmax = 10e0

#-----------------------------------------------------------------------
n = 101
x = [0]*(2*n+1); y = [0]*(2*n+1)

h = (xmax-xmin) / (n-1)
for i in range(1,n+1):
   xi = xmin + (i-1)*h
   x[i  ] = xi; y[i  ] = Intens(xi)
   x[i+n] = xi; y[i+n] = dIntens(xi)

nn = [0, n, 2*n]; col = ["", "red", "blue"]; sty = [0, 1,-1]
MultiPlot(x,y,y,nn,col,sty,2,10,
          0e0,0e0,0,0e0,0e0,0,0.15,0.85,0.15,0.85,
          "x","I, I'","Fraunhofer diffraction")

#-----------------------------------------------------------------------
xi = pi/2e0
(xi,ierr) = Secant(func,0e0,pi,xi)                  # half-width of major peak
print("Half-width of major peak = {0:8.5f}".format(xi))

#-----------------------------------------------------------------------
print("\nPositions of intensity maxima:")
h = 0.5e0
a = xmin
while (a < xmax):                                    # root separation loop
   b = a + h                                        # new search interval [a,b]
```

```
  (xi,ierr) = FalsPos(dIntens,a,b)
  if ((ierr == 0) and (xi != b) and (Intens(xi) > 1e-10)):
    print("{0:8.5f}  in   ({1:5.2f},{2:5.2f})".format(xi,a,b))
  a = b                                      # shift left boundary

MainLoop()
```

PROBLEM 6.5

Planck's law expresses the spectral energy density of the electromagnetic radiation emitted by a black body in thermal equilibrium (Toda et al. 1998):

$$u(\lambda, T) = \frac{8\pi hc}{\lambda^5} \frac{1}{\exp\left(hc/\left(\lambda k_B T\right)\right) - 1},$$

where T is the equilibrium temperature, λ is the wavelength, $h = 6.62606957 \times 10^{-34}$ J·s is Planck's constant, $k_B = 1.3806488 \times 10^{-23}$ J·K^{-1} is the Boltzmann constant, and $c = 2.99792458 \times 10^8$ m·s^{-1} is the speed of light in vacuum. $u(\lambda, T)d\lambda$ represents the energy radiated in the spectral interval $(\lambda, \lambda + d\lambda)$.

Introducing the dimensionless variable

$$x = \frac{\lambda}{\lambda_0},$$

and the parameters depending on the temperature

$$\lambda_0 = \frac{hc}{k_B T}, \quad K = \frac{8\pi k_B T}{\lambda_0^4},$$

Planck's law becomes

$$u(\lambda, T) = K \frac{1}{x^5 \left(e^{1/x} - 1\right)}. \tag{6.68}$$

The derivative of the spectral energy density,

$$\frac{\partial}{\partial \lambda} u(\lambda, T) = K\lambda_0 \frac{5x + (1 - 5x) e^{1/x}}{x^7 \left(e^{1/x} - 1\right)^2}, \tag{6.69}$$

provides by its zero the position of the maximum of $u(\lambda, T)$. *Wien's displacement law* states that for a black body, the wavelength λ_{max} corresponding to the maximum of $u(\lambda, T)$ is inversely proportional to the temperature T, or, equivalently, $\lambda_{max} T = 2.897768 \times 10^{-3}$ m·K.

a. Plot the spectral energy density $u(\lambda, T)$ and the corresponding derivative $\partial u(\lambda, T)/\partial t$ for the equilibrium temperatures $T = 4000, 5000$, and 6000 K.

b. Find the maximum of $u(\lambda, T)$ by solving the equation

$$\frac{\partial}{\partial \lambda} u(\lambda, T) = 0,$$

for the electromagnetic radiation of the sun, considering the effective photosphere temperature $T = 5778$ K (Williams 2013). Use Newton's method with numerical derivative, and check the validity of Wien's law.

Solution

The Python implementation is shown in Listing 6.19, the C/C++ version is available as supplementary material (P06-Planck.cpp). The graphical output is presented in Figure 6.10.

Listing 6.19   Planck's Law of Black-Body Radiation (Python Coding)

```python
# Planck's law of black body radiation
from math import *
from roots import *
from graphlib import *

hP = 6.62606957e-34                                # Planck's constant (J s)
kB = 1.3806488e-23                                 # Boltzmann constant (J/K)
c  = 2.99792458e8                                   # speed of light (m/s)

def u(lam):                                        # spectral energy density u(lam)
   global lam0, K                          # reference wavelength, factor of u(lam)

   if (lam == 0e0): return 0e0
   x = lam / lam0
   return K / (pow(x,5) * (exp(1e0/x) - 1e0))

def du(lam):                                                    # du/dlam
   global lam0, K                          # reference wavelength, factor of u(lam)

   if (lam == 0e0): return 0e0
   x = lam / lam0
   expx = exp(1e0/x)
   exp1 = expx - 1e0
   return K * lam0 * (5e0*x + (1e0-5e0*x)*expx) / (pow(x,7) * exp1*exp1)

# main

GraphInit(1200,600)

lam_plot = 3e-6                                    # maximum wavelength for plots
h = 1e-8                                           # step-size for wavelength
n = int(lam_plot/h + 0.5) + 1                      # number of points / plot
nmax = 3 * n                                        # total number of points

x = [0]*(nmax+1); y = [0]*(nmax+1); z = [0]*(nmax+1)

for j in range(0,3):
   T = 4000e0 + j * 1000e0                              # current temperature
   lam0 = hP*c / (kB*T)                                 # reference wavelength
   K = 8e0 * pi * kB * T / pow(lam0,4)                   # factor of u(lam)
   for i in range(1,n+1):
      lam = (i-1)*h
      x[i+j*n] = lam * 1e6                                      # microns
      y[i+j*n] = u(lam)
      z[i+j*n] = du(lam)

fmax = 0e0                                             # normalize profiles
for i in range(1,nmax+1):
   if (y[i] > fmax): fmax = y[i]
for i in range(1,nmax+1): y[i] /= fmax
fmax = 0e0
for i in range(1,nmax+1):
   if (z[i] > fmax): fmax = z[i]
for i in range(1,nmax+1): z[i] /= fmax
```

```
nn = [0, n, 2*n, 3*n]
col = ["", "blue", "green", "red"]
sty = [0, 1, 1, 1]
MultiPlot(x,y,y,nn,col,sty,3,10,
          0e0,0e0,0,0e0,0e0,0,0.10,0.45,0.15,0.85,
          "lambda (micron)","u","Spectral energy density")

MultiPlot(x,z,z,nn,col,sty,3,10,
          0e0,0e0,0,0e0,0e0,0,0.60,0.95,0.15,0.85,
          "lambda (micron)","du","Derivative of energy density")

#---------------------------------------------------------------------
T = 5778e0                              # temperature of sun's photosphere
lam0 = hP*c / (kB*T)                         # reference wavelength
K = 8e0 * pi * kB * T / pow(lam0,4)             # factor of u(lam)

fmin = 1e10; fmax = -1e10          # positions of maximum and minimum of du
for i in range(1,n+1):
   lam = (i-1)*h; f = du(lam)
   if (f > fmax): fmax = f; lam_1 = lam                   # maximum
   if (f < fmin): fmin = f; lam_2 = lam                   # minimum

                                   # find zero of du in [lam_1,lam_2]
lam_max = 0.5e0 * (lam_1 + lam_2)                 # initial approximation
(lam_max,ierr) = NewtonNumDrv(du,lam_1,lam_2,lam_max)
print("Wien's law:")
print("T = {0:4.0f} K".format(T))
print("lambda_max = {0:e} m".format(lam_max))
print("lambda_max * T = {0:e} m K".format(lam_max * T))      # 2.897768e-3 m K

MainLoop()
```

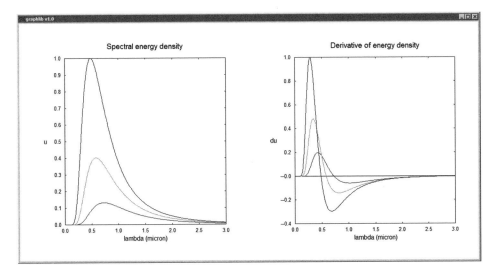

FIGURE 6.10 Planck's law of black-body radiation: spectral energy density $u(\lambda, T)$ (left panel) and derivative $\partial u(\lambda, T)/\partial \lambda$, for $T = 4000$, 5000, and 6000 K, corresponding to major peaks of increasing magnitude.

PROBLEM 6.6

Find the intersection points of the following circle and parabola:

$$(x - 1)^2 + y^2 = 4^2,$$

$$y = x^2 - 3,$$

by using the pair of routines Jacobian and NewtonSys from the modules roots.py and, respectively, roots.h. Use Figure 6.11 to find initial guesses for the coordinates of the intersection points.

Solution

Rewrite the system of equations as

$$\begin{cases} f_1(x_1, x_2) \equiv (x_1 - 1)^2 + x_2^2 - 4^2 = 0, \\ f_2(x_1, x_2) \equiv x_1^2 - x_2 - 3 = 0. \end{cases}$$

The Python implementation is given in Listing 6.20, the C/C++ version is available as supplementary material (P06-NewtonSys1.cpp), and the graphical output is depicted in Figure 6.11.

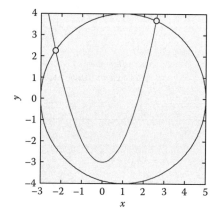

FIGURE 6.11 Intersection points of the circle $(x - 1)^2 + y^2 = 4^2$ and the parabola $y = x^2 - 3$, to be determined by Program 6.20.

Listing 6.20   Intersection Points of a Circle and a Parabola (Python Coding)

```python
# Intersection of circle and parabola
from math import *
from roots import *

def Func(f, x, n):                      # zeros: (-2.295, 2.267), (2.583, 3.673)
    f[1] = pow(x[1]-1,2) + x[2]*x[2] - 16e0
    f[2] = x[1]*x[1] - x[2] - 3e0

# main

n = 2
f = [0]*(n+1)
```

```
x = [0]*(n+1)
dx = [0]*(n+1)

x[1] = -5e0; x[2] = 5e0       # 1st initial approximation -> (-2.295, 2.267)
(dx,ierr) = NewtonSys(Func,x,n)
Func(f,x,n)

print("Solution 1:")
print("          x            dx        f")
for i in range(1,n+1):
   print("{0:d}   {1:15.7e}   {2:7.0e}   {3:7.0}".format(i,x[i],dx[i],f[i]))

x[1] = 5e0; x[2] = 5e0            # 2nd initial approximation -> (2.583, 3.673)
(dx,ierr) = NewtonSys(Func,x,n)
Func(f,x,n)

print("\nSolution 2:")
print("            x           dx        f")
for i in range(1,n+1):
   print("{0:d}   {1:15.7e}   {2:7.0e}   {3:7.0}".format(i,x[i],dx[i],f[i]))
```

PROBLEM 6.7

Solve the system of nonlinear equations (S. M. Tang and W. C. Kok, *J. Phys. A: Math. Gen*, **18**, 2691 (1985)):

$$\begin{cases} f_1(x_1, x_2, x_3) \equiv x_3 - x_1 x_2 x_3 & = 0, \\ f_2(x_1, x_2, x_3) \equiv x_1 + x_2 + x_3 & = 0, \\ f_3(x_1, x_2, x_3) \equiv x_1 x_2 + x_2 x_3 + x_3 x_1 = 0. \end{cases}$$

Solution

The Python implementation is given in Listing 6.21 and the C/C++ version is available as supplementary material (P06-NewtonSys2.cpp).

Listing 6.21   System of Three Nonlinear Equations (Python Coding)

```
# Newton-Raphson method for systems of non-linear equations
from math import *
from roots import *

def Func(f, x, n):                                    # zero: (0, 0, 0)
   f[1] = x[3] - x[1]*x[2]*x[3]
   f[2] = x[1] + x[2] + x[3]
   f[3] = x[1]*x[2] + x[2]*x[3] + x[3]*x[1]

# main

n = 3
f = [0]*(n+1)
x = [0]*(n+1)
dx = [0]*(n+1)

x[1] = 1e0; x[2] = 2e0; x[3] = 3e0              # initial approximation

(dx,ierr) = NewtonSys(Func,x,n)
Func(f,x,n)
```

```
print("Solution:")
print("             x              dx        f")
for i in range(1,n+1):
    print("{0:d}  {1:15.7e}  {2:7.0e}  {3:7.0}".format(i,x[i],dx[i],f[i]))
```

References and Suggested Further Reading

Bakhvalov, N. S. 1977. *Numerical Methods: Analysis, Algebra, Ordinary Differential Equations.* Moscow: MIR Publishers.

Beu, T. A. 2004. *Numerical Calculus in C* (3rd ed., in Romanian). Cluj-Napoca: MicroInformatica.

Burden, R. and J. Faires. 2010. *Numerical Analysis* (9th ed.). Boston: Brooks/Cole, Cengage Learning.

Demidovich, B. and I. Maron. 1987. *Computational Mathematics* (4th ed.). Moscow: MIR Publishers.

Hecht, E. 2002. *Optics* (4th ed.). San Francisco: Addison-Wesley.

Mottmann, J. 1986. Locating an orbiting object. *American Journal of Physics 54*, 838–842.

Ortega, J. M. and W. C. Rheinboldt. 2000. *Iterative Solution of Nonlinear Equations in Several Variables.* New York: Society for Industrial and Applied Mathematics.

Press, W. H., S. A. Teukolsky, W. T. Vetterling, and B. P. Flannery. 2007. *Numerical Recipes* (3rd ed.): *The Art of Scientific Computing.* Cambridge: Cambridge University Press.

Toda, M., R. Kubo, and N. Saitô. 1998. *Statistical Physics I, Equilibrium Statistical Mechanics* (2nd ed.). Heidelberg: Springer.

Toma, M. and I. Odăgescu. 1980. *Numerical Methods and Subroutines* (in Romanian). Bucharest: Editura Tehnică.

Williams, D. R. 2013. NASA, Lunar and Planetary Science, Sun Fact Sheet. http://nssdc.gsfc.nasa.gov/planetary/factsheet/sunfact.html.

7

Systems of Linear Equations

7.1 Introduction

Linear systems of equations are almost ubiquitous in applied mathematics. They may appear naturally in the mathematical formulation of the problem or, as a result of applying certain linear approximation methods for solving the original problem. Discretization of boundary value problems for differential equations by finite difference schemes or interpolation with piecewise smooth functions are just two typical examples of numerical methods leading to linear systems.

The methods for solving linear systems are certainly central to numerical analysis and can be essentially divided into two broad classes: *direct methods* and *iterative methods*.

Direct methods are *finite* algorithms for which the number of elementary operations required to obtain the solution directly depends on the size of the linear system. Direct methods are affected only by *round-off errors*, feature which makes them preferable when the system size and the number of elementary operations involved remain moderate. Despite its conceptual simplicity among the direct methods, the implementation of Cramer's rule (based on determinants) is not particularly efficient in terms of operations and, therefore, it is not recommendable in real-world applications. Examples of frequently used direct methods are Gaussian and Gauss–Jordan elimination, LU factorization, Cholesky decomposition, and so on.

Iterative methods enable finding solutions of linear systems as limits of virtually infinite iterative processes. Since only a finite number of iterations can be performed, the unavoidable roundoff errors are also accompanied by *truncation errors*. Among the advantages of iterative methods, the extreme simplicity and straightforward implementation outweigh the necessity of a more elaborate error control and make them suitable in cases that are intractable by direct methods. In addition, iterative methods can be used to refine solutions obtained by other methods. The most common examples of iterative algorithms are the Jacobi, Gauss–Seidel, and relaxation methods.

7.2 Gaussian Elimination with Backward Substitution

One of the classical direct methods for solving linear systems is known as Gaussian elimination, being named after Carl Friedrich Gauss, to whom it is attributed. Apparently, however, the method was already known in China over 2000 years ago, while Isaac Newton was using it at the beginning of the eighteenth century, almost 100 years before Gauss (Grcar 2011).

The basic concept of the Gaussian elimination implies two distinct phases: (1) transformation of the system by elementary row operations into an equivalent system with *upper triangular coefficient matrix* (with all entries below the main diagonal equal to 0); and (2) solution of the equivalent triangular system by specific recurrent techniques. The initial transformation of the system is equivalent to eliminating in sequence the unknowns from the successive equations and is therefore named *forward elimination*. The subsequent recurrent solution of the equivalent system with triangular matrix consists of expressing the solution components in reversed order, starting with the last component, and substituting them in the remaining equations, and is generically termed *backward substitution*.

The *elementary row operations* that can be applied to the equations so as to transform a linear system into an equivalent one having the same solution but a triangular coefficient matrix are

1. Interchanging two rows,
2. Multiplying a row by a nonzero scalar, and
3. Replacing a row by a linear combination of itself and another row.

Let us consider for illustration the following 3×3 system:

$$
\begin{cases}
3x_1 + 4x_2 + 5x_3 = 10 \\
4x_1 + 2x_2 + 3x_3 = 9 \\
-2x_1 - 3x_2 + x_3 = 7
\end{cases}
\quad \text{or} \quad
\begin{bmatrix} 3 & 4 & 5 \\ 4 & 2 & 3 \\ -2 & -3 & 1 \end{bmatrix}
\begin{bmatrix} x_1 \\ x_2 \\ x_3 \end{bmatrix}
=
\begin{bmatrix} 10 \\ 9 \\ 7 \end{bmatrix}.
$$

Forward elimination starts by eliminating x_1 from the second and third equations. Traditionally, this is done by solving for x_1 in the first equation and substituting it into the second and third equations to eliminate the x_1 dependence. To simplify calculations, we actually divide the first equation by the leading coefficient 3:

$$
\begin{bmatrix} 1 & 4/3 & 5/3 \\ 4 & 2 & 3 \\ -2 & -3 & 1 \end{bmatrix}
\begin{bmatrix} x_1 \\ x_2 \\ x_3 \end{bmatrix}
=
\begin{bmatrix} 10/3 \\ 9 \\ 7 \end{bmatrix}.
$$

Then, we subtract the transformed first equation multiplied by the leading coefficient 4 from the second equation and, respectively, multiplied by the leading coefficient -2 from the third equation,

$$
\begin{bmatrix} 1 & 4/3 & 5/3 \\ 0 & -10/3 & -11/3 \\ 0 & -1/3 & 13/3 \end{bmatrix}
\begin{bmatrix} x_1 \\ x_2 \\ x_3 \end{bmatrix}
=
\begin{bmatrix} 10/3 \\ -13/3 \\ 41/3 \end{bmatrix},
$$

thus effectively eliminating x_1 from the last two equations. Now, seeking to eliminate x_2 from the last equation, we first divide the second equation by the coefficient $-10/3$ of x_2,

$$
\begin{bmatrix} 1 & 4/3 & 5/3 \\ 0 & 1 & 11/10 \\ 0 & -1/3 & 13/3 \end{bmatrix}
\begin{bmatrix} x_1 \\ x_2 \\ x_3 \end{bmatrix}
=
\begin{bmatrix} 10/3 \\ 13/10 \\ 41/3 \end{bmatrix},
$$

and then multiply the transformed second equation by the coefficient of x_2 from the last equation, $-1/3$, and subtract it from the latter:

$$
\begin{bmatrix} 1 & 4/3 & 5/3 \\ 0 & 1 & 11/10 \\ 0 & 0 & 1 \end{bmatrix}
\begin{bmatrix} x_1 \\ x_2 \\ x_3 \end{bmatrix}
=
\begin{bmatrix} 10/3 \\ 13/10 \\ 3 \end{bmatrix}.
$$

Consequently, the equivalent system has, indeed, a triangular matrix.

The *backward substitution* is initiated simply by expressing the last solution component, x_3, from the last equation. From the second equation, one can then express x_2 using the already known x_3, while x_1

results from the first equation using x_2 and x_3:

$$x_3 = 3,$$
$$x_2 = 13/10 - 11/10x_3 = -2,$$
$$x_1 = 10/3 - 4/3x_2 - 5/3x_3 = 1.$$

Thus, the solution of the considered linear system is

$$x_1 = 1, \quad x_2 = -1, \quad x_3 = 3.$$

Taking a step forward toward deriving the general algorithm of the Gaussian elimination, we consider now the case of a 3×3 linear system with symbolic coefficients:

$$\begin{cases} a_{11}x_1 + a_{12}x_2 + a_{13}x_3 &= b_1 \\ a_{21}x_1 + a_{22}x_2 + a_{23}x_3 &= b_2 \\ a_{31}x_1 + a_{32}x_2 + a_{33}x_3 &= b_3. \end{cases}$$

The matrix form of this system,

$$\begin{bmatrix} a_{11} & a_{12} & a_{13} \\ a_{21} & a_{22} & a_{23} \\ a_{31} & a_{32} & a_{33} \end{bmatrix} \begin{bmatrix} x_1 \\ x_2 \\ x_3 \end{bmatrix} = \begin{bmatrix} b_1 \\ b_2 \\ b_3 \end{bmatrix}, \tag{7.1}$$

can be cast as matrix equation,

$$\mathbf{A} \cdot \mathbf{x} = \mathbf{b}, \tag{7.2}$$

where $\mathbf{A} = [a_{ij}]$ is the *coefficient matrix*, $\mathbf{x} = [x_i]$ is the column vector of the unknowns, and $\mathbf{b} = [b_i]$ is the column vector of the constant terms.

At the first step of the forward elimination (marked by an upper index (1)), we aim to eliminate x_1 from all equations, except for the first one. To this end, we start by dividing the first row by the *pivot element* a_{11}, which is assumed to be nonzero:

$$\begin{cases} x_1 + a_{12}^{(1)}x_2 + a_{13}^{(1)}x_3 = b_1^{(1)} \\ a_{21}x_1 + a_{22}x_2 + a_{23}x_3 = b_2 \\ a_{31}x_1 + a_{32}x_2 + a_{33}x_3 = b_3. \end{cases}$$

If, on the contrary, a_{11} is zero, the first equation has to be interchanged with another one that has a nonzero leading coefficient. We then subtract the first equation, multiplied by the coefficient a_{21} of x_1, from the second equation and, respectively, multiplied by a_{31}, from the third equation. Thus, we obtain the transformed system:

$$\begin{cases} x_1 + a_{12}^{(1)}x_2 + a_{13}^{(1)}x_3 &= b_1^{(1)} \\ a_{22}^{(1)}x_2 + a_{23}^{(1)}x_3 &= b_2^{(1)} \\ a_{32}^{(1)}x_2 + a_{33}^{(1)}x_3 &= b_3^{(1)}, \end{cases}$$

where the upper index (1) denotes the first step of the forward elimination. The new coefficients are given by

$$
\begin{cases}
a_{1j}^{(1)} &= a_{1j}/a_{11}, \quad j = 1, 2, 3 \\
b_1^{(1)} &= b_1/a_{11}. \\
a_{ij}^{(1)} &= a_{ij} - a_{i1}a_{1j}^{(1)}, \quad j = 1, 2, 3, \; i = 2, 3 \\
b_i^{(1)} &= b_i - a_{i1}b_1^{(1)}.
\end{cases}
\tag{7.3}
$$

In matrix notation, the first step of the Gaussian elimination leads to

$$
\begin{bmatrix}
1 & a_{12}^{(1)} & a_{13}^{(1)} \\
0 & a_{22}^{(1)} & a_{23}^{(1)} \\
0 & a_{32}^{(1)} & a_{33}^{(1)}
\end{bmatrix}
\begin{bmatrix}
x_1 \\ x_2 \\ x_3
\end{bmatrix}
=
\begin{bmatrix}
b_1^{(1)} \\ b_2^{(1)} \\ b_3^{(1)}
\end{bmatrix}.
\tag{7.4}
$$

The purpose of the second step is to eliminate the unknown x_2 from the last equation. We first divide the second equation by the pivot element $a_{22}^{(1)}$, assumed to be nonzero (otherwise, we interchange the second and third equations). We then subtract the obtained equation, multiplied by $a_{32}^{(1)}$, from the third one. The result is the matrix equation:

$$
\begin{bmatrix}
1 & a_{12}^{(1)} & a_{13}^{(1)} \\
0 & 1 & a_{23}^{(2)} \\
0 & 0 & a_{33}^{(2)}
\end{bmatrix}
\begin{bmatrix}
x_1 \\ x_2 \\ x_3
\end{bmatrix}
=
\begin{bmatrix}
b_1^{(1)} \\ b_2^{(2)} \\ b_3^{(2)}
\end{bmatrix},
\tag{7.5}
$$

with the coefficients

$$
\begin{cases}
a_{2j}^{(2)} &= a_{2j}^{(1)}/a_{22}^{(1)}, \quad j = 2, 3 \\
b_2^{(2)} &= b_2^{(1)}/a_{22}^{(1)}. \\
a_{ij}^{(2)} &= a_{ij}^{(1)} - a_{i2}^{(1)}a_{2j}^{(2)}, \quad j = 2, 3, \; i = 3 \\
b_i^{(2)} &= b_i^{(1)} - a_{i2}^{(1)}b_2^{(2)}
\end{cases}
\tag{7.6}
$$

It is readily noticeable that these relations are perfectly equivalent with Formulas 7.3 of the first elimination step and could have been obtained from the latter by simply replacing index "1" with "2." Within the second step, one no longer operates on the first line and first column.

We finalize the forward elimination by dividing the third equation by its pivot element $a_{33}^{(2)}$ (which has to be nonzero for a system with nonsingular matrix) and we get

$$
\begin{bmatrix}
1 & a_{12}^{(1)} & a_{13}^{(1)} \\
0 & 1 & a_{23}^{(2)} \\
0 & 0 & 1
\end{bmatrix}
\begin{bmatrix}
x_1 \\ x_2 \\ x_3
\end{bmatrix}
=
\begin{bmatrix}
b_1^{(1)} \\ b_2^{(2)} \\ b_3^{(3)}
\end{bmatrix},
\tag{7.7}
$$

with

$$
b_3^{(3)} = b_3^{(2)}/a_{33}^{(2)}.
$$

The *backward substitution* implies processing in reversed order the rows of the transformed system with triangular matrix 7.7—resulted from the elimination phase—and calculating the solution components by the recursive procedure:

$$\begin{cases} x_3 &= b_3^{(3)} \\ x_2 &= b_2^{(2)} - a_{23}^{(2)} x_3 \\ x_1 &= b_1^{(1)} - (a_{12}^{(1)} x_2 + a_{13}^{(1)} x_3). \end{cases} \tag{7.8}$$

The determination of the solution components proceeds from high to low indexes, each new component solely depending on the previously determined components.

A beneficial feature of the Gaussian elimination is that it also enables the calculation of the determinant of matrix $\mathbf{A}$ on the fly. Indeed, since the final matrix of the transformed system $\mathbf{A}^{(3)}$ (7.7) is triangular, its determinant amounts to the product of the diagonal elements

$$\det \mathbf{A}^{(3)} = 1.$$

On the other hand, having in view that the divisions of the rows by the pivot elements finally lead to a coefficient matrix having the determinant equal to the one of the initial matrix divided by the product of the successive pivots, there results

$$\det \mathbf{A}^{(3)} = \frac{\det \mathbf{A}}{a_{11} a_{22}^{(1)} a_{33}^{(2)}} = 1,$$

and hence

$$\det \mathbf{A} = a_{11} a_{22}^{(1)} a_{33}^{(2)}. \tag{7.9}$$

Let us now generalize the Gaussian elimination for the case of an $n \times n$ linear system (with n equations and n unknowns). Written as a matrix equation, the system reads

$$\mathbf{A} \cdot \mathbf{x} = \mathbf{b}, \tag{7.10}$$

and features the coefficient matrix $\mathbf{A} = [a_{ij}]_{nn}$, the column vector $\mathbf{b} = [b_i]_n$ of the constant terms, and the column vector $\mathbf{x} = [x_i]_n$ of the solution components.

As a result of the first $k-1$ steps of the *forward elimination*, zeros have been created below the main diagonal on the first $k-1$ columns of the coefficient matrix, and the system has acquired the equivalent form:

$$\begin{bmatrix} 1 & a_{12}^{(1)} & \cdots & a_{1k}^{(1)} & a_{1k+1}^{(1)} & \cdots & a_{1n}^{(1)} \\ 0 & 1 & \cdots & a_{2k}^{(2)} & a_{2k+1}^{(2)} & \cdots & a_{2n}^{(2)} \\ \vdots & \vdots & \ddots & \vdots & \vdots & & \vdots \\ 0 & 0 & \cdots & a_{kk}^{(k-1)} & a_{kk+1}^{(k-1)} & \cdots & a_{kn}^{(k-1)} \\ 0 & 0 & \cdots & a_{k+1k}^{(k-1)} & a_{k+1k+1}^{(k-1)} & \cdots & a_{k+1n}^{(k-1)} \\ \vdots & \vdots & & \vdots & \vdots & \ddots & \vdots \\ 0 & 0 & \cdots & a_{nk}^{(k-1)} & a_{nk+1}^{(k-1)} & \cdots & a_{nn}^{(k-1)} \end{bmatrix} \begin{bmatrix} x_1 \\ x_2 \\ \vdots \\ x_k \\ x_{k+1} \\ \vdots \\ x_n \end{bmatrix} = \begin{bmatrix} b_1^{(1)} \\ b_2^{(2)} \\ \vdots \\ b_k^{(k-1)} \\ b_{k+1}^{(k-1)} \\ \vdots \\ b_n^{(k-1)} \end{bmatrix}.$$

Obviously, equations $k, k+1, \ldots, n$ couple only the solution components $x_k, x_{k+1}, \ldots, x_n$, the rest having been successively eliminated.

The goal of step k ($k = 1, 2, \ldots, n - 1$) is to eliminate the unknown x_k from the last $n - k$ equations and to bring the system to the form:

$$
\begin{bmatrix}
1 & a_{12}^{(1)} & \cdots & a_{1k}^{(1)} & a_{1k+1}^{(1)} & \cdots & a_{1n}^{(1)} \\
0 & 1 & \cdots & a_{2k}^{(2)} & a_{2k+1}^{(2)} & \cdots & a_{2n}^{(2)} \\
\vdots & \vdots & \ddots & \vdots & \vdots & & \vdots \\
0 & 0 & \cdots & 1 & a_{kk+1}^{(k)} & \cdots & a_{kn}^{(k)} \\
0 & 0 & \cdots & 0 & a_{k+1k+1}^{(k)} & \cdots & a_{k+1n}^{(k)} \\
\vdots & \vdots & & \vdots & \vdots & \ddots & \vdots \\
0 & 0 & \cdots & 0 & a_{nk+1}^{(k)} & \cdots & a_{nn}^{(k)}
\end{bmatrix}
\begin{bmatrix}
x_1 \\ x_2 \\ \vdots \\ x_k \\ x_{k+1} \\ \vdots \\ x_n
\end{bmatrix}
=
\begin{bmatrix}
b_1^{(1)} \\ b_2^{(2)} \\ \vdots \\ b_k^{(k)} \\ b_{k+1}^{(k)} \\ \vdots \\ b_n^{(k)}
\end{bmatrix}.
\tag{7.11}
$$

The updated coefficients of the *pivot row* k are given by

$$
\begin{cases}
a_{kk}^{(k)} &= 1, \\
a_{kj}^{(k)} &= a_{kj}^{(k-1)}/a_{kk}^{(k-1)}, \quad j = k+1, \ldots, n, \\
b_k^{(k)} &= b_k^{(k-1)}/a_{kk}^{(k-1)},
\end{cases}
\tag{7.12}
$$

while the new elements of the *nonpivot rows* below the pivot line read:

$$
\begin{cases}
a_{ik}^{(k)} &= 0, \\
a_{ij}^{(k)} &= a_{ij}^{(k-1)} - a_{ik}^{(k-1)} a_{kj}^{(k)}, \quad j = k+1, \ldots, n, \; i = k+1, \ldots, n, \\
b_i^{(k)} &= b_i^{(k-1)} - a_{ik}^{(k-1)} b_k^{(k)}.
\end{cases}
\tag{7.13}
$$

It can be noticed that, at step k, one actually modifies only the elements delimited by the row and column k in the bottom-right part of the coefficient matrix, the rest of the elements remaining unchanged as compared to the previous step.

At the last elimination step, $k = n$, one only divides the last equation by the last pivot element $a_{nn}^{(n-1)}$ to create a 1 on the diagonal:

$$
\begin{cases}
a_{nn}^{(n)} &= 1, \\
b_n^{(n)} &= b_n^{(n-1)}/a_{nn}^{(n-1)}.
\end{cases}
\tag{7.14}
$$

Finally, the system takes the form

$$
\begin{bmatrix}
1 & a_{12}^{(1)} & \cdots & a_{1k}^{(1)} & a_{1k+1}^{(1)} & \cdots & a_{1n}^{(1)} \\
0 & 1 & \cdots & a_{2k}^{(2)} & a_{2k+1}^{(2)} & \cdots & a_{2n}^{(2)} \\
\vdots & \vdots & \ddots & \vdots & \vdots & & \vdots \\
0 & 0 & \cdots & 1 & a_{kk+1}^{(k)} & \cdots & a_{kn}^{(k)} \\
0 & 0 & \cdots & 0 & 1 & \cdots & a_{k+1n}^{(k+1)} \\
\vdots & \vdots & & \vdots & \vdots & \ddots & \vdots \\
0 & 0 & \cdots & 0 & 0 & \cdots & 1
\end{bmatrix}
\begin{bmatrix}
x_1 \\ x_2 \\ \vdots \\ x_k \\ x_{k+1} \\ \vdots \\ x_n
\end{bmatrix}
=
\begin{bmatrix}
b_1^{(1)} \\ b_2^{(2)} \\ \vdots \\ b_k^{(k)} \\ b_{k+1}^{(k+1)} \\ \vdots \\ b_n^{(n)}
\end{bmatrix},
\tag{7.15}
$$

or, in matrix notation:

$$\mathbf{A}^{(n)} \cdot \mathbf{x} = \mathbf{b}^{(n)}. \tag{7.16}$$

Matrix $\mathbf{A}^{(n)}$ shows indeed upper triangular structure, and the system is equivalent to the initial one, $\mathbf{A} \cdot \mathbf{x} = \mathbf{b}$, having the same solution $(x_1, x_2, \ldots, x_n)$.

In the *backward substitution* phase, one solves the transformed system (7.15), determining recursively the components of the solution in reversed order, based on the formulas:

$$\begin{cases} x_n &= b_n^{(n)}, \\ x_k &= b_k^{(k)} - \displaystyle\sum_{i=k+1}^{n} a_{ki}^{(k)} x_i, \quad k = n-1, \ldots, 1. \end{cases} \tag{7.17}$$

The determinant of the triangular matrix $\mathbf{A}^{(n)}$ equals the product of the diagonal elements and amounts to 1. Bearing in mind that the divisions of the successive pivot lines by the corresponding pivot elements lead to a matrix having a determinant equal to the one of the initial matrix $\mathbf{A}$ divided by the product of the pivot elements, we have:

$$\det \mathbf{A}^{(n)} = \frac{\det \mathbf{A}}{a_{11} a_{22}^{(1)} \cdots a_{nn}^{(n-1)}} = 1.$$

Hence, the determinant of the initial coefficient matrix $\mathbf{A}$ equals the product of the successive pivot elements:

$$\det \mathbf{A} = a_{11} a_{22}^{(1)} \cdots a_{nn}^{(n-1)}. \tag{7.18}$$

A more general problem consists of solving an arbitrary number m of *simultaneous systems*, each having the same coefficient matrix $\mathbf{A} = [a_{ij}]_{nn}$, but different column vectors of constant terms. In the equivalent matrix equation,

$$\mathbf{A} \cdot \mathbf{X} = \mathbf{B}, \tag{7.19}$$

each of the m columns of the matrices $\mathbf{B} = [b_{ij}]_{nm}$ of constant terms and $\mathbf{X} = [x_{ij}]_{nm}$ of solution components corresponds to one of the linear systems. Since, for all m systems, the transformations undergone by the matrix $\mathbf{A}$ during the forward elimination are the same, they need to be done only once. On the other hand, they determine the transformations of the individual column vectors according to the same Relations 7.12 through 7.18, which can be thus generalized by merely adding a column index j to the unknowns x_{kj} and constant terms b_{kj}.

Summarizing the operations to be carried out on the matrices $\mathbf{A}$ and $\mathbf{B}$ at step $k = 1, \ldots, n$ of the *forward elimination*, the pivot row k is divided by the pivot $a_{kk}^{(k-1)}$:

$$\begin{cases} a_{kj}^{(k)} &= a_{kj}^{(k-1)} / a_{kk}^{(k-1)}, \quad j = k+1, \ldots, n, \\ b_{kj}^{(k)} &= b_{kj}^{(k-1)} / a_{kk}^{(k-1)}, \quad j = 1, \ldots, m, \end{cases} \tag{7.20}$$

while the nonpivot rows situated *below* the pivot row are reduced according to

$$\begin{cases} a_{ij}^{(k)} &= a_{ij}^{(k-1)} - a_{ik}^{(k-1)} a_{kj}^{(k)}, \quad i = k+1, \ldots, n, \quad j = k+1, \ldots, n, \\ b_{ij}^{(k)} &= b_{ij}^{(k-1)} - a_{ik}^{(k-1)} b_{kj}^{(k)}, \quad i = k+1, \ldots, n, \quad j = 1, \ldots, m. \end{cases} \tag{7.21}$$

The subsequent *backward substitution* operates based on the relations:

$$\begin{cases} x_{nj} = b_{nj}^{(n)}, \quad j = 1, \ldots, m, \\ x_{kj} = b_{kj}^{(k)} - \displaystyle\sum_{i=k+1}^{n} a_{ki}^{(k)} x_{ij}, \quad j = 1, \ldots, m, \; k = n-1, \ldots, 1, \end{cases} \tag{7.22}$$

and yields the components of the solution matrix $\mathbf{X}$ column wise (for each of the m simultaneous systems), in reversed order.

Taking into account that the entire relevant information on the final matrix $\mathbf{A}^{(n)}$ is contained at the end of the elimination phase exclusively in the upper triangle above the main diagonal, at the elimination step k, the elements $a_{kj}^{(k)}$ of the pivot row and, respectively, $a_{ij}^{(k)}$ of the nonpivot rows underneath do not need to be explicitly evaluated on column $j = k$, since they amount by definition to 1 or 0 and they are not reused anyway. This implies that one has to operate in matrix $\mathbf{A}^{(k)}$ only on columns $j = k+1, \ldots, n$. As for the backward substitution, in implementing Relations 7.22, one can use a single array for storing both the constant terms b_{kj} and the components x_{kj} of the solution.

Algorithms capable of solving matrix equations corresponding to an arbitrary number of simultaneous linear systems open the possibility of treating a problem of great practical importance—*inversion of square matrices*. Indeed, considering as matrix of constant terms the *unit matrix*,

$$\mathbf{B} = \mathbf{E}_n \equiv [\delta_{ij}]_{nn},$$

the solution of the system $\mathbf{A} \cdot \mathbf{X} = \mathbf{B}$ is precisely the inverse of the coefficient matrix, that is, $\mathbf{X} = \mathbf{A}^{-1}$. This is, obviously, the nonspecific way in which any algorithm for solving matrix equations can be used to invert matrices.

Of a special interest, not in the least for the sake of comparisons with other solvers for linear matrix equations, is the number of floating point operations (flops) required by the Gaussian elimination. By analyzing Equations 7.20 through 7.22, one gets the following operation counts:

- Transformation of the coefficient matrix $\mathbf{A}$ into upper triangular form:
  — Summations and subtractions (Equation 7.21):

$$S_A(n) = \sum_{k=1}^{n} \sum_{j=k+1}^{n} \sum_{i=k+1}^{n} 1 = \frac{1}{3}n^3 - \frac{1}{2}n^2 + \frac{1}{6}n,$$

  — Multiplications and divisions (Equations 7.20 and 7.21):

$$M_A(n) = \sum_{k=1}^{n} \sum_{j=k+1}^{n} 1 + \sum_{k=1}^{n} \sum_{j=k+1}^{n} \sum_{i=k+1}^{n} 1 = \frac{1}{3}n^3 - \frac{1}{3}n.$$

- Transformation of a single vector of constant terms **b** ($m = 1$):
  — Summations and subtractions (Equations 7.21 and 7.22):

$$S_b(n) = \sum_{k=1}^{n} \sum_{i=k+1}^{n} 1 + \sum_{k=1}^{n-1} \left(1 + \sum_{i=k+1}^{n-1} 1 \right) = n^2 - n,$$

  — Multiplications and divisions (Equations 7.20 through 7.22):

$$M_b(n) = \sum_{k=1}^{n} 1 + \sum_{k=1}^{n} \sum_{i=k+1}^{n} 1 + \sum_{k=1}^{n-1} \sum_{i=k+1}^{n} 1 = n^2.$$

- Solution of a single system of linear equations:

$$S_A(n) + M_A(n) + S_b(n) + M_b(n) = \frac{2}{3}n^3 + \frac{3}{2}n^2 - \frac{7}{6}n \sim \frac{2}{3}n^3. \qquad (7.23)$$

- Inversion of the coefficient matrix **A** or solution of an $n \times n$ matrix equation:

$$S_A(n) + M_A(n) + n\left[S_b(n) + M_b(n)\right] = \frac{8}{3}n^3 - \frac{3}{2}n^2 - \frac{1}{6}n \sim \frac{8}{3}n^3. \qquad (7.24)$$

Despite its perfect mathematical consistency, there are still situations in which the algorithm based on Equations 7.20 through 7.22 can fail. Specifically, the implementation of the Gaussian elimination involves at every step k the division of the pivot row k by the diagonal pivot element $a_{kk}^{(k-1)}$. Besides the well-known potential of divisions to amplify roundoff errors, it is possible to encounter pivots effectively equal to zero, in which case, the algorithm cannot be continued without additional provisions. Moreover, even in cases of nonzero values, larger pivots can be shown to induce smaller roundoff errors in the transformed matrix $\mathbf{A}^{(k)}$ and, implicitly, in the final triangular matrix $\mathbf{A}^{(n)}$.

With a view to counteracting the detrimental side effects of the divisions by the pivot elements, one option is to perform the so-called *partial pivoting*. At the elimination step k in particular, pivoting implies locating the maximum element $a_{lk}^{(k-1)}$ on column k, on and below the pivot row ($l \geq k$), and placing this element on the main diagonal (to become the actual pivot $a_{kk}^{(k-1)}$) by interchanging rows k and l. Nevertheless, in calculating the determinant, as product of the successive pivots, one has to account by a change of sign for any interchange of rows. Overall, pivoting adds computational costs to the Gaussian elimination, but ensures the stability of the algorithm and the accuracy of the solution.

There are in use more effective pivoting algorithms, such as *complete pivoting*, in which the pivot is chosen from all columns and rows of the matrix $\mathbf{A}^{(k-1)}$ on which pivoting has not already been carried out. The roundoff errors are hence reduced even more, at the expense, however, of more elaborate coding and additional storage. Complete pivoting is illustrated in conjunction with the Gauss–Jordan method in the next section.

The function `Gauss` from Listing 7.1 closely implements the algorithm of the Gaussian elimination described above, and the operation of the routine is stabilized by using partial pivoting on columns. Array a contains on entry the elements of the coefficient matrix **A** and is destroyed during the elimination procedure. To be usable after the operation of the routine, matrix **A** should also be saved in a backup array. Array b contains on entry the elements of the constant term **B** and returns on exit the corresponding components of the solution matrix **X**. The determinant of the coefficient matrix is returned by the

Listing 7.1 Solves Matrix Equation by Gaussian Elimination with Partial Pivoting (Python Coding)

```python
#===========================================================================
def Gauss(a, b, n, m):
#---------------------------------------------------------------------------
#  Solves matrix equation a x = b by Gaussian elimination with partial
#  pivoting on columns.
#  a    - coefficient matrix (n x n); destroyed on exit
#  b    - matrix of constant terms (n x m); solution x on exit
#  det  - determinant of coefficient matrix (output).
#---------------------------------------------------------------------------
   det = 1e0
   for k in range(1,n+1):                              # FORWARD ELIMINATION
      amax = 0e0                                 # determine pivot row with
      for i in range(k,n+1):                     # max. element on column k
         if (amax < fabs(a[i][k])): amax = fabs(a[i][k]); imax = i
      if (amax == 0e0):
         print("Gauss: singular matrix !"); return 0e0

      if (imax != k):                            # interchange rows imax and k
         det = -det                              # to put pivot on diagonal
         for j in range(k,n+1): (a[imax][j],a[k][j]) = (a[k][j],a[imax][j])
         for j in range(1,m+1): (b[imax][j],b[k][j]) = (b[k][j],b[imax][j])

      det *= a[k][k]                          # multiply determinant with pivot

      t = 1e0/a[k][k]                             # divide pivot row by pivot
      for j in range(k+1,n+1): a[k][j] *= t
      for j in range(  1,m+1): b[k][j] *= t

      for i in range(k+1,n+1):                          # reduce non-pivot rows
         t = a[i][k]
         for j in range(k+1,n+1): a[i][j] -= a[k][j]*t
         for j in range(  1,m+1): b[i][j] -= b[k][j]*t

   for k in range(n-1,0,-1):                        # BACKWARD SUBSTITUTION
      for j in range(1,m+1):
         sum = b[k][j]
         for i in range(k+1,n+1): sum -= a[k][i]*b[i][j]
         b[k][j] = sum

   return det
```

parameter `det`. The case in which, at some stage in the elimination procedure, no nonzero pivot can be identified indicates that the coefficient matrix **A** is *singular*, and the routine exits issuing an error message.

A simple test program illustrates in Listing 7.2 the usage of the routine `Gauss`. In accordance with the basic programming techniques discussed in Section 2.5, in Python we define the 2D arrays as collections of row vectors. Concretely, the particular syntax used in Listing 7.2 allows for the last usable elements of arrays a and b to be a[n][n] and b[n][m], respectively. The C/C++ counterpart of this program, provided as supplementary material (file P07-Gauss.cpp), performs the dynamic allocation of the arrays a and b with the help of function `Matrix` from the include file `memalloc.h` (see Chapter 2). For displaying formatted output, the program calls the function `MatPrint`, defined in the module `matutil.py`.

Listing 7.2   Solution of a Linear System by Gaussian Elimination (Python Coding)

```
# Solves linear system by Gauss elimination
from linsys import *
from matutil import *

n = 4                                                     # order of system
m = 1                                          # number of constant vectors
a = [[0]*(n+1) for i in range(n+1)]                       # system matrix
b = [[0]*(m+1) for i in range(n+1)]                       # constant terms

a[1][1] = 1; a[1][2] = 2; a[1][3] = 3; a[1][4] = 4; b[1][1] = 30
a[2][1] = 2; a[2][2] = 1; a[2][3] = 2; a[2][4] = 3; b[2][1] = 22
a[3][1] = 3; a[3][2] = 2; a[3][3] = 1; a[3][4] = 2; b[3][1] = 18
a[4][1] = 4; a[4][2] = 3; a[4][3] = 2; a[4][4] = 1; b[4][1] = 20
                      # Solution: 1.0, 2.0, 3.0, 4.0; det = -20.0
print("A:")
MatPrint(a,n,n)
print("B:")
MatPrint(b,n,m)

det = Gauss(a,b,n,m)                                     # solve system

print("det A = ",det)
print("Solution:")
MatPrint(b,n,m)
```

7.3 Gauss–Jordan Elimination

The Gauss–Jordan method can be regarded as a variant of the Gaussian elimination, in which, instead of transforming the coefficient matrix **A** by elementary row operations into upper triangular form, the elimination process is carried out at each step within the entire matrix, which is finally transformed into the unit matrix. As such, even though the forward elimination phase is more elaborate, the backward substitution is eliminated altogether. In addition, by efficient coding and memory usage, the Gauss–Jordan elimination allows for the inverse of the coefficient matrix $\mathbf{A}^{-1}$ to be calculated in place, simultaneously with the solution of a matrix equation.

Let us consider again the simple numerical example that was solved in the previous section using the standard Gaussian elimination with backward substitution:

$$\begin{bmatrix} 3 & 4 & 5 \\ 4 & 2 & 3 \\ -2 & -3 & 1 \end{bmatrix} \begin{bmatrix} x_1 \\ x_2 \\ x_3 \end{bmatrix} = \begin{bmatrix} 10 \\ 9 \\ 7 \end{bmatrix}.$$

At the first step of the *elimination*, the unknown x_1 is to be eliminated from all equations (second and third), except for the first one. To this end, we first divide the first equation by the leading coefficient 3:

$$\begin{bmatrix} 1 & 4/3 & 5/3 \\ 4 & 2 & 3 \\ -2 & -3 & 1 \end{bmatrix} \begin{bmatrix} x_1 \\ x_2 \\ x_3 \end{bmatrix} = \begin{bmatrix} 10/3 \\ 9 \\ 7 \end{bmatrix}.$$

We then subtract the transformed first equation multiplied by the leading coefficient 4 from the second equation and, respectively, multiplied the leading coefficient -2 from the third equation:

$$\begin{bmatrix} 1 & 4/3 & 5/3 \\ 0 & -10/3 & -11/3 \\ 0 & -1/3 & 13/3 \end{bmatrix} \begin{bmatrix} x_1 \\ x_2 \\ x_3 \end{bmatrix} = \begin{bmatrix} 10/3 \\ -13/3 \\ 41/3 \end{bmatrix}.$$

At the second elimination step, we eliminate x_2 from all equations (first and third), except for the second one. Therefore, we first divide the second equation by the coefficient $-10/3$ of x_2,

$$\begin{bmatrix} 1 & 4/3 & 5/3 \\ 0 & 1 & 11/10 \\ 0 & -1/3 & 13/3 \end{bmatrix} \begin{bmatrix} x_1 \\ x_2 \\ x_3 \end{bmatrix} = \begin{bmatrix} 10/3 \\ 13/10 \\ 41/3 \end{bmatrix},$$

and then we subtract the transformed second equation multiplied by the coefficient $4/3$ of x_2 from the first equation and, respectively, multiplied by $-1/3$ from the third equation:

$$\begin{bmatrix} 1 & 0 & 1/5 \\ 0 & 1 & 11/10 \\ 0 & 0 & 1 \end{bmatrix} \begin{bmatrix} x_1 \\ x_2 \\ x_3 \end{bmatrix} = \begin{bmatrix} 8/5 \\ 13/10 \\ 3 \end{bmatrix}.$$

At the last elimination step, we eliminate x_3 from all equations (first and second), but the second one. For this, we subtract the third equation multiplied by the coefficients of x_3, respectively, from the first and second equations:

$$\begin{bmatrix} 1 & 0 & 0 \\ 0 & 1 & 0 \\ 0 & 0 & 1 \end{bmatrix} \begin{bmatrix} x_1 \\ x_2 \\ x_3 \end{bmatrix} = \begin{bmatrix} 1 \\ -2 \\ 3 \end{bmatrix}.$$

There is, obviously, no need for backward substitution, as for the standard Gaussian elimination, since the components of the solution are already explicit in the transformed system having the unit matrix as coefficient matrix.

To devise the general algorithm, we note that, at step k of the *forward elimination*, the unknown x_k is eliminated from all equations of the system, except for the pivot equation k, and the transformed system thus takes the form:

$$\begin{bmatrix} 1 & 0 & \cdots & 0 & a_{1k+1}^{(k)} & \cdots & a_{1n}^{(k)} \\ 0 & 1 & \cdots & 0 & a_{2k+1}^{(k)} & \cdots & a_{2n}^{(k)} \\ \vdots & \vdots & \ddots & \vdots & \vdots & & \vdots \\ 0 & 0 & \cdots & 1 & a_{kk+1}^{(k)} & \cdots & a_{kn}^{(k)} \\ 0 & 0 & \cdots & 0 & a_{k+1k+1}^{(k)} & \cdots & a_{k+1n}^{(k)} \\ \vdots & \vdots & & \vdots & \vdots & \ddots & \vdots \\ 0 & 0 & \cdots & 0 & a_{nk+1}^{(k)} & \cdots & a_{nn}^{(k)} \end{bmatrix} \begin{bmatrix} x_1 \\ x_2 \\ \vdots \\ x_k \\ x_{k+1} \\ \vdots \\ x_n \end{bmatrix} = \begin{bmatrix} b_1^{(k)} \\ b_2^{(k)} \\ \vdots \\ b_k^{(k)} \\ b_{k+1}^{(k)} \\ \vdots \\ b_n^{(k)} \end{bmatrix}. \qquad (7.25)$$

The elements of the pivot row are updated according to

$$
\begin{cases}
a_{kk}^{(k)} = 1, \\
a_{kj}^{(k)} = a_{kj}^{(k-1)}/a_{kk}^{(k-1)}, \quad j = k+1, \ldots, n, \\
b_{k}^{(k)} = b_{k}^{(k-1)}/a_{kk}^{(k-1)},
\end{cases}
\tag{7.26}
$$

and the new elements of the nonpivot rows are given by

$$
\begin{cases}
a_{ik}^{(k)} &= 0, \\
a_{ij}^{(k)} &= a_{ij}^{(k-1)} - a_{ik}^{(k-1)}a_{kj}^{(k)}, \quad j = k+1, \ldots, n, \; i = 1, \ldots, n, \; i \neq k \\
b_{i}^{(k)} &= b_{i}^{(k-1)} - a_{ik}^{(k-1)}b_{k}^{(k)}.
\end{cases}
\tag{7.27}
$$

It is apparent that at the elimination step k, only the columns *on the right* of the pivot column k are modified, the pivot column itself becoming identical with the corresponding column of the unit matrix. As for the constant terms, they are altered at all elimination steps.

At the end of the elimination phase, the matrix equation acquires the desired structure,

$$
\mathbf{E}_n \cdot \mathbf{x} = \mathbf{b}^{(n)},
$$

where $\mathbf{E}_n$ is the unit matrix of order n. It is obvious that a backward substitution phase (similar to the one featured by the Gauss elimination) is no longer necessary, since the above form allows for the straightforward identification of the solution components:

$$
x_k = b_k^{(n)}, \quad k = 1, \ldots, n.
\tag{7.28}
$$

The on-the-fly calculation of the determinant of the coefficient matrix is accomplished exactly like for the Gaussian elimination and is based on the fact that $\det \mathbf{A}^{(n)} = \det \mathbf{E}_n = 1$. Since matrix $\mathbf{A}^{(n)}$ was obtained by successive divisions of the pivot rows by the pivot elements, it follows that

$$
\det \mathbf{A}^{(n)} = \frac{\det \mathbf{A}}{a_{11}a_{22}^{(1)} \cdots a_{nn}^{(n-1)}} = 1,
$$

and from here:

$$
\det \mathbf{A} = a_{11}a_{22}^{(1)} \cdots a_{nn}^{(n-1)}.
\tag{7.29}
$$

The recurrence relations of the Gauss–Jordan method can be directly generalized for solving matrix equations by the mere addition of a supplementary column index to the constant terms b_{kj} and the solution components x_{kj}. At step $k = 1, \ldots, n$, the relations for the pivot row k are

$$
\begin{cases}
a_{kj}^{(k)} &= a_{kj}^{(k-1)}/a_{kk}^{(k-1)}, \quad j = k+1, \ldots, n, \\
b_{kj}^{(k)} &= b_{kj}^{(k-1)}/a_{kk}^{(k-1)}, \quad j = 1, \ldots, m,
\end{cases}
\tag{7.30}
$$

while for the nonpivot rows, they read:

$$
\begin{cases}
a_{ij}^{(k)} &= a_{ij}^{(k-1)} - a_{ik}^{(k-1)}a_{kj}^{(k)}, \quad i = 1, \ldots, n, \; i \neq k, \; j = k+1, \ldots, n, \\
b_{ij}^{(k)} &= b_{ij}^{(k-1)} - a_{ik}^{(k-1)}b_{kj}^{(k)}, \quad i = 1, \ldots, n, \; i \neq k, \; j = 1, \ldots, m.
\end{cases}
\tag{7.31}
$$

Finally, the solution components can be simply identified with the last values of the transformed constant terms:

$$x_{kj} = b_{kj}^{(n)}, \quad j = 1, \ldots, m, \ k = 1, \ldots, n. \tag{7.32}$$

Judging from these relations, there are two notable differences between the Gauss and Gauss–Jordan methods. First, at the step k in the Gauss method (Equations 7.20 through 7.22), only the nonpivot rows that are located *below* the pivot row k are reduced ($i = k + 1, \ldots, n$) and the unknown x_k is eliminated only from these, while in the Gauss–Jordan method *all* nonpivot rows are processed ($i = 1, \ldots, k - 1, k + 1, \ldots n$) and the unknown x_k is eliminated from all these. The second major difference consists of the total absence of the backward substitution in the case of the Gauss–Jordan method. As a technical detail, assignments (7.32) do not need to be actually coded if the same array is used both for the constant terms $\mathbf{B} = [b_{kj}]$ and the solution components $\mathbf{X} = [x_{kj}]$.

One of the most beneficial features of the Gauss–Jordan elimination in comparison with other methods is its capability of *in-place inversion* of the coefficient matrix $\mathbf{A}$. This feature relies on the fact that, since the transformed coefficient matrix $\mathbf{A}^{(k)}$ contains relevant information only in columns $j = k + 1, \ldots, n$, one has, in principle, to effectively update only these. Indeed, while "sweeping" the array from left to right during elimination, the pivot column k implicitly leaves behind columns of the unit matrix, which do not need to be actually stored. On the other hand, inverting matrix $\mathbf{A}$ essentially amounts to solving the matrix equation $\mathbf{A} \cdot \mathbf{X} = \mathbf{E}$, with the unit matrix $\mathbf{E}$ formally acting as the matrix of constant terms. By carrying out the elementary transformations of the Gauss–Jordan elimination on both matrices $\mathbf{A}$ and $\mathbf{E}$ simultaneously, the number of relevant columns of the successive matrices $\mathbf{A}^{(k)}$ decreases by 1, while the number of relevant columns of the transformed unit matrix $\mathbf{E}^{(k)}$ increases by 1. Since the total number of relevant columns of matrices $\mathbf{A}^{(k)}$ and $\mathbf{E}^{(k)}$ remains constant, it follows that the implementation of the algorithm can be accomplished in principle by using a single array for both matrices. The columns of the array are gradually replaced, from left to right, along with the reduction of the relevant columns of $\mathbf{A}^{(k)}$, by the new relevant columns of $\mathbf{E}^{(k)}$. One of the tricks of the trade is that the columns of the initial unit matrix can be generated one by one in the same columns just before their transformation. Finally, the array which contained initially the coefficient matrix $\mathbf{A}$ will contain its inverse $\mathbf{E}^{(n)} \equiv \mathbf{A}^{-1}$. Thus, the Gauss–Jordan method can be used to simultaneously solve a matrix equation and to invert its coefficient matrix.

The flops count for the Gauss–Jordan elimination with *in-place matrix inversion* can be similarly determined with the standard Gaussian elimination:

- Transformation of the coefficient matrix $\mathbf{A}$ into the unit matrix:
  — Summations and subtractions (Equation 7.31):

$$S_A(n) = \sum_{k=1}^{n} \sum_{j=1}^{n} \left(\sum_{i=1}^{k-1} 1 + \sum_{i=k+1}^{n} 1 \right) = n^3 - n^2,$$

  — Multiplications and divisions (Equations 7.30 and 7.31):

$$M_A(n) = \sum_{k=1}^{n} \sum_{j=1}^{n} 1 + \sum_{k=1}^{n} \sum_{j=1}^{n} \left(\sum_{i=1}^{k-1} 1 + \sum_{i=k+1}^{n} 1 \right) = n^3.$$

- Transformation of a single vector of constant terms **b** ($m = 1$):
  — Summations and subtractions (Equation 7.31):

$$S_b(n) = \sum_{k=1}^{n} \left(\sum_{i=1}^{k-1} 1 + \sum_{i=k+1}^{n} 1 \right) = n^2 - n,$$

  — Multiplications and divisions (Equations 7.30 and 7.31):

$$M_b(n) = \sum_{k=1}^{n} 1 + \sum_{k=1}^{n} \left(\sum_{i=1}^{k-1} 1 + \sum_{i=k+1}^{n} 1 \right) = n^2.$$

- Solution of a single system of linear equations:

$$S_A(n) + M_A(n) + S_b(n) + M_b(n) = 2n^3 + n^2 - n \sim 2n^3. \tag{7.33}$$

- Inversion of the coefficient matrix **A**:

$$S_A(n) + M_A(n) = 2n^3 - n^2 \sim 2n^3. \tag{7.34}$$

By comparing the number of flops required by the Gaussian elimination with backward substitution (Equations 7.23 and 7.24) and, respectively, the Gauss–Jordan elimination (Equations 7.33 and 7.34), it results that, while in solving systems of linear equations, the Gaussian elimination is preferable, being about 3 times faster irrespective of the system size n ($N_{\text{Gauss}} \sim (2/3)\, n^3$, $N_{\text{Gauss–Jordan}} \sim 2n^3$), for inverting matrices, the Gauss–Jordan method performs roughly 4/3 times faster ($N_{\text{Gauss}} \sim (8/3)\, n^3$, $N_{\text{Gauss–Jordan}} \sim 2n^3$) and, in addition, requires just a single array (not two) for storing both the coefficient matrix and its inverse.

The `GaussJordan0` routine from Listing 7.3 highlights the technical differences between the Gauss and Gauss–Jordan methods and follows in as much as possible the structure of the function `Gauss`, without inverting the coefficient matrix. For reducing the roundoff errors and stabilizing the algorithm, the same technique of partial pivoting on columns is used. Even though, by the absence of the backward substitution, the Gauss–Jordan algorithm appears more compact, the number of implied elementary operations exceeds the one for the Gaussian elimination due to the more elaborate elimination phase.

The function `GaussJordan1`, presented in Listing 7.4, implements in addition to the routine `GaussJordan0` the in-place inversion discussed above. It replaces on exit the array of coefficients a by its inverse and the array of constant terms b by the solution of the matrix equation.

Unlike the Gaussian elimination, the Gauss–Jordan algorithm operates at every elimination step on all columns ($j = 1, \ldots, n$), since both the coefficient matrix and the unit matrix must undergo the sequence of elementary transformations. The initialization of the columns of the unit matrix (a[k][k] = 1e0 and a[i][k] = 0e0) gradually overwrites the columns of the transformed coefficient matrix that have become redundant.

The supplementary array `ipivot` stores the indexes of the successive pivot rows. In fact, taking into account that pivoting implies *reordering of the rows* of the initial unit matrix, to obtain the correct inverse of the system matrix, a final *reordering of the columns* of array a, in reversed order of the indexes stored in `ipivot`, needs to be performed.

Listing 7.3   Gauss–Jordan Elimination with Partial Pivoting (Python Coding)

```
#===========================================================================
def GaussJordan0 (a, b, n, m):
#---------------------------------------------------------------------------
#  Solves matrix equation a x = b by Gauss-Jordan elimination with partial
#  pivoting on columns.
#  a   - coefficient matrix (n x n); destroyed on exit
#  b   - matrix of constant terms (n x m); solution x on exit
#  det - determinant of coefficient matrix (output).
#---------------------------------------------------------------------------
   det = 1e0
   for k in range(1,n+1):                              # FORWARD ELIMINATION
      amax = 0e0                               # determine pivot row with
      for i in range(k,n+1):                  # max. element on column k
         if (amax < fabs(a[i][k])): amax = fabs(a[i][k]); imax = i
      if (amax == 0e0):
         print("GaussJordan0: singular matrix !"); return 0e0

      if (imax != k):                          # interchange rows imax and k
         det = -det                            # to put pivot on diagonal
         for j in range(k,n+1): (a[imax][j],a[k][j]) = (a[k][j],a[imax][j])
         for j in range(1,m+1): (b[imax][j],b[k][j]) = (b[k][j],b[imax][j])

      det *= a[k][k]                      # multiply determinant with pivot

      t = 1e0/a[k][k]                          # divide pivot row by pivot
      for j in range(k+1,n+1): a[k][j] *= t
      for j in range(  1,m+1): b[k][j] *= t

      for i in range(1,n+1):                           # reduce non-pivot rows
         if (i != k):
            t = a[i][k]
            for j in range(1,n+1): a[i][j] -= a[k][j]*t
            for j in range(1,m+1): b[i][j] -= b[k][j]*t

   return det
```

The pivoting method achieving the most significant reduction of the roundoff errors is *complete pivoting*, which, in contrast to the partial pivoting used so far, searches for the pivot element on *all rows and columns* on which pivoting has not been already performed. Listing 7.5 exemplifies the coding of the Gauss–Jordan elimination with complete pivoting.

To obtain the correct inverse, the rows and columns of array a have to be rearranged at the end of the procedure in reversed order of the one in which they have been interchanged during pivoting. The arrays ipivot and jpivot store the row and, respectively, column indexes of the pivot elements in the sequence of their usage and the array npivot flags the rows that have been already used for pivoting, preventing them from being reused. Owing to its stability, accuracy, and versatility, among the solvers for matrix equations based on elimination, GaussJordan is the routine of choice in applications in which both solutions of systems and matrix inversions are required.

Illustrating the application of the Gauss–Jordan elimination, Listing 7.6 is essentially similar to Listing 7.2 in which the standard Gaussian elimination is used. The supplementary capability of the Gauss–Jordan algorithm to produce the inverse of the coefficient matrix is tested by evaluating the product of the initial coefficient matrix, stored in the backup array c, with the inverse returned by the function

Listing 7.4 Gauss–Jordan Elimination with Partial Pivoting and Matrix Inversion (Python Coding)

```
#===========================================================================
def GaussJordan1(a, b, n, m):
#---------------------------------------------------------------------------
#  Solves matrix equation a x = b by Gauss-Jordan elimination with partial
#  pivoting on columns.
#  a   - coefficient matrix (n x n); a^(-1) on exit
#  b   - matrix of constant terms (n x m); solution x on exit
#  det - determinant of coefficient matrix (output).
#---------------------------------------------------------------------------
   ipivot = [0]*(n+1)                                 # stores pivot rows

   det = 1e0
   for k in range(1,n+1):                             # FORWARD ELIMINATION
      amax = 0e0                               # determine pivot row with
      for i in range(k,n+1):                   # max. element on column k
         if (amax < fabs(a[i][k])): amax = fabs(a[i][k]); imax = i
      if (amax == 0e0):
         print("GaussJordan1: singular matrix !"); return 0e0
      ipivot[k] = imax                            # store pivot row index

      if (imax != k):                           # interchange rows imax and k
         det = -det                             # to put pivot on diagonal
         for j in range(1,n+1): (a[imax][j],a[k][j]) = (a[k][j],a[imax][j])
         for j in range(1,m+1): (b[imax][j],b[k][j]) = (b[k][j],b[imax][j])

      det *= a[k][k]                         # multiply determinant with pivot

      t = 1e0/a[k][k]                            # divide pivot row by pivot
      a[k][k] = 1e0                       # diagonal element of unit matrix
      for j in range(1,n+1): a[k][j] *= t
      for j in range(1,m+1): b[k][j] *= t

      for i in range(1,n+1):                         # reduce non-pivot rows
         if (i != k):
            t = a[i][k]
            a[i][k] = 0e0                   # non-diagonal element of unit matrix
            for j in range(1,n+1): a[i][j] -= a[k][j]*t
            for j in range(1,m+1): b[i][j] -= b[k][j]*t

   for k in range(n,0,-1):                      # rearrange columns of inverse
      imax = ipivot[k]
      if (imax != k):
         for i in range(1,n+1): (a[i][imax],a[i][k]) = (a[i][k],a[i][imax])

   return det
```

GaussJordan in the array a. Owing to the unavoidable roundoff errors, the result of the check is a "numerical" unit matrix, with the diagonal elements equal to 1 and the nondiagonal elements of the order $10^{-16} - 10^{-17}$.

The auxiliary functions for operations with 2D arrays—MatCopy (for copying the first array into the second one), MatPrint (for printing an array formatted onto the screen), and MatProd (returning the product of the first two arrays in the third one)—are defined, respectively, in the libraries matutil.py and matutil.h (discussed in Chapter 2).

Listing 7.5 Gauss–Jordan Elimination with Complete Pivoting (Python Coding)

```
#===========================================================================
def GaussJordan(a, b, n, m):
#---------------------------------------------------------------------------
#  Solves matrix equation a x = b by Gauss-Jordan elimination with complete
#  pivoting.
#  a   - coefficient matrix (n x n); a^(-1) on exit
#  b   - matrix of constant terms (n x m); solution x on exit
#  det - determinant of coefficient matrix (output).
#---------------------------------------------------------------------------
   ipivot = [0]*(n+1); jpivot = [0]*(n+1)     # stores pivot rows and columns
   npivot = [0]*(n+1)                            # marks used pivot columns

   det = 1e0
   for k in range(1,n+1):                                # FORWARD ELIMINATION
      amax = 0e0                                          # determine pivot
      for i in range(1,n+1):                                # loop over rows
         if (npivot[i] != 1):
            for j in range(1,n+1):                        # loop over columns
               if (npivot[j] == 0):                # pivoting not yet done?
                  if (amax < fabs(a[i][j])):
                     amax = fabs(a[i][j]); imax = i; jmax = j
               if (npivot[j] > 1):
                  print("GaussJordan: singular matrix 1 !"); return 0e0
      if (amax == 0e0): print("GaussJordan: singular matrix 2 !"); return 0e0

      ipivot[k] = imax; jpivot[k] = jmax         # store pivot row and column
      npivot[jmax] = npivot[jmax] + 1              # mark used pivot column

      if (imax != jmax):                         # interchange rows imax and jmax
         det = -det                              # to put pivot on diagonal
         for j in range(1,n+1):
            (a[imax][j],a[jmax][j]) = (a[jmax][j],a[imax][j])
         for j in range(1,m+1):
            (b[imax][j],b[jmax][j]) = (b[jmax][j],b[imax][j])

      det *= a[jmax][jmax]                       # multiply determinant with pivot

      t = 1e0/a[jmax][jmax]                        # divide pivot row by pivot
      a[jmax][jmax] = 1e0                        # diagonal element of unit matrix
      for j in range(1,n+1): a[jmax][j] *= t
      for j in range(1,m+1): b[jmax][j] *= t

      for i in range(1,n+1):                               # reduce non-pivot rows
         if (i != jmax):
            t = a[i][jmax]
            a[i][jmax] = 0e0                        # non-diagonal element of unit matrix
            for j in range(1,n+1): a[i][j] -= a[jmax][j]*t
            for j in range(1,m+1): b[i][j] -= b[jmax][j]*t

   for k in range(n,0,-1):                       # rearrange columns of inverse
      imax = ipivot[k]; jmax = jpivot[k]
      if (imax != jmax):
         for i in range(1,n+1):
            (a[i][imax],a[i][jmax]) = (a[i][jmax],a[i][imax])

   return det
```

Listing 7.6 Solution of a Matrix Equation by Gauss–Jordan Elimination (Python Coding)

```
# Solves matrix equation by the Gauss-Jordan method
from linsys import *
from matutil import *

n = 4                                              # order of system
m = 1                                    # number of constant vectors
a = [[0]*(n+1) for i in range(n+1)]              # system matrix
b = [[0]*(m+1) for i in range(n+1)]    # constant terms and solution
c = [[0]*(n+1) for i in range(n+1)]         # copy of system matrix
d = [[0]*(n+1) for i in range(n+1)]              # product matrix

a[1][1] = 1; a[1][2] = 2; a[1][3] = 3; a[1][4] = 4; b[1][1] = 30
a[2][1] = 2; a[2][2] = 1; a[2][3] = 2; a[2][4] = 3; b[2][1] = 22
a[3][1] = 3; a[3][2] = 2; a[3][3] = 1; a[3][4] = 2; b[3][1] = 18
a[4][1] = 4; a[4][2] = 3; a[4][3] = 2; a[4][4] = 1; b[4][1] = 20
                     # Solution: 1.0, 2.0, 3.0, 4.0; det = -20.0
print("A:")
MatPrint(a,n,n)
print("B:")
MatPrint(b,n,m)

MatCopy(a,c,n,n)                              # save system matrix

det = GaussJordan(a,b,n,m)            # solve system, inverse in a

print("det A = ",det)
print("Solution:")
MatPrint(b,n,m)

print("Check A^(-1)A = I:")
MatProd(a,c,d,n,n,n)                  # multiply inverse with original
MatPrint(d,n,n)
```

7.4 LU Factorization

Despite its intuitive nature, a drawback of the Gaussian elimination is that the vector of constant terms **b** has to be available concomitantly with the coefficient matrix **A**, since both have to be transformed *simultaneously*. Therefore, the solution of another system with the same coefficient matrix but different right-hand side terms again implies the entire process of transforming the matrix into upper triangular form. In certain applications, it is desirable for the details of the transformations of the system matrix to be stored for subsequent use.

A very efficient method, having at its core the idea of preprocessing the coefficient matrix and storing the details of the transformations, is the *LU factorization (decomposition)*. Considering the system of linear equations

$$\mathbf{A} \cdot \mathbf{x} = \mathbf{b}, \tag{7.35}$$

the basic idea is to decompose the coefficient matrix as a product of two matrices,

$$\mathbf{A} = \mathbf{L} \cdot \mathbf{U}, \tag{7.36}$$

whereby matrix $\mathbf{L}$ is *lower triangular* and matrix $\mathbf{U}$ is *upper triangular*. Correspondingly, system (7.35) may be rewritten as $\mathbf{L} \cdot (\mathbf{U} \cdot \mathbf{x}) = \mathbf{b}$ and is equivalent with the following set of two systems:

$$\mathbf{L} \cdot \mathbf{y} = \mathbf{b}, \tag{7.37}$$

$$\mathbf{U} \cdot \mathbf{x} = \mathbf{y}, \tag{7.38}$$

which are to be solved in sequence. It can be noted right away that the solution $\mathbf{y}$ of the first system serves as a vector of constant terms for the second one, which actually provides the solution of the initial system (7.35).

Factorization (7.36) of the coefficient matrix plays a similar role to the forward elimination in the Gaussian methods. The enhancement consists of the fact that the LU factorization can be realized independently of the actual solution of a particular system and, as such, the vector of constant terms $\mathbf{b}$ does not need to be available at this stage. Once the LU decomposition is accomplished, solving the resulting linear triangular systems (7.37) and (7.38) can be carried out in single sweeps, by very efficient recursive procedures.

In principle, by formally identifying the elements of matrix $\mathbf{A} = [a_{ij}]_{nn}$ with those of the product matrix $\mathbf{L} \cdot \mathbf{U}$, one can univocally determine only n^2 of the $2 \times n(n+1)/2 = n^2 + n$ nonzero elements of the matrices $\mathbf{L}$ and $\mathbf{U}$. Therefore, for the complete definition of their elements, still, n supplementary conditions need to be specified. The choice of these conditions benefits from a certain freedom, reflected by the existence of several factorization methods, such as the Doolittle, Crout, and Cholesky methods. In the following, *Doolittle's LU factorization* is employed, in which one considers, without any loss of generality, the diagonal elements of the matrix $\mathbf{L}$ to be equal to 1. With these considerations in mind, the matrices $\mathbf{L}$ and $\mathbf{U}$ have the form

$$\mathbf{L} = \begin{bmatrix} 1 & & & & & \\ \alpha_{21} & 1 & & & 0 & \\ \vdots & \ddots & \ddots & & & \\ \alpha_{i1} & \cdots & \alpha_{i,i-1} & 1 & & \\ \vdots & & & \ddots & \ddots & \\ \alpha_{n1} & \cdots & \cdots & \cdots & \alpha_{n,n-1} & 1 \end{bmatrix}, \tag{7.39}$$

and, respectively,

$$\mathbf{U} = \begin{bmatrix} \beta_{11} & \beta_{12} & \cdots & \beta_{1j} & \cdots & \beta_{1n} \\ & \beta_{22} & & \vdots & & \vdots \\ & & \ddots & \vdots & & \vdots \\ & & & \beta_{jj} & & \vdots \\ & 0 & & & \ddots & \vdots \\ & & & & & \beta_{nn} \end{bmatrix}. \tag{7.40}$$

Technically, a given entry a_{ij} of matrix $\mathbf{A}$ is identified with the element resulting by multiplying row i of matrix $\mathbf{L}$ with column j of matrix $\mathbf{U}$. The number of possibly nonvanishing terms in the corresponding sum is given by the minimum between i and j:

$$a_{ij} = \begin{cases} \sum_{k=1}^{i-1} \alpha_{ik}\beta_{kj} + \beta_{ij}, & i \leq j, \\ \\ \sum_{k=1}^{j-1} \alpha_{ik}\beta_{kj} + \alpha_{ij}\beta_{jj}, & i > j, \end{cases} \tag{7.41}$$

and the form of the last term depends, as evidenced by the two expressions, on the order relation between i and j. The first form of a_{ij} corresponds to the upper triangle of $\mathbf{A}$ and enables expressing the elements β_{ij} of the upper triangular matrix $\mathbf{U}$, whereas the second form corresponds to the lower triangle of $\mathbf{A}$ and enables expressing the elements α_{ij} of the lower triangular matrix $\mathbf{L}$:

$$\beta_{ij} = a_{ij} - \sum_{k=1}^{i-1} \alpha_{ik}\beta_{kj}, \quad i = 1, 2, \ldots, j, \quad j = 1, \ldots, n, \tag{7.42}$$

$$\alpha_{ij} = \frac{1}{\beta_{jj}} \left[a_{ij} - \sum_{k=1}^{j-1} \alpha_{ik}\beta_{kj} \right], \quad i = j+1, \ldots, n. \tag{7.43}$$

It is readily noticeable that if one fills the matrices $\mathbf{U}$ and $\mathbf{L}$ *column wise*, calculating for each column $j = 1, \ldots, n$ first all elements β_{ij} with $i = 1, \ldots, j$ and then all elements α_{ij} with $i = j+1, \ldots, n$, Relations 7.42 and 7.43 are algorithmically self-consistent, that is, all elements intervening in the right-hand side terms are available as they become necessary.

Furthermore, it can be seen from Formulas 7.42 and 7.43 that a given element a_{ij} contributes to determining just a single element of just one of the factorization matrices: either $\beta_{ij} \in \mathbf{U}$ if $i \leq j$, or $\alpha_{ij} \in \mathbf{L}$ if $i > j$. This useful finding enables actual implementations of the LU factorization to use a single array, which initially contains the matrix $\mathbf{A}$ and, finally, the relevant elements of matrices $\mathbf{L}$ and $\mathbf{U}$:

$$\begin{bmatrix} \beta_{11} & \cdots & \beta_{1j} & \cdots & \cdots & \beta_{1n} \\ \alpha_{21} & \ddots & \vdots & & & \vdots \\ \vdots & \ddots & \beta_{jj} & & & \vdots \\ \vdots & & \alpha_{j+1,j} & \ddots & & \vdots \\ \vdots & & \vdots & \ddots & \ddots & \vdots \\ \alpha_{n1} & \cdots & \alpha_{nj} & \cdots & \alpha_{n,n-1} & \beta_{nn} \end{bmatrix}. \tag{7.44}$$

Pivoting is one of the subtle aspects of the LU factorization. However, it affects exclusively the elements α_{ij} in the lower triangle, the only ones implying divisions (Press et al. 2007). From Equation 7.43, it can be seen that on every column j, all elements α_{ij} with $i = j+1, \ldots, n$ are divided by the pivot β_{jj}, which is the last of the elements β_{ij} on that particular column. With a view to avoiding divisions by zero and, moreover, to reducing the roundoff errors, it is desirable to maximize the pivots.

One robust and efficient strategy of *partial pivoting on columns* is based on running through the LU factorization column wise ($j = 1, \ldots, n$) and defining for each particular column j the following quantities (suggested by Equation 7.43):

$$\tilde{\alpha}_{ij} = a_{ij} - \sum_{k=1}^{j-1} \alpha_{ik}\beta_{kj}, \quad i = j, \ldots, n. \tag{7.45}$$

In terms of these, the entries β_{jj} and α_{ij}, located on column j on and below the main diagonal, can be expressed in a unifying way:

$$\beta_{jj} = \tilde{\alpha}_{jj}, \tag{7.46}$$

$$\alpha_{ij} = \tilde{\alpha}_{ij}/\beta_{jj}, \quad i = j+1, \ldots, n. \tag{7.47}$$

The most advantageous pivot β_{jj} on column j corresponds, evidently, to the largest of the elements $\tilde{\alpha}_{ij}$ and, to ensure the stability of the algorithm, it should be placed on the diagonal.

Pivoting on column j actually starts with calculating in sequence the entries β_{ij} for $i = 1, \ldots, j-1$ (using Equation 7.42) and the entries $\tilde{\alpha}_{ij}$ for $i = j, \ldots, n$ (from Equation 7.45). It should be noted that the elements $\tilde{\alpha}_{ij}$ involve only already-calculated values of α_{ik} (located on the left, on the same row) and of β_{kj} (located above, on the same column).

In parallel with calculating the elements $\tilde{\alpha}_{ij}$, their maximum also needs to be determined. Assuming the maximum to be $\tilde{\alpha}_{lj}$, one has to interchange rows l and j to relocate it on the diagonal. According to Equation 7.46, the new entry $\tilde{\alpha}_{jj}$ takes on the role of the pivot β_{jj}, while the rest of the elements $\tilde{\alpha}_{ij}$ provide by divisions by β_{jj} the elements α_{ij} of matrix $\mathbf{L}$ (according to Equation 7.47). It is noteworthy that at this stage, the entries on the right side of column j are still the initial ones of matrix $\mathbf{A}$ and the working array has the structure:

$$\begin{bmatrix}
\beta_{11} & \cdots & \cdots & \beta_{1j} & a_{1,j+1} & \cdots & \cdots & \cdots & a_{1n} \\
\alpha_{21} & \ddots & & \vdots & & & & & \vdots \\
\vdots & \ddots & \ddots & \beta_{j-1,j} & & & & & \\
\alpha_{j1} & \cdots & \ddots & \boxed{\tilde{\alpha}_{jj}} & a_{j,j+1} & \cdots & \cdots & \cdots & a_{jn} \\
\vdots & & & \tilde{\alpha}_{j+1,j} & \ddots & & & & \vdots \\
\vdots & & & \vdots & \ddots & \ddots & & & \\
\alpha_{l1} & \cdots & \cdots & \boxed{\tilde{\alpha}_{lj}} & a_{l,j+1} & \ddots & \ddots & \cdots & a_{ln} \\
\vdots & & & \vdots & & \ddots & \ddots & & \vdots \\
\alpha_{n1} & \cdots & \cdots & \tilde{\alpha}_{nj} & a_{n,j+1} & \cdots & \cdots & a_{n,n-1} & a_{nn}
\end{bmatrix}. \tag{7.48}$$

We conclude this technical discussion by stressing that pivoting causes in general not the original matrix $\mathbf{A}$ being factorized, but rather a *row-wise permutation* thereof, and this essential aspect should be accounted for when using the LU decomposition to solve systems of linear equations or invert matrices.

The determinant of matrix $\mathbf{A}$ can be calculated as the product of the determinants of matrices $\mathbf{L}$ and $\mathbf{U}$, which, owing to their triangular shape, amounts to the product of their diagonal elements:

$$\det \mathbf{A} = \prod_{j=1}^{n} \beta_{jj}. \tag{7.49}$$

The LU factorization based on Doolittle's method, regarded as a distinct task, is coded as the function LUFactor in Listing 7.7. The array a should contain on entry the elements of the original matrix $\mathbf{A}$ and it returns on exit the corresponding LU decomposition (7.44). Since, due to the applied pivoting, the decomposition results by factoring a row-wise permutation of $\mathbf{A}$, the information is incomplete without the specification of the sequence of pivot rows. This information is essential when solving systems of linear equations, because the vector of constant terms $\mathbf{b}$ must undergo, albeit subsequently, the same permutations as the coefficient matrix $\mathbf{A}$. For storing the indexes of the pivot rows, the routine uses the array ipivot. On the other hand, any permutation of rows is accounted for by changing the sign of the determinant.

Listing 7.7 LU Factorization of a Matrix by Doolittle's Method (Python Coding)

```python
#===============================================================================
def LUFactor(a, ipivot, n):
#-------------------------------------------------------------------------------
#  Performs LU factorization of (n x n) matrix a (diag(L) = 1). On exit,
#  replaces upper triangle and diagonal with U, and lower triangle, with L.
#  Uses partial pivoting on columns.
#  a      - coefficient matrix (n x n); LU decomposition on exit
#  ipivot - array of pivot row indexes (output)
#  det    - determinant of coefficient matrix (output).
#-------------------------------------------------------------------------------
   det = 1e0
   for j in range(1,n+1):                            # loop over columns
      for i in range(1,j):                           # elements of matrix U
         sum = a[i][j]
         for k in range(1,i): sum -= a[i][k]*a[k][j]
         a[i][j] = sum

      amax = 0e0
      for i in range(j,n+1):                         # elements of matrix L
         sum = a[i][j]                               # undivided by pivot
         for k in range(1,j): sum -= a[i][k]*a[k][j]
         a[i][j] = sum
                                                      # determine pivot
         if (amax < fabs(a[i][j])): amax = fabs(a[i][j]); imax = i

      if (amax == 0e0): print("LUFactor: singular matrix !"); return 0e0

      ipivot[j] = imax                               # store pivot row index
                                         # interchange rows imax and j
      if (imax != j):                    # to put pivot on diagonal
         det = -det
         for k in range(1,n+1):
            t = a[imax][k]; a[imax][k] = a[j][k]; a[j][k] = t

      det *= a[j][j]                     # multiply determinant with pivot

      t = 1e0/a[j][j]                    # divide elements of L by pivot
      for i in range(j+1,n+1): a[i][j] *= t

   return det
```

Once the LU decomposition $\mathbf{A} = \mathbf{L} \cdot \mathbf{U}$ is completed, as discussed above, solving a particular system of linear equations $\mathbf{A} \cdot \mathbf{x} = \mathbf{b}$ assumes in the first stage solving the lower triangular system $\mathbf{L} \cdot \mathbf{y} = \mathbf{b}$. This system can be detailed as

$$
\begin{bmatrix}
1 & & & & & \\
\alpha_{21} & 1 & & & 0 & \\
\vdots & \ddots & \ddots & & & \\
\alpha_{i1} & \cdots & \alpha_{i,i-1} & 1 & & \\
\vdots & & & \ddots & \ddots & \\
\alpha_{n1} & \cdots & \cdots & \cdots & \alpha_{n,n-1} & 1
\end{bmatrix}
\begin{bmatrix}
y_1 \\ y_2 \\ \vdots \\ y_i \\ \vdots \\ y_n
\end{bmatrix}
=
\begin{bmatrix}
b_1 \\ b_2 \\ \vdots \\ b_i \\ \vdots \\ b_n
\end{bmatrix},
\tag{7.50}
$$

and its solution can be expressed recurrently, starting with the lowest rank component:

$$
\begin{cases}
y_1 &= b_1, \\
y_i &= b_i - \displaystyle\sum_{j=1}^{i-1} \alpha_{ij} y_j, \quad i = 2, 3, \ldots, n.
\end{cases}
\tag{7.51}
$$

In the second stage, the components of the intermediate solution (7.51) play the role of right-hand side terms for the upper triangular system $\mathbf{U} \cdot \mathbf{x} = \mathbf{y}$, which can be detailed as

$$
\begin{bmatrix}
\beta_{11} & \cdots & \cdots & \cdots & \beta_{1n} \\
 & \ddots & & & \vdots \\
 & & \beta_{ii} & \cdots & \beta_{in} \\
 & 0 & & \ddots & \vdots \\
 & & & & \beta_{nn}
\end{bmatrix}
\begin{bmatrix}
x_1 \\ \vdots \\ x_i \\ \vdots \\ x_n
\end{bmatrix}
=
\begin{bmatrix}
y_1 \\ \vdots \\ y_i \\ \vdots \\ y_n
\end{bmatrix}.
\tag{7.52}
$$

Solving this system yields the solution $\mathbf{x}$ of the initial system $\mathbf{A} \cdot \mathbf{x} = \mathbf{b}$, whose components can be expressed recursively backward, starting with the last one:

$$
\begin{cases}
x_n &= \dfrac{y_n}{\beta_{nn}}, \\
x_i &= \dfrac{1}{\beta_{ii}} \left[y_i - \displaystyle\sum_{j=i+1}^{n} \beta_{ij} x_j \right], \quad i = n-1, n-2, \ldots, 1.
\end{cases}
\tag{7.53}
$$

The routine LUSystem presented in Listing 7.8 implements the recursive processes described by Relations 7.51 and 7.53 and solves a system of linear equations by receiving in the array a, instead of the coefficient matrix $\mathbf{A}$, its LU decomposition, as returned by the function LUFactor. In the course of solving the first system, $\mathbf{L} \cdot \mathbf{y} = \mathbf{b}$, the constant terms stored in array b are rearranged according to the components of array ipivot (returned along with the LU decomposition by the routine LUFactor) to

Listing 7.8   Solution of a Linear System Using the LU Decomposition of its Matrix (Python Coding)

```
#===========================================================================
def LUSystem(a, ipivot, b, n):
#---------------------------------------------------------------------------
#  Solves linear system a x = b of order n by LU factorization.
#  a       - LU decomposition of coefficient matrix (returned by LUFactor)
#  ipivot  - array of pivot row indexes (input)
#  b       - vector of constant terms (input); solution x (on exit)
#---------------------------------------------------------------------------
    for i in range(1,n+1):                               # solves Ly = b
        sum = b[ipivot[i]]
        b[ipivot[i]] = b[i]
        for j in range(1,i): sum -= a[i][j]*b[j]
        b[i] = sum

    for i in range(n,0,-1):                              # solves Ux = y
        sum = b[i]
        for j in range(i+1,n+1): sum -= a[i][j]*b[j]
        b[i] = sum/a[i][i]
```

Listing 7.9   Typical Sequence for Solving a Linear System by LU Factorization (Python Coding)

```
a = [[0]*(n+1) for i in range(n+1)]              # coefficient matrix
b = [0]*(n+1)                               # vector of constant terms
ipivot = [0]*(n+1)                                  # vector of pivots
...
det = LUFactor(a,ipivot,n)
LUSystem(a,ipivot,b,n)
```

reflect the row permutations performed by the pivoting algorithm during the factorization. Specifically, the components of `ipivot` are just the indexes of the successive pivot rows and, accordingly, before being used, the constant terms `b[i]` and `b[ipivot[i]]` are interchanged. In solving the second system, $U \cdot x = y$, the array b plays at first the role of the vector of constant terms y and, thereafter, on exit, it represents the final solution x.

A typical code sequence for solving a system of linear equations by LU decomposition is provided by Listing 7.9. It is assumed that arrays a, b, and `ipivot` are allocated with offset 1 and n components. The call to routine `LUFactor` replaces the coefficient matrix stored in a by its LU decomposition, returning at the same time the list of pivot rows in `ipivot`. Using the LU decomposition stored in a and the constant terms stored in b, the call to routine `LUSystem` returns in b the solution of the system.

The pair of routines `LUFactor` and `LUSystem` is particularly useful for calculating the inverse of a matrix and this is implemented in the routine `MatInv` included in Listing 7.10. Essentially, inverting a matrix implies first performing its LU factorization with the help of `LUFactor` and then the solution

Listing 7.10   Matrix Inversion Using LU Factorization (Python Coding)

```
#===========================================================================
def MatInv(a, n):
#---------------------------------------------------------------------------
#  Calculates inverse of (n x n) matrix a by LU factorization.
#  a   - (n x n) matrix (input); a^(-1) (output)
#  det - determinant of coefficient matrix (output).
#  Calls: LUFactor, LUSystem.
#---------------------------------------------------------------------------
   ainv = [[0]*(n+1) for i in range(n+1)]     # temporary storage for inverse
   b = [0]*(n+1)
   ipivot = [0]*(n+1)                              # stores pivot rows

   det = LUFactor(a,ipivot,n)                     # LU factorization of a
   if (det == 0e0):                                 # singular matrix
      print("MatInv: singular matrix !"); return 0e0

   for j in range(1,n+1):                   # loop over columns of unit matrix
      for i in range(1,n+1): b[i] = 0e0                         # column j
      b[j] = 1e0
      LUSystem(a,ipivot,b,n)                               # solve system
      for i in range(1,n+1): ainv[i][j] = b[i]      # column j of inverse

   for j in range(1,n+1):                              # copy inverse in a
      for i in range(1,n+1): a[i][j] = ainv[i][j]

   return det
```

Listing 7.11   Application of Matrix Inversion Based on LU Factorization (Python Coding)

```
# Check of matrix inversion using LU decomposition
from random import *
from linsys import *
from matutil import *

n = 5                                                   # order of matrix
a = [[0]*(n+1) for i in range(n+1)]          # original matrix and inverse
b = [[0]*(n+1) for i in range(n+1)]                  # backup of original
c = [[0]*(n+1) for i in range(n+1)]                      # check matrix

for i in range(1,n+1):                              # generate random matrix
   for j in range(1,n+1): a[i][j] = random()

print("Original matrix:")
MatPrint(a,n,n)

MatCopy(a,b,n,n)                                  # backup original matrix

det = MatInv(a,n)                                       # invert original
if (det == 0e0): print("Singular matrix"); exit(1)

print("Inverse matrix:")
MatPrint(a,n,n)

print("Check A^(-1)A = I:")
MatProd(a,b,c,n,n,n)                        # multiply inverse with original
MatPrint(c,n,n)                                       # print check matrix
```

of a sequence of linear systems having as constant vectors the columns of the unit matrix by means of LUSystem. The successive solutions precisely represent the columns of the inverse matrix and, for their temporary storage, the routine allocates a supplementary array, ainv, which is finally copied in array a. In this way, the routine returns the inverse by the same array by which the original matrix is initially conveyed.

An example program designed for testing the matrix inversion performed by the routine MatInv is presented in Listing 7.11. The program generates a random matrix, creates a copy, inverts the copy, and multiplies the original array with its inverse. Owing to the roundoff errors, the result of the check is a numerical approximation of the unit matrix, having nondiagonal elements of the order $10^{-15} - 10^{-17}$. The invoked auxiliary functions for operations with 2D arrays are defined in the files matutil.py and matutil.h, respectively.

The flops count involved by applying the LU factorization can be established similarly to the Gaussian elimination:

- LU factorization of the coefficient matrix **A** (Equations 7.42 and 7.43):
  — Summations and subtractions:

$$S_A(n) = \sum_{j=1}^{n}\sum_{i=1}^{j}\left(1 + \sum_{k=1}^{i-2} 1\right) + \sum_{j=1}^{n}\sum_{i=j+1}^{n}\left(1 + \sum_{k=1}^{j-2} 1\right) = \frac{1}{3}n^3 - \frac{1}{2}n^2 + \frac{1}{6}n,$$

— Multiplications and divisions:

$$M_A(n) = \sum_{j=1}^{n}\sum_{i=1}^{j}\sum_{k=1}^{i-1} 1 + \sum_{j=1}^{n}\sum_{i=j+1}^{n}\left(1 + \sum_{k=1}^{j-1} 1\right) = \frac{1}{3}n^3 - \frac{1}{3}n.$$

- Transformation of the vector of constant terms **b** (Equations 7.51 and 7.53):
  — Summations and subtractions:

$$S_b(n) = \sum_{i=2}^{n}\left(1 + \sum_{j=1}^{i-2}\right) + \sum_{i=1}^{n-1}\left(1 + \sum_{j=i+1}^{n-1}\right) = n^2 - n,$$

  — Multiplications and divisions:

$$M_b(n) = \sum_{i=2}^{n}\sum_{j=1}^{i-1} 1 + 1 + \sum_{i=1}^{n-1}\left(1 + \sum_{j=i+1}^{n}\right) = n^2.$$

- Solution of a single system of linear equations:

$$S_A(n) + M_A(n) + S_b(n) + M_b(n) = \frac{2}{3}n^3 + \frac{3}{2}n^2 - \frac{7}{6}n \sim \frac{2}{3}n^3. \tag{7.54}$$

- Inversion of the coefficient matrix **A**:

$$S_A(n) + M_A(n) + n\left(S_b(n) + M_b(n)\right) = \frac{8}{3}n^3 - \frac{3}{2}n^2 - \frac{1}{6}n \sim \frac{8}{3}n^3. \tag{7.55}$$

It can be noticed right away that the costs for all operations based on the LU decomposition perfectly coincide with the ones for the standard Gaussian elimination, proving that the two methods are *perfectly equivalent*. However, due to the possibility to store the details of the forward elimination, as contained in the matrices **L** and **U**, the LU factorization may be preferable in situations in which repeated solutions of a system are required for several vectors of constant terms that are not available at the same time.

7.5 Inversion of Triangular Matrices

Sometimes, even though rarely directly from the primary formulation of practical problems, the necessity arises for inverting *triangular* matrices. Two such examples are the solution of generalized eigenvalue problems for symmetric positive-definite matrices and the inversion thereof. In both cases, the standard procedure implies the Cholesky factorization of the matrix (see Section 7.6), followed by the inversion of the resulting lower triangular matrix.

It is easily provable that the inverse of a nonsingular (invertible) triangular matrix is equally a triangular matrix. From the detailed equality of the product of the lower triangular matrices $\mathbf{A} = [a_{ij}]_{nn}$ and

$\mathbf{A}^{-1} = [a'_{ij}]_{nn}$ with the unit matrix,

$$
\begin{bmatrix}
a_{11} & & & & & \\
\vdots & \ddots & & & 0 & \\
& & \ddots & & & \\
\cdots & a_{ij} & \cdots & a_{ii} & & \\
& & & & \ddots & \\
a_{n1} & \cdots & \cdots & \cdots & \cdots & a_{nn}
\end{bmatrix}
\cdot
\begin{bmatrix}
a'_{11} & & & & & \\
\vdots & \ddots & & & 0 & \\
\vdots & & a'_{jj} & & & \\
\vdots & & \vdots & \ddots & & \\
\vdots & & a'_{ij} & & \ddots & \\
a'_{n1} & & \vdots & & \cdots & a'_{nn}
\end{bmatrix}
= \mathbf{E}_n, \qquad (7.56)
$$

the following relations can be established by identifying the corresponding elements:

$$
\sum_{k=j}^{i} a_{ik} a'_{kj} = \delta_{ij}, \quad i = j, \dots, n, \quad j = 1, 2, \dots, n. \qquad (7.57)
$$

The diagonal equation (for $i = j$) has the simplest form, since the sum reduces to a single term, $a_{jj} a'_{jj} = 1$, and it allows for the diagonal element of the inverse to be expressed:

$$
a'_{jj} = \frac{1}{a_{jj}}.
$$

Descending on the same column j, the equation corresponding to $i = j+1$ reads $a_{j+1,j} a'_{jj} + a_{j+1,j+1} a'_{j+1,j} = 0$, and enables a new element of the inverse situated on the same column to be expressed in terms of the already-calculated a'_{jj}:

$$
a'_{j+1,j} = -\frac{1}{a_{j+1,j+1}} a_{j+1,j} a'_{jj}.
$$

Generalizing the procedure and expressing from the last terms of the sums in Equation 7.57 the elements a'_{ij}, we obtain the recurrence relations:

$$
\begin{cases}
a'_{jj} = \dfrac{1}{a_{jj}}, & j = 1, 2, \dots, n, \\[2ex]
a'_{ij} = -\dfrac{1}{a_{ii}} \displaystyle\sum_{k=j}^{i-1} a_{ik} a'_{kj}, & i = j+1, \dots, n.
\end{cases} \qquad (7.58)
$$

Scanning these relations column wise and evaluating on each column j first the diagonal element a'_{jj} and then the elements a'_{ij} located on the rows $i = j+1, \dots, n$ underneath, as indicated by Equations 7.58, the entries a'_{kj} on the right-hand sides are already calculated as they become necessary and Equations 7.58 prove to be algorithmically self-consistent.

The function `MatTriInv` from Listing 7.12 exactly translates Relations 7.58 and it is conceived so as to return the inverse through the same lower triangle of array a by which the initial matrix is introduced. The algorithm does not require pivoting, being perfectly stable for any invertible lower triangular matrix.

Listing 7.12   Inversion of a Triangular Matrix (Python Coding)

```
#===============================================================================
def MatTriInv(a, n):
#-------------------------------------------------------------------------------
#  Calculates inverse of lower triangular matrix L of order n contained in
#  (n x n) array a and replaces on exit lower triangle of a by L^(-1).
#  Error flag = 0 - normal execution
#               1 - singular matrix.
#-------------------------------------------------------------------------------
   for j in range(1,n+1):                                  # loop over columns
      if (a[j][j] <= 0e0): print("MatTriInv: singular matrix !"); return 1
      a[j][j] = 1e0/a[j][j]                                # diagonal element

      for i in range(j+1,n+1):                       # non-diagonal elements
         sum = 0e0
         for k in range(j,i): sum -= a[i][k]*a[k][j]
         a[i][j] = sum/a[i][i]

   return 0
```

The flops count for inverting a triangular matrix as a function of the size n amounts to

$$\underbrace{\sum_{j=1}^{n}\sum_{i=j+1}^{n}\sum_{k=j}^{i-2}1 + \sum_{j=1}^{n}1}_{\text{Summations}} + \underbrace{\sum_{j=1}^{n}\sum_{i=j+1}^{n}\left(1 + \sum_{k=j}^{i-1}1\right)}_{\text{Multiplications and divisions}} = \frac{1}{3}n^3 + \frac{2}{3}n, \qquad (7.59)$$

and it turns out to be approximately 8 times lower than for the standard Gaussian elimination or the LU factorization and 6 times lower than for the Gauss–Jordan elimination. In conclusion, in addition to the optimal usage of the storage, practically reduced to a single triangular array, and the absence of pivoting, the efficiency of the presented algorithm for inverting triangular matrices is by almost one order superior to other general methods.

7.6 Cholesky Factorization

Owing to their special features, *symmetric positive-definite matrices* arise in many areas of science and engineering. For instance, quantum mechanics and theories based on its formalism offer extensive examples of such matrices. A symmetric matrix **A** is said to be *positive definite* if the condition

$$\mathbf{x}^T \cdot \mathbf{A} \cdot \mathbf{x} > 0$$

holds true for any column vector $\mathbf{x} \neq 0$ from the space in which the matrix **A** is defined, $\mathbf{x}^T$ being the corresponding transposed row vector. Equivalently, a positive-definite matrix has only positive eigenvalues.

It can be shown that a symmetric positive-definite matrix can always be factorized under the particular form

$$\mathbf{A} = \mathbf{L} \cdot \mathbf{L}^T, \qquad (7.60)$$

where **L** is a *lower triangular* matrix and $\mathbf{L}^T$ is its transpose (upper triangular). This decomposition, known as *Cholesky factorization*, is an important particular case of the LU factorization and leads to a significant reduction of the computational cost. As a matter of fact, the Cholesky factorization can be shown to perform in the case of linear systems with positive-definite matrices roughly 2 times faster than the general methods based on Gaussian elimination or LU factorization.

From the explicit form of decomposition (7.60),

$$
\mathbf{A} =
\begin{bmatrix}
l_{11} & & & & \\
\vdots & \ddots & & 0 & \\
l_{i1} & \cdots & l_{ii} & & \\
\vdots & & & \ddots & \\
l_{n1} & \cdots & \cdots & \cdots & l_{nn}
\end{bmatrix}
\cdot
\begin{bmatrix}
l_{11} & \cdots & l_{j1} & \cdots & l_{n1} \\
& \ddots & \vdots & & \vdots \\
& & l_{jj} & & \vdots \\
0 & & & \ddots & \vdots \\
& & & & l_{nn}
\end{bmatrix},
\qquad (7.61)
$$

we can establish the following relations between the elements of matrices $\mathbf{A}$ and $\mathbf{L}$:

$$
a_{ij} = a_{ji} = \sum_{k=1}^{j} l_{ik} l_{jk}, \quad i \geq j. \qquad (7.62)
$$

Expressing from the last term of the sum the entries on the column j of matrix $\mathbf{L}$ and employing for reasons of storage efficiency just the *lower triangle* of matrix $\mathbf{A}$, we get:

$$
\begin{cases}
l_{jj} = \left[a_{jj} - \displaystyle\sum_{k=1}^{j-1} l_{jk}^2 \right]^{1/2}, & j = 1, 2, \ldots, n, \\[4mm]
l_{ij} = \dfrac{1}{l_{jj}} \left[a_{ij} - \displaystyle\sum_{k=1}^{j-1} l_{ik} l_{jk} \right], & i = j+1, \ldots, n.
\end{cases} \qquad (7.63)
$$

Analyzing the algorithmic consistency of these relations, we find that if matrix $\mathbf{L}$ is filled *column wise*, calculating on each column j first the diagonal element l_{jj} and then the elements located underneath, l_{ij} with $i = j + 1, \ldots, n$, all the implied quantities are available as they become necessary. It can be also noticed that a given element a_{ij} is used for determining just a single l_{ij} and, consequently, the implementation of the Cholesky decomposition can employ a single array, whose lower triangle initially contains the defining entries of matrix $\mathbf{A}$ and finally, the relevant elements of matrix $\mathbf{L}$.

The determinant of matrix $\mathbf{A}$ is given by the product of the determinants of matrices $\mathbf{L}$ and $\mathbf{L}^T$, namely, given their triangular structure, by the product of their diagonal elements:

$$
\det \mathbf{A} = \prod_{j=1}^{n} l_{jj}^2. \qquad (7.64)
$$

The implementation of the Cholesky factorization based on Formulas 7.63 is presented in Listing 7.13. The lower triangle of the array a (including the diagonal) is expected to contain on entry the relevant elements of matrix $\mathbf{A}$, and, on exit, it returns the matrix $\mathbf{L}$. The upper triangle is practically unused and thus remains unchanged.

A crucial aspect, which greatly simplifies the coding and distinguishes the Cholesky factorization from any other method based on elimination or factorization, is that *no pivoting* is necessary, since the positive definiteness of the matrices to which the Cholesky decomposition is applicable ensures the *unconditional stability* of the algorithm. Having in view that the only situation in which the factorization fails

Listing 7.13   Cholesky Factorization of Symmetric Positive-Definite Matrix (Python Coding)

```
#===============================================================================
def Cholesky(a, n):
#-------------------------------------------------------------------------------
#   Performs Cholesky factorization L * LT of real symmetric positive-definite
#   (n x n) matrix a. On exit, replaces lower triangle of a with L.
#   a   - coefficient matrix (n x n); L in lower triangle and diagonal on exit
#   det - determinant of coefficient matrix (output).
#   Error flag = 0 - normal execution
#                1 - a is not positive-definite.
#-------------------------------------------------------------------------------
    det = 1e0
    for j in range(1,n+1):                               # loop over columns
        sum = a[j][j]                                    # diagonal element
        for k in range(1,j): sum -= a[j][k]*a[j][k]
        if (sum <= 0e0):
            print("Cholesky: matrix b is not positive-definite !"); return (0,1)
        a[j][j] = sqrt(sum)

        for i in range(j+1,n+1):                         # non-diagonal elements
            sum = a[i][j]
            for k in range(1,j): sum -= a[i][k]*a[j][k]
            a[i][j] = sum / a[j][j]

        det *= a[j][j]*a[j][j]

    return (det,0)
```

is the one in which the matrix is not positive definite; the routine `Cholesky` also provides a convenient way for checking the property of positive definiteness, returning in case of failure the error index 1.

Once the Cholesky decomposition of a matrix **A** is accomplished, the composing triangular matrices can be used to solve an associated linear system

$$\mathbf{A} \cdot \mathbf{x} = \mathbf{b} \tag{7.65}$$

by backward substitution. By virtue of the factorization, this system is equivalent with the following two systems:

$$\mathbf{L} \cdot \mathbf{y} = \mathbf{b}, \tag{7.66}$$

$$\mathbf{L}^T \cdot \mathbf{x} = \mathbf{y}. \tag{7.67}$$

Explicitly, the first triangular system (7.66) has the structure

$$\begin{bmatrix} l_{11} & & & & \\ \vdots & \ddots & & & \\ l_{i1} & \cdots & l_{ii} & & \\ \vdots & & & \ddots & \\ l_{n1} & \cdots & \cdots & \cdots & l_{nn} \end{bmatrix} \begin{bmatrix} y_1 \\ \vdots \\ y_i \\ \vdots \\ y_n \end{bmatrix} = \begin{bmatrix} b_1 \\ \vdots \\ b_i \\ \vdots \\ b_n \end{bmatrix}, \tag{7.68}$$

and its solution can be determined recursively starting with its first component, which is actually explicit:

$$
\begin{cases}
y_1 = \dfrac{b_1}{l_{11}}, \\[2ex]
y_i = \dfrac{1}{l_{ii}}\left[b_i - \displaystyle\sum_{j=1}^{i-1} l_{ij} y_j \right], \quad i = 2, 3, \ldots, n.
\end{cases}
\tag{7.69}
$$

The intermediate solution $\mathbf{y}$ provides the vector of constant terms to the second triangular system (7.67):

$$
\begin{bmatrix}
l_{11} & \cdots & \cdots & \cdots & l_{n1} \\
 & \ddots & & & \vdots \\
 & & l_{ii} & \cdots & l_{ni} \\
 & 0 & & \ddots & \vdots \\
 & & & & l_{nn}
\end{bmatrix}
\begin{bmatrix}
x_1 \\ \vdots \\ x_i \\ \vdots \\ x_n
\end{bmatrix}
=
\begin{bmatrix}
y_1 \\ \vdots \\ y_i \\ \vdots \\ y_n
\end{bmatrix}.
\tag{7.70}
$$

The vector $\mathbf{x}$ is the sought solution of the original system $\mathbf{A} \cdot \mathbf{x} = \mathbf{b}$ and its components can be expressed recursively by backward substitution in decreasing order of the indexes, starting with the last component, which is explicit:

$$
\begin{cases}
x_n = \dfrac{y_n}{l_{nn}}, \\[2ex]
x_i = \dfrac{1}{l_{ii}}\left[y_i - \displaystyle\sum_{j=i+1}^{n} l_{ji} x_j \right], \quad i = n-1, n-2, \ldots, 1.
\end{cases}
\tag{7.71}
$$

For reasons of algorithmic efficiency, the elements of $\mathbf{L}^T$ are expressed by the equivalent entries of $\mathbf{L}$, with a view to working consistently only with the lower triangles of all implied matrices.

The routine for solving systems of linear equations with symmetric positive-definite matrices based on the Cholesky factorization, according to the recursive relations 7.69 and 7.71, is presented in Listing 7.14. The array a is expected to contain on entry into routine `CholeskySys` the matrix $\mathbf{L}$ resulting from a previous factorization of the coefficient matrix $\mathbf{A}$, as returned by function `Cholesky`. The constant terms are conveyed to the routine by way of vector b, which returns on exit the corresponding solution.

A typical program sequence for solving a system of linear equations by Cholesky factorization is given in Listing 7.15.

Inverting a symmetric positive-definite matrix using the Cholesky decomposition is straightforward, being based on the fact that the inverse of an invertible lower triangular matrix is lower triangular and the transpose of an invertible matrix is the transpose of the inverse of the original matrix. Indeed, the inverse of matrix $\mathbf{A}$ can be expressed as

$$
\mathbf{A}^{-1} = \left(\mathbf{L} \cdot \mathbf{L}^T\right)^{-1} = \left(\mathbf{L}^T\right)^{-1} \cdot \mathbf{L}^{-1} = \left(\mathbf{L}^{-1}\right)^T \cdot \mathbf{L}^{-1},
\tag{7.72}
$$

is obviously symmetric positive definite, like the original matrix $\mathbf{A}$ itself, and it can be calculated very efficiently by a single inversion of the factor matrix $\mathbf{L}$. Denoting the elements of the inverses by primes,

Listing 7.14   Solution of Positive-Definite System by Cholesky Factorization (Python Coding)

```
#===============================================================================
def CholeskySys(a, b, n):
#-------------------------------------------------------------------------------
#  Solves linear system a x = b with real symmetric positive-definite (n x n)
#  matrix by Cholesky factorization L * LT.
#  a - matrix L in lower triangle and on diagonal (returned by Cholesky)
#  b - vector of constant terms (input); solution x (on exit)
#-------------------------------------------------------------------------------
   for i in range(1,n+1):                                   # solves Ly = b
      sum = b[i]
      for j in range(1,i): sum -= a[i][j]*b[j]
      b[i] = sum/a[i][i]

   for i in range(n,0,-1):                                  # solves LT x = y
      sum = b[i]
      for j in range(i+1,n+1): sum -= a[j][i]*b[j]
      b[i] = sum/a[i][i]
```

Listing 7.15   Typical Sequence for Solving a Linear System by Cholesky Factorization (Python Coding)

```
a = [[0]*(n+1) for i in range(n+1)]                  # coefficient matrix
b = [0]*(n+1)                                         # vector of constant terms
...
(det,ierr) = Cholesky(a,n)
CholeskySys(a,b,n)
```

the detailed identity

$$
\mathbf{A}^{-1} =
\begin{bmatrix}
l'_{11} & \cdots & \cdots & \cdots & l'_{n1} \\
 & \ddots & & & \vdots \\
 & & l'_{ii} & \cdots & l'_{ni} \\
 & 0 & & \ddots & \vdots \\
 & & & & l'_{nn}
\end{bmatrix}
\begin{bmatrix}
l'_{11} & & & & \\
\vdots & \ddots & & 0 & \\
\vdots & & l'_{jj} & & \\
\vdots & & \vdots & \ddots & \\
l'_{n1} & \cdots & l_{nj}; & \cdots & l'_{nn}
\end{bmatrix}
\tag{7.73}
$$

enables us to express the elements of $\mathbf{A}^{-1}$:

$$
a'_{ij} = a'_{ji} = \sum_{k=i}^{n} l'_{ik} l'_{kj} = \sum_{k=i}^{n} l'_{ki} l'_{kj},
\tag{7.74}
$$

$$
i = j+1, \ldots, n, \ j = 1, 2, \ldots, n.
$$

To allow for an optimized implementation, the entries l'_{ik} of the transposed matrix $\left(\mathbf{L}^{-1}\right)^T$ are expressed by the corresponding entries l'_{ki} of the lower triangular matrix $\mathbf{L}^{-1}$.

The implementation presented in Listing 7.16 emphasizes that the inversion based on the Cholesky decomposition may be accomplished using just a single array a, containing initially the lower triangle of matrix **A** and finally its inverse. The upper triangle of a is used by the routine as work space and does not

Listing 7.16   Inversion of Positive-Definite Matrix Using Cholesky Factorization (Python Coding)

```
#========================================================================
def MatSymInv(a, n):
#------------------------------------------------------------------------
#  Calculates inverse of symmetric positive-definite (n x n) matrix by
#  Cholesky factorization a = L * LT as a^(-1) = (L^(-1))T * L^(-1).
#  L^(-1) is calculated using MatTriInv.
#  a    - (n x n) matrix (input); a^(-1) (output)
#  det - determinant of coefficient matrix (output).
#  Calls: Cholesky, MatTriInv.
#------------------------------------------------------------------------
   (det,ierr) = Cholesky(a,n)                         # L * LT factorization of a
   if (ierr == 1):
      print("MatSymInv: matrix not positive-definite !"); return det
   if (det == 0e0): print("MatSymInv: singular matrix !"); return det

   MatTriInv(a,n)                                      # L^(-1) in lower triangle

   for i in range(1,n+1):                                # (L^(-1))T * L^(-1)
      for j in range(1,i+1):
         sum = 0e0
         for k in range(i,n+1): sum += a[k][i] * a[k][j]
         a[j][i] = sum                                 # store in upper triangle

   for i in range(1,n+1):                              # complete lower triangle
      for j in range(1,i+1): a[i][j] = a[j][i]

   return det
```

need to be initialized. The lower triangle successively accommodates the matrix $\mathbf{L}$ after the call to function Cholesky, and the matrix $\mathbf{L}^{-1}$ after the call to function MatTriInv, which is specifically designed for inverting triangular matrices, as described in Section 7.5. The final multiplication $\left(\mathbf{L}^{-1}\right)^T \cdot \mathbf{L}^{-1}$ produces in the upper triangle of a the relevant elements of $\mathbf{A}^{-1}$, which are then copied into the lower triangle, before the routine terminates the execution.

The flop counts involved in the Cholesky decomposition are as follows:

- $\mathbf{L} \cdot \mathbf{L}^T$ factorization of the coefficient matrix $\mathbf{A}$ (Equation 7.63):
  — Summations and subtractions:

$$S_A(n) = \sum_{j=1}^{n}\left(1 + \sum_{k=1}^{j-2} 1\right) + \sum_{j=1}^{n}\sum_{i=j+1}^{n}\left(1 + \sum_{k=1}^{j-2} 1\right) = \frac{1}{6}n^3 - \frac{1}{6}n,$$

  — Multiplications and divisions:

$$M_A(n) = \sum_{j=1}^{n}\sum_{k=1}^{j-1} 1 + \sum_{j=1}^{n}\sum_{i=j+1}^{n}\left(1 + \sum_{k=1}^{j-1} 1\right) = \frac{1}{6}n^3 + \frac{1}{2}n^2 - \frac{2}{3}n,$$

  — Square-root operations: n.

- Multiplication $\mathbf{A}^{-1} = \left(\mathbf{L}^{-1}\right)^T \cdot \mathbf{L}^{-1}$ (Equation 7.74):
  — Summations and subtractions:

$$S_I(n) = \sum_{j=1}^{n} \sum_{i=j+1}^{n} \sum_{k=i}^{n-1} 1 = \frac{1}{6}n^3 - \frac{1}{2}n^2 + \frac{1}{3}n,$$

  — Multiplications and divisions:

$$M_I(n) = \sum_{j=1}^{n} \sum_{i=j+1}^{n} \sum_{k=i}^{n} 1 = \frac{1}{6}n^3 - \frac{1}{6}n.$$

- Transformation of the vector of constant terms $\mathbf{b}$ (Equations 7.69 and 7.71):
  — Summations and subtractions:

$$S_b(n) = \sum_{i=2}^{n} \left(1 + \sum_{j=1}^{i-2}\right) + \sum_{i=1}^{n-1} \left(1 + \sum_{j=i+1}^{n-1}\right) = n^2 - n,$$

  — Multiplications and divisions:

$$M_b(n) = 1 + \sum_{i=2}^{n} \left(1 + \sum_{j=1}^{i-1}\right) + 1 + \sum_{i=1}^{n-1} \left(1 + \sum_{j=i+1}^{n}\right) = n^2 + n.$$

- Solution of a single system of linear equations (plus n square-root operations):

$$S_A(n) + M_A(n) + S_b(n) + M_b(n) = \frac{1}{3}n^3 + \frac{5}{2}n^2 - \frac{5}{6}n \sim \frac{1}{3}n^3, \tag{7.75}$$

- Inversion of a positive-definite matrix $\mathbf{A}$ (plus n square-root operations):

$$S_A(n) + M_A(n) + \frac{1}{3}n^3 + \frac{2}{3}n + S_I(n) + M_I(n) = n^3. \tag{7.76}$$

Apart from n supplementary square-root operations, the Cholesky decomposition is roughly 2 times faster in inverting positive-definite matrices ($\sim n^3$) than the Gauss–Jordan elimination ($\sim 2n^3$).

7.7 Tridiagonal Systems of Linear Equations

In many numerical applications in science and engineering, the necessity arises for solving systems of linear equations with matrices having special structures. Quite often, these are *sparse matrices* and their elements are primarily equal to zero. Sometimes, again, the nonzero elements are structured in bands comprising the main diagonal and, possibly, additional diagonals on either side, and these are called *band matrices*.

There are cases in which the coefficient matrix is *tridiagonal*, the relevant elements being confined to the main diagonal and the two neighboring diagonals. Systems with tridiagonal matrices result, for example, when boundary value problems for differential equations are discretized by finite difference schemes, or when interpolating tables of values with piecewise smooth functions, as, for example, with spline functions (treated in Chapter 9). Even though one can apply for the solution of such systems any of the general methods (Gaussian elimination, LU factorization, etc.), taking into account the particularities of the coefficient matrix leads to algorithms incomparable more efficient in terms of elementary operations and storage necessities.

Let us consider the tridiagonal system of linear equations

$$\mathbf{A} \cdot \mathbf{x} = \mathbf{d}, \tag{7.77}$$

which can be detailed as

$$\begin{bmatrix} b_1 & c_1 & & & & & & \\ a_2 & b_2 & c_2 & & & & 0 & \\ & \ddots & \ddots & \ddots & & & & \\ & & a_{i-1} & b_{i-1} & c_{i-1} & & & \\ & & & a_i & b_i & c_i & & \\ & & & & \ddots & \ddots & \ddots & \\ & 0 & & & & a_{n-1} & b_{n-1} & c_{n-1} \\ & & & & & & a_n & b_n \end{bmatrix} \begin{bmatrix} x_1 \\ x_2 \\ \vdots \\ x_{i-1} \\ x_i \\ \vdots \\ x_{n-1} \\ x_n \end{bmatrix} = \begin{bmatrix} d_1 \\ d_2 \\ \vdots \\ d_{i-1} \\ d_i \\ \vdots \\ d_{n-1} \\ d_n \end{bmatrix}. \tag{7.78}$$

Relying on the philosophy of the LU factorization described in Section 7.4, the most efficient way of solving this system is by decomposing the matrix $\mathbf{A}$ as a product of a *lower bidiagonal* matrix $\mathbf{L}$ and an *upper bidiagonal* matrix $\mathbf{U}$:

$$\mathbf{A} = \mathbf{L} \cdot \mathbf{U}. \tag{7.79}$$

The matrices $\mathbf{L}$ and $\mathbf{U}$ can involve as many relevant (in principle, nonzero) entries as present in matrix $\mathbf{A}$, namely $3n - 2$ (the two codiagonals of $\mathbf{A}$ have only $n - 1$ entries). In other words, from the four diagonals of $\mathbf{L}$ and $\mathbf{U}$, only three can be determined univocally from the elements of $\mathbf{A}$, and thus, one has the freedom of arbitrarily specifying the elements of the fourth diagonal. As mentioned also in the context of the general LU factorization, one can impose a unitary diagonal to either $\mathbf{L}$ or $\mathbf{U}$, the two choices being algebraically perfectly equivalent.

In what follows, we choose the matrix $\mathbf{L}$ to have diagonal elements equal to 1, so that the two factorization matrices have the structures:

$$\mathbf{L} = \begin{bmatrix} 1 & & & & & & & \\ \alpha_2 & 1 & & & & & 0 & \\ & \ddots & \ddots & & & & & \\ & & \alpha_{i-1} & 1 & & & & \\ & & & \alpha_i & 1 & & & \\ & & & & \ddots & \ddots & & \\ & 0 & & & & \alpha_{n-1} & 1 & \\ & & & & & & \alpha_n & 1 \end{bmatrix}, \tag{7.80}$$

$$\mathbf{U} = \begin{bmatrix} \beta_1 & \gamma_1 & & & & & & \\ & \beta_2 & \gamma_2 & & & & 0 & \\ & & \ddots & \ddots & & & & \\ & & & \beta_{i-1} & \gamma_{i-1} & & & \\ & & & & \beta_i & \gamma_i & & \\ & & & & & \ddots & \ddots & \\ & 0 & & & & & \beta_{n-1} & \gamma_{n-1} \\ & & & & & & & \beta_n \end{bmatrix}. \tag{7.81}$$

Identifying the corresponding elements of matrices $\mathbf{A}$ and $\mathbf{L} \cdot \mathbf{U}$ results in:

$$
\begin{cases}
& b_1 = \beta_1, \qquad\qquad c_1 = \gamma_1, \\
a_i = \alpha_i \beta_{i-1}, & b_i = \alpha_i \gamma_{i-1} + \beta_i, \qquad c_i = \gamma_i, \quad i = 2, 3, \ldots, n-1, \\
a_n = \alpha_n \beta_{n-1}, & b_n = \alpha_n \gamma_{n-1} + \beta_n.
\end{cases}
$$

By inverting these relations, one can express the entries of matrices $\mathbf{L}$ and $\mathbf{U}$:

$$
\begin{cases}
& \beta_1 = b_1, \\
\alpha_i = a_i / \beta_{i-1}, & \beta_i = b_i - \alpha_i c_{i-1}, \quad i = 2, 3, \ldots, n,
\end{cases}
\tag{7.82}
$$

$$
\gamma_i = c_i, \quad i = 1, 2, \ldots, n-1.
$$

As one may notice, the elements γ_i of $\mathbf{U}$ coincide with the elements c_i of $\mathbf{A}$ and do not imply any calculations at all. Provided $\beta_i \neq 0$ for $i = 1, 2, \ldots, n$, the matrix $\mathbf{A}$ is nonsingular and its factorization according to Formulas 7.82 is always possible.

Having at hand the LU factorization, the original system (7.78) becomes

$$
\mathbf{L} \cdot (\mathbf{U} \cdot \mathbf{x}) = \mathbf{d},
$$

and finding its solution amounts to solving two bidiagonal systems:

$$
\begin{cases}
\mathbf{L} \cdot \mathbf{y} = \mathbf{d}, \\
\mathbf{U} \cdot \mathbf{x} = \mathbf{y}.
\end{cases}
\tag{7.83}
$$

These matrix equations can be detailed as

$$
\begin{bmatrix}
1 & & & & & & & \\
\alpha_2 & 1 & & & & 0 & & \\
& \ddots & \ddots & & & & & \\
& & \alpha_{i-1} & 1 & & & & \\
& & & \alpha_i & 1 & & & \\
& & & & \ddots & \ddots & & \\
& 0 & & & & \alpha_{n-1} & 1 & \\
& & & & & & \alpha_n & 1
\end{bmatrix}
\begin{bmatrix}
y_1 \\ y_2 \\ \vdots \\ y_{i-1} \\ y_i \\ \vdots \\ y_{n-1} \\ y_n
\end{bmatrix}
=
\begin{bmatrix}
d_1 \\ d_2 \\ \vdots \\ d_{i-1} \\ d_i \\ \vdots \\ d_{n-1} \\ d_n
\end{bmatrix},
\tag{7.84}
$$

$$
\begin{bmatrix}
\beta_1 & c_1 & & & & & & \\
& \beta_2 & c_2 & & & 0 & & \\
& & \ddots & \ddots & & & & \\
& & & \beta_{i-1} & c_{i-1} & & & \\
& & & & \beta_i & c_i & & \\
& & & & & \ddots & \ddots & \\
& 0 & & & & & \beta_{n-1} & c_{n-1} \\
& & & & & & & \beta_n
\end{bmatrix}
\begin{bmatrix}
x_1 \\ x_2 \\ \vdots \\ x_{i-1} \\ x_i \\ \vdots \\ x_{n-1} \\ x_n
\end{bmatrix}
=
\begin{bmatrix}
y_1 \\ y_2 \\ \vdots \\ y_{i-1} \\ y_i \\ \vdots \\ y_{n-1} \\ y_n
\end{bmatrix}.
\tag{7.85}
$$

Identifying the corresponding entries from Equation 7.84, one can express the components of the intermediate solution **y** in ascending order of the indexes:

$$\begin{cases} y_1 = d_1, \\ y_i = d_i - \alpha_i y_{i-1}, \quad i = 2, 3, \dots, n. \end{cases} \quad (7.86)$$

To improve the efficiency of the calculations, we group the above relations with Equations 7.82, for which the index equally runs in ascending order, to define the *factorization phase* of the algorithm:

$$\begin{cases} \beta_1 = b_1, \quad\quad\quad\quad y_1 = d_1, \\ \alpha_i = a_i/\beta_{i-1}, \quad \beta_i = b_i - \alpha_i c_{i-1}, \quad y_i = d_i - \alpha_i y_{i-1}, \quad i = 2, 3, \dots, n. \end{cases} \quad (7.87)$$

Through the same technique of identifying the corresponding elements from both sides of Equation 7.85, we express the final solution **x** by *backward substitution*:

$$\begin{cases} x_n &= y_n/\beta_n, \\ x_i &= \left(y_i - c_i x_{i+1}\right)/\beta_i, \quad i = n-1, \dots, 1. \end{cases} \quad (7.88)$$

The analysis of the data flow involved in Equations 7.87 and 7.88 reveals that, for the concrete implementation of the algorithm, four arrays, a, b, c, and d, associated, respectively, with the quantities a_i, b_i, c_i, and d_i, are sufficient. The coefficients α_i can be stored in array a, overwriting the values a_i that have been used to calculate them. Similarly, coefficients β_i replace the values b_i in array b. The entries of array c remain unchanged, since each element γ_i coincides with the corresponding c_i. Having in view their logical overlap, the quantities d_i, y_i, and x_i can be successively stored in the same array d, which finally contains the solution of the system.

Algorithm 7.87–7.88 and the above considerations on the optimal storage usage are coded within the function `TriDiagSys` shown in Listing 7.17. Since no pivoting is carried out, `TriDiagSys` can fail, in principle, even if the coefficient matrix is nonsingular, but one of the coefficients β_i vanishes. Typically, however, the practical problems leading to tridiagonal systems are by nature well behaved and

Listing 7.17   Solver for Linear Systems with Tridiagonal Matrix (Python Coding)

```
#============================================================================
def TriDiagSys(a, b, c, d, n):
#----------------------------------------------------------------------------
#   Solves a system with tridiagonal matrix by LU factorization (diag(L) = 1).
#   a - lower codiagonal (i=2,n)
#   b - main diagonal (i=1,n)
#   c - upper codiagonal (i=1,n-1)
#   d - constant terms (i=1,n); solution on exit
#   n - order of system.
#----------------------------------------------------------------------------
   if (b[1] == 0e0): print("TriDiagSys: singular matrix !"); return
   for i in range(2,n+1):                              # factorization
      a[i] /= b[i-1]
      b[i] -= a[i]*c[i-1]
      if (b[i] == 0e0): print("TriDiagSys: singular matrix !"); return
      d[i] -= a[i]*d[i-1]

   d[n] /= b[n]                                     # backward substitution
   for i in range(n-1,0,-1): d[i] = (d[i] - c[i]*d[i+1])/b[i]
```

Listing 7.18 Solution of Linear System with Tridiagonal Matrix (Python Coding)

```
# Solves system with tridiagonal matrix by LU factorization
from linsys import *

n = 4                                              # order of system
a = [0]*(n+1)                                      # lower diagonal
b = [0]*(n+1)                                      # main diagonal
c = [0]*(n+1)                                      # upper diagonal
d = [0]*(n+1)                          # constant terms and solution

a[1] = 0; b[1] = 1; c[1] = 2; d[1] = 1
a[2] = 2; b[2] = 1; c[2] = 2; d[2] = 2
a[3] = 2; b[3] = 1; c[3] = 2; d[3] = 3
a[4] = 2; b[4] = 1; c[4] = 0; d[4] = 4    # Solution: -3.0, 2.0, 3.0, -2.0

TriDiagSys(a,b,c,d,n)                              # solve tridiagonal system

print("Solution:")
for i in range (1,n+1): print('{0:10.3f}'.format(d[i]),end="")
print()
```

this is reflected from a mathematical perspective by *diagonally dominant* coefficient matrices and from a numerical perspective, by the stability of the presented factorization method.

The typical usage of function `TriDiagSys`, assumed to be included in module `linsys.py`, is indicated in Listing 7.18. The system is specified just by its three nonzero diagonals, a, b, and c, while the vector d is used both to convey the constant terms to the routine and to return the calculated solution.

7.8 Block Tridiagonal Systems of Linear Equations

By applying discretization methods for solving systems of ordinary differential equations or differential equations with partial derivatives, sometimes, systems of linear equations with *block tridiagonal matrix* can result. Essentially, such matrix equations reflect some sort of intrinsic 2D coupling within the modeled system and their structure is

$$
\begin{bmatrix}
\mathbf{B}_1 & \mathbf{C}_1 & & & & & \\
\mathbf{A}_2 & \mathbf{B}_2 & \mathbf{C}_2 & & & \mathbf{0} & \\
& \ddots & \ddots & \ddots & & & \\
& & \mathbf{A}_i & \mathbf{B}_i & \mathbf{C}_i & & \\
& & & \ddots & \ddots & \ddots & \\
& \mathbf{0} & & & \mathbf{A}_{n-1} & \mathbf{B}_{n-1} & \mathbf{C}_{n-1} \\
& & & & & \mathbf{A}_n & \mathbf{B}_n
\end{bmatrix}
\begin{bmatrix}
\mathbf{x}_1 \\ \mathbf{x}_2 \\ \vdots \\ \mathbf{x}_i \\ \vdots \\ \mathbf{x}_{n-1} \\ \mathbf{x}_n
\end{bmatrix}
=
\begin{bmatrix}
\mathbf{d}_1 \\ \mathbf{d}_2 \\ \vdots \\ \mathbf{d}_i \\ \vdots \\ \mathbf{d}_{n-1} \\ \mathbf{d}_n
\end{bmatrix},
\tag{7.89}
$$

where $\mathbf{A}_i$, $\mathbf{B}_i$, and $\mathbf{C}_i$ are ($m \times m$) square matrices, while $\mathbf{x}_i$ and $\mathbf{d}_i$ are column vectors with m components. Obviously, the order of the coefficient matrix is ($m \times n$).

In many problems of practical interest, due to the excessive size of the coefficient matrix, the solution of blocked systems by general methods, such as Gaussian elimination or LU factorization, is intractable and, therefore, out of question. Instead, a block tridiagonal matrix equation such as Equation 7.89 can be solved efficiently and with minimized storage requirements by a procedure analogous to the factorization method developed for tridiagonal systems. In fact, the recurrence relations 7.87 and 7.88 can be generalized by simply replacing the scalar operations with matrix operations. In particular, the scalar division is replaced by the multiplication with the inverse matrix. In addition, the algorithm does not require for the whole coefficient matrix to be available from the outset, but the blocks $\mathbf{A}_i$, $\mathbf{B}_i$, $\mathbf{C}_i$, and $\mathbf{d}_i$ can be rather generated as they become necessary and, consequently, only working arrays for them need to be allocated, not for the whole coefficient matrix.

The forward elimination phase of the algorithm for the solution of a matrix equation with block tridiagonal structure is synthesized by the relations:

$$\begin{cases} \qquad\qquad \beta_1 = \mathbf{B}_1, \qquad\qquad \mathbf{y}_1 = \mathbf{d}_1, \\ \alpha_i = \beta_{i-1}^{-1} \cdot \mathbf{A}_i, \quad \beta_i = \mathbf{B}_i - \alpha_i \cdot \mathbf{C}_{i-1}, \quad \mathbf{y}_i = \mathbf{d}_i - \alpha_i \cdot \mathbf{y}_{i-1}, \quad i = 2, 3, \ldots, n, \end{cases} \tag{7.90}$$

while the backward substitution is accomplished according to

$$\begin{cases} \mathbf{x}_n = \beta_n^{-1} \cdot \mathbf{y}_n. \\ \mathbf{x}_i = \beta_i^{-1} \cdot \left(\mathbf{y}_i - \mathbf{C}_i \cdot \mathbf{x}_{i+1} \right), \quad i = n-1, \ldots, 1. \end{cases} \tag{7.91}$$

Here, α_i and β_i ($i = 1, \ldots, n-1$) are square matrices of order m and $\mathbf{y}_i$ are column vectors with m components. The inversions of the blocks β_i can be performed in place with the help of a general method for dense matrices, overwriting the blocks $\mathbf{B}_i$.

A concrete implementation of the above algorithm as a general routine with a well-defined list of arguments is of little relevance, due to the difficulties incurred by communicating the large amounts of data involved. Much more efficient in practice is to adapt Algorithm 7.89–7.90 to the particularities of the studied problem, optimizing the calculation of the individual blocks, the memory usage, and the data transfer between the main program and the routine (possibly by taking advantage of the regularities in the appearance of vanishing entries and by using global arrays).

7.9 Complex Matrix Equations

Complex matrix equations have in general complex solutions and may be written as

$$(\mathbf{A} + i\mathbf{A}') \cdot (\mathbf{X} + i\mathbf{X}') = (\mathbf{B} + i\mathbf{B}'), \tag{7.92}$$

where $\mathbf{A}$ and $\mathbf{A}'$ are square matrices of order n and $\mathbf{X}$, $\mathbf{X}'$, $\mathbf{B}$, and $\mathbf{B}'$ are $(n \times m)$ matrices. The unprimed and primed matrices represent, respectively, the real and imaginary parts.

By formally carrying out the multiplications and by separating the real from the imaginary parts on both sides of Equation 7.92, one obtains the following system of two matrix equations, coupling the real and imaginary parts of the solution, $\mathbf{X}$ and $\mathbf{X}'$:

$$\begin{cases} \mathbf{A} \cdot \mathbf{X} - \mathbf{A}' \cdot \mathbf{X}' = \mathbf{B}, \\ \mathbf{A}' \cdot \mathbf{X} + \mathbf{A} \cdot \mathbf{X}' = \mathbf{B}'. \end{cases}$$

This system may be rewritten as a single matrix equation by concatenating column wise the solution matrix and, respectively, the matrix of constant terms:

$$\begin{bmatrix} \mathbf{A} & -\mathbf{A}' \\ \mathbf{A}' & \mathbf{A} \end{bmatrix} \begin{bmatrix} \mathbf{X} \\ \mathbf{X}' \end{bmatrix} = \begin{bmatrix} \mathbf{B} \\ \mathbf{B}' \end{bmatrix}. \tag{7.93}$$

Consequently, the problem of solving a complex matrix equation of order n amounts to solving a real matrix equation of order $2n$. The only complication is to build the extended real matrices, but, otherwise, any of the general methods presented so far can be employed.

7.10 Jacobi and Gauss–Seidel Iterative Methods

The direct methods presented in this chapter (Gaussian elimination, LU factorization, etc.) enable the solution of a system of linear equations of order n at an approximate cost of n^3 floating point operations. However, with increasing system size, the roundoff errors can propagate in the final solution to an unacceptable degree. As extensively discussed, minimizing these errors can be achieved by pivoting, that is, by rearranging at each elimination step the equations in such a way as to have the largest possible pivot on the diagonal. Obviously, any pivoting technique involves an operation overhead and, altogether, the computation cost for large systems can become prohibitive, preventing the direct methods from being used for such systems.

The class of iterative methods, mentioned in the introduction to this chapter, enables under certain conditions circumventing the typical n^3-dependence of the operation cost featured by direct methods, by determining the solution on the basis of an iterative process started from an initial approximation. If the matrix of the system is well conditioned, in a sense that will be quantified below, the iterative process converges to the exact solution. Even though the process is theoretically infinite, practically, it must be terminated after a finite number of iterations, yielding the solution with a certain precision. In addition to the smaller roundoff errors than for direct methods, iterative methods are also affected by truncation errors, which can, nevertheless, be controlled by the number of performed iterations. In fact, iterative methods can be even employed to refine solutions produced by other methods. If the known initial approximation of the solution is close to the exact solution, the iterative methods converge fast, with a lower operation count than the direct methods. Not in the least, iterative methods stand out by the simplicity of their coding.

One of the oldest and well-known iterative methods for solving systems of linear equations is the Jacobi method. To expound this method, let us consider the linear system

$$\mathbf{A} \cdot \mathbf{x} = \mathbf{b}, \tag{7.94}$$

with $\mathbf{A} = [a_{ij}]_{nn}$, $\mathbf{b} = [b_i]_n$, and $\mathbf{x} = [x_i]_n$ or, in detailed form:

$$\begin{cases} a_{11}x_1 + a_{12}x_2 + \cdots + a_{1n}x_n &= b_1 \\ a_{21}x_1 + a_{22}x_2 + \cdots + a_{2n}x_n &= b_2 \\ &\vdots \\ a_{n1}x_1 + a_{n2}x_2 + \cdots + a_{nn}x_n &= b_n. \end{cases} \tag{7.95}$$

We assume the diagonal elements a_{ii} to be nonzero, or, otherwise, we rearrange the equations to meet this condition. We express from each equation i the diagonal unknown x_i and obtain the equivalent system:

$$\begin{cases} x_1 &= t_1 + s_{12}x_2 + \cdots + s_{1n}x_n \\ x_2 &= s_{21}x_1 + t_2 + \cdots + s_{2n}x_n \\ &\vdots \\ x_n &= s_{n1}x_1 + s_{n2}x_2 + \cdots + t_n, \end{cases} \tag{7.96}$$

where the new coefficients are defined by

$$\begin{cases} s_{ii} &= 0, \quad i = 1, 2, \ldots, n, \\ s_{ij} &= -a_{ij}/a_{ii}, \quad j = 1, 2, \ldots, n, \, j \neq i, \\ t_i &= b_i/a_{ii}. \end{cases}$$

By introducing the matrix $\mathbf{S} = [s_{ij}]_{nn}$ and the column vector $\mathbf{t} = [t_i]_n$, System 7.96 becomes in matrix form:

$$\mathbf{x} = \mathbf{S} \cdot \mathbf{x} + \mathbf{t}. \tag{7.97}$$

We solve this so-called *reduced system* by the *method of successive approximations* starting from the initial approximation

$$\mathbf{x}^{(0)} = \mathbf{t}, \tag{7.98}$$

and build the kth-order approximation based on the $(k - 1)$th approximation, by using the recurrence relation

$$\mathbf{x}^{(k)} = \mathbf{S} \cdot \mathbf{x}^{(k-1)} + \mathbf{t}, \quad k = 1, 2, \ldots. \tag{7.99}$$

If the sequence of successive approximations $\mathbf{x}^{(0)}, \mathbf{x}^{(1)}, \ldots, \mathbf{x}^{(k)}, \ldots$ has a limit,

$$\mathbf{x} = \lim_{k \to \infty} \mathbf{x}^{(k)},$$

then this is the solution of system (7.97), and by passing to the limit as $k \to \infty$ in Equation 7.99, there results Equation 7.97. This means that the limit vector $\mathbf{x}$ is the solution of the equivalent system (7.97) and, consequently, also of the original system.

Explicitly, the formulas describing the iterative procedure read:

$$x_i^{(k)} = \sum_{\substack{j=1 \\ j \neq i}}^{n} s_{ij} x_j^{(k-1)} + t_i, \quad i = 1, 2, \ldots, n. \tag{7.100}$$

By introducing the absolute corrections of the solution components,

$$\Delta_i^{(k)} = x_i^{(k)} - x_i^{(k-1)}, \quad i = 1, 2, \ldots, n, \tag{7.101}$$

which can be regarded as estimates of the corresponding absolute errors, and, by redefining the diagonal elements of $\mathbf{S}$ to be $s_{ii} = -1$, the recurrence relations defining the *Jacobi iterative method* take the form:

$$\begin{cases} \Delta_i^{(k)} = \sum_{j=1}^{n} s_{ij} x_j^{(k-1)} + t_i, \\ x_i^{(k)} = x_i^{(k-1)} + \Delta_i^{(k)}, \quad i = 1, 2, \ldots, n, \end{cases} \tag{7.102}$$

where

$$\begin{cases} s_{ij} = -a_{ij}/a_{ii}, \quad i, j = 1, 2, \ldots, n, \\ t_i = b_i/a_{ii}. \end{cases} \tag{7.103}$$

Relations 7.102 point to the fact that the Jacobi method requires in implementations two arrays for the consecutive approximations of the solution, $\mathbf{x}^{(k-1)}$ and $\mathbf{x}^{(k)}$. However, since by simple changes, both the convergence speed and the memory usage can be improved significantly, the concrete implementation of the Jacobi method remains of little practical interest.

The *Gauss–Seidel method* represents an improved variant of the Jacobi method. The basic idea is to use in Recurrence 7.102 the most recently updated solution components $x_i^{(k)}$ instead of those from the previous iteration, $x_i^{(k-1)}$, as soon as they become available, not only after completing the iteration.

Correspondingly, the recurrence relations characterizing the Gauss–Seidel method can be represented as

$$\begin{cases} \Delta_i^{(k)} = \sum_{j=1}^{i-1} s_{ij} x_j^{(k)} + \sum_{j=i}^{n} s_{ij} x_j^{(k-1)} + t_i, \\ x_i^{(k)} = x_i^{(k-1)} + \Delta_i^{(k)}, \quad i = 1, 2, \ldots, n. \end{cases} \tag{7.104}$$

It can be noticed that, indeed, in evaluating the solution component $x_i^{(k)}$, instead of just components from the previous iteration, the updated ones $(x_1^{(k)}, \ldots, x_{i-1}^{(k)})$ are also used.

The iterative process has to be continued in principle until the maximum relative correction of the solution components drops under a given tolerance ε, and the *stopping criterion* can be written as

$$\max_i |\Delta_i^{(k)}/x_i^{(k)}| \leq \varepsilon. \tag{7.105}$$

In the case of vanishing components, the absolute error estimate $\Delta_i^{(k)}$ should be used instead of the relative error estimate $\Delta_i^{(k)}/x_i^{(k)}$.

Regarding the conditions under which the iterative process (either Equation 7.99 or 7.104) is convergent, it can be rigorously demonstrated (Demidovich and Maron (1987)) that if at least one of the following conditions

$$\sum_{j=1}^{n} |s_{ij}| < 1, \quad i = 1, 2, \ldots, n, \tag{7.106}$$

$$\sum_{i=1}^{n} |s_{ij}| < 1, \quad j = 1, 2, \ldots, n, \tag{7.107}$$

applies to the reduced system (7.97), then the recursive process converges to the unique solution of the system, provided it exists, independently of the initial approximation. A more practical consequence requires that the coefficient matrix **A** must be *strictly diagonally dominant*, that is,

$$|a_{ii}| > \sum_{j \neq i} |a_{ij}|, \quad i = 1, 2, \ldots, n, \tag{7.108}$$

or, in other words, that on every row of the system, the absolute value of the diagonal coefficient a_{ii} exceeds the sum of the absolute values of the nondiagonal coefficients. However, in practice, a sufficient condition for convergence is the *diagonal dominance* of **A**, that is,

$$|a_{ii}| > \max_{j \neq i} |a_{ij}|, \quad i = 1, 2, \ldots, n. \tag{7.109}$$

The iterative Gauss–Seidel process converges in general significantly faster than the Jacobi method. Moreover, the first process may converge even in cases in which the latter diverges. Nonetheless, situations are possible in which even the Gauss–Seidel algorithm may not converge, unless an additional transformation of the initial system is operated.

It can be proven that for a so-called *normal system*, that is, having a *symmetric positive-definite* coefficient matrix, the Gauss–Seidel process is always converging, irrespective of the initial approximation (Demidovich and Maron (1987)). In practice, there is a simple way of transforming a general system with a nonsingular matrix **A** into a normal system, and that is by left-multiplying both sides of the original matrix equation with $\mathbf{A}^T$:

$$\mathbf{A}^T \cdot \mathbf{A} \cdot \mathbf{x} = \mathbf{A}^T \cdot \mathbf{b}.$$

From Equation 7.104, it follows that the Gauss–Seidel algorithm can be implemented using *a single array* to store the successive approximations of the solution. The array can hold at the same time components from consecutive iterations, $(k-1)$ and k, which are updated continuously as the new components $x_i^{(k)}$ are calculated. In addition to optimized storage, a single array also enables more compact coding, as can be seen in Listing 7.19.

Listing 7.19   Solution of System of Linear Equations by the Gauss–Seidel Method (Python Coding)

```python
#===========================================================================
def GaussSeidel(a, b, x, n, init):
#---------------------------------------------------------------------------
#  Solves linear system a x = b by the Gauss-Seidel method.
#  Convergence is ensure by left-multiplying the system with a^T.
#  a    - coefficient matrix (n x n)
#  b    - vector of constant terms
#  x    - initial approximation of solution (input); solution (output)
#  n    - order of system
#  err  - maximum relative error of the solution components
#  init - initialization option: 0 - refines initial approximation
#                                1 - initializes solution
#---------------------------------------------------------------------------
   eps = 1e-10                                 # relative precision criterion
   itmax = 1000                                # max no. of iterations

   s = [[0]*(n+1) for i in range(n+1)]         # matrices of reduced system
   t = [0]*(n+1)

   for i in range(1,n+1):                      # matrices of normal system
      for j in range(1,i+1):                   # by multiplication with aT
         s[i][j] = 0e0                         # store result in s and t
         for k in range(1,n+1): s[i][j] += a[k][i]*a[k][j]
         s[j][i] = s[i][j]

      t[i] = 0e0
      for j in range(1,n+1): t[i] += a[j][i]*b[j]

   for i in range(1,n+1):                 # matrices s and t of reduced system
      f = 1e0/s[i][i]; t[i] /= s[i][i]
      for j in range(1,n+1): s[i][j] *= f

   if (init):
      for i in range(1,n+1): x[i] = t[i]       # initialize solution

   for k in range(1,itmax+1):                  # loop of iterations
      err = 0e0
      for i in range(1,n+1):
         delta = t[i]                                        # correction
         for j in range(1,n+1): delta += s[i][j]*x[j]
         x[i] += delta                    # new approximation to solution
         if (x[i]): delta /= x[i]                       # relative error
         if (fabs(delta) > err): err = fabs(delta)       # maximum error

      if (err <= eps): break                         # check convergence

   if (k > itmax): printf("GaussSeidel: max. no. of iterations exceeded !")

   return err
```

The routine's input consists of the coefficient matrix a, the vector of constant terms b, the vector x holding the initial approximation of the solution, the order n of the system, and the initialization option init. If init is set to 0, the routine refines the initial approximation received via x, or else, the solution is initialized with the constant terms of the reduced system. On exit from the routine, x returns the refined solution, while err represents the maximum relative error of its components.

The convergence of the algorithm is ensured, as explained above, by multiplying both sides of the original system with the transpose of the coefficient matrix. Array s is thus used at first to store the matrix $\mathbf{A}^T \cdot \mathbf{A}$ of the normal system and, then, the matrix $\mathbf{S}$ of the reduced system. Analogously, array t contains at first the vector of constant terms $\mathbf{A}^T \cdot \mathbf{b}$ of the normal system and, then, the vector $\mathbf{t}$ of the reduced system.

Except for the additional requirement of specifying an initial approximation for the solution, the usage of the routine GaussSeidel is similar to the one of direct methods.

7.11 Implementations in C/C++

Listing 7.20 shows the content of the file linsys.h, which contains equivalent C/C++ implementations of the Python functions developed in the main text and included in the module linsys.py. The corresponding routines have identical names and functionalities.

Listing 7.20 Solvers for Systems of Linear Equations (linsys.h)

```
//--------------------------------- linsys.h ---------------------------------
// Contains routines for solving systems of linear equations.
// Part of the numxlib numerics library. Author: Titus Beu, 2013
//----------------------------------------------------------------------------
#ifndef _LINSYS_
#define _LINSYS_

#include <stdio.h>
#include <math.h>
#include "memalloc.h"


//============================================================================
void Gauss(double **a, double **b, int n, int m, double &det)
//----------------------------------------------------------------------------
// Solves matrix equation a x = b by Gaussian elimination with partial
// pivoting on columns.
// a   - coefficient matrix (n x n); destroyed on exit
// b   - matrix of constant terms (n x m); solution x on exit
// det - determinant of coefficient matrix (output).
//----------------------------------------------------------------------------
{
#define Swap(a,b) { t = a; a = b; b = t; }
   double amax, sum, t;
   int i, imax, j, k;

   det = 1e0;
   for (k=1; k<=n; k++) {                          // FORWARD ELIMINATION
      amax = 0e0;                              // determine pivot row with
      for (i=k; i<=n; i++)                     // max. element on column k
         if (amax < fabs(a[i][k])) { amax = fabs(a[i][k]); imax = i; }
      if (amax == 0e0)
         { printf("Gauss: singular matrix !\n"); det = 0e0; return; }

      if (imax != k) {                         // interchange rows imax and k
```

```
              det = -det;                              // to put pivot on diagonal
              for (j=k; j<=n; j++) Swap(a[imax][j],a[k][j])
              for (j=1; j<=m; j++) Swap(b[imax][j],b[k][j])
           }

           det *= a[k][k];                             // multiply determinant with pivot

           t = 1e0/a[k][k];                            // divide pivot row by pivot
           for (j=k+1; j<=n; j++) a[k][j] *= t;
           for (j=  1; j<=m; j++) b[k][j] *= t;

           for (i=k+1; i<=n; i++) {                     // reduce non-pivot rows
              t = a[i][k];
              for (j=k+1; j<=n; j++) a[i][j] -= a[k][j]*t;
              for (j=  1; j<=m; j++) b[i][j] -= b[k][j]*t;
           }
       }

       for (k=n-1; k>=1; k--)                           // BACKWARD SUBSTITUTION
          for (j=1; j<=m; j++) {
             sum = b[k][j];
             for (i=k+1; i<=n; i++) sum -= a[k][i]*b[i][j];
             b[k][j] = sum;
          }
}

//============================================================================
void GaussJordan0(double **a, double **b, int n, int m, double &det)
//----------------------------------------------------------------------------
// Solves matrix equation a x = b by Gauss-Jordan elimination with partial
// pivoting on columns.
// a   - coefficient matrix (n x n); destroyed on exit
// b   - matrix of constant terms (n x m); solution x on exit
// det - determinant of coefficient matrix (output).
//----------------------------------------------------------------------------
{
#define Swap(a,b) { t = a; a = b; b = t; }
   double amax, t;
   int i, imax, j, k;

   det = 1e0;
   for (k=1; k<=n; k++) {                               // FORWARD ELIMINATION
      amax = 0e0;                                       // determine pivot row with
      for (i=k; i<=n; i++)                              // max. element on column k
         if (amax < fabs(a[i][k])) { amax = fabs(a[i][k]); imax = i; }
      if (amax == 0e0)
         { printf("GaussJordan0: singular matrix !\n"); det = 0e0; return; }

      if (imax != k) {                                 // interchange rows imax and k
         det = -det;                                    // to put pivot on diagonal
         for (j=k; j<=n; j++) Swap(a[imax][j],a[k][j])
         for (j=1; j<=m; j++) Swap(b[imax][j],b[k][j])
      }

      det *= a[k][k];                                   // multiply determinant with pivot

      t = 1e0/a[k][k];                                  // divide pivot row by pivot
      for (j=k+1; j<=n; j++) a[k][j] *= t;
      for (j=  1; j<=m; j++) b[k][j] *= t;
```

```
      for (i=1; i<=n; i++)                            // reduce non-pivot rows
         if (i != k) {
            t = a[i][k];
            for (j=1; j<=n; j++) a[i][j] -= a[k][j]*t;
            for (j=1; j<=m; j++) b[i][j] -= b[k][j]*t;
         }
   }
}

//===========================================================================
void GaussJordan1(double **a, double **b, int n, int m, double &det)
//---------------------------------------------------------------------------
// Solves matrix equation a x = b by Gauss-Jordan elimination with partial
// pivoting on columns.
// a   - coefficient matrix (n x n); a^(-1) on exit
// b   - matrix of constant terms (n x m); solution x on exit
// det - determinant of coefficient matrix (output).
//---------------------------------------------------------------------------
{
#define Swap(a,b) { t = a; a = b; b = t; }
   double amax, t;
   int i, imax, j, k;
   int *ipivot;

   ipivot = IVector(1,n);                             // stores pivot rows

   det = 1e0;
   for (k=1; k<=n; k++) {                             // FORWARD ELIMINATION
      amax = 0e0;                                     // determine pivot row with
      for (i=k; i<=n; i++)                            // max. element on column k
         if (amax < fabs(a[i][k])) { amax = fabs(a[i][k]); imax = i; }
      if (amax == 0e0)
         { printf("GaussJordan1: singular matrix !\n"); det = 0e0; return; }
      ipivot[k] = imax;                               // store pivot row index

      if (imax != k) {                                // interchange rows imax and k
         det = -det;                                  // to put pivot on diagonal
         for (j=1; j<=n; j++) Swap(a[imax][j],a[k][j])
         for (j=1; j<=m; j++) Swap(b[imax][j],b[k][j])
      }

      det *= a[k][k];                         // multiply determinant with pivot

      t = 1e0/a[k][k];                                // divide pivot row by pivot
      a[k][k] = 1e0;                          // diagonal element of unit matrix
      for (j=1; j<=n; j++) a[k][j] *= t;
      for (j=1; j<=m; j++) b[k][j] *= t;

      for (i=1; i<=n; i++)                            // reduce non-pivot rows
         if (i != k) {
            t = a[i][k];
            a[i][k] = 0e0;                    // non-diagonal element of unit matrix
            for (j=1; j<=n; j++) a[i][j] -= a[k][j]*t;
            for (j=1; j<=m; j++) b[i][j] -= b[k][j]*t;
         }
   }

   for (k=n; k>=1; k--) {                             // rearrange columns of inverse
```

```
         imax = ipivot[k];
         if (imax != k)
            for (i=1; i<=n; i++) Swap(a[i][imax],a[i][k])
   }

   FreeIVector(ipivot,1);
}

//==========================================================================
void GaussJordan(double **a, double **b, int n, int m, double &det)
//--------------------------------------------------------------------------
// Solves matrix equation a x = b by Gauss-Jordan elimination with complete
// pivoting.
// a    - coefficient matrix (n x n); a^(-1) on exit
// b    - matrix of constant terms (n x m); solution x on exit
// det - determinant of coefficient matrix (output).
//--------------------------------------------------------------------------
{
#define Swap(a,b) { t = a; a = b; b = t; }
   double amax, t;
   int i, imax, j, jmax, k;
   int *ipivot, *jpivot, *npivot;

   ipivot = IVector(1,n); jpivot = IVector(1,n);      // pivot rows and columns
   npivot = IVector(1,n);                             // marks used pivot columns

   for (i=1; i<=n; i++) npivot[i] = 0;

   det = 1e0;
   for (k=1; k<=n; k++) {                           // FORWARD ELIMINATION
      amax = 0e0;                                         // determine pivot
      for (i=1; i<=n; i++) {                             // loop over rows
         if (npivot[i] != 1)
            for (j=1; j<=n; j++) {                     // loop over columns
               if (npivot[j] == 0)                  // pivoting not yet done?
                  if (amax < fabs(a[i][j]))
                     { amax = fabs(a[i][j]); imax = i; jmax = j; }
               if (npivot[j] > 1) {
                  printf("GaussJordan: singular matrix 1 !\n");
                  det = 0e0; return;
               }
            }
      }
      if (amax == 0e0)
         { printf("GaussJordan: singular matrix 2 !\n"); det = 0e0; return; }

      ipivot[k] = imax; jpivot[k] = jmax;       // store pivot row and column
      ++(npivot[jmax]);                             // mark used pivot column

      if (imax != jmax) {                       // interchange rows imax and jmax
         det = -det;                                // to put pivot on diagonal
         for (j=1; j<=n; j++) Swap(a[imax][j],a[jmax][j])
         for (j=1; j<=m; j++) Swap(b[imax][j],b[jmax][j])
      }
      det *= a[jmax][jmax];                      // multiply determinant with pivot

      t = 1e0/a[jmax][jmax];                          // divide pivot row by pivot
      a[jmax][jmax] = 1e0;                      // diagonal element of unit matrix
      for (j=1; j<=n; j++) a[jmax][j] *= t;
```

```
         for (j=1; j<=m; j++) b[jmax][j] *= t;

         for (i=1; i<=n; i++)                          // reduce non-pivot rows
            if (i != jmax) {
               t = a[i][jmax];
               a[i][jmax] = 0e0;           // non-diagonal element of unit matrix
               for (j=1; j<=n; j++) a[i][j] -= a[jmax][j]*t;
               for (j=1; j<=m; j++) b[i][j] -= b[jmax][j]*t;
            }
      }

      for (k=n; k>=1; k--) {                          // rearrange columns of inverse
         imax = ipivot[k]; jmax = jpivot[k];
         if (imax != jmax)
            for (i=1; i<=n; i++) Swap(a[i][imax],a[i][jmax])
      }

      FreeIVector(ipivot,1); FreeIVector(jpivot,1);
      FreeIVector(npivot,1);
}

//===========================================================================
void LUFactor(double **a, int ipivot[], int n, double &det)
//---------------------------------------------------------------------------
// Performs LU factorization of (n x n) matrix a (diag(L) = 1). On exit,
// replaces upper triangle and diagonal with U, and lower triangle, with L.
// Uses partial pivoting on columns.
// a      - coefficient matrix (n x n); LU decomposition on exit
// ipivot - array of pivot row indexes (output)
// det    - determinant of coefficient matrix (output).
//---------------------------------------------------------------------------
{
   double amax, sum, t;
   int i, imax, j, k;

   det = 1e0;
   for (j=1; j<=n; j++) {                              // loop over columns
      for (i=1; i<=(j-1); i++) {                       // elements of matrix U
         sum = a[i][j];
         for (k=1; k<=(i-1); k++) sum -= a[i][k]*a[k][j];
         a[i][j] = sum;
      }
      amax = 0e0;
      for (i=j; i<=n; i++) {                           // elements of matrix L
         sum = a[i][j];                                // undivided by pivot
         for (k=1; k<=(j-1); k++) sum -= a[i][k]*a[k][j];
         a[i][j] = sum;
                                                       // determine pivot
         if (amax < fabs(a[i][j])) {amax = fabs(a[i][j]); imax = i;}
      }
      if (amax == 0e0)
         { printf("LUFactor: singular matrix !\n"); det = 0e0; return; }

      ipivot[j] = imax;                                // store pivot row index
                                                       // interchange rows imax and j
      if (imax != j) {                                 // to put pivot on diagonal
         det = -det;
         for (k=1; k<=n; k++)
            { t = a[imax][k]; a[imax][k] = a[j][k]; a[j][k] = t; }
```

```
      }

      det *= a[j][j];                            // multiply determinant with pivot

      t = 1e0/a[j][j];                           // divide elements of L by pivot
      for (i=j+1; i<=n; i++) a[i][j] *= t;
   }
}

//============================================================================
void LUSystem(double **a, int ipivot[], double b[], int n)
//----------------------------------------------------------------------------
// Solves linear system a x = b of order n by LU factorization.
// a      - LU decomposition of coefficient matrix (returned by LUFactor)
// ipivot - array of pivot row indexes (input)
// b      - vector of constant terms (input); solution x (on exit)
//----------------------------------------------------------------------------
{
   double sum;
   int i, j;

   for (i=1; i<=n; i++) {                              // solves Ly = b
      sum = b[ipivot[i]];
      b[ipivot[i]] = b[i];
      for (j=1; j<=(i-1); j++) sum -= a[i][j]*b[j];
      b[i] = sum;
   }

   for (i=n; i>=1; i--) {                              // solves Ux = y
      sum = b[i];
      for (j=i+1; j<=n; j++) sum -= a[i][j]*b[j];
      b[i] = sum/a[i][i];
   }
}

//============================================================================
void MatInv(double **a, int n, double &det)
//----------------------------------------------------------------------------
// Calculates inverse of real matrix by LU factorization.
// a   - (n x n) matrix (input); a^(-1) (output)
// det - determinant of coefficient matrix (output).
// Calls: LUFactor, LUSystem.
//----------------------------------------------------------------------------
{
   double **ainv, *b;
   int *ipivot, i, j;

   ainv = Matrix(1,n,1,n);              // temporary storage for inverse matrix
   b = Vector(1,n);
   ipivot = IVector(1,n);                             // stores pivot rows

   LUFactor(a,ipivot,n,det);                          // LU factorization of a
   if (det == 0e0)                                    // singular matrix
      { printf("MatInv: singular matrix !\n"); return; }

   for (j=1; j<=n; j++) {                      // loop over columns of unit matrix
      for (i=1; i<=n; i++) b[i] = 0e0;                // column j
      b[j] = 1e0;
      LUSystem(a,ipivot,b,n);                         // solve system
```

```
      for (i=1; i<=n; i++) ainv[i][j] = b[i];              // column j of inverse
   }

   for (j=1; j<=n; j++)                                     // copy inverse in a
      for (i=1; i<=n; i++) a[i][j] = ainv[i][j];

   FreeMatrix(ainv,1,1); FreeVector(b,1);
   FreeIVector(ipivot,1);
}

//===========================================================================
int MatTriInv(double **a, int n)
//---------------------------------------------------------------------------
// Calculates inverse of lower triangular matrix L of order n contained in
// (n x n) array a and replaces on exit lower triangle of a by L^(-1).
// Error flag = 0 - normal execution
//              1 - singular matrix.
//---------------------------------------------------------------------------
{
   double sum;
   int i, j, k;

   for (j=1; j<=n; j++) {                                   // loop over columns
      if (a[j][j] <= 0e0)
         { printf("MatTriInv: singular matrix !\n"); return 1; }
      a[j][j] = 1e0/a[j][j];                                // diagonal element

      for (i=j+1; i<=n; i++) {                              // non-diagonal elements
         sum = 0e0;
         for (k=j; k<=(i-1); k++) sum -= a[i][k]*a[k][j];
         a[i][j] = sum/a[i][i];
      }
   }
   return 0;
}


//===========================================================================
int Cholesky(double **a, int n, double &det)
//---------------------------------------------------------------------------
// Performs Cholesky factorization L * LT of real symmetric positive-definite
// (n x n) matrix a. On exit, replaces lower triangle of a with L.
// a   - coefficient matrix (n x n); L in lower triangle and diagonal on exit
// det - determinant of coefficient matrix (output).
// Error flag = 0 - normal execution
//              1 - a is not positive-definite.
//---------------------------------------------------------------------------
{
   double sum;
   int i, j, k;

   det = 1e0;
   for (j=1; j<=n; j++) {                                   // loop over columns
      sum = a[j][j];                                        // diagonal element
      for (k=1; k<=(j-1); k++) sum -= a[j][k]*a[j][k];
      if (sum <= 0e0)
         { printf("Cholesky: matrix not positive-definite !\n"); return 1; }
      a[j][j] = sqrt(sum);

      for (i=j+1; i<=n; i++) {                              // non-diagonal elements
```

```
            sum = a[i][j];
            for (k=1; k<=(j-1); k++) sum -= a[i][k]*a[j][k];
            a[i][j] = sum / a[j][j];
         }

         det *= a[j][j]*a[j][j];
      }
      return 0;
}


//===============================================================================
void CholeskySys(double **a, double b[], int n)
//-------------------------------------------------------------------------------
// Solves linear system a x = b with real symmetric positive-definite (n x n)
// matrix by Cholesky factorization L * LT.
// a - matrix L in lower triangle and on diagonal (returned by Cholesky)
// b - vector of constant terms (input); solution x (on exit)
//-------------------------------------------------------------------------------
{
   double sum;
   int i, j;

   for (i=1; i<=n; i++) {                                       // solves Ly = b
      sum = b[i];
      for (j=1; j<=(i-1); j++) sum -= a[i][j]*b[j];
      b[i] = sum/a[i][i];
   }

   for (i=n; i>=1; i--) {                                       // solves LT x = y
      sum = b[i];
      for (j=i+1; j<=n; j++) sum -= a[j][i]*b[j];
      b[i] = sum/a[i][i];
   }
}


//===============================================================================
void MatSymInv(double **a, int n, double &det)
//-------------------------------------------------------------------------------
// Calculates inverse of symmetric positive-definite (n x n) matrix by
// Cholesky factorization a = L * LT as a^(-1) = (L^(-1))T * L^(-1).
// L^(-1) is calculated using MatTriInv.
// a   - (n x n) matrix (input); a^(-1) (output)
// det - determinant of coefficient matrix (output).
// Calls: Cholesky, MatTriInv.
//-------------------------------------------------------------------------------
{
   double sum;
   int ierr, i, j, k;

   ierr = Cholesky(a,n,det);                          // L * LT factorization of a
   if (ierr == 1)
      { printf("MatSymInv: matrix not positive-definite !\n"); return; }
   if (det == 0e0) { printf("MatSymInv: singular matrix !\n"); return; }

   MatTriInv(a,n);                                        // L^(-1) in lower triangle

   for (i=1; i<=n; i++)                                   // (L^(-1))T * L^(-1)
      for (j=1; j<=i; j++) {
         sum = 0e0;
```

```
              for (k=i; k<=n; k++) sum += a[k][i] * a[k][j];
              a[j][i] = sum;                          // store in upper triangle
          }

      for (i=1; i<=n; i++)                            // complete lower triangle
          for (j=1; j<=i; j++) a[i][j] = a[j][i];
}

//===========================================================================
void TriDiagSys(double a[], double b[], double c[], double d[], int n)
//---------------------------------------------------------------------------
// Solves a system with tridiagonal matrix by LU factorization (diag(L) = 1).
// a - lower codiagonal (i=2,n)
// b - main diagonal (i=1,n)
// c - upper codiagonal (i=1,n-1)
// d - constant terms (i=1,n); solution on exit
// n - order of system.
//---------------------------------------------------------------------------
{
   int i;

   if (b[1] == 0e0) { printf("TriDiagSys: singular matrix !\n"); return; }
   for (i=2; i<=n; i++) {                                      // factorization
      a[i] /= b[i-1];
      b[i] -= a[i]*c[i-1];
      if (b[i] == 0e0) { printf("TriDiagSys: singular matrix !\n"); return; }
      d[i] -= a[i]*d[i-1];
   }

   d[n] /= b[n];                                        // backward substitution
   for (i=(n-1); i>=1; i--) d[i] = (d[i] - c[i]*d[i+1])/b[i];
}

//===========================================================================
void TriDiagSys(dcmplx a[], dcmplx b[], dcmplx c[], dcmplx d[], int n)
//---------------------------------------------------------------------------
// Solves a complex system with tridiagonal matrix by LU factorization.
// a - lower codiagonal (i=2,n)
// b - main diagonal (i=1,n)
// c - upper codiagonal (i=1,n-1)
// d - constant terms (i=1,n); solution on exit
// n - order of system.
//---------------------------------------------------------------------------
{
   int i;

   if (b[1] == 0e0) { printf("TriDiagSys: singular matrix !\n"); return; }
   for (i=2; i<=n; i++) {                                      // factorization
      a[i] /= b[i-1];
      b[i] -= a[i]*c[i-1];
      if (b[i] == 0e0) { printf("TriDiagSys: singular matrix !\n"); return; }
      d[i] -= a[i]*d[i-1];
   }

   d[n] /= b[n];                                        // backward substitution
   for (i=(n-1); i>=1; i--) d[i] = (d[i] - c[i]*d[i+1])/b[i];
}

//===========================================================================
```

```
void GaussSeidel(double **a, double b[], double x[], int n, int init,
                 double &err)
//-------------------------------------------------------------------------
// Solves linear system a x = b by the Gauss-Seidel method.
// To ensure convergence, the system is left-multiplied with a^T.
// a    - coefficient matrix (n x n)
// b    - vector of constant terms
// x    - initial approximation of solution (input); solution (output)
// n    - order of system
// err  - maximum relative error of the solution components
// init - initialization option: 0 - refines initial approximation
//                               1 - initializes solution
//-------------------------------------------------------------------------
{
   const double eps = 1e-10;                  // relative precision criterion
   const int itmax = 1000;                    // max no. of iterations
   double del, f, **s, *t;
   int i, j, k;

   s = Matrix(1,n,1,n);                       // matrices of reduced system
   t = Vector(1,n);

   for (i=1; i<=n; i++) {                      // matrices of normal system
      for (j=1; j<=i; j++) {                   // by multiplication with aT
         s[i][j] = 0e0;                        // store result in s and t
         for (k=1; k<=n; k++) s[i][j] += a[k][i]*a[k][j];
         s[j][i] = s[i][j];
      }
      t[i] = 0e0;
      for (j=1; j<=n; j++) t[i] += a[j][i]*b[j];
   }

   for (i=1; i<=n; i++) {                      // matrices s and t of reduced system
      f = -1e0/s[i][i]; t[i] /= s[i][i];
      for (j=1; j<=n; j++) s[i][j] *= f;
   }

   if (init) for (i=1; i<=n; i++) x[i] = t[i];  // initialize solution

   for (k=1; k<=itmax; k++) {                  // loop of iterations
      err = 0e0;
      for (i=1; i<=n; i++) {
         del = t[i];                           // correction
         for (j=1; j<=n; j++) del += s[i][j]*x[j];
         x[i] += del;                          // new approximation to solution
         if (x[i]) del /= x[i];                // relative error
         if (fabs(del) > err) err = fabs(del);  // maximum error
      }
      if (err <= eps) break;                   // check convergence
   }
   if (k > itmax) printf("GaussSeidel: max. no. of iterations exceeded !\n");

   FreeMatrix(s,1,1); FreeVector(t,1);
}

#endif
```

7.12 Problems

The Python and C/C++ programs for the following problems may import the functions developed in this chapter from the modules `linsys.py` and, respectively, `linsys.h`, which are available as supplementary material. In addition, the utility routines contained in files `matutil.py` and, respectively, `matutil.h` can be used to perform basic operations with one- and 2D arrays.

PROBLEM 7.1

Use the routine `GaussJordan` to write Python and C/C++ programs for solving the matrix equation $AX = B$. Consider matrices of order $n = 5$, and generate a random coefficient matrix A and a unit matrix B.

Verify that, on exit from the solver, the inverse replacing matrix A coincides with the solution that replaces matrix B. Check the inverse also by showing that $A^{-1}A$ equals the unit matrix.

Solution

The Python and C/C++ implementations are given in Listings 7.21 and 7.22.

Listing 7.21   Solution of a Matrix Equation by Gauss–Jordan Elimination (Python Coding)

```
# Solves matrix equation by the Gauss-Jordan method
from random import *
from linsys import *
from matutil import *

n = 5                                                      # order of system
a = [[0]*(n+1) for i in range(n+1)]                        # system matrix
b = [[0]*(n+1) for i in range(n+1)]            # constant terms and solution
c = [[0]*(n+1) for i in range(n+1)]                # copy of system matrix
d = [[0]*(n+1) for i in range(n+1)]                        # work array

for i in range(1,n+1):                                     # random matrix
    for j in range(1,n+1): a[i][j] = random()
print("A:")
MatPrint(a,n,n)

for i in range(1,n+1):                                     # unit matrix
    for j in range(1,n+1): b[i][j] = 0e0
    b[i][i] = 1e0
print("B:")
MatPrint(b,n,n)

MatCopy(a,c,n,n)                                           # save system matrix

det = GaussJordan(a,b,n,n)                         # solve system, inverse in a

print("Check A^(-1) - X = 0:")
MatDiff(a,b,d,n,n)                          # difference between inverse and solution
MatPrint(d,n,n)

print("Check A^(-1)A = I:")
MatProd(a,c,d,n,n,n)                               # multiply inverse with original
MatPrint(d,n,n)
```

Listing 7.22   Solution of a Matrix Equation by Gauss–Jordan Elimination (C/C++ Coding)

```
// Solves matrix equation by the Gauss-Jordan method
#include <stdio.h>
#include "memalloc.h"
#include "linsys.h"
#include "matutil.h"

int main()
{
   double **a, **b, **c, **d, det;
   int i, j, n;

   n = 5;                                            // order of system
   a = Matrix(1,n,1,n);                              // system matrix
   b = Matrix(1,n,1,n);               // constant terms and solution
   c = Matrix(1,n,1,n);                     // copy of system matrix
   d = Matrix(1,n,1,n);                             // work array

   for (i=1; i<=n; i++)                             // random matrix
      for (j=1; j<=n; j++) a[i][j] = rand();
   printf("A:\n");
   MatPrint(a,n,n);

   for (i=1; i<=n; i++) {                           // unit matrix
      for (j=1; j<=n; j++) b[i][j] = 0e0;
      b[i][i] = 1e0;
   }
   printf("B:\n");
   MatPrint(b,n,n);

   MatCopy(a,c,n,n);                          // save system matrix

   GaussJordan(a,b,n,n,det);            // solve system, inverse in a

   printf("Check A^(-1) - X = 0:\n");
   MatDiff(a,b,d,n,n);          // difference between inverse and solution
   MatPrint(d,n,n);

   printf("Check A^(-1)A = I:\n");
   MatProd(a,c,d,n,n,n);                  // multiply inverse with original
   MatPrint(d,n,n);
}
```

PROBLEM 7.2

Use the pair of routines LUFactor and LUSystem to write Python and C/C++ programs for solving multiple linear systems $\mathbf{Ax} = \mathbf{b}$ for the same coefficient matrix $\mathbf{A}$ and n different vectors of constant terms $\mathbf{b}$. Consider systems of order $n = 5$ and, for simplicity, generate the matrices randomly. Perform the LU factorization once, and use the resulting decomposition to solve the successive systems. Check each solution synthetically, by verifying the numerical vanishing of the maximum residual error

$$\Delta = \max_i \left| \sum_{j=1}^{n} A_{ij} x_j - b_i \right|.$$

Solution

The Python and C/C++ implementations are given in Listings 7.23 and 7.24.

Listing 7.23 Solution of Multiple Linear Systems by LU Factorization (Python Coding)

```python
# Solves multiple linear systems by LU factorization
from random import *
from linsys import *
from matutil import *

n = 5                                                    # order of system
a  = [[0]*(n+1) for i in range(n+1)]                     # system matrix
a0 = [[0]*(n+1) for i in range(n+1)]                     # backup matrix
b = [0]*(n+1)                                            # constant terms
x = [0]*(n+1)                                            # solution
ipivot = [0]*(n+1)                                       # pivots

for i in range(1,n+1):                                   # random matrix
   for j in range(1,n+1): a[i][j] = a0[i][j] = random()
print("A:")
MatPrint(a,n,n)

det = LUFactor(a,ipivot,n)                               # LU decomposition of a
print("LU decomposition:")
MatPrint(a,n,n)

for k in range(1,n+1):                                   # loop over constant vectors
   for i in range(1,n+1): b[i] = x[i] = random()         # random constant terms
   print("b:")
   VecPrint(b,n)

   LUSystem(a,ipivot,x,n)                                # solve linear system

   err = 0e0                                             # check max(Ax-b) = 0
   for i in range(1,n+1):
      erri = -b[i]
      for j in range (1,n+1): erri += a0[i][j] * x[j]    # element i of (Ax-b)
      err = max(err,fabs(erri))
   print("   max(Ax-b) = {0:.1e}".format(err))
```

Listing 7.24 Solution of Multiple Linear Systems by LU Factorization (C/C++ Coding)

```c
// Solves multiple linear systems by LU factorization
#include <stdio.h>
#include "memalloc.h"
#include "linsys.h"
#include "matutil.h"

int main()
{
   double **a, **a0, *b, *x, det, err, erri;
   int *ipivot, i, j, k, n;

   n = 5;                                                // order of system
   a  = Matrix(1,n,1,n);                                 // system matrix
   a0 = Matrix(1,n,1,n);                                 // backup matrix
   b = Vector(1,n);                                      // constant terms
   x = Vector(1,n);                                      // solution
   ipivot = IVector(1,n);                                // pivots
```

```
    for (i=1; i<=n; i++)                                    // random matrix
        for (j=1; j<=n; j++) a[i][j] = a0[i][j] = rand();
    printf("A:\n");
    MatPrint(a,n,n);

    LUFactor(a,ipivot,n,det);                               // LU decomposition of a
    printf("LU decomposition:\n");
    MatPrint(a,n,n);

    for (k=1; k<=n; k++) {                         // loop over constant vectors
        for (i=1; i<=n; i++) b[i] = x[i] = rand();     // random constant terms
        printf("b:\n");
        VecPrint(b,n);

        LUSystem(a,ipivot,x,n);                            // solve linear system

        err = 0e0;                                         // check max(Ax-b) = 0
        for (i=1; i<=n; i++) {
            erri = -b[i];
            for (j=1; j<=n; j++) erri += a0[i][j] * x[j];  // element i of (Ax-b
                )
            err = max(err,fabs(erri));
        }
        printf("    max(Ax-b) = %.1e\n",err);
    }
}
```

PROBLEM 7.3

Write Python and C/C++ programs employing routine `TriDiagSys` for solving a tridiagonal linear system (see Equation 7.78) of order $n = 100$ with the matrix elements generated randomly. Check the solution synthetically, by verifying the vanishing of the maximum residual error $\Delta = \max_i |a_i x_{i-1} + b_i x_i + c_i x_{i+1} - d_i|$.

Solution

The Python implementation is given in Listing 7.25 and the C/C++ version is available as supplementary material (`P07-TriDiagSys1.cpp`).

Listing 7.25 Solution of System with Tridiagonal Matrix by LU Factorization (Python Coding)

```
# Solves system with tridiagonal matrix by LU factorization
from random import *
from linsys import *

n = 100                                          # order of system
a = [0]*(n+1); a0 = [0]*(n+1)                     # lower diagonal
b = [0]*(n+1); b0 = [0]*(n+1)                     # main diagonal
c = [0]*(n+1); c0 = [0]*(n+1)                     # upper diagonal
d = [0]*(n+1); x  = [0]*(n+1)              # constant terms and solution

for i in range(1,n+1):                           # generate system
    if (i > 1): a[i] = a0[i] = random()          # a[i], i=2,n
    if (i < n): c[i] = c0[i] = random()          # c[i], i=1,n-1
    b[i] = b0[i] = random()
    d[i] = x[i]  - random()

TriDiagSys(a,b,c,x,n)                             # solve tridiagonal system
```

```
err = 0e0                                          # check max(Ax-d) = 0
for i in range(1,n+1):
    erri = b0[i]*x[i] - d[i]                        # element i of (Ax-d)
    if (i > 1): erri += a0[i]*x[i-1]
    if (i < n): erri += c0[i]*x[i+1]
    err = max(err,fabs(erri))
print("max(Ax-d) = {0:.1e}".format(err))
```

PROBLEM 7.4

For two nonsingular matrices **A** and **B**, check numerically the identity:

$$\det (\mathbf{A} \cdot \mathbf{B}) = \det (\mathbf{A}) \det (\mathbf{B}) .$$

For this purpose, (a) generate two *random* 2D matrices **A** and **B** of order $n = 100$; (b) use routine Gauss to calculate $\det (\mathbf{A} \cdot \mathbf{B})$ and $\det (\mathbf{A}) \det (\mathbf{B})$; and (c) check the vanishing of the relative difference between the two expressions.

Solution

Owing to its simple elimination phase, the routine Gauss is particularly efficient for calculating determinants. The backward substitution is inhibited by considering $m = 0$ columns of constant terms. The Python implementation is given in Listing 7.26 and the C/C++ version is available as supplementary material (P07-MatDetIdent.cpp).

Listing 7.26   Check of Identity for Determinant of Matrix Product (Python Coding)

```
# Check matrix identity det(A*B) = det(A)*det(B)
from random import *
from linsys import *
from matutil import *

n = 100
a = [[0]*(n+1) for i in range(n+1)]
b = [[0]*(n+1) for i in range(n+1)]
c = [[0]*(n+1) for i in range(n+1)]

for i in range(1,n+1):                      # generate matrices a and b randomly
    for j in range(1,n+1):
        a[i][j] = random()
        b[i][j] = random()

MatProd(a,b,c,n,n,n)                                    # C = A * B
detA  = Gauss(a,b,n,0)                       # m=0: performs only elimination
detB  = Gauss(b,b,n,0)                       # and calculates the determinant
detAB = Gauss(c,b,n,0)

print("det(A)  = {0:9.2e}".format(detA))
print("det(B)  = {0:9.2e}".format(detB))
print("det(AB) = {0:9.2e}".format(detAB))
print("Error   = {0:9.2e}".format(1e0 - detA*detB/detAB))
```

PROBLEM 7.5

Check numerically the matrix identity

$$(\mathbf{A} \cdot \mathbf{B})^{-1} = \mathbf{B}^{-1} \cdot \mathbf{A}^{-1}.$$

To this end, (a) generate two *random* 2D matrices **A** and **B** of order $n = 100$; (b) use routine GaussJordan or MatInv to calculate the inverses of **A** and **B** and evaluate the difference matrix

$$\mathbf{D} = (\mathbf{A} \cdot \mathbf{B})^{-1} - \mathbf{B}^{-1} \cdot \mathbf{A}^{-1},$$

as a numerical approximation of the zero matrix; and (c) determine the maximum absolute element of **D**, as an estimate of the maximum error.

Solution

The Python implementation is given in Listing 7.27 and the C/C++ version is available as supplementary material (P07-MatInvIdent.cpp).

Listing 7.27   Check of Identity for Inverse of Matrix Product (Python Coding)

```
# Check matrix identity: (A*B)^(-1) = B^(-1)*A^(-1)
from random import *
from linsys import *
from matutil import *

n = 100
a = [[0]*(n+1) for i in range(n+1)]
b = [[0]*(n+1) for i in range(n+1)]
c = [[0]*(n+1) for i in range(n+1)]
d = [[0]*(n+1) for i in range(n+1)]

for i in range(1,n+1):                           # generate matrices a and b randomly
    for j in range(1,n+1):
        a[i][j] = random()
        b[i][j] = random()

MatProd(a,b,c,n,n,n)                                          # c = a * b
det = GaussJordan(c,d,n,0)   # det = MatInv(c,n)              # c = (a b)^(-1)

det = GaussJordan(a,d,n,0)   # det = MatInv(a,n)              # a -> a^(-1)
det = GaussJordan(b,d,n,0)   # det = MatInv(b,n)              # b -> b^(-1)
MatProd(b,a,d,n,n,n)                                          # d = b^(-1) a^(-1)

MatDiff(c,d,d,n,n)                               # d = (a b)^(-1) - b^(-1)a^(-1)
print("(A B)^(-1) - B^(-1)A^(-1) (sample):\n")
MatPrint(d,5,5)                                  # print sample of "zero" matrix

err = MatNorm(d,n,n)                             # max(abs(d[i][j]))
print("\nMaximum error = ","{0:8.2e}".format(err))
```

Listing 7.28 Matrix Inversion by Cholesky Factorization (Python Coding)

```python
# Inversion of symmetric positive-definite matrix by Cholesky factorization
from random import *
from linsys import *
from matutil import *

n = 100                                      # order of matrices
a = [[0]*(n+1) for i in range(n+1)]
b = [[0]*(n+1) for i in range(n+1)]
c = [[0]*(n+1) for i in range(n+1)]

for i in range(1,n+1):                       # generate random lower triangular matrix
    for j in range(1  ,i+1): a[i][j] = random()           # lower triangle
    for j in range(i+1,n+1): a[i][j] = 0e0                # upper triangle
    a[i][i] += 1e0                                        # increase diagonal

MatCopy(a,b,n,n)
MatTrans(b,n)                        # create upper triangular matrix b = a^T
MatProd(a,b,c,n,n,n)                 # c = a a^T symmetric positive-definite
MatCopy(c,a,n,n)
MatCopy(c,b,n,n)

det = MatSymInv(a,n)                          # inverse by Cholesky factorization

MatProd(a,b,c,n,n,n)                              # c = a^(-1)a  "unit" matrix
print("A^(-1)A (sample):\n")
MatPrint(c,5,5)                                  # print sample of "unit" matrix

for i in range(1,n+1): c[i][i] -= 1e0            # transform to "zero" matrix
err = MatNorm(c,n,n)                                 # max(abs(c[i][j]))
print("\nMaximum error = ","{0:8.2e}".format(err))
```

PROBLEM 7.6

Calculate the inverse of a symmetric positive-definite matrix using the Cholesky factorization $\mathbf{A} = \mathbf{L} \cdot \mathbf{L}^T$. To this end, (a) generate a *random* 2D lower triangular matrix $\mathbf{L}$ of order $n = 100$; (b) generate the positive-definite matrix $\mathbf{A}$ by "inverse" Cholesky factorization $\mathbf{A} = \mathbf{L} \cdot \mathbf{L}^T$; (c) use routine MatSymInv to calculate the inverse $\mathbf{A}^{-1}$; and (d) print out the maximum absolute value of the elements of the "zero" matrix $\mathbf{A}^{-1}\mathbf{A} - \mathbf{I}$, as an estimate of the maximum error (typically, of the order 10^{-12}).

Solution

The Python implementation is given in Listing 7.28 and the C/C++ version is available as supplementary material (P07-MatSymInv.cpp).

PROBLEM 7.7

The Daubechies D4 wavelet transform (named after the mathematician Ingrid Daubechies) is characterized by four wavelet and scaling function coefficients:

$$c_0 = \frac{1+\sqrt{3}}{4\sqrt{2}}, \quad c_1 = \frac{3+\sqrt{3}}{4\sqrt{2}}, \quad c_2 = \frac{3-\sqrt{3}}{4\sqrt{2}}, \quad c_3 = \frac{1-\sqrt{3}}{4\sqrt{2}}.$$

For a particular signal sampled at eight points $y_0, y_1, \ldots, y_7$, with a view to finding the inverse wavelet transform, the following system of linear equations has to be solved (Bordeianu et al. 2008, 2009):

$$
\begin{bmatrix}
c_0 & c_1 & c_2 & c_3 & 0 & 0 & 0 & 0 \\
c_3 & -c_2 & c_1 & -c_0 & 0 & 0 & 0 & 0 \\
0 & 0 & c_0 & c_1 & c_2 & c_3 & 0 & 0 \\
0 & 0 & c_3 & -c_2 & c_1 & -c_0 & 0 & 0 \\
0 & 0 & 0 & 0 & c_0 & c_1 & c_2 & c_3 \\
0 & 0 & 0 & 0 & c_3 & -c_2 & c_1 & -c_0 \\
c_2 & c_3 & 0 & 0 & 0 & 0 & c_0 & c_1 \\
c_1 & -c_0 & 0 & 0 & 0 & 0 & c_3 & -c_2
\end{bmatrix}
\begin{bmatrix}
y_0 \\ y_1 \\ y_2 \\ y_3 \\ y_4 \\ y_5 \\ y_6 \\ y_7
\end{bmatrix}
=
\begin{bmatrix}
143 \\ 1543 \\ 220 \\ -403 \\ 591 \\ 50 \\ 68 \\ -58
\end{bmatrix}.
$$

Solve this system of equations using the routine `GaussSeidel`.

Listing 7.29   Solves Linear System for Daubechies D4 Wavelet (Python Coding)

```python
# Solves linear system for Daubechies D4 wavelet by Gauss-Seidel iteration
from math import *
from linsys import *
from matutil import *

n = 8
a = [[0]*(n+1) for i in range(n+1)]                    # coefficient matrix
b = [0]*(n+1)                                          # constant terms
x = [0]*(n+1)                                          # solution

sqrt3 = sqrt(3e0)
f = 1e0/(4*sqrt(2e0))
c0 = (1e0 + sqrt3) * f                    # Daubechies D4 wavelet coefficients
c1 = (3e0 + sqrt3) * f
c2 = (3e0 - sqrt3) * f
c3 = (1e0 - sqrt3) * f

for i in range(1,n+1):
   for j in range(1,n+1): a[i][j] = 0e0

a[1][1] = c0; a[1][2] = c1; a[1][3] = c2; a[1][4] = c3; b[1] =  143e0
a[2][1] = c3; a[2][2] =-c2; a[2][3] = c1; a[2][4] =-c0; b[2] = 1543e0
a[3][3] = c0; a[3][4] = c1; a[3][5] = c2; a[3][6] = c3; b[3] = 2200e0
a[4][3] = c3; a[4][4] =-c2; a[4][5] = c1; a[4][6] =-c0; b[4] = -403e0
a[5][5] = c0; a[5][6] = c1; a[5][7] = c2; a[5][8] = c3; b[5] =  591e0
a[6][5] = c3; a[6][6] =-c2; a[6][7] = c1; a[6][8] =-c0; b[6] =   50e0
a[7][1] = c2; a[7][2] = c3; a[7][7] = c0; a[7][8] = c1; b[7] =   68e0
a[8][1] = c1; a[8][2] =-c0; a[8][7] = c3; a[8][8] =-c2; b[8] =  -58e0

for i in range(1,n+1): x[i] = 1e0         # initial approximation of solution

err = GaussSeidel(a,b,x,n,0)

print("Solution:")
for i in range (1,n+1): print('{0:10.3f}'.format(x[i]),end="")
print()

err = 0e0                                              # check solution
for i in range(1,n+1):
   erri = -b[i]
   for j in range(1,n+1): erri += a[i][j] * x[j]        # element i of (Ax-b)
   err = max(err,fabs(erri))                            # err = max(Ax-b)
print("\nMaximum error = ","{0:8.2e}".format(err))
```

Solution

The Python implementation is given in Listing 7.29 and the C/C++ version is available as supplementary material (`P07-Wavelet.cpp`).

References and Suggested Further Reading

Beu, T. A. 2004. *Numerical Calculus in C* (3rd ed., in Romanian). Cluj-Napoca: MicroInformatica.

Bordeianu, C. C., R. H. Landau, and M. J. Pàez. 2008. *A Survey of Computational Physics*. Princeton, NJ: Princeton University Press.

Bordeianu, C. C., R. H. Landau, and M. J. Pàez. 2009. Wavelet analyses and applications. *European Journal of Physics* 30(5), 1049–1062.

Burden, R. and J. Faires. 2010. *Numerical Analysis* (9th ed.). Boston: Brooks/Cole, Cengage Learning.

Demidovich, B. and I. Maron. 1987. *Computational Mathematics* (4th ed.). Moscow: MIR Publishers.

Grcar, J. F. 2011. Mathematicians of Gaussian elimination. *Notices of the AMS* 58(6), 782–792.

Press, W. H., S. A. Teukolsky, W. T. Vetterling, and B. P. Flannery. 2007. *Numerical Recipes: The Art of Scientific Computing* (3rd ed.). Cambridge: Cambridge University Press.

Toma, M. and I. Odăgescu. 1980. *Numerical Methods and Subroutines* (in Romanian). Bucharest: Editura tehnică.

8

Eigenvalue Problems

8.1 Introduction

Many fundamental problems of applied sciences and engineering are formulated naturally as eigenvalue problems. In general terms, an eigenvalue equation is a homogeneous operator or matrix equation, featuring a *parameter*, whose admissible values—the *eigenvalues*—need to be determined simultaneously with the corresponding solutions, termed *eigenvectors* or *eigenfunctions*.

For instance, the eigenvectors of the inertia tensor enable placing a rigid body in its system of *principal axes*, which greatly simplifies the description of its dynamics. The characterization of the *vibration eigenmodes* of complex structures plays a key role in diverse areas, ranging from structural engineering to molecular spectroscopy, and implies solving the eigenvalue problem of the so-called *stiffness or dynamic matrix*. Even phonon dispersion relations for crystal lattices result from similar eigenvalue problems. The *Schrödinger equation*, which is central to quantum mechanics, is in its time-independent form a differential eigenvalue equation. Upon expanding the eigenfunctions in terms of complete sets of basis functions, it transforms into an eigenvalue problem for the real symmetric Hamiltonian matrix. The electronic structure of molecular systems is calculated in quantum chemistry precisely from the solutions of such matrix eigenvalue problems.

In general, matrix eigenvalue problems are representations of operator equations. A nonzero vector ξ is called *eigenvector* of a linear operator $\mathcal{A}$ if, by the action of the operator, it transforms into a collinear vector:

$$\mathcal{A}\xi = \lambda\xi. \tag{8.1}$$

The scalar λ is named *eigenvalue* of the operator $\mathcal{A}$ corresponding to the eigenvector ξ. Obviously, apart from their magnitude, the eigenvectors are invariant to the operator's action.

Using a complete set of basis vectors $\{v_i, i = 1, \ldots, n\}$, one can represent in $\mathbb{R}^n$ the operator $\mathcal{A}$ by the associated $(n \times n)$ matrix $\mathbf{A} = [a_{ij}]_{nn}$, having as elements the scalar products $a_{ij} = \langle v_i, \mathcal{A}v_j \rangle$, and the eigenvector ξ, by the column vector $\mathbf{x} = [x_i]_n$, with $x_i = \langle v_i, \xi \rangle$. The operator equation (8.1) is thus represented by the matrix eigenvalue problem

$$\mathbf{A} \cdot \mathbf{x} = \lambda\mathbf{x}, \tag{8.2}$$

or, equivalently, by the linear system

$$\begin{bmatrix} a_{11} & a_{12} & \cdots & a_{1n} \\ a_{21} & a_{22} & \cdots & a_{2n} \\ \vdots & \vdots & \ddots & \vdots \\ a_{n1} & a_{n2} & \cdots & a_{nn} \end{bmatrix} \begin{bmatrix} x_1 \\ x_2 \\ \vdots \\ x_n \end{bmatrix} = \lambda \begin{bmatrix} x_1 \\ x_2 \\ \vdots \\ x_n \end{bmatrix}. \tag{8.3}$$

Rewriting Equation 8.2 as

$$(\mathbf{A} - \lambda\mathbf{E}) \cdot \mathbf{x} = 0 \tag{8.4}$$

emphasizes the *characteristic matrix* $\mathbf{A} - \lambda\mathbf{E}$ and the fact that the eigenvalue problem is actually a *homogeneous* linear system, which has nontrivial solutions if and only if its determinant, referred to as *characteristic (secular) determinant*, is equal to zero:

$$\det(\mathbf{A} - \lambda\mathbf{E}) = 0. \tag{8.5}$$

This so-called *characteristic (secular) equation* has the detailed form:

$$\begin{vmatrix} a_{11} - \lambda & a_{12} & \cdots & a_{1n} \\ a_{21} & a_{22} - \lambda & \cdots & a_{2n} \\ \vdots & \vdots & \ddots & \vdots \\ a_{n1} & a_{n2} & \cdots & a_{nn} - \lambda \end{vmatrix} = 0. \tag{8.6}$$

Being equivalent to a polynomial equation of order n in λ, the secular equation has in principle n real or complex roots $\lambda_1, \lambda_2, \ldots, \lambda_n$, distinct or not, which are precisely the eigenvalues of the matrix $\mathbf{A}$. Calculating the roots of the expanded characteristic determinant can be accomplished by any general method for polynomial equations. The ensemble of all the eigenvalues forms the *spectrum* of the matrix $\mathbf{A}$.

Once the eigenvalues are known, determining the eigenvectors can be accomplished in the most elementary manner by replacing in sequence each eigenvalue λ_j in the homogeneous linear system (8.3), the obtained solution (expressed in terms of one arbitrary component) representing the corresponding eigenvector $\mathbf{x}^{(j)}$ of the matrix $\mathbf{A}$. One can proceed analogously for all the eigenvalues, calculating the respective eigenvectors. In practice, however, there are much more efficient numerical approaches.

Denoting by $\mathbf{X}$ the matrix having as columns the eigenvectors $\mathbf{x}^{(j)}$, the n matrix equations (8.3) resulting by the concrete substitution of the eigenvalues λ_j and eigenvectors $\mathbf{x}^{(j)}$ $(j = 1, 2, \ldots, n)$ can be written synthetically as

$$\mathbf{A} \cdot \mathbf{X} = \mathbf{X} \cdot \Lambda, \tag{8.7}$$

where

$$\mathbf{X} = \begin{bmatrix} x_1^{(1)} & x_1^{(2)} & x_1^{(n)} \\ x_2^{(1)} & x_2^{(2)} & x_2^{(n)} \\ & & \\ x_n^{(1)} & x_n^{(2)} & x_n^{(n)} \end{bmatrix}, \quad \Lambda = \begin{bmatrix} \lambda_1 & 0 & 0 \\ 0 & \lambda_2 & 0 \\ 0 & 0 & \lambda_n \end{bmatrix}. \tag{8.8}$$

Matrix $\mathbf{X}$ is called *modal matrix* and Equation 8.7 is known as *modal equation*.

For a well-behaved matrix $\mathbf{A}$, it is straightforward to show that the modal matrix $\mathbf{X}$ is composed of linearly independent columns (as eigenvectors of $\mathbf{A}$), is therefore nonsingular ($\det \mathbf{X} \neq 0$) and admits an inverse, $\mathbf{X}^{-1}$. Consequently, the modal equation can be reshaped as

$$\mathbf{X}^{-1} \cdot \mathbf{A} \cdot \mathbf{X} = \Lambda, \tag{8.9}$$

and, provided the eigenvectors of $\mathbf{A}$ are effectively known, the corresponding eigenvalues are the diagonal elements of the matrix $\mathbf{X}^{-1} \cdot \mathbf{A} \cdot \mathbf{X}$.

8.2 Diagonalization of Matrices by Similarity Transformations

Matrices associated with the same linear operator $\mathcal{A}$ relative to different bases are called *similar*. In particular, it can be shown that:

Theorem 8.1

Two real ($n \times n$) matrices, $\mathbf{A}, \mathbf{A}' \in \mathbf{M}_{\mathbb{R}}^{n \times n}$, are similar if and only if they are connected by a *similarity transformation*,

$$\mathbf{A}' = \mathbf{S}^{-1} \cdot \mathbf{A} \cdot \mathbf{S}, \tag{8.10}$$

where $\mathbf{S} \in \mathbf{M}_{\mathbb{R}}^{n \times n}$ is an invertible matrix.

Indeed, assuming that the action of the operator is described relative to a first basis by the matrix $\mathbf{A}$ and with respect to a second basis by another matrix $\mathbf{A}'$, we have

$$\mathbf{y} = \mathbf{A} \cdot \mathbf{x}, \quad \mathbf{y}' = \mathbf{A}' \cdot \mathbf{x}', \tag{8.11}$$

where $\mathbf{x}$ and $\mathbf{x}'$, on the one hand, and $\mathbf{y}$ and $\mathbf{y}'$, on the other, are the column matrix representations of the same vectors relative to the first basis and, respectively, to the second one. If $\mathbf{S}$ is the nonsingular matrix performing the basis change, then

$$\mathbf{x} = \mathbf{S} \cdot \mathbf{x}', \quad \mathbf{y} = \mathbf{S} \cdot \mathbf{y}'. \tag{8.12}$$

Plugging Relations 8.12 into the first of Equations 8.11, it follows $\mathbf{y}' = (\mathbf{S}^{-1} \cdot \mathbf{A} \cdot \mathbf{S}) \cdot \mathbf{x}'$, and, by comparing this result with the second of Equations 8.11, one obtains just Equation 8.10, which provides the necessary and sufficient condition for two matrices to be similar.

Theorem 8.2

Two similar matrices have identical eigenvalues.

Indeed, let λ be an eigenvalue of matrix $\mathbf{A}$, and $\mathbf{x}$, the corresponding eigenvector. By performing a basis change with the nonsingular matrix $\mathbf{S}$, we have $\mathbf{x} = \mathbf{S} \cdot \mathbf{x}'$ and the eigenvalue equation may be reformulated as $(\mathbf{S}^{-1} \cdot \mathbf{A} \cdot \mathbf{S}) \cdot \mathbf{x}' = \lambda \mathbf{x}'$, which proves the assertion.

By way of consequence, two similar matrices have identical eigenvalues and the corresponding eigenvectors can be deduced from one another by means of a similarity transformation. Specifically, from the reformulation (8.9) of the modal equation it follows that the matrices $\mathbf{A}$ and $\mathbf{\Lambda}$ are similar. Matrices similar with diagonal matrices are termed *diagonalizable*.

The essence of the *diagonalization of matrices* consists in the fact that, provided a transformation matrix can be found which diagonalizes the considered matrix, the sought eigenvalues will reside on the main diagonal of the transformed matrix, while the columns of the transformation matrix (which can be identified with the modal matrix) will represent precisely the corresponding eigenvectors of the original matrix.

8.3 Jacobi Method

As mentioned in the introduction to this chapter, the solution of eigenvalue problems for *real symmetric matrices* is of utmost interest in many areas of applied sciences and engineering. Moreover, the effective utilization of the symmetry of the involved matrices leads to the considerable simplification of the developed numerical methods, which thus gain stability, accuracy, and efficiency (both in terms of operation cost and memory usage).

An essential property of the eigenvectors of real symmetric matrices is that they are real and orthogonal. As a consequence, the modal matrix $\mathbf{X}$ having them as columns is orthogonal and, implicitly, the diagonalization of a real symmetric matrix can be accomplished by means of an *orthogonal transformation matrix* $\mathbf{R}$, defined by

$$\mathbf{R} \cdot \mathbf{R}^T = \mathbf{R}^T \cdot \mathbf{R} = \mathbf{E}.$$

The main benefit of this orthogonality property is that it allows for the expensive inversion operation of matrix $\mathbf{R}$ to be replaced in the modal equation (8.9) by the significantly simpler transposition operation, $\mathbf{R}^{-1} = \mathbf{R}^T$. As such, the diagonalization of matrix $\mathbf{A}$ is achieved by the *orthogonal similarity transformation*

$$\mathbf{R}^T \cdot \mathbf{A} \cdot \mathbf{R} = \Lambda. \tag{8.13}$$

The Jacobi method is conceived as an infinite iterative algorithm, which essentially consists in applying a *sequence* of orthogonal transformations to matrix $\mathbf{A}$, with a view to gradually reducing its off-diagonal elements until they vanish numerically. Suggested by the two-dimensional rotations (discussed below), which are the simplest orthogonal transformations, the elementary transformations have the form of plane rotations involving pairs of coordinates and aim at zeroing two symmetric off-diagonal elements. As a detrimental side effect, the action of the elementary rotations cannot be restricted to the two targeted elements and they actually destroy previously created off-diagonal zeros. However, there results a gradual overall reduction of the off-diagonal elements, until the transformed matrix becomes diagonal in the limit of roundoff errors and can be identified with matrix Λ. In the end, the product of the successive rotation matrices provides the modal matrix, having on the columns the eigenvectors of matrix $\mathbf{A}$, which in turn correspond to the eigenvalues located on the diagonal of matrix Λ.

Considering first, for simplicity, the case of a (2×2) matrix $\mathbf{A}$, its orthogonal transformation can be accomplished by means of the rotation matrix

$$\mathbf{R} = \begin{bmatrix} \cos\varphi & -\sin\varphi \\ \sin\varphi & \cos\varphi \end{bmatrix}, \tag{8.14}$$

which performs a rotation of angle φ of the system of basis vectors (the unit vectors of the Cartesian axes). The thus defined rotation matrix $\mathbf{R}$ is orthogonal irrespective of the angle φ and, therefore, there can be imposed the additional condition on φ that the transformation $\mathbf{R}$ diagonalizes the matrix $\mathbf{A}$. The resulting matrix may be formally written:

$$\mathbf{A}' = \mathbf{R}^T \cdot \mathbf{A} \cdot \mathbf{R},$$

and its elements amount to

$$\begin{cases} a'_{11} = a_{11}\cos^2\varphi + 2a_{21}\sin\varphi\cos\varphi + a_{22}\sin^2\varphi, \\ a'_{22} = a_{11}\sin^2\varphi - 2a_{21}\sin\varphi\cos\varphi + a_{22}\cos^2\varphi, \\ a'_{21} = a_{21}(\cos^2\varphi - \sin^2\varphi) + (a_{22} - a_{11})\sin\varphi\cos\varphi = a'_{12}. \end{cases}$$

Here we have made use of the fact that the orthogonal transformation conserves the symmetric character of the matrix.

By imposing now explicitly the vanishing of the symmetric off-diagonal elements of $\mathbf{A}'$ ($a'_{21} = a'_{12} = 0$), the following algebraic equation results for $\cot\varphi$:

$$\cot^2\varphi + \frac{a_{22} - a_{11}}{a_{21}}\cot\varphi - 1 = 0.$$

One of the mathematically equivalent ways of expressing the solution, which will reveal its virtues later on, is

$$\tan \varphi = \left[\frac{a_{11} - a_{22}}{2a_{21}} \pm \sqrt{\left(\frac{a_{11} - a_{22}}{2a_{21}}\right)^2 + 1} \right]^{-1}.$$

Rather than expressing herefrom the angle φ itself, one can determine numerically most efficiently the function values $\cos \varphi = (1 + \tan \varphi)^{-1/2}$ and $\sin \varphi = \tan \varphi \cos \varphi$, which are necessary for specifying both the matrices $\mathbf{R}$ and $\mathbf{A}'$.

Finally, the eigenvalues of $\mathbf{A}$ are obtained by evaluating the diagonal elements of the transformed matrix $\mathbf{A}'$ using the determined $\cos \varphi$ and $\sin \varphi$, while the corresponding eigenvectors are the columns of the rotation matrix $\mathbf{R}$:

$$\lambda_1 = a'_{11}, \quad \mathbf{x}^{(1)} = \begin{bmatrix} \cos \varphi \\ \sin \varphi \end{bmatrix},$$

$$\lambda_2 = a'_{22}, \quad \mathbf{x}^{(2)} = \begin{bmatrix} -\sin \varphi \\ \cos \varphi \end{bmatrix}.$$

To sum up this case, the diagonalization of (2×2) matrices can be performed *exactly*, by a single orthogonal transformation.

In the n-dimensional space, the rotation matrix can be generalized to operate with respect to the axis defined by the pair of coordinates i and j,

$$\mathbf{R}(i,j) = \begin{bmatrix} 1 & \vdots & & \vdots & 0 \\ \cdots & \cos \varphi & \cdots & -\sin \varphi & \cdots \\ & \vdots & \ddots & \vdots & \\ \cdots & \sin \varphi & \cdots & \cos \varphi & \cdots \\ 0 & \vdots & & \vdots & 1 \end{bmatrix} \begin{matrix} \\ \text{row } i \\ \\ \text{row } j \\ \\ \end{matrix} \tag{8.15}$$

$$\text{column } i \qquad \text{column } j$$

differing from the unit matrix only by the four nonzero elements $a_{ii} = a_{jj} = \cos \varphi$ and $a_{ij} = -a_{ji} = -\sin \varphi$, located at the intersection of rows and columns i and j. The rotation matrix $\mathbf{R}(i,j)$ is meant to carry out the orthogonal transformation

$$\mathbf{A}' = \mathbf{R}^T(i,j) \cdot \mathbf{A} \cdot \mathbf{R}(i,j), \tag{8.16}$$

which makes the off-diagonal elements a'_{ij} and a'_{ji} vanish.

The only elements of $\mathbf{A}'$ which differ from those of $\mathbf{A}$ are located on the rows and columns i and j, since, by left-multiplying $\mathbf{A}$ with $\mathbf{R}^T(i,j)$, *only rows* i and j get modified, while by the subsequent right-multiplication with $\mathbf{R}(i,j)$, *only columns* i and j are affected. In fact, defining the intermediate matrix $\tilde{\mathbf{A}}$, we can rewrite the transformations (8.16) as

$$\tilde{\mathbf{A}} = \mathbf{A} \cdot \mathbf{R}(i,j), \quad \mathbf{A}' = \mathbf{R}^T(i,j) \cdot \tilde{\mathbf{A}}. \tag{8.17}$$

The detailed calculations for the matrix $\tilde{\mathbf{A}}$,

$$
\tilde{\mathbf{A}} \equiv
\begin{bmatrix}
& \tilde{a}_{1i} & & \tilde{a}_{1j} & \\
& \vdots & & \vdots & \\
\cdots & \tilde{a}_{ki} & \cdots & \tilde{a}_{kj} & \cdots \\
& \vdots & & \vdots & \\
& \tilde{a}_{ni} & & \tilde{a}_{nj} &
\end{bmatrix}
$$

$$
=
\begin{bmatrix}
& a_{1i} & & a_{1j} & \\
& \vdots & & \vdots & \\
\cdots & a_{ki} & \cdots & a_{kj} & \cdots \\
& \vdots & & \vdots & \\
& a_{ni} & & a_{nj} &
\end{bmatrix}
\cdot
\begin{bmatrix}
1 & \vdots & & \vdots & 0 \\
\cdots & \cos\varphi & \cdots & -\sin\varphi & \cdots \\
& \vdots & \ddots & \vdots & \\
\cdots & \sin\varphi & \cdots & \cos\varphi & \cdots \\
0 & \vdots & & \vdots & 1
\end{bmatrix},
$$

yield relevant elements differing from those of $\mathbf{A}$ located solely on columns i and j:

$$
\begin{cases}
\tilde{a}_{ki} = & a_{ki}\cos\varphi + a_{kj}\sin\varphi, \quad k = 1, 2, \ldots, n, \\
\tilde{a}_{kj} = & -a_{ki}\sin\varphi + a_{kj}\cos\varphi.
\end{cases}
\tag{8.18}
$$

Furthermore, the matrix $\mathbf{A}'$ may be expressed as

$$
\mathbf{A}' \equiv
\begin{bmatrix}
& & \vdots & & \\
a'_{i1} & \cdots & a'_{ik} & \cdots & a'_{in} \\
& & \vdots & & \\
a'_{j1} & \cdots & a'_{jk} & \cdots & a'_{jn} \\
& & \vdots & &
\end{bmatrix}
$$

$$
=
\begin{bmatrix}
1 & \vdots & & \vdots & 0 \\
\cdots & \cos\varphi & \cdots & \sin\varphi & \cdots \\
& \vdots & \ddots & \vdots & \\
\cdots & -\sin\varphi & \cdots & \cos\varphi & \cdots \\
0 & \vdots & & \vdots & 1
\end{bmatrix}
\cdot
\begin{bmatrix}
& & \vdots & & \\
\tilde{a}_{i1} & \cdots & \tilde{a}_{ik} & \cdots & \tilde{a}_{in} \\
& & \vdots & & \\
\tilde{a}_{j1} & \cdots & \tilde{a}_{jk} & \cdots & \tilde{a}_{jn} \\
& & \vdots & &
\end{bmatrix},
$$

and has as relevant elements differing from those of $\tilde{\mathbf{A}}$ situated only on rows i and j:

$$
\begin{cases}
a'_{ik} = & \tilde{a}_{ik}\cos\varphi + \tilde{a}_{jk}\sin\varphi, \quad k = 1, 2, \ldots, n, \\
a'_{jk} = & -\tilde{a}_{ik}\sin\varphi + \tilde{a}_{jk}\cos\varphi.
\end{cases}
\tag{8.19}
$$

To express the modified elements of the transformed matrix $\mathbf{A}'$ in terms of those of the initial matrix $\mathbf{A}$, one has to bear in mind that the elements at the intersections of rows and columns i and j result by applying both transformations (8.18) and (8.19), whereas the rest of the elements of these rows and columns are obtained by applying just one of the transformations. Considering also the symmetry of $\mathbf{A}$

and $\mathbf{A}'$, we find that Relations 8.18 and 8.19 actually share the same structure. Finally, the elements of the matrix $\mathbf{A}'$ modified as a result of the transformation $\mathbf{R}(i, j)$ may be cast under the form:

$$
\begin{cases}
a'_{ik} = a'_{ki} = \quad a_{ik} \cos \varphi + a_{jk} \sin \varphi, \quad k = 1, 2, \ldots, n, \\
a'_{jk} = a'_{kj} = -a_{ik} \sin \varphi + a_{jk} \cos \varphi, \quad k \neq i, j, \\
a'_{ii} = a_{ii} \cos^2 \varphi + 2a_{ji} \sin \varphi \cos \varphi + a_{jj} \sin^2 \varphi, \\
a'_{jj} = a_{ii} \sin^2 \varphi - 2a_{ji} \sin \varphi \cos \varphi + a_{jj} \cos^2 \varphi, \\
a'_{ij} = a'_{ji} = a_{ji}(\cos^2 \varphi - \sin^2 \varphi) + (a_{jj} - a_{ii}) \sin \varphi \cos \varphi.
\end{cases} \tag{8.20}
$$

Imposing the condition that the elements a'_{ij} and a'_{ji} given by the last relation vanish, results in a second-order algebraic equation in $\cot \varphi$:

$$
\cot^2 \varphi + \frac{a_{jj} - a_{ii}}{a_{ji}} \cot \varphi - 1 = 0. \tag{8.21}
$$

Its solution can be inverted to give

$$
\tan \varphi = \left[\frac{a_{ii} - a_{jj}}{2a_{ji}} \pm \sqrt{\left(\frac{a_{ii} - a_{jj}}{2a_{ji}} \right)^2 + 1} \right]^{-1}, \tag{8.22}
$$

by the use of which one can evaluate the functions

$$
\cos \varphi = \frac{1}{\sqrt{1 + \tan^2 \varphi}}, \quad \sin \varphi = \cos \varphi \tan \varphi, \tag{8.23}
$$

necessary for specifying the transformed elements from Formulas 8.20. Still, one is faced with the problem of justifying the choice of one of the signs in Equation 8.22.

On the basis of the pattern defined by Equation 8.16, one can define a sequence of similar matrices,

$$
\mathbf{A}_0 = \mathbf{A}, \quad \mathbf{A}_1 = \mathbf{R}_1^T \cdot \mathbf{A}_0 \cdot \mathbf{R}_1, \ldots, \quad \mathbf{A}_l = \mathbf{R}_l^T \cdot \mathbf{A}_{l-1} \cdot \mathbf{R}_l, \ldots, \tag{8.24}
$$

which, by virtue of Theorem 8.2, have the same set of eigenvalues. At each stage, the rotation $\mathbf{R}_l \equiv \mathbf{R}(i_l, j_l)$ corresponds to a sequentially chosen combination of indexes ($i_l \neq j_l$) and its purpose is to make the two symmetric off-diagonal elements located at the intersection of the rows and columns i_l and j_l vanish. Even though, according to Equations 8.20, possible zeros created by previous transformations are partially destroyed, they are nevertheless replaced by linear combinations of off-diagonal elements, weighted with subunitary functions ($\sin \varphi_l$ and $\cos \varphi_l$) and, hence, an overall reduction of the off-diagonal elements results.

By defining the orthogonal matrices

$$
\mathbf{X}_l = \mathbf{R}_0 \cdot \mathbf{R}_1 \cdots \mathbf{R}_l, \quad l = 0, 1, 2, \ldots, \tag{8.25}
$$

with $\mathbf{X}_0 \equiv \mathbf{R}_0 \equiv \mathbf{E}$, the sequence of orthogonal transformations (8.24) may be rewritten:

$$
\mathbf{A}_0 = \mathbf{A}, \quad \mathbf{A}_1 = \mathbf{X}_1^T \cdot \mathbf{A} \cdot \mathbf{X}_1, \ldots, \quad \mathbf{A}_l = \mathbf{X}_l^T \cdot \mathbf{A} \cdot \mathbf{X}_l, \ldots \tag{8.26}
$$

At the limit, $\mathbf{A}_l$ is identified with the matrix Λ, having on the diagonal the eigenvalues of the original matrix $\mathbf{A}$, while $\mathbf{X}_l$ is identified with the modal matrix $\mathbf{X}$, having on the columns the corresponding eigenvectors:

$$
\lim_{l \to \infty} \mathbf{A}_l = \Lambda, \quad \lim_{l \to \infty} \mathbf{X}_l = \mathbf{X}.
$$

Establishing the recurrence relation

$$\mathbf{X}_l = \mathbf{X}_{l-1} \cdot \mathbf{R}_l(i,j),$$

the modified elements of $\mathbf{X}_l$ can be obtained in terms of the elements of $\mathbf{X}_{l-1}$ from equations similar to (8.18):

$$\begin{cases} x'_{ki} = x_{ki}\cos\varphi + x_{kj}\sin\varphi, & k = 1,2,\ldots,n, \\ x'_{kj} = -x_{ki}\sin\varphi + x_{kj}\cos\varphi. \end{cases} \tag{8.27}$$

In practical implementations, the sequence of approximations (8.26) is interrupted at some step l, for which the absolute values of the off-diagonal elements of matrix $\mathbf{A}_l$ have dropped in absolute value below a predefined tolerance $\varepsilon > 0$. The stopping condition of this iterative process can be thus formulated as

$$\max_{i \neq j} |a'_{ij}| \leq \varepsilon, \tag{8.28}$$

and, when satisfied, one can consider the diagonal elements of $\mathbf{A}_l$ to approximate "sufficiently well" the eigenvalues of matrix $\mathbf{A}$, and the columns of $\mathbf{X}_l$ to approximate the corresponding eigenvectors.

The discussion about the appropriate sign in Expression 8.22 of $\tan\varphi$ is nontrivial. In fact, the convergence of the Sequence 8.26 and, at the same time, the numerical stability of the method is ensured by choosing at each step in Formula 8.22 the minimal solution ($\tan\varphi \leq 1$), implying a rotation of the basis vectors by a minimal angle ($\varphi \leq \pi/4$). This choice implies maximizing the absolute value of the denominator in Equation 8.22 by considering the "+" sign, if $(a_{ii} - a_{jj})/a_{ji} > 0$, and the "−" sign in the opposite case. Herewith, the tangent of the rotation angle takes the effective form:

$$\tan\varphi = \text{sign}\left(\frac{a_{ii} - a_{jj}}{2a_{ji}}\right) \left[\left|\frac{a_{ii} - a_{jj}}{2a_{ji}}\right| + \sqrt{\left(\frac{a_{ii} - a_{jj}}{2a_{ji}}\right)^2 + 1} \right]^{-1}. \tag{8.29}$$

Regarding the stability of the Jacobi method, interesting insights can be gained by taking a different route to solving Equation 8.21. Rewriting it first as an equation in $\tan\varphi$, it leads directly to the alternative solution

$$\tan\varphi = \frac{a_{jj} - a_{ii}}{2a_{ji}} \pm \sqrt{\left(\frac{a_{jj} - a_{ii}}{2a_{ji}}\right)^2 + 1},$$

which is perfectly equivalent from a pure mathematical perspective with the previous Result 8.22. However, since minimizing this expression implies rewriting it under the form

$$\tan\varphi = \text{sign}\left(\frac{a_{jj} - a_{ii}}{2a_{ji}}\right) \left[\left|\frac{a_{jj} - a_{ii}}{2a_{ji}}\right| - \sqrt{\left(\frac{a_{jj} - a_{ii}}{2a_{ji}}\right)^2 + 1} \right]$$

and thus performing subtractions with operands close in absolute value, using the above relation instead of Equation 8.29 favors the onset of instabilities due to the propagation of the roundoff errors.

Listing 8.1 presents the coding of the Jacobi method for calculating the eigenvalues and eigenvectors of real symmetric matrices according to the algorithm described above.

To reduce the storage requirements, the implementation makes use of the symmetry of the matrices $\mathbf{A}_l$. Specifically, for the $2n(n + 1)/2 = n^2 + n$ relevant elements of both matrices $\mathbf{A}$ and $\mathbf{A}_l$, the routine employs just a single array a with n rows and n columns, supplemented by a vector d with n components. With a view to subsequent calculations, the elements of the original matrix $\mathbf{A}$ are secured in the upper triangle of array a, including the diagonal. The off-diagonal elements of the matrices $\mathbf{A}_l$ are temporarily

Listing 8.1 Solver for Eigenvalue Problems for Symmetric Matrices (Python Coding)

```python
#===========================================================================
def Jacobi(a, x, d, n):
#---------------------------------------------------------------------------
#  Solves the eigenvalue problem of a real symmetric matrix using the
#  Jacobi method
#  a  - real symmetric matrix (lower triangle is destroyed)
#  x  - modal matrix: eigenvectors on columns (output)
#  d  - vector of eigenvalues
#  n  - order of matrix a
#  Error flag: 0 - normal execution, 1 - exceeded max. no. of iterations
#---------------------------------------------------------------------------
   eps = 1e-30                                          # precision criterion
   itmax = 50                                           # max no. of iterations

   for i in range(1,n+1):                               # initialization
      for j in range(1,n+1): x[i][j] = 0e0             # modal matrix = unit matrix
      x[i][i] = 1e0
      d[i] = a[i][i]                                    # eigenvalues = diagonal elements

   for it in range(1,itmax+1):                          # loop of iterations
      amax = 0e0
      for i in range(2,n+1):                            # lower triangle: i > j
         for j in range(1,i):
            aii = d[i]; ajj = d[j]                      # diagonal elements
            aij = fabs(a[i][j])
            if (aij > amax): amax = aij                 # max. non-diagonal element
            if (aij > eps):                             # perform rotation
               c = 0.5e0*(aii-ajj)/a[i][j]             # tangent
               t = 1e0/(fabs(c) + sqrt(1e0+c*c))
               if (c < 0e0): t = -t                     # sign
               c = 1e0/sqrt(1e0+t*t); s = c*t           # cos, sin
               for k in range(1,j):                     # columns k < j
                  t       = a[j][k]*c - a[i][k]*s
                  a[i][k] = a[i][k]*c + a[j][k]*s
                  a[j][k] = t
               for k in range(j+1,i):                   # columns k > j
                  t       = a[k][j]*c - a[i][k]*s        # interchange j <> k
                  a[i][k] = a[i][k]*c + a[k][j]*s
                  a[k][j] = t
               for k in range(i+1,n+1):                  # columns k > i
                  t       = a[k][j]*c - a[k][i]*s        # interchange i <> k
                  a[k][i] = a[k][i]*c + a[k][j]*s        # interchange j <> k
                  a[k][j] = t

               for k in range(1,n+1):                    # transform modal matrix
                  t       = x[k][j]*c - x[k][i]*s
                  x[k][i] = x[k][i]*c + x[k][j]*s
                  x[k][j] = t

               t = 2e0 * s * c * a[i][j]
               d[i] = aii*c*c + ajj*s*s + t              # update eigenvalues
               d[j] = ajj*c*c + aii*s*s - t
               a[i][j] = 0e0

      if (amax<=eps): break                              # check convergence

   if (it > itmax):
      print("Jacobi: max. no. of iterations exceeded !"); return 1
   return 0
```

stored in the lower triangle of array a, while their diagonal elements are placed in vector d. Since the limit of the matrices $\mathbf{A}_l$ approximates the diagonal matrix Λ, the vector d contains on exit the calculated eigenvalues of matrix $\mathbf{A}$.

The array x returns on its columns the eigenvectors of $\mathbf{A}$. During the operation of the routine, array x is used for storing the elements of the successive rotation matrices, being initialized right after the entry into the routine by the unit matrix ($\mathbf{X}_0 = \mathbf{E}$). The justification for this approach is that the modal matrix $\mathbf{X}$ is obtained as the limit of the sequence of transformation matrices $\mathbf{X}_l$.

The local variables t, c, and s store, respectively, the tangent, cosine, and sine of the rotation angle φ characterizing the transformation. eps represents the predefined tolerance, the created off-diagonal numerical "zeros" are compared with, and amax contains after each orthogonal transformation the maximum of their absolute values.

Since, within the adopted strategy of economic storage usage, the upper triangle of the array a is not meant to take part in updating the elements of $\mathbf{A}_l$ based on formulas 8.20, instead of a[j][k] with k > j (from the upper triangle), the symmetric element a[k][j] has to be used, while instead of a[i][k] with k > i, the symmetric element a[k][i], equally from the lower triangle, needs to be considered. However, this methodology requires decomposing the double loop with i > j, in which the elements a'_{ik} and a'_{jk} are evaluated, into three cycles, corresponding to the three possible order relations of the index k with i and j.

Listing 8.2 presents a useful auxiliary routine, EigSort, which reorders by the *selection sort* method (similar to insertion sort; see Chapter 4) the eigenvalues stored in d and the corresponding eigenvectors stored in x according to the eigenvalues—in ascending order if isort>0 and, respectively, in descending order if isort<0.

A typical usage of the functions Jacobi and EigSort, assumed to be included in the module eigsys.py, is illustrated in Listing 8.3. For the correct operation of the routine Jacobi, the specification of just the lower triangle of the matrix $\mathbf{A}$ is sufficient. Nevertheless, with the intention of subsequently checking the accuracy of the calculated solution, the main program symmetrizes the array a, using the upper triangle as backup copy. As an error indicator, the *max norm* of the eigenvalue equation is used,

Listing 8.2   Sorting Eigenvalues and Eigenvectors (Python Coding)

```
#=============================================================================
def EigSort(x, d, n, isort):
#-----------------------------------------------------------------------------
#  Sorts the eigenvalues d and eigenvectors x according to the eigenvalues
#  x   - eigenvectors on columns
#  d   - vector of eigenvalues
#  n   - dimension of eigenvalue problem
#  isort = 0: no sorting; > 0: ascending sorting; < 0: descending sorting
#-----------------------------------------------------------------------------
   if (isort == 0): return

   for j in range(1,n):
      jmax = j; dmax = d[j]                     # find column of maximum eigenvalue
      for i in range(j+1,n+1):
         if (isort * (dmax - d[i]) > 0e0):
            jmax = i; dmax = d[i]

      if (jmax != j):
         d[jmax] = d[j]; d[j] = dmax     # swap current component with maximum
         for i in range(1,n+1):
            t = x[i][j]; x[i][j] = x[i][jmax]; x[i][jmax] = t
```

Listing 8.3   Eigenvalues and Eigenvectors of Symmetric Matrices (Python Coding)

```
# Eigenvalues and eigenvectors of symmetric matrices by the Jacobi method
from eigsys import *

n = 4
a = [[0]*(n+1) for i in range(n+1)]
x = [[0]*(n+1) for i in range(n+1)]
d = [0]*(n+1)

a[1][1] = 1                                              # lower triangle
a[2][1] = 2; a[2][2] = 1
a[3][1] = 3; a[3][2] = 2; a[3][3] = 1
a[4][1] = 4; a[4][2] = 3; a[4][3] = 2; a[4][4] = 1
# Eigenvalues: -3.414214, -1.099019, -0.585786, 9.099020

for i in range(2,n+1):                          # complete upper triangle
    for j in range(1,i): a[j][i] = a[i][j]      # - not altered by Jacobi

Jacobi(a,x,d,n)
EigSort(x,d,n,1)                         # sort eigenvalues and eigenvectors

print("Eigenvalues:")
for i in range(1,n+1): print("{0:10.5f}".format(d[i]),end="")
print()

for i in range(2,n+1):                          # restore original matrix
    for j in range(1,i): a[i][j] = a[j][i]      # from upper triangle

err = 0e0                                        # accuracy check
for i in range(1,n+1):
    for j in range(1,n+1):
        t = -x[i][j] * d[j]
        for k in range(1,n+1): t += a[i][k] * x[k][j]
        if (err < fabs(t)): err = fabs(t)        # err = max|a x - lambda x|

print("\nMaximum error = ","{0:8.2e}".format(err))
```

namely $\|\mathbf{A} \cdot \mathbf{x} - \lambda \mathbf{x}\| = \max \left|(\mathbf{A} \cdot \mathbf{x} - \lambda \mathbf{x})_{ij}\right|$, which is expected to match, in principle, the accuracy of the internal double-precision floating-point representation.

8.4 Generalized Eigenvalue Problems for Symmetric Matrices

In certain problems of applied mathematics, stemming, for example, from the quantum mechanical formalism, the so-called *generalized eigenvalue problems* can arise, which are defined not by a single matrix, but by two square matrices **A** and **B**:

$$\mathbf{A} \cdot \mathbf{x} = \lambda \mathbf{B} \cdot \mathbf{x}. \tag{8.30}$$

In most cases, such problems are matrix representations of operator equations with respect to *nonorthogonal bases* in a finite dimensional space, such as $\mathbb{R}^n$.

Assuming the matrix **B** to be nonsingular, a straightforward possibility of transforming the generalized equation (8.30) into a standard one ($\mathbf{A} \cdot \mathbf{x} = \lambda \mathbf{x}$) consists in left-multiplying the equation with the inverse $\mathbf{B}^{-1}$:

$$(\mathbf{B}^{-1} \cdot \mathbf{A}) \cdot \mathbf{x} = \lambda \mathbf{x}.$$

Unfortunately, even in the particular case of symmetric matrices **A** and **B**, the product $\mathbf{B}^{-1} \cdot \mathbf{A}$ no longer preserves the symmetry and, consequently, the resulting standard eigenvalue problem can no longer be solved by methods specific to symmetric matrices (such as the Jacobi method), and a series of otherwise ensuing advantages are thereby lost.

If the matrices **A** and **B** are *symmetric* and, in addition, **B** is positive definite, the transformation of the generalized problem (8.30) into a standard problem can be accomplished by first performing the *Cholesky factorization* of matrix **B** (see Section 7.6):

$$\mathbf{B} = \mathbf{L} \cdot \mathbf{L}^T, \tag{8.31}$$

where **L** is a lower triangular matrix and $\mathbf{L}^T$ is its transpose. By left-multiplication with $\mathbf{L}^{-1}$ and replacement of the factorization (8.31), Equation 8.30 becomes

$$\mathbf{L}^{-1} \cdot \mathbf{A} \cdot \mathbf{x} = \lambda \mathbf{L}^T \cdot \mathbf{x}.$$

Furthermore, by incorporating the trivial decomposition of the unit matrix

$$\mathbf{E} = \left(\mathbf{L} \cdot \mathbf{L}^{-1}\right)^T = \left(\mathbf{L}^{-1}\right)^T \cdot \mathbf{L}^T$$

into the product $\mathbf{A} \cdot \mathbf{x}$, the eigenvalue problem takes the form:

$$\mathbf{L}^{-1} \cdot \mathbf{A} \cdot \left(\mathbf{L}^{-1}\right)^T \cdot \left(\mathbf{L}^T \cdot \mathbf{x}\right) = \lambda \left(\mathbf{L}^T \cdot \mathbf{x}\right).$$

Making the self-evident notations

$$\widetilde{\mathbf{x}} = \mathbf{L}^T \cdot \mathbf{x} \tag{8.32}$$

and

$$\widetilde{\mathbf{A}} = \mathbf{L}^{-1} \cdot \mathbf{A} \cdot \left(\mathbf{L}^{-1}\right)^T, \tag{8.33}$$

one obtains finally for the *symmetric* matrix $\widetilde{\mathbf{A}}$ the standard eigenvalue problem:

$$\widetilde{\mathbf{A}} \cdot \widetilde{\mathbf{x}} = \lambda \widetilde{\mathbf{x}}. \tag{8.34}$$

The eigenvalues of $\widetilde{\mathbf{A}}$ obviously coincide with those of the original matrix **A**, while the eigenvectors of **A** can be recomposed from those of $\widetilde{\mathbf{A}}$ by the linear transformation

$$\mathbf{x} = \left(\mathbf{L}^{-1}\right)^T \cdot \widetilde{\mathbf{x}}. \tag{8.35}$$

From the standpoint of a practical implementation, the solution of the generalized eigenvalue problem (8.30) can be broken down into the following steps:

1. Perform the Cholesky decomposition $\mathbf{B} = \mathbf{L} \cdot \mathbf{L}^T$.
2. Calculate the inverse $\mathbf{L}^{-1} = [l'_{ij}]_{nn}$.
3. Build the transformed matrix $\widetilde{\mathbf{A}} = \mathbf{L}^{-1} \cdot \mathbf{A} \cdot (\mathbf{L}^{-1})^T$ according to

$$\tilde{a}_{ij} = \sum_{k=1}^{i} \sum_{m=1}^{j} l'_{ik} a_{km} l'_{jm}. \tag{8.36}$$

4. Solve the standard eigenvalue problem $\widetilde{\mathbf{A}} \cdot \widetilde{\mathbf{x}} = \lambda \widetilde{\mathbf{x}}$ by a method suited for symmetric matrices (such as the Jacobi method).

5. Recompose the eigenvectors $\mathbf{x} = (\mathbf{L}^{-1})^T \cdot \tilde{\mathbf{x}}$ of the initial eigenvalue problem (8.30) based on

$$x_{ij} = \sum_{k=i}^{n} l'_{ki} \tilde{x}_{kj}. \qquad (8.37)$$

In the implementation of the above algorithm given in Listing 8.4, the Cholesky factorization of matrix **B** is carried out by calling the function `Cholesky`, while the inversion of the yielded lower triangular matrix **L** is performed by the routine `MatTriInv` (both routines are developed in Chapter 7 and included in the module `linsys.py`). The actual solution of the resulting standard eigenvalue problem (8.34) is performed by calling the function `Jacobi`, described in Section 8.3.

Listing 8.4   Solver for Generalized Eigenvalue Problems for Symmetric Matrices (Python Coding)

```
#===========================================================================
def EigSym(a, b, x, d, n):
#---------------------------------------------------------------------------
#  Solves the generalized eigenvalue problem a x = lambda b x for the real
#  symmetric matrix a and the real positive-definite symmetric matrix b,
#  reducing it to standard form at xt = lambda xt by Cholesky factorization
#  of matrix b = L * LT.
#  a   - real symmetric matrix
#  b   - real positive-definite symmetric matrix (L replaces lower triangle)
#  x   - modal matrix: eigenvectors on columns (output)
#  d   - vector of eigenvalues
#  n   - order of matrix a
#  Error flag: 0 - normal execution
#              1 - matrix b not positive-definite
#  Calls: Cholesky, MatTriInv, Jacobi.
#---------------------------------------------------------------------------
   (det,ierr) = Cholesky(b,n)              # Cholesky factorization b = L * LT
   if (ierr):
      print("EigSym: matrix b is mot positive-definite !"); return 1

   MatTriInv(b,n)                                           # b = L^(-1)

   for i in range(1,n):                             # fill upper triangle of a
      for j in range(i+1,n+1): a[i][j] = a[j][i]

   at = [[0]*(n+1) for i in range(n+1)]
   xt = [[0]*(n+1) for i in range(n+1)]

   for i in range(1,n+1):        # transformed matrix at = L^(-1) * a * LT^(-1)
      for j in range(1,i+1):
         sum = 0e0
         for k in range(1,i+1):
            for m in range(1,j+1): sum += b[i][k] * a[k][m] * b[j][m]
         at[i][j] = sum

   Jacobi(at,xt,d,n)              # solve transformed problem at xt = lambda xt

   for j in range(1,n+1):                              # recompose eigenvectors of
      for i in range(1,n+1):                           # initial eigenvalue problem
         sum = 0e0
         for k in range(i,n+1): sum += b[k][i] * xt[k][j]
         x[i][j] = sum

   return 0
```

The elements of matrices **A** and **B** are passed to the routine `EigSym` by the lower triangles of the arrays a and b. The lower triangle of b is then used to store successively the elements of the factorization matrix **L** and those of the inverse $\mathbf{L}^{-1}$. The allocated arrays at and xt correspond, respectively, to the matrices $\widetilde{\mathbf{A}}$ and $\widetilde{\mathbf{x}}$. On exit from the routine, the array d contains the calculated eigenvalues, while the array x returns on its columns the corresponding eigenvectors.

8.5 Implementations in C/C++

Listing 8.5 shows the content of the file `eigsys.h`, which contains equivalent C/C++ implementations of the Python functions developed in the main text and included in the module `eigsys.py`. The corresponding routines have identical names, parameters, and functionalities.

Listing 8.5   Solvers for Eigenvalue Problems (`eigsys.h`)

```
//--------------------------------- eigsys.h ---------------------------------
// Contains routines for solving eigenvalue problems.
// Part of the numxlib numerics library. Author: Titus Beu, 2013
//----------------------------------------------------------------------------
#ifndef _EIGSYS_
#define _EIGSYS_

#include <stdio.h>
#include <math.h>
#include "memalloc.h"
#include "linsys.h"


//============================================================================
int Jacobi(double **a, double **x, double d[], int n)
//----------------------------------------------------------------------------
// Solves the eigenvalue problem of a real symmetric matrix using the
// Jacobi method
// a  - real symmetric matrix (lower triangle is destroyed)
// x  - modal matrix: eigenvectors on columns (output)
// d  - vector of eigenvalues
// n  - order of matrix a
// Error flag: 0 - normal execution, 1 - exceeded max. no. of iterations
//----------------------------------------------------------------------------
{
   const double eps = 1e-30;                      // precision criterion
   const int itmax = 50;                          // max no. of iterations
   double aii, aij, ajj, amax, c, s, t;
   int i, it, j, k;

   for (i=1; i<=n; i++) {                         // initialization
      for (j=1; j<=n; j++) x[i][j] = 0e0;         // modal matrix = unit matrix
      x[i][i] = 1e0;
      d[i] = a[i][i];                             // eigenvalues = diagonal elements
   }

   for (it=1; it<=itmax; it++) {                  // loop of iterations
      amax = 0e0;
      for (i=2; i<=n; i++)                        // lower triangle: i > j
         for (j=1; j<=(i-1); j++) {
            aii = d[i]; ajj = d[j];               // diagonal elements
            aij = fabs(a[i][j]);
```

```
                 if (aij > amax) amax = aij;           // max. non-diagonal element
                 if (aij > eps) {                               // perform rotation
                    c = 0.5e0*(aii-ajj)/a[i][j];
                    t = 1e0/(fabs(c) + sqrt(1e0+c*c));                    // tangent
                    if (c < 0e0) t = -t;                                    // sign
                    c = 1e0/sqrt(1e0+t*t); s = c*t;                      // cos, sin
                    for (k=1; k<=(j-1); k++) {                    // columns k < j
                       t       = a[j][k]*c - a[i][k]*s;
                       a[i][k] = a[i][k]*c + a[j][k]*s;
                       a[j][k] = t;
                    }
                    for (k=(j+1); k<=(i-1); k++) {                // columns k > j
                       t       = a[k][j]*c - a[i][k]*s;     // interchange j <> k
                       a[i][k] = a[i][k]*c + a[k][j]*s;
                       a[k][j] = t;
                    }
                    for (k=(i+1); k<=n; k++) {                    // columns k > i
                       t       = a[k][j]*c - a[k][i]*s;     // interchange i <> k
                       a[k][i] = a[k][i]*c + a[k][j]*s;     // interchange j <> k
                       a[k][j] = t;
                    }
                    for (k=1; k<=n; k++) {              // transform modal matrix
                       t       = x[k][j]*c - x[k][i]*s;
                       x[k][i] = x[k][i]*c + x[k][j]*s;
                       x[k][j] = t;
                    }
                    t = 2e0 * s * c * a[i][j];
                    d[i] = aii*c*c + ajj*s*s + t;              // update eigenvalues
                    d[j] = ajj*c*c + aii*s*s - t;
                    a[i][j] = 0e0;
                 }
              }
        if (amax<=eps) break;                           // check convergence
     }

     if (it > itmax) {
        printf("Jacobi: max. no. of iterations exceeded !\n"); return 1;
     }
     return 0;
}

//===========================================================================
int EigSym(double **a, double **b, double **x, double d[], int n)
//---------------------------------------------------------------------------
// Solves the generalized eigenvalue problem a x = lambda b x for the real
// symmetric matrix a and the real positive-definite symmetric matrix b,
// reducing it to standard form at xt = lambda xt by Cholesky factorization
// of matrix b = L * LT.
// a  - real symmetric matrix
// b  - real positive-definite symmetric matrix (L replaces lower triangle)
// x  - modal matrix: eigenvectors on columns (output)
// d  - vector of eigenvalues
// n  - order of matrix a
// Error flag: 0 - normal execution
//             1 - matrix b not positive-definite
// Calls: Cholesky, MatTriInv, Jacobi.
//---------------------------------------------------------------------------
{
   double **at, **xt, det, sum;
```

```
   int i, j, k, m;

   if (Cholesky(b,n,det))                        // Cholesky factorization b = L * LT
      { printf("EigSym: matrix b is mot positive-definite !\n"); return 1; }

   MatTriInv(b,n);                                                      // b = L^(-1)

   for (i=1; i<=(n-1); i++)                            // fill upper triangle of a
      for (j=i+1; j<=n; j++) a[i][j] = a[j][i];

   at = Matrix(1,n,1,n);
   xt = Matrix(1,n,1,n);

   for (i=1; i<=n; i++)          // transformed matrix at = L^(-1) * a * LT^(-1)
      for (j=1; j<=i; j++) {
         sum = 0e0;
         for (k=1; k<=i; k++)
            for (m=1; m<=j; m++) sum += b[i][k] * a[k][m] * b[j][m];
         at[i][j] = sum;
      }

   Jacobi(at,xt,d,n);                  // solve transformed problem at xt = lambda xt

   for (j=1; j<=n; j++)                                 // recompose eigenvectors of
      for (i=1; i<=n; i++) {                            // initial eigenvalue problem
         sum = 0e0;
         for (k=i; k<=n; k++) sum += b[k][i] * xt[k][j];
         x[i][j] = sum;
      }

   FreeMatrix(at,1,1);
   FreeMatrix(xt,1,1);
   return 0;
}

//===========================================================================
void EigSort(double **x, double d[], int n, int isort)
//---------------------------------------------------------------------------
// Sorts the eigenvalues d and eigenvectors x according to the eigenvalues
// x - eigenvectors on columns
// d - vector of eigenvalues
// n - dimension of eigenvalue problem
// isort = 0: no sorting; > 0: ascending sorting; < 0: descending sorting
//---------------------------------------------------------------------------
{
   double dmax, t;
   int i, j, jmax;

   if (isort == 0) return;

   for (j=1; j<=n-1; j++) {
      jmax = j; dmax = d[j];                    // find column of maximum eigenvalue
      for (i=j+1; i<=n; i++) {
         if (isort * (dmax - d[i]) > 0e0) {
            jmax = i; dmax = d[i];
         }
      }
      it (jmax != j) {
         d[jmax] = d[j]; d[j] = dmax;   // swap current component with maximum
```

```
            for (i=1; i<=n; i++) {
                t = x[i][j]; x[i][j] = x[i][jmax]; x[i][jmax] = t;
            }
        }
    }
}

#endif
```

8.6 Problems

The Python and C/C++ programs for the following problems may import the functions developed in this chapter from the modules `eigsys.py` and, respectively, `eigsys.h`, which are available as supplementary material. In addition, the utility routines contained in files `matutil.py` and `matutil.h` can be used to perform basic operations with one- and two-dimensional arrays. For creating runtime plots, the graphical routines contained in the libraries `graphlib.py` and `graphlib.h` may be employed.

PROBLEM 8.1

Use the function `Jacobi` to prove for a random invertible symmetric matrix **A** of order $n = 100$ the general property that the eigenvalues of the inverse $\mathbf{A}^{-1}$ are the inverses $1/\lambda_i$ of the eigenvalues λ_i of the original matrix, while the eigenvectors coincide.

Invert the matrix **A** using the routine `MatInv` from the module `linsys.py` (respectively, `linsys.h`), and, employing the function `EigSort`, sort the eigenvectors of **A** and $\mathbf{A}^{-1}$ by the corresponding eigenvalues before comparing them.

Solution

The Python and C/C++ implementations are given in Listings 8.6 and 8.7.

Listing 8.6   Eigenvalues and Eigenvectors of the Inverse of a Symmetric Matrix (Python Coding)

```
# Eigenvalues and eigenvectors of inverse of symmetric matrix
from random import *
from linsys import *
from eigsys import *
from matutil import *

n = 100                                             # order of matrix
a  = [[0]*(n+1) for i in range(n+1)]           # coefficient matrix
a1 = [[0]*(n+1) for i in range(n+1)]                # inverse of a
x  = [[0]*(n+1) for i in range(n+1)]            # eigenvectors of a
x1 = [[0]*(n+1) for i in range(n+1)]           # eigenvectors of a1
d = [0]*(n+1); d1 = [0]*(n+1)              # eigenvalues of a and a1

seed()                              # initialize random number generator
for i in range(1,n+1):              # generate random coefficient matrix
    for j in range(1,i+1):
        a[i][j] = a[j][i] = random()
        a1[i][j] = a1[j][i] = a[i][j]                  # copy a in a1

Jacobi(a,x,d,n)                         # solve eigenvalue problem for a
EigSort(x,d,n,1)                     # sort eigenvalues and eigenvectors

MatInv(a1,n)                                          # a1 = a^(-1)
```

```
Jacobi(a1,x1,d1,n)                              # solve eigenvalue problem for a^(-1)
for i in range(1,n+1): d1[i] = 1e0/d1[i]   # invert eigenvalues before sorting
EigSort(x1,d1,n,1)                              # sort eigenvalues and eigenvectors

for j in range(1,n+1):                                      # loop over eigenvectors
   if (x[1][j] * x1[1][j] < 0):                 # eigenvectors have different sign?
      for i in range(1,n+1): x1[i][j] = -x1[i][j]                      # match sign

VecDiff(d,d1,d1,n)                                  # difference of eigenvalues in d1
norm = VecNorm(d1,n)
print("Norm of eigenvalue difference = {0:8.2e}".format(norm))

MatDiff(x,x1,x1,n,n)                               # difference of eigenvectors in x1
norm = MatNorm(x1,n,n)
print("Norm of eigenvector difference = {0:8.2e}".format(norm))
```

Listing 8.7   Eigenvalues and Eigenvectors of the Inverse of a Symmetric Matrix (C/C++ Coding)

```
// Eigenvalues and eigenvectors of inverse of symmetric matrix
#include <stdio.h>
#include <time.h>
#include "linsys.h"
#include "eigsys.h"
#include "matutil.h"

int main()
{
   double **a, **a1, **x, **x1, *d, *d1;
   double det, norm;
   int i, j, n;

   n = 100;                                              // order of matrix
   a  = Matrix(1,n,1,n);                           // coefficient matrix
   a1 = Matrix(1,n,1,n);                                  // inverse of a
   x  = Matrix(1,n,1,n);                            // eigenvectors of a
   x1 = Matrix(1,n,1,n);                           // eigenvectors of a1
   d = Vector(1,n); d1 = Vector(1,n);        // eigenvalues of a and a1

   srand(time(NULL));                     // initialize random number generator
   for (i=1; i<=n; i++)                   // generate random coefficient matrix
      for (j=1; j<=n; j++) {
         a[i][j] = a[j][i] = (double)rand()/RAND_MAX;
         a1[i][j] = a1[j][i] = a[i][j];              // copy a in a1
      }

   Jacobi(a,x,d,n);                            // solve eigenvalue problem for a
   EigSort(x,d,n,1);                        // sort eigenvalues and eigenvectors

   MatInv(a1,n,det);                                        // a1 = a^(-1)
   Jacobi(a1,x1,d1,n);                       // solve eigenvalue problem of a^(-1)
   for (i=1; i<=n; i++) d1[i] = 1e0/d1[i]; // invert eigvalues before sorting
   EigSort(x1,d1,n,1);                       // sort eigenvalues and eigenvectors

   for (j=1; j<=n; j++)                                // loop over eigenvectors
      if (x[1][j] * x1[1][j] < 0)          // eigenvectors have different sign?
         for (i=1; i<=n; i++) x1[i][j] = -x1[i][j];            // match sign

   VecDiff(d,d1,d1,n);                           // difference of eigenvalues in d1
```

```
        norm = VecNorm(d1,n);
        printf("Norm of eigenvalue difference = %9.2e\n",norm);

        MatDiff(x,x1,x1,n,n);                    // difference of eigenvectors in x1
        norm = MatNorm(x1,n,n);
        printf("Norm of eigenvector difference = %9.2e\n",norm);
}
```

PROBLEM 8.2

Consider a massless elastic vibrating string, loaded with n identical point masses equal to m, distributed at equal distance d under constant tension T. Assuming small transversal displacements of the string, determine the frequencies and displacements for its normal modes of vibration. Consider, in particular, $n = 100$ and $dm/T = 1$ and check that the dimensionless eigenfrequencies are multiples of π.

Solution

Denoting by $u_i = u_i(t)$, $i = 1, 2, \ldots, n$, the instantaneous transverse displacements of the points, the transverse force exerted by the string on the point mass i can be approximated as (Maor 1975)

$$F_i = T\frac{u_{i-1} - 2u_i + u_{i+1}}{d},$$

and the corresponding equation of motion is

$$m\ddot{u}_i = (T/d)\,(u_{i-1} - 2u_i + u_{i+1}), \quad i = 2, \ldots, n-1.$$

This set of coupled equations must be supplemented with boundary conditions, which, in the simplest case of a string fixed at both ends, read:

$$u_1 = u_n = 0.$$

The normal modes of the string are characterized by the in-phase vibration of all the points with a common angular frequency ω. Consequently, the law of motion for each point will have the form:

$$u_i = x_i \exp(j\omega t),$$

where x_i is the corresponding *transverse displacement* (amplitude) and j is the imaginary unit. The discretized equations of motion together with the boundary conditions finally lead to the following eigenvalue problem with tridiagonal matrix:

$$
\begin{bmatrix}
1 & 0 & & & & & 0 \\
0 & 2 & -1 & & & & \\
& -1 & 2 & -1 & & & \\
& & \ddots & \ddots & \ddots & & \\
& & & -1 & 2 & -1 & \\
& & & & -1 & 2 & 0 \\
0 & & & & & 0 & 1
\end{bmatrix}
\begin{bmatrix}
x_1 \\ x_2 \\ x_3 \\ \vdots \\ x_{n-2} \\ x_{n-2} \\ x_n
\end{bmatrix}
= \lambda
\begin{bmatrix}
x_1 \\ x_2 \\ x_3 \\ \vdots \\ x_{n-2} \\ x_{n-2} \\ x_n
\end{bmatrix},
$$

where the eigenvalue λ relates to the angular frequency ω by

$$\lambda = (dm/T)\omega^2.$$

The implementations are given in Listings 8.8 and 8.9, and the plots for the four lowest-frequency normal modes are depicted in Figure 8.1.

Listing 8.8   Normal Modes of Loaded String (Python Coding)

```
# Normal modes of a loaded string with fixed ends
from eigsys import *
from graphlib import *

n = 100                                              # number of point masses
a = [[0]*(n+1) for i in range(n+1)]                  # coefficient matrix
x = [[0]*(n+1) for i in range(n+1)]          # eigenvectors = displacements
d = [0]*(n+1)                                 # eigenvalues ~ frequencies^2
xp = [0]*(n+1)                                      # mesh points for plotting
yp = [0]*(n+1)                                # function values for plotting

for i in range(1,n+1): a[i][i] = 2e0                  # build coefficient matrix
for i in range(2,n-1): a[i+1][i] = -1e0; a[i][i+1] = -1e0

Jacobi(a,x,d,n)                                      # solve eigenvalue problem
EigSort(x,d,n,1)                             # sort eigenvalues and eigenvectors

GraphInit(1200,600)

h = 1e0/(n-1)
for i in range(1,n+1): xp[i] = (i-1) * h              # mesh points for plotting

mode = 1                                             # normal mode 1
omega = sqrt(d[mode])*(n-1)                                   # frequency
for i in range(1,n+1): yp[i] = x[i][mode]                   # displacements
title = "omega({0:d}) = {1:6.2f}".format(mode,omega)
Plot(xp,yp,n,"blue",3,0.10,0.45,0.60,0.90,"x","y",title)

mode = 2                                             # normal mode 2
omega = sqrt(d[mode])*(n-1)                                   # frequency
for i in range(1,n+1): yp[i] = x[i][mode]                   # displacements
title = "omega({0:d}) = {1:6.2f}".format(mode,omega)
Plot(xp,yp,n,"blue",3,0.60,0.95,0.60,0.90,"x","y",title)

mode = 3                                             # normal mode 3
omega = sqrt(d[mode])*(n-1)                                   # frequency
for i in range(1,n+1): yp[i] = x[i][mode]                   # displacements
title = "omega({0:d}) = {1:6.2f}".format(mode,omega)
Plot(xp,yp,n,"blue",3,0.10,0.45,0.10,0.40,"x","y",title)

mode = 4                                             # normal mode 4
omega = sqrt(d[mode])*(n-1)                                   # frequency
for i in range(1,n+1): yp[i] = x[i][mode]                   # displacements
title = "omega({0:d}) = {1:6.2f}".format(mode,omega)
Plot(xp,yp,n,"blue",3,0.60,0.95,0.10,0.40,"x","y",title)

MainLoop()
```

Listing 8.9 Normal Modes of Loaded String (C/C++ Coding)

```
// Normal modes of a loaded string with fixed ends
#include <math.h>
#include "memalloc.h"
#include "eigsys.h"
#include "graphlib.h"

int main(int argc, wchar_t** argv)
{
   double **a, **x, *d, *xp, *yp;
   double h, omega;
   int i, mode, n;
   char title[20];

   n = 100;                                        // number of point masses
   a = Matrix(1,n,1,n);                            // coefficient matrix
   x = Matrix(1,n,1,n);                   // eigenvectors = displacements
   d = Vector(1,n);                       // eigenvalues ~ frequencies^2
   xp = Vector(1,n);                      // mesh points for plotting
   yp = Vector(1,n);                      // function values for plotting

   for (i=1; i<=n; i++) a[i][i] = 2e0;            // build coefficient matrix
   for (i=2; i<=n-2; i++) { a[i+1][i] = -1e0; a[i][i+1] = -1e0; }

   Jacobi(a,x,d,n);                               // solve eigenvalue problem
   EigSort(x,d,n,1);                      // sort eigenvalues and eigenvectors

   h = 1e0/(n-1);
   for (i=1; i<=n; i++) xp[i] = (i-1) * h;         // mesh points for plotting

   PyGraph w(argc, argv);
   w.GraphInit(1200,600);

   mode = 1;                                       // normal mode 1
   omega = sqrt(d[mode])*(n-1);                    // frequency
   for (i=1; i<=n; i++) yp[i] = x[i][mode];        // displacements
   sprintf(title,"omega(%i) = %6.2f",mode,omega);
   w.Plot(xp,yp,n,"blue",3,0.10,0.45,0.60,0.90,"x","y",title);

   mode = 2;                                       // normal mode 2
   omega = sqrt(d[mode])*(n-1);                    // frequency
   for (i=1; i<=n; i++) yp[i] = x[i][mode];        // displacements
   sprintf(title,"omega(%i) = %6.2f",mode,omega);
   w.Plot(xp,yp,n,"blue",3,0.60,0.95,0.60,0.90,"x","y",title);

   mode = 3;                                       // normal mode 3
   omega = sqrt(d[mode])*(n-1);                    // frequency
   for (i=1; i<=n; i++) yp[i] = x[i][mode];        // displacements
   sprintf(title,"omega(%i) = %6.2f",mode,omega);
   w.Plot(xp,yp,n,"blue",3,0.10,0.45,0.10,0.40,"x","y",title);

   mode = 4;                                       // normal mode 4
   omega = sqrt(d[mode])*(n-1);                    // frequency
   for (i=1; i<=n; i++) yp[i] = x[i][mode];        // displacements
   sprintf(title,"omega(%i) = %6.2f",mode,omega);
   w.Plot(xp,yp,n,"blue",3,0.60,0.95,0.10,0.40,"x","y",title);

   w.MainLoop();
}
```

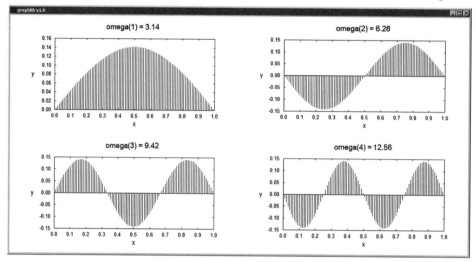

FIGURE 8.1   The lowest four normal modes of a loaded string with fixed ends.

PROBLEM 8.3

Consider the tetragonal methane molecule CH_4, characterized by the relative atomic masses $m_C = 12$ and $m_H = 1$, and by the following Cartesian positions of the 5 atoms in a body-fixed reference frame (in Å):

$$x = \begin{bmatrix} 0.000 & 0.635 & -0.635 & -0.635 & 0.635 \end{bmatrix},$$

$$y = \begin{bmatrix} 0.000 & 0.635 & -0.635 & 0.635 & -0.635 \end{bmatrix},$$

$$z = \begin{bmatrix} 0.000 & 0.635 & 0.635 & -0.635 & -0.635 \end{bmatrix},$$

$$r = \begin{bmatrix} 1.700 & 1.200 & 1.200 & 1.200 & 1.200 \end{bmatrix}.$$

The C atom is located at the origin and the quantities r represent the van der Waals radii of the atoms (Bondi 1964).

a. Place the molecule in its *center-of-mass system* by subtracting from the individual coordinates the coordinates of the mass center:

$$x_{CM} = \frac{1}{m_{CM}} \sum_i m_i x_i, \quad y_{CM} = \frac{1}{m_{CM}} \sum_i m_i y_i, \quad z_{CM} = \frac{1}{m_{CM}} \sum_i m_i z_i,$$

where

$$m_{CM} = \sum_i m_i$$

is the total mass of the system of atoms.

b. Calculate the *inertia tensor* of the molecule in its Cartesian center-of-mass system:

$$I = \begin{bmatrix} I_{xx} & I_{xy} & I_{xz} \\ I_{yx} & I_{yy} & I_{yz} \\ I_{zx} & I_{zy} & I_{zz} \end{bmatrix},$$

where

$$I_{xx} = \sum_i m_i \left(y_i^2 + z_i^2 \right), \quad I_{xy} = I_{yx} = -\sum_i m_i x_i y_i,$$

$$I_{yy} = \sum_i m_i \left(z_i^2 + x_i^2 \right), \quad I_{xz} = I_{zx} = -\sum_i m_i z_i x_i,$$

$$I_{zz} = \sum_i m_i \left(x_i^2 + y_i^2 \right), \quad I_{yz} = I_{zy} = -\sum_i m_i y_i z_i.$$

c. Solve the eigenvalue problem of the inertia tensor,

$$\begin{bmatrix} I_{xx} & I_{xy} & I_{xz} \\ I_{yx} & I_{yy} & I_{yz} \\ I_{zx} & I_{zy} & I_{zz} \end{bmatrix} \begin{bmatrix} R_{xx} & R_{xy} & R_{xz} \\ R_{yx} & R_{yy} & R_{yz} \\ R_{zx} & R_{zy} & R_{zz} \end{bmatrix} = \begin{bmatrix} R_{xx} & R_{xy} & R_{xz} \\ R_{yx} & R_{yy} & R_{yz} \\ R_{zx} & R_{zy} & R_{zz} \end{bmatrix} \begin{bmatrix} I_x & 0 & 0 \\ 0 & I_y & 0 \\ 0 & 0 & I_z \end{bmatrix},$$

to determine the *principal moments of inertia*—I_1, I_2, and I_3—as eigenvalues, and the rotation matrix **R**, as modal matrix, containing the eigenvectors on the columns.

d. Use the transposed transformation matrix $\mathbf{R}^T$ to rotate the atomic coordinates to the system of principal axes:

$$\begin{bmatrix} x_i' \\ y_i' \\ z_i' \end{bmatrix} = \begin{bmatrix} R_{xx} & R_{yx} & R_{zx} \\ R_{xy} & R_{yy} & R_{zy} \\ R_{xz} & R_{yz} & R_{zz} \end{bmatrix} \begin{bmatrix} x_i \\ y_i \\ z_i \end{bmatrix}, \quad i = 1, 2, \ldots, n.$$

e. Visualize the initial configuration, and the configurations with the *x*-axis and, respectively, the *z*-axis as principal axes of symmetry using the routine `PlotParticles` included in the libraries `graphlib.py` and `graphlib.h` (see Chapter 3 for details).

Solution

Python functions for placing a system of particles into the center-of-mass system and for rotating them to the system of principal axes are included in the module `coords.py` and are presented in Listing 8.10. C/C++ counterparts are available as supplementary material in the file `coords.h`.

The Python main program is shown in Listing 8.11 and the C/C++ version is accessible as supplementary material (`P08-MomInert.cpp`). The produced graphical output is shown in Figure 8.2.

Listing 8.10 Transformation of the Coordinates of a System of Particles (`coords.py`)

```
#------------------------------ coords.py -------------------------------
#  Contains routines for transforming the coordinates of systems of particles
#  Author: Titus Beu, 2013
#------------------------------------------------------------------------
from eigsys import *

#========================================================================
def MovetoCM(m, x, y, z, n):
#------------------------------------------------------------------------
#  Moves a system of n particles to the CM system
#------------------------------------------------------------------------
   mCM = 0e0; xCM = 0e0; yCM = 0e0; zCM = 0e0
   for i in range(1,n+1):
      mCM += m[i]
```

```
      xCM += m[i] * x[i]
      yCM += m[i] * y[i]
      zCM += m[i] * z[i]
   xCM /= mCM; yCM /= mCM; zCM /= mCM

   for i in range(1,n+1):
      x[i] -= xCM; y[i] -= yCM; z[i] -= zCM

#===============================================================================
def PrincipalAxes(m, x, y, z, n, isort):
#-------------------------------------------------------------------------------
#  Rotates a set of n particles to the system of principal axes
#  isort =  1 - highest symmetry axis along x-axis
#  isort = -1 - highest symmetry axis along z-axis
#-------------------------------------------------------------------------------
   Inert = [[0]*4 for i in range(4)]
   Rot   = [[0]*4 for i in range(4)]
   MomInert = [0]*4

   for i in range(1,4):
      for j in range(1,i+1): Inert[i][j] = 0e0

   for i in range(1,n+1):
      mi = m[i]; xi = x[i]; yi = y[i]; zi = z[i]
      Inert[1][1] += mi * (yi*yi + zi*zi)                 # inertia tensor
      Inert[2][2] += mi * (zi*zi + xi*xi)
      Inert[3][3] += mi * (xi*xi + yi*yi)
      Inert[2][1] -= mi * xi*yi
      Inert[3][1] -= mi * zi*xi
      Inert[3][2] -= mi * yi*zi

   Jacobi(Inert,Rot,MomInert,3)                     # diagonalize inertia tensor
   EigSort(Rot,MomInert,3,isort)            # sort eigenvalues and eigenvectors

   for i in range(1,n+1):                       # rotate system to principal axes
      xi = x[i]; yi = y[i]; zi = z[i]
      x[i] = Rot[1][1] * xi + Rot[2][1] * yi + Rot[3][1] * zi
      y[i] = Rot[1][2] * xi + Rot[2][2] * yi + Rot[3][2] * zi
      z[i] = Rot[1][3] * xi + Rot[2][3] * yi + Rot[3][3] * zi

   return MomInert
```

Listing 8.11   Rotation of a Molecule to the System of Principal Axes (Python Coding)

```
# Rotation of a molecule to the system of principal axes
from math import *
from eigsys import *
from coords import *
from matutil import *
from graphlib import *

n = 5                                                  # number of particles
m = [0,12.000, 1.000, 1.000, 1.000, 1.000]                     # masses
x = [0, 0.000, 0.635,-0.635,-0.635, 0.635]                   # positions
y = [0, 0.000, 0.635,-0.635, 0.635,-0.635]              # [0] not used
z = [0, 0.000, 0.635, 0.635, 0.635,-0.635]
r = [0, 0.300, 0.200, 0.200, 0.200, 0.200]                      # radii
col = ["", "red", "blue", "blue", "blue", "gray"]              # colors
```

```
dmax = 1.5e0                                    # cutoff distance for bonds

m[5] += 1e-10          # "mark" last particle to set principal symmetry axis

GraphInit(1200,600)

title = "Initial configuration"
PlotParticles(x,y,z,r,col,n,dmax,0e0,0e0,0,0e0,0e0,0,0.05,0.25,0.2,0.8,title)

MovetoCM(m,x,y,z,n)                                       # move to CM system

MomInert = PrincipalAxes(m,x,y,z,n,1)          # align main symmetry axis to x
print("Structure aligned to x-axis:")
VecPrint(x,n); VecPrint(y,n); VecPrint(z,n)
print("Principal moments of inertia:")
VecPrint(MomInert,3)
title = "System of principal axes - main x"
PlotParticles(x,y,z,r,col,n,dmax,0e0,0e0,0,0e0,0e0,0,0.45,0.65,0.2,0.8,title)

MomInert = PrincipalAxes(m,x,y,z,n,-1)         # align main symmetry axis to z
print("\nStructure aligned to z-axis:")
VecPrint(x,n); VecPrint(y,n); VecPrint(z,n)
print("Principal moments of inertia:")
VecPrint(MomInert,3)
title = "System of principal axes - main z"
PlotParticles(x,y,z,r,col,n,dmax,0e0,0e0,0,0e0,0e0,0,0.75,0.95,0.2,0.8,title)

MainLoop()
```

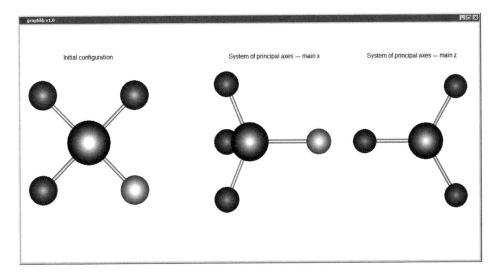

FIGURE 8.2 Methane molecule in the initial configuration (left), rotated to the system of principal axes with the main symmetry axis along *x* (center), and, respectively, with the main symmetry axis along *z* (right).

PROBLEM 8.4

Consider an airplane described by 10 significant points, with the associated masses and positions:

$$m = \begin{bmatrix} 1 & 1 & 1 & 0 & 0 & 1 & 0 & 0 & 0 & 0 \end{bmatrix},$$
$$x = \begin{bmatrix} -2 & -6 & 2 & 0 & 2 & 6 & 1 & 2 & -2 & 2 \end{bmatrix},$$
$$y = \begin{bmatrix} -2 & 2 & -6 & 2 & 0 & 6 & 2 & 1 & -2 & 2 \end{bmatrix},$$
$$z = \begin{bmatrix} 0 & 0 & 0 & 0 & 0 & 0 & 1 & 1 & 4 & 1 \end{bmatrix}.$$

The fuselage is considered to be defined by nine triangles, with the corner points specified by the indexes:

$$i_1 = \begin{bmatrix} 1 & 1 & 1 & 1 & 1 & 1 & 1 & 6 & 1 \end{bmatrix},$$
$$i_2 = \begin{bmatrix} 2 & 3 & 4 & 5 & 4 & 5 & 7 & 7 & 9 \end{bmatrix},$$
$$i_3 = \begin{bmatrix} 4 & 5 & 6 & 6 & 7 & 8 & 8 & 8 & 10 \end{bmatrix}.$$

Following the steps and using the formulas of the preceding problem, carry out the following tasks:

a. Place the airplane in its *center-of-mass system* by subtracting from the individual coordinates the coordinates of the mass center.
b. Calculate the *inertia tensor* of the airplane in its Cartesian center-of-mass system.
c. Solve the eigenvalue problem of the inertia tensor to determine the *principal moments of inertia*— I_1, I_2, and I_3—as eigenvalues, and the transformation matrix $\mathbf{R}$, as modal matrix, containing the eigenvectors on the columns.
d. Use the transposed transformation matrix, $\mathbf{R}^T$, to rotate the coordinates of the airplane to the system of principal axes.
e. Apply to the airplane a "yaw" (rotation about the vertical z-axis) of 20° and, respectively, a "roll" (rotation about the longitudinal x-axis) of $-35°$. The rotation matrices for arbitrary angles φ are

$$\mathbf{R}_x(\varphi) = \begin{bmatrix} 1 & 0 & 0 \\ 0 & \cos\varphi & -\sin\varphi \\ 0 & \sin\varphi & \cos\varphi \end{bmatrix}, \quad \mathbf{R}_z(\varphi) = \begin{bmatrix} \cos\varphi & -\sin\varphi & 0 \\ \sin\varphi & \cos\varphi & 0 \\ 0 & 0 & 1 \end{bmatrix}.$$

f. Visualize the airplane in the initial configuration, in the system of principal axes, and in the two rotated configurations using the routine PlotStruct included in the libraries graphlib.py and graphlib.h (see Chapter 3 for details).

Solution

The main program uses the functions developed for placing a configuration of point-like particles into the center-of-mass system and for rotating them to the system of principal axes, as included in the files coords.py and coords.h, which have been developed for the previous problem.

The Python implementation is shown in Listing 8.12, while the C/C++ version is provided as supplementary material (P08-Airplane.cpp). The graphical output is shown in Figure 8.3.

Listing 8.12 Rotation of an Airplane (Python Coding)

```python
# Rotation of an airplane
from math import *
from eigsys import *
from coords import *
from matutil import *
from graphlib import *

n = 10                                          # number of corner points
m = [0, 1, 1, 1, 0, 0, 1, 0, 0, 0, 0]                      # masses
x = [0,-2,-6, 2, 0, 2, 6, 1, 2,-2, 2]                   # positions
y = [0,-2, 2,-6, 2, 0, 6, 2, 1,-2, 2]                   # [0] not used
z = [0, 0, 0, 0, 0, 0, 0, 1, 1, 4, 1]

n3 = 9                                     # number of defining triangles
ind1 = [0, 1, 1, 1, 1, 1, 1, 1, 6, 1]        # indexes of triangle corners
ind2 = [0, 2, 3, 4, 5, 4, 5, 7, 7, 9]
ind3 = [0, 4, 5, 6, 6, 7, 8, 8, 8,10]

GraphInit(800,800)

title = "Initial configuration"
PlotStruct(x,y,z,n,ind1,ind2,ind3,n3,
          0e0,0e0,0,0e0,0e0,0,0.05,0.45,0.55,0.92,title)

MovetoCM(m,x,y,z,n)                                   # move structure to CM

PrincipalAxes(m,x,y,z,n,1)                   # rotate airplane to principal axes
print("Airplane in system of principal axes:")
VecPrint(x,6); VecPrint(y,6); VecPrint(z,6)
title = "System of principal axes"
PlotStruct(x,y,z,n,ind1,ind2,ind3,n3,
          0e0,0e0,0,0e0,0e0,0,0.55,0.95,0.55,0.92,title)

f = pi/180e0
phi = 20                                                        # yaw
cosp = cos(phi*f); sinp = sin(phi*f)
for i in range(1,n+1):                          # rotate airplane about z-axis
   xi = x[i]
   x[i] = cosp * xi - sinp * y[i]
   y[i] = sinp * xi + cosp * y[i]
print("\nRotated airplane yaw = {0:4.1f}".format(phi))
VecPrint(x,6); VecPrint(y,6); VecPrint(z,6)
title = "yaw = {0:4.1f}".format(phi)
PlotStruct(x,y,z,n,ind1,ind2,ind3,n3,
          0e0,0e0,0,0e0,0e0,0,0.05,0.45,0.05,0.42,title)

phi = -35                                                       # roll
cosp = cos(phi*f); sinp = sin(phi*f)
for i in range(1,n+1):                          # rotate airplane about x-axis
   yi = y[i]
   y[i] = cosp * yi - sinp * z[i]
   z[i] = sinp * yi + cosp * z[i]
print("\nRotated airplane roll = {0:4.1f}".format(phi))
VecPrint(x,6); VecPrint(y,6); VecPrint(z,6)
title = "roll = {0:4.1f}".format(phi)
PlotStruct(x,y,z,n,ind1,ind2,ind3,n3,
          0e0,0e0,0,0e0,0e0,0,0.55,0.95,0.05,0.42,title)

MainLoop()
```

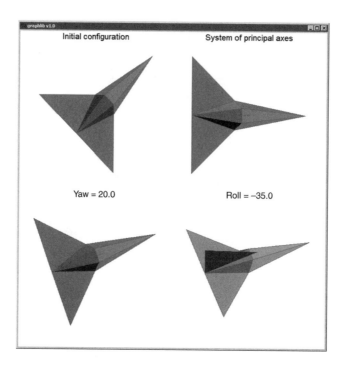

FIGURE 8.3 Airplane in the initial configuration (top left), rotated to the system of principal axes with the main symmetry axis along x (top right), rotated around the vertical z axis (yaw) (bottom left), and rotated about the x-axis (roll) (bottom right).

PROBLEM 8.5

Find the eigenenergies and eigenvectors for the cyclic benzene molecule C_6H_6 in the framework of the *Hückel theory* by solving the standard eigenvalue problem:

$$\mathbf{H} \cdot \mathbf{C} = \mathbf{C} \cdot \mathbf{E}.$$

The Hamiltonian matrix can be expressed in terms of the *Coulomb integral*, $\alpha = -6.9\,\text{eV}$, and the *resonance energy*, $\beta = -3.1\,\text{eV}$ (Brogli and Heilbronner 1972):

$$\mathbf{H} = \begin{bmatrix} \alpha & \beta & 0 & 0 & 0 & \beta \\ \beta & \alpha & \beta & 0 & 0 & 0 \\ 0 & \beta & \alpha & \beta & 0 & 0 \\ 0 & 0 & \beta & \alpha & \beta & 0 \\ 0 & 0 & 0 & \beta & \alpha & \beta \\ \beta & 0 & 0 & 0 & \beta & \alpha \end{bmatrix}.$$

The exact eigenvalues (diagonal elements of $\mathbf{E}$) and eigenvectors are

$$E_1 = \alpha + 2\beta, \quad E_2 = E_3 = \alpha + \beta, \quad E_4 = E_5 = \alpha - \beta, \quad E_6 = \alpha - 2\beta,$$

$$\mathbf{C}_1 = \frac{1}{\sqrt{6}} \begin{bmatrix} 1 \\ 1 \\ 1 \\ 1 \\ 1 \\ 1 \end{bmatrix}, \quad \mathbf{C}_2 = \frac{1}{2} \begin{bmatrix} 1 \\ 1 \\ 0 \\ -1 \\ -1 \\ 0 \end{bmatrix}, \quad \mathbf{C}_3 = \frac{1}{2} \begin{bmatrix} 1 \\ 0 \\ -1 \\ -1 \\ 0 \\ 1 \end{bmatrix},$$

$$\mathbf{C}_4 = \frac{1}{2} \begin{bmatrix} 1 \\ -1 \\ 0 \\ 1 \\ -1 \\ 0 \end{bmatrix}, \quad \mathbf{C}_5 = \frac{1}{2} \begin{bmatrix} 1 \\ 0 \\ -1 \\ 1 \\ 0 \\ -1 \end{bmatrix}, \quad \mathbf{C}_6 = \frac{1}{\sqrt{6}} \begin{bmatrix} 1 \\ -1 \\ 1 \\ -1 \\ 1 \\ -1 \end{bmatrix}.$$

Using the routine `EigSym`, prove that the exact eigenvalues and eigenvectors for the nondegenerate states E_1 and E_6 are correctly reproduced and explain why the eigenvectors for the degenerate states differ from the exact ones, being yet correct. Estimate the accuracy of the solution by means of the *max norm*:

$$\|\mathbf{H} \cdot \mathbf{C} - \mathbf{C} \cdot \mathbf{E}\| = \max_{i,j} \left| (\mathbf{H} \cdot \mathbf{C} - \mathbf{C} \cdot \mathbf{E})_{ij} \right|,$$

which is expected to vanish.

Solution

It should be noted, as a general property, that any linear combination of eigenvectors corresponding to degenerate eigenvalues is still an eigenvector corresponding to the same eigenvalue. Correspondingly, the energy levels $E_2 = E_3$ and $E_4 = E_5$ are *two-fold degenerate* and can be described properly by any linear combination $\gamma_2 \mathbf{C}_2 + \gamma_3 \mathbf{C}_3$ and, respectively, $\gamma_4 \mathbf{C}_4 + \gamma_5 \mathbf{C}_5$.

The Python implementation is shown in Listing 8.13 and the C/C++ version is available as supplementary material (`P08-Hueckel.cpp`).

Listing 8.13   Hueckel Method for Benzene (Python Coding)

```
# Hueckel method for benzene
from eigsys import *
from matutil import *

n = 6

H = [[0]*(n+1) for i in range(n+1)]                    # Hamiltonian matrix
C = [[0]*(n+1) for i in range(n+1)]                   # molecular coefficients
E = [0]*(n+1)                                              # eigenenergies

alpha = -6.9                        # Coulomb & resonance integrals: F. Brogli &
beta = -3.2                         # E. Heilbronner, Theoret. Chim. Acta 1972

for i in range(1,n+1): H[i][i] = alpha                    # diagonal elements
for i in range(1,n):
   H[i+1][i] = beta; H[i][i+1] = beta                # off-diagonal elements
H[1][6] = beta; H[6][1] = beta               # cyclic boundary conditions
                                                 # exact eigenenergies
X = (0, alpha+2*beta, alpha+beta, alpha+beta, \
        alpha-beta, alpha-beta, alpha-2*beta )
print("Eigenenergies (exact):")
VecPrint(X,n)
```

```
Jacobi(H,C,E,n)
EigSort(C,E,n,1)                                 # sort eigenvalues and eigenvectors

print("Eigenenergies (eV):")
VecPrint(E,n)

print("Eigenvextors:")
MatPrint(C,n,n)

for i in range(2,n+1):                           # restore Hamiltonian
   for j in range(1,i): H[i][j] = H[j][i]        # from upper triangle

err = 0e0
for i in range(1,n+1):
   for j in range(1,n+1):
      t = - E[j] * C[i][j]
      for k in range(1,n+1): t += H[i][k] * C[k][j]
      if (err < fabs(t)): err = fabs(t)          # err = max|a x - lambda x|

print("\nMaximum error = {0:8.2e}".format(err))
```

PROBLEM 8.6

In a density functional theory (DFT) calculation of the electronic structure of the H_2O molecule, the *Fock matrix* **F** and the *overlap matrix* **S** have the elements listed below and compose the data file H2O.dat. The Fock matrix is specified in the commonly used atomic units, the so-called Hartrees (1 Hartree = 27.21 eV).

Read in the file H2O.dat (Listing 8.14) and solve the generalized eigenvalue problem of the matrices **F** and **S**,

$$FC = SC\varepsilon,$$

known as the *Roothaan equation* (Szabo and Ostlund 1996), by using the routine EigSym.

Listing 8.14 File H2O.dat: Fock and Overlap Matrices for the H_2O Molecule

```
F: Fock matrix
-18.74
-4.74 -1.74
-3.34 -1.65 -1.26
-0.01 -0.07 -0.07 -0.10
-0.02 -0.10 -0.10  0.02 -0.08
 0.00  0.00  0.00  0.00  0.00 -0.02
 0.00 -0.05 -0.10 -0.44  0.02  0.00 -0.06
 0.00 -0.06 -0.14  0.02 -0.43  0.00  0.04 -0.03
 0.00  0.00  0.00  0.00  0.00 -0.37  0.00  0.00  0.11
-0.63 -0.56 -0.51 -0.38 -0.02  0.00 -0.29 -0.02  0.00 -0.17
-1.34 -0.84 -0.70 -0.22 -0.05  0.00 -0.16 -0.07  0.00 -0.48 -0.44
-0.63 -0.56 -0.51  0.11 -0.36  0.00  0.08 -0.28  0.00 -0.11 -0.26 -0.17
-1.34 -0.84 -0.70  0.03 -0.22  0.00 -0.02 -0.18  0.00 -0.26 -0.39 -0.48 -0.44
S: overlap matrix
 1.00
 0.23  1.00
 0.17  0.76  1.00
 0.00  0.00  0.00  1.00
 0.00  0.00  0.00  0.00  1.00
 0.00  0.00  0.00  0.00  0.00  1.00
 0.00  0.00  0.00  0.50  0.00  0.00  1.00
 0.00  0.00  0.00  0.00  0.50  0.00  0.00  1.00
```

```
0.00   0.00   0.00   0.00   0.00   0.50   0.00   0.00   1.00
0.03   0.23   0.41   0.27   0.01   0.00   0.58   0.01   0.00   1.00
0.07   0.37   0.67   0.15   0.00   0.00   0.49   0.01   0.00   0.66   1.00
0.03   0.23   0.41  -0.08   0.26   0.00  -0.18   0.55   0.00   0.04   0.20   1.00
0.07   0.37   0.67  -0.05   0.14   0.00  -0.15   0.46   0.00   0.20   0.49   0.66   1.00
```

Taking into account that the H_2O molecule has 10 electrons, which, in the ground state, occupy in spin-up–spin-down pairs the lowest 5 energy levels ε_i, the highest occupied molecular orbital (HOMO) is ε_5 and the lowest unoccupied molecular orbital (LUMO) is ε_6. Determine the HOMO–LUMO gap, $\varepsilon_6 - \varepsilon_5$, and check the accuracy of the calculation by estimating the *max norm* $\|\mathbf{FC} - \mathbf{SC}\varepsilon\| = \max_{i,j} |(\mathbf{FC} - \mathbf{SC}\varepsilon)_{ij}|$.

Solution

The Python implementation is shown in Listing 8.15 and the C/C++ version is available as supplementary material (P08-Roothaan.cpp).

Listing 8.15   Solution of the Roothaan Equation for the H_2O Molecule (Python Coding)

```python
# Solves the Roothaan equation F C = S C E for the H2O molecule
from eigsys import *
from matutil import *

n = 13

C = [[0]*(n+1) for i in range(n+1)]        # molecular coefficients
F = [[0]*(n+1) for i in range(n+1)]                  # Fock matrix
S = [[0]*(n+1) for i in range(n+1)]               # overlap matrix
S0 = [[0]*(n+1) for i in range(n+1)]   # backup copy of overlap matrix
E = [0]*(n+1)                                      # energy levels

f = open('H2O.dat','r')                         # read in F and S
line = f.readline()                               # skip header
for i in range(1,n+1):                   # read in lower triangle of F
    line = f.readline()                     # F not altered by EigSym
    for j in range(1,i+1):
        F[i][j] = float(line.split()[j-1])
        F[j][i] = F[i][j]                            # symmetrize

line = f.readline()                               # skip header
for i in range(1,n+1):
    line = f.readline()
    for j in range(1,i+1):                 # read in lower triangle of S
        S[i][j] = float(line.split()[j-1])       # S destroyed by EigSym
        S0[i][j] = S0[j][i] = S[i][j]                    # backup S
f.close()

print("F: Fock matrix (sample)")
MatPrint(F,5,5)
print("S: overlap matrix (sample)")
MatPrint(S,5,5)

EigSym(F,S,C,E,n)                          # solve Roothaan equation
EigSort(C,E,n,1)                  # sort eigenvalues and eigenvectors

print("\nEigenenergies:")
VecPrint(E,n)
```

```
print("\nHOMO-LUMO gap E[6]-E[5] = {0:4.3f} Hartrees".format(E[6]-E[5]))

err = 0e0
for i in range(1,n+1):
    for j in range(1,n+1):
        t = 0e0
        for k in range(1,n+1): t += (F[i][k] - E[j]*S0[i][k]) * C[k][j]
        if (err < fabs(t)): err = fabs(t)          # err = max|a x - lambda x|

print("\nMaximum error = {0:8.2e}".format(err))
```

References and Suggested Further Reading

Beu, T. A. 2004. *Numerical Calculus in C* (3rd ed., in Romanian). Cluj-Napoca: MicroInformatica.

Bondi, A. 1964. van der Waals Volumes and Radii. *Journal of Physical Chemistry 68*(3), 441–451.

Brogli, F. and E. Heilbronner. 1972. Correlation of Hückel molecular orbital energies with π-ionization potentials. *Theoretica Chimica Acta (Berlin) 26*, 289–299.

Burden, R. and J. Faires. 2010. *Numerical Analysis* (9th ed.). Boston: Brooks/Cole, Cengage Learning.

Demidovich, B. and I. Maron. 1987. *Computational Mathematics* (4th ed.). Mockow: MIR Publishers.

Maor, E. 1975. A discrete model for the vibrating string. *International Journal of Mathematical Education in Science and Technology 6*(3), 345–352.

Szabo, A. and N. S. Ostlund. 1996. *Modern Quantum Chemistry: Introduction to Advanced Electronic Structure Theory*. Mineola, NY: Dover Publications.

9

Modeling of Tabulated Functions

9.1 Interpolation and Regression

The necessity of modeling incompletely known or biased functional dependences occurs in rather diverse numerical contexts, the variety of developed formulations and methods being a direct reflection thereof.

Quite often in numerical applications, functions need to be used for which the mere values on a discrete mesh of points are known. Such discontinuous dependences can represent, for instance, tabulations of mathematical functions or measured physical quantities. For the sake of simplicity, we confine the discussion to real-valued dependences of a single real variable. In such cases, a function defined on some real interval,

$$f : [\alpha, \beta] \to \mathbb{R}, \quad [\alpha, \beta] \subset \mathbb{R}, \tag{9.1}$$

is specified by its values y_i at a certain number of *mesh points* $x_i \in [\alpha, \beta]$:

$$f(x_i) = y_i, \quad i = 1, 2, \ldots, n. \tag{9.2}$$

The numerical processing of such dependences requires rather frequently the use of supplementary values corresponding to arguments differing from those of the tabulation. Alas, obtaining additional information on the dependence is in many situations difficult, if at all possible. In particular, even if the algorithm is known, its complexity may render the direct computation of certain functions too expensive in large-scale computations. On the other hand, if the tabulated values stem from observations, they are typically affected by *measurement (observational) errors*.

With a view to synthesizing the information on a tabulated dependence f and using it efficiently in further calculations, it is a common practice to approximate it by a continuous function, called *model function*, $F = F(x; a_1, \ldots, a_m)$, which generally depends on a certain number of adjustable *model parameters*, a_j. The functional form of the model F and the values of the parameters a_j enjoy a certain freedom, but have to be determined, anyway, so as to approximate the discrete function f *optimally* in a predefined sense.

The concrete choice of the model function F is, indeed, a nontrivial task and it should be based, in as much as possible, on a rigorous reasoning. Broadly speaking, the following options are available:

- The model function is selected from a *convenient class* (such as polynomials, Gaussian functions, etc.), which ensures simplicity and efficiency in the postprocessing.
- The model function is derived from a *theoretical framework*, within which the parameters have well-defined significances.
- The model function results from a *compromise* between the constraints and advantages of the previous options.

To determine optimal model parameters, one has to define first a functional reflecting the degree to which the model function F approximates the tabulated dependence f. Admitting that, as commonly encountered, both f and F are *square integrable functions* on $[\alpha, \beta]$ (the integral of their squares is finite),

265

a useful synthetic measure of the deviation of the model function from the modeled function is provided by the *distance*

$$d(f, F) = \left\{ \int_\alpha^\beta \left[f(x) - F(x; \mathbf{a}) \right]^2 dx \right\}^{1/2},$$

where $\mathbf{a} \equiv (a_1, a_2, \ldots)$ denotes the ensemble of model parameters. From the general statement of the modeling problem, it is obvious that this definition is actually inoperable, since it implies the knowledge of the analytical expression of f. A less rigorous definition of the distance, however applicable only to discrete functions, is

$$d(f, F) = \left\{ \sum_{i=1}^n \left[f(x_i) - F(x_i; \mathbf{a}) \right]^2 \right\}^{1/2}. \tag{9.3}$$

The approach of *adjusting* the model parameters or *fitting* the model function to observed data sets based on the distance (9.3) is known as *least-squares approximation* and lies at the core of two basic methods, namely, *interpolation* and *regression*.

Interpolation. If the tabulated values $y_i = f(x_i)$ are considered *error-free* or numerically *exact*, it is natural to impose a *vanishing distance* between the functions f and F:

$$d(f, F) = 0. \tag{9.4}$$

This amounts to constructing a model function of a certain continuity class, which takes on *the same values* at the data nodes x_i as the modeled function $f(x)$:

$$F(x_i; \mathbf{a}) = y_i, \quad i = 1, 2, \ldots, n. \tag{9.5}$$

The model function satisfying these conditions is normally used to approximate the function $f(x)$ for interior points of the interval $[x_1, x_n]$ and the procedure is called *interpolation*. $F(x; \mathbf{a})$ is called *interpolation function* or *interpolant*, while the discrete arguments x_i are also known as *interpolation nodes*. If the arguments for which the interpolant is evaluated lie outside the tabulation interval $[x_1, x_n]$, the approach is called *extrapolation*.

In geometrical terms, interpolation implies finding the curve $y = F(x; \mathbf{a})$ which passes through *all* the data points $M_i(x_i, y_i)$ (see Figure 9.1a). With such a general statement, however, the modeling problem can have an infinity of solutions or none, depending on the continuity class to which the model function belongs. Quite in contrast, the solution becomes unambiguous if, for instance, the model function is

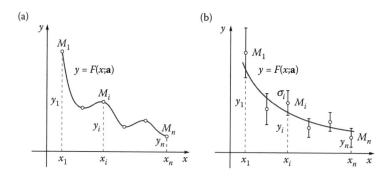

FIGURE 9.1 The model function passes in the case of the interpolation through the tabulation nodes (a), while in the case of the regression, it minimizes the merit function and passes between the nodes (b). For the same data set, the interpolation and the regression curves can be significantly different.

chosen to be a polynomial $P_m(x)$, with the number of coefficients not exceeding the number of tabulation nodes ($m + 1 \leq n$) and satisfying the interpolation conditions:

$$P_m(x_i) = y_i, \quad i = 1, 2, \ldots, n.$$

From this equation set, one can determine the polynomial coefficients and the approach is called *polynomial interpolation*.

Predominating in practice are the polynomial interpolants and, from among them, the most frequent are the *Lagrange polynomial* and the *cubic splines*. Using trigonometric functions, instead of the polynomial basis set 1, x, x^2, ..., leads to the so-called *trigonometric interpolation* and, also, to the related Fourier transform methods.

Regression. If the tabulated values $y_i = f(x_i)$ originate from observations, they are typically affected by *measurement errors*, which are usually specified as *standard deviations* σ_i (see Figure 9.1b). In such cases, the less empirical the considerations by which the model function $F = F(x; \mathbf{a})$ is established, the higher the statistical significance of the obtained model parameters. Having in view the inherent inaccuracies of the observed data, the model function is not actually expected to interpolate the data points, that is, to pass rigorously through them (even though this may, in principle, happen). It is rather natural to impose the *minimization* of the distance with respect to the set of parameters $\{a_j\}$,

$$d(f, F) = \min_{\{a_j\}}, \tag{9.6}$$

not its rigorous vanishing (like for interpolation). Concretely, this amounts to considering a model function belonging to a certain continuity class (with continuous derivatives up to a certain order) and determining the m parameters a_j that minimize the sum of squared deviations of its values from the data points:

$$S = \sum_{i=1}^{n} \left[y_i - F(x_i; \mathbf{a}) \right]^2. \tag{9.7}$$

The approach of adjusting (optimizing) the model parameters is called *regression* or *curve fitting* by the *method of least squares*. The functional (9.7) represents in a broader sense a so-called *function of merit*. By convention, merit functions are defined in such a way that lower values indicate a better agreement between the model and observations. Regression can be regarded in this respect as a *multidimensional minimization* procedure.

The merit function used most commonly in practice is the *chi-square* functional:

$$\chi^2 = \sum_{i=1}^{n} \frac{1}{\sigma_i^2} \left[y_i - F(x_i; \mathbf{a}) \right]^2. \tag{9.8}$$

Unlike the simple sum of squared deviations S (Equation 9.7), the observed data enter the χ^2-merit function with different weights, namely, inverse proportional to the individual standard deviations σ_i. The relative weights of the data points affected by lower errors are thereby increased, "forcing" the model function $F(x; \mathbf{a})$ to minimize predominantly the distances with respect to these, at the expense of the less accurate observations (with larger σ_i's). In particular, in the case of equal or unknown imprecisions σ_i, the formalism simplifies and falls back to the regression based on the S merit function.

The minimum of the χ^2 functional with respect to the model parameters a_j is characterized by the conditions:

$$\frac{\partial \chi^2}{\partial a_j} = 0, \quad j = 1, 2, \ldots, m. \tag{9.9}$$

By considering the definition (9.8) of χ^2, this leads to

$$\sum_{i=1}^{n} \frac{1}{\sigma_i^2} \left[y_i - F(x_i; \mathbf{a}) \right] \frac{\partial F(x_i; \mathbf{a})}{\partial a_j} = 0, \quad j = 1, 2, \ldots, m, \tag{9.10}$$

wherefrom the optimal parameters a_j can be determined. The curve fitting based on χ^2 is also known as *chi-square regression*.

As can be seen from Figure 9.1b, the regression curve $y = F(x; \mathbf{a})$ is not expected to pass necessarily through any of the data points $M_i(x_i, y_i)$, but only through their error bars of amplitude σ_i, thereby minimizing the functional χ^2.

Obviously, the observational errors induce uncertainties (probable errors) of the adjusted model parameters and the information on the fitting procedure is incomplete without estimating them. Assuming the observed data to be independent, the *variance* (squared probable imprecision) $\sigma_{a_j}^2$ associated with the parameter a_j can be shown to accumulate the squared imprecisions of the individual data points σ_i^2, weighted by the squared "sensitivity" $\left(\partial a_j / \partial y_i \right)^2$ of the parameter to the variation of the observations y_i (Press et al. 2007):

$$\sigma_{a_j}^2 = \sum_{i=1}^{n} \sigma_i^2 \left(\frac{\partial a_j}{\partial y_i} \right)^2. \tag{9.11}$$

In an attempt to "improve" the model function $F(x; \mathbf{a})$, to make it more amenable and able to approximate the observations (ideally, even interpolating them!), it is a misleading option to arbitrarily increase the number of model parameters without a sound theoretical justification. Not only that the model function and the parameters lack significance, but they may even lead to numerical artifacts. From this perspective, interpolation should be regarded by no means as an ideal and desirable limit of regression, since the two approaches have different significances and are applicable in qualitatively distinct situations. Indeed, Figure 9.1a and b illustrates for the same data set how different the model functions can be depending on the nature of the applied approximation, namely, interpolation or regression.

9.2 Lagrange Interpolation Polynomial

One of the oldest and most widely applied interpolation formulas is the one first published by Joseph Louis Lagrange. More than in its practical utility, the importance of the Lagrange interpolating polynomial resides in its theoretical consequences. Distinct classes of integration formulas and differentiation schemes of various orders are based on approximating the function by the Lagrange polynomial.

Assuming that, for the function $f(x)$ that needs to be modeled, there are known n discrete values corresponding to the arguments $x_1, x_2, \ldots, x_n$,

$$f(x_i) = y_i, \quad i = 1, 2, \ldots, n,$$

we tackle the problem of constructing the polynomial $P_m(x)$ of the *least degree* that interpolates all the data points (x_i, y_i), that is,

$$P_m(x_i) = y_i, \quad i = 1, 2, \ldots, n. \tag{9.12}$$

To this end, we first define a set of n auxiliary polynomials $p_i(x)$, each of which being supposed to be nonzero at just one of the tabulation nodes, x_i, and to vanish at all the rest:

$$p_i(x_j) = \delta_{ij} = \begin{cases} 1 & \text{if } j = i, \\ 0 & \text{if } j \neq i, \end{cases} \tag{9.13}$$

where δ_{ij} is the Kronecker delta. In order that $p_i(x)$ vanish at the points x_j with $j \neq i$, it can be written as a product of factors of the form $(x - x_j)$,

$$p_i(x) = C_i \prod_{j \neq i} (x - x_j),$$

from which, in particular, the factor $(x - x_i)$ is missing. Thus defined, $p_i(x)$ is a polynomial of degree $(n - 1)$ and, taking into account the imposed condition that $p_i(x_i) = 1$, one finds for the normalization constant $C_i = 1/\prod_{j \neq i}(x_i - x_j)$. Therewith, the generic member of the family of auxiliary polynomials can be written as

$$p_i(x) = \frac{\prod_{j \neq i}(x - x_j)}{\prod_{j \neq i}(x_i - x_j)}, \quad i = 1, 2, \ldots, n. \tag{9.14}$$

Owing to the property that the auxiliary polynomials $p_i(x)$ vanish at all data points except for one (namely, x_i), the polynomial $P_m(x)$ satisfying the interpolation conditions (9.12) can be defined as a linear combination of these polynomials. In addition, since the polynomials $p_i(x)$ are all of degree $(n-1)$, $P_m(x)$ is of degree $(n - 1)$ itself:

$$P_{n-1}(x) = \sum_{i=1}^{n} p_i(x) y_i. \tag{9.15}$$

Using the Properties 9.13 of the auxiliary polynomials, it can be readily shown that $P_{n-1}(x)$ satisfies, indeed, the interpolation conditions (9.12):

$$P_{n-1}(x_j) = \sum_{i=1}^{n} p_i(x_j) y_i = \sum_{i=1}^{n} \delta_{ij} y_i = y_j, \quad j = 1, 2, \ldots, n.$$

Replacing the Expressions 9.14 of the auxiliary polynomials $p_i(x)$ in (9.15), one finds the *Lagrange interpolation polynomial*:

$$P_{n-1}(x) = \sum_{i=1}^{n} \frac{\prod_{j \neq i}^{n}(x - x_j)}{\prod_{j \neq i}^{n}(x_i - x_j)} y_i. \tag{9.16}$$

By convention, the number of data points used in an interpolation scheme decreased by one is called *interpolation order* and, in the case of the Lagrange interpolant, to which *all* data points contribute, it equals the degree of the interpolation polynomial.

To demonstrate the *uniqueness of the Lagrange polynomial* as interpolating polynomial of *least degree* for a given set of n data points, we assume the contrary, namely, that there exists another polynomial $\tilde{P}_{n-1}(x)$, differing from $P_{n-1}(x)$, that equally satisfies the interpolation conditions:

$$\tilde{P}_{n-1}(x_i) = y_i, \quad i = 1, 2, \ldots, n.$$

As a consequence, the polynomial defined as the difference between $\tilde{P}_{n-1}(x)$ and $P_{n-1}(x)$,

$$Q_{n-1}(x) = \tilde{P}_{n-1}(x) - P_{n-1}(x),$$

vanishes at all n interpolation points x_i. However, since its degree is at most $(n-1)$, $Q_{n-1}(x)$ has only $(n-1)$ zeros and therefore results to be identically zero, $Q_{n-1}(x) \equiv 0$. Consequently, $\tilde{P}_{n-1}(x)$ is identical with $P_{n-1}(x)$, which proves the uniqueness of the Lagrange polynomial. It is, however, self-evident that an infinite number of polynomials of higher degree than $n-1$, or, equivalently, having a higher number of adjustable parameters than the number of interpolation conditions, can be fitted to interpolate the data points.

Formula 9.16 indicates that the Lagrange polynomial essentially provides a *nonlocal approximation*, meaning that all the data points contribute to the interpolant, not only those in the close vicinity of the argument. The same nonlocality guarantees, on the one hand, the smoothness of the approximation by the continuity of the high-order derivatives, but, on the other hand, may induce artificial oscillations of the interpolation polynomial.

In the particular case of $n = 2$ interpolation points, (x_1, y_1) and (x_2, y_2), the Lagrange polynomial reduces to

$$P_1(x) = \frac{x - x_2}{x_1 - x_2} y_1 + \frac{x - x_1}{x_2 - x_1} y_2 \tag{9.17}$$

and, obviously describes the straight line passing through the two points.

In the case $n = 3$, the Lagrange interpolant corresponds to the parabola defined by the three data points (x_1, y_1), (x_2, y_2), and (x_3, y_3):

$$P_2(x) = \frac{(x - x_2)(x - x_3)}{(x_1 - x_2)(x_1 - x_3)} y_1 + \frac{(x - x_1)(x - x_3)}{(x_2 - x_1)(x_2 - x_3)} y_2 + \frac{(x - x_1)(x - x_2)}{(x_3 - x_1)(x_3 - x_2)} y_3. \tag{9.18}$$

The Lagrange interpolant (9.16) is coded in Listing 9.1. Upon receiving n data points in the arrays `x` and `y`, the function `Lagrange` returns the interpolant value for the argument `xi`. The variant `Lagrange1`, presented in Listing 9.2, receives additionally an array `xi` of `ni` interpolation arguments, for which it returns the interpolant values via the array `yi`. For efficiency reasons, the constant factors $y_i / \prod_{j \neq i}^{n} (x_i - x_j)$ are calculated in advance.

In the sample program from Listing 9.3, the routine `Lagrange` (assumed to be contained in the module `modfunc.py`) interpolates a set of data points extracted from the function $f(x) = 1/x$ and evaluates the interpolant for several arguments.

Listing 9.1 Lagrange Interpolant (Python Coding)

```
#===========================================================================
def Lagrange(x, y, n, xi):
#---------------------------------------------------------------------------
#  Evaluates the Lagrange interpolating polynomial of n data points at xi
#  x[] - x-coordinates of data points
#  y[] - y-coordinates of data points
#  n   - number of data points
#  xi  - interpolation argument
#---------------------------------------------------------------------------
   yi = 0e0
   for i in range(1,n+1):
      p = 1e0
      for j in range(1,n+1):
         if (j != i): p *= (xi - x[j])/(x[i] - x[j])
      yi += p * y[i]

   return yi
```

Listing 9.2 Lagrange Interpolation for Multiple Arguments (Python Coding)

```
#===========================================================================
def Lagrange1(x, y, n, xi, yi, ni):
#---------------------------------------------------------------------------
#  Evaluates the Lagrange interpolating polynomial of n data points on a mesh
#  of ni interpolation points
#  x[]  - x-coordinates of data points
#  y[]  - y-coordinates of data points
#  n    - number of data points
#  xi[] - interpolation arguments
#  yi[] - interpolant values (output)
#  ni   - number of interpolation points
#---------------------------------------------------------------------------
   yf = [[0]*(n+1) for i in range(n+1)]          # factors of the interpolant

   for i in range(1,n+1):        # prepare invariant factors of the interpolant
      p = 1e0
      for j in range(1,n+1):
         if (j != i): p *= (x[i] - x[j])
      yf[i] = y[i] / p

   for k in range(1,ni+1):                       #  loop over interpolation points
      xk = xi[k]; yk = 0e0
      for i in range(1,n+1):
         p = 1e0
         for j in range(1,n+1):
            if (j != i): p *= (xk - x[j])
         yk += p * yf[i]

      yi[k] = yk
```

One of the frequent causes for *spurious oscillations* of the Lagrange polynomial is the inappropriate distribution of the data points. This phenomenon is clearly illustrated for the function $f(x) = 1/x$ (having, indeed, a slowly converging power series representation), by considering the hand-picked arguments 0.15, 0.2, 0.3, 0.5, 0.8, 1.1, 1.4, and 1.7 (see Figure 9.2). Despite the rather uniform distribution of the interpolated points *along* the curve, the relatively broad intervals between the larger arguments cause the Lagrange interpolant to fail to reproduce the modeled dependence and to feature increasingly ample oscillations. On the contrary, by uniformly distributing the tabulation abscissas, the oscillations are reduced significantly. Moreover, increasing by just one the number of equally spaced mesh points, the oscillations vanish altogether. Consequently, if not backed by a judicious distribution, the mere increase of the data set does not necessarily enhance the interpolation, since the concomitant increase of the polynomial order causes additional zeros and, consequently, also supplementary oscillations to appear.

In contrast with the behavior of the Lagrange polynomial illustrated in Figure 9.2, the piecewise continuous *cubic splines* (discussed in detail in Section 9.4) can be seen to provide a clearly superior interpolation, devoid of artificial oscillations. Qualitatively, the improved performance of the cubic splines can be traced back to their lower order, resulting in a low number of zeros and a reduced oscillation tendency.

In spite of the detrimental side effects of its nonlocality, the Lagrange polynomial remains an important and widespread interpolation tool. Based on it, yet another practical problem can be solved, namely, the *inverse interpolation*, which essentially amounts to finding the argument for which the Lagrange

Listing 9.3   Interpolation Using the Lagrange Polynomial (Python Coding)

```python
# Lagrange interpolation
from modfunc import *

# main

n  = 8                                            # number of data points
ni = 100                                  # number of interpolation points

x = [0]*(n+1)                                            # data points
y = [0]*(n+1)

x[1] = 0.15; x[2] = 0.2; x[3] = 0.3; x[4] = 0.5          # data points:
x[5] = 0.8 ; x[6] = 1.1; x[7] = 1.4; x[8] = 1.7          # f(x) = 1/x
for i in range(1,n+1): y[i] = 1e0/x[i]

out = open("interpol.txt","w")                       # open output file
out.write("       x             y            xi           yi          f\n")
h = (x[n]-x[1])/(ni-1)
for i in range(1,ni+1):
   xi = x[1] + (i-1)*h                           # interpolation argument
   yi = Lagrange(x,y,n,xi)                           # interpolant value
   if (i <= n):
      out.write("{0:12.3e}{1:12.3e}{2:12.3e}{3:12.3e}{4:12.3e}\n".
               format(x[i],y[i],xi,yi,1e0/xi))
   else:
      out.write("{0:36.3e}{1:12.3e}{2:12.3e}\n".format(xi,yi,1e0/xi))
out.close()
```

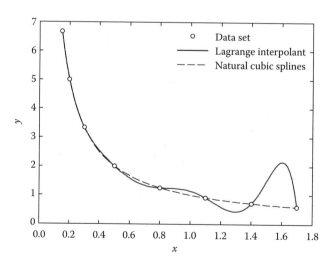

FIGURE 9.2   Case in which the Lagrange interpolant features artificial oscillations. The interpolated data sample the function $f(x) = 1/x$.

polynomial takes a given value. This can be accomplished by simply inverting the roles of x and y in Equation 9.16, respectively, considering y as independent variable and expressing x as interpolation outcome:

$$x = \sum_{i=1}^{n} \frac{\prod\limits_{j \neq i}(y - y_j)}{\prod\limits_{j \neq i}(y_i - y_j)} x_i. \tag{9.19}$$

In particular, this formula can be applied to determine an approximate root of the equation $f(x) = 0$. To this end, it is sufficient to calculate function values $y_i = f(x_i)$ for n arguments x_i close to the root. By setting $y = 0$ in (9.19), one finds as x an approximation to the root. In the ideal case, if $f(x) \equiv P_{n-1}(x)$ is a polynomial of degree $(n - 1)$, the root determined by this approach is *numerically exact*.

9.3 Neville's Interpolation Method

A shortcoming of the direct application of the Lagrange polynomial (9.16) is the absence of an usable error estimate associated with the interpolant values. Indeed, the error could be estimated by comparing the interpolant for successively embedded data sets. Such an approach does not allow though for the efficient, recurrent reuse of the interpolant of a given order in evaluating the higher-order interpolants and is, therefore, expensive.

Neville's method (named after the English mathematician Eric Harold Neville) is not a distinct interpolation approach in its own right, being rather an optimized algorithm for constructing the *unique* (Lagrange) interpolation polynomial for a given number of data points, which provides an error estimate at the same time. The algorithm involves essentially an efficient iterative evaluation of interpolants of increasing order, constructed on successively extended data sets, along with the full reuse of the previous results.

Let function $f(x)$ be specified by n values:

$$f(x_i) = y_i, \quad i = 1, 2, \ldots, n.$$

With a view to efficiently constructing the Lagrange polynomial that interpolates the n data points (x_i, y_i), we start by defining a family of auxiliary interpolation polynomials. Let $P_{i,j}(x)$ be the unique (Lagrange) polynomial that interpolates the data points with the abscissas comprised between x_i and x_j, thus satisfying the interpolation conditions:

$$P_{i,j}(x_k) = y_k, \quad i \leq k \leq j. \tag{9.20}$$

Since it interpolates $j - i + 1$ data points, the polynomial $P_{i,j}(x)$ is of degree $j - i$. In particular, $P_{i,i}(x)$ are polynomials of degree 0, which interpolate only the individual data points (x_i, y_i), and are therefore constant functions taking, respectively, the values y_i, independent of their arguments:

$$P_{i,i}(x) \equiv y_i, \quad i = 1, 2, \ldots, n. \tag{9.21}$$

Obviously, the searched-for Lagrange polynomial may be identified with $P_{1,n}(x)$, which is the unique interpolation polynomial of least degree with respect to the entire data set:

$$P_{n-1}(x) \equiv P_{1,n}(x). \tag{9.22}$$

The efficient evaluation of $P_{1,n}(x)$ for a given argument x can be accomplished by arranging the values $P_{i,j}(x)$ in the following table with "predecessors" and "descendants," the interpolation order on each column being constant and equal to $m = j - i$:

	$m = 0$	1	2	$\cdots$	$n - 2$	$n - 1$
x_1	$P_{1,1} = y_1$					
		$P_{1,2}$				
x_2	$P_{2,2} = y_2$		$P_{1,3}$			
		$P_{2,3}$		$\cdots$		
x_3	$P_{3,3} = y_3$		$\cdots$		$P_{1,n-1}$	
		$\cdots$		$P_{2,n-1}$		$P_{1,n} \equiv P_{n-1}(x)$
$\cdots$	$\cdots$		$\cdots$		$P_{2,n}$	
		$P_{n-2,n-1}$		$\cdots$		
x_{n-1}	$P_{n-1,n-1}$		$P_{n-2,n}$			
		$P_{n-1,n}$				
x_n	$P_{n,n} = y_n$					

Neville's algorithm actually prescribes the recursive filling of the columns, from left to right, starting with the polynomials $P_{i,i}(x)$ of the lowest degree (0), and leading on the last column to a single descendant that can be identified with the interpolant value, $P_{1,n} \equiv P_{n-1}(x)$. The propagation can be carried out based on the recurrence relation between a given descendant $P_{i,j}(x)$ (of degree $j - i$) and its two predecessors $P_{i,j-1}(x)$ and $P_{i+1,j}$ (of degree $j - i - 1$), located one underneath the other on the column on the left:

$$P_{i,j}(x) = \frac{(x - x_j)P_{i,j-1}(x) + (x_i - x)P_{i+1,j}(x)}{x_i - x_j}. \tag{9.23}$$

In fact, $P_{i,j}(x)$ satisfies the interpolation conditions (9.20) for any x_k with $i < k < j$, since the predecessors $P_{i,j-1}(x)$ and $P_{i+1,j}(x)$ satisfy themselves these conditions for the common subset of data points $x_{i+1}, \ldots, x_{j-1}$:

$$P_{i,j}(x_k) = \frac{(x_k - x_j)y_k + (x_i - x_k)y_k}{x_i - x_j} = y_k, \quad x_i < x_k < x_j.$$

In particular, for the boundaries x_i and x_j, the interpolation condition is met due to the alternative cancellation of the binomials $(x - x_j)$ and, respectively, $(x_i - x)$:

$$P_{i,j}(x_i) = \frac{(x_i - x_j)y_i + (x_i - x_i)P_{i+1,j}(x_i)}{x_i - x_j} = y_i,$$

$$P_{i,j}(x_j) = \frac{(x_j - x_j)P_{i,j-1}(x_j) + (x_i - x_j)y_j}{x_i - x_j} = y_j.$$

Consequently, the polynomial $P_{i,j}(x)$ defined by the recurrence relation (9.23) interpolates all the $j - i + 1$ data points $(x_i, y_i), \ldots, (x_j, y_j)$ and, being of degree $j - i$, it represents in fact the unique Lagrange interpolation polynomial.

Since the table of values $P_{i,j}(x)$ has to be filled *columnwise* and the indexes i and j are linked by $j = m + i$ to the interpolation order (column index) m, the latter appears to be more suitable than j to be used along with the data index i in identifying the table elements. Correspondingly, the recurrence (9.23) may be rewritten under an algorithmically more suitable form by introducing the notations $P_i^{(m)}(x) \equiv P_{i,j}(x)$ and $x_{m+i} \equiv x_j$:

$$P_i^{(m)}(x) = \frac{(x - x_{m+i})P_i^{(m-1)}(x) + (x_i - x)P_{i+1}^{(m-1)}(x)}{x_i - x_{m+i}}, \tag{9.24}$$

$$m = 1, 2, \ldots, n - 1, \quad i = 1, 2, \ldots, n - m.$$

The row index i on every successive column m takes $n - m$ values, one less than on the preceding column. Every pair $P_i^{(m-1)}$ and $P_{i+1}^{(m-1)}(x)$ results in a new value $P_i^{(m)}(x)$, and $P_i^{(m-1)}$ becomes unnecessary thereafter. As a consequence, in concrete implementations, instead of a two-dimensional array, a single vector for storing the polynomial values $P_i^{(m)}(x)$ of the current column is sufficient, with $P_i^{(m-1)}(x)$ being replaced by $P_i^{(m)}(x)$.

The difference between the final interpolant value, $P_{n-1}(x) \equiv P_1^{(n-1)}(x)$, and one of its two predecessors, for example, $P_2^{(n-2)}(x)$, provides a useful and conveniently implementable estimate of the absolute error associated with the approximation of the tabulated dependence by the Lagrange polynomial:

$$\Delta = \left| P_1^{(n-1)}(x) - P_2^{(n-2)}(x) \right|.$$

The implementation of Neville's method given in Listing 9.4 maintains the same significance of the arguments as for the routine `Lagrange` (discussed in the previous section). x and y are arrays containing on entry the n coordinates of the data points, while xi is the argument for which the interpolant is evaluated. In addition to the interpolant value returned by the function's name, the parameter err is used to return the associated error estimate.

In the usage example of the function `Neville`, as shown in Listing 9.5, the interpolated data set is the same as for Program 9.3 from the previous section. The modeled function, $f(x) = 1/x$, can be seen in Figure 9.3 to pass right through the error bars associated with the interpolated values. It is worth noting, however, that the estimated errors are appreciable in the region of artificial oscillations of the interpolant.

Listing 9.4 Interpolant Based on Neville's Algorithm (Python Coding)

```
#===============================================================================
def Neville(x, y, n, xi):
#-------------------------------------------------------------------------------
#  Evaluates the Lagrange interpolating polynomial of n data points at xi by
#  Neville's method and returns an estimate of the absolute error
#  x[] - x-coordinates of data points
#  y[] - y-coordinates of data points
#  n   - number of data points
#  xi  - interpolation argument
#  err - estimate of the absolute error (output)
#-------------------------------------------------------------------------------
   p = [[0]*(n+1) for i in range(n+1)]      # values of successive interpolants

   for i in range(1,n+1): p[i] = y[i]      # initialize with 0-order polynomial

   for m in range(1,n):                         # loop over columns of scheme array
      for i in range(1,n-m+1):                      # increase polynomial order
         p[i] = ((xi-x[m+i])*p[i] + (x[i]-xi)*p[i+1]) / (x[i]-x[m+i])

   yi = p[1]                                        # polynomial value
   err = fabs(p[1] - p[2])                          # error estimate

   return (yi, err)
```

Listing 9.5   Interpolation Using Neville's Method (Python Coding)

```
# Interpolation by Neville's algorithm
from modfunc import *

# main

n  = 8                                              # number of data points
ni = 50                                      # number of interpolation points

x = [0]*(n+1)                                                 # data points
y = [0]*(n+1)

x[1] = 0.15; x[2] = 0.2; x[3] = 0.3; x[4] = 0.5          # data points:
x[5] = 0.8 ; x[6] = 1.1; x[7] = 1.4; x[8] = 1.7         # f(x) = 1/x
for i in range(1,n+1): y[i] = 1e0/x[i]

out = open("interpol.txt","w")                         # open output file
out.write("        x                y            xi            yi            err\n")
h = (x[n]-x[1])/(ni-1)
for i in range(1,ni+1):
    xi = x[1] + (i-1)*h                            # interpolation argument
    (yi,err) = Neville(x,y,n,xi)                        # interpolant value
    if (i <= n):
        out.write("{0:12.3e}{1:12.3e}{2:12.3e}{3:12.3e}{4:12.3e}\n".
                  format(x[i],y[i],xi,yi,err))
    else:
        out.write("{0:36.3e}{1:12.3e}{2:12.3e}\n".format(xi,yi,err))
out.close()
```

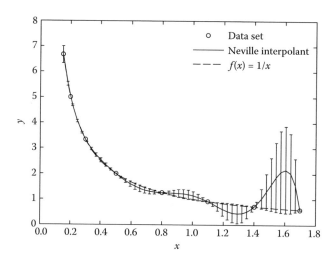

FIGURE 9.3   Plot of the Neville interpolant and the associated error bars for a case of spurious oscillations. The interpolated data sample the function $f(x) = 1/x$.

9.4 Cubic Spline Interpolation

As illustrated by the applications presented in the previous two sections, the major drawback of the Lagrange polynomial, employed as single interpolant over the entire range of tabulated data, is the possible appearance of spurious oscillations. The phenomenon can be traced back to the typically large degree and, correspondingly, large number of zeros of the interpolant, which are directly determined by the number of interpolated data points. The desirable reduction of the polynomial degree down to a reasonable degree of smoothness can be obviously achieved by considering separate lower-degree interpolation polynomials on adjacent subintervals (subsets of data points) of the total interval. Such an approach is referred to as *piecewise-polynomial interpolation.*

The separate *restrictions* of the interpolant on the different subintervals can be in principle independent. However, under such conditions, the differentiability of the interpolant is not guaranteed at the subinterval end points. If the overall smoothness (expressed by the maximum order of the continuous derivatives) is essential, then some additional constraints have to be placed on the restrictions in order to ensure the continuity of their derivatives up to a certain order at the subinterval boundaries. Such a piecewise-polynomial approximation, with coupled interpolant restrictions, is called *spline interpolation* and the restrictions, themselves, are called *splines.*

The "spline" functions were originally inspired by the "draftsman's spline," which used to be the tool employed in technical drawing before the computer era to shape smooth flat curves passing through given points. The instrument consisted of a flexible strip placed on a horizontal drawing board, sometimes strained by fixing its ends, and held in the desired curved conformation by pinning or by the frictional resistance experienced by weights attached to certain points. The spline would assume a shape corresponding to the minimum strain energy and, implicitly, to the minimum curvature (which is proportional to the former). The spline functions, aimed at imitating the draftsman's spline, benefit from the remarkable property of minimal curvature, reducing thus the adverse phenomenon of oscillation between the interpolation nodes, specific to higher-degree polynomial interpolants.

Let Δ be a division of the interval $[\alpha, \beta]$,

$$\Delta : \alpha = x_1 < x_2 < \cdots < x_n = \beta,$$

and $f : [\alpha, \beta] \to \mathbb{R}$ be a tabular dependence:

$$y_i = f(x_i), \quad i = 1, 2, \ldots, n.$$

To model $f(x)$ for arguments differing from those of the division, we construct a piecewise spline interpolant. We term *mth-order spline interpolant* relative to the division Δ a function $S : [\alpha, \beta] \to \mathbb{R}$ whose restrictions $S_i(x)$ on the subintervals $[x_i, x_{i+1}]$ are *m*th-degree polynomials (splines). The piecewise interpolant thus defined belongs to the class $C^{m-1}[\alpha, \beta]$, having the first $(m-1)$ derivatives continuous over the entire interval $[\alpha, \beta]$ and the *m*th derivative discontinuous at the nodes x_i.

The most widely used spline restrictions are third-degree polynomials, called *cubic splines*:

$$S_i(x) = A_i x^3 + B_i x^2 + C_i x + D_i, \tag{9.25}$$

$$x \in [x_i, x_{i+1}], \quad i = 1, 2, \ldots, n - 1.$$

The interpolant S having these restrictions obviously belongs to the class $C^2[\alpha, \beta]$, namely, it is *twice continuously differentiable.* The cubic splines ensure, on the one hand, sufficient overall smoothness for most practical applications, not favoring, on the other hand, the appearance of oscillations due to their low order. Moreover, owing to the simplicity of their functional form, they provide the basis for versatile integration and differentiation schemes.

The coefficients A_i, B_i, C_i, and D_i of the splines can be determined, in principle, from the continuity and interpolation conditions at the interpolation nodes x_i:

$$S(x_i) = y_i, \quad i = 1, 2, \ldots, n. \tag{9.26}$$

Unlike the first and second derivatives, the third derivatives (equal to $6A_i$) are discontinuous at the subinterval boundaries.

The second spline derivative, $S_i''(x)$, is a linear function defined in the interval $[x_i, x_{i+1}]$ by the equation

$$\frac{S_i''(x) - \tilde{D}_i}{x - x_i} = \frac{\tilde{D}_{i+1} - \tilde{D}_i}{x_{i+1} - x_i}, \tag{9.27}$$

where $\tilde{D}_i \equiv S''(x_i)$ and $\tilde{D}_{i+1} \equiv S''(x_{i+1})$ are the boundary values of the second derivative. From this equation, one may express

$$S_i''(x) = \frac{\tilde{D}_{i+1}(x - x_i) + \tilde{D}_i(x_{i+1} - x)}{h_i}, \tag{9.28}$$

with

$$h_i = x_{i+1} - x_i, \quad i = 1, 2, \ldots, n - 1. \tag{9.29}$$

Integrating Equation 9.28 twice, one obtains successively the first derivative and the spline itself:

$$S_i'(x) = \frac{\tilde{D}_{i+1}(x - x_i)^2 - \tilde{D}_i(x_{i+1} - x)^2}{2h_i} + C_i', \tag{9.30}$$

$$S_i(x) = \frac{\tilde{D}_{i+1}(x - x_i)^3 + \tilde{D}_i(x_{i+1} - x)^3}{6h_i} + C_i'x + C_i''. \tag{9.31}$$

The integration constants C_i' and C_i'' may be determined by imposing on the spline the interpolation conditions at the nodes x_i and x_{i+1}, that is,

$$S_i(x_i) = y_i, \quad S_i(x_{i+1}) = y_{i+1},$$

which result in the system:

$$\begin{cases} \dfrac{\tilde{D}_i h_i^2}{6} + C_i'x_i + C_i'' = y_i \\[2mm] \dfrac{\tilde{D}_{i+1} h_i^2}{6} + C_i'x_{i+1} + C_i'' = y_{i+1}. \end{cases}$$

Solving this system for C_i' and C_i'' yields

$$C_i' = \frac{y_{i+1} - y_i}{h_i} - \frac{(\tilde{D}_{i+1} - \tilde{D}_i)h_i}{6}, \tag{9.32}$$

$$C_i'' = \frac{x_{i+1}y_i - x_iy_{i+1}}{h_i} + \frac{(x_i\tilde{D}_{i+1} - x_{i+1}\tilde{D}_i)h_i}{6}. \tag{9.33}$$

Upon replacement of the integration constants back into Equation 9.31 and collection of the equal powers of x, one obtains the polynomial coefficients of the spline $S_i(x)$ in terms of the second derivatives $\tilde{D}_i$ at the interpolation mesh points:

$$\begin{cases} A_i = \dfrac{\tilde{D}_{i+1} - \tilde{D}_i}{6h_i}, \\[2mm] B_i = \dfrac{\tilde{D}_i x_{i+1} - \tilde{D}_{i+1} x_i}{2h_i}, \\[2mm] C_i = \dfrac{\tilde{D}_{i+1} x_i^2 - \tilde{D}_i x_{i+1}^2}{2h_i} + \dfrac{y_{i+1} - y_i}{h_i} - A_i h_i^2, \\[2mm] D_i = \dfrac{\tilde{D}_i x_{i+1}^3 - \tilde{D}_{i+1} x_i^3}{6h_i} + \dfrac{y_i x_{i+1} - y_{i+1} x_i}{h_i} - \dfrac{B_i h_i^2}{3}. \end{cases} \qquad (9.34)$$

For the complete definition of the splines, one still needs to determine the values $\tilde{D}_i$ of the second derivatives and this can be done by imposing the continuity of the first derivative at the interpolation nodes:

$$S'_{i-1}(x_i) = S'_i(x_i), \quad i = 2, 3, \ldots, n - 1. \qquad (9.35)$$

Using the Expression 9.30 of the first derivative and the form (9.32) of the coefficients C_i, the above continuity conditions result in

$$\frac{h_{i-1}}{6} \tilde{D}_{i-1} + \frac{h_{i-1} + h_i}{3} \tilde{D}_i + \frac{h_i}{6} \tilde{D}_{i+1} = \frac{y_{i+1} - y_i}{h_i} - \frac{y_i - y_{i-1}}{h_{i-1}}, \qquad (9.36)$$

$$i = 2, 3, \ldots, n - 1.$$

The System 9.36 comprises only $(n - 2)$ equations for the n second derivatives $\tilde{D}_i$ and, consequently, for their unequivocal determination, two additional equations need to be supplied. These supplementary equations can be explicitly derived for two important types of boundaries associated with the spline interpolants: *free* or *natural boundaries* and *clamped boundaries*.

The *clamped boundaries* require the first derivatives at the end points, y'_1 and y'_n, to be specified and, correspondingly, the two additional relations are

$$S'_1(x_1) = y'_1, \quad S'_{n-1}(x_n) = y'_n,$$

or, explicitly,

$$\frac{h_1}{3} \tilde{D}_1 + \frac{h_1}{6} \tilde{D}_2 = \frac{y_2 - y_1}{h_1} - y'_1$$

$$\frac{h_{n-1}}{6} \tilde{D}_{n-1} + \frac{h_{n-1}}{3} \tilde{D}_n = y'_n - \frac{y_n - y_{n-1}}{h_{n-1}}.$$

Therewith, the following tridiagonal system results, having as solution the second derivatives $\tilde{D}_i$ of the splines at the interpolation nodes:

$$\begin{cases} b_1 \tilde{D}_1 + c_1 \tilde{D}_2 = d_1 \\ a_i \tilde{D}_{i-1} + b_i \tilde{D}_i + c_i \tilde{D}_{i+1} = d_i, \quad i = 2, 3, \ldots, n - 1 \\ a_n \tilde{D}_{n-1} + b_n \tilde{D}_n = d_n, \end{cases} \qquad (9.37)$$

where the coefficients read:

$$
\begin{cases}
a_1 = 0, & b_1 = 2h_1, & c_1 = h_1, & d_1 = 6\left(\dfrac{y_2 - y_1}{h_1} - y_1'\right), \\[3mm]
a_i = h_{i-1}, & b_i = 2(h_{i-1} + h_i), & c_i = h_i, & d_i = 6\left(\dfrac{y_{i+1} - y_i}{h_i} - \dfrac{y_i - y_{i-1}}{h_{i-1}}\right), \\[3mm]
a_n = h_{n-1}, & b_n = 2h_{n-1}, & c_n = 0, & d_n = 6\left(y_n' - \dfrac{y_n - y_{n-1}}{h_{n-1}}\right).
\end{cases}
\qquad (9.38)
$$

The *free boundaries* are characteristic for the so-called *natural cubic splines* and assume the second derivatives to be identically zero at the end points of the tabulation interval:

$$
\tilde{D}_1 = 0, \quad \tilde{D}_n = 0. \qquad (9.39)
$$

Instead of programming them separately, the above conditions can be treated in a unifying manner by considering the particular coefficients in the System 9.37:

$$
c_1 = 0, \quad d_1 = 0,
$$
$$
a_n = 0, \quad d_n = 0.
$$

Essentially, the concrete implementation of the cubic spline interpolation involves:

1. Evaluation of the second derivatives $\tilde{D}_i$ by solving the System 9.37 and 9.38,
2. Calculation of the spline coefficients A_i, B_i, C_i, and D_i by using Relations 9.34,
3. Evaluation of the interpolant for a particular set of arguments.

The routine Spline given in Listing 9.6 implements the above formalism. The input is represented by the arrays x and y, conveying the n coordinates of the interpolated points, as well as the scalars d1 and dn, which contain the values of the first derivatives y_1' and y_n' at the end points of the tabulation interval. A value of 0 passed to the control parameter iopt determines the use of *natural splines*, with vanishing second derivatives at the boundaries, whereas a nonzero value determines the use of *clamped splines*, with the derivatives d1 and dn provided as input.

The main output of the routine is represented by the coefficients a, b, c, and d of the $n-1$ splines. They are particularly useful for performing differentiation or integration operations based on compact and easily programmable analytical expressions. Optionally, the routine can evaluate the interpolant internally for ni arguments received in the array xi and returns the corresponding values by the array yi.

To solve the tridiagonal system (9.37) for the second derivatives of the splines in the interpolation nodes, Spline calls the specialized routine TriDiagSys, presented in Chapter 7 and included in the library linsys.py (respectively, linsys.h, for C/C++ coding). On exit from TriDiagSys, the array d contains the second derivatives, which replace the initial constant terms of the system.

The final sequence of the routine Spline evaluates the interpolant for the ni arguments contained in the array xi. The specific spline to be evaluated for each particular argument xx is determined by incrementing the restriction index ip from 1 until a subinterval is reached whose right boundary exceeds the argument. Once the restriction index is determined, Horner's scheme is applied for the efficient evaluation of the respective spline.

Listing 9.7 presents a program which uses the routine Spline (assumed to be contained in the file modfunc.py) to interpolate with natural splines a set of data points sampled from the function $f(x) = \sin x / x$. The code also illustrates how the first two spline derivatives are calculated from their analytic expression, using the determined coefficients.

Listing 9.6 Cubic Spline Interpolant (Python Coding)

```
#===========================================================================
def Spline(x, y, n, d1, dn, iopt, a, b, c, d, xi, yi, ni):
#---------------------------------------------------------------------------
# Calculates the coefficients of the cubic splines for n data points and
# returns interpolated values on a mesh of ni points
# x[]     - x-coordinates of data points
# y[]     - y-coordinates of data points
# n       - number of data points
# d1, dn - 1st derivatives at the endpoints x[1] and x[n]
# iopt    - iopt == 0, natural splines: 2nd derivative = 0 at x[1] and x[n]
#           iopt == 1, clamped splines: uses the provided d1 and dn
# a[], b[], c[], d[], i=1..n
#         - coefficients of cubic splines i=1..n-1 (output)
# xi[]    - interpolation arguments
# yi[]    - interpolant values (output)
# ni      - number of interpolation points
# Calls: TriDiagSys (linsys.py)
#---------------------------------------------------------------------------
   if (iopt == 0): d1 = dn = 0e0           # initialization for natural splines

   hi = 0e0; di = d1                  # coefficients for system of 2nd derivatives
   for i in range(1,n):
      hm = hi; hi = x[i+1] - x[i]
      dm = di; di = (y[i+1] - y[i])/hi
      a[i] = hm; b[i] = 2e0*(hm + hi); c[i] = hi; d[i] = 6e0*(di - dm)

   a[n] = hi; b[n] = 2e0*hi; c[n] = 0e0; d[n] = 6e0*(dn - di)

   if (iopt == 0): c[1] = d[1] = a[n] = d[n] = 0e0          # natural splines

   TriDiagSys(a,b,c,d,n)                          # solve tridiagonal system
                                                  # 2nd derivatives in d
   for i in range(1,n):                           # spline coefficients
      ip = i + 1
      xx = x[i]; xp = x[ip]; hi = xp - xx
      a[i] = (d[ip] - d[i]) / (6e0*hi)
      b[i] = (d[i]*xp - d[ip]*xx) / (2e0*hi)
      c[i] = (d[ip]*xx*xx - d[i]*xp*xp) / (2e0*hi) \
                                   + (y[ip] - y[i]) / hi - a[i]*hi*hi
      d[i] = (d[i]*xp*xp*xp - d[ip]*xx*xx*xx) / (6e0*hi) \
                          + (y[i]*xp - y[ip]*xx) / hi - b[i]*hi*hi/3e0

   for i in range(1,ni):                          # interpolation loop
      xx = xi[i]
      ip = 1
      while (ip < n-1 and xx > x[ip+1]): ip += 1           # index of spline
      yi[i] = ((a[ip]*xx + b[ip])*xx + c[ip])*xx + d[ip]   # evaluate spline
```

Owing to their nonoscillatory behavior, spline interpolants provide, in general, improved approximations as compared to Lagrange polynomials. As also pointed out in Section 9.2 (see Figure 9.2), the cubic splines can smoothly model dependences for which the Lagrange polynomial fails and features significant artificial oscillations. Moreover, the increase of the number of interpolated data points merely improves the spline approximation, but does not also entail the unwanted increase of the interpolation order.

Listing 9.7   Interpolation Using Cubic Splines (Python Coding)

```python
# Interpolation with cubic splines
from modfunc import *

def sinc(x): return (sin(x)/x if x else 1e0)

# main

xmin = 0e0; xmax = 5*asin(1e0)          # tabulation interval: [0,5*pi/2]
n = 6                                       # number of data points
ni = 100                                # number of interpolation points

a = [0]*(n+1); b = [0]*(n+1)                   # spline coefficients
c = [0]*(n+1); d = [0]*(n+1)
x = [0]*(n+1); y = [0]*(n+1)                       # data points

h = (xmax-xmin)/(n-1)                          # generate data points
for i in range(1,n+1):
    xi = xmin + (i-1)*h
    x[i] = xi; y[i] = sinc(xi)

Spline(x,y,n,0e0,0e0,0,a,b,c,d,x,y,0)                  # natural splines

out = open("interpol.txt","w")                    # open output file
out.write("       x             y             xi            yi      " +
          "     y1            y2            f\n")
h = (xmax-xmin)/(ni-1)
for i in range(1,ni+1):
    xi = xmin + (i-1)*h                         # interpolation argument
    ip = 1
    while (ip < n-1 and xi > x[ip+1]): ip += 1           # index of spline

    yi = ((a[ip]*xi + b[ip])*xi + c[ip])*xi + d[ip]           # spline
    y1 = (3*a[ip]*xi + 2*b[ip])*xi + c[ip]               # 1st derivative
    y2 = (6*a[ip]*xi + 2*b[ip])                           # 2nd derivative

    if (i <= n):
        out.write(
        "{0:12.3e}{1:12.3e}{2:12.3e}{3:12.3e}{4:12.3e}{5:12.3e}{6:12.3e}\n".
                   format(x[i],y[i],xi,yi,y1,y2,sinc(xi)))
    else:
        out.write("{0:36.3e}{1:12.3e}{2:12.3e}{3:12.3e}{4:12.3e}\n".
                   format(xi,yi,y1,y2,sinc(xi)))
out.close()
```

On the other hand, however, insufficient interpolated data points, especially in regions of pronounced variations, may lead even for spline interpolants to deficient approximations. One such example is depicted in Figure 9.4 and shows in the case of the function $f(x) = 1/x^{12} - 10^4/x^6$ that, due to the sparse tabulation mesh, the spline cannot reproduce in the left-most subinterval both the appreciable slope and the curvature.

Besides the density of the interpolated data points, the quality of the spline interpolation is also directly affected by the way in which the first derivatives at the end points of the tabulation interval are modeled. The output produced by Program 9.7 is plotted in the left panel of Figure 9.5 and shows that *natural boundaries* (second derivatives implicitly set to 0) lead to visible deviations of the interpolant from

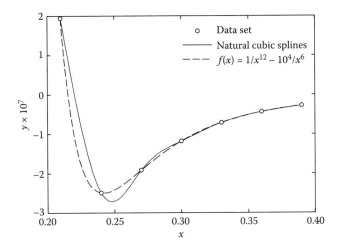

FIGURE 9.4 Case in which the natural cubic spline deviates significantly from the modeled function due to insufficient interpolated points. The interpolated data samples the function $f(x) = 1/x^{12} - 10^4/x^6$.

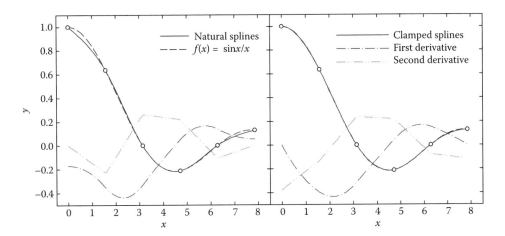

FIGURE 9.5 Interpolation with natural cubic splines (left) and clamped cubic splines (right). The modeled dependence, $f(x) = \sin x / x$, is sampled in the interval $[0, 5\pi/2]$ with a $\pi/2$ spacing.

the true behavior of the modeled function. Quite in contrast, *clamped boundaries* clearly improve the matching between the interpolant and the modeled dependence (see the right panel of Figure 9.5).

9.5 Linear Regression

In quite a few practical situations, the plot of the observations or the theoretical framework on which the data collection is based suggests a linear functional dependence of the modeled function on the independent variable (see Figure 9.6):

$$F(x; a, b) = ax + b. \tag{9.40}$$

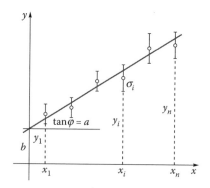

FIGURE 9.6 Linear regression based on the least-squares approximation. The parameter a of the model function represents the slope of the regression line, while b is the corresponding y-intercept.

Linear regression actually represents the simplest application of the general concepts regarding the least-squares fitting of model functions to tabulated dependences, which have been presented in Section 9.1.

Assuming the uncertainties σ_i associated with the observed values y_i to be available, the quality of the model function $F(x; a, b)$ adjusted to the observed data may be quantified by the χ^2-merit function (Equation 9.8), taking in this case the form:

$$\chi^2(a, b) = \sum_{i=1}^{n} \frac{1}{\sigma_i^2} (y_i - ax_i - b)^2. \tag{9.41}$$

The optimal model parameters a and b, which minimize χ^2, result from the condition of vanishing partial derivatives of χ^2 with respect to a and b:

$$\frac{\partial \chi^2}{\partial a} \equiv -2 \sum_{i=1}^{n} \frac{1}{\sigma_i^2} (y_i - ax_i - b)x_i = 0,$$

$$\frac{\partial \chi^2}{\partial b} \equiv -2 \sum_{i=1}^{n} \frac{1}{\sigma_i^2} (y_i - ax_i - b) = 0. \tag{9.42}$$

By introducing the notations for the involved sums

$$s = \sum_{i=1}^{n} \frac{1}{\sigma_i^2}, \quad s_x = \sum_{i=1}^{n} \frac{x_i}{\sigma_i^2}, \quad s_{xx} = \sum_{i=1}^{n} \frac{x_i^2}{\sigma_i^2}, \quad s_y = \sum_{i=1}^{n} \frac{y_i}{\sigma_i^2}, \quad s_{xy} = \sum_{i=1}^{n} \frac{x_i y_i}{\sigma_i^2}, \tag{9.43}$$

one obtains the simplified system:

$$\begin{cases} s_{xx}a + s_x b = s_{xy} \\ s_x a + s\, b = s_y. \end{cases} \tag{9.44}$$

With the additional notation

$$\Delta = s\, s_{xx} - s_x^2, \tag{9.45}$$

the solution of the system provides the values of the optimized model parameters:

$$a = \frac{s\, s_{xy} - s_x s_y}{\Delta}, \quad b = \frac{s_y s_{xx} - s_x s_{xy}}{\Delta}. \tag{9.46}$$

To evaluate the probable imprecisions associated with the optimal model parameters (9.46), one can apply the general definition (9.11) of the variance:

$$\sigma_a^2 = \sum_{i=1}^{n} \sigma_i^2 \left(\frac{\partial a}{\partial y_i} \right)^2, \quad \sigma_b^2 = \sum_{i=1}^{n} \sigma_i^2 \left(\frac{\partial b}{\partial y_i} \right)^2.$$

By using the specific expressions (9.46) of the parameters, one obtains the derivatives:

$$\frac{\partial a}{\partial y_i} = \frac{s\, x_i - s_x}{\sigma_i^2 \Delta}, \quad \frac{\partial b}{\partial y_i} = \frac{s_{xx} - s_x x_i}{\sigma_i^2 \Delta}.$$

Summation over all data points yields, for example, for the variance of the parameter a:

$$\sigma_a^2 = \sum_{i=1}^{n} \frac{s^2 x_i^2 - 2s\, x_i s_x + s_x^2}{\sigma_i^2 \Delta^2} = \frac{s^2 s_{xx} - 2s\, s_x^2 + s\, s_x^2}{\Delta^2} = \frac{s(s\, s_{xx} - s_x^2)}{\Delta^2} = \frac{s}{\Delta}.$$

Proceeding similarly for the parameter b, one finally obtains the usable expressions of the probable imprecisions of the two model parameters:

$$\sigma_a = \sqrt{\frac{s}{\Delta}}, \quad \sigma_b = \sqrt{\frac{s_{xx}}{\Delta}}. \tag{9.47}$$

The developed formalism of linear regression can be also applied to cases of nonlinear functional dependences which admit analytic inverses. For example, the exponential model

$$y = b e^{ax}$$

can be obviously linearized by applying the logarithm and naming $y' = \ln y$ and $b' = \ln b$:

$$y' = ax + b'.$$

The adjustment of the parameters a and b' of the linearized model is to be carried out with respect to the transformed observations $y_i' = \ln y_i$ and the optimized parameter b of the initial nonlinear model finally results as $b = \exp(b')$.

Based on the n observed coordinate pairs, received by way of the arrays x and y, and the associated standard deviations, `sigmy`, the routine `LinFit` shown in Listing 9.8 returns the adjusted parameters a and b of the linear model, as well as the associated probable imprecisions, `sigma` and `sigmb`. As a synthetic measure of the fit quality, the routine also returns by `chi2` the value of the χ^2-merit function calculated with the optimized model parameters.

If the control parameter `iopt` is set to 0, the routine initializes all the components of `sigmy` with 1 and the procedure falls back to the least-squares approximation based on the simple sum of squared deviations as a merit function:

$$S = \sum_{i=1}^{n} (y_i - a x_i - b)^2. \tag{9.48}$$

Listing 9.8   Linear Regression (Python Coding)

```
#===========================================================================
def LinFit(x, y, sigmy, n, iopt):
#---------------------------------------------------------------------------
#  Determines the coefficients a and b of a linear model P(x;a,b) = a * x + b
#  and the associated uncertainties, sigma and sigmb, for a set of observed
#  data points by "Chi-square" regression
#  x[]      - x-coordinates of observed data, i = 1,..,n
#  y[]      - y-coordinates of observed data
#  sigmy[]  - standard deviations of observed data
#  n        - number of observed data points
#  iopt     - iopt == 0 - initializes sigmy[i] = 1 (equal weights)
#           - iopt != 0 - uses the received values of sigmy[i]
#  a, b     - parameters of the linear model (output)
#  sigma    - uncertainties associated with the model parameters (output)
#  sigmb
#  chi2     - value of Chi-square merit function for the output parameters
#---------------------------------------------------------------------------
   if (iopt == 0):
      for i in range(1,n+1): sigmy[i] = 1e0        # iopt = 0: equall weights

   s = sx = sy = sxx = sxy = 0e0                            # prepare sums
   for i in range(1,n+1):
      f = 1e0/(sigmy[i]*sigmy[i])
      s  += f
      sx += x[i] * f; sxx += x[i] * x[i] * f
      sy += y[i] * f; sxy += x[i] * y[i] * f

   f = 1e0/(s*sxx - sx*sx)
   a = (s *sxy - sx*sy ) * f; sigma = sqrt(s*f)          # model parameters
   b = (sy*sxx - sx*sxy) * f; sigmb = sqrt(sxx*f)        # and uncertainties

   chi2 = 0e0                                   # value of Chi-square function
   for i in range(1,n+1):
      f = (y[i] - a*x[i] - b)/sigmy[i]
      chi2 += f*f

   return (a, b, sigma, sigmb, chi2)
```

This option is typically useful in situations in which the standard deviations σ_i of the observed data are unknown. Anyway, it should be noted that the comparative use of the two merit functions, χ^2 and S, may lead to rather different regression lines.

In the example shown in Figure 9.7, the merit function S, not making use of observational errors σ_i and assigning equal weights to all observed data, yields a regression line not passing through all the error bars. In contrast, by applying the χ^2-regression and considering, for the sake of simplicity, errors σ_i proportional to the observed values y_i (a somewhat typical situation), the first data point, bearing the lowest imprecision, forces the regression line to pass also through its error bar.

Specifically, the coordinates of the tabulated data points are $(1, 0.8)$, $(2, 2.1)$, $(3, 2.8)$, $(4, 4.0)$, and $(5, 4.4)$, and the corresponding standard deviations are generated simply as $\sigma_i = 0.15\, y_i$. By using the functional S one obtains the model parameters $a = 0.9100$ and $b = 0.0900$, while for the χ^2-merit function, the values $a = 0.9983$ and $b = -0.1681$ result.

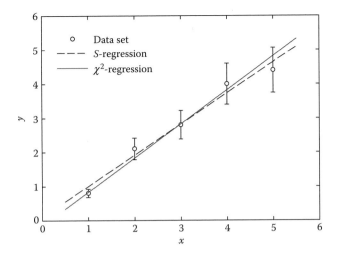

FIGURE 9.7 Example of linear regression evidencing the qualitative improvement of the fit based on the χ^2-merit function as compared to the simple sum of squared deviations, S.

9.6 Multilinear Regression Models

As a generalization of the linear regression, presented in the previous section, we consider in the following the *multilinear regression*, having as an important particular case the *polynomial regression*.

The task in the case of the multilinear regression is to fit a model represented by a *linear combination* of arbitrary functions of x,

$$F(x; \mathbf{a}) = \sum_{j=1}^{m} a_j F_j(x), \tag{9.49}$$

to a set of n observed data points, (x_i, y_i). The linear nature of the approach refers solely to the way in which the m parameters a_j enter the model. As for the *basis functions $F_j(x)$*, they have a fixed form not depending on the model parameters and can be arbitrarily nonlinear with respect to x.

The most widely used basis sets are the family of integer powers, $1, x, x^2, \ldots, x^{m-1}$, on which the polynomial regression is based, and the trigonometric functions $\sin kx$ and $\cos kx$ composing the Fourier series. Regarded as a particular case of polynomial regression, the linear regression involves only the basis functions 1 and x.

The χ^2-merit function for the multilinear regression has the form

$$\chi^2(\mathbf{a}) = \sum_{i=1}^{n} \frac{1}{\sigma_i^2} \left[y_i - \sum_{j=1}^{m} a_j F_j(x_i) \right]^2, \tag{9.50}$$

where σ_i represent, as previously, the observational errors (standard deviations) associated with the observations y_i. The following formalism remains applicable also in the case of unknown errors by setting all σ_i equal, for example, $\sigma_i = 1$.

The optimal model parameters minimize the χ^2-merit function or, equivalently, zero its partial derivatives:

$$\frac{\partial \chi^2}{\partial a_k} \equiv -2 \sum_{i=1}^{n} \frac{1}{\sigma_i^2} \left[y_i - \sum_{j=1}^{m} a_j F_j(x_i) \right] F_k(x_i) = 0, \quad k = 1, 2, \ldots, m. \tag{9.51}$$

By changing the summation order and rearranging the terms, one obtains the so-called *system of normal equations* of the multilinear fitting problem, namely,

$$\sum_{j=1}^{m} c_{kj} a_j = b_k, \quad k = 1, 2, \ldots, m, \tag{9.52}$$

whose solutions are the searched-for optimal model parameters. The system matrix $\mathbf{C} = [c_{kj}]_{mm}$ and the vector of constant terms $\mathbf{b} = [b_k]_m$ are defined by

$$c_{kj} = \sum_{i=1}^{n} \frac{1}{\sigma_i^2} F_k(x_i) F_j(x_i), \tag{9.53}$$

$$b_k = \sum_{i=1}^{n} \frac{y_i}{\sigma_i^2} F_k(x_i). \tag{9.54}$$

The estimates of the squared uncertainties (variances) associated with the optimized parameters can be obtained by applying the general formula

$$\sigma_{a_j}^2 = \sum_{i=1}^{n} \sigma_i^2 \left(\frac{\partial a_j}{\partial y_i} \right)^2. \tag{9.55}$$

To calculate the involved derivatives, we formally express the parameters a_j from the system of normal equations (9.52),

$$a_j = \sum_{k=1}^{m} c'_{jk} b_k = \sum_{k=1}^{m} c'_{jk} \sum_{i=1}^{n} \frac{y_i}{\sigma_i^2} F_k(x_i),$$

where c'_{jk} are the elements of the inverse system matrix, $\mathbf{C}^{-1} = [c'_{jk}]_{mm}$. The derivatives of the parameters are hence

$$\frac{\partial a_j}{\partial y_i} = \sum_{k=1}^{m} c'_{jk} \frac{1}{\sigma_i^2} F_k(x_i)$$

and, upon replacing them into the Formula 9.55 of the variance, inverting the summation order, and considering the definition (9.53) of the matrix elements c_{kj}, one successively obtains

$$\sigma_{a_j}^2 = \sum_{k,l=1}^{m} c'_{jk} c'_{jl} \sum_{i=1}^{n} \frac{1}{\sigma_i^2} F_k(x_i) F_l(x_i) = \sum_{k,l=1}^{m} c'_{jk} c'_{jl} c_{kl} = \sum_{l=1}^{m} c'_{jl} \delta_{jl}.$$

Here, we have also used the identity $\sum_{k=1}^{m} c'_{jk} c_{kl} = \delta_{jl}$, where δ_{jl} is Kronecker's delta. Consequently, the squared uncertainty of the optimized parameter a_j is given by

$$\sigma_{a_j}^2 = c'_{jj}, \tag{9.56}$$

that is, by the corresponding diagonal element c'_{jj} of the inverse matrix $\mathbf{C}^{-1}$ of the system of normal equations. In a broader sense, c'_{jk} represents the *covariance* associated with the pair of parameters (a_j, a_k) (Press et al. 2007).

The implementation of the described method of normal equations for the problem of multilinear regression is given in Listing 9.9. The function `MultiFit` receives the n coordinates of the observed

Listing 9.9 Multilinear Regression (Python Coding)

```python
#===============================================================================
def MultiFit(x, y, sigmy, n, iopt, a, sigma, npar, Func):
#-------------------------------------------------------------------------------
#  Determines the coefficients a[i], i=1,..,npar of a multilinear model
#     F(x;a) = a[1] * func[1](x) + ... + a[npar] * func[npar](x)
#  and the associated uncertainties, sigma[i], for a set of observed data
#  points by "Chi-square" regression
#  x[]      - x-coordinates of observed data, i = 1,..,n
#  y[]      - y-coordinates of observed data
#  sigmy[]  - standard deviations of observed data
#  n        - number of observed data points
#  iopt     - iopt == 0 - initializes sigmy[i] = 1 (equal weights)
#           - iopt != 0 - uses the received values of sigmy[i]
#  a[]      - parameters of the multilinear model (output)
#  sigma[]  - uncertainties associated with the model parameters (output)
#  npar     - number of model parameters
#  chi2     - value of Chi-square merit function for the output parameters
#  Func()   - user function returning for a given argument x the values of the
#             basis functions func[i](x):
#                Func(x, func, npar)
#  Calls: GaussJordan (linsys.py)
#-------------------------------------------------------------------------------
   c = [[0]*(n+1) for i in range(n+1)]
   b = [[0]*2 for i in range(n+1)]
   func = [0]*(npar+1)

   if (iopt -- 0):
      for i in range(1,n+1): sigmy[i] = 1e0          # iopt = 0: equall weights

   for i in range(1,npar+1):                                 # initialization
      for j in range(1,npar+1): c[i][j] = 0e0
      b[i][1] = 0e0

   for i in range(1,n+1):                            # generate the system matrices
      yi = y[i]
      Func(x[i],func,npar)                           # evaluate the basis functions
      f = 1e0/(sigmy[i]*sigmy[i])
      for j in range(1,npar+1):
         fj = f * func[j]
         for k in range(1,npar+1): c[j][k] += fj * func[k]
         b[j][1] += fj * yi

   det = GaussJordan(c,b,npar,1)                             # solve the system

   for i in range(1,npar+1):
      a[i] = b[i][1]                                         # model parameters
      sigma[i] = sqrt(c[i][i])                               # parameter uncertainties

   chi2 = 0e0                                        # value of Chi-square function
   for i in range(1,n+1):
      Func(x[i],func,npar)                           # evaluate the model function
      f = 0e0
      for j in range(1,npar+1): f += a[j]*func[j]
      f = (y[i] - f)/sigmy[i]
      chi2 += f*f
   return chi2
```

data points in the arrays x and y, along with the standard deviations of the y-coordinates, in the array sigmy. The number of model parameters is specified by the argument npar. If the control parameter iopt is set to 0, all the components of the array sigmy are initialized internally with 1 and the approximation reduces to the least-squares regression based on the merit function

$$S = \sum_{i=1}^{n} \left[y_i - \sum_{j=1}^{m} a_j F_j(x_i) \right]^2 .$$

The function MultiFit returns the adjusted model parameters and the associated uncertainties in the array a and, respectively, sigma. As a synthetic quality indicator for the performed fit, the routine also returns via chi2 the value of the χ^2-merit function calculated with the optimal parameters.

The basis functions $F_j(x)$ composing the model have to be evaluated by a separate user function, which is recognized internally by MultiFit under the generic name Func and whose actual name is conveyed by way of the parameter list. For a given argument x, the user function returns the npar values of the basis functions in the components of the array func.

The system of normal equations (9.52) may be solved, in principle, by any general-purpose method. However, since for the evaluation of the parameter uncertainties also the inverse of the system matrix is required, the most convenient turns out to be the Gauss–Jordan method, discussed in Chapter 7, which enables the simultaneous solution of the system and the in-place inversion of the system matrix. The developed routine GaussJordan solves a matrix equation $\mathbf{A} \cdot \mathbf{X} = \mathbf{B}$, where $\mathbf{A}$ is a square $(n \times n)$ matrix, while $\mathbf{B}$ and $\mathbf{X}$ are $(n \times m)$ matrices, replacing on exit the constant terms $\mathbf{B}$ by the solution $\mathbf{X}$ and the system matrix $\mathbf{A}$ by its inverse $\mathbf{A}^{-1}$. By the specific call to GaussJordan from the routine MultiFit, the array c, containing initially the fitting matrix, returns its inverse (the covariance matrix), whereas the array b of constant terms is replaced by the system's solution (the model parameters). As a technical detail, although in the present case the matrix equation reduces to an ordinary linear system with just a vector of constant terms $(m = 1)$, for reasons of compatibility with the arguments of GaussJordan, b has to be declared as a two-dimensional array, with npar rows and a single column.

Figure 9.8 illustrates the method of multilinear regression by a simple example, in which the "observed" data points (depicted with circles) are generated by sampling the function $f(x) = \sin x + \cos x$

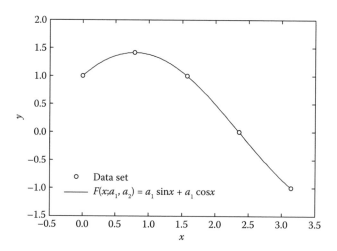

FIGURE 9.8   Multilinear regression performed with the help of the routine MultiFit. The observed data are sampled from the function $f(x) = \sin x + \cos x$.

for the hand-picked arguments 0, $\pi/2$, π, $3\pi/2$, and π. In the sample program from Listing 9.10, the model function is chosen to have two adjustable parameters,

$$F(x; a_1, a_2) = a_1 \sin x + a_2 \cos x,$$

and the evaluation of the basis functions $F_1(x) = \sin x$ and $F_2(x) = \cos x$ is carried out by the user function Func. The adjusted model parameters returned by MultiFit are $a_1 = 0.999929$ and $a_2 = 1.000212$, with the merit function $\chi^2 = 0.000000$. It is noteworthy that the fitting procedure of the model function recovers within roundoff errors the coefficients of the original function $f(x)$, which has provided the "observed" data.

Generalizing the model function to include more basis functions (1, $\sin x$, $\cos x$, $\sin 2x$, $\cos 2x$, etc.), the routine MultiFit can be used, in principle, to extract Fourier series coefficients from tabulated dependences (see Problem 9.9).

In the particular case of the polynomial regression, the model function having m adjustable parameters is a polynomial of degree $(m - 1)$:

$$F(x; \mathbf{a}) = \sum_{j=1}^{m} a_j x^{j-1}. \tag{9.57}$$

Listing 9.10 Application of Multilinear Regression (Python Coding)

```
# Multilinear fit
from math import *
from modfunc import *

def Func(x, func, npar):
    func[1] = sin(x)              # returns the basis functions sin(x) and cos(x)
    func[2] = cos(x)

# main

n = 5                                       # number of observed data
npar = 2                                    # number of model parameters

x     = [0]*(n+1)                           # x-coordinates of observed data
y     = [0]*(n+1)                           # y-coordinates of observed data
sigmy = [0]*(n+1)                       # standard deviations of observed data
func  = [0]*(npar+1)                        # values of basis functions
a     = [0]*(npar+1)                        # model parameters
sigma = [0]*(npar+1)                        # uncertainties of parameters

x[1] =  0e000; y[1] =  1e000               # observed data generated from:
x[2] =  0.785; y[2] =  1.414               # f(x) = sin(x) + cos(x)
x[3] =  1.571; y[3] =  1e000
x[4] =  2.356; y[4] =  0e000
x[5] =  3.141; y[5] = -1e000

iopt = 0                                # least squares fit: equal errors sigmy
chi2 = MultiFit(x,y,sigmy,n,iopt,a,sigma,npar,Func)

print("Multilinear fit:")
for i in range(1,npar+1):
    print(("a[{0:d}] = {1:8.4f} +/- {2:8.4f}").format(i,a[i],sqrt(sigma[i])))
print("Chi^2 = {0:8.4f}".format(chi2))
```

Listing 9.11 Polynomial Regression (Python Coding)

```
#===========================================================================
def PolFit(x, y, sigmy, n, iopt, a, sigma, npar):
#---------------------------------------------------------------------------
#  Determines the coefficients a[i], i=1,..,npar of a polynomial model
#     F(x;a) = a[1] * x^(npar-1) + a[2] * x^(npar-2) + ... + a[npar]
#  and the associated uncertainties, sigma[i], for a set of observed data
#  points by "Chi-square" regression
#  x[]     - x-coordinates of observed data, i = 1,..,n
#  y[]     - y-coordinates of observed data
#  sigmy[] - standard deviations of observed data
#  n       - number of observed data points
#  iopt    - iopt == 0 - initializes sigmy[i] = 1 (equal weights)
#          - iopt != 0 - uses the received values of sigmy[i]
#  a[]     - parameters of the polynomial model (output)
#  sigma[] - uncertainties associated with the model parameters (output)
#  npar    - number of model parameters (polynomial order + 1)
#  chi2    - value of Chi-square merit function for the output parameters
#  Calls: GaussJordan (linsys.py)
#---------------------------------------------------------------------------
   c = [[0]*(n+1) for i in range(n+1)]
   b = [[0]*2 for i in range(n+1)]
   func = [0]*(npar+1)

   if (iopt == 0):
      for i in range(1,n+1): sigmy[i] = 1e0        # iopt = 0: equall weights

   for i in range(1,npar+1):                                # initialization
      for j in range(1,npar+1): c[i][j] = 0e0
      b[i][1] = 0e0

   for i in range(1,n+1):                          # generate the system matrices
      xi = x[i]; yi = y[i]
      func[npar] = 1e0                             # basis functions 1, x, x^2,...
      for j in range(npar-1,0,-1): func[j] = xi * func[j+1]
      f = 1e0/(sigmy[i]*sigmy[i])
      for j in range(1,npar+1):
         fj = f * func[j]
         for k in range(1,npar+1): c[j][k] += fj * func[k]
         b[j][1] += fj * yi

   det = GaussJordan(c,b,npar,1)                    # solution: corrections da

   for i in range(1,npar+1):
      a[i] = b[i][1]                                         # model parameters
      sigma[i] = sqrt(c[i][i])                       # parameter uncertainties

   chi2 = 0e0                                       # value of Chi-square function
   for i in range(1,n+1):
      xi = x[i]                                      # evaluate the model function
      f = a[1]
      for j in range(2,npar+1): f = f*xi + a[j]
      f = (y[i] - f)/sigmy[i]
      chi2 += f*f

   return chi2
```

The corresponding user function can be coded in the most efficient way by taking advantage of the simple recurrence relation between the successive basis functions $1, x, x^2, \ldots$:

```
def Func(x, func, npar):    # returns basis functions 1, x, x^2,...,x^(npar-1)
    func[npar] = 1e0                        # basis functions 1, x, x^2,...
    for i in range(npar-1,0,-1): func[i] = x * func[i+1]
```

Yet another option to implement the polynomial regression is to rewrite the routine `MultiFit`, as in Listing 9.11, so as to evaluate the basis functions in-place and not to call any user function. The routine `PolFit` is useful, for example, for extracting power series coefficients from tabulated dependences.

9.7 Nonlinear Regression: The Levenberg–Marquardt Method

The least-squares method can be applied, in principle, for fitting any arbitrary model function to observed data points, but requires, in general, the solution of systems of nonlinear equations for determining the optimal model parameters. The procedure is called in such cases *nonlinear regression* and the corresponding algorithms typically feature a considerable degree of complexity. Unlike the multilinear regression, in which the model parameters are determined in a single step by solving a linear system of equations, the nonlinear regression refines the model parameters iteratively, starting from initial guesses.

The merit function is considered to have the general form:

$$\chi^2(\mathbf{a}) = \sum_{i=1}^{n} \frac{1}{\sigma_i^2} \left[y_i - F(x_i; \mathbf{a}) \right]^2. \tag{9.58}$$

With a view to characterizing the minimum of χ^2, it is useful to consider its series expansion with respect to the parameter values:

$$\chi^2(\mathbf{a} + \delta \mathbf{a}) = \chi^2(\mathbf{a}) + \sum_{j=1}^{m} \frac{\partial \chi^2}{\partial a_j} \delta a_j + \frac{1}{2} \sum_{j,l=1}^{m} \frac{\partial \chi^2}{\partial a_j \partial a_l} \delta a_j \delta a_l + \cdots, \tag{9.59}$$

where $\delta \mathbf{a}$ is the vector of parameter corrections. The merit function shows an extremum for the set of corrections $\delta \mathbf{a}$ which make its gradient vanish, or, equivalently, its partial derivatives with respect to all the parameters:

$$\left. \frac{\partial \chi^2}{\partial a_k} \right|_{\mathbf{a} + \delta \mathbf{a}} = 0, \quad k = 1, 2, \ldots, m. \tag{9.60}$$

The derivatives of χ^2 for the corrected parameters result from the Expansion 9.59,

$$\left. \frac{\partial \chi^2}{\partial a_k} \right|_{\mathbf{a} + \delta \mathbf{a}} = \sum_{j=1}^{m} \frac{\partial \chi^2}{\partial a_j} \delta_{jk} + \frac{1}{2} \sum_{j,l=1}^{m} \frac{\partial \chi^2}{\partial a_j \partial a_l} (\delta_{kj} \delta a_l + \delta a_j \delta_{kl}) + \cdots$$

$$= \frac{\partial \chi^2}{\partial a_k} + \sum_{j=1}^{m} \frac{\partial \chi^2}{\partial a_k \partial a_j} \delta a_j + \cdots \tag{9.61}$$

The first-order derivatives for the reference parameters entering the above expression can be derived from the definition (9.58) of the merit function:

$$\frac{\partial \chi^2}{\partial a_k} = -2 \sum_{i=1}^{n} \frac{1}{\sigma_i^2} \left[y_i - F(x_i; \mathbf{a}) \right] \frac{\partial F(x_i; \mathbf{a})}{\partial a_k}. \tag{9.62}$$

Analogously, for the mixed second derivatives, forming the so-called *Hessian matrix* of the merit function, one obtains:

$$\frac{\partial \chi^2}{\partial a_k \partial a_j} = 2 \sum_{i=1}^{n} \frac{1}{\sigma_i^2} \left\{ \frac{\partial F(x_i; \mathbf{a})}{\partial a_k} \frac{\partial F(x_i; \mathbf{a})}{\partial a_j} - \left[y_i - F(x_i; \mathbf{a}) \right] \frac{\partial^2 F(x_i; \mathbf{a})}{\partial a_k \partial a_j} \right\}$$

$$\approx 2 \sum_{i=1}^{n} \frac{1}{\sigma_i^2} \frac{\partial F(x_i; \mathbf{a})}{\partial a_k} \frac{\partial F(x_i; \mathbf{a})}{\partial a_j}. \tag{9.63}$$

In the vicinity of the minimum of χ^2, the mixed second derivatives $\partial^2 F(x_i; \mathbf{a})/\partial a_k \partial a_j$ are negligible in comparison with the first-order ones and, being also weighted with the quantities $[y_i - F(x_i; \mathbf{a})]$, which are randomly signed and distributed, they tend to cancel out statistically by summation. Furthermore, by imposing the vanishing conditions (9.60) on the derivatives of χ^2 and confining their Expansions 9.61 to the linear approximation, there results the following system of linear equations for the parameter corrections (Press et al. 2007):

$$\sum_{j=1}^{m} c_{kj} \delta a_j = b_k, \quad k = 1, 2, \ldots, m, \tag{9.64}$$

with the matrix elements defined by

$$c_{kj} \equiv \frac{1}{2} \frac{\partial \chi^2}{\partial a_k \partial a_j} = \sum_{i=1}^{n} \frac{1}{\sigma_i^2} \frac{\partial F(x_i; \mathbf{a})}{\partial a_k} \frac{\partial F(x_i; \mathbf{a})}{\partial a_j}, \tag{9.65}$$

$$b_k \equiv -\frac{1}{2} \frac{\partial \chi^2}{\partial a_k} = \sum_{i=1}^{n} \frac{1}{\sigma_i^2} \left[y_i - F(x_i; \mathbf{a}) \right] \frac{\partial F(x_i; \mathbf{a})}{\partial a_k}. \tag{9.66}$$

c_{kj} equal half the elements of the Hessian matrix and $\mathbf{C} = [c_{kj}]$ is called *curvature matrix*.

Far from the minimum of χ^2, the Expansion 9.59 may represent a poor local approximation to $\chi^2(\mathbf{a} + \delta\mathbf{a})$. In such a case, in order to accelerate the convergence of the process toward the minimum, one can retain only the diagonal elements of the Hessian matrix and, consequently, the components of the gradient of the merit function (9.61) become

$$\left. \frac{\partial \chi^2}{\partial a_k} \right|_{\mathbf{a}+\delta\mathbf{a}} \approx \frac{\partial \chi^2}{\partial a_k} + \frac{\partial \chi^2}{\partial a_k^2} \delta a_j. \tag{9.67}$$

Making use of the notations (9.65) and (9.66), the conditions of vanishing χ^2 derivatives (9.60) lead to the following simplified equations for the parameter corrections:

$$c_{kk} \delta a_k = b_k, \quad k = 1, 2, \ldots, m. \tag{9.68}$$

The Levenberg–Marquardt method (Marquardt (1963)) has become in practice the standard approach for nonlinear regression. It provides an efficient strategy for determining the optimal corrections of the model parameters by smoothly switching between Equations 9.68, which are applicable far from the minimum of χ^2, and Equations 9.64, which are applicable close to the minimum. Technically, by introducing a *dimensionless scale factor* λ, the two sets of equations take a unique form:

$$\sum_{j=1}^{m}(1 + \lambda\delta_{kj})c_{kj}\delta a_j = b_k, \quad k = 1, 2, \ldots, m, \tag{9.69}$$

where δ_{kj} is the Kronecker delta. For large values of λ, the system matrix is *diagonally dominant* and there results the uncoupled set of equations (9.68), which is applicable far from the minimum. On the contrary, for small values of λ, the system takes the form (9.64) and allows for the parameter corrections to be determined close to the minimum.

Analogously to the multilinear regression, the variance (squared probable imprecision) associated with the adjusted parameter a_j is given by the diagonal element c'_{jj} of the inverse $\mathbf{C}^{-1}$ of the curvature matrix, that is,

$$\sigma_{a_j}^2 = c'_{jj}. \tag{9.70}$$

The nondiagonal elements c'_{jk}, in general, represent the *covariances* corresponding to the pairs of model parameters (a_j, a_k).

The corrections to the model parameters have to be calculated iteratively from the linear system (9.69), with the coefficients specified by Equations 9.65 and 9.66, until the maximum relative change of the individual model parameters drops under a predefined tolerance ε:

$$\max_{j}|\delta a_j/(a_j + \delta a_j)| \leq \varepsilon. \tag{9.71}$$

If there exist vanishing components $a_j + \delta a_j$, rather the absolute corrections have to be employed in expressing this criterion. Iterating up to the machine representation limit is, in general, neither necessary nor useful, since the minimum of the merit function provides at best statistical estimates of the parameters a_j. Another stopping criterion, complementary to the one for the parameter values, checks the degree to which the merit function has converged:

$$|1 - \chi^2(\mathbf{a})/\chi^2(\mathbf{a} + \delta\mathbf{a})| \leq \varepsilon. \tag{9.72}$$

Synthetically, the Levenberg–Marquardt algorithm implies iterating the following steps (Press et al. 2007):

1. Calculate the merit function $\chi^2(\mathbf{a})$ for an initial set of parameters.
2. Consider a relatively small initial value for λ, such as 10^{-3}.
3. Calculate the coefficients c_{kj} and b_k from Equations 9.65 and 9.66, and solve the linear system (9.69) for the parameter corrections $\delta\mathbf{a}$.
4. Evaluate the merit function $\chi^2(\mathbf{a} + \delta\mathbf{a})$ for the corrected parameters.
5. If $\chi^2(\mathbf{a} + \delta\mathbf{a}) \geq \chi^2(\mathbf{a})$, *increase* λ by a factor of 10 and go to 3.
6. If $\chi^2(\mathbf{a} + \delta\mathbf{a}) < \chi^2(\mathbf{a})$, *decrease* λ by a factor of 10 and update $\mathbf{a} + \delta\mathbf{a} \to \mathbf{a}$.
7. If $\max_j |\delta a_j/a_j| > \varepsilon$ or $|1 - \chi^2(\mathbf{a})/\chi^2(\mathbf{a} + \delta\mathbf{a})| > \varepsilon$ go to 3.

In any attempt to implement the Levenberg–Marquardt method efficiently, one has to evaluate repeatedly the χ^2-merit function and the linearized fitting matrices $\mathbf{C}$ and $\mathbf{b}$, which depend themselves on the derivatives of the model function. To this end, the routine `Deriv` from Listing 9.12 calls the user-defined model function (with the local alias `Func`), calculates by forward finite differences its first derivatives $\partial F(x_i; \mathbf{a})/\partial a_j$ with respect to the parameters a_j, and returns these via the array `dFda` to the calling function `Chi2`. The latter builds from the model function derivatives the fitting matrices $\mathbf{C}$ and $\mathbf{b}$ according to Relations 9.65 and 9.66, and returns them together with the χ^2-merit function to the calling routine `MarqFit` (Listing 9.13).

Listing 9.12   Auxiliary Functions for the Levenberg–Marquardt Method (Python Coding)

```
#==========================================================================
def Deriv(x, a, dFda, npar, Func):
#--------------------------------------------------------------------------
#  Evaluates the derivatives of model function Func(x,a,npar) at point x with
#  respect to parameters a[j], j=1,..,npar and returns them in dFda[j]
#--------------------------------------------------------------------------
   h0 = 1e-4                            # scale factor for the derivation step

   F0 = Func(x,a,npar)                  # function value for original parameters
   for j in range(1,npar+1):
      h = h0 * fabs(a[j]) if (a[j]) else h0              # derivation step
      temp = a[j]                                        # save parameter
      a[j] += h                                   # increment parameter
      dFda[j] = (Func(x,a,npar) - F0)/h                    # derivative
      a[j] = temp                                   # restore parameter

#==========================================================================
def Chi2(x, y, sigmy, n, a, b, c, npar, Func):
#--------------------------------------------------------------------------
#  Evaluates the Chi2 merit function and the fitting matrices c and b for the
#  model function Func(x,a,npar) and the Levenberg-Marquardt algorithm
#  Calls: Deriv
#--------------------------------------------------------------------------
   dFda = [0]*(npar+1)                        # model function derivatives

   for k in range(1,npar+1):                              # initialization
      for j in range(1,npar+1): c[k][j] = 0e0
      b[k] = 0e0

   chi2 = 0e0
   for i in range(1,n+1):
      dy = y[i] - Func(x[i],a,npar)                # deviation of model from data
      fsig = 1e0/(sigmy[i]*sigmy[i])
      chi2 += fsig * dy * dy

      Deriv(x[i],a,dFda,npar,Func)    # derivatives with respect to parameters
      for k in range(1,npar+1):                       # build fitting matrices
         fact = fsig * dFda[k]
         for j in range(1,npar+1): c[k][j] += fact * dFda[j]
         b[k] += fact * dy

   return chi2
```

Even though more elaborate than the function MultiFit presented in Section 9.6, the nonlinear fitting routine MarqFit (Listing 9.13) is based on the same conventions and notations. In particular, the formal parameters have the same significance. Specifically, MarqFit receives the coordinates and the associated standard deviations for the n observed data points in the arrays x, y, and sigmy. In situations in which the data uncertainties are unknown, by setting the input parameter iopt to 0, the routine MarqFit is instructed to initialize all the components of sigmy with 1.

MarqFit returns in the array a the npar optimal model parameters and in the array sigma, the associated probable imprecisions. As a synthetic quality indicator of the fit, the parameter chi2 returns the χ^2-merit function corresponding to the adjusted parameters.

Listing 9.13 Routine for Nonlinear Fit Based on the Levenberg–Marquardt Method (Python Coding)

```
#===========================================================================
def MarqFit(x, y, sigmy, n, iopt, a, sigma, npar, Func):
#---------------------------------------------------------------------------
#  Minimizes the "Chi-square" merit function by the Levenberg-Marquardt
#  method, determining the coefficients a[i], i=1,..,npar of a non-linear
#  model F(x,a,npar) and the associated uncertainties, sigma[i], for a set of
#  observed data points
#  x[]      - x-coordinates of the observed data, i=1,..,n
#  y[]      - y-coordinates of the observed data
#  sigmy[]  - standard deviations of the observed data
#  n        - number of observed data points
#  iopt     - iopt == 0 - initializes sigmy[i] = 1 (equal weights)
#           - iopt != 0 - uses the received values of sigmy[i]
#  a[]      - parameters of the non-linear model (output)
#  sigma[]  - uncertainties of the model parameters (output)
#  npar     - number of model parameters
#  chi2     - value of Chi-square merit function for the output parameters
#  Func()   - user function returning for a given argument x the values of the
#             basis functions func[i](x):
#                Func(f, x, n)
#  Calls: GaussJordan (linsys.py)
#---------------------------------------------------------------------------
   eps = 1e-4                                    # relative precision criterion
   itmax = 1000                                  # max. no. of iterations

   c   = [[0]*(n+1) for i in range(n+1)]
   c0  = [[0]*(n+1) for i in range(n+1)]
   cov = [[0]*(n+1) for i in range(n+1)]
   da  = [[0]*2 for i in range(n+1)]
   a1 = [0]*(n+1); b = [0]*(n+1); b0 = [0]*(n+1)
   func = [0]*(npar+1)

   if (iopt == 0):
      for i in range(1,n+1): sigmy[i] = 1e0      # iopt = 0: equall weights

   lam = 1e-3                                     # initial guess for lambda
   chi2 = Chi2(x,y,sigmy,n,a,b,c,npar,Func)

   it = 0
   while (it < itmax):                           # loop of parameter approximations
      chi20 = chi2
      for k in range(1,npar+1):                  # store fitting matrices c0 and b0
         for j in range(1,npar+1): c0[k][j] = c[k][j]
         b0[k] = b[k]

      while (it < itmax):                        # loop of lambda increase
         it += 1
         for k in range(1,npar+1):               # covariance matrix
            for j in range(1,npar+1): cov[k][j] = c0[k][j]
            cov[k][k] *= (1e0 + lam)             # scale diagonal
            da[k][1] = b0[k]

         det = GaussJordan(cov,da,npar,1)        # solution: corrections da
         for j in range(1,npar+1): a1[j] = a[j] + da[j][1]   # trial params
         chi2 = Chi2(x,y,sigmy,n,a1,b,c,npar,Func)  # new linearized c and b
         if (chi2 <= chi20): break               # stop increasing lambda
         lam *= 10e0                              # increase lambda
```

```
        err = 0e0                                  # maximum relative error of parameters
        for j in range(1,npar+1):
            a[j] += da[j][1]                                       # update parameters
            errj = fabs(da[j][1]/a[j]) if a[j] else fabs(da[j][1])
            if (err < errj): err = errj

        lam *= 0.1e0                                                    # reduce lambda
                                                            # check convergence
        if ((err <= eps) and (fabs(1e0 - chi2/chi20) <= eps)): break

    if (it >= itmax): print("MarqFit: max. # of iterations exceeded !")
                                                    # uncertainties of parameters
    for j in range(1,npar+1): sigma[j] = sqrt(cov[j][j])

    return chi2
```

Besides the user function `Func`, which evaluates the model function $F(x_i; \mathbf{a})$ for particular values of the independent variable x and of the model parameters `a[j]`, the routine `MarqFit` also calls the auxiliary functions `Chi2` and `GaussJordan`. Analogously to routine `MultiFit`, upon the call to `GaussJordan`, the array `cov`, containing initially the curvature matrix $\mathbf{C}$ as system matrix, returns the inverse thereof (the covariance matrix). The array `del` of the constant terms is replaced, in its turn, by the corrections of the model parameters, as solutions of the linearized fitting equation. Here too, for reasons of compatibility with the arguments of `GaussJordan`, the array `del` is declared *two-dimensional* (not simply a vector), with npar rows and a single column.

The routine `MarqFit` is composed, essentially, of an *outer loop*, controlling the successive sets of parameter approximations, and an *inner loop*, controlling the increase (decrease) of the scale factor λ. The updated fitting matrices $\mathbf{C}$ and $\mathbf{b}$ are stored at the beginning of each outer step and are used consistently inside the inner loop for determining the parameter corrections δa_j and the χ^2-merit function for the current value of λ. The execution exits from the inner loop only if χ^2 starts decreasing, sign that the process has reached the vicinity of the merit function minimum, and it is completed by the reduction of λ by an order of magnitude.

The use of the routine `MarqFit` (assumed to be included in the module `modfunc.py`) for fitting nonlinear models is exemplified in Listing 9.14. The input data are read in from the text file `fit.dat`

Listing 9.14 Nonlinear Regression Using the Levenberg–Marquardt Algorithm (Python Coding)

```
# Non-linear fit by the Levenberg-Marquardt method
from math import *
from modfunc import *

def Func(x, a, npar):                                          # model function
    return a[1] * exp(-pow(x-a[2],2)/a[3])

# main

inp = open("fit.dat","r")                                      # open data file

line = inp.readline()                     # number of observed data and parameters
n = int(line.split()[0])
npar = int(line.split()[1])

                                                        # allocate arrays
x     = [0]*(n+1)                          # x-coordinates of observed data
y     = [0]*(n+1)                          # y-coordinates of observed data
sigmy = [0]*(n+1)                          # standard deviations of observed data
a     = [0]*(npar+1)                                        # model parameters
```

```
sigma = [0]*(npar+1)                              # uncertainties of parameters

iopt = int(inp.readline())                        # initialization option for sigmy[i]
line = inp.readline()
for i in range(1,npar+1): a[i] = float(line.split()[i-1]) # parameter guesses
for i in range(1,n+1):
   line = inp.readline()
   x[i] = float(line.split()[0])                            # observed data
   y[i] = float(line.split()[1])
   if (iopt): sigmy[i] = float(line.split()[2])                # uncertainty
inp.close()

out = open("fit.txt","w")                            # open output file
out.write(("n = {0:2d}    npar = {1:2d}\n").format(n,npar))
out.write("Initial parameters:\n")
for i in range(1,npar+1): out.write(("a[{0:d}] = {1:6.3f}\n").format(i,a[i]))

chi2 = MarqFit(x,y,sigmy,n,iopt,a,sigma,npar,Func)    # Levenberg-Marquart fit

out.write("\nFinal parameters:\n")
for i in range(1,npar+1):
   out.write(("a[{0:d}] = {1:6.3f}   sigma[{2:d}] = {3:7.1e}\n").
             format(i,a[i],i,sigma[i]))
out.write("\nChi^2 = {0:7.1e}\n".format(chi2))
out.write("\n i        x            y            sigma       yfit      y-yfit\n")
for i in range(1,n+1):
   yfit = Func(x[i],a,npar)                    # model values for adjusted parameters
   out.write(("{0:2d}{1:10.5f}{2:10.5f}{3:10.5f}{4:10.5f}{5:10.1e}\n").
             format(i,x[i],y[i],sigmy[i],yfit,y[i]-yfit))
out.close()
```

and, based on the number of observations, n, and the number of model parameters, npar, the program allocates the necessary arrays. Following the call to the function MarqFit, the produced results are written to the file fit.txt.

To illustrate the operation of the program, we consider the input data in Listing 9.15, which represent the content of the file fit.dat. The first row provides the values of the parameters n and npar, and the second specifies the initialization option iopt. The third row contains initial guesses for the model

Listing 9.15   File fit.dat: Sample Input for Program 9.14

```
11   3
1
  0.5   0.5   1.0
 -2.0   0.0003   0.02
 -1.6   0.0060   0.02
 -1.2   0.0600   0.02
 -0.8   0.3000   0.03
 -0.4   0.7000   0.07
  0.0   1.0000   0.10
  0.4   0.7000   0.07
  0.8   0.3000   0.03
  1.2   0.0600   0.02
  1.6   0.0060   0.02
  2.0   0.0003   0.02
```

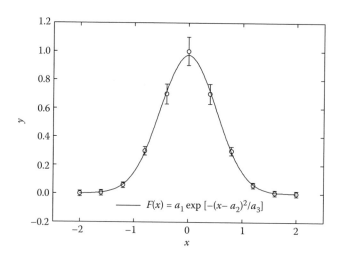

FIGURE 9.9   Example of nonlinear regression using the routine `LevMarqFit`.

parameters. The following tabular data represent (x, y) coordinates resulting by sampling the function $f(x) = \exp(-2x^2)$, along with the associated standard deviations σ_i. Aiming to introduce deviations, the function values are truncated to a single significant digit, considering as standard deviations 10% of the ordinates, but not less than 0.02.

The model function coded in the program is the general Gaussian:

$$F(x; a_1, a_2, a_3) = a_1 \exp\left[-(x - a_2)^2/a_3\right].$$

Listing 9.16   File `fit.txt`: Output from Program 9.14

```
n = 11    npar =  3
Initial parameters:
a[1] =  0.500
a[2] =  0.500
a[3] =  1.000

Final parameters:
a[1] =  0.974   sigma[1] = 4.3e-007
a[2] = -0.000   sigma[2] = 1.9e-007
a[3] =  0.531   sigma[3] = 2.5e-007

Chi^2 = 5.2e-001

 i      x          y        sigma      yfit       y-yfit
 1   -2.00000   0.00030   0.02000    0.00052  -2.2e-004
 2   -1.60000   0.00600   0.02000    0.00785  -1.8e-003
 3   -1.20000   0.06000   0.02000    0.06470  -4.7e-003
 4   -0.80000   0.30000   0.03000    0.29191   8.1e-003
 5   -0.40000   0.70000   0.07000    0.72090  -2.1e-002
 6    0.00000   1.00000   0.10000    0.97442   2.6e-002
 7    0.40000   0.70000   0.07000    0.72090  -2.1e-002
 8    0.80000   0.30000   0.03000    0.29191   8.1e-003
 9    1.20000   0.06000   0.02000    0.06470  -4.7e-003
10    1.60000   0.00600   0.02000    0.00785  -1.8c-003
11    2.00000   0.00030   0.02000    0.00052  -2.2e-004
```

The results produced by Program 9.14 are plotted in Figure 9.9 and compose the file `fit.txt` (see Listing 9.16). The nonlinear fit yields as optimal model parameter 0.974, 0.000, and 0.531, which are close, however, not identical with the original parameters, 1.0, 0.0, and 0.5, of the function $f(x)$ from which the "observed" data points are sampled. The differences are obviously caused by the intentional truncation of the exact function values in the input data set. The last two columns of the printed output list the model function and the corresponding deviations from the tabulated input data.

9.8 Implementations in C/C++

Listing 9.17 shows the content of the file `modfunc.h`, which contains equivalent C/C++ implementations of the Python functions developed in the main text and included in the module `modfunc.py`. The corresponding routines have identical names, parameters, and functionalities.

Listing 9.17 Routines for Modeling Tabulated Dependences (`modfunc.h`)

```
//---------------------------------- modfunc.h ----------------------------------
// Routines for modeling tabular dependences by interpolation/regression.
// Part of the numxlib numerics library. Author: Titus Beu, 2013
//-------------------------------------------------------------------------------
#ifndef _MODFUNC_
#define _MODFUNC_

#include <stdio.h>
#include <math.h>
#include "memalloc.h"
#include "linsys.h"

//===============================================================================
double Lagrange(double x[], double y[], int n, double xi)
//-------------------------------------------------------------------------------
// Evaluates the Lagrange interpolating polynomial of n data points at xi
// x[] - x-coordinates of data points
// y[] - y-coordinates of data points
// n   - number of data points
// xi  - interpolation argument
//-------------------------------------------------------------------------------
{
   double p, yi;
   int i, j;

   yi = 0e0;
   for (i=1; i<=n; i++) {
      p = 1e0;
      for (j=1; j<=n; j++)
         if (j != i) p *= (xi - x[j])/(x[i] - x[j]);
      yi += p * y[i];
   }
   return yi;
}

//===============================================================================
void Lagrange1(double x[], double y[], int n,
               double xi[], double yi[], int ni)
//-------------------------------------------------------------------------------
// Evaluates the Lagrange interpolating polynomial of n data points on a mesh
// of ni interpolation points
// x[]  - x-coordinates of data points
```

```
// y[]   - y-coordinates of data points
// n     - number of data points
// xi[] - interpolation arguments
// yi[] - interpolant values (output)
// ni   - number of interpolation points
//--------------------------------------------------------------------------
{
   double p, xk, yk, *yf;
   int i, j, k;

   yf = Vector(1,n);                                 // factors of the interpolant

   for (i=1; i<=n; i++) {        // prepare invariant factors of the interpolant
      p = 1e0;
      for (j=1; j<=n; j++)
         if (j != i) p *= (x[i] - x[j]);
      yf[i] = y[i] / p;
   }

   for (k=1; k<=ni; k++) {                      // loop over the interpolation points
      xk = xi[k]; yk = 0e0;
      for (i=1; i<=n; i++) {
         p = 1e0;
         for (j=1; j<=n; j++)
            if (j != i) p *= (xk - x[j]);
         yk += p * yf[i];
      }
      yi[k] = yk;
   }
   FreeVector(yf,1);
}


//============================================================================
double Neville(double x[], double y[], int n, double xi, double &err)
//--------------------------------------------------------------------------
// Evaluates the Lagrange interpolating polynomial of n data points at xi by
// Neville's method and returns an estimate of the absolute error
// x[] - x-coordinates of data points
// y[] - y-coordinates of data points
// n   - number of data points
// xi  - interpolation argument
// err - estimate of the absolute error (output)
//--------------------------------------------------------------------------
{
   double *p, yi;
   int i, m;

   p = Vector(1,n);                             // values of successive interpolants

   for (i=1; i<=n; i++) p[i] = y[i];    // initialize with 0-order polynomial

   for (m=1; m<=n-1; m++)                        // loop over columns of scheme array
      for (i=1; i<=n-m; i++)                           // increase polynomial order
         p[i] = ((xi-x[m+i])*p[i] + (x[i]-xi)*p[i+1]) / (x[i]-x[m+i]);

   yi = p[1];                                                   // polynomial value
   err = fabs(p[1] - p[2]);                                       // error estimate

   FreeVector(p,1);
```

```
      return yi;
}

//============================================================================
void Spline(double x[], double y[], int n, double d1, double dn, int iopt,
            double a[], double b[], double c[], double d[],
            double xi[], double yi[], int ni)
//----------------------------------------------------------------------------
// Calculates the coefficients of the cubic splines for n data points and
// returns interpolated values on a mesh of ni points
// x[]     - x-coordinates of data points
// y[]     - y-coordinates of data points
// n       - number of data points
// d1, dn - 1st derivatives at the endpoints x[1] and x[n]
// iopt   - iopt == 0, natural splines: 2nd derivative = 0 at x[1] and x[n]
//          iopt == 1, clamped splines: uses the provided d1 and dn
// a[], b[], c[], d[], i=1..n
//         - coefficients of cubic splines i=1..n-1 (output)
// xi[]    - interpolation arguments
// yi[]    - interpolant values (output)
// ni      - number of interpolation points
// Calls: TriDiagSys (linsys.h)
//----------------------------------------------------------------------------
{
   double di, dm, hi, hm, xx, xp;
   int i, ip;

   if (iopt == 0) d1 = dn = 0e0;          // initialization for natural splines

   hi = 0e0; di = d1;             // coefficients for system of 2nd derivatives
   for (i=1; i<=(n-1); i++) {
      hm = hi; hi = x[i+1] - x[i];
      dm = di; di = (y[i+1] - y[i])/hi;
      a[i] = hm; b[i] = 2e0*(hm + hi); c[i] = hi; d[i] = 6e0*(di - dm);
   }
   a[n] = hi; b[n] = 2e0*hi; c[n] = 0e0; d[n] = 6e0*(dn - di);

   if (iopt == 0) c[1] = d[1] = a[n] = d[n] = 0e0;       // natural splines

   TriDiagSys(a,b,c,d,n);                          // solve tridiagonal system
                                                   // 2nd derivatives in d
   for (i=1; i<=(n-1); i++) {                      // spline coefficients
      ip = i + 1;
      xx = x[i]; xp = x[ip]; hi = xp - xx;
      a[i] = (d[ip] - d[i]) / (6e0*hi);
      b[i] = (d[i]*xp - d[ip]*xx) / (2e0*hi);
      c[i] = (d[ip]*xx*xx - d[i]*xp*xp) / (2e0*hi)
                                        + (y[ip] - y[i]) / hi - a[i]*hi*hi;
      d[i] = (d[i]*xp*xp*xp - d[ip]*xx*xx*xx) / (6e0*hi)
                                  + (y[i]*xp - y[ip]*xx) / hi - b[i]*hi*hi/3e0;
   }

   for (i=1; i<=ni; i++) {                           // interpolation loop
      xx = xi[i];
      ip = 1; while (ip < n-1 && xx > x[ip+1]) ip++;       // index of spline
      yi[i] = ((a[ip]*xx + b[ip])*xx + c[ip])*xx + d[ip];  // evaluate spline
   }
}
```

```
//==============================================================================
void LinFit(double x[], double y[], double sigmy[], int n, int iopt,
            double &a, double &b, double &sigma, double &sigmb, double &chi2)
//------------------------------------------------------------------------------
// Determines the coefficients a and b of a linear model P(x;a,b) = a * x + b
// and the associated uncertainties, sigma and sigmb, for a set of observed
// data points by "Chi-square" regression
// x[]      - x-coordinates of observed data, i = 1,..,n
// y[]      - y-coordinates of observed data
// sigmy[] - standard deviations of observed data
// n        - number of observed data points
// iopt     - iopt == 0 - initializes sigmy[i] = 1 (equal weights)
//          - iopt != 0 - uses the received values of sigmy[i]
// a, b     - parameters of the linear model (output)
// sigma    - uncertainties associated with the model parameters (output)
// sigmb
// chi2     - value of Chi-square merit function for the output parameters
//------------------------------------------------------------------------------
{
   double f, s, sx, sy, sxx, sxy;
   int i;

   if (iopt == 0)
      for (i=1; i<=n; i++) sigmy[i] = 1e0;            // iopt = 0: equall weights

   s = sx = sy = sxx = sxy = 0e0;                                // prepare sums
   for (i=1; i<=n; i++) {
      f = 1e0/(sigmy[i]*sigmy[i]);
      s  += f;
      sx += x[i] * f; sxx += x[i] * x[i] * f;
      sy += y[i] * f; sxy += x[i] * y[i] * f;
   }

   f = 1e0/(s*sxx - sx*sx);
   a = (s *sxy - sx*sy ) * f; sigma = sqrt(s*f);         // model parameters
   b = (sy*sxx - sx*sxy) * f; sigmb = sqrt(sxx*f);       // and uncertainties

   chi2 = 0e0;                                    // value of Chi-square function
   for (i=1; i<=n; i++) {
      f = (y[i] - a*x[i] - b)/sigmy[i];
      chi2 += f*f;
   }
}


//==============================================================================
void MultiFit(double x[], double y[], double sigmy[], int n, int iopt,
              double a[], double sigma[], int npar, double &chi2,
              void Func(double, double [], int))
//------------------------------------------------------------------------------
// Determines the coefficients a[i], i=1,..,npar of a multilinear model
//      F(x;a) = a[1] * func[1](x) + ... + a[npar] * func[npar](x)
// and the associated uncertainties, sigma[i], for a set of observed data
// points by "Chi-square" regression
// x[]      - x-coordinates of observed data, i = 1,..,n
// y[]      - y-coordinates of observed data
// sigmy[] - standard deviations of observed data
// n        - number of observed data points
// iopt     - iopt == 0 - initializes sigmy[i] = 1 (equal weights)
//          - iopt != 0 - uses the received values of sigmy[i]
```

```
// a[]      - parameters of the multilinear model (output)
// sigma[] - uncertainties associated with the model parameters (output)
// npar     - number of model parameters
// chi2     - value of Chi-square merit function for the output parameters
// Func()   - user function returning for a given argument x the values of the
//            basis functions func[i](x):
//                void Func(double x, double func[], int npar);
// Calls: GaussJordan (linsys.h)
//-----------------------------------------------------------------------------
{
   double det, f, fj, yi;
   double **b, **c, *func;
   int i, j, k;

   c = Matrix(1,npar,1,npar);
   b = Matrix(1,npar,1,1);
   func = Vector(1,npar);

   if (iopt == 0)
      for (i=1; i<=n; i++) sigmy[i] = 1e0;        // iopt = 0: equall weights

   for (i=1; i<=npar; i++) {                                   // initialization
      for (j=1; j<=npar; j++) c[i][j] = 0e0;
      b[i][1] = 0e0;
   }

   for (i=1; i<=n; i++) {                          // generate the system matrices
      yi = y[i];
      Func(x[i],func,npar);                        // evaluate the basis functions
      f = 1e0/(sigmy[i]*sigmy[i]);
      for (j=1; j<=npar; j++) {
         fj = f * func[j];
         for (k=1; k<=npar; k++) c[j][k] += fj * func[k];
         b[j][1] += fj * yi;
      }
   }

   GaussJordan(c,b,npar,1,det);                                // solve the system

   for (i=1; i<=npar; i++) {
      a[i] = b[i][1];                                          // model parameters
      sigma[i] = sqrt(c[i][i]);                          // parameter uncertainties
   }

   chi2 = 0e0;                                        // value of Chi-square function
   for (i=1; i<=n; i++) {
      Func(x[i],func,npar);                            // evaluate the model function
      f = 0e0;
      for (j=1; j<=npar; j++) f += a[j]*func[j];
      f = (y[i] - f)/sigmy[i];
      chi2 += f*f;
   }

   FreeMatrix(c,1,1);
   FreeMatrix(b,1,1);
   FreeVector(func,1);
}

//=============================================================================
```

```
void PolFit(double x[], double y[], double sigmy[], int n, int iopt,
            double a[], double sigma[], int npar, double &chi2)
//-----------------------------------------------------------------------------
// Determines the coefficients a[i], i=1,..,npar of a polynomial model
//     F(x;a) = a[1] * x^(npar-1) + a[2] * x^(npar-2) + ... + a[npar]
// and the associated uncertainties, sigma[i], for a set of observed data
// points by "Chi-square" regression
// x[]      - x-coordinates of observed data, i = 1,..,n
// y[]      - y-coordinates of observed data
// sigmy[]  - standard deviations of observed data
// n        - number of observed data points
// iopt     - iopt == 0 - initializes sigmy[i] = 1 (equal weights)
//          - iopt != 0 - uses the received values of sigmy[i]
// a[]      - parameters of the polynomial model (output)
// sigma[]  - uncertainties associated with the model parameters (output)
// npar     - number of model parameters (polynomial order + 1)
// chi2     - value of Chi-square merit function for the output parameters
// Calls: GaussJordan (linsys.h)
//-----------------------------------------------------------------------------
{
   double det, f, fj, xi, yi;
   double **b, **c, *func;
   int i, j, k;

   c = Matrix(1,npar,1,npar);
   b = Matrix(1,npar,1,1);
   func = Vector(1,npar);

   if (iopt == 0)
      for (i=1; i<=n; i++) sigmy[i] = 1e0;            // iopt = 0: equall weights

   for (i=1; i<=npar; i++) {                                      // initialization
      for (j=1; j<=npar; j++) c[i][j] = 0e0;
      b[i][1] = 0e0;
   }

   for (i=1; i<=n; i++) {                             // generate the system matrices
      xi = x[i]; yi = y[i];
      func[npar] = 1e0;                               // basis functions 1, x, x^2,...
      for (j=npar-1; j>=1; j--) func[j] = xi * func[j+1];
      f = 1e0/(sigmy[i]*sigmy[i]);
      for (j=1; j<=npar; j++) {
         fj = f * func[j];
         for (k=1; k<=npar; k++) c[j][k] += fj * func[k];
         b[j][1] += fj * yi;
      }
   }

   GaussJordan(c,b,npar,1,det);                                   // solve the system

   for (i=1; i<=npar; i++) {
      a[i] = b[i][1];                                             // model parameters
      sigma[i] = sqrt(c[i][i]);                            // parameter uncertainties
   }

   chi2 = 0e0;                                     // value of Chi-square function
   for (i=1; i<=n; i++) {
      xi = x[i];                                        // evaluate the model function
      f = a[1];
```

```
      for (j=2; j<=npar; j++) f = f*xi + a[j];
      f = (y[i] - f)/sigmy[i];
      chi2 += f*f;
   }

   FreeMatrix(c,1,1);
   FreeMatrix(b,1,1);
   FreeVector(func,1);
}

//===========================================================================
void Deriv(double x, double a[], double dFda[], int npar,
           double Func(double, double [], int))
//---------------------------------------------------------------------------
// Evaluates the derivatives of model function Func(x,a,npar) at point x with
// respect to parameters a[j], j=1,..,npar and returns them in dFda[j]
//---------------------------------------------------------------------------
{
   const double h0 = 1e-4;                  // scale factor for the derivation step
   double a0, F0, h;
   int j;

   F0 = Func(x,a,npar);                     // function value for original parameters
   for (j=1; j<=npar; j++) {
      h = a[j] ? h0 * fabs(a[j]) : h0;                     // derivation step
      a0 = a[j];                                           // save parameter
      a[j] += h;                                    // increment parameter
      dFda[j] = (Func(x,a,npar) - F0)/h;                      // derivative
      a[j] = a0;                                        // restore parameter
   }
}


//===========================================================================
double Chi2(double x[], double y[], double sigmy[], int n,
            double a[], double b[], double **c, int npar,
            double Func(double, double [], int))
//---------------------------------------------------------------------------
// Evaluates the Chi2 merit function and the fitting matrices c and b for the
// model function Func(x,a,npar) and the Levenberg-Marquardt algorithm
// Calls: Deriv
//---------------------------------------------------------------------------
{
   int i, j, k;
   double chi2, dy, *dFda, fact, fsig;

   dFda = Vector(1,npar);                          // model function derivatives

   for (k=1; k<=npar; k++) {                                // initialization
      for (j=1; j<=npar; j++) c[k][j] = 0e0;
      b[k] = 0e0;
   }

   chi2 = 0e0;
   for (i=1; i<=n; i++) {
      dy = y[i] - Func(x[i],a,npar);          // deviation of model from data
      fsig = 1e0/(sigmy[i]*sigmy[i]);
      chi2 += fsig * dy * dy;                      // evaluate merit function

      Deriv(x[i],a,dFda,npar,Func); // derivatives with respect to parameters
```

```
      for (k=1; k<=npar; k++) {                          // build fitting matrices
         fact = fsig * dFda[k];
         for (j=1; j<=npar; j++) c[k][j] += fact * dFda[j];
         b[k] += fact * dy;
      }
   }

   FreeVector(dFda,1);
   return chi2;
}


//===========================================================================
void MarqFit(double x[], double y[], double sigmy[], int n, int iopt,
             double a[], double sigma[], int npar, double &chi2,
             double Func(double, double [], int))
//---------------------------------------------------------------------------
// Minimizes the "Chi-square" merit function by the Levenberg-Marquardt
// method, determining the coefficients a[i], i=1,..,npar of a non-linear
// model F(x,a,npar) and the associated uncertainties, sigma[i], for a set of
// observed data points
// x[]     - x-coordinates of the observed data, i=1,..,n
// y[]     - y-coordinates of the observed data
// sigmy[] - standard deviations of the observed data
// n       - number of observed data points
// iopt    - iopt == 0 - initializes sigmy[i] = 1 (equal weights)
//         - iopt != 0 - uses the received values of sigmy[i]
// a[]     - parameters of the non-linear model (output)
// sigma[] - uncertainties of the model parameters (output)
// npar    - number of model parameters
// chi2    - value of Chi-square merit function for the output parameters
// Func()  - user function returning for a given argument x the values of the
//           basis functions func[i](x):
//              void Func(double x, double func[], int npar);
// Calls: GaussJordan (linsys.h)
//---------------------------------------------------------------------------
{
   const double eps = 1e-4;                       // relative precision criterion
   const int itmax = 1000;                        // max. no. of iterations
   double *a1, *b, *b0, **c, **c0, **cov, **da;
   double chi20, det, err, errj, lam;
   int i, it, j, k;

   c   = Matrix(1,npar,1,npar); c0 = Matrix(1,npar,1,npar);
   cov = Matrix(1,npar,1,npar); da = Matrix(1,npar,1,1);
   a1 = Vector(1,npar); b = Vector(1,npar); b0 = Vector(1,npar);

   if (iopt == 0)
      for (i=1; i<=n; i++) sigmy[i] = 1e0;         // iopt = 0: equall weights

   lam = 1e-3;                                     // initial guess for lambda
   chi2 = Chi2(x,y,sigmy,n,a,b,c,npar,Func);

   it = 0;
   while (it < itmax) {                            // loop of parameter approximations
      chi20 = chi2;
      for (k=1; k<=npar; k++) {                    // store fitting matrices c0 and b0
         for (j=1; j<=npar; j++) c0[k][j] = c[k][j];
         b0[k] = b[k];
      }
```

```
      while (it < itmax) {                        // loop of lambda increase
         it++;
         for (k=1; k<=npar; k++) {                       // covariance matrix
            for (j=1; j<=npar; j++) cov[k][j] = c0[k][j];
            cov[k][k] *= (1e0 + lam);                       // scale diagonal
            da[k][1] = b0[k];
         }
         GaussJordan(cov,da,npar,1,det);           // solution: corrections da
         for (j=1; j<=npar; j++) a1[j] = a[j] + da[j][1];      // trial params
         chi2 = Chi2(x,y,sigmy,n,a1,b,c,npar,Func); // new linearized c and b
         if (chi2 <= chi20) break;                     // stop increasing lambda
         lam *= 10e0;                                       // increase lambda
      }

      err = 0e0;                       // maximum relative error of parameters
      for (j=1; j<=npar; j++) {
         a[j] += da[j][1];                               // update parameters
         errj = a[j] ? fabs(da[j][1]/a[j]) : fabs(da[j][1]);
         if (err < errj) err = errj;
      }
      lam *= 0.1e0;                                         // reduce lambda
                                                         // check convergence
      if ((err <= eps) && (fabs(1e0 - chi2/chi20) <= eps)) break;
   }
   if (it >= itmax) printf("MarqFit: max. # of iterations exceeded !\n");
                                            // uncertainties of parameters
   for (j=1; j<=npar; j++) sigma[j] = sqrt(cov[j][j]);

   FreeMatrix(c  ,1,1); FreeMatrix(c0,1,1);
   FreeMatrix(cov,1,1); FreeMatrix(da,1,1);
   FreeVector(a1,1); FreeVector(b,1); FreeVector(b0,1);
}

#endif
```

9.9 Problems

The Python and C/C++ programs for the following problems may import the functions developed in this chapter from the modules `modfunc.py` and, respectively, `modfunc.h`, which are available as supplementary material. For creating runtime plots, the graphical routines contained in the libraries `graphlib.py` and `graphlib.h` may be employed.

PROBLEM 9.1

Write Python and C/C++ programs using the `Lagrange` and `Neville` routines for interpolating $n = 8$ data points sampled from the function $f(x) = 1/x$ for the particular arguments 0.15, 0.2, 0.3, 0.5, 0.8, 1.1, 1.4, and 1.7.

    a. Evaluate and plot the interpolants for 50 equally spaced arguments from the tabulation interval. Notice the spurious oscillations that arise despite the uniformly distributed data points *along* the original dependence. Check whether the original dependence remains bounded by the error bars provided by Neville's method.

    b. Consider another set of $n = 8$ data points, with *equally spaced abscissas* in the same tabulation interval. Notice the significant reduction of the artificial oscillations and find reasons for this changed behavior.

Solution

The implementations are provided in Listings 9.18 and 9.19 and the graphical output, in Figure 9.10.

Listing 9.18 Lagrange Interpolation and Neville's Method (Python Coding)

```
# Lagrange interpolation and Neville's algorithm
from modfunc import *
from graphlib import *

# main

nn  = [0]*4                                              # end indexes of plots
col = [""]*4                                             # colors of plots
sty = [0]*4                                              # styles of plots

n  = 8                                                   # number of data points
ni = 50                                        # number of interpolation points
n1 = n + ni; n2 = n + 2*ni                                       # end indexes

x  = [0]*(n +1); y  = [0]*(n +1)                               # data points
xp = [0]*(n2+1); yp = [0]*(n2+1)                              # plotting arrays
err = [0]*(n2+1)                                         # interpolation errors

x[1] = 0.15; x[2] = 0.2; x[3] = 0.3; x[4] = 0.5              # uneven abscissas
x[5] = 0.8 ; x[6] = 1.1; x[7] = 1.4; x[8] = 1.7
# h = (x[n]-x[1])/(n-1)
# for i in range(1,n+1): x[i] = x[1] + (i-1)*h      # equally spaced abscissas
for i in range(1,n+1):
   xp[i] = x[i]
   yp[i] = y[i] = 1e0/x[i]

GraphInit(1200,600)

#-------------------------------------------------------- Lagrange interpolation
h = (x[n]-x[1])/(ni-1)
for i in range(1,ni+1):                                 # fill in plotting arrays
   xi = x[1] + (i-1)*h                                  # interpolation argument
   xp[n +i] = xi; yp[n +i] = Lagrange(x,y,n,xi)                  # interpolant
   xp[n1+i] = xi; yp[n1+i] = 1e0/xi                         # original function

nn[1] = n        ; col[1] = "red"  ; sty[1] =  0             # observed data
nn[2] = n +   ni; col[2] = "blue" ; sty[2] =  1               # interpolant
nn[3] = n + 2*ni; col[3] = "black"; sty[3] = -1                     # model
MultiPlot(xp,yp,err,nn,col,sty,3,10,0e0,0e0,0,0e0,0e0,0,
        0.07,0.47,0.15,0.85,"x","P(x)","Lagrange interpolation")

#-------------------------------------------------------- Neville interpolation
h = (x[n]-x[1])/(ni-1)
for i in range(1,ni+1):                                 # fill in plotting arrays
   xi = x[1] + (i-1)*h                                  # interpolation argument
   xp[n +i] = xi; yp[n +i],err[n+i] = Neville(x,y,n,xi)          # interpolant
   xp[n1+i] = xi; yp[n1+i] = 1e0/xi                         # original function

nn[1] = n        ; col[1] = "red"  ; sty[1] =  0             # observed data
nn[2] = n +   ni; col[2] = "blue" ; sty[2] = -4               # interpolant
nn[3] = n + 2*ni; col[3] = "black"; sty[3] = -1                     # model
MultiPlot(xp,yp,err,nn,col,sty,3,10,0e0,0e0,0,0e0,0e0,0,
        0.57,0.97,0.15,0.85,"x","P(x)","Neville interpolation")

MainLoop()
```

Listing 9.19 Lagrange Interpolation and Neville's Method (C/C++ Coding)

```
// Lagrange interpolation and Neville's algorithm
#include <stdio.h>
#include "memalloc.h"
#include "modfunc.h"
#include "graphlib.h"

int main(int argc, wchar_t** argv)
{
   double *err, *x, *xp, *y, *yp;
   double h, xi;
   int i, n, ni, n1, n2;
   int nn[4], sty[4];                      // ending indexes and styles of plots
   const char* col[4];                          // colors of plots

   n  = 8;                                    // number of data points
   ni = 50;                                 // number of interpolation points
   n1 = n + ni; n2 = n + 2*ni;                    // end indexes

   x  = Vector(1,n ); y  = Vector(1,n );          // data points
   xp = Vector(1,n2); yp = Vector(1,n2);          // plotting arrays
   err = Vector(1,n2);                       // interpolation errors

   x[1] = 0.15; x[2] = 0.2; x[3] = 0.3; x[4] = 0.5;     // uneven abscissas
   x[5] = 0.8 ; x[6] = 1.1; x[7] = 1.4; x[8] = 1.7;
// h = (x[n]-x[1])/(n-1);
// for (i=1; i<=n; i++) x[i] = x[1] + (i-1)*h;   // equally spaced abscissas
   for (i=1; i<=n; i++) {
      xp[i] = x[i];
      yp[i] = y[i] = 1e0/x[i];
   }

   PyGraph w(argc, argv);
   w.GraphInit(1200,600);

//---------------------------------------------------- Lagrange interpolation
   h = (x[n]-x[1])/(ni-1);
   for (i=1; i<=ni; i++) {                       // fill in plotting arrays
      xi = x[1] + (i-1)*h;                       // interpolation argument
      xp[n +i] = xi; yp[n +i] = Lagrange(x,y,n,xi);       // interpolant
      xp[n1+i] = xi; yp[n1+i] = 1e0/xi;              // original function
   }
   nn[1] = n ; col[1] = "red"  ; sty[1] =  0;          // data points
   nn[2] = n1; col[2] = "blue" ; sty[2] =  1;          // interpolant
   nn[3] = n2; col[3] = "black"; sty[3] = -1;       // original function
   w.MultiPlot(xp,yp,err,nn,col,sty,3,10,0e0,0e0,0,0e0,0e0,0,
               0.07,0.47,0.15,0.85,"x","P(x)","Lagrange interpolation");

//---------------------------------------------------- Neville interpolation
   h = (x[n]-x[1])/(ni-1);
   for (i=1; i<=ni; i++) {                       // fill in plotting arrays
      xi = x[1] + (i-1)*h;                       // interpolation argument
      xp[n +i] = xi; yp[n +i] = Neville(x,y,n,xi,err[n+i]);   // interpolant
      xp[n1+i] = xi; yp[n1+i] = 1e0/xi;              // original function
   }
   nn[1] = n ; col[1] = "red"  ; sty[1] =  0;          // data points
   nn[2] = n1; col[2] = "blue" ; sty[2] = -4;          // interpolant
   nn[3] = n2; col[3] = "black"; sty[3] = -1;       // original function
   w.MultiPlot(xp,yp,err,nn,col,sty,3,10,0e0,0e0,0,0e0,0e0,0,
```

```
                    0.57,0.97,0.15,0.85,"x","P(x)","Neville interpolation");

    w.MainLoop();
}
```

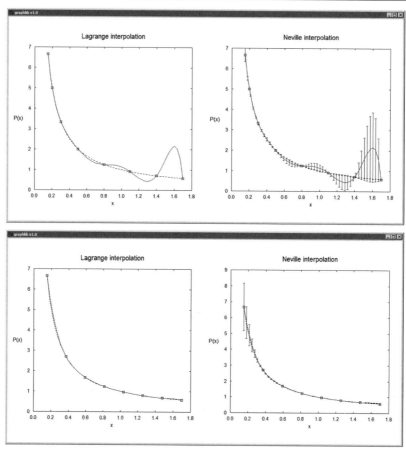

FIGURE 9.10 Lagrange and Neville interpolants (with continuous line) calculated by Programs 9.18 and 9.19. The data points (with squares) sample the function $f(x) = 1/x$ (with dashed line). The error bars represent the estimates provided by Neville's method. Upper panel uniformly distributed data points *along* the modeled function; lower panel: equally spaced *abscissas* of the data points.

PROBLEM 9.2

Write Python and C/C++ programs using the routine Spline to interpolate a set of $n = 6$ equally spaced data points sampling the function $f(x) = \sin x/x$ in the interval $[0, 5\pi/2]$.

Evaluate the splines and their first and second derivatives in each of the tabulation subintervals $[x_i, x_{i+1}]$, $i = 1, 2, \ldots, n - 1$, using their analytical expressions:

$$S_i(x) = ((A_i x + B_i) x + C_i) x + D_i,$$

$$S_i'(x) = (3A_i x + 2B_i) x + C_i,$$

$$S_i''(x) = 6A_i x + 2B_i,$$

and the coefficients A_i, B_i, C_i, and D_i returned by Spline.

Use comparatively *natural* and *clamped boundaries* (which make use of the provided boundary derivatives), by setting `iopt=0` and, respectively, nonzero in the call to `Spline`. In each case, plot the interpolant along with its first and second derivatives.

For clamped splines, use the exact first derivatives at the tabulation end points ($S'_1(0) = 0$, $S'_5(5\pi/2) = 0$) and notice the sensible improvement of the interpolant in the boundary intervals. Verify that, for the natural splines, the second derivative vanishes at the tabulation end points, whereas for the clamped splines, it is the first derivative that vanishes.

Solution

The Python implementation is given in Listing 9.20, the equivalent C/C++ version is available as supplementary material (`P09-Spline.cpp`), and the resulting plots are shown in Figure 9.11.

Listing 9.20 Interpolation Using Cubic Splines (Python Coding)

```
# Interpolation with cubic splines
from modfunc import *
from graphlib import *

def sinc(x): return (sin(x)/x if x else 1e0)

# main

xmin = 0e0; xmax = 5*asin(1e0)            # tabulation interval: [0,5*pi/2]
n   = 6                                          # number of data points
ni = 100                                      # number of interpolation points
n1 = n + ni; n2 = n + 2*ni; n3 = n + 3*ni; n4 = n + 4*ni        # end indexes
nn  = [0]*6                                        # end indexes of plots
col = [""]*6                                          # colors of plots
sty = [0]*6                                          # styles of plots

a = [0]*(n+1); b = [0]*(n+1)                          # spline coefficients
c = [0]*(n+1); d = [0]*(n+1)
x = [0]*(n+1); y = [0]*(n+1)                      # coordinates of data points
xp = [0]*(n4+1); yp = [0]*(n4+1)                      # plotting arrays

nn[1] = n ; col[1] = "black"; sty[1] =  0                  # data points
nn[2] = n1; col[2] = "black"; sty[2] = -1            # original function
nn[3] = n2; col[3] = "blue" ; sty[3] =  1                  # interpolant
nn[4] = n3; col[4] = "red"  ; sty[4] = -1              # 1st derivative
nn[5] = n4; col[5] = "green"; sty[5] = -1              # 1st derivative

h = (xmax-xmin)/(n-1)                          # generate data points
for i in range(1,n+1):
    xi = xmin + (i-1)*h
    xp[i] = x[i] = xi
    yp[i] = y[i] = sinc(xi)

h = (xmax-xmin)/(ni-1)
for i in range(1,ni+1):
    xi = xmin + (i-1)*h                          # interpolation arguments
    xp[n3+i] = xp[n2+i] = xp[n1+i] = xp[n+i] = xi          # plotting arguments

GraphInit(1200,600)

#-------------------------------------------------------------- natural splines
Spline(x,y,n,0e0,0e0,0,a,b,c,d,x,y,0)

for i in range(1,ni+1):
```

```
    xi = xp[n+i]                                        # interpolation argument
    ip = 1
    while (ip < n-1 and xi > x[ip+1]): ip += 1              # index of spline

    yp[n +i] = sinc(xi)                                # original function
    yp[n1+i] = ((a[ip]*xi + b[ip])*xi + c[ip])*xi + d[ip]          # spline
    yp[n2+i] = (3*a[ip]*xi + 2*b[ip])*xi + c[ip]              # 1st derivative
    yp[n3+i] = 6*a[ip]*xi + 2*b[ip]                        # 2nd derivative

MultiPlot(xp,yp,yp,nn,col,sty,5,10,0e0,0e0,0,0e0,0e0,0,0.07,0.47,0.15,0.85,
          "x","S","Spline interpolation - natural splines")

#------------------------------------------------------------------ clamped splines
Spline(x,y,n,0e0,0e0,1,a,b,c,d,x,y,0)

for i in range(1,ni+1):
    xi = xp[n+i]                                        # interpolation argument
    ip = 1
    while (ip < n-1 and xi > x[ip+1]): ip += 1              # index of spline

    yp[n +i] = sinc(xi)                                # original function
    yp[n1+i] = ((a[ip]*xi + b[ip])*xi + c[ip])*xi + d[ip]          # spline
    yp[n2+i] = (3*a[ip]*xi + 2*b[ip])*xi + c[ip]              # 1st derivative
    yp[n3+i] = 6*a[ip]*xi + 2*b[ip]                        # 2nd derivative

MultiPlot(xp,yp,yp,nn,col,sty,5,10,0e0,0e0,0,0e0,0e0,0,0.57,0.97,0.15,0.85,
          "x","S","Spline interpolation - clamped splines")

MainLoop()
```

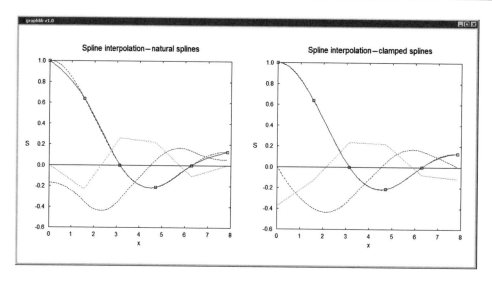

FIGURE 9.11 Interpolation with natural cubic splines (left) and clamped cubic splines (right). The modeled function, $f(x) = \sin x/x$, is sampled in the interval $[0, 5\pi/2]$ at six equally spaced points. The interpolant is plotted with continuous line, while its derivatives and the original function with dashed lines.

PROBLEM 9.3

Perform linear regression on the data set $(1, 0.8)$, $(2, 2.1)$, $(3, 2.8)$, $(4, 4.0)$ and $(5, 4.4)$, and associate with each observation y_i a proportional standard deviation $\sigma_i = 0.15\, y_i$. Write Python and C/C++ programs to analyze the data set by calling the function `LinFit` and apply comparatively the merit functions χ^2 (Equation 9.41) and S (Equation 9.48).

Plot the data set along with the two regression lines and judge the quality of the fitting procedure based on the uncertainties of the adjusted model parameters a and b.

Solution

The Python implementation is given in Listing 9.21 and the equivalent C/C++ version is available as supplementary material (`P09-LinFit.cpp`). The graphical output produced by the program is shown in Figure 9.12.

Listing 9.21   Linear Fit by the Least-Squares Method (Python Coding)

```python
# Linear fit of a model to observed data points
from modfunc import *
from graphlib import *

# main

nn  = [0]*4                                    # end-indexes of the 3 plots
col = [""]*4                                         # colors of plots
sty = [0]*4                                          # styles of plots

n = 5                                          # number of observed data
nfit = 2                              # number of points plotted from model
n1 = n + nfit; n2 = n + 2*nfit                           # end indexes

x = [0]*(n2+1); y = [0]*(n2+1)                          # observed data
sigmy = [0]*(n+1)                       # standard deviations of observed data

x[1] = 1e0; y[1] = 0.8e0                                   # data points
x[2] = 2e0; y[2] = 2.1e0
x[3] = 3e0; y[3] = 2.8e0
x[4] = 4e0; y[4] = 4.0e0
x[5] = 5e0; y[5] = 4.4e0

iopt = 0                             # least squares fit: equal errors sigmy
(a, b, sigma, sigmb, chi2) = LinFit(x,y,sigmy,n,iopt)

print("Least squares fit:")
print("a = {0:8.4f} +/- {1:8.4f}".format(a,sigma))
print("b = {0:8.4f} +/- {1:8.4f}".format(b,sigmb))
print("Chi^2 = {0:8.4f}".format(chi2))

h = (x[n]-x[1])/(nfit-1)
for i in range(1,nfit+1):                           # append model points
   x[n+i] = x[1] + (i-1)*h
   y[n+i] = a*x[n+i] + b                              # regression line

for i in range(1,n+1): sigmy[i] = 0.15*y[i]      # generate standard deviations

iopt = 1                              # Chi-square fit: different errors sigmy
(a, b, sigma, sigmb, chi2) = LinFit(x,y,sigmy,n,iopt)

print("\nChi-square fit:")
print("a = {0:8.4f} +/- {1:8.4f}".format(a,sigma))
print("b = {0:8.4f} +/- {1:8.4f}".format(b,sigmb))
print("Chi^2 = {0:8.4f}".format(chi2))
```

```
for i in range(1,nfit+1):                              # append model points
    x[n1+i] = x[n+i]
    y[n1+i] = a*x[n+i] + b                           # Chi-square regression line

GraphInit(800,600)

nn[1] = n ; col[1] = "black"; sty[1] =  4                  # data points
nn[2] = n1; col[2] = "red"  ; sty[2] = -1              # least squares fit
nn[3] = n2; col[3] = "blue" ; sty[3] =  1                # Chi-square fit
MultiPlot(x,y,sigmy,nn,col,sty,3,10,0.5e0,5.5e0,1,0e0,0e0,0,
          0.15,0.95,0.15,0.85,"x","y","Linear fit")

MainLoop()
```

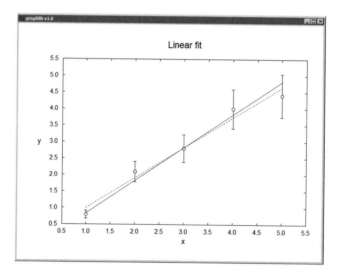

FIGURE 9.12   Linear regression showing the difference between the fit based on the χ^2-merit function and, respectively, the simple sum of squared deviations, S, as performed by Program 9.21.

PROBLEM 9.4

Based on data compiled by H. F. Ivey (*Phys. Rev.* **72**, 341, 1947), Table 9.1 lists average spectral *absorption maxima* of alkali halide crystals, λ_{max}, in terms of the interionic distance d. Ivey showed that the data can be well fitted by power laws ($x \equiv d, y \equiv \lambda_{max}$),

$$y = ax^b,$$

and found, in particular, for the so-called F-band $a = 703$ Å and $b = 1.84$.

Analyze the data in Table 9.1 both by linear regression (using the routine `LinFit`) and nonlinear regression (using the routine `MarqFit`). For the linear fit, linearize first the observational data and the model function by applying the logarithm,

$$y' = a'x' + b',$$

TABLE 9.1   Spectral Absorption Maxima for the F-Bands of Alkali Halide Crystals

Crystal	d (Å)	λ_{max} (Å)	σ (Å)	Crystal	d (Å)	λ_{max} (Å)	σ (Å)
LiF	2.01	2540	100	KF	2.67	4550	230
LiCl	2.57	3850	190	KCl	3.14	5600	100
NaF	2.31	3390	100	KBr	3.29	6280	100
NaCl	2.81	4640	120	KI	3.53	6980	350
NaBr	2.98	5400	270	RbCl	3.27	6170	150
NaI	3.23	5880	290	RbBr	3.43	7070	300
				RbI	3.66	7660	190

Note: d is the interionic salt distance and λ_{max} is the average wavelength of the absorption maxima from different references. The uncertainties σ represent the maximal difference between the experimental data sets.
Source: H. F. Ivey. 1947. *Phys. Rev. 72*, 341.

where $x' = \log x$, $y' = \log y$, $a' = b$, $b' = \log a$. The transformed standard deviations are given by $\sigma' = \log(y + \sigma) - \log y = \log(1 + \sigma/y)$. The original model parameters may be expressed finally as $a = \exp(b')$ and $b = a'$.

Consider as input file `halides.dat`, whose structure is outlined in Listing 9.22. The first row provides the number of observables n and the number of parameters npar. The value 1, specified for iopt on the second row, determines the use of the observation uncertainties provided further down. The third row specifies initial guesses for the model parameters. The following tabular data represent the observations (x, y) and the associated variances σ.

Solution

The Python implementation is given in Listing 9.23 and the equivalent C/C++ version is available as supplementary material (`P09-Halides.cpp`). The graphical output produced by the program is shown in Figure 9.13.

Listing 9.22   File `halides.dat`: Sample Input for Program 9.23

```
13   2
1
   1000.   1.0
   2.01    2540.    100.
   2.57    3850.    190.
   2.31    3390.    100.
   2.81    4640.    120.
   2.98    5400.    270.
   3.23    5880.    290.
   2.67    4550.    230.
...
```

Listing 9.23   Absorption Maximum of Halides by Linear and Nonlinear Regression (Python Coding)

```
# Absorption maximum of halides by linear and non-linear regression
from modfunc import *
from graphlib import *

def Func(x, a, npar):                       # model function for MarqFit
    return a[1] * pow(x,a[2])

# main

nfit = 100                        # number of points plotted from model
nn   = [0]*3                            # ending indexes of plots
```

```
col = [""]*3                                             # colors of plots
sty = [0]*3                                              # styles of plots

inp = open("halides.dat","r")                            # open data file

line = inp.readline()                        # number of observed data and parameters
n = int(line.split()[0])
npar = int(line.split()[1])
                                                         # allocate arrays
x    = [0]*(n+nfit+1)                         # x-coordinates of observed data
y    = [0]*(n+nfit+1)                         # y-coordinates of observed data
sigmy = [0]*(n+1)                         # standard deviations of observed data
a    = [0]*(npar+1)                                      # model parameters
sigma = [0]*(npar+1)                            # uncertainties of parameters

iopt = int(inp.readline())                      # initialization option for sigmy[i]
line = inp.readline()
for i in range(1,npar+1): a[i] = float(line.split()[i-1]) # parameter guesses
for i in range(1,n+1):
   line = inp.readline()
   x[i] = float(line.split()[0])                         # observed data
   y[i] = float(line.split()[1])
   if (iopt): sigmy[i] = float(line.split()[2])          # uncertainty
inp.close()

GraphInit(1200,600)

nn[1] = n        ; col[1] = "red" ; sty[1] = 4           # observed data
nn[2] = n + nfit; col[2] = "blue"; sty[2] = 1           # fitted model

#-------------------------------------------------------- Levenberg-Marquart fit
a[1] = 1e3; a[2] = 1e0                            # guess for model parameters
chi2 = MarqFit(x,y,sigmy,n,iopt,a,sigma,npar,Func)

print("Levenberg-Marquardt fit\n")
print("a = {0:6.2f} +/- {1:6.2f}".format(a[1],sigma[1]))
print("b = {0:6.2f} +/- {1:6.2f}".format(a[2],sigma[2]))
print("Chi^2 = {0:6.2f}".format(chi2))

h = (x[n]-x[1])/(nfit-1)
for i in range(1,nfit+1):                                # append model points
   x[n+i] = x[1] + (i-1)*h
   y[n+i] = Func(x[n+i],a,npar)                          # evaluate model

MultiPlot(x,y,sigmy,nn,col,sty,2,10,0e0,0e0,0,0e0,0e0,0,
          0.10,0.45,0.15,0.85,"d (A)","lam (A)",
          "Absorption maximum of halides - Levenberg-Marquardt fit")

#------------------------------------------------------------------- Linear fit
for i in range(1,n+1): sigmy[i] = log(1e0+sigmy[i]/y[i])   # transform data
for i in range(1,n+nfit+1): x[i] = log(x[i]); y[i] = log(y[i])

(a[1], a[2], sigma[1], sigma[2], chi2) = LinFit(x,y,sigmy,n,iopt)

print("\nLinear fit\n")
print("a = {0:6.2f} +/- {1:6.2f}".format(exp(a[2]),exp(sigma[2])))
print("b = {0:6.2f} +/- {1:6.2f}".format(a[1],sigma[1]))

for i in range(1,nfit+1): y[n+i] = a[1] * x[n+i] + a[2]   # evaluate model
```

```
MultiPlot(x,y,sigmy,nn,col,sty,2,10,0e0,0e0,0,0e0,0e0,0,
          0.60,0.95,0.15,0.85,"log(d)","log(lam)",
          "Absorption maximum of halides - Linear fit")

MainLoop()
```

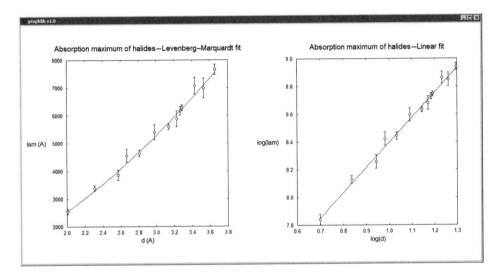

FIGURE 9.13   Absorption maxima of alkali halide crystals, λ_{max}, as functions of the interionic distance, d. The regression curves are yielded by Program 9.23 by using the Levenberg–Marquardt method (left plot), and, respectively, linear regression (right plot).

PROBLEM 9.5

Consider $n = 6$ equally spaced tabulation points of the polynomial $P(x) = 0.1x^3 - x^2 + 0.5x$ in the interval $[0, 10]$ and attempt to recover the monomial coefficients by a polynomial fit using the routine PolFit. Assign equal weights to the observables by setting the input parameter iopt to 0, case in which there is no need to specify associated standard deviations. Evaluate the polynomial model for 100 equidistant arguments and plot it along with the observed data. Comment on the accuracy of the adjusted parameters.

Solution

The Python and C/C++ implementations are available as supplementary material in files P09-PolFit.py and P09-PolFit.cpp. The graphical output produced by the programs is shown in Figure 9.14.

PROBLEM 9.6

Consider $n = 9$ equally spaced tabulation points sampling the function $f(x) = 1 - 2\sin x + \cos 2x$ in the interval $[0, 2\pi]$ and recover its Fourier coefficients by a multilinear fit using the routine MultiFit. To this end, define a user function returning as basis functions 1, $\cos x$, $\sin x$, $\cos 2x$, and $\sin 2x$. Assign equal weights to the observables by setting the input parameter iopt to 0. Evaluate the model function for 100 equidistant arguments and plot it along with the observed data. Comment on the accuracy of the adjusted parameters.

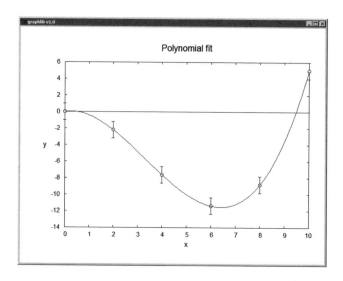

FIGURE 9.14 Polynomial regression performed by the programs `P09-PolFit.py` and `P09-PolFit.cpp`.

Solution

The Python implementation is given in Listing 9.24 and the equivalent C/C++ version is available as supplementary material (`P09-MultiFit.cpp`). The graphical output produced by the program is shown in Figure 9.15.

Listing 9.24 Multilinear Fit by the Least-Squares Method (Python Coding)

```python
# Multilinear fit
from math import *
from modfunc import *
from graphlib import *

def Func(x, func, npar):      # returns the basis functions sin(x) and cos(x)
   func[1] = 1e0
   func[2] = cos(x);   func[3] = sin(x)
   func[4] = cos(2*x); func[5] = sin(2*x)

# main

nn  = [0]*3                                  # end-indexes of the 3 plots
col = [""]*3                                        # colors of plots
sty = [0]*3                                         # styles of plots

n = 9                                       # number of observed data
npar = 5                                    # number of model parameters
nfit = 100                          # number of points plotted from model

x     = [0]*(n+nfit+1)                        # x-coordinates of observed data
y     = [0]*(n+nfit+1)                        # y-coordinates of observed data
sigmy = [0]*(n+1)                      # standard deviations of observed data
func  = [0]*(npar+1)                           # values of basis functions
a     = [0]*(npar+1)                              # model parameters
sigma = [0]*(npar+1)                         # uncertainties of parameters

                 # observed data generated from f(x) = 1 - 2sin(x) + cos(2x)
```

```
x[1] = 0.000; y[1] =  2.000; x[2] = 0.785; y[2] = -0.414
x[3] = 1.571; y[3] = -2.000; x[4] = 2.356; y[4] = -0.414
x[5] = 3.142; y[5] =  2.000; x[6] = 3.927; y[6] =  2.414
x[7] = 4.712; y[7] =  2.000; x[8] = 5.498; y[8] =  2.414
x[9] = 6.283; y[9] =  2.000

iopt = 0                             # least squares fit: equal errors sigmy
chi2 = MultiFit(x,y,sigmy,n,iopt,a,sigma,npar,Func)

print("Multilinear fit:")
for i in range(1,npar+1):
   print(("a[{0:d}] = {1:8.4f} +/- {2:8.4f}").format(i,a[i],sqrt(sigma[i])))
print("Chi^2 = {0:8.4f}".format(chi2))

h = (x[n]-x[1])/(nfit-1)
for i in range(1,nfit+1):                        # append model points
   xi = x[1] + (i-1)*h
   Func(xi,func,npar)                            # evaluate basis functions
   f = 0e0
   for j in range(1,npar+1): f += a[j]*func[j]           # evaluate model
   x[n+i] = xi; y[n+i] = f

GraphInit(800,600)

nn[1] = n         ; col[1] = "red" ; sty[1] = 0        # observed data
nn[2] = n + nfit; col[2] = "blue"; sty[2] = 1          # fitted model

MultiPlot(x,y,sigmy,nn,col,sty,2,10,0e0,0e0,0,0e0,0e0,0,
          0.15,0.95,0.15,0.85,"x","y","Multilinear fit")

MainLoop()
```

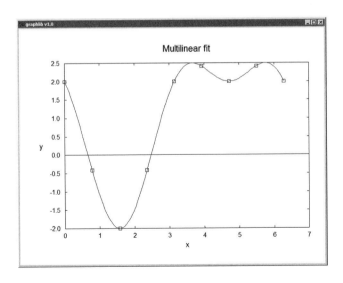

FIGURE 9.15 Least-squares multilinear regression performed by Program 9.24, using as model function $f(x) = a_1 + a_2 \cos x + a_3 \sin x + a_4 \cos 2x + a_5 \sin 2x$.

PROBLEM 9.7

Using the routine `MarqFit` (based on the Levenberg–Marquardt method), write a program to fit linear combinations of parametrized Gaussians,

$$F(x; \mathbf{a}) = \sum_{i=1,4}^{n} a_i \exp\left[-(x - a_{i+1})^2 / a_{i+2}\right],$$

to observed data. The number of model parameters is, of course, an integer multiple of 3.

Consider as input the data contained in the file `fit2.dat` and shown in Listing 9.25. The first row of `fit2.dat` provides the number of observables n and the number of parameters `npar`. The value 1, specified for the control parameter `iopt` on the second row, determines the regression routine to make use of the observation uncertainties provided further down. The third row specifies initial guesses for the model parameters. The following tabular data represent (x, y) coordinates extracted from the function:

$$f(x) = \exp\left[-(x + 0.8)^2 / 0.5\right] + 2 \exp\left[-(x - 0.8)^2 / 0.5\right],$$

with the ordinates distorted randomly by $\pm 5\%$, along with standard deviations σ, representing 10% of the ordinates, but not less than 0.02.

Plot the observed data and the fitted model and compare the optimized model parameters with those of the original dependence $f(x)$.

Solution

The Python implementation is provided in Listing 9.26 and the equivalent C/C++ version is available as supplementary material (`P09-MarqFit.cpp`). The graphical output produced by the program is shown in Figure 9.16.

Listing 9.25   File `fit2.dat`: Sample Input for the Nonlinear Fitting Program 9.26

```
 15   6
 1
    0.5  -1.0   1.0   1.0   1.0   1.0
   -2.8    0.0004   0.02
   -2.4    0.0063   0.02
   -2.0    0.0589   0.02
   -1.6    0.2642   0.03
   -1.2    0.6905   0.07
   -0.8    1.0625   0.11
   -0.4    0.8803   0.09
    0.0    0.7924   0.08
    0.4    1.4330   0.14
 . . .
```

Listing 9.26   Levenberg–Marquardt Fit (Python Coding)

```
# Non-linear fit by the Levenberg-Marquardt method
from math import *
from modfunc import *
from graphlib import *

def Func(x, a, npar):                                    # f = sum of Gaussians
```

```
   f = 0e0                                          # npar = multiple of 3
   for i in range(1,npar+1,3): f += a[i] * exp(-pow(x-a[i+1],2)/a[i+2])
   return f

# main

nfit = 100                               # number of points plotted from model
nn  = [0]*3                                      # ending indexes of plots
col = [""]*3                                           # colors of plots
sty = [0]*3                                           # styles of plots

inp = open("fit2.dat","r")                               # open data file

line = inp.readline()                    # number of observed data and parameters
n = int(line.split()[0])
npar = int(line.split()[1])

                                                    # allocate arrays
x     = [0]*(n+nfit+1)                     # x-coordinates of observed data
y     = [0]*(n+nfit+1)                     # y-coordinates of observed data
sigmy = [0]*(n+1)                     # standard deviations of observed data
a     = [0]*(npar+1)                              # model parameters
sigma = [0]*(npar+1)                         # uncertainties of parameters

iopt = int(inp.readline())                  # initialization option for sigmy[i]
line = inp.readline()
for i in range(1,npar+1): a[i] = float(line.split()[i-1]) # parameter guesses
for i in range(1,n+1):
   line = inp.readline()
   x[i] = float(line.split()[0])                             # observed data
   y[i] = float(line.split()[1])
   if (iopt): sigmy[i] = float(line.split()[2])              # uncertainty
inp.close()

out = open("fit.txt","w")                                # open output file
out.write(("n = {0:2d}   npar = {1:2d}\n").format(n,npar))
out.write("Initial parameters:\n")
for i in range(1,npar+1): out.write(("a[{0:d}] = {1:6.3f}\n").format(i,a[i]))

chi2 = MarqFit(x,y,sigmy,n,iopt,a,sigma,npar,Func)    # Levenberg-Marquart fit

out.write("\nFinal parameters:\n")
for i in range(1,npar+1):
   out.write(("a[{0:d}] = {1:6.3f}   sigma[{2:d}] = {3:7.1e}\n").
             format(i,a[i],i,sigma[i]))
out.write("\nChi^2 = {0:7.1e}\n".format(chi2))
out.write("\n i      x          y         sigma     yfit      y-yfit\n")
for i in range(1,n+1):
   yfit = Func(x[i],a,npar)                 # model values for adjusted parameters
   out.write(("{0:2d}{1:10.5f}{2:10.5f}{3:10.5f}{4:10.5f}{5:10.1e}\n").
             format(i,x[i],y[i],sigmy[i],yfit,y[i]-yfit))
out.close()

GraphInit(800,600)
                                                         # prepare plots:
nn[1] = n        ; col[1] = "red" ; sty[1] = 4         # observed data
nn[2] = n + nfit; col[2] = "blue"; sty[2] = 1          # fitted model
h = (x[n]-x[1])/(nfit-1)
for i in range(1,nfit+1):
   x[n+i] = x[1] + (i-1)*h                          # append data for 2nd plot
```

```
   y[n+i] = Func(x[n+i],a,npar)

MultiPlot(x,y,sigmy,nn,col,sty,2,10,0e0,0e0,0,0e0,0e0,0,
          0.15,0.95,0.15,0.85,"x","y","Non-linear fit")

MainLoop()
```

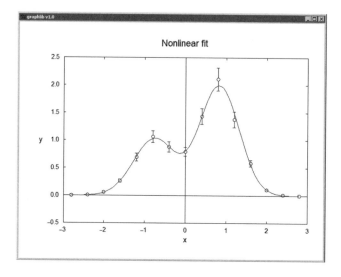

FIGURE 9.16   Nonlinear regression based on the Levenberg–Marquardt method, carried out by Program 9.26 for a model function defined as a linear combination of parametrized Gaussians.

PROBLEM 9.8

The explanation of the tidal cycles in connection with the daily cycles of the earth and moon has always awakened interest. Using the hourly tidal levels for Eastport, Maine, provided by the Center for Operational Oceanographic Products and Services (NOAA, `http://co-ops.nos.noaa.gov/`), R. de Levie has succeeded (*Am. J. Phys.* **72**, 644, 2003) to model the tidal cycles by a linear combination of two shifted sines, corresponding to the lunar and solar cycles:

$$F(x; A_1, A_2, A_3, \delta_1, \delta_2) = A_1 \sin(\omega_1 t + \delta_1) + A_2 \sin(\omega_2 t + \delta_2) + A_3.$$

The lunar and solar angular frequencies are given, respectively, by $\omega_1 = 2\pi/(T_1/2)$ and $\omega_2 = 2\pi/(T_2/2)$, where $T_1 = 24.84$ h and $T_2 = 24$ h. The adjustable model parameters are the amplitudes A_1, A_2, and A_3, along with the phase shifts δ_1 and δ_2.

To take advantage of the *intrinsically multilinear* character of the model function, rewrite it as a linear combination of nonparametrized sines and cosines of ω_1 and ω_2:

$$F(x; \mathbf{a}) = a_1 + a_2 \cos \omega_1 t + a_3 \sin \omega_1 t + a_4 \cos \omega_2 t + a_5 \sin \omega_2 t. \tag{9.73}$$

Use comparatively the routines `MarqFit` (for nonlinear regression by the Levenberg–Marquardt method) and `MultiFit` (for multilinear regression) to fit the model (9.73) to the tidal data for the same port, but the more recent period from December 1 to December 15, 2012. The data is compiled in Listing 9.27 and composes the input file `tides.dat`.

The first row of `tides.dat` specifies the number of observables n and the number of model parameters npar. The second row contains the initialization option iopt, which is set to 0 in this case to instruct the regression routines to use equal uncertainties for all observations. The third row specifies initial guesses for the five model parameters, which, due to the robustness of the multilinear fit, can be simply set to 0. The following 360 rows contain the observations, where the first entry is the time (in hours) and the second the tide level (in meters).

Elaborate on the match between the optimized model parameters provided by the two applied methods and assess the correctness of the fitted model by plotting it along with the observations.

Solution

The Python and C/C++ implementations are given, respectively, in Listing 9.28 and as supplementary material (`P09-Tides.cpp`). The graphical output produced by the programs is shown in Figure 9.17.

Listing 9.27   File `tides.dat`: Sample Input for Program 9.28

```
360   5
0
   0.0   0.0   0.0   0.0   0.0
     0.0    0.363
     1.0    1.329
     2.0    2.591
     3.0    3.870
     4.0    4.935
     5.0    5.358
     6.0    5.014
     7.0    4.232
     8.0    3.165
  . . .
```

Listing 9.28   Tidal Analysis Based on Levenberg–Marquardt and Multilinear Fits (Python Coding)

```python
# Tidal analysis
from math import *
from modfunc import *
from graphlib import *

T1 = 24.8412e0                                          # lunar period
T2 = 24e0                                               # solar period
omg1 = 2e0 * pi / (T1 / 2e0)               # lunar frequency (1/h)
omg2 = 2e0 * pi / (T2 / 2e0)               # solar frequency (1/h)

def Func1(x, a, npar):                          # model function for MarqFit
    return a[1] + a[2] * cos(omg1 * x) + a[3] * sin(omg1 * x) \
                + a[4] * cos(omg2 * x) + a[5] * sin(omg2 * x)

def Func2(x, func, npar):                       # model function for MultiFit
    func[1] = 1e0
    func[2] = cos(omg1 * x); func[3] = sin(omg1 * x)
    func[4] = cos(omg2 * x); func[5] = sin(omg2 * x)

# main

nn  = [0]*3                                     # ending indexes of plots
col = [""]*3                                    # colors of plots
sty = [0]*3                                     # styles of plots
```

```
inp = open("tides.dat","r")                                    # open data file

line = inp.readline()                          # number of observed data and parameters
n = int(line.split()[0])
npar = int(line.split()[1])
nfit = n                                       # number of points plotted from model
                                                               # allocate arrays
x     = [0]*(n+nfit+1)                              # x-coordinates of observed data
y     = [0]*(n+nfit+1)                              # y-coordinates of observed data
sigmy = [0]*(n+1)                              # standard deviations of observed data
func  = [0]*(npar+1)                                # values of basis functions
a     = [0]*(npar+1)                                    # model parameters
sigma = [0]*(npar+1)                                # uncertainties of parameters

iopt = int(inp.readline())                     # initialization option for sigmy[i]
line = inp.readline()
for i in range(1,npar+1): a[i] = float(line.split()[i-1]) # parameter guesses
for i in range(1,n+1):
   line = inp.readline()
   x[i] = float(line.split()[0])                               # observed data
   y[i] = float(line.split()[1])
   if (iopt): sigmy[i] = float(line.split()[2])                  # uncertainty
inp.close()

GraphInit(1200,800)

nn[1] = n        ; col[1] = "red" ; sty[1] = 1                # observed data
nn[2] = n + nfit; col[2] = "blue"; sty[2] = 1                # fitted model

#-------------------------------------------------- Levenberg-Marquart fit
chi2 = MarqFit(x,y,sigmy,n,iopt,a,sigma,npar,Func1)

print("Levenberg-Marquardt fit\n")
for i in range(1,npar+1):
   print(("a[{0:d}] = {1:6.3f}  sigma[{2:d}] = {3:7.1e}").
         format(i,a[i],i,sigma[i]))

h = (x[n]-x[1])/(nfit-1)
for i in range(1,nfit+1):                                  # append model points
   x[n+i] = x[1] + (i-1)*h
   y[n+i] = Func1(x[n+i],a,npar)

MultiPlot(x,y,sigmy,nn,col,sty,2,10,0e0,0e0,0,0e0,0e0,0,0.10,0.95,0.60,0.90,
          "t (h)","z (m)","Tidal analysis - Levenberg-Marquardt fit")

#-------------------------------------------------------- Multilinear fit
chi2 = MultiFit(x,y,sigmy,n,iopt,a,sigma,npar,Func2)

print("\nMultilinear fit\n")
for i in range(1,npar+1):
   print(("a[{0:d}] = {1:6.3f}  sigma[{2:d}] = {3:7.1e}").
         format(i,a[i],i,sigma[i]))

h = (x[n]-x[1])/(nfit-1)
for i in range(1,nfit+1):                                  # append model points
   xi = x[1] + (i-1)*h
   Func2(xi,func,npar)                              # evaluate basis functions
   f = 0e0
   for j in range(1,npar+1): f += a[j]*func[j]                # evaluate model
```

```
    x[n+i] = xi; y[n+i] = f

MultiPlot(x,y,sigmy,nn,col,sty,2,10,0e0,0e0,0,0e0,0e0,0,0.10,0.95,0.10,0.40,
          "t (h)","z (m)","Tidal analysis - Multilinear fit")

MainLoop()
```

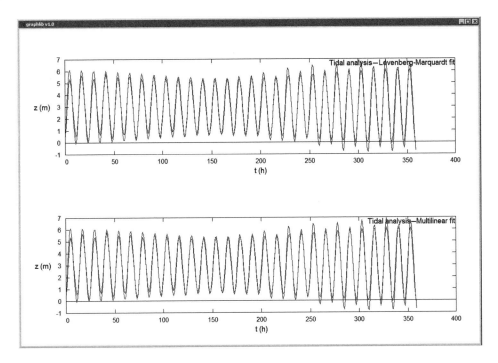

FIGURE 9.17 Tidal analysis of water levels in Eastport, Maine, over a 15-day period based on a Levenberg–Marquardt fit (upper plot) and a multilinear fit (lower plot).

PROBLEM 9.9

Build a program calling the routine `MultiFit` to perform the real Fourier analysis of observational data by multilinear regression.

The Fourier series for an arbitrary real function may be written as

$$f(x) = a_1 + \sum_{i=1} (a_{2i} \cos ix + a_{2i+1} \sin ix).$$

The involved cosines and sines can be calculated most efficiently by using the recurrences

$$\cos ix = \cos x \cos(i-1)x - \sin x \sin(i-1)x,$$

$$\sin ix = \sin x \cos(i-1)x + \cos x \sin(i-1)x,$$

started with $\cos 0 = 1$ and $\sin 0 = 0$. Indeed, the procedure involves the effective evaluation of a single $\cos x$ and a single $\sin x$ at the beginning of the recurrence.

In the case of odd functions, $f(-x) = -f(x)$, the Fourier series reduces to the *Fourier sine series*, $f(x) = \sum_i a_i \sin ix$, whereas for even functions, $f(-x) = f(x)$, the so-called *Fourier cosine series* is applied, $f(x) = a_1 + \sum_i a_i \cos ix$.

Consider as data sources the *periodic step function*, the *sawtooth function*, and the *shifted sawtooth function*, defined, respectively, by

$$f_1(x) = \begin{cases} -1 & \text{for } -\pi \le x < 0, \\ 1 & \text{for } x \ge 0 \ge \pi, \end{cases}$$

$$f_2(x) = x, \quad x \in [-\pi, \pi],$$

$$f_3(x) = x - \frac{\pi}{2}, \quad x \in [-\pi, \pi].$$

Generate for each of these functions a number of $n = 500$ uniformly distributed "observed" data points and perform multilinear regressions using $11, 21, 51, \ldots$ Fourier basis functions. The odd number of basis functions is required to have matching cosines and sines in the general Fourier series. Since the functions $f_1(x)$ and $f_2(x)$ are odd, their Fourier series actually contain merely sines. On the other hand, since $f_3(x)$ has no particular symmetry, both cosines and sines have to be considered in its Fourier representation.

Use the adjusted coefficients to construct the Fourier series for the three functions and plot these in pairs. Plot, equally, the Fourier coefficients a_i against their index i and note the qualitative difference between the resulting power spectra.

Solution

The Python implementation is provided in Listing 9.29 and the equivalent C/C++ version is available as supplementary material (P09-Fourier.cpp). The sine basis functions necessary for expanding the antisymmetric functions $f_1(x)$ and $f_2(x)$ are evaluated by the routine Func0, while in the case of $f_3(x)$, lacking a particular symmetry, the required sine and cosine basis functions are returned by the routine Func.

The graphical output produced by the program in the three cases is shown in Figures 9.18 through 9.20.

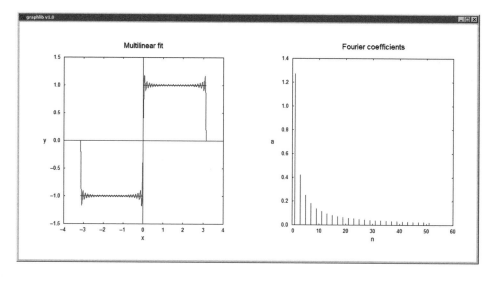

FIGURE 9.18 Fourier analysis of a periodic step function by multilinear regression and sine basis functions. Left: original function and Fourier representation. Right: Fourier coefficients of sines.

Listing 9.29 Fourier Analysis Based on Multilinear Fit (Python Coding)

```python
# Fourier analysis using multilinear fit
from math import *
from modfunc import *
from graphlib import *

def Func0(x, func, npar):                                   # sine basis functions
   for i in range(1,npar+1): func[i] = sin(i*x)


def Func(x, func, npar):                              # Fourier basis functions
   cosx = cos(x); cos1 = 1e0                                  # npar = odd
   sinx = sin(x); sin1 = 0e0

   func[1] = 1e0
   for i in range(2,npar,2):
      cosi = cosx * cos1 - sinx * sin1
      sini = sinx * cos1 + cosx * sin1
      cos1 = cosi; sin1 = sini

      func[i] = cosi; func[i+1] = sini

# main

nn  = [0]*3                                  # end-indexes of the 3 plots
col = [""]*3                                       # colors of plots
sty = [0]*3                                        # styles of plots

n = 500                                      # number of observed data
npar = 51                                   # number of model parameters
nfit = n                               # number of points plotted from model

x     = [0]*(n+nfit+1)                       # x-coordinates of observed data
y     = [0]*(n+nfit+1)                       # y-coordinates of observed data
sigmy = [0]*(n+1)                        # standard deviations of observed data
func  = [0]*(npar+1)                              # values of basis functions
a     = [0]*(npar+1)                                   # model parameters
sigma = [0]*(npar+1)                          # uncertainties of parameters

xmin = -pi; xmax = pi                                 # generate data points
h = (xmax-xmin)/(n-1)
for i in range(1,n+1):
   x[i] = xmin + (i-1)*h
   y[i] = -1e0 if x[i] < 0e0 else 1e0               # periodic step function
#  y[i] = x[i]                                      # "sawtooth" function
#  y[i] = x[i] - pi/2                            # shifted "sawtooth" function

iopt = 0                                # least squares fit: equal errors sigmy
chi2 = MultiFit(x,y,sigmy,n,iopt,a,sigma,npar,Func0)

h = (x[n]-x[1])/(nfit-1)
for i in range(1,nfit+1):                              # append model points
   xi = x[1] + (i-1)*h
   Func0(xi,func,npar)                                # evaluate basis functions
   f = 0e0
   for j in range(1,npar+1): f += a[j]*func[j]                # evaluate model
   x[n+i] = xi; y[n+i] = f

GraphInit(1200,600)
```

```
nn[1] = n          ; col[1] = "red" ; sty[1] = 1          # observed data
nn[2] = n + nfit; col[2] = "blue"; sty[2] = 1             # fitted model

MultiPlot(x,y,sigmy,nn,col,sty,2,10,0e0,0e0,0,0e0,0e0,0,
          0.10,0.45,0.15,0.85,"x","y","Multilinear fit")

for i in range(1,npar+1): x[i] = float(i)
Plot(x,a,npar,"red",3,0.60,0.95,0.15,0.85,"n","a","Fourier coefficients")

MainLoop()
```

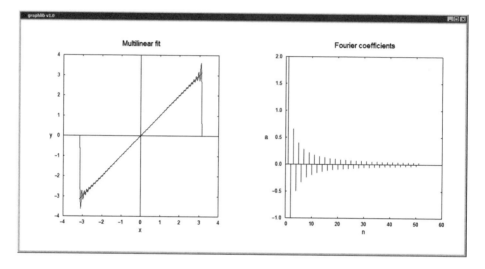

FIGURE 9.19 Fourier analysis of a "sawtooth" function using multilinear regression and sine basis functions. Left: original function and Fourier representation. Right: Fourier coefficients of sines.

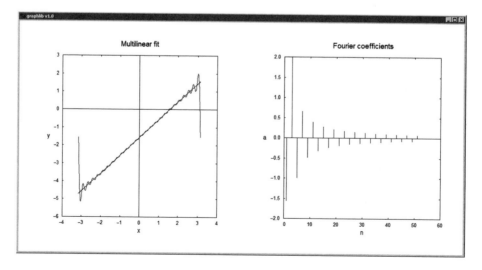

FIGURE 9.20 Fourier analysis of a shifted "sawtooth" function by multilinear regression. Left: original function and Fourier representation. Right: Fourier coefficients of cosines (even n) and sines (odd n).

References and Suggested Further Reading

Beu, T. A. 2004. *Numerical Calculus in C* (3rd ed., in Romanian). Cluj-Napoca: MicroInformatica.

Burden, R. and J. Faires. 2010. *Numerical Analysis* (9th ed.). Boston: Brooks/Cole, Cengage Learning.

Demidovich, B. and I. Maron. 1987. *Computational Mathematics* (4th ed.). Moscow: MIR Publishers.

Iacob, C., D. Homentcovschi, N. Marcov, and A. Nicolau. 1983. *Classical and Modern Mathematics* (in Romanian). Bucharest: Editura tehnică.

Ivey, H. F. 1947. *Phys. Rev. 72*, 341.

Lawson, C. L. and R. Hanson. 1974. *Solving Least Squares Problems*. Englewood Cliffs, NJ: Prentice-Hall.

Marquardt, D. W. 1963. An Algorithm for Least-Squares Estimation of Nonlinear Parameters. *Journal of the Society for Industrial and Applied Mathematics 11*(2), 431–441.

Press, W. H., S. A. Teukolsky, W. T. Vetterling, and B. P. Flannery. 2007. *Numerical Recipes 3rd Edition: The Art of Scientific Computing*. Cambridge: Cambridge University Press.

10
Integration of Functions

10.1 Introduction

The definite integral of a real-valued continuous function $f(x)$, defined on some real interval $[a, b]$ and whose *primitive* (antiderivative) $F(x)$ can be determined, may be evaluated using the Newton–Leibniz formula:

$$I = \int_a^b f(x)dx = F(b) - F(a).$$

However, in practical applications, quite often, the primitive $F(x)$ cannot be found or, determining it by elementary methods, involves a tremendous analytical effort. The integrand $f(x)$ itself is sometimes specified merely as a table of numerical values, rendering the very concept of the primitive inoperable. Furthermore, the inherent complications arising from the coupling of variables in many multidimensional cases rule out the analytical integration altogether. These and many other practical arguments entirely justify the importance of developing numerical methods to calculate definite integrals.

Numerical integration is also referred to as *quadrature*. Even though the term typically applies to one-dimensional integrals, in a broader sense, it is used along with the term *cubature* for higher-dimensional integration, too.

In general, any quadrature formula replaces the integral by a weighted sum of integrand values for particular points from the integration domain, $x_i \in [a, b]$, and features a specific *truncation error* (error term) R_n:

$$\int_a^b f(x)dx = \sum_{i=1}^n w_i f(x_i) + R_n.$$

A quadrature formula employing only information from interior points of the integration domain is called *open*, while a formula based also on the integrand values on the boundaries is called *closed*.

10.2 Trapezoidal Rule; A Heuristic Approach

Starting from the geometrical interpretation of the definite integral, as the net signed area bounded by the integrand $f(x)$, the x-axis, and the vertical lines $x = a$ and $x = b$, the simplest approach to a practical quadrature algorithm is to replace the graph of the integrand by the polygonal line defined by a finite number of integrand values (as shown in Figure 10.1) and to sum the trapezoidal areas formed thereby.

To sample the integration interval $[a, b]$ uniformly, one can choose equidistant points,

$$x_i = a + (i - 1)h, \quad i = 1, 2, \ldots, n, \tag{10.1}$$

defined by the spacing

$$h = (b - a)/(n - 1). \tag{10.2}$$

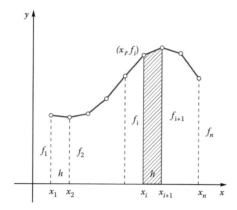

FIGURE 10.1 The integrand's graph is approximated by the line segments connecting the points (x_i, f_i).

Replacing the partial integrals on each of the $(n-1)$ subintervals (x_i, x_{i+1}) by the corresponding trapezoidal areas, the integral is approximated by

$$\int_a^b f(x)dx \approx \frac{h}{2}(f_1 + f_2) + \cdots + \frac{h}{2}(f_i + f_{i+1}) + \cdots + \frac{h}{2}(f_{n-1} + f_n),$$

and, by further grouping the terms, one obtains the *trapezoidal rule*:

$$\int_a^b f(x)dx \approx h\left[\frac{f_1}{2} + \sum_{i=2}^{n-1} f_i + \frac{f_n}{2}\right]. \tag{10.3}$$

Obviously, the smaller the mesh spacing h, the more accurate becomes the quadrature, so that, in the limit $h \to 0$, one regains the Riemann integral.

Listing 10.1 Function Integration Using the Trapezoidal Rule (Python Coding)

```python
# Evaluates an integral using the trapezoidal rule
from math import *

def Func(x): return (x*x*x) * exp(-x)

#===========================================================================
def qTrapz(Func, a, b, n):
#---------------------------------------------------------------------------
#  Integrates function Func on interval [a,b] using the trapezoidal rule
#  with n integration points
#---------------------------------------------------------------------------
   h = (b-a)/(n-1)
   s = 0.5*(Func(a) + Func(b))
   for i in range(1,n-1): s += Func(a+i*h)

   return h*s

# main

a = 0e0; b = 1e0; n = 100
print("I = ",qTrapz(Func,a,b,n))
```

To evaluate the test integral

$$\int_0^1 x^3 e^{-x} dx = 0.1139289\ldots,$$

the sample program 10.1 calls function qTrapz, which implements the trapezoidal rule.

Along with the integration limits, a and b, and the number of integration points, n, the function qTrapz receives the actual name of the user function that codes the integrand by the procedural argument Func. In generating the integration nodes, instead of the recursive incrementation $x_{i+1} = x_i + h$, the nonrecursive expression $a + ih$ is preferable, since it reduces roundoff errors (while the shifted index spares a subtraction) and can be directly passed as an argument in calls to function Func.

10.3 The Newton–Cotes Quadrature Formulas

The trapezoidal rule can be shown to be just the lowest-order member of an entire family, called *Newton–Cotes quadrature formulas*, devised for integrands specified on regular integration meshes. The general underlying idea is to replace the integrand by a suitable interpolating polynomial based on the equidistant integrand values.

With a view to evaluating the definite integral

$$I = \int_a^b f(x)dx, \tag{10.4}$$

we start by dividing the interval $[a, b]$ into $(n - 1)$ equal subintervals of length

$$h = (b - a)/(n - 1), \tag{10.5}$$

by the equally spaced abscissas

$$x_i = a + (i - 1)h, \quad i = 1, 2, \ldots, n. \tag{10.6}$$

Assuming that the integrand values $f_i \equiv f(x_i)$ at the integration nodes x_i are known, we approximate the integrand by the *Lagrange interpolating polynomial* (see Section 9.2):

$$P_{n-1}(x) = \sum_{i=1}^n \frac{\prod_{j \neq i}^n (x - x_j)}{\prod_{j \neq i}^n (x_i - x_j)} f_i.$$

By introducing the dimensionless variable

$$q = (x - a)/h, \quad q \in [0, n - 1],$$

the function arguments may be expressed as $x = a + qh$ and the products occurring in the coefficients of the Lagrange polynomial can be rewritten as

$$\prod_{\substack{j \neq i}}^{n} (x - x_j) = h^{n-1} \prod_{\substack{j \neq i}}^{n} [q - (j-1)],$$

$$\prod_{\substack{j \neq i}}^{n} (x_i - x_j) = h^{n-1} \prod_{\substack{j \neq i}}^{n} (i - j) = (-1)^{n-i} h^{n-1} \prod_{j=1}^{i-1} (i - j) \prod_{j=i+1}^{n} (j - i)$$

$$= (-1)^{n-i} h^{n-1} (i-1)! (n-i)!$$

The Lagrange polynomial thus takes the form

$$P_{n-1}(x) = \sum_{i=1}^{n} \frac{\prod_{j \neq i}^{n} [q - (j-1)]}{(-1)^{n-i}(i-1)!(n-i)!} f_i, \tag{10.7}$$

and the following approximation to the integral is obtained:

$$\int_a^b f(x)dx \approx \int_a^b P_{n-1}(x)dx = \sum_{i=1}^{n} A_i f_i. \tag{10.8}$$

The coefficients A_i weighting the integrand values can be successively transformed,

$$A_i = \int_a^b \frac{\prod_{j \neq i}^{n} [q - (j-1)]dx}{(-1)^{n-i}(i-1)!(n-i)!} = \frac{h \int_0^{n-1} \prod_{j \neq i}^{n} [q - (j-1)]dq}{(-1)^{n-i}(i-1)!(n-i)!}, \tag{10.9}$$

and, by factoring out explicitly the extent of the integration interval, $A_i = (b-a)H_i$, the classical *Newton–Cotes quadrature formula* results:

$$\int_a^b f(x)dx \approx (b - a) \sum_{i=1}^{n} H_i f_i. \tag{10.10}$$

The corresponding *Cotes coefficients* are given by

$$H_i = \frac{\int_0^{n-1} \prod_{j \neq i}^{n} [q - (j-1)]dq}{(-1)^{n-i}(i-1)!(n-i)!(n-1)}, \quad i = 1, 2, \ldots, n. \tag{10.11}$$

It should be noted that, as per definition, the Cotes coefficients do neither depend on the integrated function, nor on the integration interval. In addition, they satisfy:

$$\sum_{i=1}^{n} H_i = 1, \quad H_i = H_{n-i+1}. \tag{10.12}$$

The first property, reflecting the *normalization* of the coefficients, may be demonstrated by simply considering the particular integrand $f(x) \equiv 1$ in the general Newton–Cotes formula 10.10, while the second property, implying the *symmetry* of the coefficient sequence with respect to its median, follows from the very definition of the coefficients, Equation 10.11, which remains invariant when replacing index i with its symmetric $n - i + 1$.

10.4 Trapezoidal Rule

Next, we consider the Newton–Cotes quadrature (10.10) for the particular case $n = 2$, which implies that the only integrand values used are the ones at the integration interval limits. The two involved Cotes coefficients (10.11) take equal values:

$$H_1 = -\int_0^1 (q - 1)dq = \frac{1}{2}, \tag{10.13}$$

$$H_2 = \int_0^1 qdq = \frac{1}{2}, \tag{10.14}$$

and therewith, the *trapezoidal formula* results:

$$\int_{x_1}^{x_2} f(x)dx \approx \frac{h}{2}(f_1 + f_2). \tag{10.15}$$

The name of the method comes from the fact that, as seen from Figure 10.2, the formula can be obtained directly by replacing the function $f(x)$ with the segment connecting the points (x_1, f_1) and (x_2, f_2), and the integral thus amounts to the area of the trapeze formed by the segment and its projection onto the x-axis.

The error associated with the trapezoidal formula 10.15, depicted as hatched in the plot of Figure 10.2, may be established imposing on $f(x)$ and its first two derivatives to be continuous on $[a, b]$ ($f(x) \in C^{(2)}[a, b]$). Aiming to establish a practical measure for the precision of a quadrature formula, one typically expresses the error term as a function of the integration step size h (Demidovich and Maron, 1987):

$$R(h) = \int_{x_1}^{x_1+h} f(x)dx - \frac{h}{2}[f(x_1) + f(x_1 + h)].$$

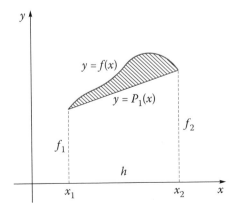

FIGURE 10.2 The trapezoidal formula approximates the integrand by the segment connecting the points (x_1, f_1) and (x_2, f_2).

By taking twice the derivative of the above expression with respect to h, we get:

$$R'(h) = \frac{1}{2}[f(x_1 + h) - f(x_1)] - \frac{h}{2}f'(x_1 + h),$$

$$R''(h) = -\frac{h}{2}f''(x_1 + h).$$

We now integrate $R''(h)$ twice, noting that $R(0) = R'(0) = 0$. Employing the *mean value theorem* (which essentially states that there is at least one point on an arc of a differentiable curve at which the derivative is equal to the average derivative of the arc), it results successively:

$$R'(h) = R'(0) + \int_0^h R''(u)du = -\frac{1}{2}f''(\xi_1)\int_0^h udu = -\frac{1}{4}h^2f''(\xi_1),$$

$$R(h) = R(0) + \int_0^h R'(u)du = -\frac{1}{4}f''(\xi)\int_0^h u^2du = -\frac{1}{12}h^3f''(\xi),$$

where $\xi, \xi_1 \in (x_1, x_1 + h)$. Finally, we obtain for the error term of the trapezoidal formula:

$$R(h) = -\frac{1}{12}h^3f''(\xi), \quad \xi \in (x_1, x_2). \tag{10.16}$$

It appears that the error term has opposite sign to the second derivative in the interval (x_1, x_2) and, consequently, the trapezoidal formula overestimates the integral if $f'' > 0$ and underestimates it otherwise.

The trapezoidal formula 10.15 is, obviously, of no practical interest as such, having rather a theoretical importance. Instead, by partitioning the integration interval $[a, b]$ by n equidistant points x_i and making use of the *additive property of integrals for subintervals*, one regains the *trapezoidal rule* (derived heuristically in Section 10.2),

$$\int_a^b f(x)dx \approx h\left[\frac{f_1}{2} + \sum_{i=2}^{n-1} f_i + \frac{f_n}{2}\right], \tag{10.17}$$

where h is the spacing of the integration points.

The error term associated with the trapezoidal rule cumulates the errors corresponding to the $(n - 1)$ integration subintervals, namely,

$$R_T(h) = -\frac{(n-1)}{12}h^3f''(\xi) = -\frac{(b-a)}{12}h^2f''(\xi), \quad \xi \in [a, b], \tag{10.18}$$

and shows the characteristic $O(h^2)$ order. Even though such an error term cannot be used as a practical estimate of the integration error, it implies that, for instance, by halving the mesh spacing h, the error roughly reduces by a factor of 4. Moreover, the error term enables by the *step-halving technique* the adaptive control of the integration mesh itself. This consists of calculating the integral for a given h (corresponding to n nodes) and comparing it with the result for $h/2$ (corresponding to $(2n - 1)$ nodes). The iterative halving of the mesh spacing is to be continued, in principle, until the relative difference between two consecutive approximations, regarded as on error estimate, drops under an admissible value ε.

10.5 Simpson's Rule

Considering the Newton–Cotes formulas 10.10 and 10.11 in particular for $n = 3$ mesh points, one obtains a quadrature method superior in precision to the trapezoidal formula. The corresponding Cotes coefficients are

$$H_1 = \frac{1}{4} \int_0^2 (q-1)(q-2)dq = \frac{1}{6}, \tag{10.19}$$

$$H_2 = -\frac{1}{2} \int_0^2 q(q-2)dq = \frac{2}{3}, \tag{10.20}$$

$$H_3 = \frac{1}{4} \int_0^2 q(q-1)dq = \frac{1}{6}, \tag{10.21}$$

and, given that $b - a \equiv x_3 - x_1 = 2h$, *Simpson's formula* results:

$$\int_{x_1}^{x_3} f(x)dx \approx \frac{h}{3}(f_1 + 4f_2 + f_3). \tag{10.22}$$

From a geometrical viewpoint, Simpson's formula implies replacing the curve $y = f(x)$ by the parabola $y = P_2(x)$, with the Lagrange polynomial $P_2(x)$ defined by the three points (x_1, f_1), (x_2, f_2), and (x_3, f_3).

Assuming the integrand $f(x)$ to be continuous together with its first four derivatives on $[a, b]$ ($f(x) \in C^{(4)}[a, b]$), one obtains an estimate of the error term for Simpson's formula by a technique similar to the one employed in the case of the trapezoidal formula (see Demidovich and Maron 1987):

$$R = -\frac{1}{90}h^5 f^{(4)}(\xi), \quad \xi \in [a, b]. \tag{10.23}$$

While the error depends on h^3 for the trapezoidal formula, it scales with h^5 for Simpson's formula and the latter can be shown to be exact not only for second-order polynomials, but also for third-order polynomials.

To establish a method of practical interest, usable with controllable precision for arbitrary integrands, one generalizes Simpson's formula 10.22 in a similar manner to the trapezoidal rule by making use of the *additive property of integrals for subintervals*. Concretely, dividing the interval $[a, b]$ by an *odd* number, $n = 2m + 1$, of equally spaced points,

$$x_i = a + (i-1)h, \quad i = 1, 2, \ldots, n,$$

separated by

$$h = \frac{b-a}{n-1} = \frac{b-a}{2m},$$

one can apply Simpson's formula 10.22 to each of the m double subintervals (of length $2h$) determined by three consecutive mesh points: $[x_1, x_3], \ldots, [x_{n-2}, x_n]$. Geometrically, this approach boils down to approximating the integrand by a piecewise parabolic function and, obviously, the generalization can be solely based on a mesh with *odd* number of points. Thus, the integral over the entire interval $[a, b]$ may be composed as

$$\int_a^b f(x)dx \approx \frac{h}{3}\left(f_1 + 4f_2 + f_3\right) + \frac{h}{3}\left(f_3 + 4f_4 + f_5\right) + \cdots$$

$$+ \frac{h}{3}\left(f_{n-4} + 4f_{n-3} + f_{n-2}\right) + \frac{h}{3}\left(f_{n-2} + 4f_{n-1} + f_n\right).$$

By regrouping the terms, one obtains *Simpson's rule*:

$$\int_a^b f(x)dx \approx \frac{h}{3}\left(f_1 + 4\sigma_2 + 2\sigma_1 + f_n\right), \qquad (10.24)$$

where

$$\sigma_1 = \sum_{i=3,5}^{n-2} f_i, \quad \sigma_2 = \sum_{i=2,4}^{n-1} f_i. \qquad (10.25)$$

Sum σ_1 is composed only of the odd-index terms, while σ_2 cumulates the even-index ones and contains one term more than σ_1.

Assuming $f(x)$ to be continuous of class $C^{(4)}[a, b]$, the error term of Simpson's rule results by summing errors (10.23) for the m double subintervals of length $2h$:

$$R_S = -\frac{m}{90}h^5 f^{(4)}(\xi) = -\frac{(b-a)}{180}h^4 f^{(4)}(\xi), \quad \xi \in [a, b]. \qquad (10.26)$$

Since it implies derivatives at unspecified interior points, the remainder is not an operational error estimate, but should be merely regarded as an indication of the *order of the global error*, that is, $O(h^4)$. Correspondingly, the *degree of precision* of Simpson's rule is three, implying that the composite formula is *exact* for polynomials up to degree three.

With a view to an efficient implementation, the indexes in the sums σ_1 and σ_2 are shifted by -1, so as to eliminate the subtraction $(i - 1)$ in the definition of the integration points:

$$\sigma_1 = \sum_{i=2,4}^{n-3} f(a + ih), \quad \sigma_2 = \sum_{i=1,3}^{n-2} f(a + ih). \qquad (10.27)$$

Listing 10.2   Function Integrator Based on Simpson's Rule (Python Coding)

```
#=============================================================================
def qSimpson(Func, a, b, n):
#-----------------------------------------------------------------------------
#  Integrates function Func on interval [a,b] using Simpson's rule with n
#  (odd) integration points
#-----------------------------------------------------------------------------
   if (n % 2 == 0): n += 1                                # increment n if even

   h = (b-a)/(n-1)
   s1 = s2 = 0e0
   for i in range(2,n-2,2): s1 += Func(a + i*h)           # odd-index sum
   for i in range(1,n-1,2): s2 += Func(a + i*h)           # even-index sum

   return (h/3)*(Func(a) + 4*s2 + 2*s1 + Func(b))
```

Function qSimpson in Listing 10.2 is a direct implementation of Simpson's rule according to Formulas 10.24 and 10.27. The significance of the arguments is the same as for function qTrapz (described in Section 10.2): the procedural argument Func receives the name of the user function coding the integrand, while a and b are the integration limits. If the received number of integration points n is even, it is automatically corrected to the next higher odd value, so that Simpson's formula can be applied to an integer number of node triplets, with the odd-index and even-index sums (10.27) calculated separately.

10.6 Adaptive Quadrature Methods

The implementations qTrapz and qSimpson presented in the previous section are of limited practical interest as such, since, besides requiring the number of integration points, they do not provide an error estimate for the integral. A straightforward strategy with automatic integration precision and mesh step control would call the elementary integrator for grids obtained by successive *step halving* and would be continued as long as the relative difference between consecutive approximations exceeds an admissible tolerance ε. Nonetheless, an obvious drawback would remain the redundant evaluation of the integrand at the common nodes of the successive meshes.

The calculation of the integral with a required precision by the *step-halving method* can be significantly optimized by establishing a recursive relation between the consecutive approximations. According to Figure 10.3, which shows the integration grids for the first four stages of the halving process, the evaluation of the integrand is necessary solely at the new integration points (indicated by circles) that divide the previous subintervals. The already-employed function values are implicitly reused by applying the recurrence relation between two consecutive approximations of the integral.

The first three approximations yielded by the *trapezoidal rule* are

$$T_0 = h_0 \left[\frac{f(a)}{2} + \frac{f(b)}{2} \right], \quad h_0 = b - a,$$

$$T_1 = h_1 \left[\frac{f(a)}{2} + f(a + h_1) + \frac{f(b)}{2} \right], \quad h_1 = h_0/2, \tag{10.28}$$

$$T_2 = h_2 \left[\frac{f(a)}{2} + f(a + h_2) + f(a + 2h_2) + f(a + 3h_2) + \frac{f(b)}{2} \right], \quad h_2 = h_1/2.$$

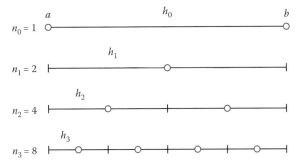

FIGURE 10.3 The subinterval-halving scheme in the adaptive trapezoidal rule. Circles indicate the new values that have to be evaluated at each halving iteration.

The above sequence can be actually described in terms of the iterative process:

$$T_0 = \frac{h_0}{2}[f(a) + f(b)], \quad h_0 = b - a, \quad n_0 = 1, \tag{10.29}$$

$$T_k = \frac{1}{2}\left[T_{k-1} + h_{k-1}\sum_{i=1}^{n_{k-1}} f\left(a + (i - 1/2)h_{k-1}\right)\right], \tag{10.30}$$

$$h_k = h_{k-1}/2, \quad n_k = 2n_{k-1}, \quad k = 1, 2, \dots,$$

where $h_k = (b - a)/n_k$ is the mesh spacing after the halving step k, with $n_k = 2^k$ representing the corresponding number of subintervals or, equivalently, the number of *new* integration points relevant for the next approximation of the integral.

The recursive process (10.29)–(10.30) is typically continued until two consecutive approximations of the integral become indistinguishable to a given accuracy, that is, until their relative difference, $|T_k - T_{k-1}|/|T_k|$, considered as an estimate of the relative error, drops under a predefined tolerance ε. To avoid the singular division in the particular situation when $T_k = 0$, the convergence criterion may be conveniently formulated as

$$\left|T_k - T_{k-1}\right| \le \varepsilon\left|T_k\right|. \tag{10.31}$$

The function qTrapzCtrl (Listing 10.3) implements the described adaptive algorithm based on the trapezoidal rule. To ensure that, during the recursive process, the current number of subintervals, n, as a power of 2, does not exceed the maximum representable value of type long ($2^{31} - 1$ for standard C/C++ implementations), the number of halving iterations is limited by kmax to 30. However, if, due to weak convergence, the kmax iterations are exhausted without the quadrature having achieved the requested accuracy eps, the step-halving loop is left with counter k incremented to kmax+1 and an error message is issued. Another, yet more versatile variant of treating the absence of convergence would be to return an error index to the calling program unit, leaving on its behalf the action to be taken.

Listing 10.3   Adaptive Integrator Based on the Trapezoidal Rule (Python Coding)

```
#===========================================================================
def qTrapzCtrl(Func, a, b, eps = 1e-6):
#---------------------------------------------------------------------------
#  Integrates function Func on interval [a,b] with relative precision eps
#  using the adaptive trapezoidal rule
#---------------------------------------------------------------------------
   kmax = 30                                 # max. no. of step halving iterations

   h = b-a; n = 1
   t0 = 0.5*h*(Func(a) + Func(b))                          # initial approximation

   for k in range(1,kmax+1):                               # step halving loop
      sumf = 0e0
      for i in range(1,n+1): sumf += Func(a+(i-0.5)*h)
      t = 0.5*(t0 + h*sumf)                                # new approximation
      if (k > 1):                                          # convergence check
         if (fabs(t-t0) <= eps*fabs(t)): break
         if (fabs(t) <= eps and fabs(t) <= fabs(t-t0)): break # integral ~= 0
      h *= 0.5; n *= 2                                     # halve integration step
      t0 = t

   if (k >= kmax): print("qTrapzCtrl: max. no. of iterations exceeded !")

   return t
```

On the basis of *Simpson's rule*, a similar, yet not stand-alone, adaptive algorithm can be devised. Indeed, while for the trapezoidal rule, a self-consistent recurrence relation between consecutive approximations obtained by dividing the mesh spacing can be established, for Simpson's rule, such a simple algorithmic relationship does not exist. Instead, starting from the first three approximations of the trapezoidal rule (10.28), consecutive approximations to Simpson's formula can be expressed in terms of the former:

$$S_1 = \frac{h_1}{3}[f(a) + 4f(a + h_1) + f(b)] = \frac{4T_1 - T_0}{3}, \tag{10.32}$$

$$S_2 = \frac{h_2}{3}[f(a) + 4f(a + h_2) + 2f(a + 2h_2) + 4f(a + 3h_2) + f(b)] = \frac{4T_2 - T_1}{3}. \tag{}$$

By straightforward generalization, the relationship is obtained:

$$S_k = \frac{4T_k - T_{k-1}}{3}. \tag{10.33}$$

Thus, it follows that the recursive algorithm with automatic step control based on Simpson's rule may be described yet in terms of trapezoidal approximations:

$$T_0 = \frac{h_0}{2}[f(a) + f(b)], \quad h_0 = b - a, \quad n_0 = 1, \tag{10.34}$$

$$T_k = \frac{1}{2}\left[T_{k-1} + h_{k-1}\sum_{i=1}^{n_{k-1}} f\left(a + (i - 1/2)h_{k-1}\right)\right], \quad k = 1, 2, \ldots, \tag{10.35}$$

$$S_k = \frac{4T_k - T_{k-1}}{3}, \quad h_k = h_{k-1}/2, \quad n_k = 2n_{k-1}. \tag{10.36}$$

Listing 10.4 Adaptive Integrator Based on Simpson's Rule (Python Coding)

```
#===========================================================================
def qSimpsonCtrl(Func, a, b, eps = 1e-6):
#---------------------------------------------------------------------------
#   Integrates function Func on interval [a,b] with relative precision eps
#   using the adaptive Simpson rule
#---------------------------------------------------------------------------
   kmax = 30                                    # max. no. of step halving iterations

   h = b-a; n = 1
   s0 = t0 = 0.5*h*(Func(a) + Func(b))                      # initial approximation

   for k in range(1,kmax+1):                                # step halving loop
      sumf = 0e0
      for i in range(1,n+1): sumf += Func(a+(i-0.5)*h)
      t = 0.5*(t0 + h*sumf)
      s = (4*t - t0)/3                                      # new approximation
      if (k > 1):                                           # convergence check
         if (fabs(s-s0) <= eps*fabs(s)): break
         if (fabs(s) <= eps and fabs(s) <= fabs(s-s0)): break # integral ~= 0
      h *= 0.5; n *= 2                                      # halve integration step
      s0 = s; t0 = t

   if (k >= kmax): print("qSimpsonCtrl: max. no. of iterations exceeded!")

   return s
```

The convergence criterion, implying the relative difference between two consecutive approximations of Simpson's formula (not of the trapezoidal ones!), can be conveniently written as

$$|S_k - S_{k-1}| \leq \varepsilon \, |S_k|. \tag{10.37}$$

The arguments, variables, and constants of function `qSimpsonCtrl`, implementing in Listing 10.4 the above ideas, maintain the significance they were assigned in function `qTrapzCtrl`.

Despite the intrinsic connection to the trapezoidal rule, the convergence rate of the process is enhanced owing to the superior approximation provided by Simpson's rule ($O(h^4)$ as compared to $O(h^2)$), requiring, in general, significantly less halving iterations for the same accuracy. A comparative example concerning the adaptive trapezoidal, Simpson's, and Romberg's methods is given in the next section.

10.7 Romberg's Method

Romberg's method (Romberg 1955) can be regarded as a generalization to higher-order Newton–Cotes formulas of the adaptive algorithm based on Simpson's formula that was developed in Section 10.6. Integration step halving is accompanied in Romberg's method by the additional recursive increase of the order of the elementary integrator and this brings about a convergence rate unmatched by any other quadrature scheme based on the Newton–Cotes formulas.

In Section 10.6, we established the simple relationship

$$S_k = \frac{4T_k - T_{k-1}}{3},$$

between the approximations T_k of the trapezoidal rule for successively halved integration meshes (Equations 10.28) and the corresponding approximations S_k given by Simpson's formula (Equation 10.32). With a view to generalizing this result to higher-order Newton–Cotes formulas, we relabel the trapezoidal approximations as $R_{k,0}$:

$$R_{0,0} = h_0 \left[\frac{f(a)}{2} + \frac{f(b)}{2} \right], \quad h_0 = b - a,$$

$$R_{1,0} = h_1 \left[\frac{f(a)}{2} + f(a + h_1) + \frac{f(b)}{2} \right], \quad h_1 = \frac{h_0}{2}, \tag{10.38}$$

$$R_{2,0} = h_2 \left[\frac{f(a)}{2} + f(a + h_2) + f(a + 2h_2) + f(a + 3h_2) + \frac{f(b)}{2} \right], \quad h_2 = \frac{h_1}{2}.$$

The number of subintervals for a given approximation $R_{k,0}$ and the corresponding mesh spacing at iteration k are given, respectively, by

$$n_k = 2^k, \quad h_k = (b - a)/n_k. \tag{10.39}$$

The recursive relation enabling the optimal evaluation of $R_{k,0}$, by implicitly reusing the previously calculated integrand values (avoiding redundant evaluations), reads:

$$R_{k,0} = \frac{1}{2}\left[R_{k-1,0} + h_{k-1}\sum_{i=1}^{n_{k-1}} f\left(a + (i-1/2)h_{k-1}\right)\right], \quad k = 1, 2, \ldots, k_{max}, \tag{10.40}$$

and it represents the exact transcription of Equation 10.30. Consistently, the first two approximations yielded by Simpson's formula are:

$$R_{1,1} = \frac{h_1}{3}[f(a) + 4f(a + h_1) + f(b)] = \frac{4R_{1,0} - R_{0,0}}{3},$$

$$R_{2,1} = \frac{h_2}{3}[f(a) + 4f(a + h_2) + 2f(a + 2h_2) + 4f(a + 3h_2) + f(b)]$$

$$= \frac{4R_{2,0} - R_{1,0}}{3}.$$

Then, the general relationship 10.33 between the approximations based on Simpson's formula and the trapezoidal rule takes the form:

$$R_{k,1} = \frac{4R_{k,0} - R_{k-1,0}}{3} = \frac{4R_{k,0} - R_{k-1,0}}{4 - 1}, \quad k = 1, 2, \ldots, k_{max}. \tag{10.41}$$

Pursuing the idea of recursively increasing the order of the basic Newton–Cotes formula, the first approximation of the next quadrature scheme is expected to rely on the nodes resulting by dividing the characteristic subintervals of the first approximation of Simpson's formula, $R_{1,1}$, or, equivalently, on the nodes of the second approximation, $R_{2,1}$. This requirement is intrinsically met by the five-point Boole's formula* (Boole and Moulton 2003), which essentially approximates the integrand by the fourth-order Lagrange interpolation polynomial:

$$\int_{x_1}^{x_5} f(x)dx = \frac{2h}{45}\left[7f_1 + 32f_2 + 12f_3 + 32f_4 + 7f_5\right] - \frac{8h^7 f^{(6)}(\xi)}{945}. \tag{10.42}$$

This first approximation based on Boole's formula can be expressed in terms of the integration step h_2 corresponding to two consecutive halving iterations of the total integration interval $[a, b]$,

$$R_{2,2} = \frac{2h_2}{45}\left[7f(a) + 32f(a + h_2) + 12f(a + 2h_2) + 32f(a + 3h_2) + 7f(b)\right], \tag{10.43}$$

and, moreover, can be related to the first two approximations provided by Simpson's formula:

$$R_{2,2} = \frac{16R_{2,1} - R_{1,1}}{15} = \frac{4^2 R_{2,1} - R_{1,1}}{4^2 - 1}. \tag{10.44}$$

It can be readily anticipated that the increase of the order of the basic quadrature scheme implies doubling the number of characteristic subintervals from 2^k (2 for Simpson's formula) to 2^{k+1} (4 for Boole's formula) and, inherently, doubling the order of the Lagrange interpolation polynomial by which the integrand is approximated. Consequently, except for the initial approximation $R_{0,0}$, all superior ones are based on integration grids with odd numbers of nodes, $(2^k + 1)$.

---

* Mistakenly referred to as "Bode's formula."

One can prove in general, though not without certain difficulty, the recurrence relation between the approximation $R_{k,j}$ (resulting by applying the Newton–Cotes formula of order 2^j to 2^k subintervals) and the approximations $R_{k,j-1}$ and $R_{k-1,j-1}$ based on the inferior quadrature of order 2^{j-1}:

$$R_{k,j} = \frac{4^j R_{k,j-1} - R_{k-1,j-1}}{4^j - 1}, \quad j = 1, 2, \ldots, k, \quad k = 1, 2, \ldots, k_{\max}. \tag{10.45}$$

For a given number of subintervals (2^k), the order of the applicable Newton–Cotes formulas (2^j) can obviously not exceed the number of subintervals and hence $j \leq k$. Conversely, the lowest approximation of the scheme of order 2^j should be $R_{j,j}$. It is, thus, natural to arrange all approximations $R_{k,j}$ of the integral in a triangular array with "predecessors" and "descendants":

$$
\begin{array}{c|ccccc}
n_k & \text{Trapezoid} & \text{Simpson} & \text{Boole} & \cdots & \cdots \\
\hline
1 & R_{0,0} & & & & \\
2 & R_{1,0} & R_{1,1} & & & \\
2^2 & R_{2,0} & R_{2,1} & R_{2,2} & & \\
\vdots & \vdots & \vdots & & \ddots & \\
2^{k_{\max}} & R_{k_{\max},0} & R_{k_{\max},1} & \cdots & \cdots & R_{k_{\max},k_{\max}}
\end{array}
\tag{10.46}
$$

Column j contains the approximations based on the Newton–Cotes formula of order 2^j for successively doubled numbers of subintervals (halved subinterval sizes).

The optimal evaluation of the integral can be achieved by completing array (10.46) recursively in a *row-wise top-down* manner, determining for each row k the approximations based on all Newton–Cotes formulas applicable on the respective mesh of 2^k integration points (starting with the trapezoidal rule and ending with the scheme of order 2^k) (Beu 2004). Specifically, the array elements are evaluated in the order $R_{0,0}$, $(R_{1,0}, R_{1,1})$, $(R_{2,0}, R_{2,1}, R_{2,2})\ldots$ Within a given number of halving iterations, $k_{\max}$, which, however, does not need to be fixed in advance, the best approximation of the integral is provided by the diagonal descendant on the last completed row, $R_{k_{\max},k_{\max}}$.

The essence of Romberg's method consists hence in recursively filling the array of approximations $R_{k,j}$ of the integral and can be cast under the form:

$$R_{0,0} = \frac{h_0}{2}[f(a) + f(b)], \quad h_0 = b - a, \quad n_0 = 1, \tag{10.47}$$

$$R_{k,0} = \frac{1}{2}\left[R_{k-1,0} + h_{k-1}\sum_{i=1}^{n_{k-1}} f\big(a + (i - 1/2)h_{k-1}\big)\right], \tag{10.48}$$

$$R_{k,j} = \frac{4^j R_{k,j-1} - R_{k-1,j-1}}{4^j - 1}, \quad j = 1, 2, \ldots, k, \tag{10.49}$$

$$h_k = h_{k-1}/2, \quad n_k = 2n_{k-1}, \quad k = 1, 2, \ldots, k_{\max}.$$

The initial element $R_{k,0}$ of each new row k is calculated according to Equation 10.48 by halving the integration mesh and using its "trapezoidal" predecessor $R_{k-1,0}$. Proceeding along a row with a new $R_{k,j}$ is carried out according to Equation 10.49 by using the estimates $R_{k,j-1}$ and $R_{k-1,j-1}$ given by the previous Newton–Cotes formula. The iterative process must be continued until the relative difference between two consecutive diagonal approximations becomes smaller than a predefined tolerance ε:

$$\left|R_{k,k} - R_{k-1,k-1}\right| \leq \varepsilon \left|R_{k,k}\right|. \tag{10.50}$$

An alternative strategy would be to fill array (10.46) in a *column-wise left-to-right* manner. This would require, however, to fix the maximum number of halving iterations k_{max} in advance and to evaluate *all* approximations on each column until convergence is met for some diagonal estimate $R_{k,k}$. At this point, if $k < k_{max}$, the already-calculated estimates located beneath row k would become unnecessary, rendering the algorithm inefficient.

Romberg's method (10.47)–(10.50) is implemented in function qRomberg (Listing 10.5), having the same parameters as routines qTrapzCtrl and qSimpsonCtrl. Instead of the triangular array of approximations $R_{k,j}$, the implementation uses just two one-dimensional arrays for the two most recent consecutive rows, r1 and r2, which are shifted at the end of each subinterval-halving iteration. Like in the case of the functions qTrapzCtrl and qSimpsonCtrl, the number of halving iterations is limited to 30, which corresponds to a maximum of 2^{30} integration points (the largest power of 2 representable by a long-type variable in standard C).

Every halving iteration of the integration spacing starts with the evaluation of the integral based on the trapezoidal formula. The order of the basic quadrature is then raised recursively and the convergence of the process is checked. If the requested precision eps is not achieved after running through all Newton–Cotes formulas applicable for the current number of mesh points, the next halving step is prepared by doubling the number of subintervals (halving the mesh spacing). If convergence is not achieved within the kmax allowed halving iterations, a warning message is issued.

Listing 10.5 Adaptive Integrator Based on Romberg's Method (Python Coding)

```
#===========================================================================
def qRomberg(Func, a, b, eps = 1e-6):
#---------------------------------------------------------------------------
#  Integrates function Func on interval [a,b] with relative precision eps
#  using the adaptive Romberg method
#---------------------------------------------------------------------------
   kmax = 30                             # max. no. of step halving iterations
   r1 = [0]*(kmax+1)                                   # two consecutive lines
   r2 = [0]*(kmax+1)                                   # from the method table

   h = b-a; n = 1
   r1[0] = 0.5*h*(Func(a) + Func(b))                    # initial approximation
   for k in range(1,kmax+1):                               # step halving loop
      sumf = 0e0
      for i in range(1,n+1): sumf += Func(a+(i-0.5)*h)
      r2[0] = 0.5*(r1[0] + h*sumf)                         # trapezoid formula
      f = 1e0
      for j in range(1,k+1):                         # increase quadrature order
         f *= 4
         r2[j] = (f*r2[j-1] - r1[j-1])/(f-1)              # new approximation

      if (k > 1):                                          # convergence check
         if (fabs(r2[k]-r1[k-1]) <= eps*fabs(r2[k])): break
         if (fabs(r2[k]) <= eps and fabs(r2[k]) <= fabs(r2[k]-r1[k-1])):break
      h *= 0.5; n *= 2                               # halve integration step
      for j in range(0,k+1): r1[j] = r2[j]                 # shift table lines

   if (k >= kmax):
      print("qRomberg: max. no. of iterations exceeded !")
      k -= 1

   return r2[k]
```

Listing 10.6   Application of Adaptive Quadratures (Python Coding)

```
# Evaluates an integral using adaptive classical quadratures
from math import *
from integral import *

def Func(x): return (x*x*x) * exp(-x)

# main

a = 0e0; b = 1e0; eps = 1e-10

print("I TrapzCtrl   = ",qTrapzCtrl(Func,a,b,eps))
print("I SimpsonCtrl = ",qSimpsonCtrl(Func,a,b,eps))
print("I Romberg     = ",qRomberg(Func,a,b,eps))
```

To assess the efficiency of the adaptive quadrature methods presented so far, we consider the integral

$$\int_0^1 x^3 \exp(-x)dx = 6 - 16e^{-1} = 0.11392894125692\ldots$$

A simple program, calling adaptive integrators (assumed to be part of module `integral.py`) to evaluate the above integral, is given in Listing 10.6. For reasons of numerical efficiency, instead of raising to the third power in the integrand, a repeated multiplication is preferred. The integral is evaluated by `qTrapzCtrl`, `qSimpsonCtrl`, and `qRomberg` with the relative precision $\varepsilon = 10^{-14}$ within 23, 13, and, respectively, six halving iterations, which clearly indicates the superior efficiency of Romberg's method.

10.8 Improper Integrals: Open Formulas

An improper integral is the one for which either the integration interval is infinite (first kind), or the integrand is undefined at certain points (second kind). Mathematically, both kinds of improper integrals imply taking the limit of the integral either at the infinite limits or at the singular points, where, for the second-kind integrals, one first splits the integration domain at the singular points. The examples below illustrate the typical features of both kinds of improper integrals:

$$\int_0^{+\infty} x \exp\left(-x^2\right) dx = \frac{1}{2}, \tag{10.51}$$

$$\int_0^\infty \frac{\sin x}{x} dx = \frac{\pi}{2}, \tag{10.52}$$

$$\int_0^1 x^{-1/2} dx = 2. \tag{10.53}$$

In the first two cases, the upper limit $(+\infty)$ cannot be implemented in a concrete numerical evaluation. In addition, the second integrand cannot be evaluated as such in 0 and needs to be replaced by its known limit $(\lim_{x\to 0} \sin x/x = 1)$. The third integrand has an integrable singularity in 0, but cannot be actually evaluated there.

A concrete implementation for first-kind improper integrals must take the *numerical* limit at the infinite boundary by gradually extending the integration domain, and this is exemplified by routine `qImprop1` given in Listing 10.7. The parameter `a` stands for the finite integration limit, the *sign of* `xinf` selects the integration interval $([a, +\infty)$ or $(-\infty, a])$, and `eps` is the required relative precision.

Listing 10.7 Adaptive Integrator for Improper Integrals of the First Kind (Python Coding)

```
#=============================================================================
def qImprop1(Func, a, xinf, eps = 1e-6):
#-----------------------------------------------------------------------------
#   Integrates function Func on interval [a,+inf) (for xinf >= 0) or (-inf,a]
#   (for xinf < 0) with relative precision eps. On output, xinf contains the
#   integration domain limit for which convergence was achieved.
#   Calls: qRomberg
#-----------------------------------------------------------------------------
    h = 1e0; x = a      # subinterval length and initial left limit for [a,+inf)
    if (xinf < 0e0): h = -h; x = a + h                        # for (-inf,a]

    s1 = 1e0
    s = 0e0
    while(fabs(s1) > eps*fabs(s) or fabs(Func(x)) > eps):
        s1 = qRomberg(Func,x,x+fabs(h),eps)          # integral for [x,x+h]
        s += s1                                      # update total integral
        x += h                                       # shift interval [x,x+h]

    xinf = x                                     # final "infinite" boundary

    return (s, xinf)
```

Technically, partial integrals `s1` are added to the integral estimate `s` for iteratively shifted subintervals, while monitoring the reduction of the relative error and the asymptotic vanishing of the integrand. In principle, any of the *adaptive* quadrature methods described so far can be employed to evaluate the partial integrals and the process is continued until the required precision is achieved. The size of the integration subinterval (`h` in the implementation) is not essential, since an overestimated value brings about at most wasted computations at the last expansion of the integration interval. On output, `xinf` contains the actual limit for which the integral has converged.

We now turn to the evaluation of improper integrals of the second kind. In particular, in the case of *interior* integrable singularities, the domain may be decomposed in such a way that the singularities become boundaries for partial integrals with continuous integrands at all interior points. Nevertheless, with a view to concrete calculations, the *closed* algorithms discussed so far prove to be inadequate due to the explicit involvement of the integration boundaries. Instead, so-called *open* the quadrature algorithms can be developed, which make use around the singular boundaries of interior points solely. It is obvious, however, that even these remain useless for *divergent* integrals, such as $\int_0^1 x^{-p} dx$ with $p > 1$.

The basic ideas are (1) to create a restricted "core" interval that excludes singularities and where one of the closed integration algorithms may be successfully applied and (2) to iteratively correct the total integral by adding contributions from the vicinities of the singular boundary points, until the requested precision is achieved.

Assuming, in general, that both a and b are singular points, the integration interval $[a, b]$ is actually split into three regions: the "left" interval $[a, a + h_0]$, the "core" interval $[a + h_0, b - h_0]$, and the "right" interval $[b - h_0, b]$, with h_0 equal to some fraction of the whole interval. In particular, assuming a to be a singular point, within its vicinity $[a, a + h_0]$, one adds contributions to the integral from successively halved subintervals, which gradually extend the core interval toward a (Figure 10.4).

The sequence of *left* boundaries $x_k^{(a)}$ of the subintervals $[x_k^{(a)}, x_k^{(a)} + h_k]$ is given by

$$h_0 = (b - a)/10,$$
$$h_k = h_{k-1}/2, \quad x_k^{(a)} = a + h_k, \quad k = 1, 2, \ldots$$

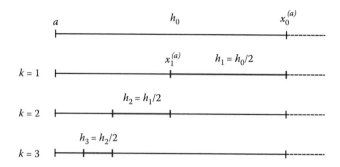

FIGURE 10.4 Improper integrals of the second kind: sequence of recursively halved intervals contributing to the partial integral in the vicinity $[a, a + h_0]$ of the singular boundary a.

Listing 10.8 Adaptive Integrator for Improper Integrals of the Second Kind (Python Coding)

```
#============================================================================
def qImprop2(Func, a, b, eps):
#----------------------------------------------------------------------------
#  Integrates function Func on interval [a,b] with a and/or b singular
#  integrable points.
#  Calls: qRomberg
#----------------------------------------------------------------------------
   h0 = 0.1e0 * (b-a)                # extent of vicinities of singular boundaries

   s = 0e0
   try: Func(a)
   except ZeroDivisionError:                          # a is singular point
      h = h0
      s1 = 1e0
      while(fabs(s1) > eps*fabs(s)):
         h *= 0.5                                      # halve interval
         x = a + h                            # left boundary of [x,x+h]
         s1 = qRomberg(Func,x,x+h,eps)        # partial integral on [x,x+h]
         s += s1                          # add contribution to total integral
      a += h0                             # new left boundary of core interval

   try: Func(b)
   except ZeroDivisionError:                          # b is singular point
      h = h0
      s1 = 1e0
      while(fabs(s1) > eps*fabs(s)):
         h *= 0.5                                      # halve interval
         x = b - h                            # right boundary of [x-h,x]
         s1 = qRomberg(Func,x-h,x,eps)        # partial integral on [x-h,x]
         s += s1                          # add contribution to total integral
      b -= h0                             # new right boundary of core interval

   s += qRomberg(Func,a,b,eps)                  # add integral on core interval

   return s
```

Listing 10.9 Improper Integrals of the First and Second Kind (Python Coding)

```python
# Evaluates improper integrals
from math import *
from integral import *

def Func1(x): return x * exp(-x*x)
def Func2(x): return sin(x)/x if x else 1e0
def Func3(x): return 1/sqrt(x)

# main

xinf = 1e10
eps = 1e-6
(s, xinf) = qImprop1(Func1,0e0,xinf,eps)
print("I1 = ",s," xinf = ",xinf)
(s, xinf) = qImprop1(Func2,0e0,xinf,eps)
print("I2 = ",s," xinf = ",xinf)
print("I3 = ",qImprop2(Func3,0e0,1e0,eps))
```

Similarly, the *right* boundaries $x_k^{(b)}$ of the sequence of subintervals $[x_k^{(b)} - h_k, x_k^{(b)}]$ within the vicinity $[b - h_0, b]$ are generated as

$$h_k = h_{k-1}/2, \quad x_k^{(b)} = b - h_k, \quad k = 1, 2, \ldots$$

The asymptotic convergence of $x_k^{(a)}$ to a and of $x_k^{(b)}$ to b is guaranteed by the convergence of the infinite series:

$$\sum_{k=1}^{\infty} (1/2)^k = 1.$$

The integrator qImprop2 in Listing 10.8 implements the exposed concepts for evaluating improper integrals of the second kind. For the sake of algorithmic efficiency, the routine first identifies the singular boundaries (a and/or b), then evaluates the integral contributions from their vicinities, establishing at the same time the boundaries of the core domain, and finally adds the core integral. Specifically, the adequate Python coding implies treating the singularities as exceptions using a try-except construct. First, the try clause attempts to evaluate the user function for the particular boundary point. For nonsingular boundaries, no exception occurs and the except clause is simply skipped. For singular points, the attempted division by 0 raises the ZeroDivisionError exception, which is properly addressed within the except block by applying the devised iterative refinement algorithm. All partial integrals are performed by means of the highly efficient qRomberg integrator.

Program 10.9 illustrates the use of functions qImprop1 and qImprop2 (assumed to be included in module integral.py). In the case of the particularly demanding integral (10.52), due to the oscillating and very slowly decaying integrand, the convergence is remarkably slow. Specifically, the relative precisions $\varepsilon = 10^{-6}$ and $\varepsilon = 10^{-8}$ are achieved by extending the integration interval to as much as $2 \cdot 10^5$ and, respectively, $2 \cdot 10^7$. Quite in contrast, for the integral (10.51), the precision $\varepsilon = 10^{-8}$ is already achieved for the upper limit extended to 6.

10.9 Midpoint Rule

In an attempt to consistently construct an open-type formula not employing explicitly the end points of the integration interval, one can start by approximating the integral of $f(x)$ on $[a, b]$ by the trapezoidal rule, respectively for the integration spacings h and $h/2$:

$$\int_a^b f(x)dx \approx h\left[\frac{f_1}{2} + f_2 + f_3 + \cdots + \frac{f_n}{2}\right], \tag{10.54}$$

$$\int_a^b f(x)dx \approx \frac{h}{2}\left[\frac{f_1}{2} + f_{3/2} + f_2 + f_{5/2} + f_3 + \cdots + \frac{f_n}{2}\right]. \tag{10.55}$$

The integrand values for semiinteger indexes correspond to the *midpoints* of the $(n-1)$ integration subintervals of length h:

$$f_{i+1/2} \equiv f(a + (i - 1/2)h).$$

Subtracting twice Equation 10.55 from 10.54, the integer-index values (including those for the domain boundaries) cancel out and one obtains the so-called *midpoint rule*:

$$\int_a^b f(x)dx \approx h\sum_{i=1}^{n-1} f_{i+1/2}. \tag{10.56}$$

Not using information from the integration limits, this is indeed an *open* formula. The remainder of the midpoint rule results from the remainders of the trapezoidal formulas 10.54 and 10.55, and it turns out to maintain the same order with respect to h:

$$R = \frac{(b-a)h^2}{24}f''(\xi), \quad \xi \in [a, b]. \tag{10.57}$$

The straightforward implementation of the midpoint rule (10.56) is of limited practical interest. On the contrary, an adaptive algorithm with automatic integration step control, similar to the one implemented in function qTrapzCtrl, is most useful for practical evaluations of improper integrals of the second kind.

The particular arrangement of the integration abscissas in the midpoint rule does not enable, however, a recurrence relation between approximations resulted by *doubling* the number of subintervals. Such an approach is possible instead by *tripling* the number of subintervals and involves only integrand values in the new mesh points, by implicitly reusing the information from previously used nodes, as can be seen

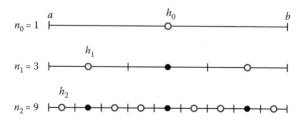

FIGURE 10.5 Algorithmic reduction of integration subintervals in the adaptive midpoint rule. Empty circles indicate new values that have to be calculated, while filled circles indicate values already calculated at the previous steps.

from Figure 10.5. The outer integration nodes tend toward the interval limits a and b, without effectively reaching them in a finite number of algorithmic steps.

The first few approximations provided by the midpoint rule are

$$S_0 = h_0 f(a + h_0/2), \quad h_0 = b - a,$$

$$S_1 = h_1[f(a + h_1/2) + f(a + 3h_1/2) + f(a + 5h_1/2)], \quad h_1 = h_0/3,$$

$$S_2 = h_2[f(a + h_2/2) + f(a + 3h_2/2) + f(a + 5h_2/2) + f(a + 7h_2/2) + f(a + 9h_2/2)$$

$$+ f(a + 11h_2/2) + f(a + 13h_2/2) + f(a + 15h_2/2) + f(a + 17h_2/2)], \quad h_2 = h_1/3.$$

In general, the approximation S_k may be expressed in terms of the previous approximation, S_{k-1}, and $2n_{k-1}$ *new* integrand values, where n_{k-1} is the number of subintervals at step $k - 1$. Denoting as $n_k = 3^k$ the number of subintervals and by h_k their size at step k, the adaptive algorithm based on the midpoint rule may be finally summarized as follows:

$$S_0 = h_0 f(a + h_0/2), \quad h_0 = b - a, \quad n_0 = 1,$$

$$S_k = \frac{1}{3} \left\{ S_{k-1} + h_{k-1} \sum_{i=1}^{n_{k-1}} [f(a + (i - 5/6)h_{k-1}) + f(a + (i - 1/6)h_{k-1})] \right\},$$

$$h_k = h_{k-1}/3, \quad n_k = 3n_{k-1}, \quad k = 1, 2, \dots \tag{10.58}$$

The recursive process is terminated when the relative difference between two consecutive approximations of the integral, regarded as a estimate for the relative error, becomes smaller than a tolerance ε:

$$\left| S_k - S_{k-1} \right| \leq \varepsilon \left| S_k \right|. \tag{10.59}$$

Listing 10.10 Adaptive Integrator Based on the Midpoint Rule (Python Coding)

```
#===============================================================================
def qMidPoint(Func, a, b, eps = 1e-6):
#-------------------------------------------------------------------------------
#  Integrates function Func on interval (a,b) with relative precision eps
#  using the adaptive midpoint rule
#-------------------------------------------------------------------------------
   kmax = 19                                    # max. no. of subdivisions
   f1p6 = 1./6.; f5p6 = 5./6.

   h = b-a; n = 1
   s0 = h * Func(a+0.5*h)                        # initial approximation

   for k in range(1,kmax+1):                     # step subdivision loop
      sumf = 0e0
      for i in range(1,n+1): sumf += Func(a+(i-f5p6)*h) + Func(a+(i-f1p6)*h)
      s = (s0 + h*sumf)/3                         # new approximation
      if (fabs(s - s0) <= eps*fabs(s)): break     # convergence check
      h /= 3; n *= 3                              # reduce step
      s0 = s

   if (k >= kmax): print("qMidPoint: max. no. of iterations exceeded !")

   return s
```

The function `qMidPoint` from Listing 10.10 closely implements Relations 10.58 and 10.59 and its formal arguments maintain the significance they have been assigned in the implementations of the other adaptive integration methods presented so far. As a difference from the operation of function `qImprop2` discussed in the previous section, which employs independent integration meshes locally adapted to the vicinities of the singular boundaries, `qMidPoint` relies on a uniform mesh, which can bring about in certain cases a slower overall convergence.

10.10 Gaussian Quadratures

The Newton–Cotes formulas imply as per construction equally spaced integration nodes. For a given order of the formula, the abscissas x_i and the weights w_i directly follow from the Lagrange interpolation polynomial based on the n integration points. By comparison, the basic idea of the so-called Gaussian quadratures consists of optimally choosing *both* the abscissas and weights, so as to attain maximum precision for certain classes of integrands (e.g., polynomials). Disposing over a double number of adjustable parameters (separately abscissas and weights), one can construct, in principle, quadrature formulas of double order as compared to the Newton–Cotes formulas with the same number of integration points.

By a linear change of variable, any integral on a finite interval $[a, b]$ may be transformed into an integral on $[-1, 1]$:

$$\int_a^b f(\xi)d\xi = \frac{b-a}{2} \int_{-1}^1 f\left(\frac{b-a}{2}x + \frac{b+a}{2}\right) dx.$$

Hence, without restricting their generality, one can develop quadratures for the standard interval $[-1, 1]$. Concretely, we approach the task of choosing the integration weights and abscissas (w_i, x_i, $i = 1, 2, \ldots, n$) in such a way that the quadrature formula

$$\int_{-1}^1 f(x)dx = \sum_{i=1}^n w_i f(x_i) + R_n \tag{10.60}$$

is *exact* ($R_n = 0$) for all polynomials of the highest possible degree. Implying $2n$ freely adjustable parameters that can be linked to an equal number of polynomial coefficients, it is obvious that the *highest degree* can be $N = 2n - 1$.

For the quadrature formula 10.60 to be exact for any polynomial up to degree $(2n - 1)$, it is necessary and sufficient for it to be identically satisfied by the *fundamental set* of functions $1, x, x^2, \ldots, x^{2n-1}$:

$$\int_{-1}^1 x^k dx = \sum_{i=1}^n w_i x_i^k, \quad k = 0, 1, \ldots, 2n - 1. \tag{10.61}$$

Indeed, any polynomial of degree up to $(2n - 1)$ may be represented as a linear combination of fundamental functions:

$$Q(x) = \sum_{k=0}^{2n-1} a_k x^k.$$

Integrating both members between -1 and 1, and using Equation 10.61 successively gives:

$$\int_{-1}^1 Q(x)dx = \sum_{k=0}^{2n-1} a_k \int_{-1}^1 x^k dx = \sum_{i=1}^n w_i \sum_{k=0}^{2n-1} a_k x_i^k = \sum_{i=1}^n w_i Q(x_i),$$

which proves the proposition.

A straightforward way to determine the nodes x_i and weights w_i is to solve the nonlinear system (10.61) with the integrals of the fundamental functions given by

$$\int_{-1}^{1} x^k dx = \begin{cases} 2/(k+1) & k = 0, 2, \ldots, 2n - 2, \\ 0 & k = 1, 3, \ldots, 2n - 1. \end{cases}$$

Nevertheless, since, for large n, the direct solution of system (10.61) in terms of (w_i, x_i) may become cumbersome, we follow a different path to determine the abscissas and weights of the quadrature formula 10.60.

Integration abscissas. The Legendre polynomials $P_n(x)$ form a widely used *orthogonal basis set* for expanding real-valued square-integrable functions defined on $[-1, 1]$. The main properties and computational methods of these polynomials have been presented in Chapter 6, along with those for other relevant special functions.

In particular, any polynomial of degree k may be expressed as a linear combination of Legendre polynomials of degrees up to k:

$$Q_k(x) = \sum_{m=0}^{k} c_m P_m(x). \tag{10.62}$$

Taking the scalar product of both members with the Legendre polynomial $P_n(x)$ of degree $n > k$ and making use of the orthogonality ($\int_{-1}^{1} P_m(x)P_n(x)dx = 0$ for $m \neq n$) yields:

$$\int_{-1}^{1} Q_k(x)P_n(x)dx = \sum_{m=0}^{k} c_m \int_{-1}^{1} P_m(x)P_n(x)dx = 0, \quad k < n.$$

Particularly, for $Q_k(x) \equiv x^k$, we have:

$$\int_{-1}^{1} x^k P_n(x)dx = 0, \quad k < n. \tag{10.63}$$

On the other hand, since the degree of the integrand $x^k P_n(x)$ does not exceed $2n - 1$, the quadrature formula 10.60 remains exact and takes the particular form

$$\sum_{i=1}^{n} w_i x_i^k P_n(x_i) = 0, \quad k = 0, 1, \ldots, n - 1. \tag{10.64}$$

The equations of this system can be satisfied simultaneously for arbitrary values w_i only provided that

$$P_n(x_i) = 0, \quad i = 1, 2, \ldots n. \tag{10.65}$$

Consequently, an n-point quadrature formula 10.60 on the interval $[-1, 1]$ having maximum precision (exact for polynomials up to degree $2n - 1$) is obtained by adopting as integration nodes the zeros of the Legendre polynomial $P_n(x)$. Since the Legendre polynomials have a *well-defined parity*, which is given by their degree n, the zeros of any given polynomial are *symmetrical* with respect to the origin (for odd polynomials, the origin is a zero itself).

Integration weights. With a view to determining the weights w_i associated with abscissas x_i, we factorize the Legendre polynomial $P_n(x)$ and its first derivative in terms of their zeros x_i:

$$P_n(x) = c \prod_{i=1}^{n}(x - x_i), \quad P'_n(x) = c \sum_{j=1}^{n} \prod_{i \neq j}^{n}(x - x_i). \qquad (10.66)$$

Additionally, we define a polynomial of degree $(n-1)$ that vanishes at all zeros of $P_n(x)$ except for x_k:

$$Q_{n-1}^{(k)}(x) = \frac{P_n(x)}{x - x_k} = c \prod_{i \neq k}^{n}(x - x_i). \qquad (10.67)$$

On the other hand, since all terms from $P'_n(x_k)$ containing the factor $(x - x_k)$ vanish, it follows that

$$Q_{n-1}^{(k)}(x_k) = P'_n(x_k). \qquad (10.68)$$

Being a polynomial of degree $(2n - 2)$, $\left[Q_{n-1}^{(k)}(x)\right]^2$ satisfies the quadrature formula 10.60 identically and one obtains, successively:

$$\int_{-1}^{1} \left[Q_{n-1}^{(k)}(x)\right]^2 dx = \sum_{i=1}^{n} w_i \left[Q_{n-1}^{(k)}(x_i)\right]^2 = w_k \left[Q_{n-1}^{(k)}(x_k)\right]^2 = w_k \left[P'_n(x_k)\right]^2, \qquad (10.69)$$

where it was used that $Q_{n-1}^{(k)}(x_i) = 0$ for all $x_i \neq x_k$. Then again, by using definition (10.67) and integrating by parts, one obtains:

$$\int_{-1}^{1} \left[Q_{n-1}^{(k)}(x)\right]^2 dx = -\frac{P_n^2(1)}{1 - x_k} - \frac{P_n^2(-1)}{1 + x_k} + 2 \int_{-1}^{1} Q_{n-1}^{(k)}(x) P'_n(x) dx. \qquad (10.70)$$

We now combine Equations 10.69 and 10.70, and apply the quadrature formula 10.60 to the last integral, whose integrand $Q_{n-1}^{(k)}(x) P'_n(x)$ is a polynomial of degree $(2n - 2)$. Using the particular values $P_n^2(1) = P_n^2(-1) = 1$ and property (10.68) leads to

$$w_k [P'_n(x_k)]^2 = -\frac{2}{1 - x_k^2} + 2 \sum_{i=1}^{n} w_i Q_{n-1}^{(k)}(x_i) P'_n(x_i)$$

$$= -\frac{2}{1 - x_k^2} + 2 w_k \left[P'_n(x_k)\right]^2.$$

Herefrom, we finally obtain the sought integration weights:

$$w_i = \frac{2}{(1 - x_i^2)[P'_n(x_i)]^2}, \quad i = 1, 2, \ldots, n, \qquad (10.71)$$

which are obviously *equal for symmetrical zeros* x_i.

Formula 10.60, with the n integration abscissas x_i coinciding with the zeros of the Legendre polynomial $P_n(x)$ and the associated weights w_i given by Equation 10.71, is called *Gauss–Legendre quadrature formula*, and it is exact for polynomials up to degree $(2n - 1)$. Other established quadrature formulas, constructed on the basis of similar concepts for other integration intervals, bear names derived from the underlying set of orthogonal polynomials (see Abramowitz and Stegun 1972; Ralston and Rabinowitz 2001).

For a finite interval $[a, b]$, the Gauss–Legendre quadrature formula reads:

$$\int_a^b f(\xi)d\xi = \frac{b-a}{2} \sum_{i=1}^n w_i f(\xi_i) + R_n, \tag{10.72}$$

with the integration nodes given in terms of those for the standard interval $[-1, 1]$ by

$$\xi_i = \frac{b-a}{2}x_i + \frac{b+a}{2}. \tag{10.73}$$

The corresponding remainder can be shown to be (Abramowitz and Stegun 1972):

$$R_n = \frac{(b-a)^{2n+1}(n!)^4}{(2n+1)[(2n)!]^3}f^{(2n)}(\xi), \quad \xi \in [a, b]. \tag{10.74}$$

The practical usage of the Gauss–Legendre quadratures still requires a methodology to compute the zeros of the Legendre polynomials. This can be most conveniently accomplished based on the inequality involving their *limiting values* (Abramowitz and Stegun 1972, Equation 22.16.6):

$$\cos\left(\frac{2i-1}{2n+1}\pi\right) \leq x_i \leq \cos\left(\frac{2i}{2n+1}\pi\right).$$

The cosine for the median argument,

$$x_i^{(0)} = \cos\left(\frac{i-0.25}{n+0.50}\pi\right), \tag{10.75}$$

can be used as an initial approximation for the zero x_i, which can be recursively refined by the Newton–Raphson method:

$$x_i^{(k+1)} = x_i^{(k)} - \frac{P_n(x_i^{(k)})}{P_n'(x_i^{(k)})}, \quad k = 0, 1, 2, \dots \tag{10.76}$$

Once the zeros x_i are determined, the weights w_i readily result from Relation 10.71.

The function xGaussLeg (Listing 10.11) returns integration abscissas x and weights w for the n-point Gauss–Legendre integration on interval [a,b] and Table 10.1 illustrates the output. The function actually calculates only the zeros and weights corresponding to the semiinterval $[-1, 0]$, the remaining points being completed as per symmetry. The Legendre polynomials are calculated using the function Legendre (presented in Chapter 5), which iterates the recurrence relation with respect to the degree and returns, in addition, also the first derivative of the polynomial. For odd n, the median integration point is just the origin $x_{n/2+1} = 0$. In the end, the function, xGaussLeg, scales the abscissas and weights to the original interval $[a, b]$.

The generation of integration points by function xGaussLeg is illustrated as part of the simple integrator qGaussLeg provided in Listing 10.12, which receives the name of the integrand by argument Func and performs the integration between a and b using the Gauss–Legendre quadrature with n points.

Listing 10.11   Abscissas and Weights for Gauss–Legendre Quadratures (Python Coding)

```
#===========================================================================
def xGaussLeg(a, b, n):
#---------------------------------------------------------------------------
#  Calculates abscissas x[] and weights w[] for the n-point Gauss-Legendre
#  quadrature on interval [a,b]
#  Calls: Legendre (from specfunc.py)
#---------------------------------------------------------------------------
   eps = 1e-14                                  # relative precision of zeros
   n2 = int(n/2)
   x = [0]*n
   w = [0]*n

   for i in range (0,n2):
      xi = cos(pi*(i+1-0.25e0)/(n+0.5e0))    # initial approximation for zeros
      f = 9e99
      while (fabs(f) > eps*fabs(xi)):             # Newton-Raphson refinement
         (f,d) = Legendre(n,xi); f /= d
         xi -= f
      x[i] = -xi; x[n-i-1] = xi                            # symmetrical zeros
      w[i] = w[n-i-1] = 2e0/((1e0-xi*xi)*d*d)                # equal weights

   if (n % 2 == 1):                              # odd no. of mesh points
      (f,d) = Legendre(n,0e0)
      x[n2] = 0e0
      w[n2] = 2e0/(d*d)

   f = 0.5e0*(b-a); xc = 0.5e0*(b+a)              # scaling to interval [a,b]
   for i in range (0,n):
      x[i] = f*x[i] + xc
      w[i] = f*w[i]

   return (x,w)
```

One of the frequently used Gaussian integration methods for exponential-like decaying integrands over infinite integration intervals is the *Gauss–Laguerre quadrature* (Abramowitz and Stegun 1972, Equation 25.4.45):

$$\int_0^\infty e^{-x} f(x)\,dx \approx \sum_{i=1}^n w_i f(x_i). \tag{10.77}$$

The n abscissas x_i are the zeros of the Laguerre polynomial $L_n(x)$ of degree n and the weights are given by

$$w_i = \frac{x_i}{(n+1)^2 [L_{n+1}(x_i)]^2} = \frac{1}{x_i [L_n'(x_i)]^2}, \quad i = 1, 2, \ldots, n. \tag{10.78}$$

The second form is computationally more convenient since the first derivative $L_n'(x_i)$ is efficiently calculated from L_n and L_{n-1} at the end of the ascending recurrence for the Laguerre polynomials with respect to the degree.

TABLE 10.1   Abscissas and Weights of the n-Point Gauss–Legendre Quadrature Formula for the Standard Interval $[-1, 1]$, Calculated by Means of Function xGaussLeg

n	i	x_i	w_i
2	1, 2	∓ 0.57735026918963	1
3	1, 3	∓ 0.77459666924148	0.55555555555555
	2	0	0.88888888888889
4	1, 4	∓ 0.86113631159405	0.34785484513745
	2, 3	∓ 0.33998104358486	0.65214515486255
5	1, 5	∓ 0.90617984593866	0.23692688505618
	2, 4	∓ 0.53846931010568	0.47862867049937
	3	0	0.56888888888889

Listing 10.12   Gauss–Legendre Integrator (Python Coding)

```
#===============================================================================
def qGaussLeg(Func, a, b, n):
#-------------------------------------------------------------------------------
#  Integrates function Func on interval [a,b] using n-point Gauss-Legendre
#  quadratures
#  Calls: xGaussLeg
#-------------------------------------------------------------------------------
   x = [0]*n
   w = [0]*n

   (x,w) = xGaussLeg(a,b,n)

   s = 0e0
   for i in range(0,n): s += w[i] * Func(x[i])

   return s
```

For the more general case of a finite but nonzero lower boundary a and an integrand not featuring explicitly an exponential factor, the Gauss–Laguerre quadrature formula can be rewritten as

$$\int_a^\infty f(x)dx \approx \sum_{i=1}^n w_i f(x_i + a), \tag{10.79}$$

$$w_i = \frac{\exp(x_i)}{x_i [L_n'(x_i)]^2}, \quad i = 1, 2, \ldots, n. \tag{10.80}$$

The function xGaussLag (Listing 10.13) returns the integration abscissas x and weights w for the n-point Gauss–Laguerre integration on the interval $[a, +\infty)$ and its overall structure is similar to the one of function xGaussLeg. The initial approximations of the zeros are taken from Stroud and Secrest (1966) and Stroud (1971), and have been commonly used in similar implementations (see, e.g., Press et al. 2007; Beu 2004).

A minimal Gauss–Laguerre integrator calling xGaussLag to integrate the user function Func on the interval $[a, +\infty)$ with n nodes is given in Listing 10.14.

Considering the test integrals defined over finite and, respectively, infinite domains

$$\int_0^1 12x^{11}dx = 1, \quad \int_0^\infty x^5 e^{-x}dx = 120,$$

Program 10.15 is conceived to enable assessing the performance of the Gauss–Legendre and Gauss–Laguerre quadratures for various numbers of nodes ($n \leq 6$).

Listing 10.13   Abscissas and Weights for Gauss–Laguerre Quadratures (Python Coding)

```
#===========================================================================
def xGaussLag(a, n):
#---------------------------------------------------------------------------
#  Calculates abscissas x[] and weights w[] for the n-point Gauss-Laguerre
#  quadrature on interval [a,+inf)
#  Calls: Laguerre (from specfunc.py)
#  Initial approximation for zeros:
#  A. Stroud & D. Secrest, Gaussian Quadrature Formulas, Prentice Hall, 1966.
#---------------------------------------------------------------------------
   eps = 1e-14                                  # relative precision of zeros
   x = [0]*n
   w = [0]*n

   for i in range(0,n):
      if (i == 0):          # initial approximation for zeros (Stroud & Secrest)
         xi = 3e0/(1e0+2.4e0*n)                               # 1st zero
      elif (i == 1):
         xi = 15e0/(1e0+2.5e0*n) + x[0]                       # 2nd zero
      else:
         f = (1e0/(i+1)+2.55e0)/1.9e0
         xi = (1e0+f)*x[i-1] - f*x[i-2]                       # recurrence

      f = 9e99
      while (fabs(f) > eps*fabs(xi)):                 # Newton-Raphson refinement
         (f,d) = Laguerre(n,xi); f /= d
         xi -= f
      x[i] = xi
      w[i] = exp(xi)/(xi*d*d)

   for i in range(0,n): x[i] += a                # scaling to interval [a,+inf)

   return (x,w)
```

Listing 10.14   Gauss–Laguerre Integrator (Python Coding)

```
#===========================================================================
def qGaussLag(Func, a, n):
#---------------------------------------------------------------------------
#  Integrates function Func on interval [a,+inf) using n-point Gauss-Laguerre
#  quadratures
#  Calls: xGaussLag
#---------------------------------------------------------------------------
   x = [0]*n
   w = [0]*n

   (x,w) = xGaussLag(a,n)

   s = 0e0
   for i in range(0,n): s += w[i] * Func(x[i])

   return s
```

Listing 10.15   Gauss–Legendre and Gauss–Laguerre Quadratures (Python Coding)

```
# Gaussian quadratures
from math import *
from integral import *

def Func1(x): return 12*pow(x,11)
def Func2(x): return x*x*x * exp(-x)

# main

nmax = 6
x = [0]*(nmax+1)
w = [0]*(nmax+1)

print("Integral of 12*x^11 on [-1,1]")
for n in range(2,nmax+1):
    print("n = ",n,"I1 = ",qGaussLeg(Func1,0e0,1e0,n))

print("\nIntegral of x^3 * exp(-x) on [0,+inf)")
for n in range(2,nmax+1):
    print("n = ",n,"I2 = ",qGaussLag(Func2,0e0,n))
```

As expected, the n-point Gaussian quadratures provide exact results only for polynomials up to degree $(2n - 1)$. The first integral is reproduced for $n = 5$ with barely four exact digits ($0.999905\ldots$), being evaluated with full double precision only for $n \geq 6$. For the second integral, due to the lower x^5 dependence implied supplementary to the decaying exponential, one obtains the exact result already for $n = 3$ integration points.

10.11 Multidimensional Integration

As compared to the one-dimensional case, the multidimensional integration is complicated by a series of factors, among which, the most significant are

- The rapid increase of the number of function evaluations with the dimension of the integral—for n function evaluations in a generic one-dimensional quadrature scheme, the number of evaluations increases as n^d in the d-dimensional case.
- The difficulty of describing the $(d - 1)$-dimensional hypersurface representing the integration boundary for a d-dimensional integral.

Most often, the multidimensional integration scheme is imposed by the complexity of the integration domain, the smoothness of the integrand, and the required precision. For instance, in the case of *complicated boundaries*, smooth integrands, and moderate precisions, *Monte Carlo integration* (described in Chapter 11) is a reasonable choice. Nevertheless, despite its simplicity and good error control, the Monte Carlo method tends to be slowly converging with increasing number of integration points. In contrast, for *simple boundaries* but high required precisions, rather, the *direct product of one-dimensional quadratures*, presented in what follows, is adequate. In particular, for integrands with marked local extrema, the integration domain may be decomposed into subdomains enclosing smooth integrands.

A convenient way of circumventing the description of complicated boundaries is the *zero-padding technique*, which implies artificially extending the integration domain up to the closest enclosing hypercuboid and setting the integrand to zero outside the real boundaries. In this manner, each integration point results by the direct product of the one-dimensional meshes and the integration is carried out over the entire extended domain at the expense of evaluating the integrand also at points where it explicitly vanishes.

All one-dimensional quadrature formulas discussed so far have been shown to comply with the general form:

$$\int_a^b f(x)dx \approx \sum_{i=1}^n w_i f(x_i).$$

(10.81)

Now, considering the integrand $f(x, y, z)$ defined on a regular *cuboid* within real space,

$$f: [a_x, b_x] \times [a_y, b_y] \times [a_z, b_z] \to \mathbb{R},$$

the corresponding 3D integral may be approximated by the direct product of three one-dimensional quadrature schemes:

$$\int_{a_x}^{b_x} \int_{a_y}^{b_y} \int_{a_z}^{b_z} f(x, y, z)dxdydz \approx \sum_{i=1}^{n_x} w_i^{(x)} \sum_{j=1}^{n_y} w_j^{(y)} \sum_{k=1}^{n_z} w_k^{(z)} f(x_i, y_j, z_k).$$

(10.82)

Since the one-dimensional integration meshes are independent, the integrand evaluations run over the $(n_x \times n_y \times n_z)$ points (x_i, y_j, z_k) of the regular composite grid.

Two functions are provided in Listings 10.16 and 10.17, which implement the direct product method in Cartesian coordinates, using as basic one-dimensional integrator Simpson's rule. Given that the algorithm for calculating the integration points is the same for all three Cartesian directions, it is conceived as a separate auxiliary function, xSimpson (Listing 10.16), which receives the integration limits, a and b, and returns the n-requested abscissas and weights via arrays x [ ] and w [ ]. If the number of integration nodes is even, it is incremented to the next higher odd integer, as appropriate for Simpson's rule. The

Listing 10.16 Integration Points and Weights for Simpson's Rule (Python Coding)

```
#=====================================================================
def xSimpson(a, b, n):
#---------------------------------------------------------------------
#  Calculates abscissas x[] and weights w[] for Simpson's rule with n
#  integration points on interval [a,b]
#---------------------------------------------------------------------
   c13 = 1e0/3e0; c23 = 2e0/3e0; c43 = 4e0/3e0

   if (n % 2 == 0): n += 1                            # increment n if even

   x = [0]*n; w = [0]*n

   h = (b-a)/(n-1)
   for i in range(0,n):
      x[i] = a + i*h; w[i] = h * (c23 if (i+1) % 2 else c43)
   w[0] = w[n-1] = h * c13

   return (x, w)
```

Listing 10.17   3D Integrator Based on Simpson's Rule (Python Coding)

```
#============================================================================
def qSimpson3D(Func, ax, bx, nx, ay, by, ny, az, bz, nz):
#----------------------------------------------------------------------------
#  Integrates function Func(x,y,z) in the cuboid [ax,bx] x [ay,by] x [az,bz]
#  using Simpson's rule with (nx x ny x nz) integration points
#----------------------------------------------------------------------------
   if (nx % 2 == 0): nx += 1                            # increment nx if even
   if (ny % 2 == 0): ny += 1                            # increment ny if even
   if (nz % 2 == 0): nz += 1                            # increment nz if even

   x = [0]*nx; wx = [0]*nx
   y = [0]*ny; wy = [0]*ny
   z = [0]*nz; wz = [0]*nz

   (x,wx) = xSimpson(ax,bx,nx)              # generate integartion points
   (y,wy) = xSimpson(ay,by,ny)
   (z,wz) = xSimpson(az,bz,nz)

   s = 0e0
   for i in range(0,nx):
      sx = 0e0
      for j in range(0,ny):
         sy = 0e0
         for k in range(0,nz):
            sy += wz[k] * Func(x[i],y[j],z[k])
         sx += wy[j] * sy
      s += wx[i] * sx

   return s
```

integration weights are assigned, depending on the index parity, the values 2/3 (even index), 4/3 (odd index), or 1/3 for the end points.

The 3D integrator qSimpson3D (Listing 10.17) effectively employing the nodes generated by xSimpson receives as parameters the name of the user function, as well as the integration limits and number of nodes for each separate Cartesian direction. If necessary, qSimpson3D corrects in a first instance the parity of the node numbers and then stores the integration points returned by xSimpson in six local arrays. The core of qSimpson3D consists of a triple loop, within which (starting with the innermost cycle) the partial sums over the z_k, y_j, and x_i points are calculated.

Except for the correction of the node number parities, which is a specific requirement of Simpson's rule, the features of routine qSimpson3D are typical for any 3D integrator. Having the arguments, structure, and functionality similar to those of qSimpson3D, the 3D Cartesian integrator qGaussLeg3D based on Gauss–Legendre quadratures provided in Listing 10.18 calls function xGaussLeg instead of xSimpson to generate the corresponding one-dimensional integration points and abscissas.

Let us consider as an example the integral over the volume of a *torus* of major radius $R = 3$ and minor (tube) radius $r = 1$ (Figure 10.6), centered at the origin and having Oz as symmetry axis. The points of the toroidal surface satisfy the equation:

$$\left(R - \sqrt{x^2 + y^2}\right)^2 + z^2 = r^2. \tag{10.83}$$

Listing 10.18   3D Integrator Based on Gaussian Quadratures (Python Coding)

```
#===============================================================================
def qGaussLeg3D(Func, ax, bx, nx, ay, by, ny, az, bz, nz):
#-------------------------------------------------------------------------------
#   Integrates function Func(x,y,z) in the cuboid [ax,bx] x [ay,by] x [az,bz]
#   using Gauss-Legendre quadratures with (nx x ny x nz) integration points
#-------------------------------------------------------------------------------
   x = [0]*nx; wx = [0]*nx
   y = [0]*ny; wy = [0]*ny
   z = [0]*nz; wz = [0]*nz

   (x,wx) = xGaussLeg(ax,bx,nx)                          # generate integartion points
   (y,wy) = xGaussLeg(ay,by,ny)
   (z,wz) = xGaussLeg(az,bz,nz)

   s = 0e0
   for i in range(0,nx):
      sx = 0e0
      for j in range(0,ny):
         sy = 0e0
         for k in range(0,nz):
            sy += wz[k] * Func(x[i],y[j],z[k])
         sx += wy[j] * sy
      s += wx[i] * sx

   return s
```

The radial distances from a particular point (x, y, z) to the revolution axis Oz and, respectively, to the tube axis (central circle) are given by

$$R' = \sqrt{x^2 + y^2}, \quad r' = \sqrt{(R - R')^2 + z^2}, \tag{10.84}$$

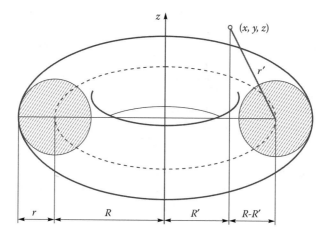

FIGURE 10.6   Toroidal integration domain revealing the radial distances R' and r' from the current point (x, y, z) to the symmetry axis Oz and, respectively, to the tube axis.

and the condition satisfied by the *interior points* of the torus takes the very simple form:

$$r' \le r. \tag{10.85}$$

We consider for numerical illustration the simple, but still realistic case of a density function $f(x, y, z)$ depending only on the distance r' to the tube axis and vanishing outside the torus:

$$f(x, y, z) = \begin{cases} (1 - r'/r)^2 & r' \le r, \\ 0 & r' > r. \end{cases} \tag{10.86}$$

Specifically, the function is chosen to parabolically decrease from 1 on the tube axis to 0 on the surface, and its integral amounts to $\frac{1}{3}\pi^2 R r^2 \approx 9.8696044$.

Employing Cartesian coordinates, the straightforward approach to describe the boundaries is to enclose the whole torus within a cuboid. However, making use of the revolution symmetry and the additional reflection symmetry with respect to the Oxy-plane, the integration domain can be reduced to $[0, R + r] \times [0, R + r] \times [0, r]$ within the first octant and the final result has to be multiplied, accordingly, by 8.

For integrating function (10.86), Program 10.19 comparatively uses the routines qSimpson3D and qGaussLeg3D. To comply with the three-parameter list of the user functions employed by all discussed 3D integrators, the geometrical details of the torus (R and r) are passed to Func by global variables. For reasons of efficiency, squaring is always replaced by multiplication.

In particular, function qGaussLeg3D yields the integral with seven exact significant digits employing $200 \times 200 \times 50$ integration points (such chosen as to ensure the same density of nodes along all three Cartesian directions). On the other hand, using the same integration mesh, function qSimpson3D barely achieves a two-order lower precision.

Listing 10.19   Integral Over Toroidal Domain Using Cartesian Coordinates (Python Coding)

```python
# Evaluates a 3D integral on a torus using Cartesian coordinates
from math import *
from integral import *

def Func(x, y, z):
   global R, r
   Rp = sqrt(x*x + y*y)                          # major radial distance
   dR = R - Rp
   rp = sqrt(dR*dR + z*z)                        # minor radial distance
   dr = 1e0 - rp/r
   return (dr*dr if rp <= r else 0e0)                     # zero-padding

# main

R = 3e0; r = 1e0                             # major & minor torus radii
ax = ay = az = 0e0                          # bounding box: 1st octant
bx = by = R+r; bz = r                       # multiply results with 8
nx = ny = 200                               # equal density of mesh
nz = int(nx * r/(R+r))                      # points along x, y, z

I = 8e0 * qSimpson3D(Func,ax,bx,nx,ay,by,ny,az,bz,nz)
print("I Simpson  = ",I)
I = 8e0 * qGaussLeg3D(Func,ax,bx,nx,ay,by,ny,az,bz,nz)
print("I GaussLeg = ",I)                            # result 9.869603...
```

Making $f(x, y, z) = 1$ at all interior points, the integral obviously represents the torus volume ($2\pi^2 R r^2 \approx 59.217626$), and this is a convenient test case for the program.

For 3D integrals defined in *spherical coordinates*, a quadrature formula similar to the one for Cartesian coordinates can be devised:

$$\int_0^a \int_0^\pi \int_0^{2\pi} f(r, \theta, \varphi) r^2 \, dr \sin\theta \, d\theta \, d\varphi \approx \sum_{i=1}^{n_r} w_i^{(r)} \sum_{j=1}^{n_\theta} w_j^{(\theta)} \sum_{k=1}^{n_\varphi} w_k^{(\varphi)} f(r_i, \theta_j, \varphi_k), \qquad (10.87)$$

however, with weights also accounting for the supplementary r^2 and $\sin\theta$ factors stemming from the volume element. In principle, the same one-dimensional schemes can be employed for the individual coordinates as for the Cartesian case. The integrator qSimpsonSph (Listing 10.20) is based on Simpson's rule and expects, besides the user function Func describing the integrand, a single geometrical parameter, namely, the outer radial limit a. The number of integration points for r, θ, and φ (nr, nt, and np) must also be provided.

After retrieving generic abscissas and weights from routine xSimpson for all three spherical coordinates, function qSimpsonSph scales the radial weights wr by r^2, and the angular weights wt for the θ integration by $\sin\theta$. With the redefined weights, the subsequent triple summation remains similar to the one for Cartesian coordinates.

Listing 10.20   Integrator in Spherical Coordinates Based on Simpson's Rule (Python Coding)

```
#===============================================================================
def qSimpsonSph(Func, a, nr, nt, np):
#-------------------------------------------------------------------------------
#   Integrates function Func(r,theta,phi) on [0,a] x [0,pi] x [0,2*pi]
#   in sperical coordinates using Simpson's rule with (nr x nt x np) points
#-------------------------------------------------------------------------------
   if (nr % 2 == 0): nr += 1                              # increment nr if even
   if (nt % 2 == 0): nt += 1                              # increment nt if even
   if (np % 2 == 0): np += 1                              # increment np if even

   r = [0]*nr; wr = [0]*nr
   t = [0]*nt; wt = [0]*nt
   p = [0]*np; wp = [0]*np

   (r,wr) = xSimpson(0e0,a,nr)                         # generate integartion points
   (t,wt) = xSimpson(0e0,pi,nt)
   (p,wp) = xSimpson(0e0,2e0*pi,np)

   for i in range(0,nr): wr[i] *= r[i] * r[i]                    # factors from
   for j in range(0,nt): wt[j] *= sin(t[j])                     # volume element

   s = 0e0
   for i in range(0,nr):
      sr = 0e0
      for j in range(0,nt):
         st = 0e0
         for k in range(0,np):
            st += wp[k] * Func(r[i],t[j],p[k])
         sr += wt[j] * st
      s += wr[i] * sr

   return s
```

For integrals in spherical coordinates extending over the entire space, Gaussian quadratures are better suited than classical formulas. In particular, the radial integration can be most conveniently performed using the Gauss–Laguerre formula 10.79, which is specifically designed for the semiinfinite interval $[0, +\infty)$, while the angular contributions can be efficiently evaluated using Gauss–Legendre quadratures. Maintaining the overall structure of qSimpsonSph, the routine qGaussSph (Listing 10.21) calls xGaussLag and xGaussLeg to generate the radial and, respectively, angular abscissas and weights.

In what follows, we consider as an example the wave function of the electron in a 2p state in the hydrogen atom (for quantum numbers $n = 2$, $l = 1$, and $m = 0$). Expressed in dimensionless radial units (see Fitts 2002), this wave function reads:

$$\Psi_{210}(r, \theta, \varphi) = \frac{1}{4\sqrt{2\pi}} r e^{-r/2} \cos\theta.$$

Its norm is given by the integral of the probability density $|\Psi_{210}(r, \theta, \varphi)|^2$,

$$\int_0^\infty \int_0^\pi \int_0^{2\pi} |\Psi_{210}(r, \theta, \varphi)|^2 r^2 dr \sin\theta d\theta d\varphi = 1,$$

and, since it represents the total probability of finding the electron somewhere in space, it naturally amounts to 1. Program 10.22 comparatively evaluates the above integral using Simpson's rule and

Listing 10.21 Integrator in Spherical Coordinates Based on Gaussian Quadratures (Python Coding)

```
#===========================================================================
def qGaussSph(Func, nr, nt, np):
#---------------------------------------------------------------------------
#  Integrates function Func(r,theta,phi) on [0,inf] x [0,pi] x [0,2*pi]
#  in spherical coordinates using Gauss-Laguerre and Gauss-Legendre formulas
#  with (nr x nt x np) points
#---------------------------------------------------------------------------
   r = [0]*nr; wr = [0]*nr
   t = [0]*nt; wt = [0]*nt
   p = [0]*np; wp = [0]*np

   (r,wr) = xGaussLag(0e0,nr)                 # Gauss-Laguerre radial quadrature
   (t,wt) = xGaussLeg(0e0,pi,nt)              # Gauss-Legendre angular quadratures
   (p,wp) = xGaussLeg(0e0,2e0*pi,np)

   for i in range(0,nr): wr[i] *= r[i] * r[i]             # factors from
   for j in range(0,nt): wt[j] *= sin(t[j])               # volume element

   s = 0e0
   for i in range(0,nr):
      sr = 0e0
      for j in range(0,nt):
         st = 0e0
         for k in range(0,np):
            st += wp[k] * Func(r[i],t[j],p[k])
         sr += wt[j] * st
      s += wr[i] * sr

   return s
```

Listing 10.22   Integral in Spherical Coordinates (Python Coding)

```
# Evaluates a 3D integral using spherical coordinates
from math import *
from integral import *

def Func(r, theta, phi):                          # probability density for 2p state
   c = 1e0/(32e0*pi)
   rcos = r * cos(theta)
   return c * rcos * rcos * exp(-r)

# main

a = 35e0                                                           # radial limit
nr = 150; nt = 60; np = 3
I = qSimpsonSph(Func,a,nr,nt,np)
print("I SimpsonSph = ",I)

nr = 3; nt = 9; np = 1
I = qGaussSph(Func,nr,nt,np)
print("I GaussSph   = ",I)
```

Listing 10.23   Integrator in Cylindrical Coordinates Based on Simpson's Rule (Python Coding)

```
#===========================================================================
def qSimpsonCyl(Func, a, az, bz, nr, np, nz):
#---------------------------------------------------------------------------
#  Integrates function Func(r,phi,z) on domain [0,a] x [0,2*pi] x [az,bz]
#  in cylindrical coordinates using Simpson's rule with (nr x np x nz) points
#---------------------------------------------------------------------------
   if (nr % 2 == 0): nr += 1                              # increment nr if even
   if (np % 2 == 0): np += 1                              # increment np if even
   if (nz % 2 == 0): nz += 1                              # increment nz if even

   r = [0]*nr; wr = [0]*nr
   p = [0]*np; wp = [0]*np
   z = [0]*nz; wz = [0]*nz

   (r,wr) = xSimpson(0e0,a,nr)                         # generate integartion points
   (p,wp) = xSimpson(0e0,2e0*pi,np)
   (z,wz) = xSimpson(az,bz,nz)

   for i in range(0,nr): wr[i] *= r[i]                # factor from volume element

   s = 0e0
   for i in range(0,nr):
      sr = 0e0
      for j in range(0,np):
         sp = 0e0
         for k in range(0,nz):
            sp += wz[k] * Func(r[i],p[j],z[k])
         sr += wp[j] * sp
      s += wr[i] * sr

   return s
```

Gaussian quadratures. Concretely, to reproduce the result with seven exact figures, function `qSimpsonSph` needs to radially integrate out to $a = 35$ using $n_r = 150$, $n_\theta = 60$, and $n_\varphi = 3$ nodes (the integrand lacks φ dependence). Given their remarkable properties, especially, the appropriateness of the Gauss–Laguerre formula for the radial contributions, the Gaussian quadratures achieve the same result with as few as $n_r = 3$, $n_\theta = 9$, and $n_\varphi = 1$ integration points.

Combining functions `qSimpson3D` and `qSimpsonSph`, one can readily build an integrator for cylindrical coordinates based on Simpson's rule (Listing 10.23).

10.12 Adaptive Multidimensional Integration

In the case of high-dimensional integrals with irregular boundaries and dispersed integrand values, the direct product technique, with zero padding on uniform meshes and regular extended boundaries, may become intractable if the grid is adjusted to the least convergent regions. In such cases, rather a methodology operating directly inside the irregular boundaries and featuring local mesh control is desirable. To this end, we will represent the integral as embedded, functionally coupled one-dimensional integrals, using the same one-dimensional adaptive integrator for each separate level of integration.

Concretely, a 2D integral in Cartesian coordinates may be regarded as an "outer" integral with respect to x, whose integrand is the "inner" integral with respect to y along a line of fixed x:

$$\int_{a_x}^{b_x} \int_{a_y(x)}^{b_y(x)} f(x,y)\,dx\,dy = \int_{a_x}^{b_x} F_x(x)\,dx, \tag{10.88}$$

$$F_x(x) = \int_{a_y(x)}^{b_y(x)} f(x,y)\,dy.$$

For each particular outer argument $x_i \in [a_x, b_x]$ (see Figure 10.7), the inner integral $F_x(x_i)$ may be evaluated on an independent grid, which is completely determined by the local behavior of the integrand and

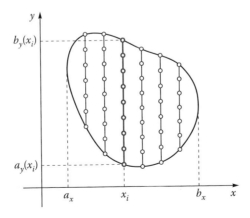

FIGURE 10.7 The "inner" integration with respect to y operates on independent adaptive grids for each single argument x_i of the "outer" integration.

the particular integration limits, $a_y(x_i)$ and $b_y(x_i)$. The corresponding quadrature may be formulated as

$$\int_{a_x}^{b_x} \int_{a_y(x)}^{b_y(x)} f(x,y)\,dxdy \approx \sum_{i=1}^{n_x} w_i^{(x)} F_x(x_i), \tag{10.89}$$

$$F_x(x_i) \approx \sum_{j=1}^{n_y(x_i)} w_j^{(y)}(x_i) f(x_i, y_j),$$

and it emphasizes the dependence of the inner integration points y_j (in terms of number, positions, and associated weights) on the outer argument x_i.

By generalizing the decomposition into embedded one-dimensional integrals to three dimensions, we may represent the quadrature by

$$\int_{a_x}^{b_x} \int_{a_y(x)}^{b_y(x)} \int_{a_z(x,y)}^{b_z(x,y)} f(x,y,z)\,dxdydz \approx \sum_{i=1}^{n_x} w_i^{(x)} F_x(x_i), \tag{10.90}$$

$$F_x(x_i) \approx \sum_{j=1}^{n_y(x_i)} w_j^{(y)}(x_i) F_y\left(x_i, y_j\right),$$

$$F_y\left(x_i, y_j\right) \approx \sum_{k=1}^{n_z(x_i,y_j)} w_k^{(z)}\left(x_i, y_j\right) f(x_i, y_j, z_k).$$

The three sums are formally one-dimensional, yet not independent. They could be evaluated, in principle, by means of the same one-dimensional adaptive integrator by making use of the built-in recursivity. There is, however, an additional complication: while $F_x(x_i)$ complies with the prototype of single-argument integrand handled by the one-dimensional integrators, $F_y\left(x_i, y_j\right)$ and $f(x_i, y_j, z_k)$ obviously do not conform.

A solution at hand, though departing from the spirit of modular programming (which requires program units to communicate only through well-defined interfaces—headers with lists of arguments), is to define two single-argument interface functions, $\widetilde{F}_y(y_j) \equiv F_y\left(x_i, y_j\right)$ and $\widetilde{F}_z(z_k) \equiv f(x_i, y_j, z_k)$, that receive the other arguments, x_i and, respectively, $\left(x_i, y_j\right)$, via different mechanisms in concrete implementations (e.g., by means of global variables).

To illustrate the above ideas, we consider the integral

$$I = \int_0^1 dx \int_0^{\sqrt{1-x^2}} dy \int_0^{\sqrt{1-x^2-y^2}} dz = \frac{\pi}{6} \approx 0.5235987\ldots,$$

whose result is the volume of the first octant of the sphere of radius 1. The implementation in Listing 10.24 uses the same one-dimensional integrator, qRomberg, assumed to be contained in the module integral.py, for all three Cartesian coordinates. Although not directly, the integrator calls itself recursively by way of interface functions, producing three levels of subordination corresponding to the x, y, and z-directions.

Listing 10.24 Adaptive 3D Integration on Spherical Domain (Python Coding)

```python
# 3D adaptive quadrature over 1st octant of unity sphere
from math import *
from integral import *

#----------------------------------------------------------- integrand
def Func(x, y, z): return 1e0

#----------------------------------------------------- integration limits
def ay(x): return 0e0
def by(x): return sqrt(1e0 - x*x)
def az(x, y): return 0.0
def bz(x, y): return sqrt(fabs(1e0 - x*x - y*y))

#----------------------------------------------------------- interfaces
def Fz(z):                                        # z-integrand
   global xvar, yvar
   return Func(xvar,yvar,z)                # (x,y) as global (xvar,yvar)
def Fy(y):                                        # y-integrand
   global xvar, yvar
   yvar = y                                # stores y in yvar
   return qRomberg(Fz,az(xvar,y),bz(xvar,y),eps) # x as global xvar
def Fx(x):                                        # x-integrand
   global xvar
   xvar = x                                # stores x in xvar
   return qRomberg(Fy,ay(x),by(x),eps)

# main

ax = 0e0; bx = 1e0; eps = 1e-7

I = qRomberg(Fx,ax,bx,eps)                        # outer integral
print("Integral = ",I)                     # result 0.5235987...
```

All particular aspects related to the specific integral are restricted to the user function Func (returning the integrand $f(x, y, z)$) and to the functions ay, by, az, bz and the scalars ax and by (describing the integration limits). The overall relative precision is set in the main program and is available through the global variable eps.

The outer integral with respect to x is calculated right in the main program by calling the integrator qRomberg for integrand Fx and the fixed integration limits ax and bx. The integrand Fx is actually called by qRomberg for various x-values, which are stored in the global variable xvar for subsequent use in functions Fy and Fz. Furthermore, Fx initiates the evaluation of the y-integral by recursively calling qRomberg for the single-argument interface Fy and the current limits ay(x) and by(x). Function Fy is called by qRomberg for various y-values, which are stored in the global variable yvar for use in Fz. Continuing, Fy starts the innermost z-integration, however, not for the "real" integrand Func (which does not comply with the required single-argument prototype), but, instead, for the single-argument interface Fz, which receives the z-value as an argument and the x- and y-values by way of the global variables xvar and yvar. The partial integrals are finalized in reversed order of their recursivity levels.

If the integral is improper relative to some of its variables, the integration domain may be partitioned in such a way as to place the singular regions on the new boundaries and an open-quadrature routine (like, for instance, function qMidPoint) may be employed to calculate the partial integrals.

10.13 Implementations in C/C++

Listing 10.25 shows the content of the file `integral.h`, which contains equivalent C/C++ implementations of the Python functions developed in the main text and included in the module `integral.py`. The corresponding routines have identical names, parameters, and functionalities.

The Gaussian integrators call the special functions `Legendre` and `Laguerre` included in the header file `specfunc.h`, which, for this reason, also needs to be accessible along with `integral.h`.

Listing 10.25   Integrators for Real Functions (`integral.h`)

```
//-------------------------------- integral.h --------------------------------
// Contains 1D and 3D integrators for real functions with real variables.
// Part of the numxlib numerics library. Author: Titus Beu, 2013
//----------------------------------------------------------------------------
#ifndef _INTEGRAL_
#define _INTEGRAL_

#include <stdio.h>
#include <math.h>
#include "memalloc.h"
#include "specfunc.h"


//============================================================================
double qTrapz(double Func(double), double a, double b, int n)
//----------------------------------------------------------------------------
// Integrates function Func on interval [a,b] using the trapezoidal rule
// with n integration points
//----------------------------------------------------------------------------
{
   double h, s;
   int i;

   h = (b-a)/(n-1);
   s = 0.5*(Func(a) + Func(b));
   for (i=1; i<=(n-2); i++) s += Func(a+i*h);

   return h*s;
}


//============================================================================
double qSimpson(double Func(double), double a, double b, int n)
//----------------------------------------------------------------------------
// Integrates function Func on interval [a,b] using Simpson's rule with n
// (odd) integration points
//----------------------------------------------------------------------------
{
   double h, s1, s2;
   int i;

   if (n % 2 == 0) n++;                                  // increment n if even

   h = (b-a)/(n-1);
   s1 = s2 = 0e0;
   for (i=2; i<=n-3; i+=2) { s1 += Func(a + i*h); }        // odd-index sum
   for (i=1; i<=n-2; i+=2) { s2 += Func(a + i*h); }        // even-index sum

   return (h/3)*(Func(a) + 4*s2 + 2*s1 + Func(b));
}
```

```
//===========================================================================
double qTrapzCtrl(double Func(double), double a, double b, double eps)
//---------------------------------------------------------------------------
// Integrates function Func on interval [a,b] with relative precision eps
// using the adaptive Trapezoidal Rule
//---------------------------------------------------------------------------
{
   const int kmax = 30;                      // max. no. of step halving iterations
   double h, sum, t, t0;
   long i, n;
   int k;

   h = b-a; n = 1;
   t0 = 0.5*h*(Func(a) + Func(b));                    // initial approximation

   for (k=1; k<=kmax; k++) {                           // step halving loop
      sum = 0e0;
      for (i=1; i<=n; i++) sum += Func(a+(i-0.5)*h);
      t = 0.5*(t0 + h*sum);                            // new approximation
      if (k > 1) {                                     // convergence check
         if (fabs(t-t0) <= eps*fabs(t)) break;
         if (fabs(t) <= eps && fabs(t) <= fabs(t-t0)) break; // integral ~= 0
      }
      h *= 0.5; n *= 2;                                // halve integration step
      t0 = t;
   }
   if (k > kmax) printf("qTrapzCtrl: max. no. of iterations exceeded !\n");

   return t;
}

//===========================================================================
double qSimpsonCtrl(double Func(double), double a, double b, double eps)
//---------------------------------------------------------------------------
// Integrates function Func on interval [a,b] with relative precision eps
// using the adaptive Simpson Rule
//---------------------------------------------------------------------------
{
   const int kmax = 30;                      // max. no. of step halving iterations
   double h, s, s0, sum, t, t0;
   long i, n;
   int k;

   h = b-a; n = 1;
   s0 = t0 = 0.5*h*(Func(a) + Func(b));               // initial approximation

   for (k=1; k<=kmax; k++) {                           // step halving loop
      sum = 0e0;
      for (i=1; i<=n; i++) sum += Func(a+(i-0.5)*h);
      t = 0.5*(t0 + h*sum);
      s = (4*t - t0)/3;                                // new approximation
      if (k > 1) {                                     // convergence check
         if (fabs(s-s0) <= eps*fabs(s)) break;
         if (fabs(s) <= eps && fabs(s) <= fabs(s-s0)) break; // integral ~= 0
      }
      h *= 0.5; n *= 2;                                // halve integration step
      s0 = s; t0 = t;
   }
   if (k > kmax) printf("qSimpsonCtrl: max. no. of iterations exceeded!\n");
```

```
      return s;
}


//==============================================================================
double qRomberg(double Func(double), double a, double b, double eps)
//------------------------------------------------------------------------------
// Integrates function Func on interval [a,b] with relative precision eps
// using the adaptive Romberg method
//------------------------------------------------------------------------------
{
   const int kmax = 30;                        // max. no. of step halving iterations
   double r1[kmax+1];                          // two consecutive lines
   double r2[kmax+1];                          // from the method table
   double f, h, sum;
   long i, n;
   int j, k;

   h = b-a; n = 1;
   r1[0] = 0.5*h*(Func(a) + Func(b));          // initial approximation
   for (k=1; k<=kmax; k++) {                    // step halving loop
      sum = 0e0;
      for (i=1; i<=n; i++) sum += Func(a+(i-0.5)*h);
      r2[0] = 0.5*(r1[0] + h*sum);             // trapezoid formula
      f = 1e0;
      for (j=1; j<=k; j++) {                    // increase quadrature order
         f *= 4;
         r2[j] = (f*r2[j-1] - r1[j-1])/(f-1);  // new approximation
      }
      if (k > 1) {                             // convergence check
         if (fabs(r2[k]-r1[k-1]) <= eps*fabs(r2[k])) break;
         if (fabs(r2[k]) <= eps && fabs(r2[k]) <= fabs(r2[k]-r1[k-1])) break;
      }
      h *= 0.5; n *= 2;                        // halve integration step
      for (j=0; j<=k; j++) r1[j] = r2[j];      // shift table lines
   }
   if (k > kmax) {
      printf("qRomberg: max. no. of iterations exceeded !\n");
      k--;
   }

   return r2[k];
}


//==============================================================================
double qImprop1(double Func(double), double a, double &xinf, double eps)
//------------------------------------------------------------------------------
// Integrates function Func on interval [a,+inf) (for xinf >= 0) or (-inf,a]
// (for xinf < 0) with relative precision eps. On output, xinf contains the
// integration domain limit for which convergence was achieved.
// Calls: qRomberg
//------------------------------------------------------------------------------
{
   double h, s, s1, x;

   h = 1e0; x = a;  // subinterval length and initial left limit for [a,+inf)
   if (xinf < 0e0) { h = -h; x = a + h; }                 // for (-inf,a]

   s1 = 1e0;
```

```
   s = 0e0;
   while(fabs(s1) > eps*fabs(s) || fabs(Func(x)) > eps) {
      s1 = qRomberg(Func,x,x+fabs(h),eps);            // integral for [x,x+h]
      s += s1;                                        // update total integral
      x += h;                                         // shift interval [x,x+h]
   }
   xinf = x;                                      // final "infinite" boundary

   return s;
}

//===========================================================================
double qImprop2(double Func(double), double a, double b, double eps)
//---------------------------------------------------------------------------
// Integrates function Func on interval [a,b] with a and/or b singular
// integrable points.
// Calls: qRomberg
//---------------------------------------------------------------------------
{
   double h, h0, s, s1, x;

   h0 = 0.1e0 * (b-a);         // extent of vicinities of singular boundaries

   s = 0e0;
   if (fabs(Func(a)) > 9e99) {                        // a is singular point?
      h = h0;
      s1 = 1e0;
      while(fabs(s1) > eps*fabs(s)) {
         h *= 0.5;                                          // halve interval
         x = a + h;                                // left boundary of [x,x+h]
         s1 = qRomberg(Func,x,x+h,eps);          // partial integral on [x,x+h]
         s += s1;                           // add contribution to total integral
      }
      a += h0;                               // new left boundary of core interval
   }

   if (fabs(Func(b)) > 9e99) {                        // b is singular point?
      h = h0;
      s1 = 1e0;
      while(fabs(s1) > eps*fabs(s)) {
         h *= 0.5;                                          // halve interval
         x = b - h;                               // right boundary of [x-h,x]
         s1 = qRomberg(Func,x-h,x,eps);          // partial integral on [x-h,x]
         s += s1;                           // add contribution to total integral
      }
      b -= h0;                              // new right boundary of core interval
   }

   s += qRomberg(Func,a,b,eps);               // add integral on core interval

   return s;
}

//===========================================================================
double qMidPoint(double Func(double), double a, double b, double eps)
//---------------------------------------------------------------------------
// Integrates function Func on interval (a,b) with relative precision eps
// using the adaptive midpoint rule
//---------------------------------------------------------------------------
```

```
{
   const int kmax = 19;                                  // max. no. of divisions
   double h, s, s0, sum;
   double f1p6 = 1e0/6e0, f5p6 = 5e0/6e0;
   long i, n;
   int k;

   h = b-a; n = 1;
   s0 = h * Func(a+0.5*h);                               // initial approximation

   for (k=1; k<=kmax; k++) {                             // step division loop
      sum = 0e0;
      for (i=1; i<=n; i++) sum += Func(a+(i-f5p6)*h) + Func(a+(i-f1p6)*h);
      s = (s0 + h*sum)/3;                                // new approximation
      if (fabs(s - s0) <= eps*fabs(s)) break;            // convergence check
      h /= 3; n *= 3;                                    // reduce step
      s0 = s;
   }
   if (k > kmax) printf("qMidPoint: max. no. of iterations exceeded !\n");

   return s;
}

//=============================================================================
void xGaussLeg(double a, double b, double x[], double w[], int n)
//-----------------------------------------------------------------------------
// Calculates abscissas x[] and weights w[] for the n-point Gauss-Legendre
// quadrature on interval [a,b]
// Calls: Legendre (from specfunc.h)
//-----------------------------------------------------------------------------
{
#define pi 3.141592653589793
   const double eps = 1e-14;                             // relative precision of zeros
   double d, f, xc, xi;
   int i, n2;

   n2 = n/2;
   for (i=1; i<=n2; i++) {
      xi = cos(pi*(i-0.25e0)/(n+0.5e0));    // initial approximation for zeros
      f = 9e99;
      while (fabs(f) > eps*fabs(xi)) {               // Newton-Raphson refinement
         f = Legendre(n,xi,d) / d;
         xi -= f;
      }
      x[i] = -xi; x[n-i+1] = xi;                        // symmetrical zeros
      w[i] = w[n-i+1] = 2e0/((1e0-xi*xi)*d*d);          // equal weights
   }

   if (n % 2 == 1) {                                    // odd no. of mesh points
      Legendre(n,0e0,d);
      x[n2+1] = 0e0;
      w[n2+1] = 2e0/(d*d);
   }

   f = 0.5e0*(b-a); xc = 0.5e0*(b+a);                   // scaling to interval [a,b]
   for (i=1; i<=n; i++) {
      x[i] = f*x[i] + xc;
      w[i] = f*w[i];
   }
```

```
}

//============================================================================
double qGaussLeg(double Func(double), double a, double b, int n)
//----------------------------------------------------------------------------
// Integrates function Func on interval [a,b] using n-point Gauss-Legendre
// quadratures
// Calls: xGaussLeg
//----------------------------------------------------------------------------
{
   double s, *x, *w;
   int i;

   x = Vector(1,n);
   w = Vector(1,n);

   xGaussLeg(a,b,x,w,n);

   s = 0e0;
   for (i=1; i<=n; i++) s += w[i] * Func(x[i]);

   FreeVector(x,1);
   FreeVector(w,1);

   return s;
}


//============================================================================
void xGaussLag(double a, double x[], double w[], int n)
//----------------------------------------------------------------------------
// Calculates abscissas x[] and weights w[] for the n-point Gauss-Laguerre
// quadrature on interval [a,+inf)
// Calls: Laguerre (from specfunc.h)
// Initial approximation for zeros:
// A. Stroud & D. Secrest, Gaussian Quadrature Formulas, Prentice Hall, 1966.
//----------------------------------------------------------------------------
{
   const double eps = 1e-14;                    // relative precision of zeros
   double d, f, xi;
   int i;

   for (i = 1; i <= n; i++) {
      if (i == 1)           // initial approximation for zeros (Stroud & Secrest)
         xi = 3e0/(1e0+2.4e0*n);                              // 1st zero
      else if (i == 2)
         xi = 15e0/(1e0+2.5e0*n) + x[1];                      // 2nd zero
      else {
         f = (1e0/i+2.55e0)/1.9e0;                            // recurrence
         xi = (1e0+f)*x[i-1] - f*x[i-2];
      }
      f = 9e99;
      while (fabs(f) > eps*fabs(xi)) {           // Newton-Raphson refinement
         f = Laguerre(n,xi,d) / d;
         xi -= f;
      }
      x[i] = xi;
      w[i] = exp(xi)/(xi*d*d);
   }
```

```
      for (i = 1; i <= n; i++) x[i] += a;          // scaling to interval [a,+inf)
}

//=================================================================================
double qGaussLag(double Func(double), double a, int n)
//---------------------------------------------------------------------------------
// Integrates function Func on interval [a,+inf) using n-point Gauss-Laguerre
// quadratures
// Calls: xGaussLag
//---------------------------------------------------------------------------------
{
   double s, *x, *w;
   int i;

   x = Vector(1,n);
   w = Vector(1,n);

   xGaussLag(a,x,w,n);

   s = 0e0;
   for (i=1; i<=n; i++) s += w[i] * Func(x[i]);

   FreeVector(x,1);
   FreeVector(w,1);

   return s;
}


//=================================================================================
double qTrapz3D(double Func(double,double,double),
                double ax, double bx, int nx,
                double ay, double by, int ny,
                double az, double bz, int nz)
//---------------------------------------------------------------------------------
// Integrates function Func(x,y,z) in the cuboid [ax,bx] x [ay,by] x [az,bz]
// using the trapezoidal rule with (nx x ny x nz) integration points
//---------------------------------------------------------------------------------
{
   double hx, hy, hz, s, sx, sy, wx, wy, wz, x, y, z;
   int i, j, k;

   hx = (bx-ax)/(nx-1);
   hy = (by-ay)/(ny-1);
   hz = (bz-az)/(nz-1);

   s = 0e0;
   for (i=1; i<=nx; i++) {
      x = ax + (i-1)*hx; wx = ((i-1)*(i-nx) ? hx : 0.5e0*hx);
      sx = 0e0;
      for (j=1; j<=ny; j++) {
         y = ay + (j-1)*hy; wy = ((j-1)*(j-ny) ? hy : 0.5e0*hy);
         sy = 0e0;
         for (k=1; k<=nz; k++) {
            z = az + (k-1)*hz; wz = ((k-1)*(k-nz) ? hz : 0.5e0*hz);
            sy += wz * Func(x,y,z);
         }
         sx += wy * sy;
      }
      s += wx * sx;
```

```
   }
   return s;
}

//===========================================================================
void xSimpson(double a, double b, double x[], double w[], int n)
//---------------------------------------------------------------------------
// Calculates abscissas x[] and weights w[] for Simpson's rule with n
// integration points on interval [a,b]
//---------------------------------------------------------------------------
{
   const double c13 = 1e0/3e0, c23 = 2e0/3e0, c43 = 4e0/3e0;
   double h;
   int i;

   if (n % 2 == 0) n++;                             // increment n if even

   h = (b-a)/(n-1);
   for (i=1; i<=n; i++) {
      x[i] = a + (i-1)*h; w[i] = h * (i % 2 ? c23 : c43);
   }
   w[1] = w[n] = h * c13;
}


//===========================================================================
double qSimpson3D(double Func(double,double,double),
                  double ax, double bx, int nx,
                  double ay, double by, int ny,
                  double az, double bz, int nz)
//---------------------------------------------------------------------------
// Integrates function Func(x,y,z) in the cuboid [ax,bx] x [ay,by] x [az,bz]
// using Simpson's rule with (nx x ny x nz) integration points
//---------------------------------------------------------------------------
{
   double *wx, *wy, *wz, *x, *y, *z;
   double s, sx, sy;
   int i, j, k;

   if (nx % 2 == 0) nx++;                           // increment nx if even
   if (ny % 2 == 0) ny++;                           // increment ny if even
   if (nz % 2 == 0) nz++;                           // increment nz if even

   x = Vector(1,nx); wx = Vector(1,nx);
   y = Vector(1,ny); wy = Vector(1,ny);
   z = Vector(1,nz); wz = Vector(1,nz);

   xSimpson(ax,bx,x,wx,nx);                         // generate integartion points
   xSimpson(ay,by,y,wy,ny);
   xSimpson(az,bz,z,wz,nz);

   s = 0e0;
   for (i=1; i<=nx; i++) {
      sx = 0e0;
      for (j=1; j<=ny; j++) {
         sy = 0e0;
         for (k=1; k<=nz; k++) {
            sy += wz[k] * Func(x[i],y[j],z[k]);
         }
         sx += wy[j] * sy;
```

```
         }
         s += wx[i] * sx;
      }

   FreeVector(x,1); FreeVector(wx,1);
   FreeVector(y,1); FreeVector(wy,1);
   FreeVector(z,1); FreeVector(wz,1);

   return s;
}

//============================================================================
double qGaussLeg3D(double Func(double,double,double),
                   double ax, double bx, int nx,
                   double ay, double by, int ny,
                   double az, double bz, int nz)
//----------------------------------------------------------------------------
// Integrates function Func(x,y,z) in the cuboid [ax,bx] x [ay,by] x [az,bz]
// using Gauss-Legendre quadratures with (nx x ny x nz) integration points
//----------------------------------------------------------------------------
{
   double *wx, *wy, *wz, *x, *y, *z;
   double s, sx, sy;
   int i, j, k;

   x = Vector(1,nx); wx = Vector(1,nx);
   y = Vector(1,ny); wy = Vector(1,ny);
   z = Vector(1,nz); wz = Vector(1,nz);

   xGaussLeg(ax,bx,x,wx,nx);                          // generate integartion points
   xGaussLeg(ay,by,y,wy,ny);
   xGaussLeg(az,bz,z,wz,nz);

   s = 0e0;
   for (i=1; i<=nx; i++) {
      sx = 0e0;
      for (j=1; j<=ny; j++) {
         sy = 0e0;
         for (k=1; k<=nz; k++) {
            sy += wz[k] * Func(x[i],y[j],z[k]);
         }
         sx += wy[j] * sy;
      }
      s += wx[i] * sx;
   }

   FreeVector(x,1); FreeVector(wx,1);
   FreeVector(y,1); FreeVector(wy,1);
   FreeVector(z,1); FreeVector(wz,1);

   return s;
}

//============================================================================
double qSimpsonAng(double Func(double,double), int nt, int np)
//----------------------------------------------------------------------------
// Integrates function Func(theta,phi) on [0,pi] x [0,2*pi] in sperical
// coordinates using Simpson's rule with (nt x np) points
//----------------------------------------------------------------------------
```

```
{
#define pi 3.141592653589793
   double *wt, *wp, *t, *p;
   double s, st;
   int i, j;

   if (nt % 2 == 0) nt++;                                // increment nt if even
   if (np % 2 == 0) np++;                                // increment np if even

   t = Vector(1,nt); wt = Vector(1,nt);
   p = Vector(1,np); wp = Vector(1,np);

   xSimpson(0e0,pi,t,wt,nt);                             // generate integartion points
   xSimpson(0e0,2e0*pi,p,wp,np);

   for (i=1; i<=nt; i++) { wt[i] *= sin(t[i]); }         // volume element

   s = 0e0;
   for (i=1; i<=nt; i++) {
      st = 0e0;
      for (j=1; j<=np; j++) {
         st += wp[j] * Func(t[i],p[j]);
      }
      s += wt[i] * st;
   }

   FreeVector(t,1); FreeVector(wt,1);
   FreeVector(p,1); FreeVector(wp,1);

   return s;
}

//===========================================================================
double qSimpsonSph(double Func(double,double,double), double a,
                   int nr, int nt, int np)
//---------------------------------------------------------------------------
// Integrates function Func(r,theta,phi) on [0,a] x [0,pi] x [0,2*pi]
// in sperical coordinates using Simpson's rule with (nr x nt x np) points
//---------------------------------------------------------------------------
{
#define pi 3.141592653589793
   double *wr, *wt, *wp, *r, *t, *p;
   double s, sr, st;
   int i, j, k;

   if (nr % 2 == 0) nr++;                                // increment nr if even
   if (nt % 2 == 0) nt++;                                // increment nt if even
   if (np % 2 == 0) np++;                                // increment np if even

   r = Vector(1,nr); wr = Vector(1,nr);
   t = Vector(1,nt); wt = Vector(1,nt);
   p = Vector(1,np); wp = Vector(1,np);

   xSimpson(0e0,a,r,wr,nr);                              // generate integartion points
   xSimpson(0e0,pi,t,wt,nt);
   xSimpson(0e0,2e0*pi,p,wp,np);

   for (i=1; i<=nr; i++) { wr[i] *= r[i] * r[i]; }       // factors from
   for (j=1; j<=nt; j++) { wt[j] *= sin(t[j]); }         // volume element
```

```
   s = 0e0;
   for (i=1; i<=nr; i++) {
      sr = 0e0;
      for (j=1; j<=nt; j++) {
         st = 0e0;
         for (k=1; k<=np; k++) {
            st += wp[k] * Func(r[i],t[j],p[k]);
         }
         sr += wt[j] * st;
      }
      s += wr[i] * sr;
   }

   FreeVector(r,1); FreeVector(wr,1);
   FreeVector(t,1); FreeVector(wt,1);
   FreeVector(p,1); FreeVector(wp,1);

   return s;
}

//===========================================================================
double qGaussSph(double Func(double,double,double), int nr, int nt, int np)
//---------------------------------------------------------------------------
// Integrates function Func(r,theta,phi) on [0,inf] x [0,pi] x [0,2*pi]
// in spherical coordinates using Gauss-Laguerre and Gauss-Legendre formulas
// with (nr x nt x np) points
//---------------------------------------------------------------------------
{
#define pi 3.141592653589793
   double *wr, *wt, *wp, *r, *t, *p;
   double s, sr, st;
   int i, j, k;

   r = Vector(1,nr); wr = Vector(1,nr);
   t = Vector(1,nt); wt = Vector(1,nt);
   p = Vector(1,np); wp = Vector(1,np);

   xGaussLag(0e0,r,wr,nr);                  // Gauss-Laguerre radial quadrature
   xGaussLeg(0e0,pi,t,wt,nt);               // Gauss-Legendre angular quadratures
   xGaussLeg(0e0,2e0*pi,p,wp,np);

   for (i=1; i<=nr; i++) { wr[i] *= r[i] * r[i]; }        // factors from
   for (j=1; j<=nt; j++) { wt[j] *= sin(t[j]); }          // volume element

   s = 0e0;
   for (i=1; i<=nr; i++) {
      sr = 0e0;
      for (j=1; j<=nt; j++) {
         st = 0e0;
         for (k=1; k<=np; k++) {
            st += wp[k] * Func(r[i],t[j],p[k]);
         }
         sr += wt[j] * st;
      }
      s += wr[i] * sr;
   }

   FreeVector(r,1); FreeVector(wr,1);
```

```
      FreeVector(t,1); FreeVector(wt,1);
      FreeVector(p,1); FreeVector(wp,1);

      return s;
}

//===========================================================================
double qSimpsonCyl(double Func(double,double,double),
                   double a, double az, double bz, int nr, int np, int nz)
//---------------------------------------------------------------------------
// Integrates function Func(r,phi,z) on domain [0,a] x [0,2*pi] x [az,bz]
// in cylindrical coordinates using Simpson's rule with (nr x np x nz) points
//---------------------------------------------------------------------------
{
#define pi 3.141592653589793
   double *wr, *wp, *wz, *r, *p, *z;
   double s, sr, sp;
   int i, j, k;

   if (nr % 2 == 0) nr++;                         // increment nr if even
   if (np % 2 == 0) np++;                         // increment np if even
   if (nz % 2 == 0) nz++;                         // increment nz if even

   r = Vector(1,nr); wr = Vector(1,nr);
   p = Vector(1,np); wp = Vector(1,np);
   z = Vector(1,nz); wz = Vector(1,nz);

   xSimpson(0e0,a,r,wr,nr);                       // generate integartion points
   xSimpson(0e0,2e0*pi,p,wp,np);
   xSimpson(az,bz,z,wz,nz);

   for (i=1; i<=nr; i++) { wr[i] *= r[i]; }       // factor from volume element

   s = 0e0;
   for (i=1; i<=nr; i++) {
      sr = 0e0;
      for (j=1; j<=np; j++) {
         sp = 0e0;
         for (k=1; k<=nz; k++) {
            sp += wz[k] * Func(r[i],p[j],z[k]);
         }
         sr += wp[j] * sp;
      }
      s += wr[i] * sr;
   }

   FreeVector(r,1); FreeVector(wr,1);
   FreeVector(p,1); FreeVector(wp,1);
   FreeVector(z,1); FreeVector(wz,1);

   return s;
}

#endif
```

10.14 Problems

The Python and C/C++ programs for the following problems may import the functions developed in this chapter from the modules `integral.py` and, respectively, `integral.h`, which are available as supplementary material. For creating runtime plots, the graphical routines contained in the libraries `graphlib.py` and `graphlib.h` may be employed.

PROBLEM 10.1

Evaluate the integral

$$I = \int_0^\pi \sin x\, dx = 2,$$

by using comparatively the composite trapezoidal and Simpson rules (routines `qTrapz` and `qSimpson`) for numbers of subintervals doubled in sequence from $n = 2$ to 1024. For each number of subintervals, evaluate the relative errors with respect to the exact result

$$\delta_T^{(n)} = |1 - I_T^{(n)}/I|, \quad \delta_S^{(n)} = |1 - I_S^{(n)}/I|,$$

as well as the error reduction factor between two consecutive estimates of the integral

$$f_T^{(n)} = I_T^{(n-1)}/I_T^{(n)}, \quad f_S^{(n)} = I_S^{(n-1)}/I_S^{(n)}.$$

Compare the numerical reduction factors with the theoretical values 4 and, respectively, 16, resulting from the orders $O(h^2)$ and $O(h^4)$ of the global errors featured by the trapezoidal and Simpson rules.

Listing 10.26   Convergence of the Trapezoidal and Simpson Rules (Python Coding)

```python
# Convergence of trapezoidal and Simpson quadratures
from math import *
from integral import *

def Func(x): return sin(x)                                      # integrand

a = 0e0; b = pi                                          # integration domain

I = cos(a) - cos(b)                                         # exact result
print("Exact result = ",I)
print("   n        IT        errT   facT     IS        errS   facS")

n = 1
errT0 = 1e0; errS0 = 1e0
for i in range(1,11):
   n = 2*n                                            # number of intervals

   IT = qTrapz(Func,a,b,n+1)                             # trapezoidal rule
   errT = fabs(1e0 - IT/I)                                # relative error
   facT = errT0/errT                                     # error reduction

   IS = qSimpson(Func,a,b,n+1)                           # Simpson's rule
   errS = fabs(1e0 - IS/I)                                # relative error
   facS = errS0/errS                                     # error reduction

   print(("{0:5d}{1:11.6f}{2:10.1e}{3:6.1f}{4:11.6f}{5:10.1e}{6:6.1f}").
         format(n,IT,errT,facT,IS,crrS,facS))

   errT0 = errT; errS0 = errS
```

Listing 10.27   Output for Problem 10.2

```
Exact result =  0.2222222222222222
   n        IT        errT    facT      IS        errS    facS
   2    1.140625    4.1e+00   0.2    0.854167    2.8e+00   0.4
   4    0.653870    1.9e+00   2.1    0.491618    1.2e+00   2.3
   8    0.400656    8.0e-01   2.4    0.316251    4.2e-01   2.9
  16    0.272359    2.3e-01   3.6    0.229593    3.3e-02  12.8
  32    0.235120    5.8e-02   3.9    0.222707    2.2e-03  15.2
  64    0.225470    1.5e-02   4.0    0.222253    1.4e-04  15.8
 128    0.223036    3.7e-03   4.0    0.222224    8.7e-06  15.9
 256    0.222426    9.2e-04   4.0    0.222222    5.4e-07  16.0
 512    0.222273    2.3e-04   4.0    0.222222    3.4e-08  16.0
1024    0.222235    5.7e-05   4.0    0.222222    2.1e-09  16.0
```

Solution

The Python implementation is given in Listing 10.26 and the C/C++ version is available as supplementary material (P10-qTrapSimpConv.cpp).

PROBLEM 10.2

Solve the requirements of the previous problem for the norm of the Legendre polynomials (see Section 6.4 for definition and numerical methods):

$$\int_{-1}^{+1} [P_N(x)]^2 dx = \frac{2}{2N+1},$$

using the function Legendre (from specfunc.py or specfunc.h) for evaluating the polynomials of the particular order $N = 4$.

Solution

The Python and C/C++ implementations are available as supplementary material (files P10-qTrapSimpConv1.py and P10-qTrapSimpConv1.cpp) and the output is shown in Listing 10.27.

PROBLEM 10.3

The Fresnel integrals (see Guenther 1990),

$$C(w) = \int_0^w \cos \frac{\pi u^2}{2} du,$$

$$S(w) = \int_0^w \sin \frac{\pi u^2}{2} du,$$

are central to the theory of optical diffraction. Calculate them using the qRomberg integrator for $-3.5 \leq w \leq 3.5$ with relative precision $\varepsilon = 10^{-6}$, and plot the curve $S(w)$ versus $C(w)$, which is known as the *Cornu spiral.*

Solution

The implementations are given in Listings 10.28 and 10.29 and the graphical output is shown in Figure 10.8.

Listing 10.28   Fresnel Integrals (Python Coding)

```python
# Fresnel integrals and Cornu spiral
from math import *
from integral import *
from graphlib import *

def CosF(u): return cos(0.5*pi*u*u)          # integrands of Fresnel integrals
def SinF(u): return sin(0.5*pi*u*u)

# main

eps = 1e-6                                   # relative integration precision
xmin = -3.5; xmax = 3.5                       # interval of upper limits
h = 0.05;                                          # plotting mesh spacing
n = int((xmax-xmin)/h) + 1                    # number of upper limits

x = [0]*(2*n+1); c = [0]*(2*n+1); s = [0]*(n+1)

for i in range(1,n+1):
   x[i] = xmin + (i-1)*h; x[i+n] = x[i]                      # upper limit
   c[i] = qRomberg(CosF,0e0,x[i],eps)               # Fresnel integrals
   s[i] = qRomberg(SinF,0e0,x[i],eps); c[i+n] = s[i]

GraphInit(1200,600)

nn = [0, n, 2*n]; col = ["", "red", "blue"]; sty = [0, 1,-1]
MultiPlot(x,c,c,nn,col,sty,2,10,
          0e0,0e0,0,0e0,0e0,0,0.10,0.45,0.15,0.85,
          "x","C, S","Fresnel integrals")

Plot(c,s,n,"green",1,0.60,0.95,0.15,0.85,"C(x)","S(x)","Cornu Spiral")

MainLoop()
```

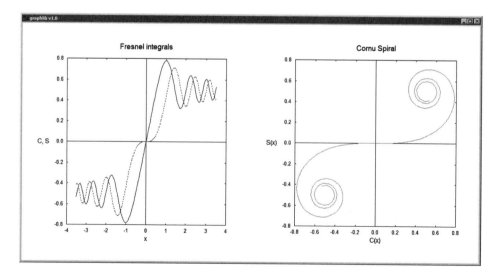

FIGURE 10.8   Fresnel integrals $C(w)$ and $S(w)$ (dashed) and Cornu spiral, as calculated by programs 10.28 and 10.29.

Listing 10.29 Fresnel Integrals (C/C++ Coding)

```
// Fresnel integrals and Cornu spiral
#include <math.h>
#include "memalloc.h"
#include "integral.h"
#include "graphlib.h"
#define pi 3.141592653589793

double CosF(double u) { return cos(0.5e0*pi*u*u); }      // integrands of
double SinF(double u) { return sin(0.5e0*pi*u*u); }      // Fresnel integrals

int main(int argc, wchar_t** argv)
{
    int i, n;
    double eps, h, xmin, xmax;
    double *c, *s, *x;
    int sty[3], nn[3];
    const char* col[3];

    eps = 1e-6;                                  // relative integration precision
    xmin = -3.5e0; xmax = 3.5e0;                      // interval of upper limits
    h = 0.05e0;                                       // plotting mesh spacing
    n = int((xmax-xmin)/h + 0.5) + 1;                 // number of upper limits

    x = Vector(1,2*n); c = Vector(1,2*n); s = Vector(1,n);

    for (i=1; i<=n; i++) {
        x[i] = xmin + (i-1)*h; x[i+n] = x[i];                    // upper limit
        c[i] = qRomberg(CosF,0e0,x[i],eps);                // Fresnel integrals
        s[i] = qRomberg(SinF,0e0,x[i],eps); c[i+n] = s[i];
    }

    PyGraph w(argc, argv);
    w.GraphInit(1200,600);

    nn[1] = n  ; col[1] = "red" ; sty[1] =  1;
    nn[2] = 2*n; col[2] = "blue"; sty[2] = -1;                   // dashed line
    w.MultiPlot(x,c,c,nn,col,sty,2,10,
                0e0,0e0,0,0e0,0e0,0,0.10,0.45,0.15,0.85,
                "x","C, S","Fresnel integrals");
    w.Plot(c,s,n,"green",1,0.60,0.95,0.15,0.85,"C(x)","S(x)","Cornu Spiral");

    w.MainLoop();
}
```

PROBLEM 10.4

The wave function of the electron in a 2s state ($n = 2$, $l = 0$, and $m = 0$) within the hydrogen atom can be defined in dimensionless radial units as (Fitts 2002):

$$\psi_{200}(r,\theta,\varphi) = \frac{1}{4\sqrt{2\pi}} (2 - r)\, e^{-r/2}.$$

The probability for the electron to be localized between the radial positions $r = 0$ and R is given by

$$\int_0^R \int_0^\pi \int_0^{2\pi} |\Psi_{210}(r,\theta,\varphi)|^2\, r^2 dr \sin\theta\, d\theta\, d\varphi = \frac{1}{8} \int_0^R r^2 (2-r)^2\, e^{-r}\, dr,$$

Listing 10.30   Radial Localization Probability of the Electron in the H Atom (Python Coding)

```
# Radial localization probability of an electron within the H atom
from math import *
from integral import *

def Func(r):                                    # probability density for 2s state
    f = r * (2-r)
    return f * f * exp(-r) / 8e0

# main

eps = 1e-8                                       # relative precision of integrals

print("Radial localization probability:")
for R in range(10,31):
    I = qRomberg(Func,0e0,R,eps)
    print("[0,{0:.1f}] = {1:10.8f}".format(R,I))

print("\nTotal localization probability:")
Rinf = 25e0                                      # guess for +infinity
(I,Rinf) = qImprop1(Func,0e0,Rinf,eps)
print("I Improp1  = {0:10.8f}   Rinf = {1:.1f}".format(I,Rinf))

n = 3                                            # number of radial integration nodes
I = qGaussLag(Func,0e0,n)
print("I GaussLag = {0:10.8f}".format(I))
```

to which the angular integrals contribute by a factor of 4π.

  a. Calculate the probabilities to find the electron between the radial positions $r = 0$ and $R \in [10, 25]$, with increments equal to 1, using Romberg's method (routine qRomberg) with a relative precision $\varepsilon = 10^{-8}$.
  b. Calculate the radial integral from $r = 0$ to $+\infty$ using the routines qImprop1 and, respectively, qGaussLag, and check that the total probability to find the electron anywhere in space is, indeed, equal to 1. Note the very small number of integration points necessary for the Gauss–Laguerre quadrature.

Solution

The Python implementation is given in Listing 10.30 and the C/C++ version is available as supplementary material (P10-ElecProbRad.cpp).

PROBLEM 10.5

The wave function of the electron in a $3d_0$ state ($n = 3, l = 2, m = 0$) within the hydrogen atom has the following form in dimensionless radial units (Fitts 2002):

$$\Psi_{210}(r, \theta, \varphi) = \frac{1}{81\sqrt{6\pi}} r^2 e^{-r/3}(3 \cos^2 \theta - 1).$$

Use functions qSimpsonSph and qGaussSph comparatively to verify that the norm of $\Psi_{210}(r, \theta, \varphi)$, as total localization probability of the electron, is, indeed, unitary:

$$\int_0^\infty \int_0^\pi \int_0^{2\pi} |\Psi_{210}(r, \theta, \varphi)|^2 r^2 dr \sin \theta \, d\theta \, d\varphi = 1.$$

Listing 10.31   Total Localization Probability of the Electron in the H Atom (Python Coding)

```python
# Total localization probability of an electron in the H atom
from math import *
from integral import *

def Func(r, theta, phi):                    # probability density for 3d_0 state
    c = 1e0/(81e0*sqrt(6e0*pi))
    cost = cos(theta)
    psi = c * r * r * exp(-r/3e0) * (3e0*cost*cost - 1e0)
    return psi * psi

# main

a = 40e0                                     # radial limit
nr = 100; nt = 70; np = 3
I = qSimpsonSph(Func,a,nr,nt,np)
print("I SimpsonSph = ",I)

nr = 10; nt = 10; np = 1
I = qGaussSph(Func,nr,nt,np)
print("I GaussSph   = ",I)
```

In conjunction with the routine qSimpsonSph, integrate radially up to $a = 40$ using $n_r = 100$, $n_\theta = 70$, and $n_\varphi = 3$ nodes. For the Gaussian quadratures, employ $n_r = 10$, $n_\theta = 10$, and $n_\varphi = 1$ integration points. Compare the number of correct significant digits in the results and determine appropriate numbers of mesh points for each of the quadrature schemes for correcting an additional significant digit. Justify that $n_\varphi = 1$ is a proper choice in the case of the Gaussian quadratures.

Solution
The Python program is shown in Listing 10.31 and the C/C++ version is available as supplementary material (P10-ElecProb.cpp).

PROBLEM 10.6
Evaluate the integral in cylindrical coordinates:

$$\int_0^\infty \exp(-r)r\,dr \int_0^{2\pi} (\cos\varphi)^2 \, d\varphi \int_0^1 z\,dz = \frac{1}{2}\pi,$$

using function qSimpsonCyl with $n_r = 200$, $n_\varphi = 20$, and $n_z = 2$ integration points. Identify the number of correct significant digits in the result and determine by trial and error the number n_r of radial mesh points reproducing eight significant digits of the exact result.

Solution
The Python implementation is given in Listing 10.32 and the C/C++ version is available as supplementary material (P10-IntegCyl.cpp).

PROBLEM 10.7
Elaborate a 3D integrator in Cartesian coordinates based on the trapezoidal rule, using as model the routine qSimpson3D. Employ the integrator in a program similar to the one in Listing 10.19 for integrating

$$f(x, y, z) = \begin{cases} exp(-r'/r) & r' \le r, \\ 0 & r' > r, \end{cases}$$

Listing 10.32   Integral in Cylindrical Coordinates (Python Coding)

```
# Evaluates a 3D integral using cylindrical coordinates
from math import *
from integral import *

def Func(r, phi, z):
    return exp(-r) * pow(cos(phi),2) * z

# main

a = 35e0                                                 # radial limit
az = 0e0; bz = 1e0                                       # axial limits
nr = 200; np = 20; nz = 2                          # numbers of mesh points

I = qSimpsonCyl(Func,a,az,bz,nr,np,nz)
print("I SimpsonCyl = ",I)
print("I exact       = ",pi/2)
```

in a toroidal domain with $R = 3$ and $r = 1$ (see Figure 10.6), where

$$R' = \sqrt{x^2 + y^2}, \quad r' = \sqrt{(R - R')^2 + z^2}.$$

Use zero padding and the condition

$$r' \leq r,$$

Listing 10.33   3D integrator Based on the Trapezoidal Rule (Python Coding)

```
#===========================================================================
def qTrapz3D(Func, ax, bx, nx, ay, by, ny, az, bz, nz):
#---------------------------------------------------------------------------
#  Integrates function Func(x,y,z) in the cuboid [ax,bx] x [ay,by] x [az,bz]
#  using the trapezoidal rule with (nx x ny x nz) integration points
#---------------------------------------------------------------------------
   hx = (bx-ax)/(nx-1)
   hy = (by-ay)/(ny-1)
   hz = (bz-az)/(nz-1)

   s = 0e0
   for i in range(0,nx):
      x = ax + i*hx; wx = (hx if i*(i+1-nx) else 0.5e0*hx)
      sx = 0e0
      for j in range(0,ny):
         y = ay + j*hy; wy = (hy if j*(j+1-ny) else 0.5e0*hy)
         sy = 0e0
         for k in range(0,nz):
            z = az + k*hz; wz = (hz if k*(k+1-nz) else 0.5e0*hz)
            sy += wz * Func(x,y,z)
         sx += wy * sy
      s += wx * sx

   return s
```

Listing 10.34 Integral over Toroidal Domain Using Cartesian Coordinates (Python Coding)

```python
# Evaluates a 3D integral on a torus using Cartesian coordinates
from math import *
from integral import *

def Func(x, y, z):
    global R, r
    Rp = sqrt(x*x + y*y)                                  # major radial distance
    dR = R - Rp
    rp = sqrt(dR*dR + z*z)                                # minor radial distance

    return (exp(-rp/r) - exp(-1e0) if rp <= r else 0e0)      # zero-padding

# main

R = 3e0; r = 1e0                              # major & minor torus radii
ax = ay = az = 0e0                            # bounding box: 1st octant
bx = by = R+r; bz = r                         # multiply results with 8
nx = ny = 200                                 # equal density of mesh
nz = int(nx * r/(R+r))                        # points along x, y, z

I = 8e0 * qTrapz3D(Func,ax,bx,nx,ay,by,ny,az,bz,nz)
print("I Trapezes = ",I)
I = 8e0 * qSimpson3D(Func,ax,bx,nx,ay,by,ny,az,bz,nz)
print("I Simpson  = ",I)
I = 8e0 * qGaussLeg3D(Func,ax,bx,nx,ay,by,ny,az,bz,nz)
print("I GaussLeg = ",I)                        # result = 9.510516...
```

to distinguish between interior and exterior points. Having in view the exact value of the integral $(4\pi^2 Rr^2 (1 - \frac{5}{2}e^{-1}) \approx 9.5105163)$ and using the same number of grid points ($200 \times 200 \times 50$), compare the obtained result in terms of exact significant digits with those provided by functions qSimpson3D and qGaussLeg3D.

Solution

The Python integrator is to be found in Listing 10.33, being part of the module integral.py. Its C/C++ counterpart is included in the header file integral.h.

The Python program performing the required integration is given in Listing 10.34 and the C/C++ version is available as supplementary material (P10-Integ3D1.py).

PROBLEM 10.8

Write Python and C/C++ codes based on the adaptive 3D technique presented in Section 10.12 to integrate over a toroidal domain of radii $R = 3$ and $r = 1$ (see Figure 10.6) the integrand

$$f(x, y, z) = \begin{cases} (1 - r'/r)^2 & r' \leq r, \\ 0 & r' > r, \end{cases}$$

depending explicitly only on the distance r' to the tube axis. Impose the relative precision $\varepsilon = 10^{-6}$.

Solution

The implementations are given in Listings 10.35 and 10.36.

Listing 10.35   Adaptive 3D Integration on Toroidal Domain (Python Coding)

```python
# 3D adaptive quadrature over 1st octant of a torus
from math import *
from integral import *

#----------------------------------------------------------------- integrand
def Func(x, y, z):
   global R, r
   Rp = sqrt(x*x + y*y)                            # major radial distance
   dR = R - Rp
   rp = sqrt(dR*dR + z*z)                          # minor radial distance
   dr = 1e0 - rp/r
   return (dr*dr if rp <= r else 0e0)                  # zero-padding

#-------------------------------------------------------- integration limits
def ay(x):
   return (sqrt((R-r)*(R-r) - x*x) if x <= R-r else 0.0)
def by(x):
   return sqrt((R+r)*(R+r) - x*x)
def az(x, y):
   return 0.0
def bz(x, y):
   r0 = R - sqrt(x*x + y*y)
   return sqrt(fabs(r*r - r0*r0))

#---------------------------------------------------------------- interfaces
def Fz(z):                                          # z-integrand
   global xvar, yvar
   return Func(xvar,yvar,z)                    # (x,y) as global (xvar,yvar)
def Fy(y):                                          # y-integrand
   global xvar, yvar
   yvar = y                                         # stores y in yvar
   return qRomberg(Fz,az(xvar,y),bz(xvar,y),eps)    # x as global xvar
def Fx(x):                                          # x-integrand
   global xvar
   xvar = x                                         # stores x in xvar
   return qRomberg(Fy,ay(x),by(x),eps)

# main

R = 3e0; r = 1e0
ax = 0e0; bx = R + r; eps = 1e-6
I = 8e0 * qRomberg(Fx,ax,bx,eps)                     # outer integral
print("Integral = ",I)                              # result 9.869605...
```

Listing 10.36   Adaptive 3D Integration on Toroidal Domain (C/C++ Coding)

```c
// 3D adaptive quadrature over 1st octant of a torus
#include <stdio.h>
#include <math.h>
#include "integral.h"

//-------------------------------------------------------------------- globals
double eps,  r, R, xvar, yvar;

//------------------------------------------------------------------ integrand
```

```
double Func(double x, double y, double z)
{
    double dR, dr, Rp, rp;
    Rp = sqrt(x*x + y*y);                          // major radial distance
    dR = R - Rp;
    rp = sqrt(dR*dR + z*z);                        // minor radial distance
    dr = 1e0 - rp/r;
    return (rp <= r ? dr*dr : 0e0);                     // zero-padding
}

//----------------------------------------------------------- integration limits
double ay(double x)
    { return (x <= R-r ? sqrt((R-r)*(R-r) - x*x) : 0.0); }
double by(double x)
    { return sqrt((R+r)*(R+r) - x*x); }
double az(double x, double y)
    { return 0.0; }
double bz(double x, double y)
    { double r0 = R - sqrt(x*x + y*y); return sqrt(fabs(r*r - r0*r0)); }

//--------------------------------------------------------------------- interfaces
double Fz(double z)                                     // z-integrand
{
    return Func(xvar,yvar,z);                      // (x,y) as global (xvar,yvar)
}
double Fy(double y)                                     // y-integrand
{
    yvar = y;                                      // stores y in yvar
    return qRomberg(Fz,az(xvar,y),bz(xvar,y),eps);      // x as global xvar
}
double Fx(double x)                                     // x-integrand
{
    xvar = x;                                      // stores x in xvar
    return qRomberg(Fy,ay(x),by(x),eps);
}

//-----------------------------------------------------------------------------
int main()
{
    double ax, bx, I;

    R = 3e0; r = 1e0;
    ax = 0e0; bx = R + r; eps = 1e-6;
    I = 8e0 * qRomberg(Fx,ax,bx,eps);                   // outer integral
    printf("Integral = %f\n",I);                        // result 9.869605...
}
```

References and Suggested Further Reading

Abramowitz, M. and I. A. Stegun (Eds.). 1972. *Handbook of Mathematical Functions: With Formulas, Graphs, and Mathematical Tables.* New York: Dover Publications.

Beu, T. A. 2004. *Numerical Calculus in C* (3rd ed., in Romanian). Cluj-Napoca: MicroInformatica.

Boole, G. and J. F. Moulton. 2003. *A Treatise on the Calculus of Finite Differences* (2nd ed.). New York: Dover Publications.

Burden, R. and J. Faires. 2010. *Numerical Analysis* (9th ed.). Boston: Brooks/Cole, Cengage Learning.

Demidovich, B. and I. Maron. 1987. *Computational Mathematics* (4th ed.). Moscow: MIR Publishers.

Fitts, D. 2002. *Principles of Quantum Mechanics: As Applied to Chemistry and Chemical Physics.* Cambridge: Cambridge University Press.

Guenther, R. D. 1990. *Modern Optics.* New York: John Wiley & Sons.

Press, W. H., S. A. Teukolsky, W. T. Vetterling, and B. P. Flannery. 2007. *Numerical Recipes: The Art of Scientific Computing* (3rd ed.). Cambridge: Cambridge University Press.

Ralston, A. and P. Rabinowitz. 2001. *A First Course in Numerical Analysis* (2nd ed.). New York: Dover Publications.

Romberg, W. 1955. Vereinfachte numerische Integration (Simplified Numerical Integration (in German)). *Det Kongelige Norske Videnskabers Selskab Forhandlinger (Trondheim) 28*, 30–36.

Stroud, A. 1971. *Approximate Calculation of Multiple Integrals.* Englewood Cliffs: Prentice-Hall.

Stroud, A. and D. Secrest. 1966. *Gaussian Quadrature Formulas.* Englewood Cliffs: Prentice-Hall.

11

Monte Carlo Method

11.1 Introduction

The classical way to solve a problem of numerical analysis relies on a rigorous algorithm, which, for a given input, provides a well-defined solution in a predetermined number of steps. Such an approach is essentially *deterministic*, in the sense that its implementation is expected to lead on repeated runs, upon completion of the same number of operations, to the same result. Still, for numerous topical problems in applied sciences and engineering, the complexity of deterministic algorithms renders the computations intractable within reasonable amounts of time.

Certain areas of modern sciences are concerned with systems composed of a huge number of coupled components, which are sometimes also prone to fluctuations. Such complex systems can be as different as collections of spins, biological populations, or galaxies. Their characterization is often accomplished by means of high-dimensional integrals. A typical example is the *classical canonical partition function* of a system of N interacting particles,

$$ Z = \int \cdots \int d^3 r_1 \cdots d^3 r_N \, \exp\left[-\frac{1}{k_B T} E(\mathbf{r}_1 \cdots \mathbf{r}_N) \right], $$

where $E(\mathbf{r}_1 \cdots \mathbf{r}_N)$ is the total energy of the system, with $\mathbf{r}_i$, the position of particle i, T is the system temperature, and k_B, the Boltzmann constant. The evaluation of this $3N$-dimensional integral by any of the classical quadrature methods can be completely ruled out even for the lowest particle numbers of practical interest. In fact, considering the modest value $N = 20$ and only 10 integration points for each dimension, the number of required elementary operations would be on the order of 10^{60}. Even using the latest petascale supercomputers, which are capable of more than 10^{16} floating point operations per second, the computation would require approximately $3 \cdot 10^{36}$ years!

A rewarding alternative to the deterministic approaches for complex high-dimensional problems are the so-called *stochastic methods*, based on the *law of large numbers* from the probability theory. Here, the quantities of interest are defined as expectation values of random variables or, in other words, the average values of large sequences of random variables are considered under certain assumptions probabilistic estimates of the sought-for quantities. Such techniques, generically referred to as *Monte Carlo methods*, have an intrinsically *nondeterministic* character, since they use the outcome of stochastic experiments and, within statistical errors, they exhibit different behaviors on different runs. Essentially, instead of deterministically covering the domains of the involved functions, the Monte Carlo methods sample these randomly. In general, stochastic techniques do not require *genuine* random numbers, but rather *pseudorandom* sequences, nevertheless with a high degree of uniformity and low sequential correlations.

The seminal Monte Carlo method was developed in the late 1940s at the Los Alamos National Laboratory by Stanislaw Ulam, Nicholas Metropolis, and John von Neumann, and it was named, due to the gambling-like underlying principles, after the famous Monte Carlo casino. The Monte Carlo method started unveiling its virtues with the advent of the high-performance computers, because the attainment of sufficiently accurate results generally implies a tremendous number of operations and conveying vast amounts of data. By comparison with the deterministic numerical methods, the success of the Monte Carlo method is mainly due to the more advantageous scaling of the computational effort with increasing problem size. In fact, while being the most inefficient quadrature method for one-dimensional integrals, with increasing dimensionality, the Monte Carlo method progressively exceeds the deterministic methods.

The main uses of the Monte Carlo method can be broadly categorized into optimization, numerical integration, and generation of samples from probability distributions. In applied mathematics, the Monte Carlo method reveals its efficiency most clearly in applications such as the evaluation of multidimensional integrals with complex boundaries, the solution of large linear systems, or the solution of Dirichlet problems for differential equations. In this chapter, we specifically focus on the integration of functions.

11.2 Integration of Functions

In the Monte Carlo method, the integral of a function f over a multidimensional domain $\mathcal{D}$ is simply estimated by the product of the arithmetic mean of the function and the domain volume V,

$$\int_{\mathcal{D}} f dV \approx V \langle f \rangle \pm \sigma, \tag{11.1}$$

where the average $\langle f \rangle$ is calculated for n random points, $\mathbf{x}_1, ..., \mathbf{x}_n$, sampling *randomly* and *uniformly* the domain $\mathcal{D}$. The statistical uncertainty associated with the result of the integration is specified by the *standard deviation* σ,

$$\sigma = V \sqrt{\frac{\langle f^2 \rangle - \langle f \rangle^2}{n}} = \frac{V}{\sqrt{n}} \sigma_f, \tag{11.2}$$

or its square, σ^2, called *variance*. The standard deviation,

$$\sigma_f = \sqrt{\langle (f - \langle f \rangle)^2 \rangle} = \sqrt{\langle f^2 \rangle - \langle f \rangle^2},$$

occurring in the above expression, measures the deviation of the integrand f from its mean $\langle f \rangle$ within the integration domain. The averages $\langle f \rangle$ and $\langle f^2 \rangle$ are given, respectively, by

$$\langle f \rangle \equiv \frac{1}{n} \sum_{i=1}^{n} f(\mathbf{x}_i), \quad \langle f^2 \rangle \equiv \frac{1}{n} \sum_{i=1}^{n} f^2(\mathbf{x}_i). \tag{11.3}$$

Expression 11.2 of the standard deviation σ illustrates two important features of the Monte Carlo integration. The first refers to the $n^{-1/2}$ scaling of the uncertainty σ. As expected, the larger the number of sampling points, the more precise the result of the integration, even if the decrease of σ is rather slow. By comparison, the error of the basic trapezoidal formula, the most rudimentary of the Newton–Cotes quadrature family, features an n^{-2} dependence and the required numerical effort for achieving the same precision in a *one-dimensional* integration proves to be definitely lower than for the Monte Carlo method. However, the advantage of the more favorable scaling of the operation count for the classical quadrature formulas vanishes in multidimensional cases. It can be easily seen that the remainder of a d-dimensional integration scheme based on a conventional one-dimensional integrator, characterized by n function evaluations for each dimension, increases as $n^{-2/d}$. On the contrary, the $n^{-1/2}$ dependence of the uncertainty σ is conserved irrespective of the dimensionality, so that, for $d > 4$, the Monte Carlo method becomes more efficient.

The second defining aspect of the Monte Carlo integration concerns the fact that, since the uncertainty σ is proportional to the standard deviation σ_f of the integrand, it is smaller, the "flatter" the function f, that is, the less f departs from its average value. In the particular case of a constant function, its evaluation at a single sampling point is sufficient for the variance to vanish and the average to be precisely defined. On the contrary, assuming that f has significant values only in a confined region, while the integration points $\mathbf{x}_i$ are generated with *equal probability* within the entire domain $\mathcal{D}$, it is likely that their majority samples regions with insignificant contributions, resulting in an inaccurate estimate of the integral.

Even though the real virtues of the Monte Carlo method emerge in the evaluation of multidimensional integrals, for the sake of clarity, we confine the discussion for the moment to the one-dimensional case. The generic integral,

$$I = \int_a^b f(X)dX,$$

(11.4)

may be represented relative to the standard integration domain $[0, 1]$ by the change of variables $X = (b - a)x + a$, namely:

$$I = (b - a)\int_0^1 f((b - a)x + a)dx.$$

(11.5)

The general Monte Carlo formula 11.1 becomes in this particular case:

$$I \approx (b - a)\langle f \rangle = \frac{b - a}{n}\sum_{i=1}^n f((b - a)x_i + a).$$

(11.6)

The average $\langle f \rangle$ is expressed here by sampling the integrand at the points $X_i = (b - a)x_i + a$ from the interval $[a, b]$, corresponding to n random points x_i distributed uniformly (with equal probability) within the standard interval $[0, 1)$.

We defer the task of generating random variables with different distributions for Section 11.5. Since computers are by essence deterministic machines, by "random" we will rather understand "pseudo-random," as these numbers will result from deterministic recurrences producing bit-wise truncations. For the time being, we just briefly mention language built-in utility functions for creating uniform distributions of random numbers.

Python provides as part of the standard module `random.py` a wealth of functions related to random sequences. The function `seed()` called without arguments uses the current system time to initialize the random number generator. The call to `random()` returns the next random floating point number of a sequence in the range $[0, 1)$.

The C/C++ built-in random number generator can be initialized by invoking the function `srand`, which generates a "seed" for a subsequent random sequence depending on the received argument. A simple initialization method is to call `srand` passing the current Central Processing Unit (CPU) time as returned by the function `time` (defined in the standard header file `time.h`), like in the following macro:

```
// Initializes the random number generator using the current system time
#define seed() srand((unsigned)time(NULL))
```

In particular, `time(NULL)` returns the number of seconds elapsed since 00:00 hours, January 1, 1970.

The standard C/C++ function `rand()` (defined in the header file `stdlib.h`) generates on repeated calls sequences of integer pseudorandom numbers uniformly distributed in the interval `[0, RAND_MAX]`. The predefined constant `RAND_MAX` is library dependent, but is guaranteed to be at least 32,767 in any standard implementation of the C/C++ compiler. To generate real random numbers in the standard interval [0, 1), the values returned by rand() have to be simply divided by (RAND_MAX+1):

```
// Generates random floating point numbers in the range [0,1)
#define random() ((double)rand()/(RAND_MAX+1))
```

The macros seed() and random() will be assumed in what follows to be included in the header file random.h.

As a first, elementary example of Monte Carlo quadrature, we consider the integral

$$\int_0^1 xe^{-x}dx = 1 - 2e^{-1} \approx 0.26424. \tag{11.7}$$

The corresponding program (Listing 11.1) does not actually implement the Monte Carlo methodology as a distinct procedure, since a suchlike routine lacks practical interest, as the whole idea of using the Monte Carlo method for performing one-dimensional integration. Instead, the main program is kept simple, with a view to emphasizing the general concept. Here, n represents the number of sampling points, s is the integral estimate, and sigma represents the associated standard deviation. The variables f1 and f2 are used to store, respectively, the average $\langle f \rangle$ and the average squared function $\langle f^2 \rangle$.

The results of the program execution for different numbers of integration points n are compiled in the first three columns of Table 11.1. Along with the increase of n, the estimates of the integral converge toward the exact value, with obviously decreasing associated uncertainty estimates σ. As will be shown in the next section, the standard deviations σ scale indeed as $n^{-1/2}$ and a comparison with any

Listing 11.1   Monte Carlo Integration of a Function (Python Coding)

```
# One-dimensional Monte Carlo quadrature
from math import *
from random import *

def func(x): return x * exp(-x)                                # integrand

# main

n = eval(input("n = "))                          # number of sampling points

seed()

f1 = f2 = 0e0                                # quadrature with uniform sampling
for i in range(1,n+1):
   x = random()                              # RNG with uniform distribution
   f = func(x)                                              # integrand
   f1 += f; f2 += f * f                                        # sums

f1 /= n; f2 /= n                                             # averages
s = f1                                                       # integral
sigma = sqrt((f2-f1*f1)/n)                            # standard deviation
print("s = ",s," +/- ",sigma)
```

TABLE 11.1 Monte Carlo Estimates of the Integral $I = \int_0^1 xe^{-x}dx$ and Associated Standard Deviations σ, Calculated Using Random Sequences with the Distributions $w(x) = 1$ and, Respectively, $w(x) = (3/2)x^{1/2}$

	$w(x) = 1$		$w(x) = (3/2)x^{1/2}$	
n	I	σ	I	σ
10	0.25108	0.03855	0.26860	0.00727
100	0.26252	0.01085	0.26397	0.00285
1000	0.26622	0.00336	0.26458	0.00089
10,000	0.26544	0.00105	0.26423	0.00027

classical one-dimensional quadrature formula proves the latter to be significantly more efficient. While the Monte Carlo integration can be yet improved, its full potential comes to light only in the evaluation of multidimensional integrals.

11.3 Importance Sampling

Having in view that the uncertainty σ (11.2) of the Monte Carlo quadrature is directly related to the variance of the integrand σ_f, which measures the deviation of the integrand from its average value, to increase the precision and efficiency of the quadrature, one can apply a general strategy, known as *variance reduction*. More precisely, we discuss a technique called *importance sampling*.

Essentially, one introduces a positive *weight function*, $w(x)$, normalized to unity over the interval $[0, 1]$,

$$\int_0^1 w(x)dx = 1, \tag{11.8}$$

by which integrand (11.5) is multiplied and divided:

$$I = (b - a) \int_0^1 \frac{f((b - a)x + a)}{w(x)} w(x)dx.$$

Performing the change of variable $d\xi = w(x)dx$ or, in integral form,

$$\xi(x) = \int_0^x w(x')dx', \tag{11.9}$$

with the boundary conditions $\xi(0) = 0$ and $\xi(1) = 1$ (to comply with normalization (11.8)), the integral becomes

$$I = (b - a) \int_0^1 \frac{f((b - a)x(\xi) + a)}{w(x(\xi))} d\xi. \tag{11.10}$$

The conservation of the integration domain upon the change of variable is a direct consequence of the normalization of $w(x)$. To evaluate this integral, one can directly apply the Monte Carlo method described in the preceding section, by averaging the new integrand, f/w, over a set of *uniformly distributed* sampling points ξ_i from the interval $[0, 1)$:

$$I \approx \frac{b - a}{n} \sum_{i=1}^n \frac{f((b - a)x(\xi_i) + a)}{w(x(\xi_i))}. \tag{11.11}$$

The performed change of variable may seem at first glance to turn the quadrature more complicated. The advantages of this technique become apparent only if one chooses the weight function $w(x)$, in as much as possible, *similar* to the function $f((b - a)x + a)$. In such a case, the integrand f/w becomes *smooth and slowly varying* (in the ideal case $w \equiv f$, even constant). As a consequence of the reduced fluctuations of the integrand with respect to its average, there results a variance reduction for the integral itself. Naturally, this benefit is directly contingent upon choosing an adequate weight function $w(x)$ and on the possibility of expressing the dependence $x = x(\xi)$ by inverting the integral relation 11.9.

A deeper insight into the usefulness of the variable change (11.9) is gained by noting that the uniform distribution of the random points ξ_i corresponds to a, generally, *nonuniform distribution* of the arguments $x_i \equiv x(\xi_i)$, namely, according to the weight function $w(x)$. In other words, the probability of a transformed abscissa x_i to be located in the interval $(x, x + dx)$ is $w(x)dx$. This means that the arguments x_i are generated with maximum probability and are, therefore, concentrated in regions where w is large. For an adequate choice of w, the sampling points are, thus, predominantly concentrated in regions with significant values of the original integrand f. One achieves in such a manner an optimal sampling of the integration domain (importance sampling), diminishing the effort "wasted" on evaluating the integrand at points with reduced statistical significance in sum (11.11).

To illustrate the operation of the variance reduction by importance sampling, we consider again the integral

$$\int_0^1 xe^{-x}dx \approx 0.26424. \tag{11.12}$$

The integrand monotonically rises in the domain $[0, 1]$ from 0 to e^{-1}, while maintaining the negative second derivative. A similar behavior is featured by the function $e^{-1}x^{1/2}$, for which, most importantly, the integral transformation (11.9) is invertible. The adequately normalized weight function is in this case

$$w(x) = (3/2)x^{1/2}. \tag{11.13}$$

By inverting the change of variable (11.9), one obtains

$$x = \xi^{2/3}. \tag{11.14}$$

Listing 11.2 shows the implementation of the Monte Carlo quadrature 11.12 with importance sampling based on Formula 11.11. The functions `seed()` and `random()` (defined in the modules `random.py` and `random.h`) are called to initialize and, respectively, generate real uniform random sequences. For variables with the distribution $w(x) = (3/2)x^{1/2}$, the program defines the routine `ranSqrt()`, which calls, in its turn, the routine `random()` for producing the underlying random numbers ξ with uniform distribution.

The results of the program execution are listed in the last two columns of Table 11.1 and evidence a clear-cut improvement over those not relying on variance reduction (for $w(x) = 1$). In fact, for the same number of sampling points, the standard deviation decreases typically more than 4 times.

The improvement brought about by the importance sampling may also be judged from the plots depicted in Figure 11.1. The left panel plots the estimates of the integral for numbers of sampling points between $n = 100$ and 30,000, and one can notice right away the significantly lower spread of the values obtained with importance sampling, even though both approaches converge to the same result. This behavior is also reflected by the standard deviations σ depicted in the right panel. While the scaling of σ (obtained by regression) is, for both approaches, the one predicted theoretically, namely $\sigma \sim n^{-1/2}$, the uncertainties resulting from the importance sampling can be seen to be shifted to lower values by a factor of approximately 3.7.

The variance reduction is particularly effective when the integrand is itself a product of two functions, $F(x) = w(x)f(x)$. In such a case, if the factor $w(x)$ can be integrated analytically according to Equation 11.9 to yield the dependence $\xi = \xi(x)$, and the latter may be furthermore inverted, then the factor $w(x)$ can be treated as a distribution function. On the basis of a sequence of uniformly distributed random

Listing 11.2 Monte Carlo integration of a Function with Importance Sampling (Python Coding)

```python
# One-dimensional Monte Carlo quadrature with variance reduction
from math import *
from random import *

def ranSqrt():
#--------------------------------------------------------------------
#  Returns a random number x in the range [0,1) with the distribution
#  w(x) = 3/2 x^(1/2), and the corresponding value w(x)
#--------------------------------------------------------------------
    x = pow(random(),2e0/3e0)
    w = 1.5e0 * sqrt(x)
    return (x, w)

def func(x): return x * exp(-x)                       # integrand

# main

n = eval(input("n = "))                      # number of sampling points

seed()

f1 = f2 = 0e0                           # quadrature with importance sampling
for i in range(1,n+1):
    (x, w) = ranSqrt()                       # RNG with distribution w(x)
    if (w):
        f = func(x) / w                              # integrand
        f1 += f; f2 += f * f                             # sums

f1 /= n; f2 /= n                                     # averages
s = f1                                              # integral
sigma = sqrt((f2-f1*f1)/n)                  # standard deviation
print("s = ",s," +/- ",sigma)
```

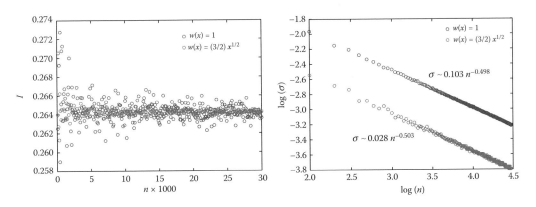

FIGURE 11.1 Monte Carlo estimates of the integral $\int_0^1 xe^{-x}dx$ (left panel) and associated standard deviations σ (right panel) using uniform sampling and, respectively, importance sampling with the distribution $w(x) = (3/2)x^{1/2}$, for numbers of integration points ranging between $n = 100$ and 30,000.

numbers ξ, the inverted relation $x = x(\xi)$ provides a sequence of random variables with the distribution $w(x)$, which is necessary for carrying out the Monte Carlo integration based on Equation 11.11.

In spite of the significant improvement achieved by using adequate weight functions, the Monte Carlo method still remains basically inefficient for one-dimensional integration, even when compared with the classical trapezoidal formula. While the accuracy of the Monte Carlo quadrature barely reaches 10^{-4} for $n = 5000$ sampling points, the trapezoidal formula performs by one order of magnitude better (10^{-5}). Nonetheless, the relative performance of the Monte Carlo integration as compared to the deterministic quadrature schemes qualitatively changes in the case of the multidimensional integration.

11.4 Multidimensional Integrals

As already mentioned, the real virtues of the Monte Carlo integration come plenary to light in multi-dimensional cases. Even though the description of the integration boundary remains one of the difficult tasks even for the Monte Carlo approach, the fact that the integration points are not generated as separate variables along each coordinate direction, but as tuples within the entire integration domain in a single cycle, renders possible the modeling of more complex frontiers than for deterministic quadratures.

Let us tackle the integration of a function f over a complex domain $\mathcal{D}$ and let us assume that, due to the complexity of the frontier, both expressing the volume V and its exclusive sampling by random points are difficult. To make the Monte Carlo quadrature formula (11.1) applicable, it is necessary to define an easy-to-sample *extended domain* $\widetilde{\mathcal{D}}$, that tightly encloses the original integration domain $\mathcal{D}$, and whose volume $\tilde{V}$ is easily expressible (usually, a hyperparallelepiped). By defining the new integrand

$$\tilde{f}(\mathbf{x}) = \begin{cases} f(\mathbf{x}) & \text{if } \mathbf{x} \in \mathcal{D}, \\ 0 & \text{if } \mathbf{x} \notin \mathcal{D}, \end{cases} \tag{11.15}$$

the quadrature formula becomes

$$\int_{\mathcal{D}} f dV \approx \tilde{V} \langle \tilde{f} \rangle \pm \tilde{\sigma}, \tag{11.16}$$

where the average $\langle \tilde{f} \rangle$ and the variance $\tilde{\sigma}$ refer this time to the extended volume.

The extended domain $\widetilde{\mathcal{D}}$ is desirable to be as similar as possible with the original domain $\mathcal{D}$ to reduce the relative contribution of the sampling points falling outside $\mathcal{D}$. Corresponding to vanishing integrand values, these points imply trivial information and indirectly deteriorate the standard deviation $\tilde{\sigma}$ by the entailed decrease of the number of effective points.

The Monte Carlo integration is not suitable for "boxed," general-purpose implementations operating with different user-supplied integrands, as in the case of classical quadratures (see Chapter 10). Even though the programs have, in principle, the same general structure, the code is susceptible to additional optimizations by taking advantage of the particular integrand features. To demonstrate the basic ideas, we discuss in what follows a couple of 2D integrals and a 3D integral.

First, we consider evaluating the area of the unit circle from the integral:

$$I = \iint_{x^2+y^2 \leq 1} dx \, dy = 4 \int_0^1 dx \int_0^{\sqrt{1-x^2}} dy = \pi. \tag{11.17}$$

Taking advantage of the problem's symmetry, the integration domain $\mathcal{D}$ is reduced to the circular sector in the first quadrant. With a view to convenient sampling, we define as extended domain the unit square $\widetilde{\mathcal{D}} = [0, 1) \times [0, 1)$ and we rewrite the integral as

$$I = 4 \int_0^1 dx \int_0^1 dy \, H[1 - (x^2 + y^2)], \tag{11.18}$$

where the points external to the domain $\mathcal{D}$ can be discriminated by using the Heaviside function

$$H(x) = \begin{cases} 0 & \text{if } x < 0, \\ 1 & \text{if } x \geq 0. \end{cases} \tag{11.19}$$

The corresponding Monte Carlo quadrature formula is thus

$$I \approx \frac{4}{n} \sum_{i=1}^{n} H[1 - (x_i^2 + y_i^2)] = 4\frac{n_i}{n}, \tag{11.20}$$

and it implies n random sampling points (x_i, y_i), uniformly distributed within the square $\widetilde{\mathcal{D}}$. In this case, the sum has obviously the significance of the number n_i of points interior to the circular sector $\mathcal{D}$ and its area is, as can be seen, proportional to the ratio between n_i and the total number of sample points n. In other words, the area of the circular sector is proportional with the probability of generating interior sampling points. A simple calculation shows that the mean integrand and the mean squared integrand are equal in this case, that is, $\langle H \rangle = \langle H^2 \rangle = n_i/n$.

Program 11.3 precisely implements the discussed ideas. The estimates for the unit circle area are listed in Table 11.2 for several numbers of sampling points. For $n = 50,000$, the expected π value is reproduced with four exact decimal digits, even though the associated standard error σ suggests a poorer estimate.

Pursuing the idea that for a uniform sampling of the extended domain $\widetilde{\mathcal{D}}$, the ratio of the volumes V and $\widetilde{V}$ is approximated by the ratio n_i/n of the corresponding numbers of sampling points; the above

Listing 11.3   Monte Carlo Evaluation of the Area of the Unit Circle (Python Coding)

```python
# Monte Carlo calculation of the unit circle area
from math import *
from random import *

# main

n = eval(input("n = "))                           # number of sampling points

seed()

ni = 0                                            # number of interior points
for i in range(1,n+1):
    x = random(); y = random()
    if (x*x + y*y <= 1e0): ni += 1                # add interior point

fi = ni/n                                         # fraction of interior points
s = 4e0 * fi                                             # integral
sigma = 4e0 * sqrt((fi - fi*fi)/n)                 # standard deviation
print("Unit circle area = ",s," +/- ",sigma)
```

TABLE 11.2   Monte Carlo Estimates of the Integral $I = \int \int_{x^2+y^2 \leq 1} dx\, dy$ and Associated Standard Deviations σ

n	I	σ	n	I	σ
100	3.20000	0.16000	5,000	3.14400	0.02320
500	3.22400	0.07074	10,000	3.15520	0.01633
1000	3.12800	0.05223	50,000	3.14128	0.00734

technique can be generalized for the evaluation of complicated multidimensional volumes based on the relation:

$$V = \lim_{n \to \infty} \frac{n_i}{n} \tilde{V}.$$

Technically, the calculation amounts to register the points interior to the original integration domain along with the total number of points generated within the entire extended domain.

The second 2D example considers the integral:

$$\frac{1}{4\pi} \int_{-\infty}^{\infty} \int_{-\infty}^{\infty} \left(x^2 + y^2\right) e^{-(x^2+y^2)/2} dy\, dx = 1. \tag{11.21}$$

Certainly, it can be evaluated using the general Monte Carlo methodology with uniform sampling. However, one may take advantage of the Gaussian factor and perform a more meaningful sampling using integration points generated by the Gaussian distribution:

$$w(x, y) = \frac{1}{2\pi} e^{-(x^2+y^2)/2}. \tag{11.22}$$

Since this probability distribution is normalized to unity, $\int_{-\infty}^{\infty} \int_{-\infty}^{\infty} w(x,y) dy\, dx = 1$, the Monte Carlo quadrature amounts to calculating the average

$$\langle \tilde{f} \rangle = \frac{1}{n} \sum_{i=1}^{n} \frac{\tilde{f}(x_i, y_i)}{w(x_i, y_i)} = \frac{1}{2n} \sum_{i=1}^{n} \left(x_i^2 + y_i^2\right), \tag{11.23}$$

for arguments sampled according to the normal probability distribution $w(x,y)$. Techniques for generating random numbers with such a distribution are presented in Section 11.5. One of the developed routines is `randNrm2`, which actually returns pairs of random variables x and y, along with the corresponding value w of the distribution function.

Program 11.4 comparatively uses uniform and Gaussian sampling. Results illustrating the variance reduction for the integral in question are depicted in Figure 11.2. The integral estimates obtained with importance sampling are seen in the left panel to feature a significantly lower spread than for uniform

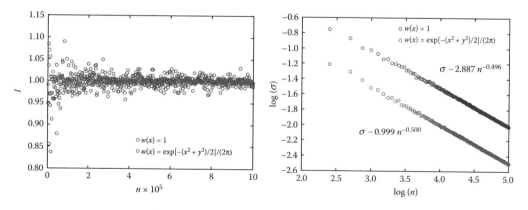

FIGURE 11.2 Monte Carlo estimates of the integral $(1/4\pi) \int_{-\infty}^{\infty} \int_{-\infty}^{\infty} (x^2 + y^2) e^{-(x^2+y^2)/2} dy\, dx = 1$ (left panel) and associated standard deviations σ (right panel) using uniform and, respectively, Gaussian sampling, for numbers of integration points ranging between $n = 250$ and $100,000$.

Listing 11.4   2D Monte Carlo Integration with Importance Sampling (Python Coding)

```
# Two-dimensional Monte Carlo quadrature with variance reduction
from math import *
from random import *
from random1 import *

def func(x,y):                                              # integrand
    return (x*x + y*y)*exp(-0.5e0*(x*x + y*y))/(4e0*pi)

# main

L = 8e0                                    # integration domain [-L,L] x [-L,L]
L2 = L * L                                           # area of sampling domain

n = 100000                                          # number of sampling points

seed()

f1 = f2 = 0e0                               # quadrature with uniform sampling
for i in range(1,n+1):
    x = L * random(); y = L * random()
    f = func(x,y)                                             # integrand
    f1 += f; f2 += f * f                                      # sums

f1 /= n; f2 /= n                                             # averages
s = 4e0 * L2 * f1                                            # integral
sigma = 4e0 * L2 * sqrt((f2-f1*f1)/n)                # standard deviation
print("Uniform sampling : s = ",s," +/- ",sigma)

f1 = f2 = 0e0                            # quadrature with importance sampling
for i in range(1,n+1):
    (w, x, y) = randNrm2()          # random numbers with normal distribution
    f = func(x,y) / w                                        # integrand
    f1 += f; f2 += f * f                                     # sums

f1 /= n; f2 /= n                                             # averages
s = f1                                                      # integral
sigma = sqrt((f2-f1*f1)/n)                          # standard deviation
print("Gaussian sampling: s = ",s," +/- ",sigma)
```

sampling. A further confirmation of the better performance of the Gaussian sampling is provided by the linear fits of the log–log plots of the standard deviations as functions of the number of sampling points, which indicate a reduction of the uncertainties roughly by a factor of 2.9 when using normally instead of uniformly distributed random numbers.

Let us consider as an application of the Monte Carlo methodology to 3D quadratures the evaluation of the mass and mass center position of a *torus* of major radius $R = 3$ and minor radius $r = 1$ (see Figure 11.3). The torus is centered at the origin and is symmetric about the z-axis. A similar problem was treated in Chapter 10 by using deterministic quadratures. The equation of the toroidal surface is

$$\left(R - \sqrt{x^2 + y^2}\right)^2 + z^2 = r^2. \tag{11.24}$$

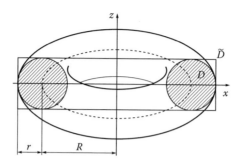

FIGURE 11.3   Integration domain and coordinate system for the Monte Carlo calculation of the mass of a torus and its mass center position.

The radial distances from a given point (x, y, z) to the z-axis and, respectively, to the tube axis are given by

$$R' = \sqrt{x^2 + y^2}, \quad r' = \sqrt{(R - R')^2 + z^2}. \tag{11.25}$$

The condition for a point to be *inside the torus* can be simply expressed as

$$r' \leq r. \tag{11.26}$$

As a density function, we consider

$$f(x, y, z) = \begin{cases} (1 - r'/r)^2 & r' \leq r, \\ 0 & r' > r, \end{cases} \tag{11.27}$$

which depends only on the minor radial distance r' and decreases parabolically from 1 at the tube axis to 0 at the torus' surface. The integral of the density yields the mass and amounts to $m = \frac{1}{3}\pi^2 R r^2 \approx 9.8696044$. As for the mass center, due to the symmetry, it is obviously located at the origin: $x_c = y_c = z_c = 0$.

As an extended integration domain, we consider the parallelepiped tightly enclosing the torus, $\widetilde{D} = [-R - r, R + r] \times [-R - r, R + r] \times [-r, r]$. The mass and the coordinates of the mass center are given, respectively, by

$$m = \int_{\widetilde{D}} \rho \, dx \, dy \, dz,$$

$$x_c = \frac{1}{m} \int_{\widetilde{D}} \rho x \, dx \, dy \, dz, \quad y_c = \frac{1}{m} \int_{\widetilde{D}} \rho y \, dx \, dy \, dz, \quad z_c = \frac{1}{m} \int_{\widetilde{D}} \rho z \, dx \, dy \, dz.$$

The Monte Carlo evaluation of these integrals can be accomplished according to the relations:

$$m \approx \tilde{V}\langle \rho \rangle \pm \tilde{V}\sqrt{\left[\langle \rho^2 \rangle - \langle \rho \rangle^2\right]/n}, \tag{11.28}$$

$$x_c \approx (\tilde{V}/m)\langle \rho x \rangle \pm (\tilde{V}/m)\sqrt{\left[\langle (\rho x)^2 \rangle - \langle \rho x \rangle^2\right]/n}, \tag{11.29}$$

$$y_c \approx (\tilde{V}/m)\langle \rho y \rangle \pm (\tilde{V}/m)\sqrt{\left[\langle (\rho y)^2 \rangle - \langle \rho y \rangle^2\right]/n}, \tag{11.30}$$

$$z_c \approx (\tilde{V}/m)\langle \rho z \rangle \pm (\tilde{V}/m)\sqrt{\left[\langle (\rho z)^2 \rangle - \langle \rho z \rangle^2\right]/n}. \tag{11.31}$$

On the basis of the random variables ξ_i, η_i, and ζ_i, distributed uniformly in the interval $[0, 1)$, one can generate random points (x_i, y_i, z_i) uniformly distributed within the domain $\widetilde{\mathcal{D}}$ as

$$x_i = (2\xi_i - 1)(R + r), \quad y_i = (2\eta_i - 1)(R + r), \quad z_i = (2\zeta_i - 1) r. \tag{11.32}$$

The program in Listing 11.5 implements relations 11.28 through 11.32, and the particular functional form of the density is coded in function `Func`.

Listing 11.5 Monte Carlo Evaluation of the Mass Center of a Torus (Python Coding)

```python
# Monte Carlo calculation of the mass center of a torus of radii R and r,
# centered at the origin and with Oz as symmetry axis
from math import *
from random import *

def Func(x, y, z):
    global R, r
    Rp = sqrt(x*x + y*y)                          # major radial distance
    dR = R - Rp
    rp = sqrt(dR*dR + z*z)                         # minor radial distance
    dr = 1e0 - rp/r
    return (dr*dr if rp <= r else 0e0)                  # zero-padding

# main

R = 3e0; r = 1e0                              # major & minor torus radii
Lx = Ly = R + r; Lz = r                     # extended domain: 1st octant
V = 8e0 * Lx * Ly * Lz                 # volume of total extended domain

n = 10000000                              # number of sampling points

seed()

sm  = sx  = sy  = sz  = 0e0
sm2 = sx2 = sy2 = sz2 = 0e0
for i in range(1,n+1):
    x = Lx * (2e0*random() - 1e0)                     # -Lx <= x <= Lx
    y = Ly * (2e0*random() - 1e0)                     # -Ly <= y <= Ly
    z = Lz * (2e0*random() - 1e0)                     # -Lz <= x <= Lz
    dens = Func(x,y,z)                                    # density
    if (dens):
        f = dens     ; sm += f; sm2 += f * f                   # sums
        f = dens * x; sx += f; sx2 += f * f
        f = dens * y; sy += f; sy2 += f * f
        f = dens * z; sz += f; sz2 += f * f

sm /= n; sx /= n; sy /= n; sz /= n                        # averages
m  = V * sm; sigm = V * sqrt((sm2/n - sm*sm)/n); f = V/m    # integrals
xc = f * sx; sigx = f * sqrt((sx2/n - sx*sx)/n)
yc = f * sy; sigy = f * sqrt((sy2/n - sy*sy)/n)
zc = f * sz; sigz = f * sqrt((sz2/n - sz*sz)/n)

print("m  = {0:8.5f} +/- {1:8.5f}".format(m ,sigm))
print("xc = {0:8.5f} +/- {1:8.5f}".format(xc,sigx))
print("yc = {0:8.5f} +/- {1:8.5f}".format(yc,sigy))
print("zc = {0:8.5f} +/- {1:8.5f}".format(zc,sigz))
```

TABLE 11.3   Monte Carlo Estimations of the Mass and Mass Center Position of a
Torus of Major Radius $R = 3$ and Minor Radius $r = 1$

n	m	x_c	y_c	z_c
100,000	9.96246	0.02386	0.02089	0.00139
1,000,000	9.87478	0.00374	0.00026	0.00024
10,000,000	9.86233	0.00114	0.00236	0.00006
100,000,000	9.86754	−0.00027	−0.00005	0.00011

As can be seen from Table 11.3, the Monte Carlo estimate of the mass features only three exact significant digits for $n = 100,000,000$ integration points, which reflects the rather slow convergence of the method. Anyway, the simplicity of the methodology and the invariant scaling law of the associated uncertainties, independent of dimensionality, makes Monte Carlo the method of choice for high-dimension integrals.

11.5 Generation of Random Numbers

The preceding sections have shown that the Monte Carlo integration of a function $f(x)$ basically implies

- Generation of sampling points with a certain distribution $w(x)$ within the integration domain.
- Evaluation of the modified integrand $f(x)/w(x)$ for the generated sampling points.

While the evaluation of the integrand does not require, in general, any techniques in addition to those presented in Chapter 5, the generation of random variables with a given distribution may require special treatment.

Randomness is not an attribute of an individual number, but rather of a *sequence* of (ideally) uncorrelated values characterized by a given distribution function.

The use of computers, which are intrinsically deterministic machines, for generating random numbers may appear deprived of sense. In fact, a sequence of numbers produced by a computer can be just *pseudorandom*, since it is completely determined by the first number, called *seed*, and the underlying algorithm. Nevertheless, the reduced sequential correlation enables one to use pseudorandom sequences successfully in most applications instead of the genuinely random sequences. Therefore, we commonly refer to computer-generated pseudorandom numbers simply by "random numbers" and, within the limits of the associated statistical errors, the applications using them should produce compatible results for different random number generators.

Truly random quantities can only result from physical processes, such as, for instance, the decay of the nuclei from a radioactive sample. Since one can neither predict which of the nuclei will decay, nor the moment of time when this will happen, a genuinely random quantity is the time interval between two consecutive radioactive decays.

The random number generator for sequences with *uniform distribution* is the basic building block for any algorithm based on random variables, even in the case of nonuniform distributions. The uniformity of the distribution implies the generation of the numbers with *equal probability* in any equal-sized subintervals of the definition domain. The generators most commonly implemented by high-level programming languages are *linear congruential generators*, which, starting from a *seed*, x_0, "grow" by repeated application of an entire sequence of random numbers x_i, using a recurrence relation of the form:

$$x_{i+1} = (ax_i + c) \mod m. \tag{11.33}$$

a, c, and m are integers, named *multiplier*, *increment*, and *modulus*, respectively, with the latter being typically large, since it determines the period length of the sequence, that is, the number of iterations after which the values repeat themselves. Obviously, the same "magic numbers" a, c, and m and the same seed always lead to the same sequence.

The random sequence generated using the recurrence relation 11.33 is confined between 0 and $m - 1$ and may repeat itself with a period inferior to m. For an optimal choice of the parameters a, c, and m, the period may be maximized and become equal to m. In such a case, *all* the integers between 0 and $m - 1$ appear in a certain order, no matter which of them was used as a seed.

In practical implementations, the value employed for the modulus m has to reconcile two contradictory requirements: on the one hand, m needs to be large enough to lead to the longest possible sequence of distinct values, and, on the other, the product of m, as the upper limit of the generated numbers, and the multiplier a should not exceed the machine representation of integers. To generate real random numbers in the unit interval $[0, 1)$, one may use the fractions x_{j+1}/m.

The function `randLCG1` given in Listing 11.6 generates real random deviates in the range $[0, 1)$ using a linear congruential generator with the coefficients a, c, and m defined in Press et al. (2007). With a view to extending the length of the generated sequence, the routine `randLCG1` employs long (32-bit) integers. The sequence may be reinitialized at any time by calling the routine with the argument `iopt` set to 0, which causes the local control variable of the sequence, `irnd`, to be reinitialized by a call to the built-in integer random number generator.

Figure 11.4 represents as histogram the distribution of 100,000 random numbers generated with the help of routine `randLCG1`. The roughly constant distribution envelope suggests that, despite its simplicity, the featured generator produces deviates with a reduced degree of sequential correlation and remarkable uniformity.

Listing 11.6   Linear Congruential Generators (Python Coding)

```python
#=============================================================================
def randLCG1(iopt=1):
#-----------------------------------------------------------------------------
#  Linear Congruential Generator for real random numbers in the range [0,1)
#  Press, Teukolsky, Vetterling, Flannery, Numerical Recipes 3rd Ed., 2007
#  iopt = 0 initializes the sequence
#-----------------------------------------------------------------------------
   global irnd                                        # conserved between calls
   a = 8121; c = 28411; m = 134456

   if (iopt == 0): irnd = randrange(0xFFFFFFFF)             # initialization

   irnd = (a * irnd + c) % m
   return irnd/m


#=============================================================================
def randLCG2(iopt=1):
#-----------------------------------------------------------------------------
#  Linear Congruential Generator for real random numbers in the range [0,1)
#  D. Rapaport, The Art of Molecular Dynamics Simulation, Cambridge, 2004
#  iopt = 0 initializes the sequence
#-----------------------------------------------------------------------------
   global irnd                                        # conserved between calls
   a = 314159269; c = 453806245; m = 2147483647

   if (iopt == 0): irnd = randrange(0xFFFFFFFF)             # initialization

   irnd = (a * irnd + c) & m
   return irnd/m
```

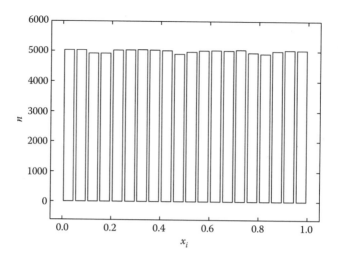

FIGURE 11.4 Distribution of 100,000 random numbers generated with the function `randLCG1`.

The binning of the successively generated random values can be carried out by calling the routine `HistoBin` presented in Chapter 3. Performing the binning in parallel with the generation of the random numbers is preferable to the prior generation and storage of the entire sequence, followed by binning, since it requires significantly less storage. To effectively plot the histogram, in the end, one can call the routine `Plot` with the style parameter `sty` set to 4.

Another, equally simple and efficient linear congruential approach uses the bit-wise AND operator & instead of the integer modulo operation (Rapaport 2004):

$$x_{i+1} = (ax_i + c) \,\&\, m, \tag{11.34}$$

and is coded as function `LCG2` in Listing 11.6.

George Marsaglia, a mathematician and computer scientist who earned his fame for outstanding contributions to the development of random number generators, described in 1999 in a post on the *Sci. Stat.*

Listing 11.7   Multiply-with-Carry Generator of George Marsaglia (Python Coding)

```
#=================================================================================
def randMCG(iopt=1):
#---------------------------------------------------------------------------------
#  Multiply-with-Carry Generator for real random numbers in the range [0,1)
#  George Marsaglia, post to Sci. Stat. Math, 1999
#  iopt = 0 initializes the sequence
#---------------------------------------------------------------------------------
   global irnd1, irnd2                                # conserved between calls

   if (iopt == 0):                                              # initialization
      irnd1 = randrange(0xFFFFFFFF); irnd2 = randrange(0xFFFFFFFF)

   irnd1 = 36969 * (irnd1 & 0xFFFF) + (irnd1 >> 0xF)
   irnd2 - 18000 * (irnd2 & 0xFFFF) + (irnd2 >> 0xF)
   return (((irnd1 << 0xF) + irnd2) & 0xFFFFFFFF)/0xFFFFFFFF
```

Math forum a series of eight high-quality 32-bit random number generators. One of these uses the so-called *multiply-with-carry* method to generate a sequence of uniformly distributed random integers based on two seed values. Essentially, this fast generator, implemented as function `randMCG` in Listing 11.7, concatenates two 16-bit multiply-with-carry generators and has an extremely long period of about 2^{60}.

The hexadecimal value `0xFFFF` (decimal 65,535) is used in bit-wise AND operations to produce from the integers `irand1` and `irand2` two 16-bit integer values, which are then combined. `0xFFFFFFFF` is the largest representable 32-bit integer value. An entry value of 0 of the parameter `iopt` causes the routine to reinitialize the random sequence by calling the built-in integer random number generator.

Quite frequently in practice, random numbers with nonuniform distributions are required. Such an example was considered in Section 11.3, where the method of importance sampling was exposed. Let us now deal in this context with the integral

$$I = \int_0^\infty f(x)dx \equiv \int_0^\infty g(x)e^{-x}dx. \tag{11.35}$$

Importance sampling can be carried out in this case by considering the normalized weight function

$$w(x) = e^{-x}, \tag{11.36}$$

and performing the change of variable

$$\xi(x) = \int_0^x e^{-x'}dx' = 1 - e^{-x}. $$

Inversion of the above relation leads to

$$x(\xi) = -ln(1 - \xi), \tag{11.37}$$

and enables one to express the integral as

$$I = \int_0^\infty g(x(\xi))d\xi. \tag{11.38}$$

Applying the Monte Carlo quadrature to the function $g(x) \equiv f(x)/w(x)$ leads to a significant variance reduction as compared to function $f(x)$, at the expense, however, of having to evaluate $g(x)$ for random arguments $x(\xi_i)$ with *exponential distribution*. The `randExp` function listed below generates numbers with such a distribution based on Relation 11.37, while producing the intermediate numbers with uniform distribution ξ_i by means of the function `random()` (presented in Section 11.2). Figure 11.5 illustrates the distribution of 100,000 random numbers generated by function `randExp`, which are binned in 20 subintervals in the range $[0, 4]$.

```
# Generates random numbers with exponential distribution
def randExp(): -log(1e0 - random())
```

One of the frequently used distribution functions in applications based on Monte Carlo techniques, also playing a primordial role in the probability theory, is the *normal* or *Gaussian distribution*:

$$w(x) = (2\pi)^{-1/2}e^{-x^2/2}. \tag{11.39}$$

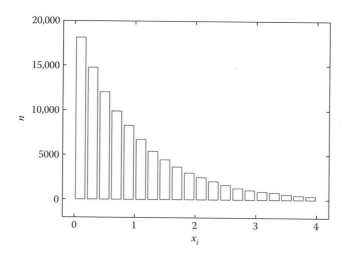

FIGURE 11.5   Distribution of 100,000 random numbers generated with the function `randExp`.

In principle, random numbers with such a distribution can be generated by inverting its incomplete integral, namely, the *error function*,

$$\xi(x) = (2\pi)^{-1/2} \int_0^x e^{-x'^2/2} dx',$$

by using, for example, a polynomial approximation thereof. However, this method lacks efficiency and we present, therefore, two alternative algorithms.

An efficient method for generating variables with normal distribution is based on the *central limit theorem*, according to which, the mean of a large number of uniformly distributed random variables approximates a normal distribution. Having in view that the mean of, for example, 12 random numbers uniformly distributed in the interval $[0, 1)$ is 6 and the corresponding standard deviation equals 1, the function `randNrm` (Listing 11.8) returns a random variable with normal distribution, featuring a vanishing mean and unitary standard deviation. As an illustration, Figure 11.6 shows the distribution of 100,000 values generated by function `randNrm`.

Another efficient technique of generating random deviates with normal distribution is known as the *Box–Muller method* (Box and Muller 1958) and is based on the 2D Gaussian distribution:

$$w(x, y) = \frac{1}{2\pi} e^{-(x^2+y^2)/2} \equiv \frac{1}{2\pi} e^{-r^2/2},$$

with $x = r\cos\theta$ and $y = r\sin\theta$ in polar coordinates. Performing the change of variable $u = r^2/2$, one may rewrite this distribution as

$$w(u, \theta) = \frac{1}{2\pi} e^{-u}.$$

Accordingly, by generating for u exponentially distributed deviates in the range $[0, \infty)$ and for θ uniformly distributed values in the interval $[0, 2\pi]$, the corresponding Cartesian projections,

$$x = (2u)^{1/2} \cos\theta, \quad y = (2u)^{1/2} \sin\theta, \tag{11.40}$$

are *normally distributed*. Function `randNrm2` (Listing 11.8) uses these relations to generate two random deviates, x and y, with normal distribution. The intermediate variable u with exponential distribution,

Listing 11.8   Random Number Generators with Normal Distribution (Python Coding)

```
#==============================================================================
def randNrm():
#------------------------------------------------------------------------------
#  Returns random numbers with normal distribution
#  w = exp(-0.5e0*x*x) / sqrt(2*pi)
#  using the central limit theorem and sums of 12 uniform random numbers
#------------------------------------------------------------------------------
   sum = 0e0
   for i in range(1,13): sum += random()
   x = sum - 6e0
   w = 0.398942280401433 * exp(-0.5e0*x*x)           # 1/sqrt(2*pi) = 0.3989...
   return (w, x)

#==============================================================================
def randNrm2():
#------------------------------------------------------------------------------
#  Returns 2 random numbers (x,y) with normal distribution
#  w = exp(-(x*x+y*y)/2) / (2*pi)
#  and the corresponding distribution value
#------------------------------------------------------------------------------
   r2 = -log(1e0 - random())             # exponential distribution for r**2/2
   w = exp(-r2) / (2e0*pi)                        # distribution function value
   r = sqrt(2e0*r2); theta = 2e0 * pi * random()              # polar coordinates
   x = r * cos(theta); y = r * sin(theta)              # Cartesian projections
   return (w, x, y)
```

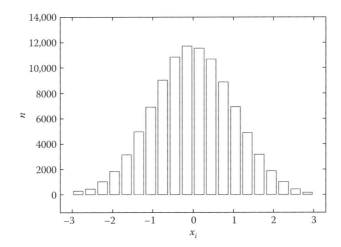

FIGURE 11.6   Distribution of 100,000 random numbers generated with the function `randNrm`.

though not coded as an explicit program variable, is generated by the technique also employed in the case of the routine `randExp()`.

The methodology of variance reduction discussed in Section 11.3 may be directly applied for generating random numbers with particular distributions only provided that the corresponding integral relations prescribing the change of variable are invertible. Unfortunately, in many applications, the inversion is difficult, if at all possible. Furthermore, if the distribution is multidimensional, the difficulties may become insurmountable. There is an obvious need for algorithms avoiding the inversion of the integral change of variable and solely implying the evaluation of the distribution function itself. One of the most representative algorithms of this category, used successfully in a wide range of topical problems of applied statistics, is the so-called *Metropolis algorithm* (Metropolis et al. 1953; Beichl and Sullivan 2000).

Let us assume that it is required to generate a sequence of random points x_i, distributed according to the probability density $w(\mathbf{x})$ in a certain (possibly multidimensional) space. The Metropolis algorithm generates these random deviates as angular points of a so-called *random walk*.

The rules according to which the random walk is performed reflect its character of Markow chain. This means that, having reached a given point x_i in the sequence, the probability density $w(\mathbf{x})$ completely determines the next point, x_{i+1}, and no memory of the previous states is involved. Specifically, to determine x_{i+1}, one first performs a *trial step* to a point $\mathbf{x}_t = \mathbf{x}_i + \delta$, randomly chosen within a hypercube of predefined size δ, centered about the starting point x_i.

Defining the ratio between the trial and the old value of the distribution function,

$$r = \frac{w(\mathbf{x}_t)}{w(\mathbf{x}_i)}, \tag{11.41}$$

if $r \geq 1$, the trial step is accepted and one sets $x_{i+1} = \mathbf{x}_t$, favoring moves toward regions with higher probability density $w(\mathbf{x})$. Otherwise, if $r < 1$, the trial step is accepted only conditionally, namely, with the probability r. Technically, this amounts to generating a random variable ρ with uniform distribution in $[0, 1)$, comparing it with r, and accepting the trial step only if

$$\rho \leq r.$$

Actually, this can be used as the sole acceptance condition of the trial step, since it combines both cases $r \geq 1$ and $r < 1$. If the trial step is rejected, one maintains the starting point, setting $x_{i+1} = x_i$. Once x_{i+1} is determined, one can pass to generating the following point of the sequence, according to the same rules. One can choose as origin of the entire random walk any point of the respective space.

Finally, we present in Listing 11.9 the routine `randMet`, which illustrates the described algorithm. The routine receives as input parameters the actual name of the distribution function w, the size `del` of the sampling hypercube, and the option parameter `iopt`, which enables one to (re)initialize the sequence if it is set to 0. Every call to the routine executes a single step of the random walk, returning based on

Listing 11.9   Random Number Generator Based on the Metropolis Method (Python Coding)

```
#===========================================================================
def randMet(w, delta, iopt=1):
#---------------------------------------------------------------------------
#  Generates random numbers x with the distribution w(x) by the Metropolis
#  method, using a maximal trial step size delta
#  iopt = 0 initializes the sequence
#---------------------------------------------------------------------------
   global xrnd                                      # conserved between calls

   if (iopt == 0): xrnd = 2e0*random() - 1e0               # initialization

   dx = delta * (2e0*random() - 1e0)                      # trial step size
   if (random() <= w(xrnd+dx)/w(xrnd)): xrnd += dx   # accept with prob. w(x)

   return xrnd
```

the starting value `xrnd` (stored locally between the calls) a new value corresponding to the distribution function $w(x)$. The acceptance condition of the trial step is checked by comparing the ratio between the trial and the old values of the distribution function with a subunitary random number generated by calling the routine `random()`.

The distribution of 100,000 random deviates generated by the routine `randMet` for the Gaussian distribution $w(x) = exp(-x^2/2)$ is similar with the one depicted in Figure 11.6.

11.6 Implementations in C/C++

Listing 11.10 shows the content of the file `random.h`, which contains equivalent C/C++ implementations of the Python functions developed in the main text and included in the module `random1.py`. The corresponding routines have identical names, parameters, and functionalities.

Listing 11.10   Routines for Generating Random Numbers with Various Distributions (`random.h`)

```
//------------------------------ random.h ------------------------------
// Contains routines for generating pseudo-random numbers.
// Part of the numxlib numerics library. Author: Titus Beu, 2013
//----------------------------------------------------------------------
#ifndef _RANDOM_
#define _RANDOM_

#include <stdlib.h>
#include <time.h>
#include <math.h>

// Initializes the random number generator using the current system time
#define seed() srand((unsigned)time(NULL))

// Generates random floating point numbers in the range [0,1)
#define random() ((double)rand()/(RAND_MAX+1))

// Generates random numbers with exponential distribution
#define randExp() -log(fabs(1e0 - random()))


//===========================================================================
double randLCG1(int iopt)
//---------------------------------------------------------------------------
// Linear Congruential Generator for real random numbers in the range [0,1)
// Press, Teukolsky, Vetterling, Flannery, Numerical Recipes 3rd Ed., 2007
// iopt = 0 initializes the sequence
//---------------------------------------------------------------------------
{
   const unsigned int a = 8121, c = 28411, m = 134456;
   static unsigned int irnd;                       // conserved between calls

   if (iopt == 0) irnd = rand();                           // initialization

   irnd = (a * irnd + c) % m;
   return (double)irnd/m;
}


//===========================================================================
double randLCG2(int iopt)
//---------------------------------------------------------------------------
// Linear Congruential Generator for real random numbers in the range [0,1)
// D. Rapaport, The Art of Molecular Dynamics Simulation, Cambridge, 2004
```

```
// iopt = 0 initializes the sequence
//-----------------------------------------------------------------------------
{
   const unsigned int a = 314159269, c = 453806245, m = 2147483647;
   static unsigned int irnd;                              // conserved between calls

   if (iopt == 0) irnd = rand();                                    // initialization

   irnd = (a * irnd + c) & m;
   return (double)irnd/m;
}

//=============================================================================
double randMCG(int iopt)
//-----------------------------------------------------------------------------
// Multiply-with-Carry Generator for real random numbers in the range [0,1)
// George Marsaglia, post to Sci. Stat. Math, 1999
// iopt = 0 initializes the sequence
//-----------------------------------------------------------------------------
{
   static unsigned int irand1, irand2;                   // conserved between calls

   if (iopt == 0) { irand1 = rand(); irand2 = rand(); }      // initialization

   irand1 = 36969 * (irand1 & 0xFFFF) + (irand1 >> 16);
   irand2 = 18000 * (irand2 & 0xFFFF) + (irand2 >> 16);
   return (double)((irand1 << 16) + irand2)/0xFFFFFFFF;
}

//=============================================================================
double randNrm(double &w)
//-----------------------------------------------------------------------------
// Returns random numbers with normal distribution
// w = exp(-0.5e0*x*x) / sqrt(2*pi)
// using the central limit theorem and sums of 12 uniform random numbers
//-----------------------------------------------------------------------------
{
   double sum, x;
   int i;

   sum = 0e0;
   for (i=1; i<=12; i++) sum += random();
   x = sum - 6e0;
   w = 0.398942280401433 * exp(-0.5e0*x*x);              // 1/sqrt(2*pi) = 0.3989...
   return x;
}

//=============================================================================
void randNrm2(double &w, double &x, double &y)
//-----------------------------------------------------------------------------
// Returns 2 random numbers (x,y) with normal distribution
// w = exp(-(x*x+y*y)/2) / (2*pi)
// and the corresponding distribution value
//-----------------------------------------------------------------------------
{
#define pi2 6.283185307179586
   double r, r2, theta;

   r2 = -log(1e0 - random());               // exponential distribution for r**2/2
```

```
   w = exp(-r2) / pi2;                           // distribution function value
   r = sqrt(2e0*r2); theta = pi2 * random();           // polar coordinates
   x = r * cos(theta); y = r * sin(theta);           // Cartesian projections
}


//=================================================================
double randMet(double w(double), double delta, int iopt)
//-----------------------------------------------------------------
// Generates random numbers x with the distribution w(x) by the Metropolis
// method, using a maximal trial step size delta
// iopt = 0 initializes the sequence
//-----------------------------------------------------------------
{
   static double dx, xrnd;                      // conserved between calls

   if (iopt == 0) xrnd = random();                       // initialization

   dx = delta * (2e0*random() - 1e0);                    // trial step size
   if (random() <= w(xrnd+dx)/w(xrnd)) xrnd += dx;   // accept with prob. w(x)

   return xrnd;
}

#endif
```

11.7 Problems

The Python and C/C++ programs for the following problems may import the functions developed in this chapter from the modules `random1.py` and, respectively, `random.h`, which are available as supplementary material. For creating runtime plots, the graphical routines contained in the libraries `graphlib.py` and `graphlib.h` may be employed.

PROBLEM 11.1

Consider the one-dimensional integrals:

$$\frac{1}{\sqrt{2\pi}} \int_0^\infty e^{-x^2/2} dx = \frac{1}{2}, \tag{11.42}$$

$$\frac{1}{\sqrt{2\pi}} \int_0^\infty x^4 e^{-x^2/2} dx = \frac{3}{2}. \tag{11.43}$$

a. Evaluate the integrals by Monte Carlo techniques, applying comparatively uniform and exponential sampling (using function `randExp`). Consider numbers of sampling points ranging from $n = 250$ to 100,000, and plot the integral estimates and the associated standard deviations.

b. Explain the significant variance reduction when using exponential sampling for the first integral and the lack of effect in the case of the second integral.

c. Use the routine `LinFit` (from `modfunc.py`) to perform linear regression and extract the scaling laws of the standard deviations for the two probability distributions from their log–log plots.

Solution

The Python implementation is provided in Listing 11.11 and the C/C++ version is available as sup-plementary material (`P11-MC-1Dx.cpp`). The plots of the Gaussian integral 11.42 and of the corresponding standard deviations are shown in Figure 11.7.

Listing 11.11   Monte Carlo Integration of a Function with Importance Sampling (Python Coding)

```python
# One-dimensional Monte Carlo quadrature with variance reduction
from math import *
from random import *
from random1 import *
from modfunc import *
from graphlib import *

wnorm = 1e0/sqrt(2e0*pi)
def func(x): return wnorm * exp(-0.5e0*x*x)                        # integrand

# main

L = 8e0                                                   # integration domain [0,L)

nn  = [0]*3                                               # ending indexes of plots
col = [""]*3                                                   # colors of plots
sty = [0]*3                                                    # styles of plots

np = 400                                                  # number of plotting points
x1 = [0]*(2*np+1); y1 = [0]*(2*np+1)                           # plotting points
x2 = [0]*(2*np+1); y2 = [0]*(2*np+1); sig = [0]*(2*np+1)

seed()

out = open("mcarlo.txt","w")                              # open output file
out.write("     n        Int       sig      Int_w     sig_w\n")

for ip in range(1,np+1):
   n = 250 * ip                                           # number of sampling points

   f1 = f2 = 0e0                                          # quadrature with uniform sampling
   for i in range(1,n+1):
      x = L * random()                          # RNG with uniform distribution in [0,L)
      f = func(x)                                             # integrand
      f1 += f; f2 += f * f                                       # sums

   f1 /= n; f2 /= n                                           # averages
   s = L * f1                                                 # integral
   sigma = L * sqrt((f2-f1*f1)/n)                         # standard deviation
   out.write(("{0:8d}{1:10.5f}{2:10.5f}").format(n,s,sigma))
   x1[ip] =      n ; y1[ip] = s
   x2[ip] = log10(n); y2[ip] = log10(sigma)

   f1 = f2 = 0e0                                # quadrature with importance sampling
   for i in range(1,n+1):
      x = randExp()                             # RNG with exponential distribution
      f = func(x) / exp(-x)                                     # integrand
      f1 += f; f2 += f * f                                       # sums

   f1 /= n; f2 /= n                                           # averages
   s = f1                                                     # integral
   sigma = sqrt((f2-f1*f1)/n)                             # standard deviation
```

```
      out.write((("{0:10.5f}{1:10.5f}\n").format(s,sigma))
      x1[np+ip] =          n ; y1[np+ip] = s
      x2[np+ip] = log10(n); y2[np+ip] = log10(sigma)

out.close()
                                                    # linear regression
(a, b, sigma, sigmb, chi2) = LinFit(x2[0:],y2[0:],sig,np,0)
print("sigma = {0:6.3f} n**({1:6.3f})  w(x) = 1".format(pow(10e0,b),a))

(a, b, sigma, sigmb, chi2) = LinFit(x2[np:],y2[np:],sig,np,0)
print("sigma = {0:6.3f} n**({1:6.3f})  w(x) = exp(-x)". \
      format(pow(10e0,b),a))

GraphInit(1200,600)

nn[1] =   np; col[1] = "blue"; sty[1] = 0           # uniform sampling
nn[2] = 2*np; col[2] = "red" ; sty[2] = 0          # importance sampling
MultiPlot(x1,y1,y1,nn,col,sty,2,10,0e0,0e0,0,0e0,0e0,0,
         0.10,0.45,0.15,0.85,"n","Int","Monte Carlo - integral")

MultiPlot(x2,y2,y2,nn,col,sty,2,10,0e0,0e0,0,0e0,0e0,0,
         0.60,0.95,0.15,0.85,"log(n)","log(sig)",
         "Monte Carlo - standard deviation")

MainLoop()
```

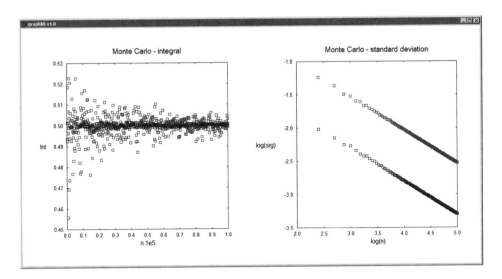

FIGURE 11.7 Monte Carlo estimates of integral 11.42 (left panel) and associated standard deviations (right panel) using uniform and, respectively, exponential sampling.

PROBLEM 11.2

Consider the 2D integral:

$$\frac{1}{4\pi} \int_{-\infty}^{\infty} \int_{-\infty}^{\infty} \left(x^2 + y^2\right) e^{-(x^2+y^2)/2} dy\, dx. \qquad (11.44)$$

a. Evaluate the integral by Monte Carlo quadratures, applying for the x and y coordinates comparatively uniform sampling, Gaussian sampling (routine `randNrm2`), and exponential sampling (routine `randExp`). Taking into account the symmetry of the problem, restrict for exponential sampling the integration domain to the first quadrant and multiply the results accordingly. Consider numbers of sampling points ranging between $n = 250$ and 100,000, and plot the integral values and the associated standard deviations.

b. Use the routine `LinFit` (from `modfunc.py`) to fit the log–log plots of the standard deviations, and verify that these scale as $n^{-1/2}$ independent of the employed sampling technique. Judging by the lower leading factor in the corresponding scaling law, explain the apparently better performance of exponential sampling.

Solution

The Python implementation is given in Listing 11.12 and the C/C++ version is available as supplementary material (`P11-MC-2Dx.cpp`). The plots of integral (11.44) and of the corresponding standard deviations for uniform, Gaussian, and exponential sampling are shown in Figure 11.8.

Listing 11.12 2D Monte Carlo Integration with Importance Sampling (Python Coding)

```
# Two-dimensional Monte Carlo quadrature with variance reduction
from math import *
from random import *
from random1 import *
from modfunc import *
from graphlib import *

pi4 = 4 * pi

def func(x,y):                                               # integrand
    r2 = x*x + y*y
    return r2*exp(-0.5e0*r2)/pi4

# main

L = 8e0                                   # integration domain [-L,L] x [-L,L]
L2 = L * L                                        # area of sampling domain

nn  = [0]*4                                       # ending indexes of plots
col = [""]*4                                              # colors of plots
sty = [0]*4                                              # styles of plots

np = 400                                          # number of plotting points
x1 = [0]*(3*np+1); y1 = [0]*(3*np+1)                     # plotting points
x2 = [0]*(3*np+1); y2 = [0]*(3*np+1); sig = [0]*(3*np+1)

seed()

out = open("mcarlo.txt","w")                             # open output file
out.write("      n         Int         sig        Int_w       sig_w\n")

for ip in range(1,np+1):
    n = 250 * ip                                  # number of sampling points

    f1 = f2 = 0.0                # quadrature with uniform sampling in [0,L] x [0,L]
    for i in range(1,n+1):
        x = L * random(); y = L * random()
        f = func(x,y)                                           # integrand
```

```
      f1 += f; f2 += f * f                                              # sums

   f1 /= n; f2 /= n                                                   # averages
   s = 4e0 * L2 * f1                                                  # integral
   sigma = 4e0 * L2 * sqrt(fabs(f2-f1*f1)/n)            # standard deviation
   out.write(("{0:8d}{1:10.5f}{2:10.5f}").format(n,s,sigma))
   x1[ip] =           n ; y1[ip] = s
   x2[ip] = log10(n); y2[ip] = log10(sigma)

   f1 = f2 = 0.0                            # quadrature with Gaussian sampling
   for i in range(1,n+1):
       (w, x, y) = randNrm2()       # random numbers with normal distribution
       f = func(x,y) / w                                            # integrand
       f1 += f; f2 += f * f                                              # sums

   f1 /= n; f2 /= n                                                   # averages
   s = f1                                                             # integral
   sigma = sqrt((f2-f1*f1)/n)                             # standard deviation
   out.write(("{0:10.5f}{1:10.5f}").format(s,sigma))
   x1[np+ip] =           n ; y1[np+ip] = s
   x2[np+ip] = log10(n); y2[np+ip] = log10(sigma)

   f1 = f2 = 0.0                         # quadrature with exponential sampling
   for i in range(1,n+1):
       x = randExp()             # random variables with exponential distribution
       y = randExp()
       w = exp(-(x+y))
       f = func(x,y) / w                                            # integrand
       f1 += f; f2 += f * f                                              # sums

   f1 /= n; f2 /= n                                                   # averages
   s = 4e0 * f1                                                       # integral
   sigma = 4e0 * sqrt((f2-f1*f1)/n)                       # standard deviation
   out.write(("{0:10.5f}{1:10.5f}\n").format(s,sigma))
   x1[2*np+ip] =           n ; y1[2*np+ip] = s
   x2[2*np+ip] = log10(n); y2[2*np+ip] = log10(sigma)

out.close()
                                                           # linear regression
(a, b, sigma, sigmb, chi2) = LinFit(x2[0:],y2[0:],sig,np,0)
print("sigma = {0:6.3f} n**({1:6.3f})  uniform sampling". \
      format(pow(10e0,b),a))

(a, b, sigma, sigmb, chi2) = LinFit(x2[np:],y2[np:],sig,np,0)
print("sigma = {0:6.3f} n**({1:6.3f})  Gaussian sampling". \
      format(pow(10e0,b),a))

(a, b, sigma, sigmb, chi2) = LinFit(x2[2*np:],y2[2*np:],sig,np,0)
print("sigma = {0:6.3f} n**({1:6.3f})  exponential sampling". \
      format(pow(10e0,b),a))

GraphInit(1200,600)

nn[1] =   np; col[1] = "blue" ; sty[1] = 0                    # uniform sampling
nn[2] = 2*np; col[2] = "red"  ; sty[2] = 0                   # Gaussian sampling
nn[3] = 3*np; col[3] = "green"; sty[3] = 0                # exponential sampling
MultiPlot(x1,y1,y1,nn,col,sty,3,10,0e0,0e0,0,0e0,0e0,0,
          0.10,0.45,0.15,0.85,"n","Int","Monte Carlo - integral")
```

```
MultiPlot(x2,y2,y2,nn,col,sty,3,10,0e0,0e0,0,0e0,0e0,0,
          0.60,0.95,0.15,0.85,"log(n)","log(sig)",
          "Monte Carlo - standard deviation")

MainLoop()
```

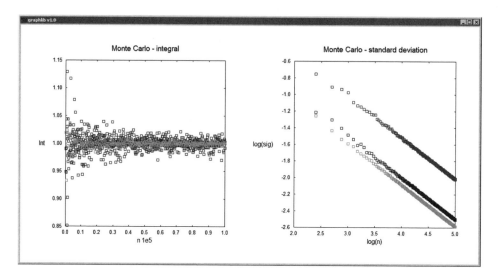

FIGURE 11.8   Monte Carlo estimates of integral (11.44) (left panel) and associated standard deviations (right panel) using uniform, Gaussian, and exponential sampling (lowest log–log dependence).

PROBLEM 11.3

Build the histograms for 10^6 random deviates, generated uniformly in the range $[0, 1)$ by the routines `random` (based on the built-in random number generator), `randLCG1` (linear congruential generator using specifications from Press et al. 2007), `randLCG2` (linear congruential generator based on specifications from Rapaport 2004), and, respectively, `randMCG` (George Marsaglia's multiply-with-carry generator), which are included in the libraries `random1.py` and, respectively, `random.h`.

For initializing, cumulating, and normalizing the data for the histograms, use the routine `HistoBin`, and for plotting the histograms, the function `Plot` with the style parameter `sty` set to 4. Both routines are part of the graphic modules `graphlib.py` and, respectively, `graphlib.h`.

Assess the quality of the random number generators by the uniformity of the corresponding histograms.

Solution

The Python implementation is given in Listing 11.13 and the C/C++ version is available as supplementary material (`P11-RandUni.cpp`). The histograms of the generated random distributions are shown in Figure 11.9.

PROBLEM 11.4

Build the histograms for 10^6 random deviates with normal (Gaussian) distribution, generated by the routines `randNrm` (based on the central limit theorem), `randNrm2` (based on polar change of variables), and `randMet` (based on the Metropolis algorithm). For `randMet` consider sampling hypercubes of sizes $\delta = 0.1$ and $\delta = 0.5$.

Listing 11.13 Generation of Random Variables with Uniform Distribution (Python Coding)

```python
# Generation of random numbers with uniform distribution
from random1 import *
from graphlib import *

# main

n = 21                                          # number of bin boundaries
nrnd = 1000000                                  # number of random numbers
x = [0]*(n+1); y = [0]*(n+1)                         # plotting points

seed()

GraphInit(1200,800)

a = 0e0; b = 1e0

HistoBin(0e0,a,b,x,y,n,0)                            # initialize histogram
for irnd in range(1,nrnd+1): HistoBin(random(),a,b,x,y,n,1)
HistoBin(0e0,a,b,x,y,n,2)                            # normalize histogram
Plot(x,y,n,"blue",4,0.10,0.45,0.60,0.90,"x","n",
     "Built-in RNG")
                                            # Linear Congruential Generator 1
HistoBin(0e0,a,b,x,y,n,0)                            # initialize histogram
randLCG1(0)                                      # initialize RNG
for irnd in range(1,nrnd+1): HistoBin(randLCG1(1),a,b,x,y,n,1)
HistoBin(0e0,a,b,x,y,n,2)                            # normalize histogram
Plot(x,y,n,"blue",4,0.60,0.95,0.60,0.90,"x","n",
     "Linear Congruential Generator 1")
                                            # Linear Congruential Generator 2
HistoBin(0e0,a,b,x,y,n,0)                            # initialize histogram
randLCG2(0)                                      # initialize RNG
for irnd in range(1,nrnd+1): HistoBin(randLCG2(1),a,b,x,y,n,1)
HistoBin(0e0,a,b,x,y,n,2)                            # normalize histogram
Plot(x,y,n,"blue",4,0.10,0.45,0.10,0.40,"x","n",
     "Linear Congruential Generator 2")
                                            # Multiply-with-Carry Generator
HistoBin(0e0,a,b,x,y,n,0)                            # initialize histogram
randMCG(0)                                       # initialize RNG
for irnd in range(1,nrnd+1): HistoBin(randMCG(),a,b,x,y,n,1)
HistoBin(0e0,a,b,x,y,n,2)                            # normalize histogram
Plot(x,y,n,"blue",4,0.60,0.95,0.10,0.40,"x","n",
     "Multiply-with-Carry Generator")

MainLoop()
```

For initializing, cumulating, and normalizing the data for the histograms, use the routine `HistoBin`, and for plotting the histograms, the function `Plot` with the style parameter `sty` set to 4. Both routines are included in `graphlib.py` and, respectively, `graphlib.h`.

Solution
The Python implementation is given in Listing 11.14 and the C/C++ version is available as supplementary material (`P11-RandNrm.cpp`). The histograms of the generated random distributions are shown in Figure 11.10.

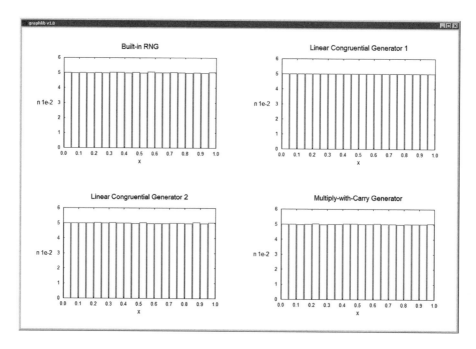

FIGURE 11.9 Histograms of 10^6 uniform random deviates generated by the routines `random`, `randLCG1`, `randLCG2`, and, respectively, `randMCG`.

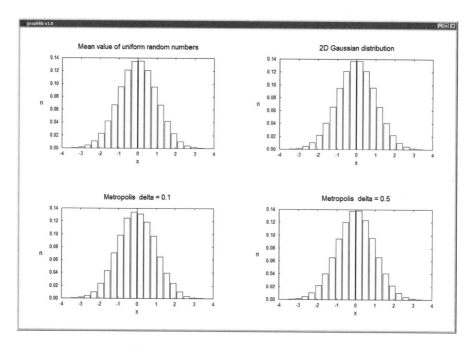

FIGURE 11.10 Histograms of 10^6 random deviates with Gaussian distribution generated by the routines `randNrm`, `randNrm2`, and `randmet`.

Listing 11.14   Generation of Random Variables with Normal Distribution (Python Coding)

```
# Generation of random numbers with normal distribution
from math import *
from random1 import *
from graphlib import *

def wMet(x): return exp(-0.5*x*x)                        # distribution for randMet

# main

n = 21                                                  # number of bin boundaries
nrnd = 1000000                                          # number of random numbers
x = [0]*(n+1); y = [0]*(n+1)                                  # plotting points

seed()

GraphInit(1200,800)

a = -3.5e0; b = 3.5e0
                                                        # Central limit theorem
HistoBin(0e0,a,b,x,y,n,0)                                 # initialize histogram
for irnd in range(1,nrnd+1):
    (w, rnd) = randNrm()
    HistoBin(rnd,a,b,x,y,n,1)
HistoBin(0e0,a,b,x,y,n,2)                                  # normalize histogram
Plot(x,y,n,"blue",4,0.10,0.45,0.60,0.90,"x","n",
     "Mean value of uniform random numbers")
                                                       # 2D Gaussian distribution
HistoBin(0e0,a,b,x,y,n,0)                                 # initialize histogram
for irnd in range(1,nrnd+1):
    (w,rnd,rnd) = randNrm2()
    HistoBin(rnd,a,b,x,y,n,1)
HistoBin(0e0,a,b,x,y,n,2)                                  # normalize histogram
Plot(x,y,n,"blue",4,0.60,0.95,0.60,0.90,"x","n",
     "2D Gaussian distribution")
                                                            # Metropolis method
HistoBin(0e0,a,b,x,y,n,0)                                 # initialize histogram
randMet(wMet,0.1e0,0)                                        # initialize RNG
for irnd in range(1,nrnd+1): HistoBin(randMet(wMet,0.1e0,1),a,b,x,y,n,1)
HistoBin(0e0,a,b,x,y,n,2)                                  # normalize histogram
Plot(x,y,n,"blue",4,0.10,0.45,0.10,0.40,"x","n",
     "Metropolis  delta = 0.1")
                                                            # Metropolis method
HistoBin(0e0,a,b,x,y,n,0)                                 # initialize histogram
randMet(wMet,0.1e0,0)                                        # initialize RNG
for irnd in range(1,nrnd+1): HistoBin(randMet(wMet,0.5e0,1),a,b,x,y,n,1)
HistoBin(0e0,a,b,x,y,n,2)                                  # normalize histogram
Plot(x,y,n,"blue",4,0.60,0.95,0.10,0.40,"x","n",
     "Metropolis  delta = 0.5")

MainLoop()
```

References and Suggested Further Reading

Beichl, I. and F. Sullivan: 2000. The Metropolis algorithm. *Computing in Science and Engineering* 2(5), 65–69.

Beu, T. A. 2004. *Numerical Calculus in C* (3rd ed., in Romanian). Cluj-Napoca: MicroInformatica.

Box, G. E. P. and M. E. Muller. 1958. A note on the generation of random normal deviates. *Annals of Mathematical Statistics* 29, 610–611.

Gould, H., J. Tobochnik, and W. Christian. 2007. *An Introduction to Computer Simulation Methods: Applications to Physical Systems* (3rd ed.). San Francisco, CA: Pearson Addison-Wesley.

Koonin, S. E. and D. C. Meredith. 1990. *Computational Physics: Fortran Version*. Redwood City, CA: Addison-Wesley.

Landau, R. H., M. J. Páez, and C. C. Bordeianu. 2007. *Computational Physics: Problem Solving with Computers* (2nd ed.). Weinheim: Wiley-VCH.

Metropolis, N., A. W. Rosenbluth, M. N. Rosenbluth, A. H. Teller, and E. Teller. 1953. Equation of state calculations by fast computing machines. *Journal of Chemical Physics* 21(6), 1087–1092.

Press, W. H., S. A. Teukolsky, W. T. Vetterling, and B. P. Flannery. 2007. *Numerical Recipes: The Art of Scientific Computing* (3rd ed.). Cambridge: Cambridge University Press.

Rapaport, D. C. 2004. *The Art of Molecular Dynamics Simulation* (2nd ed.). Cambridge: Cambridge University Press.

Ordinary Differential Equations

12.1 Introduction

The particular importance of the numerical methods for solving ordinary differential equations (ODEs) or systems thereof is due to the fact that many of the laws of nature are conveniently expressed in differential form. The classical equations of particle motion, the equations of mass diffusion or heat transport, or Schrödinger's wave equation are just a few examples illustrating the extraordinary diversity of essentially different physical phenomena, which are modeled by way of differential equations.

A multitude of phenomena are described by partial differential equations (PDEs), implying derivatives of various orders of the unknown function with respect to *several independent variables*. Again, in many situations, one has to deal with ODEs depending on the derivatives of various orders of the unknown function with respect to a *single independent variable*.

The ODEs can be *intrinsically* ordinary, that is, they can result as such from the very modeling of the concerned phenomenon, or they may result by a prior process of separation of variables from an initial PDE, usually by taking advantage of certain symmetry properties featured by the modeled system. In a more complex situation, the solution may comprise several unknown functions, satisfying a *system of coupled ODEs*.

To illustrate the close connection between higher-order ODEs and systems of ODEs, let us consider as a prototypical example the classical 1D Newton's equation of motion of a particle of mass m, experiencing a force field $F(x)$:

$$m\frac{d^2x}{dt^2} = F(x). \tag{12.1}$$

Defining the particle momentum, $p = m\left(dx/dt\right)$, this second-order equation may be decomposed into a set of two first-order equations:

$$\frac{dx}{dt} = \frac{p}{m}, \quad \frac{dp}{dt} = F(x), \tag{12.2}$$

which are the so-called *Hamilton's canonical equations of motion*. Emphasizing rather the particle's velocity $v = p/m$ than its momentum, the above equations become

$$\frac{dx}{dt} = v, \quad \frac{dv}{dt} = \frac{1}{m}F(x). \tag{12.3}$$

By the equivalent functional form of the first-order differential equations composing this coupled system, the two unknown functions, x and v, are set on equal footing. Moreover, the methodology of transforming the initial equation of motion (12.1) into system (12.3) reveals that it is sufficient, in principle, to develop only methods for solving first-order differential equations or systems of such equations. Nevertheless, due to the particular practical importance of the second-order equations of motion (e.g., in quantum or classical molecular dynamics), in the course of this chapter, there will also be developed specific numerical methods for solving second-order ODEs or systems of such equations.

With a view to generalizing the above technique, let us consider now a general nth-order differential equation for the unknown function $y(t)$:

$$y^{(n)} = f(t, y, y', \ldots, y^{(n-1)}). \tag{12.4}$$

One can easily show that this equation is equivalent with the system of n first-order differential equations

$$y'_i(t) = f_i(t, y_1(t), y_2(t), \ldots, y_n(t)), \quad i = 1, 2, \ldots, n, \tag{12.5}$$

by considering as unknown functions besides y, also its first $(n - 1)$ derivatives,

$$y_i \equiv y^{(i-1)}, \quad i = 1, 2, \ldots, n,$$

and identifying the original unknown with the first of the new unknown functions, $y \equiv y_1$. The resulting system thus takes the form:

$$\begin{cases} y'_i(t) = y_{i+1}(t), \quad i = 1, 2, \ldots, n-1, \\ y'_n(t) = f(t, y_1(t), y_2(t), \ldots, y_n(t)), \end{cases}$$

where the first $n-1$ equations result from bare notations and are nonspecific, while only the last equation contains the particular right-hand-side function f, being problem specific.

To simplify the notations and the derivation of numerical solution methods, the general system of first-order ODEs (12.5) may be rewritten with vector notation as

$$y'(t) = f(t, y), \tag{12.6}$$

where

$$y = \begin{bmatrix} y_1(t) \\ \vdots \\ y_n(t) \end{bmatrix}, \quad f(t, y) = \begin{bmatrix} f_1(t, y_1(t), \ldots, y_n(t)) \\ \vdots \\ f_n(t, y_1(t), \ldots, y_n(t)) \end{bmatrix}.$$

The following sections do not explicitly discuss methods for solving *systems* of ODEs, since these may be obtained simply from those for single equations, by replacing the scalar with vector notations.

A problem for an ODE is not completely specified solely by the equation itself. To obtain a particular solution from the family of compatible solutions, one has to attach *supplementary conditions* concerning the solution values at certain points in the definition domain. More than by the specific form of the ODE, the type of numerical strategy that is suitable to solve the problem is determined by these supplementary conditions.

Essentially, the supplementary conditions that can be associated to ODEs fall into two broad categories:

- *Initial-value (Cauchy) problems*
- *Two-point (bilocal) boundary-value problems*

An initial-value problem results by attaching to system (12.6) n supplementary conditions at a "starting" point t_0,

$$y_i(t_0) = y_{i0}, \quad i = 1, 2, \ldots, n, \tag{12.7}$$

or, in vector form,

$$y(t_0) = y_0. \tag{12.8}$$

Solving this problem implies the progressive, step-by-step calculation of the solution at a sequence of points $t_0, t_1, t_2, \ldots$

Initial-value problems are generally, but not always, depending on time as an independent variable and describe the evolution of the modeled system. A typical example is the oblique throw of a projectile experiencing gravitation and drag. Newton's equations of motion solely provide a family of solutions parametrized with arbitrary integration constants. A particular solution, though, describing a concrete physical situation, can be obtained only by also specifying initial values for the position and velocity vectors of the projectile.

In their turn, the methods for solving initial-value problems can be divided into two classes. The first class comprises the so-called *direct* or *one-step methods*, which provide the solution at some point t_{m+1} based only on information (solution and derivatives) from the preceding point, t_m, and, possibly, from the interval $[t_m, t_{m+1}]$. The number of initial conditions associated has to match for a single ODE the order of the equation, while for a system of ODEs, the cumulated orders of the involved equations. We mention from this category the general Euler and Runge–Kutta methods, as well as the more problem-adapted Euler–Cromer, Euler–Richardson, and Verlet methods.

The second class of algorithms for initial-value problems gathers the so-called *indirect* or *multistep methods*, which enable the calculation of the solution at some point t_{m+1} based on information from several preceding points $t_m, t_{m-1}, \ldots$ From this category, we mention the Adams–Moulton, Milne, Fox–Goodwin, and Numerov methods. The methods of each of the two classes feature specific advantages and disadvantages, being typically effective under different circumstances and subject to different requirements.

By associating to an ODE supplementary conditions regarding the solution and, possibly, its derivatives at the ends of the definition range $[x_1, x_2]$,

$$\alpha_1 y(x_1) + \beta_1 y'(x_1) = \gamma_1, \tag{12.9}$$

$$\alpha_2 y(x_2) + \beta_2 y'(x_2) = \gamma_2, \tag{12.10}$$

there results a so-called *two-point problem*. As in the case of the Cauchy problems, the number of supplementary conditions has to match, here too, the cumulated orders of the equations. The aim of any numerical approach is in this case the calculation of the solution at some interior mesh points of the domain $[x_1, x_2]$.

Typical physical problems that can be modeled as two-point problems concern the deformation of a clamped string or a supported beam. In the closing of this chapter, the *shooting method* and the *finite-difference method* are applied to solve such bilocal problems associated to linear second-order ODEs.

12.2 Taylor Series Method

One of the oldest and most general methods of solving initial-value problems for ODEs is based on the expansion of the solution in a Taylor series. The Taylor series method allows, in principle, the solution of any differential equation, but its practical application is quite often difficult and inefficient. The importance of this method resides rather in its ability to provide criteria for evaluating the accuracy of methods of practical interest. Basically, the underlying idea is to replace differentiations by finite-difference operations.

Let us consider the initial-value problem:

$$y' = f(t, y), \tag{12.11}$$

$$y(t_0) = y_0. \tag{12.12}$$

We assume the solution to exist, to be unique, finite, and expandable in a Taylor series in the vicinity of the initial point t_0:

$$y(t) = y(t_0) + \frac{t - t_0}{1!} y'(t_0) + \frac{(t - t_0)^2}{2!} y''(t_0) + \cdots . \tag{12.13}$$

The Taylor expansion (12.13) provides a step-by-step procedure for evaluating the solution $y(t)$ at given mesh points $t_0, t_1, \ldots, t_m, \ldots$ as a sequence of solutions of elementary Cauchy problems. Specifically, propagating the solution from t_m to t_{m+1} is equivalent to solving the differential equation (12.11) subject to the initial condition

$$y(t_m) = y_m,$$

and it is accomplished based on the expansion

$$y_{m+1} = y_m + h y'_m + \frac{h^2}{2} y''_m + \cdots , \tag{12.14}$$

where $y_m \equiv y(t_m)$, $y'_m \equiv y'(t_m)$, $y''_m \equiv y''(t_m)$, and the mesh spacing is given by

$$h = t_{m+1} - t_m.$$

The smaller the mesh spacing h in comparison to the convergence radius of the series, and the larger the number of terms kept in series (12.14), the closer the estimate y_{m+1} to the exact solution $y(t_{m+1})$.

Determining the propagated solution y_{m+1} requires the derivatives at t_m, namely $y'_m, y''_m, \ldots$. As a matter of fact, the first derivative is already available from ODE (12.11) itself and it just needs to be evaluated at t_m:

$$y'_m = f(t_m, y_m).$$

On the other hand, by taking the derivatives of both members of Equation 12.11 with respect to t (bearing in mind that f is a function of t both explicitly and through y), one finds for the second derivative at t_m:

$$y''_m = \left[\frac{\partial f}{\partial t} + \frac{\partial f}{\partial y} \frac{\partial y}{\partial t} \right]_{t_m, y_m} = \left[\frac{\partial f}{\partial t} + f \frac{\partial f}{\partial y} \right]_{t_m, y_m} .$$

Now, plugging the expressions of y'_m and y''_m into Expansion 12.14, we get the propagation relation:

$$y_{m+1} = y_m + h f(t_m, y_m) + \frac{h^2}{2} \left[\frac{\partial f}{\partial t} + f \frac{\partial f}{\partial y} \right]_{t_m, y_m} + O(h^3). \tag{12.15}$$

The notation $O(h^3)$ indicates that the remainder (truncation error) is of the order h^3, that is, all the neglected terms involve h at powers greater or equal to three, while the retained terms imply at most h^2. The Taylor series method is obviously a *one-step method*, since determining y_{m+1} requires only information from the preceding point t_m.

For higher-order approximations of the solution, one needs to consider additional terms in series (12.14). Anyhow, the implied higher-order derivatives become more and more intricate, both to derive and to program, leading to inefficient propagation formulas. In certain situations, the bare evaluation of the derivatives becomes impossible.

Notwithstanding the practical deficiencies, the Taylor series method retains, as already mentioned, a special theoretical importance, since it provides a unifying criterion for assessing the accuracy of other methods for Cauchy problems. The criterion refers to the degree in which the approximate solution coincides with the Taylor series expansion of the exact solution. More precisely, a method is said to be *of order p* if the approximate solution is equivalent with the exact Taylor series up to terms of the order

h^p, or, equivalently, if the method's truncation error is $O(h^{p+1})$. This way of assessing the precision class also remains applicable for methods that do not even imply the effective evaluation of derivatives of the right-hand-side function $f(t, y)$ of the ODE.

12.3 Euler's Method

Euler's method, also called *forward Euler* or *explicit Euler method*, is the simplest implementable *one-step solver* for initial-value problems. Referring again to the general Cauchy problem 12.11–12.12, Euler's method is simply the *linear approximation* of the Taylor series method and results by neglecting the $O(h^2)$ terms:

$$y_{m+1} = y_m + hf(t_m, y_m), \quad m = 0, 1, 2, \ldots, \tag{12.16}$$

with $h = t_{m+1} - t_m$. We point again to the fact that y_m may represent here a single value (in the case of a single ODE), or may stand for a vector of solution components (in the case of a system of ODEs).

The recurrence relation 12.16 *explicitly* provides the propagated solution y_{m+1} in terms of the preceding value y_m (regarded as initial value of an elementary Cauchy problem). Hence, starting from an initial value $y_0 = y(t_0)$, one can determine, step by step, the sequence of approximations $y_1, y_2, \ldots$ of the exact solution values $y(t_1), y(t_2), \ldots$

Figure 12.1 shows that repeatedly applying Euler's method amounts to replacing the curve $y = y(t)$ with the piecewise linear function connecting the points $(t_0, y_0), (t_1, y_1), (t_2, y_2), \ldots$ The initial segment represents the line tangent to the exact solution that passes through the initial point (t_0, y_0). In general, the propagated solution y_{m+1} can be regarded as the straight-line extrapolation of the particular solution of the ODE passing through the previous point (t_m, y_m), however, not necessarily satisfying the initial condition $y(t_0) = (y_0)$.

Euler's method is an $O(h)$ algorithm, since the propagated solution coincides with the Taylor series up to terms proportional to h. The *local error* is dominated by the $O(h^2)$ truncation error and characterizes the deviation of y_{m+1} from the particular solution passing through (t_m, y_m). The *global error* cumulates the local errors along the propagation and measures the total deviation $\Delta_{m+1} = y_{m+1} - y(t_{m+1})$ of the approximate solution y_{m+1} from the exact solution. The global error obviously increases with the number of propagation steps and, according to a general convergence theorem, if the local error of a one-step ODE solver is $O(h^{p+1})$, then the global error is $O(h^p)$. The global truncation error of the Euler method is hence $O(h)$.

At this point, it is important to stress that decreasing the step size h of any ODE solver indefinitely, with the aim of reducing the truncation error and obtaining at the limit the "exact" solution, is not a practical approach. Indeed, the accumulation of roundoff errors, brought about by the inherent increase of the

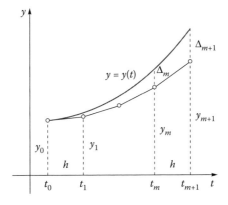

FIGURE 12.1 The approximate solutions y_{m+1} propagated by Euler's method lie on the line tangents to the exact solutions of the ODE passing through (x_m, y_m), $m = 0, 1, 2, \ldots$

number of propagation steps, starts dominating the solution, which ceases to improve by the further reduction of h below a certain value.

One practical way of overcoming the low accuracy of Euler's method is to raise the order of the approximation by applying its variant called *predictor–corrector Euler method*, which operates with two solution estimates at each propagation step:

$$\tilde{y}_{m+1} = y_m + h f(t_m, y_m), \quad m = 0, 1, 2, \ldots, \tag{12.17}$$

$$y_{m+1} = y_m + \frac{h}{2} \left[f(t_m, y_m) + f(t_m + h, \tilde{y}_{m+1}) \right]. \tag{12.18}$$

The *predicted solution*, $\tilde{y}_{m+1}$, is calculated as the basic Euler method, from the *predictor formula* 12.17, and it is used to predict the propagated derivative $f(t_m + h, \tilde{y}_{m+1})$. The *corrected solution*, y_{m+1}, is then obtained based on the *corrector formula* 12.18, which employs the predicted average derivative on the interval $[x_m, x_{m+1}]$, namely $\left[f(t_m, y_m) + f(t_m + h, \tilde{y}_{m+1}) \right]/2$.

In Section 12.4, both the basic and the predictor–corrector Euler methods are shown to be particular cases of an entire family of algorithms, known as Runge–Kutta methods. The basic Euler method is equivalent with the first-order Runge–Kutta method, while the predictor–corrector Euler algorithm is $O(h^2)$, being equivalent with the second-order Runge–Kutta method. The superior precision of the predictor–corrector Euler algorithm obviously comes about at the expense of a double number of evaluations of the right-hand-side functions $f(x, y)$ of the ODEs.

The Euler method in its basic and predictor–corrector variants is implemented in functions `Euler` and, respectively, `EulerPC` in Listing 12.1.

Listing 12.1   Euler and Euler Predictor–Corrector ODE Solvers (Python Coding)

```
#===============================================================================
def Euler(t, ht, y, n, Func):
#-------------------------------------------------------------------------------
#  Propagates the solution y of a system of 1st order ODEs
#     y'[i] = f[i](t,y[]), i = 1..n
#  from t to t+ht using Euler's method
#  Calls: Func(t, y, f) - RHS of ODEs
#-------------------------------------------------------------------------------
   f = [0]*(n+1)                                             # RHS of ODEs

   Func(t,y,f)                                               # get RHS of ODEs
   for i in range(1,n+1): y[i] += ht * f[i]              # propagate solution


#===============================================================================
def EulerPC(t, ht, y, n, Func):
#-------------------------------------------------------------------------------
#  Propagates the solution y of a system of 1st order ODEs
#     y'[i] = f[i](t,y[]), i = 1..n
#  from t to t+ht using Euler's predictor-corrector method
#  Calls: Func(t, y, f) - RHS of ODEs
#-------------------------------------------------------------------------------
   f1 = [0]*(n+1); f2 = [0]*(n+1)                            # RHS of ODEs
   yt = [0]*(n+1)                                         # predicted solution

   Func(t,y,f1)                                            # RHS of ODEs at t
   for i in range(1,n+1): yt[i] = y[i] + ht * f1[i]             # predictor
   Func(t+ht,yt,f2)                                       # RHS of ODEs at t+ht

   ht2 = ht/2e0
   for i in range(1,n+1): y[i] += ht2 * (f1[i] + f2[i])          # corrector
```

To illustrate the accuracy and stability of the discussed Euler methods, let us consider the Cauchy problem for the linear second-order ODE:

$$y'' + y = 0, \tag{12.19}$$

$$y(0) = y_0, \quad y'(0) = y_0'. \tag{12.20}$$

The general solution of this problem is $y(t) = A \sin t + B \cos t$ and the specific constants depend on the particular initial values y_0 and y_0'. According to the methodology described in the introduction to this chapter, by making the notations

$$y_1 \equiv y, \quad y_2 \equiv y',$$

the considered Cauchy problem may be cast under the form:

$$\begin{cases} y_1' = y_2, \\ y_2' = -y_1, \end{cases} \quad \begin{cases} y_1(0) = y_0, \\ y_2(0) = y_0'. \end{cases} \tag{12.21}$$

The program designed to solve Problem 12.21 (Listing 12.2) defines the right-hand sides of the ODEs in the user function Func. The employed solvers, Euler and EulerPC, are imported from

Listing 12.2 Solution of Cauchy Problem for System of ODEs by Euler's Method (Python Coding)

```
#   Solves a Cauchy problem for a 2nd order ODE by Euler's method
#      y" + y = 0,     y(0) = y0, y'(0) = y0'
#   Equivalent problem: y[1] = y, y[2] = y'
#      y[1]' =  y[2],     y[1](0) = y0
#      y[2]' = -y[1],     y[2](0) = y0'
#----------------------------------------------------------------------
from math import *
from ode import *

def Func(t, y, f):                             # Right-hand sides of ODEs
   f[1] =  y[2]
   f[2] = -y[1]

# main

y0 = 0e0; dy0 = 1e0                    # initial values => y(t) = sin(t)
tmax = 100e0                                           # time span
ht = 0.05e0                                            # step size

n = 2                                    # number of 1st order ODEs
y = [0]*(n+1)                               # solution components

out = open("ode.txt","w")                          # open output file
out.write("        t          y1          y2        check\n")

t = 0e0
y[1] = y0; y[2] = dy0                                  # initial values
out.write(("{0:10.5f}{1:10.5f}{2:10.5f}{3:10.5f}\n"). \
          format(t,y[1],y[2],y[1]*y[1]+y[2]*y[2]))

while (t+ht <= tmax):                               # propagation loop
   Euler(t,ht,y,n,Func)
   t += ht

   out.write(("{0:10.5f}{1:10.5f}{2:10.5f}{3:10.5f}\n"). \
             format(t,y[1],y[2],y[1]*y[1]+y[2]*y[2]))
out.close()
```

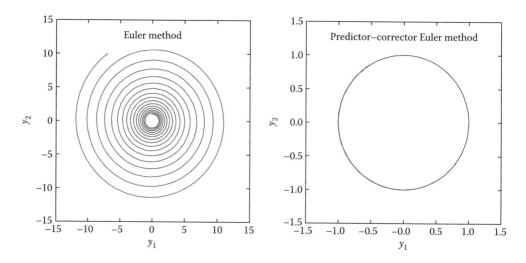

FIGURE 12.2 Numerical solution yielded by the Euler method (left panel) and the predictor–corrector Euler method (right panel) for the Cauchy problem 12.21–12.22 for a propagation step size $h = 0.05$.

the library ode.py (respectively, ode.h, for C/C++ coding). The program implements a characteristic while-do propagation loop, which repeatedly calls the solver to propagate the solution over elementary time intervals ht, also being responsible for advancing the current time.

The exact solution to Problem 12.21 in the particular case of the initial values

$$y_1(0) = 0, \quad y_2(0) = 1, \tag{12.22}$$

is

$$y_1(t) = \sin t, \quad y_2(t) = \cos t,$$

and can be readily compared with the numerical output produced by the program. An even more eloquent indication on the numerical stability of the solution is provided by the dependence of one solution component on the other, concretely, by the plot of y_1 versus y_2. In a case of stable propagation, the curve $y_1 = y_1(y_2)$ is expected to be a closed circle of unit radius.

As can be noticed from Figure 12.2, which depicts the evolution of the profiles over a time span $t_{max} = 100$, obtained with a propagation step size $h = 0.05$, the solution yielded by the basic Euler method (left panel) progressively departs from the expected circular behavior, whereas, judging by the closed curve of radius close to 1, the predictor–corrector Euler method (right panel) exhibits fair stability and accuracy. The basic Euler method achieves a stability degree comparable to the one of the predictor–corrector algorithm only for a 50 times smaller step size, that is, for $h = 0.001$.

12.4 Runge–Kutta Methods

The Runge–Kutta methods represent a whole family of algorithms for the numerical solution of initial-value problems for first-order ODEs or systems thereof. They have been initially developed by C. Runge, who published the first schemes of this type in a seminal paper in 1895 (Runge 1895), while W. Kutta extended this class of methods up to order 5 (Kutta 1901). Owing to their reliability and robustness, the Runge–Kutta methods have acquired the status of standard, general-purpose methods for numerical integration of ODEs.

The Runge–Kutta algorithms are self-starting *one-step methods*, which require only the evaluation of the right-hand sides of the ODEs, but not of their derivatives. According to the general convention, the pth-order method provides solutions coinciding with the $O(h^p)$ approximations of the corresponding Taylor series. The increase of the order is accompanied naturally by superior precision, yet also by increased implementation complexity.

Let us consider the generic Cauchy problem for a first-order ODE or system of ODEs:

$$y' = f(t, y), \quad y(t_0) = y_0. \tag{12.23}$$

The propagation of the solution $y(t)$ from t_m to t_{m+1} may be accomplished, as discussed in Section 12.2, by applying the Taylor expansion

$$y_{m+1} = y_m + hf(t_m, y_m) + \frac{h^2}{2} \left[\frac{\partial f}{\partial t} + f \frac{\partial f}{\partial y} \right]_{t_m, y_m} + \cdots \tag{12.24}$$

Retaining a large number of terms in this series, with a view to increasing the accuracy, is impractical, due to the rapid increase of the complexity of the involved derivatives of the right-hand-side function $f(t, y)$. The Runge–Kutta methods follow instead another leading idea: to construct solvers depending on the values of $f(t, y)$, but not on derivatives thereof, and whose power-series representations in h should be equivalent with the exact expansion (12.24) up to highest possible terms, irrespective of the particular function $f(t, y)$.

We represent y_{m+1} as a linear combination:

$$y_{m+1} = y_m + \sum_{i=1}^{p} w_i k_i, \tag{12.25}$$

where w_i are weights and k_i take values proportional to $f(t, y)$,

$$k_i = hf(\xi_i, \eta_i),$$

for p particular argument pairs from the vicinity of t_m and y_m:

$$\begin{cases} \xi_i = t_m + \alpha_i h, \\ \eta_i = y_m + \sum_{j=1}^{i-1} \beta_{ij} k_j. \end{cases} \tag{12.26}$$

α_i, β_{ij}, and w_i are adjustable parameters, which need to be determined so as to maximize the agreement between the linear combination (12.25) and the Taylor series (12.24). Without restriction of generality, one can consider in all cases:

$$\alpha_1 = 0, \quad \beta_{11} = 0, \quad \xi_1 = t_m, \quad \eta_1 = y_m, \tag{12.27}$$

which guarantees that $hf(t_m, y_m)$ is always the first term in any Runge–Kutta propagator (12.25). We then have:

$$\begin{cases} k_1 = hf(t_m, y_m), \\ k_2 = hf(t_m + \alpha_2 h, y_m + \beta_{21} k_1), \\ k_3 = hf(t_m + \alpha_3 h, y_m + \beta_{31} k_1 + \beta_{32} k_2), \\ \cdots\cdots\cdots \end{cases} \tag{12.28}$$

The above relations reveal the *explicit* algorithmic character of the Runge–Kutta methods, meaning that the various quantities are calculated in a straightforward manner based only on already-available quantities.

Let us determine in the following the parameters α_i, β_{ij}, and w_i for methods of various orders p, in such a way that the expansion of the propagation formula 12.25 in powers of h should be equivalent with the $O(h^p)$ approximation of the solution's Taylor series 12.24.

For $p = 1$, there results:

$$y_{m+1} = y_m + hf(t_m, y_m), \tag{12.29}$$

and one regains Euler's formula, which turns out to be identical with the first-order Runge–Kutta method.

For $p = 2$, the general propagation formula 12.25 and coefficients 12.28 take on the particular forms:

$$y_{m+1} = y_m + w_1 k_1 + w_2 k_2,$$

and, respectively,

$$\begin{cases} k_1 = hf(t_m, y_m), \\ k_2 = hf(t_m + \alpha_2 h, y_m + \beta_{21} k_1). \end{cases}$$

Combining these relations yields:

$$y_{m+1} = y_m + w_1 hf(t_m, y_m) + w_2 hf(t_m + \alpha_2 h, y_m + \beta_{21} hf(t_m, y_m)),$$

and, by expanding the last term as a power series in h, one obtains:

$$y_{m+1} = y_m + (w_1 + w_2) hf(t_m, y_m) + h^2 \left[w_2 \alpha_2 \frac{\partial f}{\partial t} + w_2 \beta_{21} f \frac{\partial f}{\partial y} \right]_{t_m, y_m}. \tag{12.30}$$

Comparing this formula with the Taylor series 12.24 and equating the coefficients of the same powers of h leads to

$$\begin{cases} w_1 + w_2 = 1 \\ w_2 \alpha_2 = 1/2 \\ w_2 \beta_{21} = 1/2. \end{cases}$$

This undetermined nonlinear system with four unknowns does not have a unique solution. Noting that it is necessary to have $w_2 \neq 0$, a simple choice at hand, which also favors symmetry, is $w_2 = 1/2$ and thereby results:

$$\alpha_2 = \beta_{21} = 1,$$

$$w_1 = w_2 = 1/2.$$

With these coefficients, the characteristic formulas of the *second-order Runge–Kutta method* become:

$$y_{m+1} = y_m + \frac{1}{2}(k_1 + k_2) + O(h^3), \tag{12.31}$$

with

$$\begin{cases} k_1 = hf(t_m, y_m), \\ k_2 = hf(t_m + h, y_m + k_1). \end{cases} \tag{12.32}$$

Or, by combining the above relations, there results:

$$y_{m+1} = y_m + \frac{h}{2} \left[f(t_m, y_m) + f(t_m + h, y_m + hf(t_m, y_m)) \right], \tag{12.33}$$

and we regain the predictor–corrector Euler method, with the predictor embedded in the corrector formula.

The most important and widely used variant is the fourth-order Runge–Kutta method ($p = 4$), which offers both a good control of the local truncation error and an easily implementable functional form. Following the same reasoning for determining the coefficients α_i, β_{ij}, and w_i as in the case $p = 2$, there results a nonlinear system of 10 equations with 13 unknowns, which, being underdetermined, does not have a unique solution. Several sets of coefficients are in use, being though preferred those either featuring simplicity or favoring efficient storage usage. The commonly employed variant of the *fourth-order Runge–Kutta method* is defined by the relations:

$$y_{m+1} = y_m + \frac{1}{6}\left(k_1 + 2k_2 + 2k_3 + k_5\right) + O(h^5),\tag{12.34}$$

$$\begin{cases} k_1 = hf(t_m, y_m), \\ k_2 = hf(t_m + h/2, y_m + k_1/2), \\ k_3 = hf(t_m + h/2, y_m + k_2/2), \\ k_4 = hf(t_m + h, y_m + k_3), \end{cases}\tag{12.35}$$

and it implies at each propagation step four evaluations of the right-hand-side function $f(t, y)$. The corresponding truncation error reflects the neglect of the $O(h^5)$ terms:

$$\Delta_{m+1} \sim Kh^5.\tag{12.36}$$

It is noteworthy that, for a right-hand-side function f independent of y, the differential equation turns into the simple integral of f over the interval $[t_m, t_m + h]$, and the fourth-order Runge–Kutta method becomes equivalent with Simpson's rule (see Chapter 10):

$$y_{m+1} = y_m + \frac{h}{6}\left[f(t_m) + 4f(t_m + h/2) + f(t_m + h)\right].$$

The double values of the standard coefficients of Simpson's rule ($1/3, 4/3, 1/3$) stem from the definition of the integration interval length as $2h$ instead of h.

In addition to the satisfactory precision and the simple implementation, being an explicit one-step algorithm, the fourth-order Runge–Kutta method stands out by its self-starting ability, as well. The Runge–Kutta solver is also useful for providing initial values for more accurate, yet not self-starting multistep methods.

Taking advantage of the simple structure of the coefficients k_1, k_2, k_3, and k_4 of the fourth-order Runge–Kutta method 12.34–12.35, for a *system* of n first-order ODEs, it is convenient to explicitly refer to the values of the right-hand-side functions f_i:

$$y_{m+1,i} = y_{m,i} + \frac{h}{6}\left(f_{1,i} + 2f_{2,i} + 2f_{3,i} + f_{4,i}\right), \quad i = 1, 2, \dots, n,\tag{12.37}$$

$$\begin{cases} f_{1,i} = f_i(t_m, \{y_{m,i}\}), \\ f_{2,i} = f_i(t_m + h/2, \{y_{m,i} + (h/2)f_{1,i}\}), \\ f_{3,i} = f_i(t_m + h/2, \{y_{m,i} + (h/2)f_{2,i}\}), \\ f_{4,i} = f_i(t_m + h, \{y_{m,i} + hf_{3,i}\}). \end{cases}\tag{12.38}$$

The curly braces indicate the whole set of arguments for all solution components (running over all values of i). Obviously, the formalism can also be applied for solving Cauchy problems for higher-order ODEs, by first transforming the equation into a system of first-order ODEs.

The routine `RungeKutta` (Listing 12.3) represents the one-to-one coding of the described algorithm for solving an "elementary" Cauchy problem for a system of first-order ODEs. Concretely, the routine performs on each call a one-step propagation of the known solution ($y_{m,i}$, $i = 1, 2, \dots n$) from t_m to $t_{m+1} = t_m + h$, producing updated components $y_{m+1,i}$ according to Equations 12.37 through

Listing 12.3 Fourth-Order Runge–Kutta Solver for Cauchy Problems (Python Coding)

```
#===============================================================================
def RungeKutta(t, ht, y, n, Func):
#-------------------------------------------------------------------------------
#  Propagates the solution y of a system of 1st order ODEs
#      y'[i] = f[i](t,y[]), i = 1..n
#  from t to t+ht using the 4th order Runge-Kutta method
#  Calls: Func(t, y, f) - RHS of ODEs
#-------------------------------------------------------------------------------
   f1 = [0]*(n+1); f2 = [0]*(n+1)                           # RHS of ODEs
   f3 = [0]*(n+1); f4 = [0]*(n+1)
   yt = [0]*(n+1)                                       # predicted solution

   ht2 = ht/2e0
   Func(t,y,f1)                                              # RHS at t
   for i in range(1,n+1): yt[i] = y[i] + ht2*f1[i]
   Func(t+ht2,yt,f2)                                       # RHS at t+ht/2
   for i in range(1,n+1): yt[i] = y[i] + ht2*f2[i]
   Func(t+ht2,yt,f3)                                       # RHS at t+ht/2
   for i in range(1,n+1): yt[i] = y[i] + ht *f3[i]
   Func(t+ht,yt,f4)                                         # RHS at t+ht

   h6 = ht/6e0                                         # propagate solution
   for i in range(1,n+1): y[i] += h6*(f1[i] + 2*(f2[i] + f3[i]) + f4[i])
```

12.38. The array y is supposed to contain on entry the n components of the initial solution, which are replaced on exit by their values propagated with a step size ht. The routine calls by the local alias Func the user-supplied function that returns the values $f_{1,i}$, $f_{2,i}$, $f_{3,i}$, and $f_{4,i}$ of the right-hand-side functions $f_i(t, y_1, \ldots, y_n)$, $i = 1, 2, \ldots n$. The corresponding work arrays f1, f2, f3, and f4 are allocated on the first entry into the routine.

As an application of the Runge–Kutta algorithm, we consider below the problem of a nonlinear pendulum. For a pendulum of constant length l, the equation of motion for the angular displacement u with respect to the vertical passing through the suspension point is a nonlinear second-order ODE,

$$u'' = -\frac{g}{l} \sin u, \tag{12.39}$$

which is typically solved subject to a given initial displacement and zero time derivative:

$$u(0) = u_0, \quad u'(0) = 0. \tag{12.40}$$

For oscillations of *small amplitude* u_0, the right-hand side of Equation 12.39 may be linearized using the approximation $\sin u \approx u$, and the corresponding *harmonic solution*,

$$u(t) = u_0 \cos\left(\frac{2\pi}{T_0} t\right), \quad T_0 = 2\pi \sqrt{\frac{l}{g}},$$

is characterized by an amplitude-independent period T_0.

For oscillations of arbitrary amplitude, the period can be shown to strongly depend on the initial angular displacement $u_0 \in [-\pi, \pi]$ (Beléndez et al. 2007):

$$T = T_0 \frac{2}{\pi} K\left(\sin^2 \frac{u_0}{2}\right), \tag{12.41}$$

where $K(m)$ is the *complete elliptic integral of the first kind*, defined as (Abramowitz and Stegun 1972)

$$K(m) = \int_0^1 \frac{dz}{\sqrt{\left(1 - z^2\right)\left(1 - mz^2\right)}}. \tag{12.42}$$

Listing 12.4   Oscillations of a Nonlinear Pendulum by the Runge–Kutta Method (Python Coding)

```python
# Angular motion of a nonlinear pendulum by the Runge-Kutta method
#    u" = -g/l * sin(u) - k * u',    u(0) = u0, u'(0) = u0'
from math import *
from ode import *

g = 9.81e0                                    # gravitational acceleration

def Func(t, u, f):                            # RHS of 1st order ODEs
   f[1] =  u[2]                               # u[1] = u, u[2] = u'
   f[2] = -g/l * sin(u[1]) - k * u[2]

# main

l = 1e0                                             # pendulum length
k = 0e0                                           # velocity coefficient
u0 = 0.5e0*pi                                    # initial displacement
du0 = 0e0                                         # initial derivative
tmax = 20e0                                             # time span
ht = 0.001e0                                        # time step size

n = 2                                        # number of 1st order ODEs
u = [0]*(n+1)                                    # solution components

out = open("pendulum.txt","w")                       # open output file
out.write("        t          u         du\n")

t = 0e0
u[1] = u0; u[2] = du0                                 # initial values
out.write(("{0:10.5f}{1:10.5f}{2:10.5f}\n").format(t,u[1],u[2]))

nT = 0                                        # number of half-periods
t1 = t2 = 0e0                                 # bounding solution zeros
us = u[1]                                             # save solution
while (t+ht <= tmax):                             # propagation loop
   RungeKutta(t,ht,u,n,Func)
   t += ht

   if (u[1]*us < 0e0):                 # count solution passages through zero
      if (t1 == 0): t1 = t                             # initial zero
      else: t2 = t; nT += 1                            # final zero
   us = u[1]                                          # save solution

   out.write(("{0:10.5f}{1:10.5f}{2:10.5f}\n").format(t,u[1],u[2]))

T = 2e0*(t2-t1) / nT                              # calculated period
T0 = 2e0*pi*sqrt(l/g)                               # harmonic period
print("u0 = {0:7.5f}   T/T0 = {1:7.5f}".format(u0,T/T0))

out.close()
```

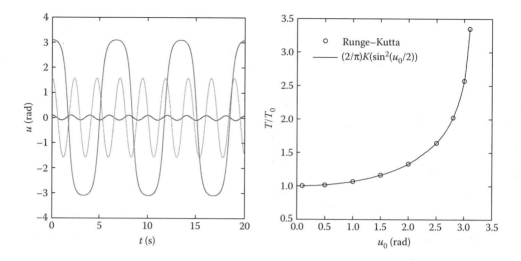

FIGURE 12.3 Solution $u(t)$ of the Cauchy problem of the nonlinear pendulum yielded by the fourth-order Runge–Kutta method for a step size $h = 0.001$ and initial angular displacements $u_0 = 0.1$, $\pi/2$, and 3.1 (left panel). Relative period T/T_0 as a function of the initial angular displacement u_0, and comparison with the theoretical dependence $(2/\pi)K(sin^2(u_0/2))$ (right panel).

To solve numerically Problem 12.39–12.40 by one of the Runge–Kutta methods, one has to transform it first into a problem for a system of first-order ODEs:

$$\begin{cases} u'_1 = u_2, \\ u'_2 = -(g/l)\sin(u_1) - ku_2, \end{cases} \qquad \begin{cases} u_1(0) = u_0, \\ u_2(0) = 0. \end{cases}$$

In addition to the initial model, here, provision is also made for a velocity-dependent damping term ($u_2 = u'$) with a velocity coefficient k.

In Program 12.4, the right-hand-side terms of the above ODEs are coded within the user function Func. Since Func has to comply with the parameter list expected by the routine RungeKutta, it gets access to the problem-specific quantities l and k through the global parameters l and k, which are assigned values in the main program.

The main program determines the oscillation period by an elementary, yet fail-safe approach. Specifically, the propagation loop identifies the first and last zeros of the solution $u_1(t)$ by the sign changes, and counts the encompassed half-periods. The period then results by dividing the time span between the extreme zeros by the number of half-periods.

Figure 12.3 shows in the left panel the tremendous differences between the time-dependent solutions obtained by running the code for the initial angular displacements $u_0 = 0.01$, $\pi/2$, and 3.1, in the absence of friction forces ($k = 0$). The right panel shows the pronounced rise with increasing initial angular displacement u_0 and the perfect match between the calculated relative periods T/T_0 and the exact analytical result (12.41).

12.5 Adaptive Step Size Control

The choice of an appropriate propagation step size h is of critical importance in numerically solving initial-value problems. Using a constant, overestimated value for h deteriorates in most applications the stability and leads to the unraveling of the solution due to aberrant propagation of the truncation errors. On the other hand, a too small step size, even though successful in overcoming domains of strong solution variation, may result in a prohibitive computational effort in regions of reduced variation. An efficient algorithm for solving Cauchy problems for ODEs is actually expected to exert an adaptive control over the step size to achieve a targeted overall precision of the solution.

The essential information required for automatic step size control is contained in the *truncation error*. In the following, we discuss the simplest way of estimating the truncation error, namely by the *step-halving method*. To this end, we compare two estimates of the solution, $\tilde{y}_{m+1}$ and y_{m+1}, obtained by propagating an initial solution $y_m \equiv y(t_m)$ in a single step of size h, and, respectively, in two steps of size $h/2$, whereby also producing the intermediate value $y_{m+1/2}$ for $t_m + h/2$. The difference between the two approximations (with y_{m+1} naturally more precise) provides useful estimates of the absolute truncation error:

$$\Delta_{m+1} = \left| y_{m+1} - \tilde{y}_{m+1} \right|, \tag{12.43}$$

and of the corresponding relative error:

$$\delta_{m+1} = \frac{\Delta_{m+1}}{\left| y_{m+1} \right|} = \left| 1 - \frac{\tilde{y}_{m+1}}{y_{m+1}} \right|. \tag{12.44}$$

The adjustment mechanism of the integration step size mainly relies on the specific *order* of the truncation error. From this perspective, there are basically two practical options: to take as reference the *local* or the *global* truncation error. As discussed in Section 12.3, the solution provided by an ODE solver of order p is equivalent with the Taylor series including terms up to $O(h^p)$ and it displays, consequently, a local truncation error $O(h^{p+1})$ and a one order lower global truncation error $O(h^p)$. Since, in applications, it appears to be generally more meaningful to control the global error, we associate in the following the relative error δ_{m+1} with the *global truncation error* and we assume it to scale with the integration step size such as:

$$\delta_{m+1} \approx K h^p. \tag{12.45}$$

Setting a maximum admissible error ε, one may assume that it is not exceeded for step sizes below a maximum value h_1, which is expected to be linked to ε by a dependence similar to the above one:

$$\varepsilon \approx K_1 h_1^p. \tag{12.46}$$

From the ratio of these two dependences, one may predict the maximum usable step size, h_1, in terms of the one effectively used, h, and the resulting error, δ_{m+1}:

$$h_1 \approx \left(\frac{\varepsilon}{\delta_{m+1}} \right)^{1/p} h. \tag{12.47}$$

If the occurring error δ_{m+1} exceeds the admissible one ($\delta_{m+1} > \varepsilon$), this formula prescribes a reduced step size, to be used in repeating the propagation step. On the contrary, if $\delta_{m+1} < \varepsilon$, the formula specifies an increased step size, which could have been used to propagate the solution without exceeding ε. In the latter case, there is actually no need to repeat the step, since the desired precision is already attained and the step size guess h_1 may be rather used at the next propagation step.

One should bear in mind, though, that Formula 12.47 is not an exact result and it merely provides a *prediction* of the step size to be used. Moreover, for rapidly varying solutions, the error estimate δ_{m+1} itself may depart significantly from its exact value. All things considered, one can further fine-tune the prediction h_1 by an ad hoc subunitary factor (see for variants Burden and Faires 2010 and Press et al. 2007), aimed at causing an overall reduction of the employed step sizes.

The scheme that we actually use to adjust the integration step size in the following implementations is

$$h_1 = 0.9 \left(\frac{\varepsilon}{\delta_{m+1}}\right)^{1/p} h, \quad h_1/h \leq 5, \tag{12.48}$$

which favors the reduction of the step size by the factor 0.9 and prevents increases by factors larger than 5.

The routine RKadapt (Listing 12.5) implements the described algorithm and employs as the basic ODE solver the fourth-order Runge–Kutta method ($p = 4$). The three calls to the RungeKutta solver (developed in Section 12.4) return in yt the initial solution propagated in a single step from t to $t + h$ and, respectively, in yt2 the initial solution propagated in a first step to $t + h/2$, and in a second one to $t + h$. The argument ht received on entry represents just a trial step size h for the current propagation process.

Listing 12.5 Adaptive ODE Solver Using Step-Halving and the Runge–Kutta Method (Python Coding)

```
#===========================================================================
def RKadapt(t, ht, eps, y, n, Func):
#---------------------------------------------------------------------------
#  Propagates the solution y of a system of 1st order ODEs
#     y'[i] = f[i](t,y[]), i = 1..n
#  from t to t+ht using 4th order Runge-Kutta and adaptive step size control
#
#  ht   - initial step size (input); final step size (output):
#         ht is unchanged if err <= eps; ht is reduced if err > eps
#  ht1  - step size guess for next propagation (output)
#  eps  - max. relative error of solution components
#  Calls: Func(t, y, f) - RHS of ODEs
#---------------------------------------------------------------------------
   p = 4                                      # order of basic ODE solver
   itmax = 10                                 # max. no. of step size reductions

   yt = [0]*(n+1); yt2 = [0]*(n+1)

   for it in range(1,itmax+1):                # loop of step size scaling
      ht2 = ht/2e0
      for i in range(1,n+1): yt2[i] = yt[i] = y[i]   # initialize trial sol.
      RungeKutta(t,ht,yt,n,Func)                      # t -> t+ht
      RungeKutta(t,ht2,yt2,n,Func)                    # t -> t+ht/2
      RungeKutta(t+ht2,ht2,yt2,n,Func)                # t+ht/2 -> t+ht

      err = 0e0                               # max. error of solution components
      for i in range(1,n+1):
         erri = abs(1e0 - yt[i]/yt2[i]) if yt2[i] else abs(yt2[i] - yt[i])
         if (err < erri): err = erri

      f = 1e0                                 # scaling factor for step size
      if (err): f = 0.9e0*pow(eps/err,1e0/p)
      if (f > 5e0): f = 5e0                             # prevent increase > 5x
      ht1 = f * ht                                      # guess for next step size
      if (err <= eps): break                            # precision attained: exit
      ht = ht1                                          # reduce step size and repeat

   if (it > itmax): print("RKadapt: max. no. of iterations exceeded !")
   for i in range(1,n+1): y[i] = yt2[i]      # update y with the best solution
   return (ht, ht1)
```

If the estimated truncation error `err` exceeds the target value `eps`, the routine iteratively reduces `ht` according to Equation 12.48 and takes repeated propagation steps from the same initial condition until the precision criterion is met. The relative truncation error `err` is evaluated as maximum relative difference between the components of the solutions `yt` and `yt2`. Once `err` drops below `eps`, the trial propagation is accepted and the process exits from the step size scaling loop. Along with the effectively used step size `ht`, the routine also returns via `ht1` a guess for the step size to be used at the next propagation step. In any case, an increase of more than 5 times of the step size is explicitly prevented.

A significantly improved efficiency is achieved in practice by using, instead of step halving for a single ODE solver, rather two embedded solvers of consecutive orders, p and $p + 1$, sharing the same defining coefficients and enabling the estimation of the truncation error in a single process. Such a scheme is the Runge–Kutta–Fehlberg method (Fehlberg 1966, 1970), in which, two particular $O(h^5)$ and $O(h^6)$ Runge–Kutta methods are used jointly:

$$\tilde{y}_{m+1} = y_m + \frac{25}{216}k_1 + \frac{1408}{2565}k_3 + \frac{2197}{4104}k_4 - \frac{1}{5}k_5, \tag{12.49}$$

$$y_{m+1} = y_m + \frac{16}{135}k_1 + \frac{6656}{12,825}k_3 + \frac{28,561}{56,430}k_4 - \frac{9}{50}k_5 + \frac{2}{55}k_6. \tag{12.50}$$

The common defining coefficients are

$$k_1 = hf(t_m, y_m), \tag{12.51}$$

$$k_2 = hf\left(t_m + \frac{1}{4}h, y_m + \frac{1}{4}k_1\right),$$

$$k_3 = hf\left(t_m + \frac{3}{8}h, y_m + \frac{3}{32}k_1 + \frac{9}{32}k_2\right),$$

$$k_4 = hf\left(t_m + \frac{12}{13}h, y_m + \frac{1932}{2197}k_1 - \frac{7200}{2197}k_2 + \frac{7296}{2197}k_3\right),$$

$$k_5 = hf\left(t_m + h, y_m + \frac{439}{216}k_1 - 8k_2 + \frac{3680}{513}k_3 - \frac{845}{4104}k_4\right),$$

$$k_6 = hf\left(t_m + \frac{1}{2}h, y_m - \frac{8}{27}k_1 + 2k_2 - \frac{3544}{2565}k_3 + \frac{1859}{4104}k_4 - \frac{11}{40}k_5\right).$$

An estimate of the truncation error results by subtracting Equation 12.49 from Equation 12.50:

$$\Delta_{m+1} = y_{m+1} - \tilde{y}_{m+1} = \frac{1}{360}k_1 - \frac{128}{4275}k_3 - \frac{2197}{75,240}k_4 + \frac{1}{50}k_5 + \frac{2}{55}k_6 + O(h^6), \tag{12.52}$$

and also provides, according to Equation 12.44, an approximation for the relative truncation error. These error estimates are obtained at each step by six evaluations of the right-hand side function f, instead of double as many (3×4) in the case of applying step halving for the fourth-order Runge–Kutta method.

The overall structure, parameters, and functionality of the routine `RKFehlberg` given in Listing 12.6 are similar with those of function `RKadapt`. The main differences concern the replacement of the three calls of the RungeKutta solver from RKadapt with the built-in calculation of the coefficients k_j (Equations 12.51), of the $O(h^6)$ approximation of the solution y_{m+1} (Equation 12.50), and of the estimate Δ_{m+1} of the absolute truncation error (Equation 12.52).

Listing 12.6   Adaptive ODE Solver Based on the Runge–Kutta–Fehlberg Method (Python Coding)

```
#===============================================================================
def RKFehlberg(t, ht, eps, y, n, Func):
#-------------------------------------------------------------------------------
#  Propagates the solution y of a system of 1st order ODEs
#     y'[i] = f[i](t,y[]), i = 1..n
#  from t to t+ht using the Runge-Kutta-Fehlberg method with stepsize control
#
#  ht    - initial step size (input); final step size (output):
#          ht is unchanged if err <= eps; ht is reduced if err > eps
#  ht1   - step size guess for next propagation (output)
#  eps   - max. relative error of solution components
#  Calls: Func(t, y, f) - RHS of ODEs
#-------------------------------------------------------------------------------
   p = 5                                            # order of basic ODE solver
   itmax = 10                                       # max. no. of step size reductions
                                                    # Runge-Kutta-Fehlberg coefficients
   a2 = 1e0/4e0; a3 = 3e0/8e0; a4 = 12e0/13e0; a5 = 1e0; a6 = 1e0/2e0
   b21 = 1e0/4e0; b31 = 3e0/32e0; b32 = 9e0/32e0
   b41 = 1932e0/2197e0; b42 = -7200e0/2197e0; b43 = 7296e0/2197e0
   b51 = 439e0/216e0; b52 = -8e0; b53 = 3680e0/513e0; b54 = -845e0/4104e0
   b61 = -8e0/27e0; b62 = 2e0; b63 = -3544e0/2565e0; b64 = 1859e0/4104e0
   b65 = -11e0/40e0
   c1 = 16e0/135e0; c3 = 6656e0/12825e0; c4 = 28561e0/56430e0
   c5 = -9e0/50e0; c6 = 2e0/55e0
   e1 = 1e0/360e0; e3 = -128e0/4275e0; e4 = -2197e0/75240e0
   e5 = 1e0/50e0; e6 = 2e0/55e0

   f1 = [0]*(n+1); f2 = [0]*(n+1); f3 = [0]*(n+1); f4 = [0]*(n+1)
   f5 = [0]*(n+1); f6 = [0]*(n+1); yt = [0]*(n+1)

   for it in range(1,itmax+1):                      # loop of step size scaling
      Func(t,y,f1)
      for i in range(1,n+1):
         yt[i] = y[i] + ht*b21*f1[i]
      Func(t+a2*ht,yt,f2)
      for i in range(1,n+1):
         yt[i] = y[i] + ht*(b31*f1[i] + b32*f2[i])
      Func(t+a3*ht,yt,f3)
      for i in range(1,n+1):
         yt[i] = y[i] + ht*(b41*f1[i] + b42*f2[i] + b43*f3[i])
      Func(t+a4*ht,yt,f4)
      for i in range(1,n+1):
         yt[i] = y[i] + ht*(b51*f1[i] + b52*f2[i] + b53*f3[i] + b54*f4[i])
      Func(t+a5*ht,yt,f5)
      for i in range(1,n+1):
         yt[i] = y[i] + ht*(b61*f1[i] + b62*f2[i] + b63*f3[i] + b64*f4[i] + \
                            b65*f5[i])
      Func(t+a6*ht,yt,f6)

      err = 0e0                                      # max. error of solution components
      for i in range(1,n+1):                         # O(h5) solution estimate
         yt[i] = y[i] + \
```

```
                    ht*(c1*f1[i] + c3*f3[i] + c4*f4[i] + c5*f5[i] + c6*f6[i])
                                                     # error estimate
         erri = ht*(e1*f1[i] + e3*f3[i] + e4*f4[i] + e5*f5[i] + e6*f6[i])
         erri = fabs(erri/yt[i])
         if (err < erri): err = erri

      f = 1e0                                 # scaling factor for step size
      if (err): f = 0.9e0*pow(eps/err,1e0/p)
      if (f > 5e0): f = 5e0                         # prevent increase > 5x
      ht1 = f * ht                                 # guess for next step size
      if (err <= eps): break                     # precision attained: exit
      ht = ht1                               # reduce step size and repeat

   if (it > itmax): print("RKFehlberg: max. no. of iterations exceeded !")
   for i in range(1,n+1): y[i] = yt[i]      # update y with the best solution
   return (ht, ht1)
```

As in the case of the RungeKutta routine, we do not actually evaluate the coefficients k_j themselves, but rather the involved values of the right-hand-side function *f*, and these are stored in the local arrays f1, f2, f3, f4, f5, and f6. For the rest, the control mechanism of the integration step size is identical with the one implemented in RKadapt.

The exit from the scaling loop of the step size is initiated once the maximum relative error of the solution components, err, drops under the desired relative tolerance eps. Anyway, increases of more than 5 times of the step size are prevented. The particular Runge–Kutta–Fehlberg coefficients coded in RKFehlberg can be readily replaced with the coefficients specific to other embedded adaptive ODE solvers.

As an illustration for the operation of the RKFehlberg solver, let us consider a Cauchy problem for a second-order ODE with a strongly discontinuous coefficient:

$$y'' = f(t)y, \quad y(0) = 0, \quad y'(0) = 1, \tag{12.53}$$

$$f(t) = \begin{cases} 1, & 0 \le t \le 5, \\ -100, & t > 5. \end{cases}$$

The exact analytic solution can be easily shown to consist of two functionally different restrictions—a rapidly growing hyperbolic sine and a combination of wildly oscillating harmonic functions:

$$y(t) = \begin{cases} \sinh t, & 0 \le t \le 5, \\ \sinh 5 \cos 10(t-5) + \frac{1}{10} \cosh 5 \sin 10(t-5), & t > 5. \end{cases}$$

It is obvious that a constant integration step size cannot ensure a uniform precision throughout the definition domain. A step size, small enough to cope with the functional discontinuity and reproduce the oscillatory behavior, results in a waste of effort in the hyperbolic part. In such cases, adaptive solvers are definitely the methods of choice. Program 12.7 implements in Func the right-hand sides of the equivalent system of first-order ODEs and repeatedly calls the routine RKFehlberg to advance the solution up to tmax. The propagation loop is supposed to update the trial step size ht with the guess ht1 returned by the preceding call to RKFehlberg. The solution $y(t)$ is shown in the left panel of Figure 12.4, while the pronounced time dependence of the employed step size is depicted in the right panel. For the same relative tolerance $\varepsilon = 10^{-8}$, the two forms of the solution can be seen to require step sizes differing by more than one order of magnitude.

Listing 12.7   Application of Adaptive Step Size Control (Python Coding)

```python
#  Solves a Cauchy problem with adaptive step size control
from math import *
from ode import *

def Func(t, y, f):                                      # RHSs of ODEs
   f[1] = y[2]
   f[2] = (1e0 if t < 5e0 else -100e0) * y[1]

# main

y0 = 0e0; dy0 = 1e0                                      # initial values
tmax = 10e0                                              # time span
ht = 0.01e0                                              # step size
eps = 1e-8                                     # relative solution precision

n = 2                                          # number of 1st order ODEs
y = [0]*(n+1)                                      # solution components

out = open("ode.txt","w")                               # open output file
out.write("       t            y           h\n")

t = 0e0
y[1] = y0; y[2] = dy0                                    # initial values
out.write(("{0:10.5f}{1:10.5f}{2:10.5f}\n").format(t,y[1],ht))

ht1 = ht                                         # initial step size guess
while (t+ht <= tmax):                                    # propagation loop
   ht = ht1                               # update initial step size with guess
   (ht, ht1) = RKFehlberg(t,ht,eps,y,n,Func)
   t += ht

   out.write(("{0:10.5f}{1:10.5f}{2:10.5f}\n").format(t,y[1],ht))

out.close()
```

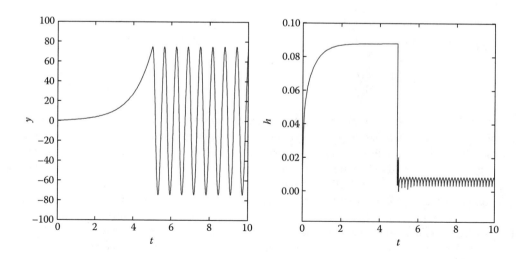

FIGURE 12.4   Solution of the Cauchy problem 12.53 using adaptive step size control based on the Runge–Kutta–Fehlberg method (left panel). Time evolution of the employed integration step size (right panel).

12.6 Methods for Second-Order ODEs

Cauchy problems for second-order ODEs arise in many areas of applied sciences and engineering, most frequently describing time or space dependences of the continuous quantities of interest. However different may seem two of the prototypical second-order ODEs, namely, Newton's second law and the 1D time-independent Schrödinger equation, the numerical methods applicable for their solution share many common features.

Let us thus consider the generic Cauchy problem:

$$\frac{d^2 y}{dt^2} = f(t, y, y'), \tag{12.54}$$

$$y(t_0) = y_0, \quad y'(t_0) = y'_0. \tag{12.55}$$

The algorithms developed for solving it typically derive from the Taylor series method (see Section 12.2), with the first derivative y' (often having a well-defined significance within the underlying formalism) being propagated explicitly along with the solution.

Considering two consecutive values of the independent variable, t_m and t_{m+1}, separated by $h = t_{m+1} - t_m$, and starting from the known initial values $y_m \equiv y(t_m)$ and $y'_m \equiv y'(t_m)$, the propagation of the solution to t_{m+1} can be performed based on the Taylor expansions of y and y', with the second derivative y'' determined directly from the ODE. The relevant equations arranged in algorithmic order read:

$$y''_m = f(t_m, y_m, y'_m),$$
$$y_{m+1} = y_m + h y'_m + \frac{1}{2} h^2 y''_m + O(h^3), \tag{12.56}$$
$$y'_{m+1} = y'_m + h y''_m + O(h^2).$$

One defining characteristic of all algorithms built on these equations is that each propagation step requires just *a single evaluation* of the second derivative from the right-hand side of the ODE. Practical implementations actually use variants of these equations, tuned alternatively for simplicity, precision, or storage efficiency.

The simplest variant is the *Euler method*, which is the equivalent for second-order ODEs of the approach described in Section 12.3. Euler's method essentially implies retaining the linear approximations in the propagators of both the position and the velocity:

$$y''_m = f(t_m, y_m, y'_m),$$
$$y_{m+1} = y_m + h y'_m + O(h^2), \tag{12.57}$$
$$y'_{m+1} = y'_m + h y''_m + O(h^2).$$

Since it uses only information from t_m (not also from t_{m+1}), this scheme is also known as *first-point approximation* and its obvious advantage is the ease of implementation.

In spite of its appealing simplicity, Euler's method is hardly recommendable due to its nonnegligible downsides. First, with its $O(h)$ global truncation error, it is the *lowest-order* solver for second-order ODEs. Second, due to the rapid propagation of the truncation errors, it often yields *unstable solutions*, especially for oscillatory systems. Third, since it only uses information from t_m, the propagation from t_m to t_{m+1} is *not time reversible* (the propagator is not symmetric in t).

The so-called *Euler–Cromer method* (Cromer 1981) uses the same $O(h)$ approximations of the Taylor series, but, very importantly, it cures the instability of the Euler method. The basic relations are formally obtained by the simple inversion of the equations for y'_{m+1} and y_{m+1}, and by using in the latter one, instead of the initial derivative y'_m, the updated value y'_{m+1}:

$$y''_m = f(t_m, y_m, y'_m),$$
$$y'_{m+1} = y'_m + hy''_m + O(h^2), \tag{12.58}$$
$$y_{m+1} = y_m + hy'_{m+1} + O(h^2).$$

In implementations, the preceding evaluation of y'_{m+1} and its subsequent use in y_{m+1} allows for a single program variable to be allocated both for y'_m and y'_{m+1}. Moreover, the use of the updated derivative, thus of information from both t_m and t_{m+1}, proves to bring about a radical improvement in the solution's stability (Cromer 1981). In terms of the solution's precision, though, the Euler–Cromer method maintains the same low $O(h)$ order as the original Euler method.

The *velocity Verlet method* (Swope et al. 1982), considered in the following, belongs to a family of algebraically equivalent $O(h^2)$ algorithms, which includes the original *Verlet method* (Verlet 1967) and the *Leapfrog method* (Hockney and Eastwood 1988). In general, these are employed for solving Newton's equations of motion in molecular dynamics simulations (Rapaport 2004), but may as well be used for solving initial-value problems for other second-order ODEs. The main difference between the Euler/Euler–Cromer and the velocity Verlet algorithms regards the additional retention within the latter of the *parabolic terms* in the Taylor expansion of the solution. Concretely, the propagated solution may be expressed successively as

$$y_{m+1} = y_m + hy'_m + \frac{1}{2}h^2 y''_m + O(h^3)$$

$$= y_m + h\left(y'_m + \frac{1}{2}hy''_m\right) + O(h^3)$$

$$= y_m + hy'_{m+1/2} + O(h^3).$$

The solution y_{m+1} appears to be propagated linearly from t_m to t_{m+1} based on the derivative $y'_{m+1/2}$ at mid-interval, that is, at $t_{m+1/2} = t_m + h/2$. From a different perspective, $y'_{m+1/2}$ can be regarded as the *average first derivative* over the interval $[t_m, t_{m+1}]$. The propagated solution y_{m+1} substituted in the ODE determines the second derivative y''_{m+1} at the interval end, which, in turn, also enables the propagation of the first derivative $y'_{m+1/2}$ up to t_{m+1}.

The algorithmic sequence of propagation formulas of the velocity Verlet method finally takes the form:

$$y'_{m+1/2} = y'_m + \frac{1}{2}hy''_m + O(h^2),$$
$$y_{m+1} = y_m + hy'_{m+1/2} + O(h^3), \tag{12.59}$$
$$y''_{m+1} = f(t_{m+1}, y_{m+1}, y'_{m+1/2}),$$
$$y'_{m+1} = y'_{m+1/2} + \frac{1}{2}hy''_{m+1} + O(h^2).$$

Besides ensuring $O(h^2)$ precision (superior by one order over the Euler and Euler–Cromer methods), the velocity Verlet method also stands out by stability and the small storage requirements facilitated by the algorithmic possibility of all quantities to overwrite themselves along the propagation. Nevertheless, an aspect requiring special attention needs to be emphasized. The second derivative y''_{m+1} determined at one step is necessary not only to propagate the first derivative $y'_{m+1/2}$ over the second interval half, from $t_{m+1/2}$ to t_m, but also to advance y'_{m+1} over the first half of the next step, from t_{m+1} to $t_{m+3/2}$. For this reason, along with y_{m+1} and y'_{m+1}, also, the second derivative y''_{m+1} needs to be saved for being used at the next step.

The implementations of the Euler, Euler–Cromer, and velocity Verlet methods for the solution of initial-value problems for single second-order ODE are presented in Listing 12.8.

Listing 12.8   Solvers for Cauchy Problems for Second-Order ODEs (Python Coding)

```
#==============================================================================
def Euler1(t, ht, y, dy, Func):
#------------------------------------------------------------------------------
#  Propagates the solution y and 1st derivative dy of a 2nd order ODE from t
#  to t+ht using Euler's method
#------------------------------------------------------------------------------
   d2y = Func(t,y,dy)                                        # d2y -> t

   y  += ht * dy                                             # y -> t+ht
   dy += ht * d2y                                            # dy -> t+ht
   return (y, dy)


#==============================================================================
def EulerCromer1(t, ht, y, dy, Func):
#------------------------------------------------------------------------------
#  Propagates the solution y and the 1st derivative dy of a 2nd order ODE
#  from t to t+ht using the Euler-Cromer method
#------------------------------------------------------------------------------
   d2y = Func(t,y,dy)                                        # d2y -> t

   dy += ht * d2y                                            # dy -> t+ht
   y  += ht * dy                                             # y -> t+ht
   return (y, dy)


#==============================================================================
def Verlet1(t, ht, y, dy, d2y, Func):
#------------------------------------------------------------------------------
#  Propagates the solution y and the 1st derivative dy of a 2nd order ODE
#  from t to t+ht using the Verlet method; returns 2nd derivative in d2y;
#  d2y needs to be initialized on first call and saved between calls
#------------------------------------------------------------------------------
   ht2 = 0.5e0 * ht
   dy += ht2 * d2y                                           # dy -> t+ht/2
   y  += ht  * dy                                            # y -> t+ht

   d2y = Func(t,y,dy)                                        # d2y -> t+ht

   dy += ht2 * d2y                                           # dy -> t+ht
   return (y, dy, d2y)
```

Having in view the particular practical importance of the initial-value problems for Newton's equation of motion,

$$m\frac{d^2\mathbf{r}}{dt^2} = \mathbf{F}(t,\mathbf{r},\mathbf{v}), \quad \mathbf{r}(t_0) = \mathbf{r}_0, \quad \mathbf{v}(t_0) = \mathbf{v}_0,$$

where m is the mass, $\mathbf{r}$ and $\mathbf{v}$ are the position and the velocity, and $\mathbf{F}$ is the force acting on the considered object, the velocity Verlet algorithm is usually formulated as

$$\mathbf{v}_{m+1/2} = \mathbf{v}_m + (h/2)\,\mathbf{a}_m.$$

$$\mathbf{r}_{m+1} = \mathbf{r}_m + h\,\mathbf{v}_{m+1/2}.$$

$$\mathbf{a}_{m+1} = (1/m)\mathbf{F}(t_{m+1},\mathbf{r}_{m+1},\mathbf{v}_{m+1/2}).$$

$$\mathbf{v}_{m+1} = \mathbf{v}_{m+1/2} + (h/2)\,\mathbf{a}_{m+1}.$$

(12.60)

Listing 12.9   2D Velocity Verlet Propagator for a Single Particle (Python Coding)

```
#===========================================================================================
def Verlet2(ht, m, x, y, vx, vy, ax, ay, Forces):
#-------------------------------------------------------------------------------------------
#   Propagates the 2D solution of Newton's equations of motion for a particle of mass m
#   over a time interval ht using the velocity Verlet method
#   x, y   - position components
#   vx, vy - velocity components
#   ax, ay - acceleration components (need to be initialized on 1st call)
#   Ekin   - kinetic energy
#   Epot   - potential energy
#-------------------------------------------------------------------------------------------
   ht2 = 0.5e0 * ht
   vx += ht2 * ax; x += ht * vx                                          # v -> t+h/2
   vy += ht2 * ay; y += ht * vy                                          # r -> t+h

   (fx, fy, Epot) = Forces(m,x,y,vx,vy)                                      # forces

   ax = fx/m; ay = fy/m                                                  # a -> t+h
   vx += ht2 * ax                                                        # v -> t+h
   vy += ht2 * ay

   Ekin = 0.5e0 * m * (vx*vx + vy*vy)                                 # kinetic energy

   return (x, y, vx, vy, ax, ay, Ekin, Epot)
```

In the first stage of each propagation step, the velocity $\mathbf{v}$ is promoted to mid-interval ($\mathbf{v}_{m+1/2}$) and the position $\mathbf{r}$, all the way to the end of the time interval ($\mathbf{r}_{m+1}$). The force $\mathbf{F}$ acting on the particle, calculated with the new position, determines the corresponding acceleration ($\mathbf{a}_{m+1}$). In the second stage, the mid-interval velocity (also considered as average velocity over $[t_m, t_{m+1}]$) is propagated up to the end of the time interval ($\mathbf{v}_{m+1}$), completing the dynamic information on the moving body for t_{m+1}.

Listing 12.10   Velocity Verlet Propagator for a System of Particles in Interaction (Python Coding)

```
#===========================================================================================
def Verlet(ht, m, x, y, z, vx, vy, vz, ax, ay, az, n, Forces):
#-------------------------------------------------------------------------------------------
#   Propagates the solution of Newton's equations of motion for a system of n particles over
#   a time interval ht using the velocity Verlet method
#   m           - masses of particles
#   x, y, z     - positions
#   vx, vy, vz  - velocities
#   ax, ay, az  - accelerations (need to be initialized on 1st call)
#   Ekin        - total kinetic energy
#   Epot        - total potential energy
#-------------------------------------------------------------------------------------------
   ht2 = 0.5e0 * ht
   for i in range(1,n+1):
      vx[i] += ht2 * ax[i]; x[i] += ht * vx[i]                           # v -> t+h/2
      vy[i] += ht2 * ay[i]; y[i] += ht * vy[i]                           # r -> t+h
      vz[i] += ht2 * az[i]; z[i] += ht * vz[i]

   Epot = Forces(m,x,y,z,ax,ay,az,n)                                         # forces

   Ekin = 0e0
   for i in range(1,n+1):                                               # corrector step
      ax[i] /= m[i]; ay[i] /= m[i]; az[i] /= m[i]                           # a -> t+ht
      vx[i] += ht2 * ax[i]                                                  # v -> t+ht
      vy[i] += ht2 * ay[i]
      vz[i] += ht2 * az[i]

      Ekin += 0.5e0 * m[i] * (vx[i]*vx[i] + vy[i]*vy[i] + vz[i]*vz[i])

   return (Ekin, Epot)
```

Listing 12.11 Oscillations of a Nonlinear Pendulum by the Euler–Cromer Method (Python Coding)

```
# Angular motion of a nonlinear pendulum by the Euler-Cromer method
#    u" = -g/l * sin(u) - k * u',    u(0) = u0, u'(0) = u0'
from math import *
from ode import *

g = 9.81e0                                          # gravitational acceleration

def Func(t, u, v):
   return -g/l * sin(u) - k * v

# main

l = 1e0                                                        # pendulum length
k = 0e0                                                    # velocity coefficient
u0 = 0.5e0*pi                                              # initial displacement
du0 = 0e0                                                   # initial derivative
tmax = 20e0                                                         # time span
ht = 0.01e0                                                    # time step size

out = open("pendulum.txt","w")                              # open output file
out.write("       t           u          du\n")

t = 0e0
u = u0; du = du0
out.write(("{0:10.5f}{1:10.5f}{2:10.5f}\n").format(t,u,du))

while (t+ht <= tmax):                                       # propagation loop
   (u, du) = EulerCromer1(t,ht,u,du,Func)
   t += ht

   out.write(("{0:10.5f}{1:10.5f}{2:10.5f}\n").format(t,u,du))

out.close()
```

The 2D implementation of the propagation relations 12.60 in routine `Verlet2` (Listing 12.9) is applicable, for instance, for describing objects evolving along planar orbits. The 3D variant `Verlet` (Listing 12.10) generalizes the algorithm to n pairwise-interacting particles and shows the main features of the full-fledged propagators currently used in large-scale molecular dynamics simulations. Besides the quantities specific to the Verlet method by construction, `Verlet2` and `Verlet` also return the total kinetic and total potential energies, `Ekin` and `Epot`, which enable assessing the stability of the trajectory by the conservation of the total energy.

The choice of implementing in both routines separate variables (arrays) instead of higher-order structures for the individual Cartesian components of the positions, velocities, and accelerations brings about, indeed, a more complex argument list, but this complication is outweighed by certain advantages. To start with, the vectorization (low-level parallelization) of the code is enhanced in operation-intensive applications with logically independent, formally similar, operation threads. Also, the allocation and manipulation of lower-level data structures can be better fine-tuned. Last but not least, operations that couple all the Cartesian components can be coded in a more compact and efficient manner.

To illustrate the stability properties of the Euler, Euler–Cromer, and velocity Verlet methods, let us consider again problem (12.39)–(12.40) of the nonlinear pendulum. Program 12.11, written to solve

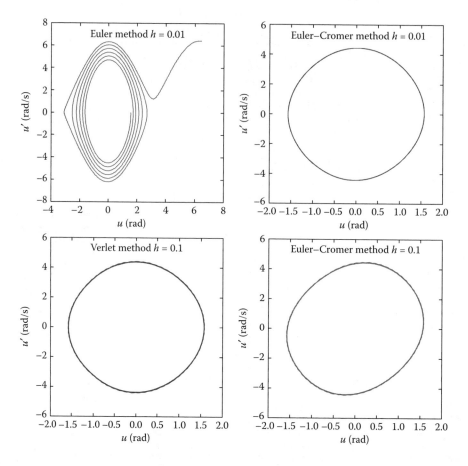

FIGURE 12.5 Trajectory of a nonlinear pendulum in phase space obtained for a step size $h = 0.01$, by using the Euler method (top left) and the Euler–Cromer method (top right); for a step size $h = 0.1$, by using the Verlet method (bottom left) and the Euler–Cromer methods (bottom right).

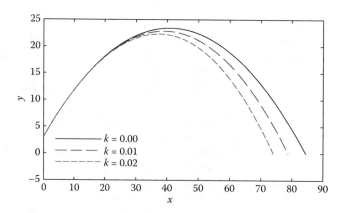

FIGURE 12.6 Trajectories of an obliquely thrown object experiencing quadratic drag, yielded by the velocity Verlet method. The medium-dash and the short-dash curves represent propagations with velocity coefficients $k = 0.01$ and $k = 0.02$, respectively.

Listing 12.12 Oblique Throw of an Object Using the Velocity Verlet Method (Python Coding)

```python
# Oblique throw of an object with drag using the velocity Verlet method
from math import *
from ode import *

g = 9.81e0                                      # gravitational acceleration

def Forces(m, x, y, vx, vy):
    fx = -k * vx*abs(vx)                                # force components
    fy = -k * vy*abs(vy) - m*g
    Epot = m*g*y                                        # potential energy
    return (fx, fy, Epot)

# main

m = 7e0                                             # mass of object
k = 0.01e0                                      # velocity coefficient
x0 = 0e0; y0 = 3e0                               # initial position
vx0 = 20e0; vy0 = 20e0                           # initial velocity
tmax = 20e0                                          # time span
ht = 0.001e0                                      # time step size

out = open("throw.txt","w")                      # open output file
out.write("      t          x          y          vx          vy\n")

t = 0e0
x = x0; vx = vx0; ax = 0e0                            # initial values
y = y0; vy = vy0; ay = 0e0
out.write(("{0:10.5f}{1:10.5f}{2:10.5f}{3:10.5f}{4:10.5f}\n"). \
          format(t,x,y,vx,vy))

while (t+ht <= tmax):                                # propagation loop
    (x,y,vx,vy,ax,ay,Ekin,Epot) = Verlet2(ht,m,x,y,vx,vy,ax,ay,Forces)
    t += ht

    out.write(("{0:10.5f}{1:10.5f}{2:10.5f}{3:10.5f}{4:10.5f}\n").
              format(t,x,y,vx,vy))
    if (y < 0.e0): break                       # stop if object hits the ground

out.close()
```

this problem, is able to use interchangeably the mentioned methods. In the case of using the `Verlet1` integrator, one needs to declare and initialize additionally in the main program a variable `d2y`, also meant to store the second derivative between calls to `Verlet1`.

In principle, in the absence of drag ($k = 0$), a stable propagator should produce a closed phase-space trajectory for the pendulum. From the top-left plot of Figure 12.5, the Euler method (routine `Euler1`) can be seen to be markedly unstable even for a step size as small as $h = 0.01$. In contrast and in spite of being equally only first-order accurate, the Euler–Cromer method (routine `EulerCromer1`) remains stable even for a 10-fold step size ($h = 0.1$) (top-right and bottom-right plots). Indeed, the trajectory is closed, thus *stable*, however considerably distorted, thus *inaccurate*. As for the velocity Verlet algorithm (routine `Verlet1`), its superior $O(h^2)$ precision brings about a stable and undistorted trajectory even for step sizes for which the Euler–Cromer method clearly fails to be accurate.

In the following, we exemplify the use of the 2D `Verlet2` integrator by the problem of the oblique throw of an object experiencing *quadratic drag* (air resistance proportional to the squared velocity). The mathematical formulation with x and y as horizontal and, respectively, vertical coordinates is

$$m\frac{d^2x}{dt^2} = -k\, v_x\, |v_x|, \quad x(0) = x_0, \quad v_x(0) = v_{x0},$$

$$m\frac{d^2y}{dt^2} = -k\, v_y\, |v_y| - mg, \quad y(0) = y_0, \quad v_y(0) = v_{y0}.$$

The particular way of expressing the drag force components, that is, by the product of the velocity components and the absolute value thereof, takes into account both the quadratic dependence and the opposed direction of the drag. The implementation is illustrated by Program 12.12 and the forces are coded in the user function `Forces`.

The above-mentioned particularity of the velocity Verlet algorithm, of using the second derivatives (accelerations) to propagate the first derivatives (velocities) both in the second half of a time step and in the first half of the subsequent step, makes it necessary for the accelerations (1) to be initialized for the first step, and (2) to be saved between the calls to the `Verlet` routine. Both requirements are solved simply by declaring the accelerations as variables of the main program, initializing them there, and exchanging them through the parameter list with the routine `Verlet`.

The output of Program 12.12 can be seen in Figure 12.6, with the impact point moving closer to the launching point with increasing drag (increasing velocity coefficient k).

12.7 Numerov's Method

Initial-value problems for second-order ODEs having the particular form

$$\frac{d^2y}{dx^2} = f(x)y + g(x), \tag{12.61}$$

that is, with absent first derivative and right-hand side linearly depending on the solution, can be solved most efficiently by using the *multistep method* developed by the astronomer B. Numerov (1923). Numerov's method owes its popularity to its high $O(h^6)$ precision, algorithmic simplicity, and direct applicability to solving Schrödinger's equation, as fundamental equation of wave mechanics. By the mere passage to vector notations, Numerov's method becomes equally suited for solving *systems* of second-order ODEs of form (12.1). It is noteworthy that a similar methodology to the one described below is also applicable to derive integrators for systems of higher-order ODEs that involve only *even-order derivatives*.

In the first stage, considering the solution y to have the necessary degree of smoothness (continuous derivatives up to the fifth order) and taking as reference its value y_m at some point x_m, one can express the values at $x_m + h$ and, respectively, at $x_m - h$ by the Taylor series:

$$y_{m+1} = y_m + hy'_m + \frac{1}{2!}h^2y''_m + \frac{1}{3!}h^3y_m^{(3)} + \frac{1}{4!}h^4y_m^{(4)} + \frac{1}{5!}h^5y_m^{(5)} + O(h^6),$$

$$y_{m-1} = y_m - hy'_m + \frac{1}{2!}h^2y''_m - \frac{1}{3!}h^3y_m^{(3)} + \frac{1}{4!}h^4y_m^{(4)} - \frac{1}{5!}h^5y_m^{(5)} + O(h^6).$$

The odd-order derivatives can be eliminated by adding these expansions,

$$y_{m+1} + y_{m-1} = 2y_m + h^2y''_m + \frac{1}{12}h^4y_m^{(4)} + O(h^6). \tag{12.62}$$

In fact, seeking an algebraic relation for the propagated solution y_{m+1} in terms of previous values, one needs to replace the remaining derivatives, y''_m and $y_m^{(4)}$, with finite-difference schemes, nevertheless preserving the overall $O(h^6)$ accuracy.

The second derivative y''_m can be readily expressed from the differential equation itself:

$$y''_m = f_m y_m + g_m, \tag{12.63}$$

where $f_m \equiv f(x_m)$ and $g_m \equiv g(x_m)$. As for $y_m^{(4)}$, it can be viewed as the second derivative of y'' and, invoking again the differential equation, it results in

$$y_m^{(4)} = \left[f(x)y + g(x) \right]''_m. \tag{12.64}$$

For second derivatives, in general, one can devise a finite-difference scheme by retaining the $O(h^4)$ approximation in Equation 12.62:

$$h^2 y''_m = y_{m+1} - 2y_m + y_{m-1} + O(h^4).$$

Applied in particular to $\left[f(x)y + g(x) \right]''_m$, this scheme transforms Equation 12.64 into

$$\begin{aligned} h^4 y_m^{(4)} = h^2 \left(f_{m+1} y_{m+1} - 2f_m y_m + f_{m-1} y_{m-1} \right) \\ + h^2 \left(g_{m+1} + 10g_m + g_{m-1} \right) + O(h^6). \end{aligned} \tag{12.65}$$

Inserting back Equations 12.63 and 12.65 into 12.62 and grouping the terms leads to *Numerov's formula*:

$$\begin{aligned} y_{m+1} = \Bigg[\left(2 + \frac{5}{6} h^2 f_m \right) y_m - \left(1 - \frac{1}{12} h^2 f_{m-1} \right) y_{m-1} \\ + \frac{1}{12} h^2 \left(g_{m+1} + 10g_m + g_{m-1} \right) \Bigg] \left(1 - \frac{1}{12} h^2 f_{m+1} \right)^{-1} + O(h^6). \end{aligned} \tag{12.66}$$

Finally, by defining the auxiliary quantities

$$u_m = 1 - \frac{1}{12} h^2 f_m, \tag{12.67}$$

Numerov's formula may be cast under the more expedient form:

$$\begin{aligned} y_{m+1} = \frac{(12 - 10u_m)\, y_m - u_{m-1} y_{m-1}}{u_{m+1}} \\ + \frac{1}{12} h^2 \frac{g_{m+1} + 10g_m + g_{m-1}}{u_{m+1}} + O(h^6). \end{aligned} \tag{12.68}$$

The downside of the Numerov method is that it is *not self-starting*, which is intrinsically linked to its multistep nature. Correspondingly, two initial solution values need to be provided before recurrence (12.68) can be initiated. This can be accomplished by an initial step performed with a self-starting algorithm (Euler–Cromer, Runge–Kutta, etc.), or, in especially demanding applications, by using the Taylor expansion of the solution, with embedded particularities of the ODE.

The Numerov method is not suited for boxed, general-purpose implementations. Aiming for efficient coding, it is rather preferable to take advantage of the particular form of the functions $f(x)$ and $g(x)$ and to adequately generate the two initial values required for starting the algorithm.

Let us consider for illustration the 1D time-independent Schrödinger equation for a particle of mass μ experiencing the potential $V(x)$:

$$-\frac{\hbar^2}{2\mu}\frac{d^2\psi}{dx^2} + V(x)\psi(x) = E\psi(x). \tag{12.69}$$

In dimensionless units, the equation can be simplified to

$$\psi'' = 2\left[V(x) - E\right]\psi, \tag{12.70}$$

and the relations defining its solution by the Numerov algorithm take the simple form:

$$u_m = 1 - \frac{1}{6}h^2\left[V(x_m) - E\right], \tag{12.71}$$

$$\psi_{m+1} = \frac{(12 - 10u_m)\,\psi_m - u_{m-1}\psi_{m-1}}{u_{m+1}}. \tag{12.72}$$

The implementation of Numerov's method for the solution of Cauchy problems for the wave equation (12.70) is shown in Listing 12.13.

Assuming the potential energy to be received tabulated in array V, the routine Numerov propagates the solution y, for a given energy E, starting from the initial values y0 and dy0 over nx equidistant mesh points. To properly start the recurrence with *two solution values*, an initial Euler–Cromer step is actually

Listing 12.13   Numerov Integrator for the 1D Schrödinger Equation (Python Coding)

```
#===============================================================================
def Numerov(E, V, x, y, nx, y0, dy0):
#-------------------------------------------------------------------------------
#  Propagates the solution of the dimensionless 1D Schrodinger equation
#      y" = 2 [V(x) - E] y, y(0) = y0, y'(0) = y'0
#  on a regular mesh x[] with nx points by the Numerov method. Receives the
#  energy in E, the tabulated potential in V[], and the initial conditions in
#  y0 and dy0. Returns the index of the divergence point (default is nx).
#-------------------------------------------------------------------------------
   hx = x[2] - x[1]                                      # propagation step size
   y[1] = y0; dy = dy0                                           # initial values

   d2y = 2e0 * (V[1] - E) * y[1]                        # initial Euler-Cromer step
   dy += hx * d2y
   y[2] = y[1] + hx * dy

   h6 = hx*hx/6e0
   um1 = 1e0 - h6 * (V[1] - E)                          # stack of auxiliary values
   um  = 1e0 - h6 * (V[2] - E)
   for m in range(2,nx):
      up1 = 1e0 - h6 * (V[m+1] - E)
      y[m+1] = ((12e0 - 10e0*um) * y[m] - um1 * y[m-1]) / up1
      um1 = um; um = up1                          # shift stack down for next step

      if (abs(y[m+1]) > 1e10): break                        # stop if y diverges

   return m                                           # index of divergence point
```

performed to calculate the solution at the second mesh point. The auxiliary values u_m (depending on the local potential values) do not need to be allocated an entire array, since they are used in triplets, and are rather stored as a stack of three variables, um1, um, and up1, corresponding to u_{m-1}, u_m, and u_{m+1}. They are downshifted at every step while updating the most advanced component, up1.

For many applications, it is essential to identify regions where the solution possibly starts diverging numerically. In fact, for infinite-range potentials and solutions expected to vanish asymptotically, it is highly likely that beyond a region where the solution becomes negligible and satisfies the asymptotic behavior, it starts growing, being overwhelmed by the accumulation of truncation and roundoff errors. Practically, in the case of diverging solutions ($|\psi(x)| > 10^{10}$), the propagation stops and the index of the divergence point, m, is returned along with the solution to the calling program for further actions to be taken.

The Numerov method is exemplified in the next section in conjunction with the so-called "shooting method" for determining bound states of Schrödinger's wave equation.

12.8 Shooting Methods for Two-Point Problems

Up to this point, we have discussed only methods for solving initial-value problems, for which the conditions complementing the ODE are specified at a single point. These relatively easy-to-use methods still prove useful when dealing with two-point boundary-value problems, provided they are embedded in a so-called "shooting" algorithm. Essentially, a shooting method transforms a two-point problem into an initial-value problem and attempts on a trial-and-error basis to reproduce the second (unused) boundary condition.

Let us consider the general second-order two-point problem with Dirichlet-type (fixed-value) boundary conditions:

$$\frac{d^2y}{dx^2} = f(x, y, y'), \tag{12.73}$$

$$y(x_a) = y_a, \quad y(x_b) = y_b, \tag{12.74}$$

whose solution $y(x)$ and first derivative $y'(x)$ are supposed to be continuous. The derivative $y'(x_a) = y'_a$ at the left interval boundary is not known *a priori*, but it is *univocally defined* by the boundary conditions. Supposing that y'_a was known from the beginning, the two-point problem could be replaced by the equivalent initial-value problem:

$$\frac{d^2y}{dx^2} = f(x, y, y'), \tag{12.75}$$

$$y(x_a) = y_a, \quad y'(x_a) = y'_a, \tag{12.76}$$

having the same solution and also satisfying the unused boundary condition $y(x_b) = y_b$.

Let us assume that we start the propagation of the solution, instead of the true (unknown) first derivative y'_a, with a pair of trial derivatives, $y'^{(1)}_a < y'^{(2)}_a$, and that we obtain, correspondingly, two different solutions of the problem, for which the values at the end point x_b are $y^{(1)}_b$ and, respectively, $y^{(2)}_b$.

If both $y^{(1)}_b$ and $y^{(2)}_b$ deviate by the same sign from the true end value y_b, by virtue of the continuity of the solution, it follows that the true derivative y'_a is not contained between the initial trial values $y'^{(1)}_a$ and

$y_a^{\prime(2)}$. On the contrary, if $y_b^{(1)}$ and $y_b^{(2)}$ fall on different sides of y_b, it follows that y_a' is located between $y_a^{\prime(1)}$ and $y_a^{\prime(2)}$, or

$$(y_b^{(1)} - y_b)(y_b^{(2)} - y_b) > 0 \Rightarrow y_a' \notin \left[y_a^{\prime(1)}, y_a^{\prime(2)} \right], \tag{12.77}$$

$$(y_b^{(1)} - y_b)(y_b^{(2)} - y_b) \leq 0 \Rightarrow y_a' \in \left[y_a^{\prime(1)}, y_a^{\prime(2)} \right]. \tag{12.78}$$

In a separation and root-refining strategy such as the ones discussed in Chapter 6, in the first case, one would drop the current interval and pass to a new trial interval $\left[y_a^{\prime(1)}, y_a^{\prime(2)} \right]$. In the second case, one would iteratively refine the derivative, and, implicitly, the solution, by bisecting the trial interval $\left[y_a^{\prime(1)}, y_a^{\prime(2)} \right]$, until the target interval $[y_b^{(1)}, y_b^{(2)}]$ bracketing the true end value y_b would decrease below a predefined tolerance.

The routine `Shoot` coded along these lines (see Listing 12.14) receives the entire x-mesh via vector `x`, along with the boundary values `ya` and `yb`, the limits `dy1` and `dy2` of the search interval for the first derivative, and the tolerance `eps`, within which the found solution is expected to meet the boundary value `yb`. For the repeated propagations from `x[1]` to `x[nx]` implied by the shooting algorithm, `Shoot` calls the auxiliary function `Propag`, which is a simple generic propagator based on the Euler–Cromer method.

`Shoot` initially performs two complete propagations for the input derivatives, `dy1` and `dy2`, and compares the corresponding solution deviations, `f1` and `f2`, from the desired boundary value `yb`. If `f1` and `f2` have the same sign, the interval between `dy1` and `dy2` certainly does not contain the sought initial derivative and the routine returns a fake `dy` and the flag `exist` set to 0. On the contrary, if the two trial solutions deviate on different sides of `yb` (`f1` and `f2` have opposite signs), the routine initiates the iterative refinement of the interval between `dy1` and `dy2` by bisection (see Section 6.2). At each halving step, `Shoot` starts the propagation with the previous average derivative, and retains, as bounding initial derivatives, those leading to final deviations of opposite signs and, which, obviously, continue to bracket the exact derivative. The ability to bracket the exact derivative, doubled by its algorithmic robustness and simplicity, are precisely the beneficial features that recommend the bisection method.

For illustration of the shooting method, we consider in the following the calculation of the Legendre polynomials, $P_n(x)$, from their differential equation. The corresponding two-point boundary-value problem, complying with the commonly used standardization $P_n(1) = 1$, can be written as

$$\frac{d^2 P_n}{dx^2} = \frac{1}{1 - x^2} \left[2x \frac{dP_n}{dx} - n(n+1)P_n \right], \tag{12.79}$$

$$P_n(-1) = (-1)^n, \quad P_n(1) = 1. \tag{12.80}$$

We make explicit use of the well-defined parity of the Legendre polynomials and, taking into account their alternatively symmetric and antisymmetric character (determined by the independence of the ODE on the sign of x), the integration domain can be reduced from $[-1, 1]$ to $[0, 1]$. In this context, the derivative at the origin, $P_n'(0)$, which may be useful for comparison with numerical results, can be shown to satisfy the simple relation:

$$P_n'(0) = nP_{n-1}(0).$$

Listing 12.14   ODE Solver Based on the Shooting Method (Python Coding)

```
#===========================================================================
def Propag(x, y, nx, y0, dy0, Func):
#---------------------------------------------------------------------------
#  Propagates the solution y[] of a Cauchy-problem for a 2nd order ODE on a
#  regular mesh x[] with nx points, starting from y0 and dy0.
#  Calls: EulerCromer1(x, hx, y, dy, Func); Func(x, y, dy) - RHS of ODE
#---------------------------------------------------------------------------
   hx = x[2] - x[1]
   y[1] = y0; dy = dy0
   for m in range(1,nx):
      (y[m+1], dy) = EulerCromer1(x[m],hx,y[m],dy,Func)


#===========================================================================
def Shoot(x, y, nx, ya, yb, dy1, dy2, eps, Func):
#---------------------------------------------------------------------------
#  Solves a two-point boundary-value problem for a 2nd order ODE
#     y" = f(x,y,y'), y(xa) = ya, y(xb) = yb
#  on a regular mesh x[] with nx points, using the shooting method with trial
#  initial derivatives dy in [dy1,dy2]. Returns the solution y[] satisfying
#  the condition y(xb) = yb within tolerance eps, the used derivative dy, and
#  an existence flag.
#  Calls: Propag(x, y, nx, y0, dy0, Func); Func(x, y, dy) - RHS of ODE
#---------------------------------------------------------------------------
   itmax = 100                                 # max. number of bisections

   Propag(x,y,nx,ya,dy1,Func)                     # propagate y for dy1
   f1 = y[nx] - yb                                      # deviation at xb
   Propag(x,y,nx,ya,dy2,Func)                     # propagate y for dy2
   f2 = y[nx] - yb                                      # deviation at xb

   if (f1*f2 < 0):                         # check if dy exists in [dy1,dy2]
      exist = 1
      for it in range(1,itmax+1):               # refine dy by bisection
         dy = 0.5e0*(dy1 + dy2)                      # new approximation
         Propag(x,y,nx,ya,dy,Func)                       # propagate y
         f = y[nx] - yb                                # deviation at xb
         if (f1*f > 0): dy1 = dy                      # new semi interval
         else:          dy2 = dy
         if (fabs(f) <= eps): break            # deviation vanishes at xb ?

      if (it >= itmax): print("Shoot: max. no. of bisections exceeded !")
   else:
      dy = 1e10; exist = 0

   return (dy, exist)
```

Hence, detailing the initial conditions to be used at $x = 0$ for even and, respectively, odd solutions, the Legendre polynomials can be regarded as solutions of the Cauchy problem:

$$y'' = \frac{1}{1 - x^2}\left[2xy' - n(n+1)y\right], \tag{12.81}$$

$$y(0) = y_a, \quad y'(0) = 0, \quad n = 0, 2, \ldots \tag{12.82}$$

$$y(0) = 0, \quad y'(0) = y'_a, \quad n = 1, 3, \ldots \tag{12.83}$$

with the additional constraint

$$y(1) = 1. \tag{12.84}$$

The homogeneity of the Legendre equation guarantees that any solution multiplied by a constant remains a solution, too. Correspondingly, for even n, the procedure amounts to the straightforward propagation from an *arbitrary initial value* $\tilde{y}(0) = y_a$, which is finalized by rescaling the found solution, $\tilde{y}(x)$, with the end value $\tilde{y}(1)$, that is, $y(x) = \tilde{y}(x)/\tilde{y}(1)$. In such a manner, the unused boundary condition, $y(1) = 1$, is clearly satisfied. As for the odd n-solutions, they require the actual application of the shooting method, which implies the determination of the correct initial derivative y'_a by iterative bisections of the interval $(-\infty, +\infty)$.

A program designed to calculate the Legendre polynomials by the shooting method is presented in Listing 12.15. In particular, for $n = 5$, an integration step size $h = 10^{-4}$, and a relative tolerance $\varepsilon = 10^{-4}$

Listing 12.15   Legendre Polynomials by the Shooting Method (Python Coding)

```
# Legendre polynomials by the shooting method
from math import *
from ode import *
from specfunc import *

global n                                            # polynomial order
def Func(x, y, dy):                                 # RHS of Legendre ODE
    return (2e0*x*dy - n*(n+1)*y) / (1e0 - x*x)

# main

n = 5                                               # order of Legendre polynomial
xa = 0e0                                             # boundary values
xb = 1e0; yb = 1e0
eps = 1e-4                                           # tolerance for solution at xb
hx = 1e-4                                            # x-mesh step size

nx = int((xb-xa)/hx + 0.5) + 1                       # number of x-mesh points

x = [0]*(nx+1); y = [0]*(nx+1)                       # x-mesh, solution

for m in range(1,nx+1): x[m] = xa + (m-1)*hx         # generate x-mesh

if (n % 2 == 0):                                     # even solutions: rescaling
    ya = 1e0; dy = 0e0
    Propag(x,y,nx,ya,dy,Func)
    for m in range(1,nx+1): y[m] /= y[nx]            # normalization
else:                                               # odd solutions: shooting
    ya = 0e0
    dy1 = -1e3; dy2 = 1e3                    # search initial derivative in [dy1,dy2]
    (dy, exist) = Shoot(x,y,nx,ya,yb,dy1,dy2,eps,Func)

out = open("shoot.txt","w")
out.write("dy = {0:8.5f}\n".format(dy))
out.write("        x           P{0:1d}            err\n".format(n))
for m in range(1,nx+1):
    (P, d) = Legendre(n,x[m])
    out.write(("{0:10.5f}{1:10.5f}{2:10.5f}\n").format(x[m],y[m],P-y[m]))
out.close()
```

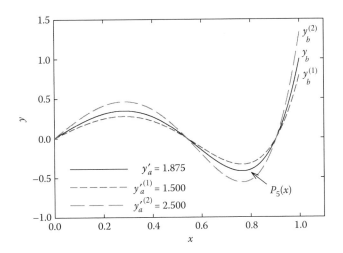

FIGURE 12.7 Solutions of the Legendre equation (12.81) for $n = 5$, yielded by the shooting method: with continuous line, the exact solution, $P_5(x)$; with dashed line, trial solutions for two values of the initial first derivative, $y_a'^{(1)} = 1.5$ and $y_a'^{(2)} = 2.5$, bracketing the exact derivative.

for the boundary value $y(1)$, the code determines the refined initial derivative $y'(0) = 1.8768$, instead of the exact value $P_5'(0) = 15/8 = 1.875$. The maximum local error with respect to the exact solution amounts to $4.3 \cdot 10^{-4}$ and is expectedly higher than ε (which concerns only the boundary value). The agreement with the exact result can be obviously improved by reducing the integration step size.

Figure 12.7 shows two trial solutions resulting for the particular initial derivatives $y_a'^{(1)} = 1.5$ and $y_a'^{(2)} = 2.5$, which bracket the exact derivative ($P_5'(0) = 1.875$). The trial solutions can be seen to border the exact solution ($P_5(x)$), similar to the way in which the corresponding end values $y_b^{(1)}$ and $y_b^{(2)}$ bracket the true boundary value y_b.

While ranking among the best-suited algorithms for boundary-value problems for ODEs, the shooting method unquestionably shows its full potential when applied to *eigenvalue problems* and, besides, when they are defined over *infinite domains*. These two additional modeling aspects call for specific numerical treatment. One of the prototypical examples of this category are the bound-state problems for the time-independent 1D Schrödinger equation.

We consider in the following the *bound states* of Schrödinger's wave equation in the presence of a *symmetric potential* $V(x)$:

$$-\frac{\hbar^2}{2\mu}\frac{d^2\psi}{dx^2} + V(x)\psi(x) = E\psi(x), \tag{12.85}$$

$$V(-x) = V(x), \quad x \in (-\infty, +\infty). \tag{12.86}$$

The corresponding eigenenergies E are contained between the limiting values of the potential

$$V_{\min} \le E \le V_{\max} \equiv V(+\infty),$$

and, owing to the symmetry, the treatment can be restricted to the spatial domain $[0, +\infty)$. The fact that the wave equation is invariant to the sign of x for a symmetric potential induces a well-defined symmetry of the eigenfunctions themselves, which have to be alternatively even ($\psi(-x) = \psi(x)$) or odd ($\psi(-x) = \psi(x)$). A practical survey on the application of the shooting method to such quantum systems is provided by Giannozzi (2013).

All things considered, the following Cauchy problem for Schrödinger's wave equation can be set up in dimensionless units:

$$\psi'' = 2\,[V(x) - E]\,\psi,\tag{12.87}$$

$$\psi(0) = \psi_0, \quad \psi'(0) = 0, \quad n = 0, 2, \ldots,\tag{12.88}$$

$$\psi(0) = 0, \quad \psi'(0) = \psi_0', \quad n = 1, 3, \ldots,\tag{12.89}$$

with the asymptotic constraint

$$\psi(x) \underset{x \to \infty}{\to} 0,\tag{12.90}$$

and the normalization condition

$$2 \int_0^\infty |\psi(x)|^2\, dx = 1.\tag{12.91}$$

In view of the formal similarities with Problem 12.81–12.84 for the Legendre polynomials, the overall numerical strategy for the quantum bound-state problem 12.87–12.90 remains essentially the same. However, instead of adjusting the solution's initial values ψ_0 and ψ_0', it is the eigenenergy E that is being adjusted as part of the shooting approach. Also, instead of meeting the second boundary condition at a precise point, rather, the asymptotic vanishing of the wave function is monitored, which actually occurs numerically at points depending on the eigenenergy E itself and has, thus, to be determined self-consistently. As for the concrete initial values of ψ_0 (for even solutions) and ψ_0' (for odd solutions) that are to be utilized in the shooting procedure, by virtue of the homogeneity of Schrödinger's equation, they can be chosen arbitrarily and the wave function can be eventually normalized according to Equation 12.91.

It is noteworthy that the practical handling of the mentioned particularities is actually based on an unavoidable and detrimental numerical phenomenon, namely, the asymptotic instability of the solution. In fact, beyond a certain point x_c, where the solution is expected to *vanish numerically* relatively to a predefined tolerance ε,

$$|\psi(x)| \le \varepsilon, \quad x = x_c,\tag{12.92}$$

the accumulation of truncation and roundoff errors starts invariably dominating the propagation and finally causes the rapid divergence of the solution at some point x_∞:

$$\psi(x) \to \pm\infty, \quad x \ge x_\infty \ge x_c.$$

Further on, x_c is called "checkpoint" and x_∞ is referred to as "divergence point." In spite of its apparent lack of relevance, the *sign of the divergence* is an essential piece of information. Indeed, as can be seen from Figure 12.8, the solutions $\psi_0^{(1)}$ and $\psi_0^{(2)}$, corresponding to the trial energies $E_0^{(1)}$ and $E_0^{(2)}$ that bracket a true eigenenergy E_0, diverge "asymptotically" in opposite directions. This numerical phenomenon can be used to practically identify intervals that contain eigenvalues—if $\psi_0^{(1)}$ and $\psi_0^{(2)}$ have divergences of opposite sign, then the interval $\left[E_0^{(1)}, E_0^{(2)}\right]$ definitely contains a true eigenvalue E:

$$\lim_{x \to +\infty} \psi_0^{(1)}(x)\psi_0^{(2)}(x) < 0 \quad \Rightarrow \quad \exists E \in \left[E_0^{(1)}, E_0^{(2)}\right].$$

The tighter the interval $\left[E_0^{(1)}, E_0^{(2)}\right]$ encloses the eigenvalue E_0, the further away from the checkpoint x_c moves the divergence point x_∞.

Listing 12.16 presents the solver ShootQM, which implements the described shooting methodology for eigenvalue problems for the 1D Schrödinger equation. On the basis of the particular potential V[], which is expected to be received tabulated on a regular mesh x[] with nx points, the routine determines,

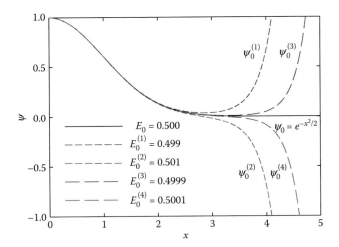

FIGURE 12.8   Solutions for the problem of the quantum harmonic oscillator: with continuous line, the exact unnormalized solution $\psi_0(x) = \exp(-x^2/2)$, with dashed lines, the solution yielded by the shooting method for the trial energies $E_0^{(1)}, E_0^{(2)}, E_0^{(3)}, E_0^{(4)}$, bracketing in pairs the true eigenvalue $E_0 = 0.5$.

Listing 12.16   Solver for the 1D Schrödinger Equation Using the Shooting Method (Python Coding)

```
#===========================================================================
def ShootQM(E1, E2, V, x, y, nx, nc, y0, dy0, eps):
#---------------------------------------------------------------------------
#  Solves the two-point eigenvalue problem for the 1D Schrodinger equation
#     y" = (2m/h^2) [V(x) - E] y, y(0) = y0, y(+inf) = 0
#  on a regular mesh x[] with nx points, using the shooting method with the
#  trial energies E in [E1,E2] and the tabulated potential in V[]. Returns
#  the solution y[] vanishing at the checkpoint x[nc] within tolerance eps,
#  and the existence flag exist.
#  Calls: Numerov(V, x, y, nx, y0, dy0, E)
#---------------------------------------------------------------------------
   itmax = 100                                   # max. number of bisections

   inf = Numerov(E1,V,x,y,nx,y0,dy0)                # propagate y for E1
   f1 = y[inf]                                          # asymptotic y
   inf = Numerov(E2,V,x,y,nx,y0,dy0)                # propagate y for E2
   f2 = y[inf]                                          # asymptotic y

   if (f1*f2 < 0):                            # check if exists E in [E1,E2]
      exist = 1
      for it in range(1,itmax+1):                   # refine E by bisection
         E = 0.5e0*(E1 + E2)                            # new approximation
         inf = Numerov(E,V,x,y,nx,y0,dy0)                  # propagate y
         f = y[inf]                                        # asymptotic y
         if (f1*f > 0): E1 = E                         # new semi interval
         else:          E2 = E
         if (fabs(y[nc]) <= eps): break     # check if y vanishes at x[nc] ?

      if (it >= itmax): print("ShootQM: max. no. of bisections exceeded !")
   else:
      E = 1e10; exist = 0

   return (E, exist)
```

if it exists, the eigenenergy E bordered by the input energies E1 and E2. Equally, the routine returns the corresponding eigenfunction y[], which is determined so as to vanish at the checkpoint x[nc] within the tolerance eps.

ShootQM works correctly if the interval between E1 and E2 separates *a single eigenenergy* E, or none. It should be called, therefore, within an external loop for energy "windows" narrow enough to guarantee the separation of the eigenvalues and that sequentially scan the entire energy domain of interest (possibly for several eigenenergies).

The global ODE integrator called by the routine ShootQM is the function Numerov developed in Section 12.7. Starting from the initial values y0 and dy0, Numerov propagates the solution y[] corresponding to the input energy E and the potential tabulated in V[] over the entire domain of nx equidistant mesh points specified in x[]. In addition, Numerov returns the index of the mesh point where the solution diverges (exceeds 10^{10}), which defaults to nx if no divergence occurs.

ShootQM initially performs two complete propagations for the limiting input energies, E1 and E2, and receives in inf in sequence the indexes of the respective points of divergence. If the corresponding divergences y[inf], stored in f1 and, respectively, f2, have the same sign, the interval between E1 and E2 certainly does not contain the sought eigenvalue and the routine returns a fake E and the flag exist set to 0. On the contrary, if the two trial solutions diverge with opposite signs (f1*f2 < 0), the routine initiates the iterative refinement of the interval between E1 and E2 by bisection. At each halving step, a propagation is carried out for the median value of the current energy subinterval, while retaining the semiinterval for whose limits the solutions diverge with opposite signs.

Again, the ability to constantly bracket the exact eigenenergy and to work in tandem with an external separation sequence for multiple eigenenergies, as explained above, is the desirable feature for which the bisection method is preferred.

Now referring to the concrete example of the quantum harmonic oscillator, we have in dimensionless units the simple parabolic potential:

$$V(x) = \frac{1}{2}x^2.$$

The exact dimensionless eigenenergies (in $\hbar\omega$ units) and the corresponding eigenfunctions are (Fitts 2002)

$$E_n = n + \frac{1}{2}, \quad n = 0, 1, 2, \ldots,$$

$$\psi_n(x) = \left(\sqrt{\pi}2^n n!\right)^{-1/2} e^{-x^2/2} H_n(x),$$

where $H_n(x)$ are the Hermite polynomials (see Section 5.4 for properties and routine). In particular, the ground-state eigenfunction is the Gaussian $\psi_0(x) = \pi^{-1/4}e^{-x^2/2}$.

Program 12.17 is designed to determine a single bound state of the quantum harmonic oscillator by the shooting method. The particular eigenstate to be calculated is determined by the parity par of the eigenfunction (0 or 1) and the lower limit Emin of the search interval for the eigenenergy E. xx specifies the limit of the *x*-mesh, xc is the location of the "checkpoint" where the solution is expected to vanish within the tolerance eps, Emax is the upper limit for searching E, dE is the width of the successive search windows for E, and hx is the spatial integration step size. Concerning the usable values of dE, they should be just low enough to separate the different eigenenergies; yet, too low values cannot harm, since they only increase the number of energy windows not containing any eigenvalue.

The program starts by determining the indexes nx and nc of the final mesh point and, respectively, of the checkpoint. It then initializes the spatial mesh x[] and tabulates the potential in V[].

Listing 12.17 Eigenstates of the Quantum Oscillator by the Shooting Method (Python Coding)

```python
#  Eigenstates of the 1D Schroedinger equation for the harmonic oscillator
#     y" = 2 [V(x) - E] y, y(0) = y0, y'(0) = y'0, y(+inf) = 0
#     V(x) = 0.5*x*x
#  using a shooting algorithm based on the Numerov method
#---------------------------------------------------------------------------
from math import *
from ode import *
from specfunc import *

def Pot(x): return 0.5e0*x*x                      # Potential for harmonic oscillator

# main

xx = 10e0                                       # limit of x-mesh
xc = 6e0                             # checkpoint for vanishing solution (xc < xx)
par = 0                                       # parity of eigenstate (0/1)
Emin = 0e0                                    # lower limit for eigenvalues
Emax = 1e2                                    # upper limit for eigenvalues
dE = 0.1e0                               # minimum separation of eigenvalues
eps = 1e-4                                    # tolerance for solution at xc
hx = 1e-4                                      # x-mesh step size

nx = int(xx/hx + 0.5) + 1                    # number of x-mesh points
nc = int(xc/hx + 0.5) + 1              # index of checkpoint for vanishing y
x = [0]*(nx+1); y = [0]*(nx+1)                 # x-mesh, solution
V = [0]*(nx+1)                                 # tabulated potential

for m in range(1,nx+1):
   x[m] = (m-1)*hx                               # integration mesh
   V[m] = Pot(x[m])                            # tabulate potential

Ew = Emin                       # lower limit of search window for E, [Ew,Ew+dE]
while (Ew < Emax):                           # loop over eigenvalue windows
                                             # initial values at x = 0
   if (par == 0): y0 = 1e0; dy0 = 0e0                 # even y
   else:          y0 = 0e0; dy0 = 1e0                 # odd y

   (E, exist) = ShootQM(Ew,Ew+dE,V,x,y,nx,nc,y0,dy0,eps)

   if (exist): break

   Ew += dE                                # shift [Ew,Ew+dE] for next shoot

if (exist):
   n = int(E)                                    # quantum number

   f = 0e0                                 # normalize y by trapezoidal rule
   for m in range(1,nc+1): f += y[m]*y[m]*hx          # norm for [0,+inf]
   f = sqrt(2e0*f)
   if (int(n/2) % 2): f = -f                        # sign correction
   for m in range(1,nx+1): y[m] /= f

   f = sqrt(pi)                             # norm of Hermite polynomials
   for i in range(1,n+1): f *= 2e0 * i
   f = 1e0/sqrt(f)

   out = open("shoot.txt","w")
   out.write("E{0:1d} = {1:8.5f}\n".format(n,E))
```

```
out.write("       x          y{0:1d}                err\n".format(n))
for m in range(1,nc+1):
    (yH, d) = Hermite(n,x[m]); yH *= f * exp(-0.5*x[m]*x[m])
    out.write((("{0:10.5f}{1:10.5f}{2:10.5f}\n").format(x[m],y[m],yH-y[m]))
out.close()
else:
    print("No solution found !")
```

The search sequence of the eigenenergy E is conceived as a while-do structure in which successive shoots using the routine ShootQM are attempted within energy windows defined by their lower boundary Ew and width dE. The initial values of the trial wave functions, y0 and dy0, are determined by the parity par according to conditions (12.88) through (12.89). If for a particular energy window, [Ew, Ew+dE], the flag exist is set by ShootQM to 0, meaning that no valid eigenvalue was found, the window is shifted by dE and a new shoot is attempted. The scanning process of the energy range between Emin and Emax is terminated either when the upper limit is reached or when the eigenenergy is found and ShootQM exits with exist set to 1.

Once the eigenstate is found, the eigenfunction is normalized (performing the integral by the trapezoidal rule) and compared with the exact result provided by the function Hermite from the library specfunc.py (or specfunc.h, for C/C++ coding).

For the particular parameters coded in Listing 12.17, that is, par=0 and Emin=0e0, the sought solution is obviously the (even) ground state ($n = 0$). The eigenenergy evaluates to 0.49997 and the largest local difference between the propagated and the exact solution amounts to $2 \cdot 10^{-5}$. By simply taking par=1 and Emin=0e0, the program yields the first excited state ($n = 1$), which is odd, or, by setting par=0 and Emin=0.6e0, there results the second excited state ($n = 2$), which is again even. This is the principle that enables the code to be generalized for determining in sequence all the eigenenergies in a given interval.

Figure 12.8 shows the unnormalized eigenfunction obtained by running program 12.17. It is graphically indistinguishable from the exact result $\psi_0(x) \sim e^{-x^2/2}$. The solutions $\psi_0^{(1)}$, $\psi_0^{(2)}$, $\psi_0^{(3)}$, and $\psi_0^{(4)}$ are obtained, respectively, for the trial energies $E_0^{(1)} = 0.499$, $E_0^{(2)} = 0.501$, $E_0^{(3)} = 0.4999$, and $E_0^{(4)} = 0.5001$, which bracket in increasingly tight pairs the true eigenenergy $E_0 = 0.5$. Even though the trial solutions reproduce comparatively well the exact solution for low arguments, they feature significant divergences for large arguments. Anyway, the closer the exact eigenvalue is bordered by the trial energies, the more shifts the onset of the divergence toward larger arguments, virtually moving to infinity.

12.9 Finite-Difference Methods for Linear Two-Point Problems

From among the numerical methods for solving boundary-value problems, finite-difference methods are known to have the best stability properties. It is true, however, that the enhanced stability comes, in general, at the expense of increased computational effort for comparable accuracy.

In fact, the shooting method (described in the previous section) is susceptible of being unstable under certain circumstances. While in the case of the Legendre equation (12.81), used for illustration, the integration proves stable when started out from the origin; if started from one of the boundaries, $x = \pm 1$, it exhibits instabilities, caused by the presence of the factor $1/(1-x^2)$ in the right-hand side of the equation.

Essentially, finite-difference methods imply approximating the derivatives in the ODE and boundary conditions by finite-difference schemes defined on a discrete set of mesh points. In any case, the particular discretization schemes used are supposed to ensure a homogeneous order of the truncation errors. The resulting system of algebraic equations can then be solved by a general-purpose method or, ideally, by a method adapted to the particular form of the system.

Let us consider the *linear two-point boundary-value problem*:

$$y'' = p(x)y' + q(x)y + r(x), \tag{12.93}$$

$$\begin{cases} \alpha_1 y(x_a) + \beta_1 y'(x_a) = 1, \\ \alpha_2 y(x_b) + \beta_2 y'(x_b) = 1, \end{cases} \tag{12.94}$$

where $p(x)$, $q(x)$, and $r(x)$ are assumed to be continuous in the interval $[x_a, x_b]$. The boundary conditions are completely defined by the four coefficients α_1, β_1, α_2, and β_2, whereby the additional natural condition should apply, that not both coefficients α and β for a given boundary can be equal to 0 ($|\alpha_i| + |\beta_i| \neq 0$, $i = 1, 2$).

In particular, for $\beta_i = 0$, there results a *Dirichlet boundary condition*, which defines the solution's *value* on the boundary, while $\alpha_i = 0$ leads to a *Neumann boundary condition*, which fixes the solution's *derivative* on the boundary. The condition types can differ on the two boundaries, depending on the modeled problem.

Let us consider a partition of the domain $[x_a, x_b]$, characterized by the equidistant mesh points:

$$x_m = x_a + (m-1)h, \quad m = 1, 2, \ldots, M, \tag{12.95}$$

with the step size equal to

$$h = \frac{x_b - x_a}{M - 1}. \tag{12.96}$$

As in the previous sections, we use the notation $y_m \equiv y(x_m)$ and, for obtaining discretized approximations for the first and second derivatives, y'_m and y''_m, we start from the Taylor expansions of the solution at $x_{m+1} = x_m + h$ and $x_{m-1} = x_m - h$:

$$y_{m+1} = y_m + hy'_m + \frac{1}{2!}h^2 y''_m + \frac{1}{3!}h^3 y_m^{(3)} + O(h^4), \tag{12.97}$$

$$y_{m-1} = y_m - hy'_m + \frac{1}{2!}h^2 y''_m - \frac{1}{3!}h^3 y_m^{(3)} + O(h^4). \tag{12.98}$$

For the first derivative y'_m, one can obtain straightforward approximations by simply neglecting all terms higher than and including $(h^2/2)\, y''_m$:

$$y'_m = \frac{y_m - y_{m-1}}{h} + O(h), \quad y'_m = \frac{y_{m+1} - y_m}{h} + O(h), \tag{12.99}$$

which, however, become only $O(h)$ by the final division by h. The first is a *backward-difference formula* (since it connects x_m with the previous point, x_{m-1}), while the second is a *forward-difference formula* (since it connects x_m with the next point, x_{m+1}).

A higher-order approximation for y'_m results from the difference of the Taylor expansions (12.97) and (12.98), by the explicit neglect of the third-order terms, $(h^3/6)\, y_m^{(3)}$, and the exact cancellation of the second-order terms, $(h^2/2)\, y''_m$:

$$y'_m = \frac{y_{m+1} - y_{m-1}}{2h} + O(h^2). \tag{12.100}$$

This *central-difference formula* employs information from mesh points located symmetrically about the point x_m where the derivative is being expressed. Again, the final division by h reduces the order of the scheme by 1.

Taking the sum of the Taylor expansions (12.97) and (12.98), the first- and third-order terms, hy'_m and $(h^3/6)\, y_m^{(3)}$, cancel out exactly and the following central-difference scheme for the second-order derivative is obtained:

$$y''_m = \frac{y_{m+1} - 2y_m + y_{m-1}}{h^2} + O(h^2), \qquad (12.101)$$

which is $O(h^2)$ due to the final division by h^2.

The application of central-difference schemes for discretized linear two-point problems leads to linear systems with symmetric band matrices, which positively impacts on the stability of the subsequent solution methods. By using the $O(h^2)$ central-difference formulas 12.100 and 12.101 for discretizing ODE 12.93 and Formulas 12.99 for approximating the boundary conditions 12.94, we obtain the following linear system:

$$\begin{cases} \dfrac{y_{m+1} - 2y_m + y_{m-1}}{h^2} = p_m \dfrac{y_{m+1} - y_{m-1}}{2h} + q_m y_m + r_m, \\ m = 2, 3, \ldots, M - 1, \\ \alpha_1 y_1 + \beta_1 \dfrac{y_2 - y_1}{h} = 1, \\ \alpha_2 y_M + \beta_2 \dfrac{y_M - y_{M-1}}{h} = 1. \end{cases}$$

Regrouping the unknown solution values y_m, the discretized linear system corresponding to the bilocal boundary-value problem 12.93–12.94 takes the form:

$$\begin{cases} (h\alpha_1 - \beta_1)y_1 + \beta_1 y_2 = h, \\ -(2 + hp_m)\,y_{m-1} + (4 + 2h^2 q_m)\,y_m - (2 - hp_m)\,y_{m+1} = -2h^2 r_m, \\ m = 2, 3, \ldots, M - 1, \\ -\beta_2 y_{M-1} + (h\alpha_2 + \beta_2)y_M = h. \end{cases} \qquad (12.102)$$

It is noticeable that this system has a tridiagonal matrix and can be represented finally as

$$\begin{bmatrix} b_1 & c_1 & & & & & & \\ a_2 & b_2 & c_2 & & & 0 & & \\ & \ddots & \ddots & \ddots & & & & \\ & & a_m & b_m & c_m & & & \\ & & & \ddots & \ddots & \ddots & & \\ & 0 & & & a_{M-1} & b_{M-1} & c_{M-1} \\ & & & & & a_M & b_M \end{bmatrix} \begin{bmatrix} y_1 \\ y_2 \\ \vdots \\ y_m \\ \vdots \\ y_{M-1} \\ y_M \end{bmatrix} = \begin{bmatrix} d_1 \\ d_2 \\ \vdots \\ d_m \\ \vdots \\ d_{M-1} \\ d_M \end{bmatrix}, \qquad (12.103)$$

with the nonzero elements having the expressions:

$$b_1 = h\alpha_1 - \beta_1, \quad c_1 = \beta_1, \quad d_1 = h,$$

$$\begin{cases} a_m = -(2 + hp_m), \quad m = 2, 3, \ldots, M - 1, \\ b_m = 4 + 2h^2 q_m, \\ c_m = -(2 - hp_m), \\ d_m = -2h^2 r_m, \end{cases} \qquad (12.104)$$

$$a_M = -\beta_2, \quad b_M = h\alpha_2 + \beta_2, \quad d_M = h.$$

The routine `Bilocal` from Listing 12.18 illustrates the implementation of the finite-difference algorithm described above. The function receives as input data the limits `xa` and `xb` of the solution's definition domain, the number `nx` of spatial mesh points, as well as the four coefficients defining the boundary conditions, `alf1`, `bet1`, `alf2`, and `bet2`. The values of the functions $p(x)$, $q(x)$, and $r(x)$ defining the ODE are returned by the user function `Func` through the arguments `p`, `q`, and, respectively,

Listing 12.18 Linear Bilocal Problem Solver Based on the Finite-Difference Method (Python Coding)

```
#=================================================================================
def Bilocal(xa, xb, y, nx, alf1, bet1, alf2, bet2, Func):
#---------------------------------------------------------------------------------
#  Solves a linear two-point boundary-value problem for a 2nd order ODE
#     y" = p(x) y' + q(x) y + r(x),    xa <= x <= xb
#     alf1 y(xa) + bet1 y'(xa) = 1
#     alf2 y(xa) + bet2 y'(xa) = 1
#  on a regular mesh with nx points, using central finite-differences.
#  Returns the solution in y[].
#  Calls: Func(x, p, q, r) - returns values of p(x), q(x), r(x)
#         TriDiagSys (linsys.py) - solves discretized tridiagonal system
#---------------------------------------------------------------------------------
   a = [0]*(nx+1); b = [0]*(nx+1); c = [0]*(nx+1)        # matrix elements

   hx = (xb-xa)/(nx-1); h2 = 2e0*hx*hx
                                                      # build system coefficients
   b[1] = hx*alf1 - bet1; c[1] = bet1; y[1] = hx
   for m in range (2,nx):
      x = xa + (m-1)*hx                                  # mesh point
      (p,q,r) = Func(x)
      a[m] = -(2e0 + hx*p); b[m] = 4e0 + h2*q; c[m] = -(2e0 - hx*p)
      y[m] = -h2*r                            # RHSs of tridiagonal system
   a[nx] = -bet2; b[nx] = hx*alf2 + bet2; y[nx] = hx

   TriDiagSys(a,b,c,y,nx)                            # solve discretized system
```

r. The most efficient way of solving system (12.103)–(12.104) relies on the LU factorization method for systems with tridiagonal matrix presented in Section 7.7 and coded as routine `TriDiagSys` (included in the libraries `linsys.py` and `linsys.h`). On the basis of the input, `Bilocal` calculates the coefficients of the discretized system, passes them to routine `TriDiagSys`, which actually solves the discretized system, and returns the solution `y[]` to the calling program.

For illustration, we consider the same bilocal problem for Legendre polynomials, which was also discussed in the context of the shooting method:

$$\frac{d^2 P_n}{dx^2} = \frac{1}{1-x^2}\left[2x\frac{dP_n}{dx} - n(n+1)P_n\right], \tag{12.105}$$

$$P_n(-1) = (-1)^n, \quad P_n(1) = 1. \tag{12.106}$$

The functions defining the right-hand side of the ODE are in this case:

$$p(x) = x/(1-x^2), \quad q(x) = -n(n+1)/(1-x^2), \quad r(x) = 0, \tag{12.107}$$

and the Dirichlet boundary conditions are characterized by the coefficients:

$$\begin{aligned} \alpha_1 &= (-1)^n, & \beta_1 &= 0, \\ \alpha_2 &= 1, & \beta_2 &= 0. \end{aligned} \tag{12.108}$$

Program 12.19, solving this problem for $n = 5$, implements the user function `Func` for returning the values of the functions $p(x)$, $q(x)$, and $r(x)$. The solutions obtained by running the code with the (rather large) spatial mesh spacings $h = 0.1$ and $h = 0.01$ are plotted in Figure 12.9 comparatively with the solutions yielded by the shooting method (Program 12.15). For $h = 0.1$, the shooting method clearly fails

Listing 12.19   Legendre Polynomials by the Finite-Difference Method (Python Coding)

```
# Legendre polynomials by the finite-difference method
from math import *
from ode import *
from specfunc import *

global n                                           # polynomial order
def Func(x):                                        # RHS of Legendre ODE
   p = 2e0*x/(1e0-x*x); q =-n*(n+1)/(1e0-x*x); r = 0e0
   return (p, q, r)

# main

n = 5                                              # order of Legendre polynomial
xa = -1e0;   xb = 1e0                                # domain boundaries
hx = 1e-3                                           # x-mesh step size

nx = int((xb-xa)/hx + 0.5) + 1                       # number of x-mesh points

x = [0]*(nx+1); y = [0]*(nx+1)                        # x-mesh, solution

for m in range(1,nx+1): x[m] = xa + (m-1)*hx          # generate x-mesh

alf1 = -1e0 if n % 2 else 1e0; bet1 = 0e0             # Dirichlet conditions
alf2 = 1e0; bet2 = 0e0

Bilocal(xa,xb,y,nx,alf1,bet1,alf2,bet2,Func)

out = open("bilocal.txt","w")
out.write("      x          P{0:1d}            err\n".format(n))
for m in range(1,nx+1):
   (P, d) = Legendre(n,x[m])
   out.write(("{0:10.5f}{1:10.5f}{2:10.5f}\n").format(x[m],y[m],P-y[m]))
out.close()
```

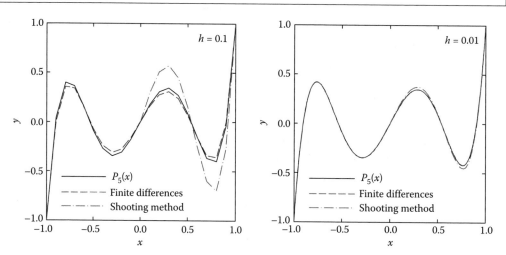

FIGURE 12.9   Solutions of the two-point problem for the Legendre polynomials of order $n = 5$, using the finite-difference method (dash line) and, respectively, the shooting method (dash-dot line), for the spatial mesh spacings $h = 0.1$ (left panel) and $h = 0.01$ (right panel). As reference, the exact Legendre polynomial $P_5(x)$ is plotted with continuous line.

to follow the analytic dependence ($P_5(x)$), even though the solution satisfies the boundary conditions. For $h = 0.01$, the solution provided by the finite-difference method becomes indistinguishable from the exact solution, while the agreement for the shooting method improves significantly.

12.10 Implementations in C/C++

Listing 12.20 shows the content of the file ode.h, which contains equivalent C/C++ implementations of the Python functions developed in the main text and included in the module ode.py. The corresponding routines have identical names, parameters, and functionalities.

Listing 12.20 Solvers for Cauchy and Two-Point Problems for Systems of ODEs (ode.h)

```
//---------------------------------ode.h------------------------------------
// Contains routines for solving systems of ordinary differential equations.
// Part of the numxlib numerics library. Author: Titus Beu, 2013
//--------------------------------------------------------------------------
#ifndef _ODE_
#define _ODE_

#include <math.h>
#include "memalloc.h"
#include "linsys.h"


//===========================================================================
void Euler(double t, double ht, double y[], int n,
           void Func(double,double[],double[]))
//---------------------------------------------------------------------------
// Propagates the solution y of a system of 1st order ODEs
//    y'[i] = f[i](t,y[]), i = 1..n
// from t to t+ht using Euler's method
// Calls: Vector (memalloc.h)
//        void Func(double t, double y[], double f[]) - RHS of ODEs
//---------------------------------------------------------------------------
{
   static double *f;
   static int init = 1;                             // initialization switch
   int i;

   if (init) { f = Vector(1,n); init = 0; }         // 1st entry allocation

   Func(t,y,f);                                     // get RHS of ODEs
   for (i=1; i<=n; i++) y[i] += ht * f[i];          // propagate solution
}


//===========================================================================
void EulerPC(double t, double ht, double y[], int n,
             void Func(double,double[],double[]))
//---------------------------------------------------------------------------
// Propagates the solution y of a system of 1st order ODEs
//    y'[i] = f[i](t,y[]), i = 1..n
// from t to t+ht using Euler's predictor-corrector method
// Calls: Vector (memalloc.h)
//        void Func(double t, double y[], double f[]) - RHS of ODEs
//---------------------------------------------------------------------------
{
   static double *f1, *f2, *yt;
   static int init = 1;                             // initialization switch
   double ht2;
   int i;
```

```
   if (init) {                                            // 1st entry allocation
      f1 = Vector(1,n); f2 = Vector(1,n);                        // RHS of ODEs
      yt = Vector(1,n);                                   // predicted solution
      init = 0;
   }

   Func(t,y,f1);                                            // RHS of ODEs at t
   for (i=1; i<=n; i++) yt[i] = y[i] + ht * f1[i];                  // predictor
   Func(t+ht,yt,f2);                                    // RHS of ODEs at t+ht

   ht2 = ht/2e0;
   for (i=1; i<=n; i++) y[i] += ht2 * (f1[i] + f2[i]);             // corrector
}

//===========================================================================
void RungeKutta(double t, double ht, double y[], int n,
                void Func(double,double[],double[]))
//---------------------------------------------------------------------------
// Propagates the solution y of a system of 1st order ODEs
//    y'[i] = f[i](t,y[]), i = 1..n
// from t to t+ht using the 4th order Runge-Kutta method
// Calls: Vector (memalloc.h)
//        void Func(double t, double y[], double f[]) - RHS of ODEs
//---------------------------------------------------------------------------
{
   static double *f1, *f2, *f3, *f4, *yt;
   static int init = 1;                                   // initialization switch
   double ht2, ht6;
   int i;

   if (init) {                                            // 1st entry allocation
      f1 = Vector(1,n); f2 = Vector(1,n); f3 = Vector(1,n); f4 = Vector(1,n);
      yt = Vector(1,n);                                   // predicted solution
      init = 0;
   }

   ht2 = ht/2e0;
   Func(t,y,f1);                                                    // RHS at t
   for (i=1; i<=n; i++) yt[i] = y[i] + ht2*f1[i];
   Func(t+ht2,yt,f2);                                          // RHS at t+ht/2
   for (i=1; i<=n; i++) yt[i] = y[i] + ht2*f2[i];
   Func(t+ht2,yt,f3);                                          // RHS at t+ht/2
   for (i=1; i<=n; i++) yt[i] = y[i] + ht *f3[i];
   Func(t+ht,yt,f4);                                             // RHS at t+ht

   ht6 = ht/6e0;                                            // propagate solution
   for (i=1; i<=n; i++) y[i] += ht6*(f1[i] + 2*(f2[i] + f3[i]) + f4[i]);
}

//===========================================================================
void RKadapt(double t, double &ht, double &ht1, double eps,
             double y[], int n, void Func(double,double[],double[]))
//---------------------------------------------------------------------------
// Propagates the solution y of a system of 1st order ODEs
//    y'[i] = f[i](t,y[]), i = 1..n
// from t to t+ht using 4th order Runge-Kutta and adaptive step size control
//
// ht   - initial step size (input); final step size (output):
```

```
//        ht is unchanged if err <= eps; ht is reduced if err > eps
// ht1  - step size guess for next propagation (output)
// eps  - max. relative error of solution components
// Calls: Vector (memalloc.h)
//        void Func(double t, double y[], double f[]) - RHS of ODEs
//-----------------------------------------------------------------------
{
   const int p = 4;                               // order of basic ODE solver
   const int itmax = 10;                     // max. no. of step size reductions
   static double *yt, *yt2;                // trial solutions for t+ht and t+ht/2
   static int init = 1;                            // initialization switch
   double err, erri, f, ht2;
   int i, it;
                                                   // 1st entry allocation
   if (init) { yt = Vector(1,n); yt2 = Vector(1,n); init = 0; }

   for (it=1; it<=itmax; it++) {                  // loop of step size scaling
      ht2 = ht/2e0;
      for (i=1; i<=n; i++) yt2[i] = yt[i] = y[i];   // initialize trial sol.
      RungeKutta(t,ht,yt,n,Func);                              // t -> t+ht
      RungeKutta(t,ht2,yt2,n,Func);                           // t -> t+ht/2
      RungeKutta(t+ht2,ht2,yt2,n,Func);                   // t+ht/2 -> t+ht

      err = 0e0;                              // max. error of solution components
      for (i=1; i<=n; i++) {
         erri = yt2[i] ? fabs(1e0 - yt[i]/yt2[i]) : fabs(yt2[i] - yt[i]);
         if (err < erri) err = erri;
      }

      f = 1e0;                                 // scaling factor for step size
      if (err) f = 0.9e0*pow(eps/err,1e0/p);
      if (f > 5e0) f = 5e0;                           // prevent increase > 5x
      ht1 = f * ht;                                   // guess for next step size
      if (err <= eps) break;                          // precision attained: exit
      ht = ht1;                                   // reduce step size and repeat
   }
   if (it > itmax) printf("RKadapt: max. no. of iterations exceeded !\n");
   for (i=1; i<=n; i++) y[i] = yt2[i];      // update y with the best solution
}

//=========================================================================
void RKFehlberg(double t, double &ht, double &ht1, double eps,
                double y[], int n, void Func(double,double[],double[]))
//-----------------------------------------------------------------------
// Propagates the solution y of a system of 1st order ODEs
//    y'[i] = f[i](t,y[]), i = 1..n
// from t to t+ht using the Runge-Kutta-Fehlberg method with stepsize control
//
// ht   - initial step size (input); final step size (output):
//        ht is unchanged if err <= eps; ht is reduced if err > eps
// ht1  - step size guess for next propagation (output)
// eps  - max. relative error of solution components
// Calls: Vector (memalloc.h)
//        void Func(double t, double y[], double f[]) - RHS of ODEs
//-----------------------------------------------------------------------
{
   const int p = 5;                                // order of basic ODE solver
   const int itmax = 10;                     // max. no. of step size reductions
   const double                            // Runge-Kutta-Fehlberg coefficients
```

```
         a2 = 1e0/4e0, a3 = 3e0/8e0, a4 = 12e0/13e0, a5 = 1e0, a6 = 1e0/2e0,
         b21 = 1e0/4e0, b31 = 3e0/32e0, b32 = 9e0/32e0,
         b41 = 1932e0/2197e0, b42 = -7200e0/2197e0, b43 = 7296e0/2197e0,
         b51 = 439e0/216e0, b52 = -8e0, b53 = 3680e0/513e0, b54 = -845e0/4104e0,
         b61 = -8e0/27e0, b62 = 2e0, b63 = -3544e0/2565e0, b64 = 1859e0/4104e0,
         b65 = -11e0/40e0,
         c1 = 16e0/135e0, c3 = 6656e0/12825e0, c4 = 28561e0/56430e0,
         c5 = -9e0/50e0, c6 = 2e0/55e0,
         e1 = 1e0/360e0, e3 = -128e0/4275e0, e4 = -2197e0/75240e0,
         e5 = 1e0/50e0, e6 = 2e0/55e0;
   static double *f1, *f2, *f3, *f4, *f5, *f6, *yt;
   static int init = 1;                                 // initialization switch
   double err, erri, f;
   int i, it;

   if (init) {                                          // 1st entry allocation
      f1 = Vector(1,n); f2 = Vector(1,n); f3 = Vector(1,n); f4 = Vector(1,n);
      f5 = Vector(1,n); f6 = Vector(1,n); yt = Vector(1,n);
      init = 0;
   }

   for (it=1; it<=itmax; it++) {                        // loop of step size scaling
      Func(t,y,f1);
      for (i=1; i<=n; i++)
         yt[i] = y[i] + ht*b21*f1[i];
      Func(t+a2*ht,yt,f2);
      for (i=1; i<=n; i++)
         yt[i] = y[i] + ht*(b31*f1[i] + b32*f2[i]);
      Func(t+a3*ht,yt,f3);
      for (i=1; i<=n; i++)
         yt[i] = y[i] + ht*(b41*f1[i] + b42*f2[i] + b43*f3[i]);
      Func(t+a4*ht,yt,f4);
      for (i=1; i<=n; i++)
         yt[i] = y[i] + ht*(b51*f1[i] + b52*f2[i] + b53*f3[i] + b54*f4[i]);
      Func(t+a5*ht,yt,f5);
      for (i=1; i<=n; i++)
         yt[i] = y[i] + ht*(b61*f1[i] + b62*f2[i] + b63*f3[i] + b64*f4[i] +
                         b65*f5[i]);
      Func(t+a6*ht,yt,f6);

      err = 0e0;                              // max. error of solution components
      for (i=1; i<=n; i++) {                           // O(h5) solution estimate
         yt[i] = y[i] +
                 ht*(c1*f1[i] + c3*f3[i] + c4*f4[i] + c5*f5[i] + c6*f6[i]);
                                                        // error estimate
         erri = ht*(e1*f1[i] + e3*f3[i] + e4*f4[i] + e5*f5[i] + e6*f6[i]);
         erri = fabs(erri/yt[i]);
         if (err < erri) err = erri;
      }

      f = 1e0;                                // scaling factor for step size
      if (err) f = 0.9e0*pow(eps/err,1e0/p);
      if (f > 5e0) f = 5e0;                            // prevent increase > 5x
      ht1 = f * ht;                                    // guess for next step size
      if (err <= eps) break;                           // precision attained: exit
      ht = ht1;                                        // reduce step size and repeat
   }

   if (it > itmax) printf("RKFehlberg: max. no. of iterations exceeded !\n");
```

```
      for (i=1; i<=n; i++) y[i] = yt[i];        // update y with the O(h5) solution
}

//==============================================================================
void Euler1(double t, double ht, double &y, double &dy,
            double Func(double,double,double))
//------------------------------------------------------------------------------
// Propagates the solution y and 1st derivative dy of a 2nd order ODE from t
// to t+ht using Euler's method
//------------------------------------------------------------------------------
{
   double d2y;

   d2y = Func(t,y,dy);                                            // d2y -> t
   y  += ht * dy;                                                 // y -> t+ht
   dy += ht * d2y;                                                // dy -> t+ht
}

//==============================================================================
void EulerCromer1(double t, double ht, double &y, double &dy,
                  double Func(double,double,double))
//------------------------------------------------------------------------------
// Propagates the solution y and the 1st derivative dy of a 2nd order ODE
// from t to t+ht using the Euler-Cromer method
//------------------------------------------------------------------------------
{
   double d2y;

   d2y = Func(t,y,dy);                                            // d2y -> t
   dy += ht * d2y;                                                // dy -> t+ht
   y  += ht * dy;                                                 // y -> t+ht
}

//==============================================================================
void Verlet1(double t, double ht, double &y, double &dy, double &d2y,
             double Func(double,double,double))
//------------------------------------------------------------------------------
// Propagates the solution y and the 1st derivative dy of a 2nd order ODE
// from t to t+ht using the Verlet method; returns 2nd derivative in d2y;
// d2y needs to be initialized on first call and saved between calls
//------------------------------------------------------------------------------
{
   double ht2;

   ht2 = 0.5e0 * ht;
   dy += ht2 * d2y;                                               // dy -> t+ht/2
   y  += ht  * dy;                                                // y -> t+ht

   d2y = Func(t,y,dy);                                            // d2y -> t+h

   dy += ht2 * d2y;                                               // dy -> t+ht
}

//==============================================================================
void Euler2(double t, double ht, double y[], double dy[], int n,
            void Func(double,double[],double[],double[]))
//------------------------------------------------------------------------------
// Propagates the solution y and the 1st derivative dy of a system of n
// 2nd order ODEs from t to t+ht using the Euler method
```

```
//-------------------------------------------------------------------------
{
   static double *d2y;
   static int init = 1;                                  // initialization switch
   int i;

   if (init) { d2y = Vector(1,n); init = 0; }           // 1st entry allocation

   Func(t,y,dy,d2y);                                                    // d2y -> t

   for (i=1; i<=n; i++) {                                   // propagate solution
      y[i]  += ht * dy[i];                                          // y -> t+ht
      dy[i] += ht * d2y[i];                                         // dy -> t+ht
   }
}

//=========================================================================
void EulerCromer(double t, double ht, double y[], double dy[], int n,
                 void Func(double,double[],double[],double[]))
//-------------------------------------------------------------------------
// Propagates the solution y and the 1st derivative dy of a system of n
// 2nd order ODEs from t to t+ht using the Euler-Cromer method
//-------------------------------------------------------------------------
{
   static double *d2y;
   static int init = 1;                                  // initialization switch
   int i;

   if (init) { d2y = Vector(1,n); init = 0; }           // 1st entry allocation

   Func(t,y,dy,d2y);                                                    // d2y -> t

   for (i=1; i<=n; i++) {                                   // propagate solution
      dy[i] += ht * d2y[i];                                         // dy -> t+ht
      y[i]  += ht * dy[i];                                          // y -> t+ht
   }
}

//=========================================================================
void Verlet2(double ht, double m,
             double &x, double &y, double &vx, double &vy,
             double &ax, double &ay, double &Ekin, double &Epot,
             void Forces(double,double,double,double,double,
                         double&,double&,double&))
//-------------------------------------------------------------------------
// Propagates the 2D solution of Newton's equations of motion for a particle
// of mass m over a time interval ht using the velocity Verlet method
// x, y   - position components
// vx, vy - velocity components
// ax, ay - acceleration components (need to be initialized on 1st call)
// Ekin   - kinetic energy
// Epot   - potential energy
//-------------------------------------------------------------------------
{
   double fx, fy, ht2;

   ht2 = 0.5e0 * ht;
   vx += ht2 * ax; x += ht * vx;                                    // v -> t+ht/2
   vy += ht2 * ay; y += ht * vy;                                    // r -> t+ht
```

```
      Forces(m,x,y,vx,vy,fx,fy,Epot);                                  // forces

      ax = fx/m; ay = fy/m;                                            // a -> t+ht
      vx += ht2 * ax;                                                  // v -> t+ht
      vy += ht2 * ay;

      Ekin = 0.5e0 * m * (vx*vx + vy*vy);                             // kinetic energy
   }

//===========================================================================
void Verlet(double ht, double m[],
            double x[], double y[], double z[],
            double vx[], double vy[], double vz[],
            double ax[], double ay[], double az[], int n,
            double &Ekin, double &Epot,
            void Forces(double[],double[],double[],double[],double[],
                        double[],double[],int,double&))
//---------------------------------------------------------------------------
// Propagates the solution of Newton's equations of motion for a system of n
// particles over a time interval ht using the velocity Verlet method
// m           - masses of particles
// x, y, z     - positions
// vx, vy, vz  - velocities
// ax, ay, az  - accelerations (need to be initialized on 1st call)
// Ekin        - total kinetic energy
// Epot        - total potential energy
//---------------------------------------------------------------------------
{
   double ht2;
   int i;

   ht2 = 0.5e0 * ht;
   for (i=1; i<=n; i++) {
      vx[i] += ht2 * ax[i]; x[i] += ht * vx[i];                       // v -> t+ht/2
      vy[i] += ht2 * ay[i]; y[i] += ht * vy[i];                       // r -> t+ht
      vz[i] += ht2 * az[i]; z[i] += ht * vz[i];
   }

   Forces(m,x,y,z,ax,ay,az,n,Epot);                                   // forces

   Ekin = 0e0;
   for (i=1; i<=n; i++) {
      ax[i] /= m[i]; ay[i] /= m[i]; az[i] /= m[i];                    // a -> t+ht
      vx[i] += ht2 * ax[i];                                           // v -> t+ht
      vy[i] += ht2 * ay[i];
      vz[i] += ht2 * az[i];

      Ekin += 0.5 * m[i] * (vx[i]*vx[i] + vy[i]*vy[i] + vz[i]*vz[i]);
   }
}

//===========================================================================
int EulerCromerQM(double E, double V[], double x[], double y[], int nx,
                  double y0, double dy0)
//---------------------------------------------------------------------------
// Propagates the solution of the dimensionless 1D Schrodinger equation
//     y" = 2 [V(x) - E] y, y(0) = y0, y'(0) = y'0
// on a regular mesh x[] with nx points by the Euler-Cromer method. Receives
```

```
// the energy in E, the tabulated potential in V[], the initial conditions in
// y0 and dy0. Returns the index of the divergence point (default is nx).
//---------------------------------------------------------------------------
{
   double dy, d2y, hx;
   int m;

   hx = x[2] - x[1];                                        // propagation step size
   y[1] = y0; dy = dy0;                                     // initial values
   for (m=1; m<=nx-1; m++) {                                // propagation loop
      d2y = 2e0 * (V[m] - E) * y[m];             // RHS of Schroedinger equation
      dy += hx * d2y;                                       // dy -> x[m]+hx
      y[m+1] = y[m] + hx * dy;                              // y -> x[m]+hx

      if (fabs(y[m+1]) > 1e10) break;                       // stop if y diverges
   }
   return m;                                        // index of divergence point
}

//============================================================================
int Numerov(double E, double V[], double x[], double y[], int nx,
            double y0, double dy0)
//---------------------------------------------------------------------------
// Propagates the solution of the dimensionless 1D Schrodinger equation
//     y" = 2 [V(x) - E] y, y(0) = y0, y'(0) = y'0
// on a regular mesh x[] with nx points by the Numerov method. Receives the
// energy in E, the tabulated potential in V[], and the initial conditions in
// y0 and dy0. Returns the index of the divergence point (default is nx).
//---------------------------------------------------------------------------
{
   double dy, d2y, hx, h6, um, um1, up1;
   int m;

   hx = x[2] - x[1];                                        // propagation step size
   y[1] = y0; dy = dy0;                                     // initial values

   d2y = 2e0 * (V[1] - E) * y[1];                    // initial Euler-Cromer step
   dy += hx * d2y;
   y[2] = y[1] + hx * dy;

   h6 = hx*hx/6e0;
   um1 = 1e0 - h6 * (V[1] - E);                      // stack of auxiliary values
   um  = 1e0 - h6 * (V[2] - E);
   for (m=2; m<=nx-1; m++) {                                // propagation loop
      up1 = 1e0 - h6 * (V[m+1] - E);
      y[m+1] = ((12e0 - 10e0*um) * y[m] - um1 * y[m-1]) / up1;
      um1 = um; um = up1;                         // shift stack down for next step

      if (fabs(y[m+1]) > 1e10) break;                       // stop if y diverges
   }
   return m;                                        // index of divergence point
}

//============================================================================
void Propag(double x[], double y[], int nx, double y0, double dy0,
            double Func(double,double,double))
//---------------------------------------------------------------------------
// Propagates the solution y[] of a Cauchy-problem for a 2nd order ODE on a
// regular mesh x[] with nx points, starting from y0 and dy0.
```

```
// Calls: EulerCromer1(x, hx, y, dy, Func); Func(x, y, dy) - RHS of ODE
//-------------------------------------------------------------------------
{
   double dy, hx;
   int m;

   hx = x[2] - x[1];                                   // propagation step size
   y[1] = y0; dy = dy0;                                     // initial values
   for (m=1; m<nx; m++)                                    // propagation loop
      { y[m+1] = y[m]; EulerCromer1(x[m],hx,y[m+1],dy,Func); }
}


//============================================================================
double Shoot(double x[], double y[], int nx, double ya, double yb,
             double dy1, double dy2, double eps, int &exist,
             double Func(double,double,double))
//-------------------------------------------------------------------------
// Solves a two-point boundary-value problem for a 2nd order ODE
//    y" = f(x,y,y'), y(xa) = ya, y(xb) = yb
// on a regular mesh x[] with nx points, using the shooting method with trial
// initial derivatives dy in [dy1,dy2]. Returns the solution y[] satisfying
// the condition y(xb) = yb within tolerance eps, the used derivative dy, and
// an existence flag.
// Calls: Propag(x, y, nx, y0, dy0, Func); Func(x, y, dy) - RHS of ODE
//-------------------------------------------------------------------------
{
   const int itmax = 100;                              // max. no. of bisections
   double dy, f, f1, f2;
   int it;

   Propag(x,y,nx,ya,dy1,Func);                             //propagate y for dy1
   f1 = y[nx] - yb;                                            //deviation at xb
   Propag(x,y,nx,ya,dy2,Func);                             //propagate y for dy2
   f2 = y[nx] - yb;                                            //deviation at xb

   if (f1*f2 < 0) {                               //check if dy exists in [dy1,dy2]
      exist = 1;
      for (it=1; it<=itmax; it++) {                         //refine dy by bisection
         dy = 0.5e0*(dy1 + dy2);                            //new approximation
         Propag(x,y,nx,ya,dy,Func);                             //propagate y
         f = y[nx] - yb;                                       //deviation at xb
         if (f1*f > 0) dy1 = dy; else dy2 = dy;            //new semi interval
         if (fabs(f) <= eps) break;                  //deviation vanishes at xb ?
      }
      if (it >= itmax) printf("Shoot: max. no. of bisections exceeded !\n");
   } else
      { dy = 1e10; exist = 0; }
   return dy;
}


//============================================================================
double ShootQM(double E1, double E2, double V[], double x[], double y[],
               int nx, int nc, double y0, double dy0, double eps, int &exist)
//-------------------------------------------------------------------------
// Solves the two-point eigenvalue problem for the 1D Schrodinger equation
//    y" = (2m/h^2) [V(x) - E] y, y(0) = y0, y(+inf) = 0
// on a regular mesh x[] with nx points, using the shooting method with the
// trial energies E in [E1,E2] and the tabulated potential in V[]. Returns
// the solution y[] vanishing at the checkpoint x[nc] within tolerance eps,
```

```
// and the existence flag exist.
// Calls: Numerov(V, x, y, nx, y0, dy0, E)
//--------------------------------------------------------------------------
{
   const int itmax = 100;                                // max. no. of bisections
   double E, f, f1, f2;
   int inf, it;

   inf = Numerov(E1,V,x,y,nx,y0,dy0);                    // propagate y for E1
   f1 = y[inf];                                                  // asymptotic y
   inf = Numerov(E2,V,x,y,nx,y0,dy0);                    // propagate y for E2
   f2 = y[inf];                                                  // asymptotic y

   if (f1*f2 < 0) {                                 //check if exists E in [E1,E2]
      exist = 1;
      for (it=1; it<=itmax; it++) {                         //refine E by bisection
         E = 0.5e0*(E1 + E2);                               //new approximation
         inf = Numerov(E,V,x,y,nx,y0,dy0);                     //propagate y
         f = y[inf];                                          // asymptotic y
         if (f1*f > 0) E1 = E; else E2 = E;               //new semi interval
         if (fabs(y[nc]) <= eps) break;         //check if y vanishes at x[nc] ?
      }
      if (it >= itmax) printf("ShootQM: max. no. of bisections exceeded!\n");
   } else
      { E = 1e10; exist = 0; }
   return E;
}


//==============================================================================
void Bilocal(double xa, double xb, double y[], int nx,
             double alf1, double bet1, double alf2, double bet2,
             void Func(double, double&, double&, double&))
//------------------------------------------------------------------------------
// Solves a linear two-point boundary-value problem for a 2nd order ODE
//     y" = p(x) y' + q(x) y + r(x),   xa <= x <= xb
//     alf1 y(xa) + bet1 y'(xa) = 1
//     alf2 y(xa) + bet2 y'(xa) = 1
// on a regular mesh with nx points, using central finite-differences.
// Returns the solution in y[].
// Calls: Func(x, p, q, r) - returns values of p(x), q(x), r(x)
//        TriDiagSys (linsys.h) - solves discretized tridiagonal system
//------------------------------------------------------------------------------
{
   double *a, *b, *c;
   double hx, h2, p, q, r, x;
   int m;

   a = Vector(1,nx); b = Vector(1,nx); c = Vector(1,nx);    // matrix elements

   hx = (xb-xa)/(nx-1); h2 = 2e0*hx*hx;
                                                        // build system coefficients
   b[1] = hx*alf1 - bet1; c[1] = bet1; y[1] = hx;
   for (m=2; m<=nx-1; m++) {
      x = xa + (m-1)*hx;                                       // mesh point
      Func(x,p,q,r);
      a[m] = -(2e0 + hx*p); b[m] = 4e0 + h2*q; c[m] = -(2e0 - hx*p);
      y[m] = -h2*r;                                    // RHSs of tridiagonal system
   }
   a[nx] = -bet2; b[nx] = hx*alf2 + bet2; y[nx] = hx;
```

```
    TriDiagSys(a,b,c,y,nx);                          // solve discretized system

    FreeVector(a,1); FreeVector(b,1); FreeVector(c,1);
}

#endif
```

12.11 Problems

The Python and C/C++ programs for the following problems may import the functions developed in this chapter from the modules ode.py and, respectively, ode.h, which are available as supplementary material. For creating runtime plots, the graphical routines contained in the libraries graphlib.py and graphlib.h may be employed.

PROBLEM 12.1

Consider the Cauchy problem:

$$y'' + y = 0, \tag{12.109}$$

$$y(0) = 0, \quad y'(0) = 1. \tag{12.110}$$

Put the ODE first under the form of a system of first-order ODEs and then solve the equivalent problem by comparatively using Euler's method and the Euler predictor–corrector method up to $t_{max} = 100$, decreasing the step size from $h = 0.05$ to $h = 0.001$. Plot the dependences $y' = y'(y)$ and estimate the value of h for which the basic Euler method starts being stable.

Solution

The Python implementation is given in Listing 12.21 and the C/C++ version is available as supplementary material (P12-ODEg.cpp). The plots resulting by applying the Euler and Euler predictor–corrector algorithms for $h = 0.05$ are shown in Figure 12.10.

Listing 12.21   Cauchy Problem for a Second-Order ODE Solved by Euler's Method (Python Coding)

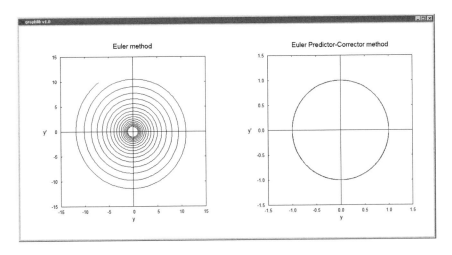

FIGURE 12.10   Solution yielded by the Euler (left) and Euler predictor–corrector methods (right) for the Cauchy problem (12.109)–(12.110) with the step size $h = 0.05$.

```
#   Solves a Cauchy problem for a 2nd order ODE by Euler's method
#       y" + y = 0,    y(0) = y0, y'(0) = y0'
from math import *
from ode import *
from graphlib import *

def Func(t, y, f):                                   # RHSs of 1st order ODEs
    f[1] =  y[2]                                       # y[1] = y, y[2] = y'
    f[2] = -y[1]

# main

y0 = 0e0; dy0 = 1e0                         # initial values => y(t) = sin(t)
tmax = 100e0                                                     # time span
ht = 0.05e0                                                      # step size

n = 2                                              # number of 1st order ODEs
nt = int(tmax/ht + 0.5) + 1                        # number of time steps
y = [0]*(n+1)                                        # solution components
y1 = [0]*(nt+1); y2 = [0]*(nt+1)                       # plotting arrays

GraphInit(1200,600)

t = 0e0; it = 1                                              # Euler method
y[1] = y0; y[2] = dy0                                     # initial values
y1[1] = y[1]; y2[1] = y[2]                            # store for plotting
while (t+ht <= tmax):                                    # propagation loop
    Euler(t,ht,y,n,Func)
    t += ht; it += 1
    y1[it] = y[1]; y2[it] = y[2]                       # store for plotting

Plot(y1,y2,nt,"red",2,0.10,0.45,0.15,0.85,"y","y'","Euler method")

t = 0e0; it = 1                             # Predictor-Corrector Euler method
y[1] = y0; y[2] = dy0                                     # initial values
y1[1] = y[1]; y2[1] = y[2]                            # store for plotting
while (t+ht <= tmax):                                    # propagation loop
    EulerPC(t,ht,y,n,Func)
    t += ht; it += 1
    y1[it] = y[1]; y2[it] = y[2]                       # store for plotting

Plot(y1,y2,nt,"blue",2,0.60,0.95,0.15,0.85,
     "y","y'","Euler Predictor-Corrector method")

MainLoop()
```

PROBLEM 12.2

The world record for men's hammer throw measures 86.74 m and has been held since 1986 by Yuriy
Sedykh.[*] The men's hammer weighs 7.26 kg, is spherical, and has a typical radius $R = 6$ cm. The drag on
the hammer can be considered proportional to the squared hammer velocity relative to the air:

$$F_D = \frac{1}{2}\rho A C_D v^2,$$

---

[*] http://hammerthrow.org

where ρ is the density of the air (1.2 kg/m^3) and $A = \pi R^2$ is the cross-sectional area of the hammer. The hammer can experience, in principle, either a "laminar-separated" flow, with a typical drag coefficient $C_d = 0.5$, or an "unsteady-oscillating" flow, with $C_d = 0.75$.*

a. Solve Newton's equation of motion for the oblique throw of the hammer using Euler's predictor–corrector method with a time step $h = 0.01$, transforming first the ODEs for the x and y motions into a system of four first-order ODEs. Consider throws from the initial position $x_0 = 0$, $y_0 = 2$ m, under the ideal angle $\varphi = 45°$ and find the initial velocity that produces the throw distance of the world record.

b. Calculate and plot the time dependence of the hammer's altitude and its trajectory $y = y(x)$ in three flow regimes: (a) frictionless, (b) laminar-separated flow, and (c) unsteady-oscillating flow. Evaluate the amount by which the throw distance is influenced by the drag.

c. Solve the same problem employing the Euler–Cromer and Verlet methods. Use the step-halving technique to estimate the error affecting the throw distance for each of the employed methods and discuss the results with respect to their known orders of precision.

Solution

The equivalent Cauchy problem for first-order equations, using the notations $y_1 = x$, $y_2 = y$, $y_3 = v_x$, and $y_4 = v_y$, reads:

$$\begin{cases} y_1' = y_3, & y_1(0) = x_0, \\ y_2' = y_4, & y_2(0) = y_0, \\ y_3' = -(k/m)\, y_3 \left| y_3 \right|, & y_3(0) = v_{x0}, \\ y_4' = -(k/m)\, y_4 \left| y_4 \right| - g, & y_4(0) = v_{y0}. \end{cases} \qquad (12.111)$$

The Python program based on the Euler predictor–corrector method is presented in Listing 12.22. The corresponding C/C++ implementation is available as supplementary material (P12-ThrowPCg.cpp), along with the variants using the Euler–Cromer method

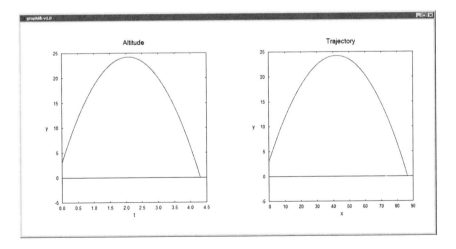

FIGURE 12.11 Time-dependent altitude and trajectory of a hammer thrown under $45°$ and experiencing quadratic drag, as yielded by Euler's predictor–corrector method (for Problem 12.2).

---

* https://www.grc.nasa.gov/www/k-12/airplane/dragsphere.html

(P12-ThrowECg.py and P12-ThrowECg.cpp), and those based on the Verlet method
(P12-ThrowVg.py and P12-ThrowVg.cpp). Typical graphical output for the laminar-separated
flow is shown in Figure 12.11.

Listing 12.22   Hammer Throw Using Euler's Predictor–Corrector Method (Python Coding)

```python
# Oblique throw of an object with drag using the Euler PC method
from math import *
from ode import *
from graphlib import *

g = 9.81e0                                    # gravitational acceleration

def Func(t, y, f):                # RHSs of 1st order ODEs for oblique throw
   f[1] = y[3]                    # y[1] = x, y[2] = y, y[3] = vx, y[4] = vy
   f[2] = y[4]
   f[3] = -k/m * y[3]*fabs(y[3])
   f[4] = -k/m * y[4]*fabs(y[4]) - g

# main

m = 7.26e0                                            # mass of hammer
R = 0.06e0                                            # radius of hammer
x0 = 0e0; y0 = 3e0                                    # initial position
v0 = 29e0                                             # initial velocity
phi = 45e0 * pi/180e0                                     # throw angle
vx0 = v0 *cos(phi); vy0 = v0*sin(phi)                 # initial velocity
rho = 1.2                                              # density of air
Cd = 0.5e0                                            # drag coefficient
k = 0.5e0*rho*(pi*R*R)*Cd                         # velocity coefficient
tmax = 20e0                                               # time span
ht = 0.001e0                                          # time step size

n = 4                                        # number of 1st order ODEs
nt = int(tmax/ht + 0.5) + 1                      # number of time steps
y = [0]*(n+1)                                    # solution components
tt = [0]*(nt+1); xt = [0]*(nt+1); yt = [0]*(nt+1)      # plotting arrays

t = 0e0; it = 1
y[1] = x0; y[3] = vx0                                   # initial values
y[2] = y0; y[4] = vy0
tt[1] = t; xt[1] = y[1]; yt[1] = y[2]               # store for plotting

while (t+ht <= tmax):                                # propagation loop
   EulerPC(t,ht,y,n,Func)
   t += ht; it += 1

   tt[it] = t; xt[it] = y[1]; yt[it] = y[2]           # store for plotting
   if (y[2] < 0.e0): break               # stop if object hits the ground

print("xmax = {0:5.2f}".format(y[1]))

GraphInit(1200,600)
Plot(tt,yt,it,"blue",1,0.10,0.45,0.15,0.85,"t","y","Altitude")
Plot(xt,yt,it,"blue",1,0.60,0.95,0.15,0.85,"x","y","Trajectory")
MainLoop()
```

PROBLEM 12.3

Consider a particle of mass $m = 1$ u (unified atomic mass units), carrying the elementary charge $q = -e$, which orbits around a fixed charge $Q = e$, located at the origin. The initial position of the particle is $x_0 = 1$ Å and $y_0 = 0$ and its velocity components are $v_{x0} = 0$ and $v_{y0} = 4.5$ Å/ps.

Listing 12.23 Orbiting Charged Particle Using Euler's Predictor–Corrector Method (Python Coding)

```
# Charged particle orbiting about a fixed charge - Euler PC method
from math import *
from ode import *
from graphlib import *

def Func(t, y, f):                              # RHS of 1st order ODEs
   factCoul = 14.3996517e0                      # e**2/(4*pi*eps0) [eV*A]
                                  # charge: e, energy: eV, force: eV/A
   r2 = y[1]*y[1] + y[2]*y[2]          # mass: u, distance: A, time: ps
   r = sqrt(r2)                            # distance from force center
   fr = factCoul * q * Q / r2                           # radial force
   f[1] = y[3]
   f[2] = y[4]
   f[3] = fr/m * y[1]/r
   f[4] = fr/m * y[2]/r

# main

m = 1e0                                         # mass of particle
q = -1.e0; Q = 1.e0                        # charges in units of e
x0 = 1e0; y0 = 0e0                              # initial position
vx0 = 0e0; vy0 = 4.5e0                          # initial velocity
tmax = 20e0                                            # time span
ht = 0.01e0                                       # time step size

n = 4                                     # number of 1st order ODEs
nt = int(tmax/ht + 0.5) + 1                   # number of time steps
y = [0]*(n+1)                                  # solution components
tt = [0]*(nt+1); xt = [0]*(nt+1); yt = [0]*(nt+1)     # plotting arrays

t = 0e0; it = 1
y[1] = x0; y[3] = vx0                              # initial values
y[2] = y0; y[4] = vy0
tt[1] = t; xt[1] = y[1]; yt[1] = y[2]          # store for plotting

while (t+ht <= tmax):                             # propagation loop
   EulerPC(t,ht,y,n,Func)
   t += ht; it += 1

   tt[it] = t; xt[it] = y[1]; yt[it] = y[2]      # store for plotting

GraphInit(1200,600)

Plot(tt,xt,it,"blue",1,0.10,0.45,0.15,0.85,
     "t","y","x-position of charged particle")
Plot(xt,yt,it,"blue",1,0.60,0.95,0.15,0.85,
     "x","y","Trajectory of charged particle")

MainLoop()
```

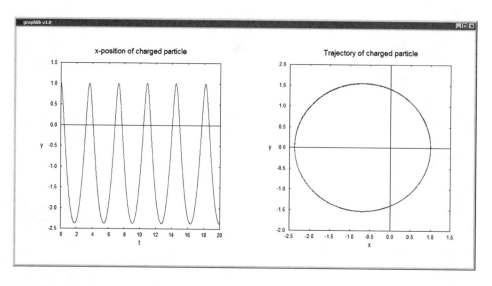

FIGURE 12.12   Time dependence of the x-position and trajectory of a charged particle orbiting about a fixed charge, calculated using the Euler predictor–corrector method (for Problem 12.3).

Propagate the trajectory of the particle up to $t_{max} = 20$ ps using the Euler predictor–corrector method with a time step $h = 0.01$. Plot the time dependence of the particle's x-position and its trajectory. Assess the stability of the orbit, and determine its defining axes and the orbital period.

Solution

The Coulomb force components are

$$F_x = K \frac{qQ}{r^2} \frac{x}{r}, \quad F_y = K \frac{qQ}{r^2} \frac{y}{r},$$

where $K = e^2/(4\pi) = 14.3996517$ eV/Å.

The Python program is given in Listing 12.23, while the C/C++ implementation is available as supplementary material (P12-ODEg.cpp). The graphical representations produced by the program for the time dependence of the particle's x-position and for its trajectory are shown in Figure 12.12.

PROBLEM 12.4

Consider the previous problem of the orbiting charged particle and use comparatively the Euler and Euler–Cromer methods for second-order ODEs with the time step $h = 0.01$. Assess the stability of the two orbits, explain the qualitative difference between them, and determine the time step for which Euler's method becomes stable, as well.

Solution

The Python implementation is shown in Listing 12.24 and the C/C++ version is available as supplementary material (P12-ChargeEC.cpp). The charged particle's trajectories are depicted in Figure 12.13.

Listing 12.24   Orbiting Charged Particle Using the Euler–Cromer Method (Python Coding)

```
# Charged particle orbiting about a fixed charge - Euler-Cromer method
from math import *
from ode import *
from graphlib import *

def Func(t, y, dy, f):                          # RHS of 2nd order ODEs
    factCoul = 14.3996517e0                     # e**2/(4*pi*eps0) [eV*A]
                                   # charge: e, energy: eV, force: eV/A
    r2 = y[1]*y[1] + y[2]*y[2]          # mass: u, distance: A, time: ps
    r = sqrt(r2)                            # distance from force center
    fr = factCoul * q * Q / r2                          # radial force
    f[1] = fr/m * y[1]/r
    f[2] = fr/m * y[2]/r

# main

m = 1e0                                             # mass of particle
q = -1.e0; Q = 1.e0                            # charges in units of e
x0 = 1e0; y0 = 0e0                                 # initial position
vx0 = 0e0; vy0 = 4.5e0                             # initial velocity
tmax = 20e0                                             # time span
ht = 0.01e0                                        # time step size

n = 2                                         # number of 2nd order ODEs
nt = int(tmax/ht + 0.5) + 1                     # number of time steps
y = [0]*(n+1); dy = [0]*(n+1)                   # solution components
xt = [0]*(nt+1); yt = [0]*(nt+1)                     # plotting arrays

GraphInit(1200,600)

t = 0e0; it = 1                                        # Euler method
y[1] = x0; dy[1] = vx0                               # initial values
y[2] = y0; dy[2] = vy0
xt[1] = y[1]; yt[1] = y[2]                         # store for plotting
while (t+ht <= tmax):                                # propagation loop
    Euler2(t,ht,y,dy,n,Func)
    t += ht; it += 1
    xt[it] = y[1]; yt[it] = y[2]                   # store for plotting

Plot(xt,yt,it,"red",1,0.10,0.45,0.15,0.85,
    "t","y","Trajectory of charged particle - Euler")

t = 0e0; it = 1                                  # Euler-Cromer method
y[1] = x0; dy[1] = vx0                               # initial values
y[2] = y0; dy[2] = vy0
xt[1] = y[1]; yt[1] = y[2]                         # store for plotting
while (t+ht <= tmax):                                # propagation loop
    EulerCromer(t,ht,y,dy,n,Func)
    t += ht; it += 1
    xt[it] = y[1]; yt[it] = y[2]                   # store for plotting

Plot(xt,yt,it,"blue",1,0.60,0.95,0.15,0.85,
    "x","y","Trajectory of charged particle - Euler-Cromer")

MainLoop()
```

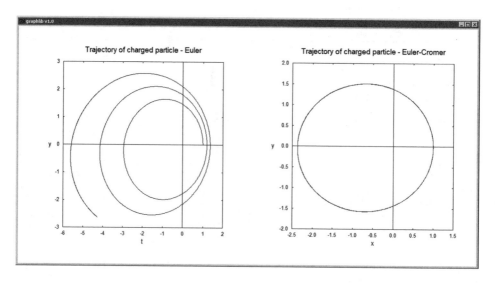

FIGURE 12.13   Trajectories of a charged particle orbiting about a fixed charge, as calculated using the Euler and Euler–Cromer methods (for Problem 12.4).

PROBLEM 12.5

Consider the previous problem of the orbiting charged particle and use for propagation the Verlet algorithm with the (rather large) time step $h = 0.1$. Plot the kinetic, potential, and total energies, as well as the particle's trajectory. Identify the aspects that accompany the occurrence of instability. Decrease the time step to $h = 0.01$ and observe the qualitative changes in the total energy and the particle's trajectory.

Solution

The Python implementation is provided in Listing 12.25 and the C/C++ version is available as supplementary material (P12-ChargeV.cpp). The charged particle's energies and trajectory are plotted in Figure 12.14.

Listing 12.25   Orbiting Charged Particle Using the Verlet Method (Python Coding)

```python
# Charged particle orbiting about a fixed charge - Verlet method
from math import *
from ode import *
from graphlib import *

def Forces(m, x, y, vx, vy):
    factCoul = 14.3996517e0                       # e**2/(4*pi*eps0) [eV*A]
                                       # charge: e, energy: eV, force: eV/A
    r2 = x*x + y*y                     # mass: u, distance: A, time: ps
    r = sqrt(r2)                            # distance from force center
    fr = factCoul * q * Q / r2                         # radial force
    fx = fr * x/r                                    # force components
    fy = fr * y/r
    Epot = fr * r                                   # potential energy
    return (fx, fy, Epot)

# main

nn  = [0]*4; sty = [0]*4                # end-indexes, styles of plots
col = [""]*4                                      # colors of plots
```

```
m = 1e0                                            # mass of particle
q = -1.e0; Q = 1.e0                                  # in units of e
x0 = 1e0; y0 = 0e0                                # initial position
vx0 = 0e0; vy0 = 4.5e0                            # initial velocity
tmax = 20e0                                              # time span
ht = 0.1e0                                          # time step size

nt = int(tmax/ht + 0.5) + 1                   # number of time steps
tt = [0]*(3*nt+1); Et = [0]*(3*nt+1)             # plotting arrays
xt = [0]*(nt+1); yt = [0]*(nt+1)

t = 0e0; it = 1
x = x0; vx = vx0                                     # initial values
y = y0; vy = vy0

(fx, fy, Epot) = Forces(m,x,y,vx,vy)    # initial forces and potential energy
ax = fx/m; ay = fy/m                        # initial acceleration
Ekin = 0.5e0 * m * (vx*vx + vy*vy)       # initial kinetic energy
xt[it] = x; yt[it] = y                        # store for plotting
tt[it] = tt[nt+it] = t; tt[2*nt+it] = t
Et[it] = Ekin; Et[nt+it] = Epot; Et[2*nt+it] = Ekin + Epot

while (t < tmax):                                 # propagation loop
   (x,y,vx,vy,ax,ay,Ekin,Epot) = Verlet2(ht,m,x,y,vx,vy,ax,ay,Forces)
   t += ht; it += 1

   xt[it] = x; yt[it] = y                        # store for plotting
   tt[it] = tt[nt+it] = t; tt[2*nt+it] = t
   Et[it] = Ekin; Et[nt+it] = Epot; Et[2*nt+it] = Ekin + Epot

GraphInit(1200,600)

nn[1] =   nt; col[1] = "blue" ; sty[1] = 1                    # Ekin
nn[2] = 2*nt; col[2] = "green"; sty[2] = 1                    # Epot
nn[3] = 3*nt; col[3] = "red"  ; sty[3] = 1                    # Etot
MultiPlot(tt,Et,Et,nn,col,sty,3,10,0e0,0e0,0,0e0,0e0,0,
          0.10,0.45,0.15,0.85,"t","Ek,Ep,Et","Energies of charged particle")
Plot(xt,yt,nt,"blue",1,0.60,0.95,0.15,0.85,
     "x","y","Trajectory of charged particle")

MainLoop()
```

PROBLEM 12.6

Consider the equation of motion for the angular displacement u of a nonlinear rigid pendulum of length l (g is the gravitational acceleration):

$$u'' = -\frac{g}{l} \sin u, \tag{12.112}$$

with the initial conditions:

$$u(0) = u_0, \quad u'(0) = 0. \tag{12.113}$$

The period of oscillations of arbitrary amplitude explicitly depends on the initial displacement $u_0 \in [-\pi, \pi]$ (Beléndez et al. 2007):

$$T = T_0 \frac{2}{\pi} K\left(\sin^2 \frac{u_0}{2}\right). \tag{12.114}$$

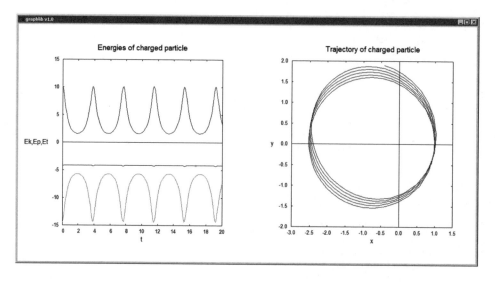

FIGURE 12.14 Time dependence of the kinetic, potential, and total energies and trajectory of a charged particle orbiting about a fixed charge, using the Verlet method (for Problem 12.5).

Here

$$T_0 = 2\pi \sqrt{\frac{l}{g}}$$

is the harmonic period, specific for the limit of vanishing amplitudes, and $K(m)$ is the *complete elliptic integral of the first kind*, defined as (Abramowitz and Stegun 1972)

$$K(m) = \int_0^1 \frac{dz}{\sqrt{(1 - z^2)(1 - mz^2)}}. \tag{12.115}$$

    a. Write a routine to calculate the elliptic integral $K(m)$ based on its definition and use the function `Improp2` (defined in `integral.py` or `integral.h`) to evaluate the involved improper integral of the second kind (having a singular integrand at one of the boundaries; see Section 10.8).

    b. Consider a pendulum of length $l = 1$ m and an initial angular displacement $u_0 = \pi/2$. Use the Runge–Kutta method to propagate the dynamic state up to $t_{max} = 20$ s, with a time step $h = 0.001$ s. Determine the number of half-periods covered by the propagation by counting the zeros (sign changes) of the solution and, from the time span between the first and the last zero, calculate the oscillation period and compare it with the exact value.

    c. Plot the time dependence of the displacement to verify the determined period qualitatively. Plot the trajectory of the pendulum in the phase space, $u' = u'(u)$, to check the stability of the solution visually.

Solution

To apply the Runge–Kutta method, the equation of the pendulum is transformed into a system of first-order ODEs:

$$\begin{cases} u_1' = u_2, \\ u_2' = -(g/l)\sin(u_1) - ku_2, \end{cases} \qquad \begin{cases} u_1(0) = u_0, \\ u_2(0) = 0. \end{cases}$$

The Python code is provided in Listing 12.26 and the C/C++ version is available as supplementary material (P12-PendulumRKg.cpp). The time dependence of the pendulum's angular displacement and the trajectory in the phase space are shown in Figure 12.15. The found period, $T/T_0 = 1.18036$, is exact to five significant digits. The program also makes provisions for taking into account the drag on the pendulum.

Listing 12.26 Oscillations of a Nonlinear Pendulum by the Runge–Kutta Method (Python Coding)

```
# Angular motion of a nonlinear pendulum by the Runge-Kutta method
#     u" = -g/l * sin(u) - k * u',    u(0) = u0, u'(0) = u0'
from math import *
from ode import *
from integral import *
from graphlib import *

g = 9.81e0                                       # gravitational acceleration

def Func(t, u, f):                               # RHS of 1st order ODEs
   f[1] = u[2]                                   # u[1] = u, u[2] = u'
   f[2] = -g/l * sin(u[1]) - k * u[2]


#==============================================================================
def Kel(m):
#------------------------------------------------------------------------------
#  Returns the complete elliptic integral of the 1st kind
#  Calls: qImprop2 (integral.py)
#------------------------------------------------------------------------------
   eps = 1e-7                                     # relative precision
   def fKel(z): return 1e0 / sqrt((1e0-z*z)*(1e0-m*z*z))      # integrand
   return qImprop2(fKel,0e0,1e0,eps)

# main

l = 1e0                                               # pendulum length
k = 0e0                                           # velocity coefficient
u0 = 0.5e0*pi                                     # initial displacement
du0 = 0e0                                          # initial derivative
tmax = 20e0                                                  # time span
ht = 0.001e0                                            # time step size

n = 2                                         # number of 1st order ODEs
nt = int(tmax/ht + 0.5) + 1                     # number of time steps
u = [0]*(n+1)                                   # solution components
tt = [0]*(nt+1); ut = [0]*(nt+1); vt = [0]*(nt+1)    # arrays for plotting

t = 0e0; it = 1
u[1] = u0; u[2] = du0                                    # initial values
tt[1] = t; ut[1] = u[1]; vt[1] = u[2]             # store for plotting

nT = 0                                         # number of half-periods
t1 = t2 = 0e0                                 # bounding solution zeros
us = u[1]                                             # save solution
while (t+ht <= tmax):                               # propagation loop
   RungeKutta(t,ht,u,n,Func)
   t += ht; it += 1

   if (u[1]*us < 0e0):                # count solution passages through zero
      if (t1 == 0): t1 = t                              # initial zero
      else: t2 = t; nT += 1                              # final zero
```

```
    us = u[1]                                                      # save solution

    tt[it] = t; ut[it] = u[1]; vt[it] = u[2]                      # store for plotting

T = 2e0*(t2-t1) / nT                                               # calculated period
T0 = 2e0*pi*sqrt(l/g)                                              # harmonic period
Tex = 2/pi * T0 * Kel(pow(sin(0.5e0*u0),2))                        # exact period
print("u0 = {0:7.5f}    T/T0 = {1:7.5f} + ({2:7.1e})". \
      format(u0,T/T0,(Tex-T)/T0))

GraphInit(1200,600)

Plot(tt,ut,it,"blue",1,0.10,0.45,0.15,0.85, \
     "t (s)","u (rad)","Displacement of nonlinear pendulum")
Plot(ut,vt,it,"blue",1,0.60,0.95,0.15,0.85, \
     "u (rad)","u' (rad/s)","Trajectory of pendulum in phase space")

MainLoop()
```

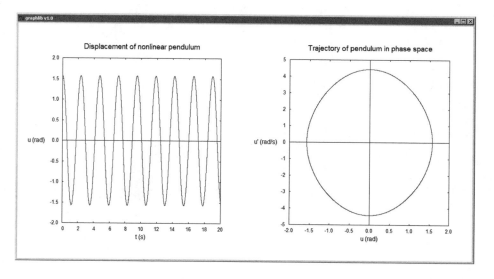

FIGURE 12.15   Time dependence of the angular displacement of a nonlinear pendulum and its trajectory in the phase space, when started from an initial displacement equal to $\pi/2$ (for Problem 12.6).

PROBLEM 12.7

Consider the previous problem for the nonlinear pendulum and propagate trajectories started from initial angular displacements $u_0 \in [0.1, 3.1]$ separated by 0.1. Plot the time dependences of the displacement, $u = u(t)$, for the lowest and the largest initial displacements and notice the significant differences both in period and in functional form. Plot the oscillation period as a function of the initial displacement, $T = T(u_0)$, and compare the dependence with the exact result (Equation 12.114).

Solution

The Python implementation is given in Listing 12.27 and the C/C++ variant is provided as supplementary material (P12-PendulumRKT.cpp). The time dependences of the pendulum's angular displacement and the dependence of the period on the initial displacement are displayed in Figure 12.16.

Listing 12.27   Oscillations of a Nonlinear Pendulum by the Runge–Kutta Method (Python Coding)

```
# Angular motion of a nonlinear pendulum by the Runge-Kutta method
#     u" = -g/l * sin(u) - k * u',    u(0) = u0, u'(0) = u0'
from math import *
from ode import *
from integral import *
from graphlib import *

g = 9.81e0                                       # gravitational acceleration

def Func(t, u, f):                                  # RHS of 1st order ODEs
   f[1] = u[2]                                      # u[1] = u, u[2] = u'
   f[2] = -g/l * sin(u[1]) - k * u[2]

#============================================================================
def Kel(m):
#----------------------------------------------------------------------------
#  Returns the complete elliptic integral of the 1st kind
#  Calls: qImprop2 (integral.py)
#----------------------------------------------------------------------------
   eps = 1e-7                                            # relative precision
   def fKel(z): return 1e0 / sqrt((1e0-z*z)*(1e0-m*z*z))       # integrand
   return qImprop2(fKel,0e0,1e0,eps)

# main

nn  = [0]*3                                          # ending indexes of plots
col = [""]*3                                              # colors of plots
sty = [0]*3                                               # styles of plots

l = 1e0                                                    # pendulum length
k = 0e0                                              # velocity coefficient
du0 = 0e0                                             # initial derivative
tmax = 20e0                                                    # time span
ht = 0.001e0                                            # time step size
u0min = 0.1e0                                                 # minimum u0
u0max = 3.1e0                                                 # maximum u0
hu = 0.1e0                                              # increment for u0

n = 2                                               # number of 1st order ODEs
nt = int(tmax/ht + 0.5) + 1                           # number of time steps
nu = int((u0max-u0min)/hu + 0.5) + 1                        # number of u0s
u = [0]*(n+1)                                         # solution components
tt = [0]*(2*nt+1); ut = [0]*(2*nt+1)               # arrays for t-dep. plots
tu = [0]*(2*nu+1); uu = [0]*(2*nu+1)              # arrays for u0-dep. plots

for iu in range(1,nu+1):
   u0 = u0min + (iu-1)*hu                               # initial displacement

   t = 0e0; it = 1
   u[1] = u0; u[2] = du0                                      # initial values
   if (iu == 1 ): tt[1    ] = t; ut[1    ] = u[1]        # for smallest u0
   if (iu == nu): tt[1+nt] = t; ut[1+nt] = u[1]         # for largest u0

   nT = 0                                            # number of half-periods
   t1 = t2 = 0e0                                  # bounding solution zeros
   us = u[1]                                               # save solution
   while (t+ht <= tmax):                                # propagation loop
      RungeKutta(t,ht,u,n,Func)
```

```
        t += ht; it += 1

        if (u[1]*us < 0e0):                    # count solution passages through zero
            if (t1 == 0): t1 = t                          # initial zero
            else: t2 = t; nT += 1                         # final zero
        us = u[1]                                    # save solution
                                                  # store for plotting
        if (iu == 1 ): tt[it   ] = t; ut[it   ] = u[1]      # for smallest u0
        if (iu == nu): tt[it+nt] = t; ut[it+nt] = u[1]      # for largest u0

    T = 2e0*(t2-t1) / nT                        # calculated period
    T0 = 2e0*pi*sqrt(1/g)                       # harmonic period
    Tex = 2/pi * T0 * Kel(pow(sin(0.5e0*u0),2))         # exact period

    uu[iu   ] = u0; tu[iu   ] = T/T0
    uu[iu+nu] = u0; tu[iu+nu] = Tex/T0

GraphInit(1200,600)

nn[1] = it   ; col[1] = "blue"; sty[1] = 1              # for smallest u0
nn[2] = it+nt; col[2] = "red" ; sty[2] = 1              # for largest u0
MultiPlot(tt,ut,ut,nn,col,sty,2,10,0e0,0e0,0,0e0,0e0,0,0.10,0.45,0.15,0.85, \
        "t (s)","u (rad)","Displacement of nonlinear pendulum")

nn[1] =   nu; col[1] = "red" ; sty[1] = 0              # calculated periods
nn[2] = 2*nu; col[2] = "blue"; sty[2] = 1              # exact periods
MultiPlot(uu,tu,tu,nn,col,sty,2,10,0e0,0e0,0,0e0,0e0,0,0.60,0.95,0.15,0.85, \
        "u0 (rad)","T/T0","Period of nonlinear pendulum")

MainLoop()
```

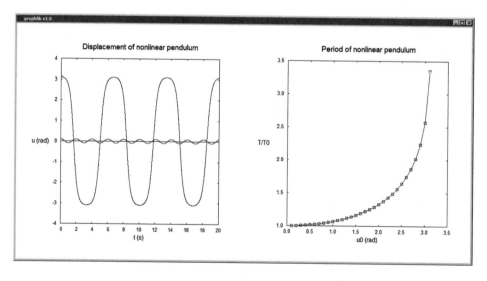

FIGURE 12.16 Time dependences of the angular displacement of a nonlinear pendulum for initial displacements equal to 0.1 and, respectively, 3.1, and dependence of the period on the initial displacement $u_0 \in [0.1, 3.1]$ (for Problem 12.7).

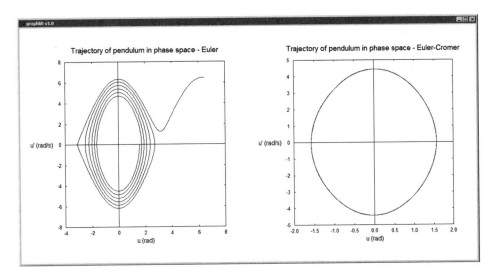

FIGURE 12.17 Phase space trajectories of a nonlinear pendulum of length $l = 1$, displaced initially by $u_0 = \pi/2$, as yielded by the Euler and Euler–Cromer methods with a time step $h = 0.01$ s (for Problem 12.8).

PROBLEM 12.8

Consider again the problem of the nonlinear pendulum of length $l = 1$ m displaced initially by $u_0 = \pi/2$. Use comparatively the Euler method for second-order ODEs (routine `Euler1`) and the Euler–Cromer method (routine `EulerCromer1`) to propagate the solution up to $t_{max} = 20$ s, with a time step $h = 0.01$ s. For both methods, plot the phase space trajectory of the pendulum and comment on the qualitative difference between their stability properties.

Solution

The Python and C/C++ implementations are provided as supplementary material (`P12-PendulumECg.py` and `P12-PendulumECg.cpp`). The phase space trajectories of the nonlinear pendulum for the Euler and Euler–Cromer methods are depicted in Figure 12.17.

PROBLEM 12.9

Consider the relative motion of the Moon around the Earth in the common center-of-mass system, under the influence of its gravitational interaction force

$$F = G\frac{m_1 m_2}{r_{12}}. \tag{12.116}$$

$G = 6.67384 \cdot 10^{-11}$ m^3/kg $\cdot$ s^2 is the gravitational constant, $m_1 = 5.97 \cdot 10^{24}$ kg is the Earth's mass, $m_2 = 7.35 \cdot 10^{22}$ kg is the Moon's mass, and r_{12} is the distance between them.

a. Place initially the Earth at rest at the origin and the Moon at the approximate apogee (largest) distance $d_0 = 4.06 \cdot 10^8$ m, with a perpendicular relative velocity $v_0 = 960$ m/s.* Use the routine `MovetoCM` (from the module `coords.py` given in Listing 8.10, or from the header file `coords.h`) to move the planets to their center-of-mass system and also to cancel their total momentum, by passing velocities instead of positions to the routine.

---

* http://solarsystem.nasa.gov/planets/profile.cfm?Object=Moon

b. Write a general routine to evaluate the pairwise gravitational interaction forces for an arbitrary number of celestial objects.

c. The *sidereal month* is the time it takes the Moon to complete one full revolution around the Earth with respect to the stars and it equals approximately 27.32 Earth days. Propagate the dynamic state of the two planets for two sidereal months using the `Verlet` propagator (Listing 12.10), with a time step $h = 1$ s.

d. Record the positions of the Earth and Moon every 10,000 steps (roughly, 3 h) and plot the Moon's trajectory to check its stability. Plot, equally, the time dependence of the Earth–Moon distance to determine its perigee (minimum) value and compare it with the observed distance of 363,104 km.

Solution

The Python code is shown in Listing 12.28 and the C/C++ version is available as supplementary material (`P12-PlanetsV.cpp`). The Earth–Moon distance as a function of time and the Moon's elliptical trajectory about the common mass center are plotted in Figure 12.18.

Listing 12.28   Orbital Motions of the Earth and Moon by the Verlet Method (Python Coding)

```python
# Relative motion of the Earth and Moon using the velocity Verlet method
from math import *
from ode import *
from coords import *
from graphlib import *

G = 6.67384e-11                                     # gravitational constant m^3 / kg s^2

#============================================================================
def Forces(m, x, y, z, fx, fy, fz, n):
#----------------------------------------------------------------------------
#   Returns gravitational forces acting on a system of n point-masses
#----------------------------------------------------------------------------
    Epot = 0e0
    for i in range(1,n+1): fx[i] = fy[i] = fz[i] = 0e0

    for i in range(1,n):                                    # loop over all pairs
        for j in range(i+1,n+1):
            dx = x[i] - x[j]                        # components of relative distance
            dy = y[i] - y[j]
            dz = z[i] - z[j]
            r2 = dx*dx + dy*dy + dz*dz                      # squared distance
            r = sqrt(r2)
            fr = G * m[i] * m[j] / r                        # |force| * r

            Epot += fr                              # total potential energy

            fr /= r2                                        # |force| / r
            fx[i] -= fr * dx; fx[j] += fr * dx      # total force components
            fy[i] -= fr * dy; fy[j] += fr * dy
            fz[i] -= fr * dz; fz[j] += fr * dz

    return Epot

# main

mEarth = 5.97e24                                    # Earth's mass (kg)
mMoon  = 7.35e22                                    # Moon's mass (kg)
d0 = 4.06e8                                 # Earth-Moon distance at apogee (m)
v0 = 969e0                                  # initial relative velocity (m/s)
```

```
km = 1e3
month = 27.32                       # sidereal month: Moon's orbital period (days)
day = 3600 * 24                                          # day length (s)
tmax = 2 * month * day                       # time span: 2 lunar months (s)
ht = 1e0                                                 # time step (s)

n = 2                                               # number of planets
m = [0]*(n+1)                                            # planet masses
x = [0]*(n+1); vx = [0]*(n+1); ax = [0]*(n+1)            # coordinates
y = [0]*(n+1); vy = [0]*(n+1); ay = [0]*(n+1)            # velocities
z = [0]*(n+1); vz = [0]*(n+1); az = [0]*(n+1)            # accelerations

nt = int(tmax/ht + 0.5) + 1                     # number of time steps
n1 = 10000                              # number of steps between records
np = int(float(nt-1)/n1) + 1              # number of plotted points
nt = (np-1) * n1 + 1                      # corrected number of time steps
tp = [0]*(np+1); dp = [0]*(np+1)                        # plotting arrays
xp = [0]*(np+1); yp = [0]*(np+1)

m[1] = mEarth                                    # initial configuration
x[1] = 0e0; vx[1] = 0e0; ax[1] = 0e0
y[1] = 0e0; vy[1] = 0e0; ay[1] = 0e0
z[1] = 0e0; vz[1] = 0e0; az[1] = 0e0
m[2] = mMoon
x[2] = d0 ; vx[2] = 0e0; ax[2] = 0e0
y[2] = 0e0; vy[2] = v0 ; ay[2] = 0e0
z[2] = 0e0; vz[2] = 0e0; az[2] = 0e0

MovetoCM(m,x,y,z,n)                                  # move to CM system
MovetoCM(m,vx,vy,vz,n)                          # cancel total momentum

t =  0e0                                               # initialization
it = ip = 1
tp[ip] = t / day
dp[ip] = sqrt(pow(x[1]-x[2],2)+pow(y[1]-y[2],2)) / km   # Earth-Moon distance
xp[ip] = x[2] / km; yp[ip] = y[2] / km               # Moon's position

GraphInit(1200,600)

while (t+ht <= tmax):                                # propagation loop
   t += ht; it += 1
   (Ekin,Epot) = Verlet(ht,m,x,y,z,vx,vy,vz,ax,ay,az,n,Forces)

   if (it % n1 == 0):
      ip += 1
      tp[ip] = t / day
      dp[ip] = sqrt(pow(x[1]-x[2],2)+pow(y[1]-y[2],2)) / km
      xp[ip] = x[2] / km; yp[ip] = y[2] / km

      GraphClear()
      Plot(tp,dp,ip,"blue",1,0.12,0.47,0.15,0.85,
           "t (days)","d (km)","Earth-Moon distance")
      Plot(xp,yp,ip,"red",2,0.62,0.97,0.15,0.85,
           "x (km)","y (km)","Moon's trajectory in the CM system")
      GraphUpdate()

MainLoop()
```

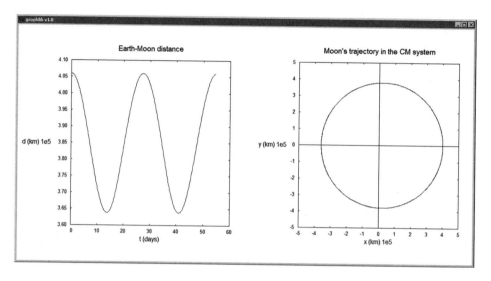

FIGURE 12.18 Time evolution of the Earth–Moon distance during two sidereal months (left) and the Moon's trajectory in the common center-of-mass system (right) (for Problem 12.9).

PROBLEM 12.10

Use the Euler–Cromer method to build a global integrator for the 1D Schrödinger equation,

$$\psi'' = 2\,[V(x) - E]\,\psi, \tag{12.117}$$

which receives the energy E, the potential $V(x)$ tabulated on a regular spatial mesh, and the initial values ψ_0 and ψ_0' at the first mesh point. The routine is supposed to identify the tendency of the wave function to diverge ($|\psi(x)| > 10^{10}$) and to return the index of the mesh point where this occurs.

Listing 12.29   Euler–Cromer Integrator for the 1D Schrödinger Equation (Python Coding)

```
#===============================================================================
def EulerCromerQM(E, V, x, y, nx, y0, dy0):
#-------------------------------------------------------------------------------
#  Propagates the solution of the dimensionless 1D Schrodinger equation
#     y" = 2 [V(x) - E] y, y(0) = y0, y'(0) = y'0
#  on a regular mesh x[] with nx points by the Euler-Cromer method. Receives
#  the energy in E, the tabulated potential in V[], the initial conditions in
#  y0 and dy0. Returns the index of the divergence point (default is nx).
#-------------------------------------------------------------------------------
   hx = x[2] - x[1]                              # propagation step size
   y[1] = y0; dy = dy0                                  # initial values
   for m in range(2,nx):                              # propagation loop
      d2y = 2e0 * (V[m] - E) * y[m]      # RHS of Schrodinger equation
      dy += hx * d2y                                  # dy -> x[m]+hx
      y[m+1] = y[m] + hx * dy                          # y -> x[m]+hx

      if (abs(y[m+1]) > 1e10): break            # stop if y diverges

   return m                                # index of divergence point
```

Solution

The Python integrator is shown in Listing 12.29, being part of the module ode.py, while the C/C++ version is contained in the header file ode.h.

PROBLEM 12.11

Solve the 1D Schrödinger equation,

$$\psi'' = 2 [V(x) - E] \psi, \tag{12.118}$$

for a particle in the symmetric finite square well

$$V(x) = \begin{cases} 0, & |x| \leq 5, \\ 10, & |x| > 5. \end{cases} \tag{12.119}$$

To this end, employ the shooting technique, as implemented in the routine ShootQM (Listing 12.16), which itself calls, as a global integrator, the function Numerov (Listing 12.13). Apply the initial conditions required by the intrinsic symmetry of the solutions and integrate the wave equation from $x_0 = 0$ out to $x_{\max} = 10$, with a step size $h = 10^{-3}$.

a. Determine the lowest eight bound states, with a relative precision $\varepsilon = 10^{-4}$ in the eigenenergies, by performing trial shoots in energy windows of width $\delta E = 0.1$, by which the energy range between the minimum and maximum potential values is scanned. Ensure the physically required vanishing of the solution at a checkpoint located at $x_c = 7$.

b. Normalize the eigenfunction for each found eigenstate, performing the integration by the trapezoidal rule, and plot it along with its square (as probability density).

c. Notice the tendency of the lower eigenvalues to be proportional to the squares of the respective quantum numbers $n = 1, 2, 3, \ldots$ Verify that the increase of the potential well depth gradually generalizes this behavior to the higher states, as well.

Solution

The Python program is provided in Listing 12.30, while the C/C++ version is available as supplementary material (P12-SqWellSHg.cpp). The eigenfunctions and the density probabilities are shown in Figure 12.19.

Listing 12.30   Bound States of a Finite Square Well by the Shooting Method (Python Coding)

```
#   Bound states of the 1D Schroedinger equation for a finite square well
#      y" = 2 [V(x) - E] y, y(0) = y0, y'(0) = y'0, y(+inf) = 0
#      V(x) = 0 for 0 < x <= 5; = 10 for x > 5
#   using a shooting algorithm based on the Numerov method
#---------------------------------------------------------------------------
from math import *
from ode import *
from graphlib import *

def Pot(x): return 0e0 if x < 5e0 else 10e0          # Square well potential

# main

xx = 10e0                                            # limit of x-mesh
xc = 7e0                          # checkpoint for vanishing solution (xc < xx)
nE = 8                              # number of eigenvalues to be calculated
dE = 0.1e0                          # minimum separation of eigenvalues
eps = 1e-4                          # tolerance for solution at xx
hx = 1e-3                           # x-mesh step size
```

```
nx = int(xx/hx + 0.5) + 1                               # number of x-mesh points
nc = int(xc/hx + 0.5) + 1                     # index of checkpoint for vanishing y
x = [0]*(nx+1); y = [0]*(nx+1); y2 = [0]*(nx+1)            # x-mesh, solution
V = [0]*(nx+1)                                             # tabulated potential

Vmin = Vmax = Pot(xx)
for m in range(1,nx+1):
   x[m] = (m-1)*hx                                      # integration mesh
   V[m] = Pot(x[m])                                      # tabulate potential
   if (Vmin > V[m]): Vmin = V[m]                         # potential minimum
   if (Vmax < V[m]): Vmax = V[m]                         # potential maximum

GraphInit(800,800)

hy = 0.92e0/nE                                  # fractional height of the plots
fy = 0.05                                 # lower fractional position of plots

iE = 0                                                   # index of found E
par = 0                                                  # parity of ground state
Ew = Vmin                       # lower limit of search window for E, [Ew,Ew+dE]
while (Ew < Vmax and iE < nE):                  # loop over eigenvalue windows
                                                         # initial values at x = 0
   if (par == 0): y0 = 1e0; dy0 = 0e0                     # even y
   else:          y0 = 0e0; dy0 = 1e0                     # odd y
                                                         # shoot in [Ew,Ew+dE]
   (E, exist) = ShootQM(Ew,Ew+dE,V,x,y,nx,nc,y0,dy0,eps)

   Ew += dE                                       # shift [Ew,Ew+dE] for next shoot

   if (exist):
      iE += 1                                             # found new E
      par = 0 if par else 1                               # parity of next y

      f = 0e0                               # normalize y by trapezoidal rule
      for m in range(1,nc+1): f += y[m]*y[m]*hx           # norm for [0,+inf]
      f = sqrt(2e0*f)
      if (int((iE-1)/2) % 2): f = -f                      # sign correction
      for m in range(1,nx+1): y[m] /= f; y2[m] = y[m]*y[m]

      if (iE == 1): E1 = E
      title = "E{0:1d}/E1 = {1:4.2f}".format(iE,E/E1)
      xtext = "x" if (iE == 1) else "None"
      Plot(x,y,nc,"blue",1,0.11,0.48,fy,fy+hy,xtext,"y",title)
      Plot(x,y2,nc,"blue",1,0.61,0.98,fy,fy+hy,xtext,"y^2",title)
      fy += hy                             # fractional y-position of next plot

MainLoop()
```

PROBLEM 12.12

Solve the 1D Schrödinger equation,

$$\psi'' = 2\left[V(x) - E\right]\psi, \tag{12.120}$$

for the quantum harmonic oscillator, having in dimensionless units the potential:

$$V(x) = \frac{1}{2}x^2. \tag{12.121}$$

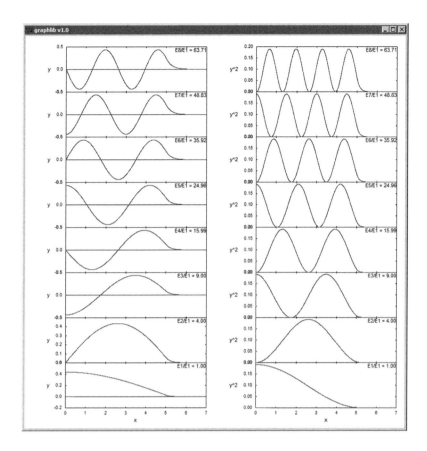

FIGURE 12.19 Lowest-energy eigenfunctions of a particle in a finite square well (left column) and the corresponding density probabilities (right column), determined by the shooting method (for Problem 12.12).

Employ for this purpose the shooting technique, as implemented in the routine `ShootQM` (Listing 12.16), which itself calls the function `Numerov` (Listing 12.13) as a global integrator. Apply the initial conditions required by the intrinsic symmetry of the solutions and integrate the wave equation from $x_0 = 0$ to $x_{max} = 10$, with a step size $h = 10^{-3}$.

a. Determine the lowest eight eigenstates, with a relative precision $\varepsilon = 10^{-4}$ in the eigenenergies, by performing trial shoots in energy windows of width $\delta E = 0.1$, by which the energy range between the minimum and maximum potential values is scanned. Ensure the physically required vanishing of the solution at a checkpoint located at $x_c = 6$.

b. Normalize the eigenfunction for each found eigenstate, performing the integration by the trapezoidal rule, and plot it along with its square (as probability density). Verify the theoretically required proportionality of the eigenenergies with $(n + 1/2)$, $n = 0, 1, 2 \ldots$

Solution

The Python and C/C++ implementations are available as supplementary material (`P12-Oscilla torSHg.py` and `P12-OscillatorSHg.cpp`). The eigenfunctions and the density probabilities are plotted in Figure 12.20.

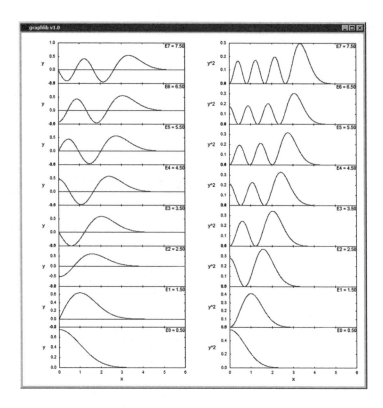

FIGURE 12.20   Lowest-energy eigenfunctions of the quantum harmonic oscillator (left column) and the corresponding density probabilities (right column), determined by the shooting method (for Problem 12.12).

PROBLEM 12.13

Solve the two-point problems for the Legendre and, respectively, Chebyshev polynomials of the first kind:

$$\frac{d^2 P_n}{dx^2} = \frac{1}{1-x^2}\left[2x\frac{dP_n}{dx} - n(n+1)P_n\right],$$

$$P_n(-1) = (-1)^n, \quad P_n(1) = 1,$$

$$\frac{d^2 T_n}{dx^2} = \frac{1}{1-x^2}\left[x\frac{dT_n}{dx} - n^2 T_n\right],$$

$$T_n(-1) = (-1)^n, \quad T_n(1) = 1.$$

For this purpose, employ the shooting method, as implemented in the routine `Shoot` (Listing 12.14), and, by virtue of the intrinsic symmetry of the solutions, reduce the integration domain to [0,1]. Use for the propagation a step size $h = 10^{-4}$, and ensure a convergence of the boundary value at $x = 1$ with a relative precision $\varepsilon = 10^{-6}$. Calculate and plot the Legendre and Chebyshev polynomials of the first kind up to the order $n_{\max} = 5$.

Solution

The two bilocal problems rewritten as Cauchy problems read:

$$y'' = \frac{1}{1-x^2}\left[2xy' - n(n+1)y\right], \tag{12.122}$$

$$y'' = \frac{1}{1-x^2}\left(xy' - n^2 y\right), \quad x \in [0,1], \tag{12.123}$$

$$y(0) = y_a, \quad y'(0) = 0, \quad n = 0, 2, 4, \ldots, \tag{12.124}$$

$$y(0) = 0, \quad y'(0) = y'_a, \quad n = 1, 3, 5, \ldots, \tag{12.125}$$

with the additional condition $y(1) = 1$.

The Python program is given in Listing 12.31 and the C/C++ version is provided as supplementary material (`P12-LegChebSH.cpp`). The Legendre and Chebyshev polynomials calculated by the shooting method are plotted in Figure 12.21.

Listing 12.31 Legendre and Chebyshev Polynomials by the Shooting Method (Python Coding)

```python
# Legendre and Chebyshev polynomials by the shooting method
from math import *
from ode import *
from graphlib import *

def Func(x, y, dy):                              # RHS of Legendre ODE
    return (2e0*x*dy - n*(n+1)*y) / (1e0 - x*x)

def Func1(x, y, dy):                             # RHS of Chebyshev ODE
    return (x*dy - n*n*y) / (1e0 - x*x)

# main

nmax = 5                                         # max. order of polynomial
xa = 0e0                                         # boundary values
xb = 1e0; yb = 1e0
eps = 1e-6                                       # tolerance for solution at xb
hx = 1e-4                                        # x-mesh step size

nx = int((xb-xa)/hx + 0.5) + 1                   # number of x-mesh points

x = [0]*(nx+1); y = [0]*(nx+1)                   # x-mesh, solution
xp = [0]*(nx*nmax+1); yp = [0]*(nx*nmax+1)       # plotting arrays
nn  = [0]*(nmax+1)                               # ending indexes of plots
sty = [0]*(nmax+1)                               # styles of plots
col = [""]*(nmax+1)                              # colors of plots
color = ["blue", "cyan", "green", "orange", "red"]

for m in range(1,nx+1): x[m] = xa + (m-1)*hx     # generate x-mesh

GraphInit(1200,600)
                                                 # Legendre polynomials
for n in range(1,nmax+1):                        # loop over polynomial order
    if (n % 2 == 0):                             # even solutions: rescaling
        ya = 1e0; dy = 0e0
        Propag(x,y,nx,ya,dy,Func)
        for m in range(1,nx+1): y[m] /= y[nx]    # normalization
    else:                                        # odd solutions: shooting
        ya = 0e0
        dy1 = -1e3; dy2 = 1e3            # search initial derivative in [dy1,dy2]
        (dy, exist) = Shoot(x,y,nx,ya,yb,dy1,dy2,eps,Func)

    nn[n] = n*nx; col[n] = color[(n-1)%5]; sty[n] = 1
    m0 = (n-1)*nx
    for m in range(1,nx+1): xp[m0+m] = x[m]; yp[m0+m] = y[m]

MultiPlot(xp,yp,yp,nn,col,sty,nmax,10,0e0,0e0,0,-1e0,1e0,1, \
          0.10,0.45,0.15,0.85,"x","Pn", \
```

```
              "Legendre polynomials - shooting method")
                                                      # Chebyshev polynomials
for n in range(1,nmax+1):                          # loop over polynomial order
    if (n % 2 == 0):                               # even solutions: rescaling
        ya = 1e0; dy = 0e0
        Propag(x,y,nx,ya,dy,Func1)
        for m in range(1,nx+1): y[m] /= y[nx]             # normalization
    else:                                          # odd solutions: shooting
        ya = 0e0
        dy1 = -1e3; dy2 = 1e3          # search initial derivative in [dy1,dy2]
        (dy, exist) = Shoot(x,y,nx,ya,yb,dy1,dy2,eps,Func1)

    nn[n] = n*nx; col[n] = color[(n-1)%5]; sty[n] = 1
    m0 = (n-1)*nx
    for m in range(1,nx+1): xp[m0+m] = x[m]; yp[m0+m] = y[m]

MultiPlot(xp,yp,yp,nn,col,sty,nmax,10,0e0,0e0,0,-1e0,1e0,1, \
          0.60,0.95,0.15,0.85,"x","Tn", \
          "Chebyshev polynomials - shooting method")

MainLoop()
```

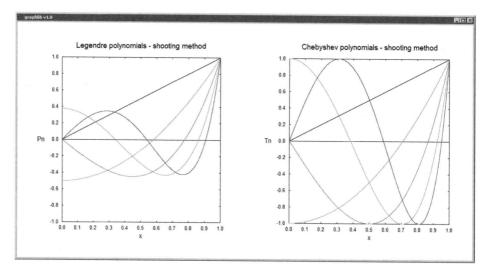

FIGURE 12.21 Legendre polynomials $P_n(x)$ (left) and Chebyshev polynomials of the first kind $T_n(x)$ (right) for $n = 1$–5, calculated by the shooting method (for Problem 12.13).

PROBLEM 12.14

Solve the two-point problems for the Legendre and, respectively, Chebyshev polynomials of the first kind:

$$\frac{d^2 P_n}{dx^2} = \frac{1}{1-x^2}\left[2x\frac{dP_n}{dx} - n(n+1)P_n\right],$$

$$P_n(-1) = (-1)^n, \quad P_n(1) = 1,$$

$$\frac{d^2 T_n}{dx^2} = \frac{1}{1-x^2}\left[x\frac{dT_n}{dx} - n^2 T_n\right],$$

$$T_n(-1) = (-1)^n, \quad T_n(1) = 1,$$

by the finite-difference method, using the routine `Bilocal` (Listing 12.18) with a step size $h = 10^{-4}$. Calculate and plot the Legendre and Chebyshev polynomials of the first kind up to the order $n_{max} = 5$.

Solution

The two bilocal problems rewritten in the standard form,

$$y'' = p(x)y' + q(x)y + r(x), \tag{12.126}$$

$$\begin{cases} \alpha_1 y(x_a) + \beta_1 y'(x_a) = 1, \\ \alpha_2 y(x_b) + \beta_2 y'(x_b) = 1, \end{cases} \tag{12.127}$$

imply the characteristic functions

$$p(x) = x/(1-x^2), \quad q(x) = -n(n+1)/(1-x^2), \quad r(x) = 0,$$

for the Legendre polynomials, and

$$p(x) = x/(1-x^2), \quad q(x) = -n^2/(1-x^2), \quad r(x) = 0,$$

for the Chebyshev polynomials. The boundary conditions are defined in both cases by the coefficients:

$$\alpha_1 = (-1)^n, \quad \beta_1 = 0,$$
$$\alpha_2 = 1, \quad \beta_2 = 0.$$

The Python program is given in Listing 12.32 and the C/C++ version is provided as supplementary material (`P12-LegChebBL.py`). The Legendre and Chebyshev polynomials calculated by the finite-difference method are plotted in Figure 12.22.

Listing 12.32 Legendre and Chebyshev Polynomials by Finite Differences (Python Coding)

```
# Legendre and Chebyshev polynomials by the finite-difference method
from math import *
from ode import *
from graphlib import *

def Func(x):                                    # RHS of Legendre ODE
    p = 2e0*x/(1e0-x*x); q =-n*(n+1)/(1e0-x*x); r = 0e0
    return (p, q, r)

def Func1(x):                                   # RHS of Chebyshev ODE
    p = x/(1e0-x*x); q =-n*n/(1e0-x*x); r = 0e0
    return (p, q, r)

# main

nmax = 5                                     # max. order of polynomial
xa = -1e0;   xb = 1e0                              # domain boundaries
hx = 1e-3                                          # x-mesh step size

nx = int((xb-xa)/hx + 0.5) + 1             # number of x-mesh points

x = [0]*(nx+1); y = [0]*(nx+1)                     # x-mesh, solution
xp = [0]*(nx*nmax+1); yp = [0]*(nx*nmax+1)           # plotting arrays
nn  = [0]*(nmax+1)                          # ending indexes of plots
sty = [0]*(nmax+1)                               # styles of plots
```

```
col = [""]*(nmax+1)                                      # colors of plots
color = ["blue", "cyan", "green", "orange", "red"]

for m in range(1,nx+1): x[m] = xa + (m-1)*hx             # generate x-mesh

GraphInit(1200,600)
                                                   # Legendre polynomials
for n in range(1,nmax+1):                          # loop over polynomial order
    alf1 = -1e0 if n % 2 else 1e0; bet1 = 0e0       # Dirichlet conditions
    alf2 = 1e0; bet2 = 0e0
    Bilocal(xa,xb,y,nx,alf1,bet1,alf2,bet2,Func)

    nn[n] = n*nx; col[n] = color[(n-1)%5]; sty[n] = 1
    m0 = (n-1)*nx
    for m in range(1,nx+1): xp[m0+m] = x[m]; yp[m0+m] = y[m]

MultiPlot(xp,yp,yp,nn,col,sty,nmax,10,0e0,0e0,0,-1e0,1e0,1, \
          0.10,0.45,0.15,0.85,"x","Pn", \
          "Legendre polynomials - finite-differences")
                                                   # Chebyshev polynomials
for n in range(1,nmax+1):                          # loop over polynomial order
    alf1 = -1e0 if n % 2 else 1e0; bet1 = 0e0       # Dirichlet conditions
    alf2 = 1e0; bet2 = 0e0
    Bilocal(xa,xb,y,nx,alf1,bet1,alf2,bet2,Func1)

    nn[n] = n*nx; col[n] = color[(n-1)%5]; sty[n] = 1
    m0 = (n-1)*nx
    for m in range(1,nx+1): xp[m0+m] = x[m]; yp[m0+m] = y[m]

MultiPlot(xp,yp,yp,nn,col,sty,nmax,10,0e0,0e0,0,-1e0,1e0,1, \
          0.60,0.95,0.15,0.85,"x","Tn", \
          "Chebyshev polynomials - finite-differences")

MainLoop()
```

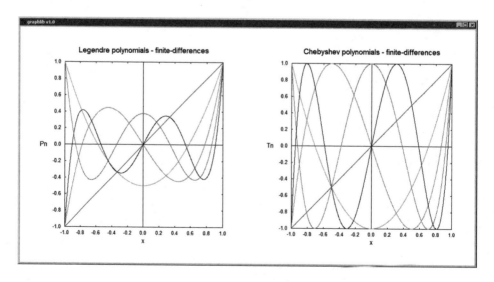

FIGURE 12.22 Legendre polynomials $P_n(x)$ (left) and Chebyshev polynomials of the first kind $T_n(x)$ (right) for $n = 1$–5, calculated by the finite-difference method (for Problem 12.14).

References and Suggested Further Reading

Abramowitz, M. and I. A. Stegun (Eds.) 1972. *Handbook of Mathematical Functions: With Formulas, Graphs, and Mathematical Tables.* New York: Dover Publications.

Beléndez, A., C. Pascual, D. I. Méndez, T. Belendez, and C. Neipp. 2007. Exact solution for the nonlinear pendulum. *Revista Brasileira de Ensino de Fisica* 29(4), 645–648.

Beu, T. A 2004. *Numerical Calculus in C* (3rd ed., in Romanian). Cluj-Napoca: MicroInformatica.

Burden, R. and J. Faires. 2010. *Numerical Analysis* (9th ed.) Boston: Brooks/Cole, Cengage Learning.

Butcher, J. C. 1996. A history of Runge–Kutta methods. *Applied Numerical Mathematics* 20, 247–260.

Cromer, A. 1981. Stable solutions using the Euler approximation. *American Journal of Physics* 49(5), 455–459.

Fehlberg, E. 1966. New high-order Runge–Kutta formulas with an arbitrarily small truncation error. *Zeitschrift für Angewandte Mathematik und Mechanik* 46(1), 1–16.

Fehlberg, E. 1970. Klassische Runge–Kutta-Formeln vierter und niedrigerer Ordnung mit Schrittweiten-Kontrolle und ihre Anwendung auf Wärmeleitungsprobleme. (Classical Runge-Kutta formulas of fourth and lower order with step size control and their application to heat conduction problems (in German)). *Computing* 6, 61–71.

Fitts, D. 2002. *Principles of Quantum Mechanics: As Applied to Chemistry and Chemical Physics.* Cambridge: Cambridge University Press.

Gear, C. W. 1971. *Numerical Initial Value Problems in Ordinary Differential Equations.* Englewood Cliffs, NJ: Prentice-Hall.

Giannozzi, P. 2013. Numerical methods in quantum mechanics. http://www.fisica.uniud.it/ giannozz/-Corsi/MQ/mq.html.

Hockney, R. W. and J. W. Eastwood. 1988. *Computer Simulation Using Particles.* Bristol: Adam Hilger.

Ixaru, L. G. 1979. *Numerical Methods for Differential Equations with Applications* (in Romanian). Bucharest: Editura Academiei.

Kutta, W. 1901. Beitrag zur näherungsweisen Integration totaler Differentialgleichungen (Contribution to the approximate integration of total differential equations (in German)). *Zeitschrift für Mathematik und Physik* 46, 435.

Landau, R. H., M. J. Páez, and C. C. Bordeianu. 2007. *Computational Physics: Problem Solving with Computers* (2nd ed.). Weinheim: Wiley-VCH.

Numerov, B. 1923. Méthode nouvelle de la détermination des orbites et le calcul des éphémérides en tenant compte des perturbations (New method for determining orbits and calculating ephemerides by taking into account perturbations (in French)). *Publications de l'Observatoire Astrophysique Central de Russie II*, 188–288.

Press, W. H., S. A. Teukolsky, W. T. Vetterling, and B. P. Flannery. 2007. *Numerical Recipes* (3rd ed.). *The Art of Scientific Computing.* Cambridge: Cambridge University Press.

Rapaport, D. C. 2004. *The Art of Molecular Dynamics Simulation* (2nd ed.). Cambridge: Cambridge University Press.

Runge, C. 1895. Über die numerische Auflösung von Differentialgleichungen (On the numerical solution of differential equations (in German)). *Mathematische Annalen* 46, 167.

Stoer, J. and R. Bulirsch. 1980. *Introduction to Numerical Analysis.* New York: Springer-Verlag.

Swope, W. C., H. C. Andersen, P. H. Berens, and K. R. Wilson. 1982. A computer simulation method for the calculation of equilibrium constants for the formation of physical clusters of molecules: Application to small water clusters. *Journal of Chemical Physics* 76(1), 637–649.

Verlet, L. 1967. Computer 'experiments' on classical fluids. I. Thermodynamical properties of Lennard–Jones molecules. *Physical Review* 159(1), 98–103.

13

Partial Differential Equations

13.1 Introduction

Partial differential equations (PDEs) arise, in general, in connection with phenomena taking place in continuous systems, in which, the quantities vary in space and time. The aimed processes show a wide diversity—heat and mass transport, propagation of mechanical, electromagnetic, or quantum mechanical waves and so on. Apart from a relatively small number of simple cases, in which the solutions can be expressed in closed form, it is necessary to resort to numerical methods to solve the underlying equations.

Typically, by using certain *discretization schemes*, the differential equation is turned into a matrix equation, having as unknowns the solution values at the nodes of a space–time grid. Even though the resulting matrix equation can be solved, in principle, by any of the general methods (Gaussian elimination, LU factorization, etc.), the considerable dimension of the system for problems of practical interest (on the order of thousands) usually makes such an approach impracticable. Fortunately, the *local character* of the considered PDEs (containing only low-order derivatives), as well as the local character of the discretization schemes applied to the differential operators (implying only neighboring mesh points), causes, most often, the discretized system of equations to have the so-called *sparse matrix*, with nonzero elements on just a few diagonals. For such matrices, there exist special inversion techniques, from which the most important ones are applied in this chapter.

Many of the PDEs of practical importance in science and engineering are *second-order equations*, containing partial derivatives of the unknown function up to the second order and typically having as independent variables either only spatial-, or space and time coordinates. Restricting the discussion, for simplicity, to linear problems in two variables, these PDEs can be cast under the generic form:

$$a(x,y)\frac{\partial^2 u}{\partial x^2} + b(x,y)\frac{\partial^2 u}{\partial x \partial y} + c(x,y)\frac{\partial^2 u}{\partial y^2}$$

$$+ d(x,y)\frac{\partial u}{\partial x} + e(x,y)\frac{\partial u}{\partial y} + g(x,y)u = f(x,y),$$

where $(x,y) \in \mathcal{D}$ (domain in the xy-plane) and the condition $a^2(x,y) + b^2(x,y) + c^2(x,y) > 0$ must hold everywhere in $\mathcal{D}$. Similarly to the conic sections, which contain powers of x and y instead of derivatives, the PDEs can be classified according to their information propagation curves into

- *Elliptic equations*, if $b^2 - 4ac < 0$ for all $(x,y) \in \mathcal{D}$,
- *Parabolic equations*, if $b^2 - 4ac = 0$ for all $(x,y) \in \mathcal{D}$,
- *Hyperbolic equations*, if $b^2 - 4ac > 0$ for all $(x,y) \in \mathcal{D}$.

The prototype for *elliptic PDEs* is *Poisson's equation*:

$$\frac{\partial^2 u}{\partial x^2} + \frac{\partial^2 u}{\partial y^2} = f(x,y),$$

509

which, in the case of an absent source term $f(x, y)$, turns into the *Laplace equation*. On the whole, elliptic equations are dominated by the homogeneous second-order derivatives, which, notably, appear in terms having the same sign. Such equations model naturally stationary phenomena, and examples thereof are the stationary heat equation, Poisson's equation for the electrostatic potential, and the time-independent Schrödinger equation.

Typical examples of *parabolic PDEs* are the *diffusion* and the *heat equations*:

$$\frac{\partial u}{\partial t} - \frac{\partial}{\partial x}\left(D\frac{\partial u}{\partial x}\right) = 0, \quad \frac{\partial u}{\partial t} - K\frac{\partial^2 u}{\partial x^2} = f(x, t),$$

where the evolution is described by first-order time derivatives, D is the *diffusion coefficient*, and $K > 0$ is called *thermal diffusivity*.

The prototypical *hyperbolic PDE* is the classical *wave equation*:

$$\frac{\partial^2 u}{\partial t^2} - v^2 \frac{\partial^2 u}{\partial x^2} = 0,$$

which implies second-order time and space derivatives of opposite signs and where c is the phase velocity of the wave.

In general, to obtain a uniquely determined solution of a PDE, corresponding to a well-defined physical situation, one also needs to specify supplementary conditions, thereby resulting in either a *boundary-value problem* or an *initial-value (Cauchy) problem*.

Boundary-value problems are typically associated with elliptic PDEs and model *equilibrium phenomena*, for which the time evolution is irrelevant. Given the absent time dependence, one seeks a steady-state solution $u(x, y)$ satisfying the differential equation in a certain domain $\mathcal{D}$, as well as the associated boundary conditions. The most frequently encountered boundary conditions may be categorized into

- *Dirichlet boundary conditions*, specifying the *solution values* on the boundary,

$$u(x, y) = u_S(x, y), \quad \text{for } (x, y) \in \mathcal{S},$$

- *Neumann boundary conditions*, defining the *normal derivatives* on the boundary,

$$\frac{\partial u}{\partial \mathbf{n}}(x, y) = v_S(x, y), \quad \text{for } (x, y) \in \mathcal{S},$$

- *Mixed boundary conditions*, prescribing *linear combinations* of solution and derivative values on the boundary,

$$\alpha(x, y)u(x, y) + \beta(x, y)\frac{\partial u}{\partial \mathbf{n}}(x, y) = \gamma(x, y), \quad \text{for } (x, y) \in \mathcal{S}.$$

A special category of boundary-value problems is the one of the *eigenproblems*, which admit stationary solutions only for particular values of a parameter, called eigenvalues.

Initial-value (Cauchy) problems are naturally associated with parabolic or hyperbolic equations and typically model *propagation phenomena*. Specifically, based on the known spatial behavior of the solution (and possibly also of its time derivative) at some initial moment t_0, the differential equation governs the propagation of the solution $u(x, t)$ in space and time.

From a numerical perspective, the classification according to the type of problem tends to prevail over the equation type, since the character of the boundary conditions is one of the determining aspects in adopting the numerical strategy. The conceptual differences between the boundary-value and initial-value problems are also suggested by Figure 13.1. Even though the main goal of both types of problems is the calculation of the solution on a spatial grid, the steady-state solution for boundary-value problems is determined by a numerical process simultaneously converging in the entire domain $\mathcal{D}$, whereas, in the

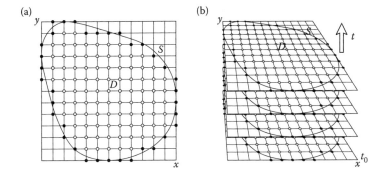

FIGURE 13.1   Integration grids for (a) boundary-value problems and (b) initial-value problems. Full circles stand for the input data, while empty circles represent the solution values that need to be calculated.

case of initial-value problems, the solution values from all the points of the spatial domain are recursively propagated in time, starting from the solution at the initial moment t_0.

13.2 Boundary-Value Problems for Elliptic Differential Equations

With a view to expounding the discretization principles for boundary-value problems, we consider in the following the 2D Poisson equation in Cartesian coordinates:

$$\nabla^2 u(x,y) \equiv \frac{\partial^2 u}{\partial x^2} + \frac{\partial^2 u}{\partial y^2} = f(x,y). \tag{13.1}$$

We choose the rectangular integration domain $\mathcal{D} \equiv [x_{\min}, x_{\max}] \times [y_{\min}, y_{\max}]$ and impose mixed boundary conditions:

$$\left[\alpha u + \beta \frac{\partial u}{\partial n} \right]_{(x,y)\in\mathcal{S}} = \gamma, \tag{13.2}$$

where $\alpha(x,y)$, $\beta(x,y)$, and $\gamma(x,y)$ are functions defined on the domain boundary $\mathcal{S}$. In particular, we have *Dirichlet conditions* for $\beta = 0$ (fixed solution values), *Neumann conditions* for $\alpha = 0$ (fixed normal derivatives), and the so-called *uniform conditions* for $\gamma = 0$.

Following the basic approach of the finite-difference methods, we restrict the solution to the $N_x \times N_y$ values $u_{ij} \equiv u(x_i, y_j)$ at the nodes of a uniform spatial grid, defined by

$$x_i = x_{\min} + (i-1)h_x, \quad i = 1, 2, \dots, N_x,$$

$$y_j = y_{\min} + (j-1)h_y, \quad j = 1, 2, \dots, N_y,$$

where h_x and h_y denote the mesh spacings along the two directions (see Figure 13.2).

Similarly to the finite-difference techniques applied in Chapter 12 to ODEs, we approximate the Laplace operator ∇^2 starting from the Taylor series of the solution at the interior points of the domain $\mathcal{D}$. In particular, along the x-direction we have, respectively, the backward and forward expansions:

$$u_{i-1,j} = u_{i,j} - \frac{h_x}{1!}\left(\frac{\partial u}{\partial x}\right)_{i,j} + \frac{h_x^2}{2!}\left(\frac{\partial^2 u}{\partial x^2}\right)_{i,j} - \frac{h_x^3}{3!}\left(\frac{\partial^3 u}{\partial x^3}\right)_{i,j} + O(h_x^4),$$

$$u_{i+1,j} = u_{i,j} + \frac{h_x}{1!}\left(\frac{\partial u}{\partial x}\right)_{i,j} + \frac{h_x^2}{2!}\left(\frac{\partial^2 u}{\partial x^2}\right)_{i,j} + \frac{h_x^3}{3!}\left(\frac{\partial^3 u}{\partial x^3}\right)_{i,j} + O(h_x^4).$$

By their addition, we get the following finite-difference scheme for the second derivative:

$$\left(\frac{\partial^2 u}{\partial x^2}\right)_{i,j} = \frac{u_{i+1,j} - 2u_{i,j} + u_{i-1,j}}{h_x^2} + O(h_x^2). \tag{13.3}$$

The $O(h_x^2)$ error term results by the exact cancellation of the third-order terms in the above $O(h_x^4)$ expansions and the subsequent division by h_x^2 for expressing the second derivative. The formula is said to be a *central-difference scheme*, since it uses information from nodes distributed symmetrically about the point where the derivative is being evaluated.

Analogously, for the second derivative along the y-direction, we have the scheme:

$$\left(\frac{\partial^2 u}{\partial y^2}\right)_{i,j} = \frac{u_{i,j+1} - 2u_{i,j} + u_{i,j-1}}{h_y^2} + O(h_y^2). \tag{13.4}$$

Thus, for the Laplacian of function u at the node (x_i, y_j), there results a five-point difference scheme, which may be correlated with the graphical representation in Figure 13.2:

$$\nabla^2 u(x, y)\big|_{i,j} = \frac{1}{h_y^2} u_{i,j-1} + \frac{1}{h_x^2} u_{i-1,j} - 2\left(\frac{1}{h_x^2} + \frac{1}{h_y^2}\right) u_{i,j}$$

$$+ \frac{1}{h_x^2} u_{i+1,j} + \frac{1}{h_y^2} u_{i,j+1} + O(h_x^2 + h_y^2). \tag{13.5}$$

Therewith, one obtains the following finite-difference representation of the Poisson equation (13.1) at the interior points of the domain $\mathcal{D}$:

$$\frac{1}{h_y^2} u_{i,j-1} + \frac{1}{h_x^2} u_{i-1,j} - 2\left(\frac{1}{h_x^2} + \frac{1}{h_y^2}\right) u_{i,j} + \frac{1}{h_x^2} u_{i+1,j} + \frac{1}{h_y^2} u_{i,j+1} = f_{i,j},$$

$$i = 2, \ldots, N_x - 1, \quad j = 2, \ldots, N_y - 1, \tag{13.6}$$

where $f_{i,j} \equiv f(x_i, y_j)$.

The linear system (13.6), having as unknowns the solution values $u_{i,j}$ at the grid nodes, does not actually contain sufficient equations for the solution to be completely determined. On the other hand, the five-point discretization scheme employed for the Laplacian prevents the direct embedding of the boundary conditions, which have thus to be treated separately.

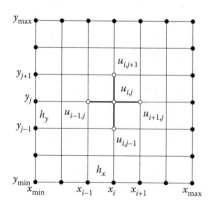

FIGURE 13.2   Discretization mesh for the finite-difference solution of the Poisson equation.

Quite often, discretizing the boundary conditions is more intricate than discretizing the PDEs themselves. For the sake of clarity, we use in the following the lowest-order approximations for the mixed boundary conditions. Concretely, considering the normal derivatives to be positive if they are oriented outward from the domain $\mathcal{D}$, the conditions on the left and, respectively, right boundary ($i = 1$ and $i = N_x$) can be expressed as

$$\alpha_j^{x_{\min}} u_{1,j} + \beta_j^{x_{\min}} \frac{u_{1,j} - u_{2,j}}{h_x} = \gamma_j^{x_{\min}}, \quad i = 1,$$

$$\alpha_j^{x_{\max}} u_{N_x,j} + \beta_j^{x_{\max}} \frac{u_{N_x,j} - u_{N_x-1,j}}{h_x} = \gamma_j^{x_{\max}}, \quad i = N_x,$$

$$j = 1, \ldots, N_y.$$

In a similar manner, the discretized conditions on the lower and, respectively, upper boundary ($j = 1$ and $j = N_y$) read:

$$\alpha_i^{y_{\min}} u_{i,1} + \beta_i^{y_{\min}} \frac{u_{i,1} - u_{i,2}}{h_y} = \gamma_i^{y_{\min}}, \quad j = 1,$$

$$\alpha_i^{y_{\max}} u_{i,N} + \beta_i^{y_{\max}} \frac{u_{i,N_y} - u_{i,N_y-1}}{h_y} = \gamma_i^{y_{\max}}, \quad j = N_y,$$

$$i = 1, \ldots, N_x.$$

To simplify the discretized equations and the boundary coefficients, we define the quantities:

$$k_x = \frac{1}{h_x^2}, \quad k_y = \frac{1}{h_y^2}, \quad k_{xy} = 2\left(\frac{1}{h_x^2} + \frac{1}{h_y^2}\right), \tag{13.7}$$

and the boundary coefficients:

$$\begin{aligned}
\bar{\beta}_i^{y_{\min}} &= \beta_i^{y_{\min}} / \left(\alpha_i^{y_{\min}} h_y + \beta_i^{y_{\min}}\right), & \bar{\gamma}_i^{y_{\min}} &= \gamma_i^{y_{\min}} / \left(\alpha_i^{y_{\min}} + \beta_i^{y_{\min}} / h_y\right), \\
\bar{\beta}_i^{y_{\max}} &= \beta_i^{y_{\max}} / \left(\alpha_i^{y_{\max}} h_y + \beta_i^{y_{\max}}\right), & \bar{\gamma}_i^{y_{\max}} &= \gamma_i^{y_{\max}} / \left(\alpha_i^{y_{\max}} + \beta_i^{y_{\max}} / h_y\right), \\
\bar{\beta}_j^{x_{\min}} &= \beta_j^{x_{\min}} / \left(\alpha_j^{x_{\min}} h_x + \beta_j^{x_{\min}}\right), & \bar{\gamma}_j^{x_{\min}} &= \gamma_j^{x_{\min}} / \left(\alpha_j^{x_{\min}} + \beta_j^{x_{\min}} / h_x\right), \\
\bar{\beta}_j^{x_{\max}} &= \beta_j^{x_{\max}} / \left(\alpha_j^{x_{\max}} h_x + \beta_j^{x_{\max}}\right), & \bar{\gamma}_j^{x_{\max}} &= \gamma_j^{x_{\max}} / \left(\alpha_j^{x_{\max}} + \beta_j^{x_{\max}} / h_x\right).
\end{aligned} \tag{13.8}$$

Therewith, the complete linear system resulting by discretizing the Poisson equation and the attached mixed boundary conditions takes the form:

$$u_{i,1} - \bar{\beta}_i^{y_{\min}} u_{i,2} = \bar{\gamma}_i^{y_{\min}}, \quad i = 1, \ldots, N_x, \quad j = 1, \quad \text{(bottom)} \tag{13.9}$$

$$u_{1,j} - \bar{\beta}_j^{x_{\min}} u_{2,j} = \bar{\gamma}_j^{x_{\min}}, \quad i = 1, \quad j = 2, \ldots, N_y - 1, \quad \text{(left)} \tag{13.10}$$

$$k_y u_{i,j-1} + k_x u_{i-1,j} - k_{xy} u_{i,j} + k_x u_{i+1,j} + k_y u_{i,j+1} = f_{i,j},$$

$$i = 2, \ldots, N_x - 1, \quad j = 2, \ldots, N_y - 1, \quad j = 1, \quad \text{(interior)} \tag{13.11}$$

$$-\bar{\beta}_j^{x_{\max}} u_{N_x-1,j} + u_{N_x,j} = \bar{\gamma}_j^{x_{\max}}, \quad i = N_x, \quad j = 2, \ldots, N_y - 1, \quad \text{(right)} \tag{13.12}$$

$$-\bar{\beta}_i^{y_{\max}} u_{i,N_y-1} + u_{i,N_y} = \bar{\gamma}_i^{y_{\max}}, \quad i = 1, \ldots, N_x, \quad j = N_y. \quad \text{(top)} \tag{13.13}$$

The entire discretized system may be formally rewritten under a general form, which helps reveal its structure:

$$d_i^j u_{i,j-1} + b_{ai}^j u_{i-1,j} + b_{bi}^j u_{i,j} + b_{ci}^j u_{i+1,j} + c_i^j u_{i,j+1} = d_i^j,$$

$$i = 1, \ldots, N_x, \quad j = 1, \ldots, N_y. \tag{13.14}$$

Considering that the discretized equation centered about the mesh node (i, j) connects not only the five neighboring values $(u_{i,j-1}, u_{i-1,j}, u_{i,j}, u_{i+1,j}, u_{i,j+1})$, but all the values on the three neighboring rows $j - 1$, j, and $j + 1$, the system can be reshaped as

$$\sum_{i'=1}^{N_x} \left[A_{ii'}^j u_{i',j-1} + B_{ii'}^j u_{i',j} + C_{ii'}^j u_{i',j+1} \right] = d_i^j, \tag{13.15}$$

$$i = 1, \ldots, N_x, \quad j = 1, \ldots, N_y,$$

where the $N_x \times N_x$ blocks $\mathbf{A}^j = [A_{ii'}^j]_{N_x N_x}$, $\mathbf{B}^j = [B_{ii'}^j]_{N_x N_x}$, and $\mathbf{C}^j = [C_{ii'}^j]_{N_x N_x}$ have the elements:

$$A_{ii'}^j = d_i^j \delta_{ii'},$$

$$B_{ii'}^j = \begin{cases} b_{ai}^j & \text{if } i' = i - 1, \\ b_{bi}^j & \text{if } i' = i, \\ b_{ci}^j & \text{if } i' = i + 1, \\ 0 & \text{if } i' \neq i, i \pm 1, \end{cases} \tag{13.16}$$

$$C_{ii'}^j = c_i^j \delta_{ii'}.$$

On the basis of these blocks, the complete system matrix can be shown to be *block tridiagonal* and its band structure can be seen in Figure 13.3. In fact, defining the vectors $\mathbf{u}^j = [u_{i',j}]_{N_x}$ and $\mathbf{d}^j = [d_i^j]_{N_x}$, with $\mathbf{u}^j$ having as components the N_x solution values on the row j and $\mathbf{d}^j$ containing the corresponding right-hand-side terms, the discretized system takes the form:

$$\mathbf{A}^j \mathbf{u}^{j-1} + \mathbf{B}^j \mathbf{u}^j + \mathbf{C}^j \mathbf{u}^{j+1} = \mathbf{d}^j, \quad j = 1, 2, \ldots, N_y, \tag{13.17}$$

and it shows itself as a set of matrix equations connecting the vectors $\mathbf{u}^{j-1}$, $\mathbf{u}^j$, and $\mathbf{u}^{j+1}$, that is, the solutions on the rows $j - 1$, j, and, respectively, $j + 1$. In matrix notations, the block tridiagonal structure may readily be observed:

$$\begin{bmatrix} \mathbf{B}^1 & \mathbf{C}^1 & & & & \\ \mathbf{A}^2 & \mathbf{B}^2 & \mathbf{C}^2 & & & \\ & \ddots & \ddots & \ddots & & \\ & & \mathbf{A}^{N_y-1} & \mathbf{B}^{N_y-1} & \mathbf{C}^{N_y-1} \\ & & & \mathbf{A}^{N_y} & \mathbf{B}^{N_y} \end{bmatrix} \begin{bmatrix} \mathbf{u}^1 \\ \mathbf{u}^2 \\ \vdots \\ \mathbf{u}^{N_y-1} \\ \mathbf{u}^{N_y} \end{bmatrix} = \begin{bmatrix} \mathbf{d}^1 \\ \mathbf{d}^2 \\ \vdots \\ \mathbf{d}^{N_y-1} \\ \mathbf{d}^{N_y} \end{bmatrix}. \tag{13.18}$$

System 13.18 can be solved, in principle, using two classes of methods that take into account the sparse structure of its matrix:

- *Direct methods*, based on the recursive inversion of the blocks composing the matrix
- *Iterative methods*, such as the Jacobi, Gauss–Seidel, and overrelaxation methods

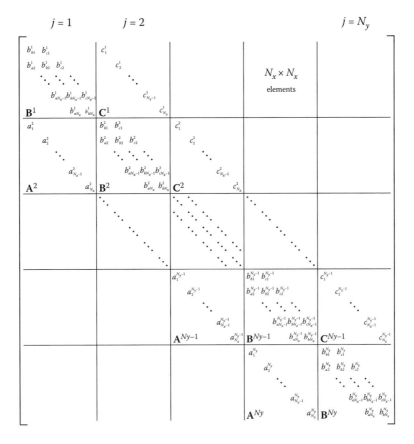

FIGURE 13.3 Structure of the matrix resulting by discretizing the Laplace operator on a rectangular domain.

Chapter 7, devoted to linear systems, presents a direct approach specifically designed for systems with block tridiagonal structure (Section 7.8). The method is, indeed, applicable in this context, but does neither take advantage of the mostly constant elements of the five nonzero diagonals, nor can it be directly generalized to higher-dimensional problems.

Owing to the less elaborate implementation and the more favorable scaling of the computations with the system size, for problems with more than two dimensions or with complex domain boundaries, iterative methods turn out to be preferable for solving the linear system obtained by discretizing boundary-value problems for PDEs. Therefore, we confine the following discussion to the *Gauss–Seidel iterative method*, which shows in practice a broader spectrum of applications, being also suitable for solving *eigenvalue problems* for PDEs.

We rewrite formally the system under the general form

$$\mathbf{A} \cdot \mathbf{u} = \mathbf{d}, \tag{13.19}$$

and consider the decomposition:

$$\mathbf{A} = \mathbf{L} + \mathbf{D} + \mathbf{U}, \tag{13.20}$$

where $\mathbf{L}$ is lower triangular, $\mathbf{D}$ is diagonal, and $\mathbf{U}$ is upper triangular. In the basic *Jacobi method* (see Section 7.10), the diagonal elements are expressed and the solution is iterated according to the relation:

$$\mathbf{D} \cdot \mathbf{u}^{(r)} = \mathbf{d} - (\mathbf{L} + \mathbf{U}) \cdot \mathbf{u}^{(r-1)},$$

or

$$\mathbf{u}^{(r)} = \mathbf{D}^{-1} \cdot \left[\mathbf{d} - (\mathbf{L} + \mathbf{U}) \cdot \mathbf{u}^{(r-1)}\right], \tag{13.21}$$

where $r = 1, 2, \ldots$ is the approximation order of the solution. Since $\mathbf{D}$ is diagonal, the inverse $\mathbf{D}^{-1}$ is likewise diagonal and contains simply the inverses of the diagonal elements of $\mathbf{D}$. The Jacobi method converges for matrices that are diagonally dominant and this condition is usually satisfied in the framework of finite-difference methodologies.

Referring specifically to Equation 13.11 for the inner mesh points from the discretized systems (13.9) through (13.13), the diagonal solution components are the ones located at the center of the star-like discretization schemes, namely $u_{i,j}$, and, correspondingly, the solution is iterated based on the recurrence relation:

$$u_{i,j}^{(r)} = \left[k_x \left(u_{i-1,j}^{(r-1)} + u_{i+1,j}^{(r-1)}\right) + k_y \left(u_{i,j-1}^{(r-1)} + u_{i,j+1}^{(r-1)}\right) - f_{i,j}\right]/k_{xy}. \tag{13.22}$$

It is defining for the Jacobi method, that all the solution values employed on the right-hand side stem from the preceding iteration, $(r-1)$.

Assuming that the algorithm runs within embedded loops, in increasing order of the indexes i and j, when calculating $u_{i,j}^{(r)}$, the updated components $u_{i,j-1}^{(r)}$ and $u_{i-1,j}^{(r)}$ are already available. Their immediate use is specific to the *Gauss–Seidel method* and the corresponding formal recurrence relation takes the form:

$$u_{i,j}^{(r)} = \left[k_x \left(u_{i-1,j}^{(r)} + u_{i+1,j}^{(r-1)}\right) + k_y \left(u_{i,j-1}^{(r)} + u_{i,j+1}^{(r-1)}\right) - f_{i,j}\right]/k_{xy}. \tag{13.23}$$

Thus, unlike the Jacobi method, $u_{i,j}^{(r)}$ is calculated using the most recent components $u_{i,j-1}^{(r)}$ and $u_{i-1,j}^{(r)}$, not their predecessors, $u_{i,j-1}^{(r-1)}$ and $u_{i-1,j}^{(r-1)}$. In addition to the enhanced convergence speed, the Gauss–Seidel method also requires just a single array (not two) for storing the solution. This array may contain at the same time components of both approximations $\mathbf{u}^{(r-1)}$ and $\mathbf{u}^{(r)}$, which are continuously updated and become usable immediately following their calculation.

The equations describing the iterative solution of the discretized system (13.9)–(13.13) with mixed boundary conditions, based on the Gauss–Seidel method, may now be compiled as follows:

$$u_{i,1}^{(r)} = \bar{\beta}_i^{y_{min}} u_{i,2}^{(r-1)} + \bar{\gamma}_i^{y_{min}}, \quad i = 1, \ldots, N_x, \quad j = 1, \quad \text{(bottom)} \tag{13.24}$$

$$u_{1,j}^{(r)} = \bar{\beta}_j^{x_{min}} u_{2,j}^{(r-1)} + \bar{\gamma}_j^{x_{min}}, \quad i = 1, \quad j = 2, \ldots, N_y - 1, \quad \text{(left)} \tag{13.25}$$

$$u_{i,j}^{(r)} = \left[k_x \left(u_{i-1,j}^{(r)} + u_{i+1,j}^{(r-1)}\right) + k_y \left(u_{i,j-1}^{(r)} + u_{i,j+1}^{(r-1)}\right) - f_{i,j}\right]/k_{xy},$$
$$i = 2, \ldots, N_x - 1, \quad j = 2, \ldots, N_y - 1, \quad j = 1, \quad \text{(interior)} \tag{13.26}$$

$$u_{N_x,j}^{(r)} = \bar{\beta}_j^{x_{max}} u_{N_x-1,j}^{(r-1)} + \bar{\gamma}_j^{x_{max}}, \quad i = N_x, \quad j = 2, \ldots, N_y - 1, \quad \text{(right)} \tag{13.27}$$

$$u_{i,N_y}^{(r)} = \bar{\beta}_i^{y_{max}} u_{i,N_y-1}^{(r-1)} + \bar{\gamma}_i^{y_{max}}, \quad i = 1, \ldots, N_x, \quad j = N_y. \quad \text{(top)} \tag{13.28}$$

The recursive process (13.24)–(13.28) is to be continued, in principle, until the *maximum relative difference* between the solution components from two consecutive iterations drops below a predefined tolerance ε:

$$\max_{i,j} |1 - u_{ij}^{(r-1)} / u_{ij}^{(r)}| \leq \varepsilon. \tag{13.29}$$

For vanishing components, this convergence criterion should rather employ the maximum *absolute* difference $\max_{i,j} |u_{ij}^{(r)} - u_{ij}^{(r-1)}|$.

The routines Poisson0 (Listing 13.1) and PoissonXY (Listing 13.2) are implementations of increasing complexity of the presented strategy for solving the linear system (13.24)–(13.28) for the discretized Poisson equation by the Gauss–Seidel method.

Listing 13.1   Solver for the 2D Poisson Equation with Dirichlet Conditions (Python Coding)

```
#===========================================================================
def Poisson0(u, x, y, nx, ny, eps, Func):
#---------------------------------------------------------------------------
#  Solves the 2D Poisson equation in Cartesian coordinates with Dirichlet
#  boundary conditions on a regular grid with (nx x ny) nodes (x[],y[]) using
#  the Gauss-Seidel method. The solution u[][] is converged with relative
#  precision eps. An error index is returned: 0 - normal execution.
#  Calls: Func(x,y) - RHS of Poisson equation
#---------------------------------------------------------------------------
   itmax = 10000                                        # max no. of iterations

   f = [[0]*(ny+1) for i in range(nx+1)]

   hx = (x[nx]-x[1])/(nx-1); kx = 1e0/(hx*hx)                    # mesh spacings
   hy = (y[ny]-y[1])/(ny-1); ky = 1e0/(hy*hy)
   kxy = 2e0*(kx + ky)

   for j in range(2,ny):                                # RHS of PDE for interior points
      for i in range(2,nx): f[i][j] = Func(x[i],y[j])

   for it in range(1,itmax+1):                          # Gauss-Seidel iteration loop
      err = 0e0
      for j in range(2,ny):
         for i in range(2,nx):                          # interior mesh points
            uij = (kx*(u[i-1][j] + u[i+1][j]) +
                   ky*(u[i][j-1] + u[i][j+1]) - f[i][j]) / kxy
            eij = 1e0 - u[i][j]/uij if uij else uij - u[i][j]   # local error
            if (fabs(eij) > err): err = fabs(eij)               # maximum error
            u[i][j] = uij

      if (err <= eps): break                            # convergence test

   if (it >= itmax):
      print("Poisson0: max. number of iterations exceeded !"); return 1
   return 0
```

Listing 13.2   Solver for the 2D Poisson Equation in Cartesian Coordinates (Python Coding)

```
#===========================================================================
def PoissonXY(u, x, y, nx, ny, eps, Func, CondX, CondY):
#---------------------------------------------------------------------------
#  Solves the 2D Poisson equation in Cartesian coordinates on a regular grid
#  with (nx x ny) nodes (x[],y[]) using the Gauss-Seidel method. The solution
#  u[][] is converged with relative precision eps.
#  An error index is returned: 0 - normal execution.
#  Calls: Func(x,y) - RHS of Poisson equation; boundary conditions:
#         CondX(y,alf_min,bet_min,gam_min,alf_max,bet_max,gam_max)
#         CondY(x,alf_min,bet_min,gam_min,alf_max,bet_max,gam_max)
#---------------------------------------------------------------------------
   itmax = 10000                                        # max no. of iterations

   f = [[0]*(ny+1) for i in range(nx+1)]
   betXmin = [0]*(ny+1); betXmax = [0]*(ny+1)
   gamXmin = [0]*(ny+1); gamXmax = [0]*(ny+1)
   betYmin = [0]*(nx+1); betYmax = [0]*(nx+1)
   gamYmin = [0]*(nx+1); gamYmax = [0]*(nx+1)
```

```
   hx = (x[nx]-x[1])/(nx-1); kx = 1e0/(hx*hx)              # mesh spacings
   hy = (y[ny]-y[1])/(ny-1); ky = 1e0/(hy*hy)
   kxy = 2e0*(kx + ky)

   for j in range(2,ny):                          # RHS of PDE for interior points
      for i in range(2,nx): f[i][j] = Func(x[i],y[j])
                                                      # boundary conditions
   for i in range(1,nx+1):                      # lower and upper boundaries
      (alf_min,bet_min,gam_min,alf_max,bet_max,gam_max) = CondY(x[i])
      betYmin[i] = bet_min/(alf_min*hy + bet_min)
      gamYmin[i] = gam_min/(alf_min + bet_min/hy)
      betYmax[i] = bet_max/(alf_max*hy + bet_max)
      gamYmax[i] = gam_max/(alf_max + bet_max/hy)

   for j in range(2,ny):                          # left and right boundaries
      (alf_min,bet_min,gam_min,alf_max,bet_max,gam_max) = CondX(y[j])
      betXmin[j] = bet_min/(alf_min*hx + bet_min)
      gamXmin[j] = gam_min/(alf_min + bet_min/hx)
      betXmax[j] = bet_max/(alf_max*hx + bet_max)
      gamXmax[j] = gam_max/(alf_max + bet_max/hx)

   for it in range(1,itmax+1):                    # Gauss-Seidel iteration loop
      err = 0e0
      j = 1                                            # lower boundary
      for i in range(1,nx+1):
         uij = betYmin[i]*u[i][2] + gamYmin[i]
         eij = 1e0 - u[i][j]/uij if uij else uij - u[i][j]
         if (fabs(eij) > err): err = fabs(eij)
         u[i][j] = uij

      for j in range(2,ny):
         i = 1                                         # left boundary
         uij = betXmin[j]*u[i+1][j] + gamXmin[j]
         eij = 1e0 - u[i][j]/uij if uij else uij - u[i][j]
         if (fabs(eij) > err): err = fabs(eij)
         u[i][j] = uij

         for i in range(2,nx):                      # interior mesh points
            uij = (kx*(u[i-1][j] + u[i+1][j]) +
                   ky*(u[i][j-1] + u[i][j+1]) - f[i][j]) / kxy
            eij = 1e0 - u[i][j]/uij if uij else uij - u[i][j]  # local error
            if (fabs(eij) > err): err = fabs(eij)           # maximum error
            u[i][j] = uij

         i = nx                                        # right boundary
         uij = betXmax[j]*u[i-1][j] + gamXmax[j]
         eij = 1e0 - u[i][j]/uij if uij else uij - u[i][j]
         if (fabs(eij) > err): err = fabs(eij)
         u[i][j] = uij

      j = ny                                           # upper boundary
      for i in range(1,nx+1):
         uij = betYmax[i]*u[i][ny-1] + gamYmax[i]
         eij = 1e0 - u[i][j]/uij if uij else uij - u[i][j]
         if (fabs(eij) > err): err = fabs(eij)
         u[i][j] = uij

      if (err <= eps): break                          # convergence test
```

```
if (it >= itmax):
    print("PoissonXY: max. number of iterations exceeded !"); return 1
return 0
```

The routine `Poisson0` is designed to operate only with Dirichlet boundary conditions. Consequently, from the entire system (13.24)–(13.28), it needs to iterate only Equation 13.26 for the interior points of the domain. The routine receives the initial approximation of the solution, including the boundary values, by way of array `u[][]`, which also returns the solution upon execution. The Cartesian coordinates of the uniformly spaced mesh points are received via arrays `x[]` and `y[]`, with the parameters `nx` and `ny` passing the corresponding numbers of nodes, and `eps`, the relative tolerance to be applied to the solution.

The `PoissonXY` solver generalizes the functionalities of the routine `Poisson0`, being able to solve problems with mixed boundary conditions, based on the full set of discretized equations (13.24) through (13.28).

The argument list of the two Poisson solvers is similar, with the array `u[][]` passing on entry an initial approximation and returning on exit the final solution. The supplementary procedural arguments `Func`, `CondX`, and `CondY` of `PoissonXY` provide the actual names of the user routines returning the values of the right-hand-side function $f(x, y)$ and, respectively, the coefficients defining the boundary conditions. Specifically, for a given `y`, the function `CondX` is expected to return the coefficients `alf_min`, `bet_min`, and `gam_min`, which define the conditions in the x-direction, normal to the left boundary, as well as `alf_max`, `bet_max`, and `gam_max`, corresponding to the right boundary. Similarly, for each given `x`, the function `CondY` is expected to return the coefficients defining the solution in the y-direction on the lower and, respectively, upper boundaries.

In the first phase, `PoissonXY` determines the auxiliary coefficients $\bar{\beta}$ and $\bar{\gamma}$ according to Equation 13.8 and stores the values of the right-hand-side function $f(x, y)$. Once the constant quantities are prepared, the routine enters the Gauss–Seidel loop, where the solution is updated at each iteration in sequence on the lower boundary, on the left boundary, at the interior mesh points, on the right boundary, and, finally, on the upper boundary. Along with the updated solution, also the maximum relative change of the solution components, as error estimate, is recorded in `err`. The Gauss–Seidel process is continued until `err` drops below the chosen tolerance `eps`. If the predefined maximum number of iterations is exceeded, an error message is issued and a nonzero error index is returned.

As an illustration for the use of the routine `PoissonXY`, we consider the toy problem associated with the Poisson equation:

$$\frac{\partial^2 u}{\partial x^2} + \frac{\partial^2 u}{\partial y^2} = \cos(x + y) - \cos(x - y), \tag{13.30}$$

which we solve in the domain $\mathcal{D} = [-\pi, \pi] \times [-\pi, \pi]$ subject to the boundary conditions:

$$\begin{aligned} u(\pm\pi, y) &= 0, \quad y \in [-\pi, \pi], \\ u(x, \pm\pi) &= 0, \quad x \in [-\pi, \pi]. \end{aligned} \tag{13.31}$$

The analytical solution is $u(x, y) = \sin x \sin y$ and can conveniently serve as reference for numerical solutions.

Program 13.3, coded to solve the above problem, defines the user functions `Func`, `CondX`, and `CondY`, which return, respectively, the values of the right-hand-side function of the PDE and the coefficients defining the boundary conditions. The program initializes the arrays `x[]` and `y[]` with the

Listing 13.3 Solution of the 2D Poisson Equation in Cartesian Coordinates (Python Coding)

```
# Solves the 2D Poisson equation in a rectangular domain
from math import *
from pde import *

def Func(x, y):                                  # RHS function for PoissonXY
   return cos(x+y) - cos(x-y)

def CondX(y):                          # Coefficients for left and right boundaries
   alf_min = 1e0; bet_min = 0e0; gam_min = 0e0
   alf_max = 1e0; bet_max = 0e0; gam_max = 0e0
   return (alf_min, bet_min, gam_min, alf_max, bet_max, gam_max)

def CondY(x):                          # Coefficients for lower and upper boundaries
   alf_min = 1e0; bet_min = 0e0; gam_min = 0e0
   alf_max = 1e0; bet_max = 0e0; gam_max = 0e0
   return (alf_min, bet_min, gam_min, alf_max, bet_max, gam_max)

# main

xmin = -pi; xmax = pi; ymin = -pi; ymax = pi         # domain boundaries
nx = 51; ny = 51                                  # number of mesh points
eps = 1e-5                                   # relative solution tolerance

u = [[0]*(ny+1) for i in range(nx+1)]                      # solution
x = [0]*(nx+1); y = [0]*(ny+1)                   # mesh point coordinates

hx = (xmax-xmin)/(nx-1)
for i in range(1,nx+1): x[i] = xmin + (i-1)*hx             # x-mesh points
hy = (ymax-ymin)/(ny-1)
for j in range(1,ny+1): y[j] = ymin + (j-1)*hy             # y-mesh points

for j in range(1,ny+1):              # initial approximation of the solution
   for i in range(1,nx+1): u[i][j] = 0e0

PoissonXY(u,x,y,nx,ny,eps,Func,CondX,CondY)

out = open("Poisson.txt","w")                         # open output file
out.write("        x          y            u\n")
for j in range(1,ny+1):
   for i in range(1,nx+1):
      out.write(("{0:10.5f}{1:10.5f}{2:14.5e}\n").format(x[i],y[j],u[i][j]))
out.close()
```

coordinates of the mesh points and starts the calculation with a trivial initial approximation, with all the solution components set to 0. The solution returned by the solver PoissonXY (or, alternatively, by Poisson0) on a 51×51-point grid (50×50 intervals), with a relative tolerance $\varepsilon = 10^{-5}$, is written to an output file and is depicted in Figure 13.4. The numerical profile is found to accurately reproduce the exact solution of the problem.

When dealing with problems for elliptic equations on *simply connected irregular domains* (without interior holes), one way of describing the boundaries is to delimit the interior nodes on a regular discretization mesh by a couple of vectors, $\mathbf{i}_{min}$ and $\mathbf{i}_{max}$, whose components, i_{min}^j and i_{max}^j, represent the limiting i-indexes on each horizontal line j. For such problems, however, defining normal derivatives to

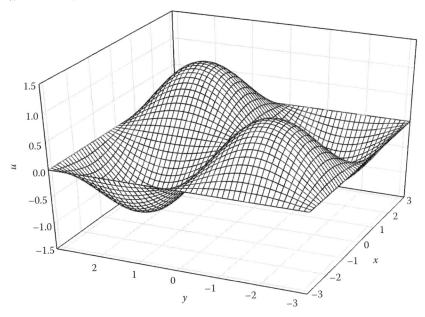

FIGURE 13.4 Solution of the boundary-value problem (13.30)–(13.31) for the Poisson equation.

the boundaries is rather intricate. For clarity reasons, we confine the discussion to Dirichlet boundary conditions, which specify the function values on the bottom, left, right, and top boundaries:

$$u_{i,1}^{(r)} = u_{i,1}^0, \quad i = i_{min}^1, \ldots, i_{max}^1, \quad j = 1,$$

$$u_{i_{min},j}^{(r)} = u_{i_{min},j}^0, \quad u_{i_{max},j}^{(r)} = u_{i_{max},j}^0, \quad j = 2, \ldots, N_y - 1, \tag{13.32}$$

$$u_{i,N_y}^{(r)} = u_{i,N_y}^0, \quad i = i_{min}^{N_y}, \ldots, i_{max}^{N_y}, \quad j = N_y.$$

Correspondingly, the only equations that need to be iterated for solving such problems are those for the interior points of the domain:

$$u_{i,j}^{(r)} = \left[k_x \left(u_{i-1,j}^{(r)} + u_{i+1,j}^{(r-1)} \right) + k_y \left(u_{i,j-1}^{(r)} + u_{i,j+1}^{(r-1)} \right) - f_{i,j} \right] / k_{xy},$$

$$i = i_{min} + 1, \ldots, i_{max} + 1, \quad j = 2, \ldots, N_y - 1. \tag{13.33}$$

Figure 13.5 exemplifies a nonrectangular integration domain by a disk-like region embedded in a Cartesian mesh, along with the discretized boundaries and the corresponding bounding line indexes.

The routine implementing the described iterative process is `Poisson2D` (Listing 13.4), whose input consists of the initial approximation of the solution, received via array u and including the appropriate boundary values, the coordinates x and y of the discretization grid, the arrays with the boundary indexes `imin` and `imax`, the number of mesh points for the two Cartesian directions, `nx` and `ny`, the required relative precision of the solution components `eps`, and the name of the user function that evaluates the right-hand side of the Poisson equation. Similarly to `PoissonXY`, `Poisson2D` performs Gauss–Seidel iterations to converge the solution u to the desired precision. The iterative refinement of the solution is terminated when the maximum relative change of the solution components between two consecutive iterations drops below the tolerance `eps`.

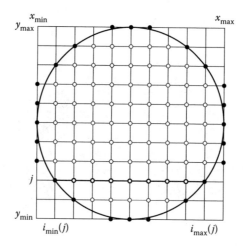

FIGURE 13.5 Integration grid and bounding line indexes, i_{min}^j and i_{max}^j, for a disk-like integration domain. Filled circles represent boundary values and empty circles indicate values that need to be calculated.

Listing 13.4   Solver for the 2D Poisson Equation on an Irregular Domain (Python Coding)

```
#============================================================================
def Poisson2D(u, x, y, imin, imax, nx, ny, eps, Func):
#----------------------------------------------------------------------------
#  Solves the 2D Poisson equation in Cartesian coordinates on a regular grid
#  with (nx x ny) nodes (x[],y[]), with limiting indexes along x specified by
#  imin[j] and imax[j] (j=1,ny), using the Gauss-Seidel method. The solution
#  u[][] is converged with relative precision eps.
#  An error index is returned: 0 - normal execution.
#  Calls: Func(x,y) - RHS of Poisson equation
#----------------------------------------------------------------------------
   itmax = 10000                                          # max no. of iterations

   f = [[0]*(ny+1) for i in range(nx+1)]

   hx = (x[nx]-x[1])/(nx-1); kx = 1e0/(hx*hx)                     # mesh spacings
   hy = (y[ny]-y[1])/(ny-1); ky = 1e0/(hy*hy)
   kxy = 2e0*(kx + ky)

   for j in range(2,ny):                       # RHS of PDE for interior points
      for i in range(2,nx): f[i][j] = Func(x[i],y[j])

   for it in range(1,itmax+1):                        # Gauss-Seidel iteration loop
      err = 0e0
      for j in range(2,ny):
         for i in range(imin[j]+1,imax[j]):                # interior mesh points
            uij = (kx*(u[i-1][j] + u[i+1][j]) - f[i][j]) / kxy
            if (i >= imin[j-1] and i <= imax[j-1]): uij += ky*u[i][j-1] / kxy
            if (i >= imin[j+1] and i <= imax[j+1]): uij += ky*u[i][j+1] / kxy

            eij = 1e0 - u[i][j]/uij if uij else uij - u[i][j]
            if (fabs(eij) > err): err = fabs(eij)          # max. component error
            u[i][j] = uij

      if (err <= eps): break                                    # convergence test
```

```
    if (it >= itmax):
        print("Poisson2D: max. number of iterations exceeded !"); return 1
    return 0
```

As an application for the static heat equation, we determine the temperature distribution on a circular conducting disk, with boundary temperatures decreasing from 100°C on the left semicircle to 0°C on the right semicricle. Concretely, we solve the Laplace equation,

$$\frac{\partial^2 u}{\partial x^2} + \frac{\partial^2 u}{\partial y^2} = 0, \tag{13.34}$$

on the disk circumscribed by the square domain $\mathcal{D} = [-a,a] \times [-a,a]$, with $a = 5$, subject to the boundary conditions:

$$
\begin{aligned}
u_{i,1}^0 &= 0, & i = i_{min}^1, \dots, i_{max}^1, & \quad j = 1, \\
u_{i_{min},j}^0 &= 100, & u_{i_{max},j}^{(r)} = 0, & \quad j = 2, \dots, N_y - 1, \\
u_{i,N_y}^0 &= 0, & i = i_{min}^{N_y}, \dots, i_{max}^{N_y}, & \quad j = N_y.
\end{aligned}
\tag{13.35}
$$

In the implementation from Listing 13.5, the user function BoundCondEllipse can actually handle the more general case of an elliptical domain, circumscribed by an $N_x \times N_y$ regular Cartesian grid. The routine determines the boundary indexes i_{min}^j and i_{max}^j on all the N_y horizontal grid lines and assigns the boundary values of the solution to the appropriate components of array u[][].

The function BoundCondEllipse does not actually receive the geometrical extent of the integration domain, and operates instead just with the dimensionless node indexes, relating all the positions to the coordinates x0 and y0 of the center of the dimensionless ellipse. On each grid line j, the routine determines from the relative y-coordinate yj the positive x-intersection with the elliptical boundary, xi, and, therewith, the bounding indexes imin[j] and imax[j]. The boundary values of the solution are then initialized based on these indexes according to Equation 13.35.

The main program is quite similar with the one from Listing 13.3 and it basically initializes the mesh coordinates, marks the exterior nodes with a distinctive value, uses the routine BoundCondEllipse to set the boundary values, calls the solver Poisson2D to calculate the interior solution values, and finally prints out the solution to a file.

The contour plot of the temperature distribution resulting by running program 13.5 is shown in Figure 13.6.

Listing 13.5 Solution of the Static Heat Problem of a Conducting Disk (Python Coding)

```
# Steady-state temperature distribution in a thin conducting disk
from math import *
from pde import *

def Func(x, y):                                  # RHS function for Poisson2D
    return 0e0

#===============================================================================
def BoundCondEllipse(u, imin, imax, nx, ny):
#-------------------------------------------------------------------------------
#  Determines the boundary indexes imin[j] and imax[j] (j=1,ny) for an
```

```
#   elliptical domain, determined by the number of mesh points nx and ny, and
#   sets values on the boundaries for the function u
#-------------------------------------------------------------------------
    x0 = 0.5e0*(nx + 1e0)                         # center of dimensionless ellipse
    y0 = 0.5e0*(ny + 1e0)
    a = x0 - 1e0                                                   # semi-axes
    b = y0 - 1e0

    for j in range(1,ny+1):                                   # boundary indexes
       yj = j - y0                                      # relative y-coordinate
       xi = a * sqrt(1e0 - yj*yj/(b*b))              # x-coordinate on ellipse
       if (xi == 0e0): xi = 0.75e0               # correction for 1-point boundary

       imin[j] = int(x0 - xi + 0.5e0)                      # left boundary index
       imax[j] = int(x0 + xi + 0.5e0)                     # right boundary index
                                                             # boundary values
    for i in range(imin[1],imax[1]+1): u[i][1] = 0e0                   # bottom
    for j in range(2,ny):
       u[imin[j]][j] = 100e0                                          # left
       u[imax[j]][j] = 0e0                                           # right
    for i in range(imin[ny],imax[ny]+1): u[i][ny] = 0e0                # top

# main

a = 5                                                           # disk radius
xmin = -a; xmax = a; ymin = -a; ymax = a                   # domain boundaries
nx = 51; ny = 51                                        # number of mesh points
eps = 1e-5                                         # relative solution tolerance

u = [[0]*(ny+1) for i in range(nx+1)]                           # solution
x = [0]*(nx+1); y = [0]*(ny+1)                         # mesh point coordinates
imin = [0]*(ny+1)                                 # indexes of y-line boundaries
imax = [0]*(ny+1)

hx = (xmax-xmin)/(nx-1)
for i in range(1,nx+1): x[i] = xmin + (i-1)*hx                  # x-mesh points
hy = (ymax-ymin)/(ny-1)
for j in range(1,ny+1): y[j] = ymin + (j-1)*hy;                # y-mesh points

for j in range(1,ny+1):                                  # initialize solution
   for i in range(1,nx+1): u[i][j] = -1e99

BoundCondEllipse(u,imin,imax,nx,ny)                         # boundary values

for j in range(2,ny):                      # initial approximation of the solution
   for i in range(imin[j]+1,imax[j]): u[i][j] = 0e0              # interior nodes

Poisson2D(u,x,y,imin,imax,nx,ny,eps,Func)

out = open("Poisson.txt","w")                              # open output file
out.write("       x           y            u\n")
for j in range(1,ny+1):
   for i in range(1,nx+1):
      out.write(("{0:10.5f}{1:10.5f}{2:14.5e}\n").format(x[i],y[j],u[i][j]))
out.close()
```

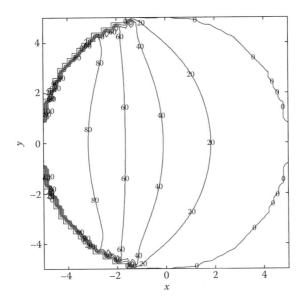

FIGURE 13.6 Temperature profile resulting by solving with Program 13.3 the static heat problem (13.34)–(13.35) for a conducting disk.

13.3 Initial-Value Problems for Parabolic Differential Equations

Science and engineering offers a broad variety of examples of problems for parabolic PDEs. As a prototype for parabolic PDEs, we deal in the following with the 1D diffusion equation, or the formally similar heat equation. Nevertheless, the applied discretization techniques and the pertaining stability analysis have general validity and are applicable to other parabolic equations, as well.

In spite of the considerable number of analytical solutions available for the diffusion equation, these are restricted to simple geometries and piecewise constant diffusion coefficients. The analytically manageable boundary conditions are equally simple; yet, the solutions are often expressed as infinite series, which are by no means trivial to evaluate. On the other hand, there is a wealth of practical problems, for which the simplifying assumptions required for exact resolvability affect to an unacceptable extent the essence of the physical model.

To obtain solutions of the diffusion equation that realistically model practical situations, one generally needs to resort to numerical algorithms. Basically, these imply restricting the solution to a discrete set of mesh points and approximating the derivatives by finite-difference schemes relative to these. Subsequently, the numerical approach boils down to solving the resulting linear system, whose unknowns are the solution values at the mesh points.

The 1D *diffusion* and *heat* equations,

$$\frac{\partial u}{\partial t} - D\frac{\partial^2 u}{\partial x^2} = 0, \tag{13.36}$$

$$\frac{\partial u}{\partial t} - K\frac{\partial^2 u}{\partial x^2} = f(x, t), \tag{13.37}$$

describe the time evolution and spatial distribution of the concentration of a diffusing species ($u(x,t) \equiv c(x,t)$) and, respectively, of the system's temperature ($u(x,t) \equiv T(x,t)$). The diffusion coefficient, D, relates the diffusive flux, J_{diff}, to the concentration gradient by Fick's law, while the thermal diffusivity, K, links the heat flux, J_{heat}, to the temperature gradient by Fourier's law:

$$J_{\text{diff}} = -D\frac{\partial c}{\partial x}, \quad J_{\text{heat}} = -K\frac{\partial T}{\partial x}.$$

The above form of the heat equation also makes provision for a heat source, $f(x, t)$.

Even though both equations appear to be spatially 1D, alternatively, they can be assumed to model a 3D planar "slab" geometry, in which there is explicit dependence only on the Cartesian coordinate perpendicular to the slab (x in the above equations) and no dependence on the other two coordinates (y and z), along which the extent of the system is implicitly considered infinite.

Modeling a particular physical situation requires, in addition to the actual PDE, provision of *initial* and *boundary conditions*. The *initial conditions* specify the solution over the whole spatial domain at some initial moment t_0:

$$u(x, t_0) = u^0(x), \quad x \in [0, L]. \tag{13.38}$$

The *boundary conditions* define the behavior of the solution on the boundaries at all times $t > t_0$. For simplicity, we consider, for the time being, Dirichlet-type boundary conditions, implying constant solution values on the boundaries:

$$u(0, t) = u_0^0, \quad u(L, t) = u_L^0. \tag{13.39}$$

There are several methods at hand for discretizing parabolic equations, and some of them have been employed in the preceding sections. Nonetheless, apparently slight modifications of the discretization schemes have a tremendous impact on the stability and accuracy of the time evolution of the solution. In the following, we address three such finite-difference approaches: the explicit, the implicit, and the Crank–Nicolson methods.

13.3.1 Explicit Finite-Difference Method

We start discretizing the diffusion equation (13.36) in conjunction with the boundary conditions (13.38) and (13.39) by establishing a regular space–time grid, characterized by the nodes (see Figure 13.7):

$$x_i = (i - 1)h_x, \quad i = 1, 2, \ldots, N_x, \tag{13.40}$$

$$t_n = nh_t, \quad n = 0, 1, 2, \ldots \tag{13.41}$$

h_t is here the time step and h_x represents the spacing between the N_x spatial nodes, being defined by

$$h_x = L/(N_x - 1). \tag{13.42}$$

The first-time derivative, $\partial u/\partial t$, at the space–time node (x_i, t_n) can be obtained in a straightforward way from the linear approximation of the Taylor series with respect to t for constant $x = x_i$:

$$u_i^{n+1} = u_i^n + h_t \left(\frac{\partial u}{\partial t}\right)_{i,n} + O(h_t^2),$$

where the notation $u_i^n \equiv u(x_i, t_n)$ is used. Indeed, expressing the time derivative, one obtains the simple *forward-difference scheme*:

$$\left(\frac{\partial u}{\partial t}\right)_{i,n} = \frac{u_i^{n+1} - u_i^n}{h_t} + O(h_t). \tag{13.43}$$

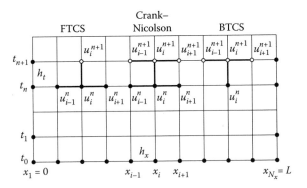

FIGURE 13.7   Space–time discretization grid for the 1D diffusion equation. The four-, six-, and four-node stencils for the explicit FTCS, Crank–Nicolson, and implicit BTCS methods are shown. Filled circles represent available data, and empty circles indicate data to be calculated.

The "forward" character is given by the presence of the propagated solution from t^{n+1} in the expression of the derivative at t^n, while the fact that the scheme remains only first-order accurate in h_t is due to the implied division by h_t.

For the second spatial derivative at the space–time node (x_i, t_n), one can use a *central-difference scheme* similar to the one used for elliptic equations in Section 13.2, namely,

$$\left(\frac{\partial^2 u}{\partial x^2} \right)_{i,n} = \frac{u_{i+1}^n - 2u_i^n + u_{i-1}^n}{h_x^2} + O(h_x^2), \tag{13.44}$$

which involves only information from t_n and from space points located symmetrically about the node where the derivative is being expressed.

Inserting the finite-difference schemes (13.43) and (13.44) in the diffusion equation (13.36), we obtain the *forward-time central-space* (FTCS) method:

$$\frac{u_i^{n+1} - u_i^n}{h_t} = D \frac{u_{i+1}^n - 2u_i^n + u_{i-1}^n}{h_x^2}, \tag{13.45}$$

which provides an $O(h_x^2 + h_t)$ approximation of the diffusion equation at the space–time node (x_i, t_n). Herefrom, we can express the solution propagated at the time step t_{n+1} for each interior spatial mesh point x_i only in terms of values from the previous time step t_n:

$$u_i^{n+1} = \lambda u_{i-1}^n + (1 - 2\lambda) u_i^n + \lambda u_{i+1}^n, \quad i = 2, 3, \dots, N_x - 1, \tag{13.46}$$

where

$$\lambda = \frac{Dh_t}{h_x^2}. \tag{13.47}$$

As for the solution values on the boundaries, they are fixed by the Dirichlet conditions (13.39) throughout the propagation and do not require any particular treatment:

$$u_1^{n+1} = u_1^n = u_0^0, \quad u_{N_x}^{n+1} = u_{N_x}^n = u_L^0. \tag{13.48}$$

Since at each time step, the propagated solution components can be expressed independently from Equation 13.46, exclusively based on data from the preceding time step, the method built on the FTCS scheme is said to be *explicit* and its essence is also apparent from Figure 13.7.

The explicit nature of the propagation process also emerges by rewriting Equations 13.46 through 13.48 in matrix notations:

$$\mathbf{u}^{n+1} = \mathbf{B} \cdot \mathbf{u}^n, \quad n = 0, 1, 2, \ldots \tag{13.49}$$

The propagation matrix $\mathbf{B}$ is tridiagonal and the vector $\mathbf{u}^n$ gathers the solution values from all the spatial mesh points at the time step t_n:

$$\mathbf{B} = \begin{bmatrix} 1 & 0 & & & 0 \\ \lambda & 1-2\lambda & \lambda & & \\ & \ddots & \ddots & \ddots & \\ & & \lambda & 1-2\lambda & \lambda \\ 0 & & & 0 & 1 \end{bmatrix}, \quad \mathbf{u}^n = \begin{bmatrix} u_1^n \\ u_2^n \\ \vdots \\ u_{N_x-1}^n \\ u_{N_x}^n \end{bmatrix}. \tag{13.50}$$

Each component of the propagated solution $\mathbf{u}^{n+1}$ results individually, by the multiplication of the matrix $\mathbf{B}$ with the previous solution vector $\mathbf{u}^n$.

The routine PropagFTCS (Listing 13.6) solves an elementary Cauchy problem for the diffusion equation (13.36), propagating the initial solution u0[] over the time interval ht according to Formulas 13.46 through 13.48 obtained by applying the explicit FTCS discretization scheme. The constant diffusion coefficient, the spatial step size, and the number of nodes are received via arguments D, hx, and, respectively, nx, and the propagated solution is returned by means of the array u[]. The driving sequence in the main program is supposed to perform any additional processing and, above all, to free the array u[] for a new propagation step by saving its content to u0[].

A typical main program is included in Listing 13.7. The code first initializes the solution u0[] by a call to the user function Init and then controls the propagation within a temporal loop by repeated calls to PropagFTCS, each followed necessarily by a backward transfer of content from u[] to u0[].

The solution of the diffusion equation (13.36) by the explicit FTCS method is prone to a particular numerical phenomenology, being confronted under certain conditions with severe *numerical instability*. The instability implies that, instead of smooth, well-behaved spatial profiles, the propagation develops spurious oscillations, which grow exponentially in time, distorting the solution irrecoverably. This critical behavior is due to the increasing domination of the roundoff errors and always arises when the time step size exceeds a certain upper bound, which correlates with the employed spatial step size.

Listing 13.6 Explicit FTCS Propagator for the Diffusion Equation (Python Coding)

```
#===========================================================================
def PropagFTCS(u0, u, nx, D, hx, ht):
#---------------------------------------------------------------------------
#  Propagates the solution u0[] of the diffusion equation
#     du/dt = D d2u/dx2,  D - diffusion coefficient (constant)
#  over the time intervat ht, using an explicit FTCS difference scheme on
#  a spatial grid with nx nodes and spacing hx. Returns the solution in u[].
#---------------------------------------------------------------------------
   lam = D * ht/(hx*hx)
   lam1 = 1e0 - 2e0*lam

   u[1] = u0[1]; u[nx] = u0[nx]
   for i in range(2,nx):
      u[i] = lam*u0[i-1] + lam1*u0[i] + lam*u0[i+1]
```

Listing 13.7 Finite-Difference Solution of the Diffusion Equation (Python Coding)

```
# Solves the 1D diffusion equation by finite-difference schemes
from math import *
from pde import *
from graphlib import *

def Init(u, x, nx):             # initial solution for the diffusion equation
    global L                                    # extent of spatial domain
    for i in range(1,nx+1): u[i] = sin(pi*x[i]/L)

# main

D    = 0.1e0                                # diffusion coefficient
L    = 1e0                                  # [0,L] spatial domain
nx   = 21                            # number of spatial mesh points
tmax = 6.0e0                                   # propagation time
ht   = 1.25e-2                                      # time step

u0 = [0]*(nx+1); u = [0]*(nx+1)                      # solution
x = [0]*(nx+1)                                  # spatial mesh

hx = L/(nx-1)
for i in range(1,nx+1): x[i] = (i-1)*hx                  # spatial mesh

Init(u0,x,nx)                                  # initial solution

t = 0e0
while (t < tmax):                                  # temporal loop
    t += ht                                       # increase time
    PropagFTCS(u0,u,nx,D,hx,ht)                  # propagate solution

    for i in range(1,nx): u0[i] = u[i]             # shift solutions

out = open("diffusion.txt","w")                # open output file
out.write(("lambda = {0:f} t = {1:f}\n").format(D*ht/(hx*hx),t))
out.write("      x        u        exact\n")
f = exp(-pi*pi*D*t/(L*L))
for i in range(1,nx+1):
    out.write(("{0:10.5f}{1:10.5f}{2:10.5f}\n"). \
            format(x[i],u[i],f*sin(pi*x[i]/L)))
out.close()

GraphInit(800,600)
Plot(x,u,nx,"blue",1,0.15,0.95,0.15,0.85,"x","u","Diffusion")
MainLoop()
```

For illustrating the critical relation between the time and spatial step sizes, we solve the 1D diffusion equation:

$$\frac{\partial u}{\partial t} = D\frac{\partial^2 u}{\partial x^2}, \quad x \in [0,L], \quad t > 0, \tag{13.51}$$

with constant diffusion coefficient, subject to the Dirichlet boundary conditions

$$u(0,t) = u(L,t) = 0, \quad t > 0, \tag{13.52}$$

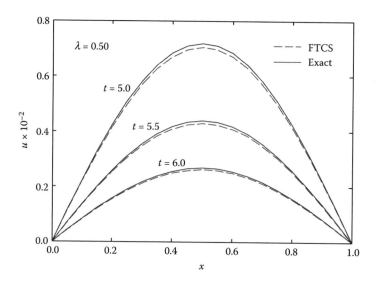

FIGURE 13.8 Numerical solutions for the diffusion problem (13.51)–(13.53) by the explicit FTCS method with the spatial step size $h_x = 0.05$ and the time step size $h_t = 0.00125$.

and with the initial solution profile

$$u(x, 0) = \sin(\pi x/L), \quad x \in [0, L]. \tag{13.53}$$

The exact solution satisfying this problem can be readily verified to be

$$u(x, t) = \exp(-\pi^2 Dt/L^2) \sin(\pi x/L). \tag{13.54}$$

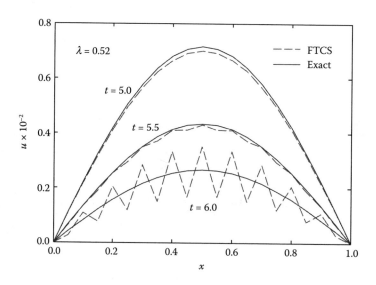

FIGURE 13.9 Numerical solutions for the diffusion problem (13.51)–(13.53) by the explicit FTCS method with the spatial step size $h_x = 0.05$ and the time step size $h_t = 0.0013$.

Specifically, we consider $D = 0.1$ and $L = 1$, and propagate the solution up to $t_{max} = 6.0$. We use comparatively two time steps, $h_t = 1.25 \cdot 10^{-2}$ and $1.30 \cdot 10^{-2}$, while maintaining the same spatial spacing $h_x = 0.05$. Correspondingly, the discretization parameter (Equation 13.47) takes the values $\lambda = 0.5$ and, respectively, 0.52, and the slight difference between these turns out to be essential.

The spatial profiles resulting for $h_t = 1.25 \cdot 10^{-2}$ ($\lambda = 0.5$) are depicted in Figure 13.8 along with the analytical solutions for the three considered times. Aside from the slight systematic underestimation caused by the rather large spatial spacing employed, the numerical profiles (with dashed lines) fairly reproduce the analytical solutions.

Similar profiles, however obtained with $h_t = 1.30 \cdot 10^{-2}$ ($\lambda = 0.52$), are shown in Figure 13.9. While at $t = 5.0$, the solutions obtained with the two time steps are barely discernible, instabilities start developing at $t = 5.5$, and completely dominate the solution at $t = 6.0$. Hence, an apparently marginal increase of the time step h_t causes a tremendous qualitative change in the solution's behavior. $\lambda = 1/2$ appears, thus, to be a critical value, separating two distinct domains of *stable* ($\lambda \leq 1/2$) and, respectively, *unstable propagation* ($\lambda > 1/2$).

13.3.2 von Neumann Stability Analysis

The standard technique for assessing the stability of finite-difference schemes as applied to *linear* (generally time dependent) PDEs is the so-called *von Neumann stability analysis* (von Neumann 1950). The stability of a numerical scheme essentially regards the way in which the errors evolve during the recursive propagation of the solution. An approach is deemed to be *unstable* if the associated errors grow by successive accumulation beyond manageable limits, producing eventually an unbounded solution.

The von Neumann analysis is based on the assumptions that (1) the PDE has *constant coefficients* (or so slowly varying as to be considered constant) and (2) the PDE is solved subject to *periodic boundary conditions* (implying linked values on opposite boundaries). Nonetheless, the analysis proves useful even in cases where the boundary conditions are actually nonperiodic. The analysis operates with the local roundoff error, defined as the difference between the *finite-precision solution* $\tilde{u}_i^n$ and the *exact discrete solution* u_i^n (resulting in the absence of roundoff errors):

$$\varepsilon_i^n \equiv \tilde{u}_i^n - u_i^n. \tag{13.55}$$

Therewith, the error must satisfy the same difference equation as the exact solution itself.

For periodic boundary conditions over the domain $[0, L]$, the error can be represented as a discrete Fourier series:

$$\varepsilon(x, t) = \sum_m \xi_m(t) e^{\imath k_m x}, \tag{13.56}$$

with the *real wave numbers* $k_m = 2\pi m/L$ defined for integer $m = 1, 2, \ldots, N_x$ ($N_x = [L/h_x + 1]$). In particular, the error associated to the spatial node $x_i = (i - 1)h_x$ and time $t_n = nh_t$ is given by

$$\varepsilon_i^n \equiv \sum_m \xi_m(nh_t) \exp\left[\imath k_m i h_x\right],$$

where $\imath = \sqrt{-1}$ is the imaginary unit (not to be mistaken for the index i) and each coefficient ξ_m also incorporates a position-independent factor $\exp(-\imath k_m h_x)$ originating from the definition of the nodes. The coefficients ξ obviously depend only on the time (index n) and the wave vector (index m), while the exponentials depend on the position (index i) and the wave vector (index m).

For linear difference equations (stemming from linear PDEs), it proves sufficient to analyze the stability of a single Fourier harmonic of the error term, with the wave number denoted simply by k. Such a generic *eigenmode* (independent solution) of the difference equation may be written under the general form:

$$\varepsilon_i^n = \xi(k)^n \exp(\imath k i h_x), \tag{13.57}$$

where the *amplification factor* ξ is in general a complex function of k. An essential feature of the von Neumann analysis is the power-law dependence of the eigenmode on time, namely, by the nth power of the amplification factor after n time steps. The temporal propagation is obviously *stable* only if the absolute value of the amplification factor is subunitary,

$$\left|\xi(k)\right| < 1, \tag{13.58}$$

since, in such a way, there cannot exist exponentially growing eigenmodes of the error term.

As a general procedure for establishing whether, or under which conditions the stability criterion (13.58) is met, one substitutes the generic eigenmode (13.57) into the difference equation resulting by discretizing the PDE, and expresses $\xi(k)$ therefrom. Specifically, in the case of the diffusion equation discretized by the explicit FCTS scheme, the eigenmodes must satisfy (see Equation 13.46)

$$\varepsilon_i^{n+1} = \lambda \varepsilon_{i-1}^n + (1 - 2\lambda)\,\varepsilon_i^n + \lambda \varepsilon_{i+1}^n, \tag{13.59}$$

and, by substituting Expression 13.57 of the generic eigenmode, one finds:

$$\xi = \lambda \exp(-\imath k h_x) + (1 - 2\lambda) + \lambda \exp(\imath k h_x).$$

Combining the exponentials to give $2\cos(kh_x)$ and using the identity $1 - \cos x = 2\sin^2(x/2)$ leads to the sought expression for the amplification factor:

$$\xi = 1 - 4\lambda \sin^2(kh_x/2). \tag{13.60}$$

Given that $0 \le \sin^2(kh_x/2) \le 1$, von Neumann's stability criterion (13.58) can only be met if the discretization parameter λ satisfies the condition:

$$0 < \lambda < 1/2. \tag{13.61}$$

Having in view Expression 13.47 of λ, the stability condition translates to

$$h_t < \frac{1}{2}\frac{h_x^2}{D}. \tag{13.62}$$

Hence, the time step size h_t ensuring the stability of the FTCS algorithm is bounded above by half the characteristic *diffusion time* across the distance h_x ($\tau_D = h_x^2/D$). Consequently, the explicit method (13.46)–(13.48) based on the FTCS forward-time discretization scheme is *conditionally stable*, whereby, the critical value $\lambda = 1/2$ separates the stable and unstable regimes.

13.3.3 Implicit Finite-Difference Method

By reference to the explicit FTCS method developed above, an apparently minor change, referring to the time step, t_n versus t_{n+1}, at which the discretized time and space derivatives are equated, brings about a qualitative change in the stability of the temporal propagation of the solution, namely, it leads to an *unconditionally stable* method. In fact, taking as reference t_{n+1}, the very same discrete approximation of the time derivative (Equation 13.43) now becomes a *backward-difference scheme*:

$$\left(\frac{\partial u}{\partial t}\right)_{i,n+1} = \frac{u_i^{n+1} - u_i^n}{h_t} + O(h_t). \tag{13.63}$$

The spatial derivative, in turn, needs to be built accordingly, based on values from time step t_{n+1}:

$$\left(\frac{\partial^2 u}{\partial x^2}\right)_{i,n+1} = \frac{u_{i+1}^{n+1} - 2u_i^{n+1} + u_{i-1}^{n+1}}{h^2} + O(h_x^2). \tag{13.64}$$

The difference equation resulting for the space–time point (x_i, t_{n+1}),

$$\frac{u_i^{n+1} - u_i^n}{h_t} = D\frac{u_{i+1}^{n+1} - 2u_i^{n+1} + u_{i-1}^{n+1}}{h_x^2}, \tag{13.65}$$

represents the so-called *backward-time central-space* (BTCS) method and is, analogously to the FTCS method, an $O(h_x^2 + h_t)$ approximation of the diffusion equation. Regrouping the terms and noting, as before, $\lambda = Dh_t/h_x^2$, one obtains for the propagated solution components u_i^{n+1} at the interior mesh points the linear system

$$-\lambda u_{i-1}^{n+1} + (1 + 2\lambda)u_i^{n+1} - \lambda u_{i+1}^{n+1} = u_i^n, \quad i = 2, 3, \ldots, N_x - 1. \tag{13.66}$$

This system is to be solved in conjunction with appropriate initial and boundary conditions. In particular, the Dirichlet conditions,

$$u_1^{n+1} = u_1^n = u_0^0, \quad u_{N_x}^{n+1} = u_{N_x}^n = u_L^0, \tag{13.67}$$

maintain the solution values constant on the boundaries.

Since, as opposed to the FTCS method, the components of the propagated solution cannot be expressed independently, but result at every time step by solving the linear system (13.66)–(13.67), the BTCS scheme is said to be an *implicit method*, and the characteristic information flow between the time steps t_n and t_{n+1} can be also learnt from Figure 13.7.

In matrix notation, the linear system of the BTCS method takes the form:

$$\mathbf{A} \cdot \mathbf{u}^{n+1} = \mathbf{u}^n, \quad n = 0, 1, 2, \ldots, \tag{13.68}$$

whereby, the preceding solution $\mathbf{u}^n$ plays the role of constant vector and the system matrix is tridiagonal:

$$\mathbf{A} = \begin{bmatrix} 1 & 0 & & & \\ -\lambda & 1 + 2\lambda & -\lambda & & \\ & \ddots & \ddots & \ddots & \\ & & -\lambda & 1 + 2\lambda & -\lambda \\ & & & 0 & 1 \end{bmatrix}. \tag{13.69}$$

Since $\lambda > 0$ by definition, matrix $\mathbf{A}$ is diagonally dominant and positive definite. Therefore, solving system (13.68)–(13.69) can be carried out most efficiently by means of the LU decomposition method developed in Section 7.7 and implemented as routine `TriDiagSys` in the libraries `linsys.py` and `linsys.h`.

Addressing the stability of the implicit BTCS method, one plugs the generic eigenmode $\varepsilon_i^n = \xi(k)^n \exp(\iota k i h_x)$ into the difference equation (13.66) and finds:

$$\xi \left[-\lambda \exp(-\iota k h_x) + (1 + 2\lambda) - \lambda \exp(\iota k h_x) \right] = 1.$$

Combining the exponentials into $2\cos(k h_x)$ and using the identity $1 - \cos x = 2\sin^2(x/2)$ then yields the expression for the amplification factor:

$$\xi = \frac{1}{1 + 4\lambda \sin^2(k h_x/2)}. \tag{13.70}$$

Since $0 \leq \sin^2(k h_x/2) \leq 1$ and $\lambda > 0$, the denominator always exceeds 1 and ξ is invariably subunitary, as required by the von Neumann stability criterion ($|\xi(k)| < 1$). Consequently, unlike the FTCS method, the implicit BTCS method is *unconditionally stable* for any time step size h_t.

It should be noted, however, that the unconditional stability of the BTCS scheme does not also guarantee a predefined degree of precision. Owing to the low $O(h_t)$ order with respect to time, the BTCS method requires rather small time steps for achieving acceptable levels of precision, and this remains a considerable weakness of this method.

The routine `PropagBTCS` presented in Listing 13.8 propagates the initial solution `u0[]` of the diffusion equation over an elementary time interval `ht` by the implicit BTCS method described above. To this end, the routine builds the elements of the three relevant diagonals of the system matrix and the constant terms in the arrays `a[]`, `b[]`, `c[]`, and, respectively, `u[]`. The discretized tridiagonal system is then solved by calling the routine `TriDiagSys` and the solution is returned via array `u[]`. The routine can be used in Program 13.7 interchangeably with the function `PropagFTCS`, without any changes.

Listing 13.8　Implicit Finite-Difference Propagator for the Diffusion Equation (Python Coding)

```
#===========================================================================
def PropagBTCS(u0, u, nx, D, hx, ht):
#---------------------------------------------------------------------------
#  Propagates the solution u0[] of the diffusion equation
#     du/dt = D d2u/dx2,   D - diffusion coefficient (constant)
#  over the time intervat ht, using an implicit BTCS difference scheme on
#  a spatial grid with nx nodes and spacing hx. Returns the solution in u[].
#---------------------------------------------------------------------------
   a = [0]*(nx+1); b = [0]*(nx+1); c = [0]*(nx+1)

   lam = D * ht/(hx*hx)
   lam1 = 1e0 + 2e0*lam

   b[1] = 1e0; c[1] = 0e0              # build coefficients of discretized system
   u[1] = u0[1]
   for i in range(2,nx):
      a[i] = -lam; b[i] = lam1; c[i] = -lam
      u[i] = u0[i]
   a[nx] = 0e0; b[nx] = 1e0
   u[nx] = u0[nx]

   TriDiagSys(a,b,c,u,nx)             # solve tridiagonal discretized system
```

13.3.4 Crank–Nicolson Method

The mentioned drawback of the implicit BTCS method of being only first-order accurate in time can be cured using as reference in discretizing the diffusion equation the intermediate time step $t_{n+1/2} \equiv t_n + h_t/2$. The thereby resulting difference scheme is second-order accurate both in space and time. Let us consider for this purpose the Taylor expansions of the solution at t_n and t_{n+1} using as reference $t_{n+1/2}$:

$$u_i^n = u_i^{n+1/2} - \left(\frac{h_t}{2}\right)\left(\frac{\partial u}{\partial t}\right)_{i,n+1/2} + \frac{1}{2}\left(\frac{h_t}{2}\right)^2 \left(\frac{\partial^2 u}{\partial t^2}\right)_{i,n+1/2} + O(h_t^3),$$

$$u_i^{n+1} = u_i^{n+1/2} + \left(\frac{h_t}{2}\right)\left(\frac{\partial u}{\partial t}\right)_{i,n+1/2} + \frac{1}{2}\left(\frac{h_t}{2}\right)^2 \left(\frac{\partial^2 u}{\partial t^2}\right)_{i,n+1/2} + O(h_t^3).$$

Upon subtraction of these expansions, both the solution $u_i^{n+1/2}$ and the second time derivative $\left(\partial^2 u/\partial t^2\right)_{i,n+1/2}$ at $t_{n+1/2}$ cancel out exactly and we obtain for the first-time derivative the *second-order central-difference scheme*:

$$\left(\frac{\partial u}{\partial t}\right)_{i,n+1/2} = \frac{u_i^{n+1} - u_i^n}{h_t} + O(h_t^2). \tag{13.71}$$

The second spatial derivative at time $t_{n+1/2}$ can be simply approximated by the mean of the difference schemes for the time steps t_n and t_{n+1}:

$$\left(\frac{\partial^2 u}{\partial x^2}\right)_{i,n+1/2} = \frac{1}{2}\left[\frac{u_{i+1}^{n+1} - 2u_i^{n+1} + u_{i-1}^{n+1}}{h_x^2} + \frac{u_{i+1}^n - 2u_i^n + u_{i-1}^n}{h_x^2}\right] + O(h_x^2). \tag{13.72}$$

Therewith, the diffusion equation can be cast under the discretized form:

$$\frac{u_i^{n+1} - u_i^n}{h_t} = \frac{D}{2}\frac{\left(u_{i+1}^{n+1} - 2u_i^{n+1} + u_{i-1}^{n+1}\right) + \left(u_{i+1}^n - 2u_i^n + u_{i-1}^n\right)}{h_x^2}. \tag{13.73}$$

This *central-time-central-space* scheme is known as the *Crank–Nicolson method*. Even though it formally represents the mean of the FTCS and BTCS schemes, the Crank–Nicolson method features an enhanced $O(h_t^2 + h_x^2)$ precision order. Separating the terms for the time steps t_n and t_{n+1} results in the following linear system, having as unknowns the components of the propagated solution:

$$-\lambda u_{i-1}^{n+1} + (1 + 2\lambda)\,u_i^{n+1} - \lambda u_{i+1}^{n+1} = \lambda u_{i-1}^n + (1 - 2\lambda)\,u_i^n + \lambda u_{i+1}^n, \tag{13.74}$$

$$i = 2, 3, \ldots, N_x - 1,$$

where

$$\lambda = \frac{1}{2}\frac{Dh_t}{h_x^2}. \tag{13.75}$$

The discretized system (13.74) along with the attached boundary conditions may be represented in matrix form as

$$\mathbf{A} \cdot \mathbf{u}^{n+1} = \mathbf{B} \cdot \mathbf{u}^n, \quad n = 0, 1, 2, \ldots, \tag{13.76}$$

where the matrices $\mathbf{A}$ and $\mathbf{B}$ are tridiagonal and, in the case of Dirichlet boundary conditions, they are respectively given by Equations 13.50 and 13.69. Matrix $\mathbf{A}$ is diagonally dominant, positive definite, and therefore nonsingular. Hence, the propagated solution $\mathbf{u}^{n+1}$ may be determined, as in the case of the implicit method, by means of the LU factorization algorithm for tridiagonal systems developed in

Section 7.7. As compared to the implicit method, there is, certainly, an operation overhead due to the additional matrix multiplication implied by the constant terms.

Following the general methodology of the von Neumann stability analysis, by substituting the generic eigenmode $\varepsilon_i^n = \xi(k)^n \exp(\imath kih_x)$ into the difference equation (13.74), one finds:

$$\xi\left[-\lambda \exp(-\imath kh_x) + (1+2\lambda) - \lambda \exp(\imath kh_x)\right] = \lambda \exp(-\imath kh_x) + (1-2\lambda) + \lambda \exp(\imath kh_x).$$

Combining on both sides the exponentials into $2\cos(kh_x)$ and making use of the identity $1 - \cos x = 2\sin^2(x/2)$, one finds for the amplification factor:

$$\xi = \frac{1 - 4\lambda \sin^2(kh_x/2)}{1 + 4\lambda \sin^2(kh_x/2)}. \tag{13.77}$$

Since $\left|1 - 4\lambda \sin^2(kh_x/2)\right| \leq \left|1 + 4\lambda \sin^2(kh_x/2)\right|$ under any circumstances, according to the von Neumann stability criterion ($\left|\xi(k)\right| < 1$), the Crank–Nicolson method is *unconditionally stable* for any time step size h_t. In addition, due to the enhanced order relative to time, the Crank–Nicolson scheme is the method of choice for solving parabolic PDEs.

The Crank–Nicolson method for the diffusion equation is implemented in the routine `PropagCN` (Listing 13.9), its list of arguments and functionalities being the same as for the functions `PropagFTCS` and `PropagBTCS`, presented in the preceding sections.

For the purpose of illustrating the beneficial effect of the higher-order accuracy with respect to time on a stable propagation, let us consider again the initial-value problem 13.51–13.53 for the diffusion equation, which we solve comparatively using the implicit BTCS and the Crank–Nicolson methods, by calling the routines `PropagBTCS` and `PropagCN` from Program 13.7.

Listing 13.9   Crank–Nicolson Propagator for the Diffusion Equation (Python Coding)

```
#===========================================================================
def PropagCN(u0, u, nx, D, hx, ht):
#---------------------------------------------------------------------------
#  Propagates the solution u0[] of the diffusion equation
#     du/dt = D d2u/dx2,  D - diffusion coefficient (constant)
#  over the time intervat ht, using the Crank-Nicolson difference scheme on
#  a spatial grid with nx nodes and spacing hx. Returns the solution in u[].
#---------------------------------------------------------------------------
   a = [0]*(nx+1); b = [0]*(nx+1); c = [0]*(nx+1)

   lam = 0.5e0 * D * ht/(hx*hx)
   lam1 = 1e0 + 2e0*lam; lam2 = 1e0 - 2e0*lam

   b[1] = 1e0; c[1] = 0e0            # build coefficients of discretized system
   u[1] = u0[1]
   for i in range(2,nx):
      a[i] = -lam; b[i] = lam1; c[i] = -lam
      u[i] = lam*u0[i-1] + lam2*u0[i] + lam*u0[i+1]

   a[nx] = 0e0; b[nx] = 1e0
   u[nx] = u0[nx]

   TriDiagSys(a,b,c,u,nx)           # solve tridiagonal discretized system
```

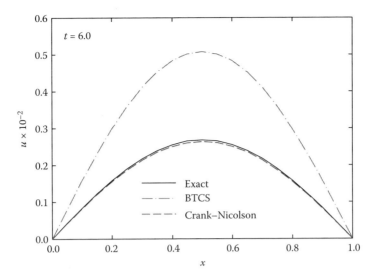

FIGURE 13.10 Numerical solutions of the diffusion problem 13.51–13.53 obtained for $t = 6.0$ by the implicit and the Crank–Nicolson methods using the spatial spacing $h_x = 0.05$ and the time step size $h_t = 0.25$.

The calculations employ the same, spatial step size, $h_x = 0.05$, as with the explicit FTCS method, while for the time step, we use $h_t = 0.25$, which is 20 times larger than the limiting value still ensuring the stability of the FTCS scheme (see Figure 13.8). As can be seen from Figure 13.10, showing profiles for $t = 6.0$, unlike the solution provided by the BTCS method, which substantially departs from the exact result, the solution yielded by the Crank–Nicolson method agrees fairly, in spite of the very large time step. The example points to the fact that the stability of a difference method (absence of spurious oscillations) does not also entail a reasonable degree of precision.

13.3.5 Spatially Variable Diffusion Coefficient

In many applications of practical importance, involving, for instance, the permeation of substances through membrane stacks, the diffusion coefficient cannot be considered by any means constant, but rather piecewise constant under the most simplifying assumptions. In such problems, the 1D diffusion equation needs to be formulated as

$$\frac{\partial u}{\partial t} = \frac{\partial}{\partial x}\left(D\frac{\partial u}{\partial x}\right). \tag{13.78}$$

Aiming to consistently maintain the second order of the discretization methods, we approximate the spatial derivative at the space–time point (x_i, t_n) by the central-difference scheme:

$$\left[\frac{\partial}{\partial x}\left(D\frac{\partial u}{\partial x}\right)\right]_{i,n} \cong \frac{1}{h_x}\left[\left(D\frac{\partial u}{\partial x}\right)_{i+1/2,n} - \left(D\frac{\partial u}{\partial x}\right)_{i-1/2,n}\right] \tag{13.79}$$

$$\cong \frac{1}{h_x}\left[D_{i+1/2}\frac{u_{i+1}^n - u_i^n}{h_x} - D_{i-1/2}\frac{u_i^n - u_{i-1}^n}{h_x}\right],$$

where the midpoints $x_{i\pm1/2} = (x_i + x_{i\pm1})/2$ ensure, on the one hand, the proper centering of the involved quantities about the reference point x_i and lead, on the other, to an $O(h_x^2)$ scheme. The median values of the diffusion coefficients, $D_{i\pm1/2}$, can be conveniently approximated assuming a linear dependence between the spatial nodes, as averages of the adjacent regular values D_{i-1}, D_i, and D_{i+1}:

$$D_{i\pm1/2} = \frac{1}{2}\left[D_i + D_{i\pm1}\right]. \tag{13.80}$$

The desired overall $O(h_t^2 + h_x^2)$ precision is ensured by applying the Crank–Nicolson method based on the spatial difference scheme (13.79), and this results in the following discretized form of the diffusion equation:

$$\frac{u_i^{n+1} - u_i^n}{h_t} = \frac{1}{2h_x}\left[D_{i+1/2}\frac{u_{i+1}^{n+1} - u_i^{n+1}}{h_x} - D_{i-1/2}\frac{u_i^{n+1} - u_{i-1}^{n+1}}{h_x}\right]$$
$$+ \frac{1}{2h_x}\left[D_{i+1/2}\frac{u_{i+1}^n - u_i^n}{h_x} - D_{i-1/2}\frac{u_i^n - u_{i-1}^n}{h_x}\right]. \tag{13.81}$$

Separating the terms corresponding to t_n and t_{n+1} gives rise to the linear system:

$$-\lambda_{i-1/2}\, u_{i-1}^{n+1} + (1 + 2\lambda_i)\, u_i^{n+1} - \lambda_{i+1/2}\, u_{i+1}^{n+1} = \tag{13.82}$$
$$\lambda_{i-1/2}\, u_{i-1}^n + (1 - 2\lambda_i)\, u_i^n + \lambda_{i+1/2}\, u_{i+1}^n, \quad i = 2, 3, \dots, N_x - 1,$$

with

$$\lambda_i = \frac{1}{2}\frac{D_i h_t}{h_x^2}, \quad \lambda_{i\pm1/2} = \frac{1}{4}\frac{(D_i + D_{i\pm1/2})h_t}{h_x^2}. \tag{13.83}$$

To account for specific physical conditions, the above system must be accompanied by appropriate boundary conditions. In particular, Dirichlet conditions fix the solution values on the boundaries:

$$u_1^{n+1} = u_1^n, \quad u_{N_x}^{n+1} = u_{N_x}^n, \tag{13.84}$$

while Neumann conditions define the boundary derivatives (fluxes):

$$-D_1\frac{u_2^{n+1} - u_1^{n+1}}{h_x} = J_{\text{diff}}^{(1)}, \quad -D_{N_x}\frac{u_{N_x}^{n+1} - u_{N_x-1}^{n+1}}{h_x} = J_{\text{diff}}^{(2)},$$

or

$$u_1^{n+1} - u_2^{n+1} = h_x J_{\text{diff}}^{(1)}/D_1, \quad -u_{N_x-1}^{n+1} + u_{N_x}^{n+1} = -h_x J_{\text{diff}}^{(2)}/D_{N_x}. \tag{13.85}$$

Finally, the relevant coefficients of the first and last rows of the tridiagonal discretized system, coding the boundary conditions, are

$$\begin{array}{llll}
b_1 = 1, & c_1 = 0, & d_1 = u_1^n, & \text{Dirichlet,}\\
b_1 = 1, & c_1 = -1, & d_1 = h_x J_{\text{diff}}^{(1)}/D_1, & \text{Neumann,}\\
a_{N_x} = 0, & b_{N_x} = 1, & d_{N_x} = u_{N_x}^n, & \text{Dirichlet,}\\
a_{N_x} = -1, & b_{N_x} = 1, & d_{N_x} = -h_x J_{\text{diff}}^{(2)}/D_{N_x}, & \text{Neumann,}
\end{array} \tag{13.86}$$

while the elements on the rest of the rows read ($i = 2, 3, \dots, N_x - 1$):

$$a_i = -\lambda_{i-1/2}, \quad b_i = 1 + 2\lambda_i, \quad c_i = -\lambda_{i+1/2}, \tag{13.87}$$

$$d_i = \lambda_{i-1/2}\, u_{i-1}^n + (1 - 2\lambda_i)\, u_i^n + \lambda_{i+1/2}\, u_{i+1}^n. \tag{13.88}$$

It is physically meaningful and it also enhances the numerical stability to impose a *global condition*, expressed in terms of the solution's integral over the entire spatial domain:

$$Q(t) = \int_0^L u(x, t)dx. \tag{13.89}$$

Such a condition results naturally by integrating the continuity equation,

$$\frac{\partial}{\partial t}u(x, t) + \frac{\partial}{\partial x}J_{\text{diff}}(x, t) = 0,$$

to give

$$\frac{\partial}{\partial t}Q(t) = J_{\text{diff}}^{(1)} - J_{\text{diff}}^{(2)}, \tag{13.90}$$

where we consider both boundary fluxes, $J_{\text{diff}}^{(1)}$ and $J_{\text{diff}}^{(2)}$, by convention, positive. The discrete form of the above condition may be used to express the expected integral of the propagated solution:

$$Q^{n+1} = Q^n + \left(J_{\text{diff}}^{(1)} - J_{\text{diff}}^{(2)}\right) h_t. \tag{13.91}$$

Assuming that (by using, e.g., the trapezoidal rule) the integral actually evaluates to the approximate value

$$\tilde{Q}^{n+1} \simeq h_x \left(\frac{\tilde{u}_1^{n+1}}{2} + \sum_{i=2}^{N_x-1} \tilde{u}_i^{n+1} + \frac{\tilde{u}_{N_x}^{n+1}}{2}\right), \tag{13.92}$$

the solution can be normalized properly with the ratio of the expected to the actual integral:

$$u_i^{n+1} = \frac{Q^{n+1}}{\tilde{Q}^{n+1}} \tilde{u}_i^{n+1}. \tag{13.93}$$

The algorithm based on Equations 13.82 through 13.93 is implemented in Listing 13.10 in the routine `PropagDiff`, which receives the diffusion coefficients at all spatial mesh points by way of array `D[]`. The type of boundary conditions (Dirichlet or Neumann) may be selected by setting the control parameters `iopBC1` and `iopBC2` to 0 or 1. In the case of Neumann conditions, boundary fluxes also have to be provided by way of the arguments `Jdiff1` and `Jdiff2`. The tridiagonal difference system is solved, as before, using the special LU factorization method for tridiagonal systems, implemented as routine `TriDiagSys` in the libraries `linsys.py` and `linsys.h`.

The use of the routine `PropagDiff` is exemplified by Program 13.11, which solves the diffusion equation:

$$\frac{\partial u}{\partial t} = \frac{\partial}{\partial x}\left(D\frac{\partial u}{\partial x}\right), \quad x \in [0, L], \quad t > 0, \tag{13.94}$$

with the diffusion coefficient varying linearly between 0.01 for $x = 0$ and 0.1 for $x = L$:

$$D(x) = 0.09(x/L) + 0.01. \tag{13.95}$$

Listing 13.10   Crank–Nicolson Propagator for Variable Diffusion Coefficient (Python Coding)

```
#===========================================================================
def PropagDiff(u0, u, D, nx, hx, ht, iopBC1, iopBC2, Jdiff1, Jdiff2):
#---------------------------------------------------------------------------
#  Propagates the solution u0[] of the diffusion equation
#     du/dt = D d2u/dx2, D - diffusion coefficient (spatially variable)
#  over the time intervat ht, using the Crank-Nicolson difference scheme on
#  a spatial grid with nx nodes and spacing hx. Returns the solution in u[].
#  iopBC1, iopBC2 - left/right boundary condition: 0 - Dirichlet, 1 - Neumann
#  Jdiff1, Jdiff2 - left/right boundary fluxes for Neumann conditions
#  ---------------------------------------------------------------------------
   a = [0]*(nx+1); b = [0]*(nx+1); c = [0]*(nx+1)

   f = 0.5e0 * ht/(hx*hx)

   b[1] = 1e0; c[1] = -iopBC1        # build coefficients of discretized system
   u[1] = hx*Jdiff1/D[1] if iopBC1 else u0[1]
   for i in range(2,nx):
      lam = D[i] * f
      lam1 = 0.5e0 * (D[i] + D[i-1]) * f
      lam2 = 0.5e0 * (D[i] + D[i+1]) * f
      a[i] = -lam1
      b[i] = 1e0 + 2e0*lam
      c[i] = -lam2
      u[i] = lam1*u0[i-1] + (1e0 - 2e0*lam)*u0[i] + lam2*u0[i+1]

   a[nx] = -iopBC2; b[nx] = 1e0
   u[nx] = -hx*Jdiff2/D[nx] if iopBC2 else u0[nx]

   TriDiagSys(a,b,c,u,nx)                      # solve tridiagonal discretized system

   for i in range(1,nx+1):                                  # keep solution >= 0
      if (u[i] < 0e0): u[i] = 0e0

   uint0 = 0.5e0*(u0[1] + u0[nx])                     # integral of old solution
   for i in range(2,nx): uint0 += u0[i]                       # trapezoidal rule
   uint0 *= hx
   if (iopBC1 == 1): uint0 += Jdiff1*ht               # contribution of left flux
   if (iopBC2 == 1): uint0 -= Jdiff2*ht              # contribution of right flux

   uint = 0.5e0*(u[1] + u[nx])                        # integral of new solution
   for i in range(2,nx): uint += u[i]                        # trapezoidal rule
   uint *= hx

   f = uint0/uint                                       # normalization factor
   if (f < 0e0): f = 0e0
   for i in range(1,nx+1): u[i] *= f                        # normalize solution
```

The initial solution is chosen to be step-like, with the entire diffusing substance concentrated in the first half of the domain:

$$u(x,0) = \begin{cases} 1, & 0 \le x < L/2, \\ 0, & L/2 \le x \le L. \end{cases} \tag{13.96}$$

The applied zero-flux Neumann boundary conditions,

$$J_{\text{diff}}^{(1)} = -\left[D\frac{\partial u}{\partial x} \right]_{x=0} = 0, \quad J_{\text{diff}}^{(2)} = -\left[D\frac{\partial u}{\partial x} \right]_{x=L} = 0, \tag{13.97}$$

Listing 13.11 Diffusion with Spatially Variable Diffusion Coefficient (Python Coding)

```python
# Solves the diffusion equation for spatially variable diffusion coefficient
from math import *
from pde import *
from graphlib import *

def Init(u, x, D, nx):                                  # initialization
    global L                                      # extent of spatial domain
    for i in range(1,nx+1):
        D[i] = 0.09e0*(x[i]/L) + 0.01e0               # diffusion coefficient
        u[i] = 1e0 if x[i] < 0.5e0*L else 0e0        # initial steplike solution

# main

L       = 1e0                                    # [0,L] spatial domain
iopBC1 = 1;  iopBC2 = 1                  # left/right boundary condition type
Jdiff1 = 0e0; Jdiff2 = 0e0                    # left/right boundary fluxes
nx      = 51                             # number of spatial mesh points
tmax    = 2e0                                    # propagation time
ht      = 1e-3                                      # time step
nout    = 100                              # output every nout steps

nt = (int)(tmax/ht + 0.5)                        # number of time steps
u0 = [0]*(nx+1); u = [0]*(nx+1)                       # solution
x = [0]*(nx+1)                                     # spatial mesh
D = [0]*(nx+1)                              # difusion coefficient

hx = L/(nx-1)
for i in range(1,nx+1): x[i] = (i-1)*hx                   # spatial mesh

Init(u0,x,D,nx)              # initialization: diffusion coefficient, solution

GraphInit(1000,700)
out = open("diffusion.txt","w")
out.write("       x    ")
for i in range (1,nx+1): out.write("{0:10.5f}".format(x[i]))   # print x-mesh
out.write("\n")
out.write("        t            u\n")

for it in range(1,nt+1):                                   # time loop
    t = it*ht                                       # propagate solution
    PropagDiff(u0,u,D,nx,hx,ht,iopBC1,iopBC2,Jdiff1,Jdiff2)
    for i in range(1,nx+1): u0[i] = u[i]                 # shift solutions

    if (it % nout == 0 or it == nt):                  # output every nout steps
        out.write("{0:10.5f}".format(t))
        for i in range (1,nx+1): out.write("{0:10.5f}".format(u[i]))
        out.write("\n")

        GraphClear()
        title = "Diffusion  t = {0:4.2f}".format(t)
        Plot(x,u,nx,"blue",1,0.15,0.95,0.50,0.90,"None","u",title)
        Plot(x,D,nx,"red",1,0.15,0.95,0.08,0.48,"x","D","")
        GraphUpdate()
out.close

MainLoop()
```

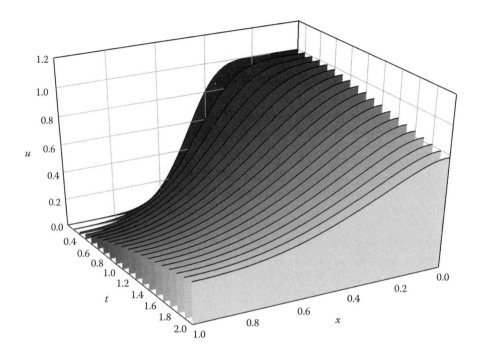

FIGURE 13.11   Time-resolved concentration profiles for the diffusion problem 13.94–13.97 with spatially variable diffusion coefficient.

prevent any substance leakage outside the domain throughout the propagation and cause the equilibrium solution to be a constant profile in the limit of infinite time:

$$u(x, t \to \infty) = 0.5.$$

Concretely, Program 13.11 uses $L = 1$ and $N_x = 51$ (resulting in $h_x = 0.02$), and $h_t = 10^{-3}$. It propagates the solution up to $t_{max} = 2$ and produces printed and graphical output every $n_{out} = 100$ time steps. Figure 13.11 shows concentration profiles with a time increment of 0.1. The solution can be seen to evolve from the initial step-like profile toward the expected uniform distribution. The equilibrium solution is indeed recovered to a fair accuracy if the program is run for times $t_{max} > 20$.

13.4 Time-Dependent Schrödinger Equation

The space and time evolution of a nonrelativistic quantum system is described by the Schrödinger equation:

$$\iota \hbar \frac{\partial \Psi}{\partial t} = \hat{H}\Psi, \tag{13.98}$$

where $\hat{H}$ is the *Hamiltonian*, $\hbar = h/2\pi$—the reduced Planck constant, and $\iota = \sqrt{-1}$—the imaginary unit. The *wave function* Ψ, as solution of the equation, is in general a complex function of position and time. In particular, for a 1D system, $\Psi = \Psi(x, t)$. From a formal perspective, the quantum wave equation is equivalent with two parabolic PDEs, coupling the real and imaginary parts of the wave function.

A central role in the quantum formalism is played by the *probability density*, $|\Psi(x, t)|^2 = \Psi^*(x, t)\Psi(x, t)$, whereby $|\Psi(x, t)|^2\, dx$ represents the probability of the system being localized in the interval $(x, x + dx)$ at time t. When solving time-dependent problems, the wave function is typically modeled as a *wave packet*, featuring both localization and wave-like behavior. The wave packet's initial shape, $\Psi(x, t = 0)$, must be specified and its localization requires that it vanishes asymptotically, that is, $\Psi \to 0$ when $x \to \pm\infty$. The norm of a *normalized wave packet* is by definition

$$\int_{-\infty}^{\infty} |\Psi|^2 dx = 1,$$

as it represents the total probability of the system being localized somewhere in space. Anyway, more than having a specific finite value, the norm must remain *constant in time*.

The solution of Schrödinger's equation (13.98) for a 1D system can be formally expressed as

$$\Psi(x, t) = e^{-\imath(t-t_0)\hat{H}}\Psi(x, t_0),$$ (13.99)

where $e^{-\imath(t-t_0)\hat{H}}$ is the so-called *evolution operator*. The evolution operator is said to be *unitary*, since, multiplied by its adjoint, $[e^{-\imath(t-t_0)\hat{H}}]^* = e^{\imath(t-t_0)\hat{H}}$, it yields the identity operator. In fact, unitarity is an essential feature of the evolution operator, as it guarantees the conservation of the wave packet's norm in time.

Considering two consecutive time steps, t_n and $t_{n+1} = t_n + h_t$, the wave packet evolves according to

$$\Psi^{n+1} = e^{-\imath h_t \hat{H}}\Psi^n.$$ (13.100)

The concrete action of the exponential operator is defined by the corresponding power series in $\hat{H}$, wherefrom, retaining the *first-order approximation*, we get the propagation formula:

$$\Psi^{n+1} = \left(1 - \imath h_t \hat{H}\right)\Psi^n.$$ (13.101)

With the Hamiltonian given by a finite-difference representation, this is an *explicit propagation scheme*, allowing for the solution Ψ^{n+1} to be determined directly in terms of the previous solution Ψ^n. Unfortunately, this method proves to be only *conditionally stable* and, thus, prone to rapid divergence. At the level of the quantum formalism, the possible instability may be linked to the fact that the linearized propagator $(1 - \imath h_t \hat{H})$ is *not unitary* for finite h_t, and therefore does not conserve the wave packet's norm. The nonunitarity results immediately by taking the norm on both sides of Equation 13.101.

The conditional stability of the explicit scheme can be cured by applying the inverse evolution operator, $e^{\imath(t-t_0)\hat{H}}$, to both sides of Equation 13.100 and retaining again the *first-order approximation*:

$$\left(1 + \imath h_t \hat{H}\right)\Psi^{n+1} = \Psi^n.$$ (13.102)

Anyway, besides being *implicit* and, therefore, entailing additional calculations for expressing the propagated solution, this scheme is likewise *nonunitary*.

The adequate way to discretize the evolution operator in Equation 13.100, leading to a both *unitary* and *second-order accurate* approximation, is based on expressing the wave function $\Psi^{n+1/2}$ at the median time step $t_{n+1/2} = t_n + h_t/2$, both by backward propagation from Ψ^{n+1} and by forward propagation from Ψ^n:

$$\Psi^{n+1/2} \equiv e^{\imath(h_t/2)\hat{H}}\Psi^{n+1} = e^{-\imath(h_t/2)\hat{H}}\Psi^n.$$ (13.103)

Approximating the two exponential operators by $O(h_t^3)$ power series results in

$$\left[1 + \imath \frac{h_t}{2}\hat{H} - \frac{1}{2}\left(\frac{h_t}{2}\hat{H}\right)^2\right]\Psi^{n+1} = \left[1 - \imath \frac{h_t}{2}\hat{H} - \frac{1}{2}\left(\frac{h_t}{2}\hat{H}\right)^2\right]\Psi^n + O(h_t^3).$$

Furthermore, relating Ψ^{n+1} and Ψ^n shows that the parabolic terms, $h_t^2\hat{H}^2\Psi^{n+1}$ and $h_t^2\hat{H}^2\Psi^n$, actually cancel out up to the h_t^2-terms, themselves. Hence, we finally obtain the desired *second-order accurate* propagation formula:

$$\left(1 + \imath \frac{h_t}{2}\hat{H}\right)\Psi^{n+1} = \left(1 - \imath \frac{h_t}{2}\hat{H}\right)\Psi^n, \tag{13.104}$$

or, equivalently,

$$\Psi^{n+1} = \left(1 + \imath \frac{h_t}{2}\hat{H}\right)^{-1}\left(1 - \imath \frac{h_t}{2}\hat{H}\right)\Psi^n. \tag{13.105}$$

The second variant introduces the so-called *Cayley form* of the evolution operator,

$$e^{-\imath h_t \hat{H}} \cong \left(1 + \imath \frac{h_t}{2}\hat{H}\right)^{-1}\left(1 - \imath \frac{h_t}{2}\hat{H}\right), \tag{13.106}$$

which is accurate to order h_t^2 according to the above argumentation.

Building the norms on both sides of Equation 13.105 and making use of the Hermitian character of the operators proves that the norm of the wave packet is conserved, that is, $\int \left|\Psi^{n+1}\right|^2 dx = \int \left|\Psi^n\right|^2 dx$, and that the Cayley form of the evolution operator (13.106) is, indeed, unitary.

Both forms of the propagation relation (Equations 13.104 and 13.105) are used in practice. Anyway, the suitability of one or the other largely depends on the way in which the wave packet is represented spatially. For example, if based on a spatial grid representation of the wave packet, Equation 13.104 is typically more adequate, since, due to the simple and regular form of the resulting coefficients, it allows for the storage of large matrices to be avoided. On the other hand, if the wave packet is more appropriately represented by a linear combination of basis functions (as in *ab initio* molecular calculations), the propagation formula 13.105, employing a matrix representation of the Cayley operator, is more likely to be useful.

Let us now consider the 1D motion of a quantum particle of mass m in the stationary potential $V(x)$. The Hamiltonian of the system is

$$\hat{H} = -\frac{\hbar^2}{2m}\frac{\partial^2}{\partial x^2} + V(x), \tag{13.107}$$

and, taking for simplicity $\hbar = 1$ and $m = 1$, the corresponding Schrödinger equation takes the form:

$$\imath \frac{\partial \Psi}{\partial t} = \left[-\frac{1}{2}\frac{\partial^2}{\partial x^2} + V(x)\right]\Psi. \tag{13.108}$$

We approximate all the functions on a regular spatial mesh defined by the nodes:

$$x_j = (j-1)h_x, \quad j = 1, 2, \ldots, N_x. \tag{13.109}$$

The use of the central-difference representation of the Hamiltonian (13.107) in Equation 13.104, with $V_j = V(x_j)$, gives rise to the following propagation formula:

$$\Psi_j^{n+1} + \imath \frac{h_t}{2}\left[-\frac{\Psi_{j+1}^{n+1} - 2\Psi_j^{n+1} + \Psi_{j-1}^{n+1}}{4h_x^2} + V_j\Psi_j^{n+1}\right]$$

$$= \Psi^n - \imath \frac{h_t}{2}\left[-\frac{\Psi_{j+1}^n - 2\Psi_j^n + \Psi_{j-1}^n}{4h_x^2} + V_j\Psi_j^n\right]. \tag{13.110}$$

Remarkably, this is precisely the same finite-difference representation as the one resulting by directly applying the general Crank–Nicholson method (see Section 13.3.4) to the Schrödinger equation (13.108):

$$\imath\frac{\Psi_j^{n+1} - \Psi_j^n}{h_t} = -\frac{\Psi_{j+1}^{n+1} - 2\Psi_j^{n+1} + \Psi_{j-1}^{n+1}}{4h_x^2} + \frac{1}{2}V_j\Psi_j^{n+1}$$

$$-\frac{\Psi_{j+1}^n - 2\Psi_j^n + \Psi_{j-1}^n}{4h_x^2} + \frac{1}{2}V_j\Psi_j^n. \tag{13.111}$$

We conclude that, when applied to the Schrödinger equation, the Crank–Nicolson method and the Cayley approach are equivalent.

We introduce the notation

$$\lambda = \frac{h_t}{4h_x^2}, \tag{13.112}$$

and separate in either of Equations 13.110 or 13.111 the terms corresponding to the time steps t_n and t_{n+1}:

$$-\lambda\Psi_{j+1}^{n+1} + \left(2\lambda + \frac{1}{2}h_tV_j - \imath\right)\Psi_j^{n+1} - \lambda\Psi_{j-1}^{n+1}$$

$$= \lambda\Psi_{j+1}^n - \left(2\lambda + \frac{1}{2}h_tV_j + \imath\right)\Psi_j^n + \lambda\Psi_{j-1}^n.$$

Finally, by still defining the quantities related to the potential values

$$W_j = 2\lambda + \frac{1}{2}h_tV_j, \tag{13.113}$$

the discretized system corresponding to the Schrödinger equation (13.108) acquires the simple *tridiagonal* form:

$$-\lambda\Psi_{j-1}^{n+1} + \left(W_j - \imath\right)\Psi_j^{n+1} - \lambda\Psi_{j+1}^{n+1} = \lambda\Psi_{j-1}^n - \left(W_j + \imath\right)\Psi_j^n + \lambda\Psi_{j+1}^n, \tag{13.114}$$

$$j = 2,\ldots,N_x - 1.$$

This system is completed by imposing the asymptotic vanishing of the wave packet, which, for an extended-enough spatial mesh, translates into the boundary conditions:

$$\Psi_1^{n+1} = \Psi_{N_x}^{n+1} = 0. \tag{13.115}$$

There are several methods at hand by which the difference system (13.114)–(13.115) can be solved to advance the state of the wave packet from t_n to t_{n+1}. Given the tridiagonal structure of the system, the most straightforward appears to be the use of the *tridiagonal solver* based on LU factorization described in Section 7.7. Correspondingly, the complex elements of the three nonzero diagonals, $\mathbf{a} = [a_i]$, $\mathbf{b} = [b_i]$, and $\mathbf{c} = [c_i]$, and the constant terms $\mathbf{d} = [d_i]$ are

$$b_1 = 1, \quad c_1 = 0, \quad d_1 = 0,$$

$$\begin{cases} a_j = -\lambda, \quad j = 2, 3, \ldots, N_x - 1, \\ b_j = W_j - \iota, \\ c_j = -\lambda, \\ d_j = \lambda \Psi_{j-1}^n - \left(W_j + \iota \right) \Psi_j^n + \lambda \Psi_{j+1}^n, \end{cases} \tag{13.116}$$

$$a_{N_x} = 0, \quad b_{N_x} = 1, \quad d_{N_x} = 0.$$

The supplementary complication of dealing with complex elements is circumvented elegantly in modern programming languages that implement function-overloading mechanisms (among which, C++ and Python). *Function overloading* is a feature that allows creating multiple functions with the same name, but with different parameters. In fact, the C++ library linsys.h provides two versions of the routine TriDiagSys—one operating with double variables, and the other, with complex variables. The compiler directs the calls to one or the other according to the type of the actual arguments. Owing to the built-in capability of dynamic typing, in Python there is no need for multiple variants, since the same implementation of a routine can be called with arguments of different types. Indeed, the library linsys.py contains a single version of TriDiagSys, which can be invoked both with real and complex parameters.

Listing 13.12   Propagator for the Schrödinger Equation Based on Tridiagonal Solver (Python Coding)

```
#===========================================================================
def PropagQTD(Psi, V, nx, hx, ht):
#---------------------------------------------------------------------------
#  Propagates the solution PSI = Psi + i Chi of the 1D Schrodinger equation
#    i d/dt PSI(x,t) = [-(1/2) d2/dx2 - V(x)] PSI(x,t),
#  over the time interval ht. Uses the Crank-Nicolson scheme on a grid with
#  nx nodes and spacing hx and solves the tridiagonal discretized system by
#  LU factorization. Uses complex arithmetic.
#---------------------------------------------------------------------------
   a = [complex(0,0)]*(nx+1)                  # diagonals of discretized system
   b = [complex(0,0)]*(nx+1)
   c = [complex(0,0)]*(nx+1)

   lam = ht/(4e0*hx*hx)

   b[1] = 1e0; c[1] = 0e0               # build coefficients of discretized system
   Psii = Psi[1]
   for i in range (2,nx):
      Psi1 = Psii; Psii = Psi[i]               # save initial wave packet values
      W = 2e0*lam + 0.5e0*ht*V[i]
      a[i] = -lam
      b[i] = complex(W,-1e0)
      c[i] = -lam
      Psi[i] = lam*Psi1 - complex(W,1e0)*Psii + lam*Psi[i+1]   # constant term

   a[nx] = 0e0; b[nx] = 1e0

                                          # solve tridiagonal discretized system
   TriDiagSys(a,b,c,Psi,nx)              # solution Psi: propagated wave packet
```

The elementary propagator built on the basis of Equations 13.112 through 13.116 is implemented as routine `PropagQTD` in Listing 13.12. The routine receives the tabular values of the initial wave packet in `Psi[]` and those of the potential in `V[]`. `PropagQTD` operates on a spatial grid with `nx` nodes, spaced regularly by `hx`, and propagates the solution over the time interval `ht`. The Python version of `PropagQTD` uses the built-in type `complex`. The C/C++ version of the routine (see Listing 13.18) employs the double–complex type `dcmplx` defined in the library `memalloc.h`, which also includes the dynamic allocation routine for complex vectors, `CVector`. The routine allocates the arrays `a[]`, `b[]`, and `c[]` for the three relevant diagonals of the discretized system, and temporarily uses the array `Psi[]` for storing the constant terms. After composing their elements, `PropagQTD` passes these arrays to the routine `TriDiagSys`, which, after solving the tridiagonal system, returns the propagated wave packet through `Psi[]`.

An alternative strategy for solving the difference system (13.114)–(13.115) is based on explicitly treating the real and imaginary parts of the wave function:

$$\Psi^n = \psi^n + \iota \chi^n. \tag{13.117}$$

We begin by inserting this expression into the difference equation (13.114), carry through some algebra, and separate the real and imaginary parts to obtain

$$\chi_j^{n+1} - \lambda \psi_{j-1}^{n+1} + W_j \psi_j^{n+1} - \lambda \psi_{j+1}^{n+1} = \chi_j^n + \lambda \psi_{j-1}^n - W_j \psi_j^n + \lambda \psi_{j+1}^n,$$

$$\psi_j^{n+1} + \lambda \chi_{j-1}^{n+1} - W_j \chi_j^{n+1} + \lambda \chi_{j+1}^{n+1} = \psi_j^n - \lambda \chi_{j-1}^n + W_j \chi_j^n - \lambda \chi_{j+1}^n.$$

Collecting in each of these equations the terms from the time step t_n as distinct quantities,

$$\psi_{j,0}^{n+1} = \psi_j^n - \lambda \left(\chi_{j-1}^n + \chi_{j-1}^n \right) + W_j \chi_j^n, \quad j = 2, 3, \dots, N_x - 1,$$

$$\chi_{j,0}^{n+1} = \chi_j^n + \lambda \left(\psi_{j-1}^n + \psi_{j-1}^n \right) - W_j \psi_j^n, \tag{13.118}$$

we actually obtain explicit zeroth-order approximations of the components of the propagated solution. Employing $\psi_{j,0}^{n+1}$ and $\chi_{j,0}^{n+1}$ as constant terms, the components ψ_j^{n+1} and χ_j^{n+1} of the propagated wave packet are expressed as

$$\psi_j^{n+1} = \psi_{j,0}^{n+1} - \lambda \left(\chi_{j-1}^{n+1} + \chi_{j+1}^{n+1} \right) + W_j \chi_j^{n+1}, \quad j = 2, 3, \dots, N_x - 1,$$

$$\chi_j^{n+1} = \chi_{j,0}^{n+1} + \lambda \left(\psi_{j-1}^{n+1} + \psi_{j+1}^{n+1} \right) - W_j \psi_j^{n+1}. \tag{13.119}$$

Given the implicit dependence of the entire set, these components may be determined numerically by Gauss–Seidel iterations, with the updated components becoming recursively right-hand-side terms until self-consistency is achieved.

The main advantage of this approach based on the Gauss–Seidel method is that it avoids complex arithmetic and significantly reduces the storage requirements, since the elements of the tridiagonal difference system do not have to be stored simultaneously for being passed to a solver routine. In addition, the overall coding is also simplified and is suitable for multidimensional generalization. Nevertheless, for the same input (essentially, the same time step h_t), this method proves slower than the previous one due to the involved Gauss–Seidel iterations.

The one-to-one implementation of the above algorithm, based on real arithmetic and the Gauss–Seidel solution of the discretized system, is the routine `PropagQGS` (Listing 13.13). Upon each call, the routine starts by building the explicit zeroth-order components `Psi0[]` and `Chi0[]` ($\psi_{j,0}^{n+1}$ and $\chi_{j,0}^{n+1}$) of the propagated solution according to Equation 13.118. These then serve as constant terms within the

Listing 13.13 Propagator for the Schrödinger Equation Using the Gauss–Seidel Method (Python Coding)

```
#===========================================================================
def PropagQGS(Psi, Chi, V, nx, hx, ht):
#---------------------------------------------------------------------------
#  Propagates the solution PSI = Psi + i Chi of the 1D Schrodinger equation
#     i d/dt PSI(x,t) = [-(1/2) d2/dx2 - V(x)] PSI(x,t),
#  over the time interval ht. Uses the Crank-Nicolson scheme on a grid with
#  nx nodes and spacing hx and solves the discretized system by Gauss-Seidel
#  iterations. Uses real arithmetic.
#---------------------------------------------------------------------------
   eps = 1e-6                                          # relative tolerance
   itmax = 100                                         # max no. of iterations

   Psi0 = [0]*(nx+1); Chi0 = [0]*(nx+1); W = [0]*(nx+1)

   lam = ht/(4e0*hx*hx)

   for i in range(2,nx):                       # explicit 0th-order approximations
      W[i] = 0.5e0*ht*V[i] + 2e0*lam
      Psi0[i] = Psi[i] - lam*(Chi[i-1] + Chi[i+1]) + W[i]*Chi[i]
      Chi0[i] = Chi[i] + lam*(Psi[i-1] + Psi[i+1]) - W[i]*Psi[i]

   for it in range(1,itmax+1):                        # Gauss-Seidel iteration loop
      err = 0e0
      for i in range (2,nx):
         Psii = Psi0[i] - lam*(Chi[i-1] + Chi[i+1]) + W[i]*Chi[i]
         Chii = Chi0[i] + lam*(Psi[i-1] + Psi[i+1]) - W[i]*Psii
                            # local error estimate based on probability density
         erri = fabs((Psii*Psii+Chii*Chii) - (Psi[i]*Psi[i]+Chi[i]*Chi[i]))
         if (erri > err): err = erri                  # maximum error estimate
         Psi[i] = Psii
         Chi[i] = Chii

      if (err <= eps): break                            # convergence test

   if (it > itmax):
      print("PropagQGS: max. number of iterations exceeded !"); return 1
   return 0
```

Gauss–Seidel solution of the discretized system (13.119) and the process is continued until the maximum local difference between the density probabilities at two consecutive iterations drops below the predefined tolerance eps.

In the following applications, we choose to represent the initial state of the quantum particle as a normalized *Gaussian wave packet*:

$$\Psi(x,0) = \frac{1}{\sqrt{(2\pi)^{1/2}\sigma}} e^{-(x-x_0)^2/(4\sigma^2)} e^{ik_0 x}. \tag{13.120}$$

x_0 represents the position of the wave packet's center, σ is the packet's half-width, and k_0 is the associated average wave number. The first exponential defines the Gaussian envelope, while the second exponential is responsible for the motion of the packet, in accordance with the sign of k_0, in the positive or negative x-direction. Gaussian wave packets are very much in use due to their special properties, among which of primary importance is that of being the wave packets of *minimum uncertainty*. In fact, the product of their position and momentum uncertainties reaches the lowest value prescribed by Heisenberg's uncertainty principle, namely, $\Delta x \Delta p = \hbar/2$.

As a first example, we consider the scattering (reflexion and transmission) of the wave packet (13.120) on a square potential barrier of width a and height V_0, extending symmetrically about the origin:

$$V(x) = \begin{cases} V_0, & |x| \leq a/2, \\ 0, & |x| > a/2. \end{cases} \qquad (13.121)$$

The numerical strategy relying on the tridiagonal solver and complex arithmetic, as implemented in the routine `PropagQTD` (assumed to be included in the module `pde.py`), is implemented in Program 13.14.

Listing 13.14   Scattering of a Quantum Wave Packet Using Tridiagonal Solver (Python Coding)

```
# Reflexion/transmission of quantum wave packet using tridiagonal solver
from math import *
from cmath import *
from pde import *

def Pot(x, a, V0):                                        # Potential
   return V0 if fabs(x) <= 0.5e0*a else 0e0


#===============================================================================
def Init(Psi, x, nx, x0, sig, k0):
#-------------------------------------------------------------------------------
#   Initial Gaussian wave packet
#       Psi(x,0) = 1/sqrt(sqrt(2*pi)*sig) * exp[-(x-x0)^2/(4*sig^2)] * exp(ikx)
#   x0 - position of center, sig - half-width, k0 - average wave number
#-------------------------------------------------------------------------------
   a = 1e0/sqrt(sqrt(2e0*pi)*sig)
   b =-1e0/(4e0*sig*sig)
   for i in range(1,nx+1):
      dx = x[i] - x0
      f = a * exp(b*dx*dx)
      if (f < 1e-10): f = 0e0
      Psi[i] = f * complex(cos(k0*x[i]),sin(k0*x[i]))


#===============================================================================
def ProbDens(Psi, Psi2, nx, hx):
#-------------------------------------------------------------------------------
#   Calculates the probability density Psi2[] of the wave function Psi[]
#-------------------------------------------------------------------------------
   for i in range(1,nx+1):
      Psi2[i] = abs(Psi[i])*abs(Psi[i])     # unnormalized probability density
      if (Psi2[i] <= 1e-10): Psi2[i] = 0e0

   PsiNorm = 0.5e0*(Psi2[1] + Psi2[nx])          # integral by trapezoidal rule
   for i in range(2,nx): PsiNorm += Psi2[i]
   PsiNorm *= hx

   for i in range(1,nx+1): Psi2[i] /= PsiNorm       # normalized prob. density
   return PsiNorm

# main

a    = 5e0                                       # width of potential barrier
V0   = 50e0                                      # height of potential barrier
x0   = -20e0                                  # initial position of wave packet
```

```
sig  = 1e0                                              # half-width of packet
k0   = 10e0                                       # average wave number of packet
xmax = 100e0                                                         # maximum x
hx   = 5e-2                                                 # spatial step size
tmax = 5e0                                          # maximum propagation time
ht   = 5e-3                                                         # time step
nout = 40                                          # output every nout steps

nx  = 2*(int)(xmax/hx + 0.5) + 1                  # odd number of spatial nodes
nt  = (int)(tmax/ht + 0.5)                                # number of time steps
nx2 = int(nx/2)

Psi  = [complex(0,0)]*(nx+1)                               # wave function
Psi2 = [0]*(nx+1)                                        # probability density
V = [0]*(nx+1)                                                    # potential
x = [0]*(nx+1)                                               # spatial mesh

for i in range(1,nx+1):                  # tabulate spatial mesh and potential
   x[i] = (i-nx2-1)*hx
   V[i] = Pot(x[i],a,V0)

Init(Psi,x,nx,x0,sig,k0)                                # initial wave packet

for it in range(1,nt+1):                                          # time loop
   t = it*ht
   PropagQTD(Psi,V,nx,hx,ht)           # propagate solution by tridiagonal solver

   PsiNorm = ProbDens(Psi,Psi2,nx,hx)                     # probability density

   if (it % nout == 0 or it == nt):                   # output every nout steps
      fname = "scatter_{0:4.2f}.txt".format(t)
      out = open(fname,"w")
      out.write("t = {0:4.2f}\n".format(t))
      out.write("      x          V         PsiR      PsiI      Psi2\n")
      for i in range(1,nx+1):
         out.write("{0:10.5f}{1:10.5f}{2:10.5f}{3:10.5f}{4:10.5f}\n".\
                format(x[i],V[i],Psi[i].real,Psi[i].imag,Psi2[i]))
      out.close
```

Program 13.14 performs the scattering calculations assuming units resulting by setting $\hbar = 1$ and $m = 1$. The potential barrier is characterized by $a = 5$ and $V_0 = 50$, while the initial wave packet is centered at $x_0 = -20$, has the half-width $\sigma = 1$, and the average wave vector $k_0 = 10$. The latter value is chosen in such a way that the total energy equals the height of the potential barrier ($\hbar^2 k^2 / 2m = V_0$). The spatial grid extends in the range $[-x_{max}, x_{max}]$, with $x_{max} = 100$ and nodes spaced regularly at $h_x = 5 \cdot 10^{-2}$. The propagation is carried out up to $t_{max} = 5$, with a time step $h_t = 5 \cdot 10^{-3}$.

The function Init initializes the wave packet and the routine ProbDens produces normalized probability densities from the wave packet Psi[] propagated by the routine PropagQTD. Output, including the real and imaginary components of the wave packet, as well as the corresponding probability density, is produced every $n_{out} = 40$ time steps to files with the current time marked in their names.

Snapshots of the probability density profiles are depicted in Figure 13.12 and emphasize the splitting of the wave packet in reflected and transmitted waves at both barrier walls.

A similar program, however, using the routine PropagQGS as propagator (based on the Gauss–Seidel method), is presented in Listing 13.15. The solved problem concerns the scattering of the wave packet from a square potential well with the bottom $V_0 = -50$. Apart from the larger spatial step size, $h_x = 10^{-1}$, the rest of the numerical parameters are the same as for Program 13.14.

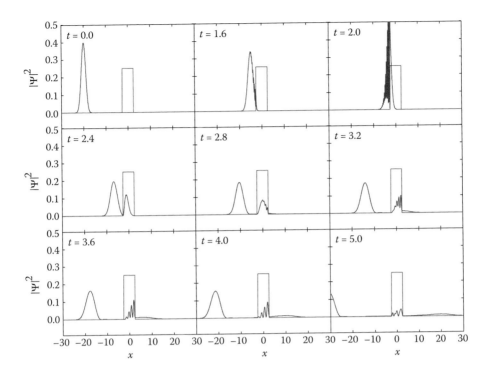

FIGURE 13.12 Scattering of a Gaussian wave packet on a square potential barrier, calculated with Program 13.14, using the tridiagonal technique for complex systems implemented in routine `PropagQTD`. The average energy is equal to the barrier height ($k_0^2/2 = V_0 = 50$).

For reasons of consistency with the overall philosophy of using real arithmetic, Program 13.15 defines a distinct routine `InitR` for initializing the real and imaginary components of the wave packet, and a routine `ProbDensR` for evaluating the normalized probability density. The rest of the functionalities are identical with those of Program 13.14. Snapshots from an actual run are depicted in Figure 13.13. One can readily notice that reflected waves are produced even though the reflection probability of a classical particle under the same conditions is zero.

Even though the strategy using Gauss–Seidel iterations and real arithmetic for solving the discretized Schrödinger equation (implemented in routine `PropagQGS`) is somewhat slower than the dedicated LU factorization technique for complex linear systems (implemented in routine `PropagQTD`), the former is more amenable in situations of higher complexity, as, for instance, higher dimensionality or more sophisticated boundary conditions.

Listing 13.15   Scattering of a Quantum Wave Packet Using a Gauss–Seidel Solver (Python Coding)

```
# Reflexion/transmission of quantum wave packet using Gauss-Seidel solver
from math import *
from pde import *

def Pot(x, a, V0):                                              # Potential
   return V0 if fabs(x) <= 0.5e0*a else 0e0

#===========================================================================
def InitR(Psi, Chi, x, nx, x0, sig, k0):
#---------------------------------------------------------------------------
#  Initial Gaussian wave packet
```

```
#       Psi(x,0) = 1/sqrt(sqrt(2*pi)*sig) * exp[-(x-x0)^2/(4*sig^2)] * exp(ikx)
#    x0 - position of center, sig - half-width, k0 - average wave number
#-----------------------------------------------------------------------------
   a = 1e0/sqrt(sqrt(2e0*pi)*sig)
   b =-1e0/(4e0*sig*sig)
   for i in range(1,nx+1):
      dx = x[i] - x0
      f = a * exp(b*dx*dx)
      if (f < 1e-10): f = 0e0
      Psi[i] = f * cos(k0*x[i]); Chi[i] = f * sin(k0*x[i])


#=============================================================================
def ProbDensR(Psi, Chi, Psi2, nx, hx):
#-----------------------------------------------------------------------------
#  Calculates the probability density Psi2[] of the wave function Psi[]
#-----------------------------------------------------------------------------
   for i in range(1,nx+1):                       # unnormalized probability density
      Psi2[i] = Psi[i]*Psi[i] + Chi[i]*Chi[i]
      if (Psi2[i] <= 1e-10): Psi2[i] = 0e0

   PsiNorm = 0.5e0*(Psi2[1] + Psi2[nx])         # integral by trapezoidal rule
   for i in range(2,nx): PsiNorm += Psi2[i]
   PsiNorm *= hx

   for i in range(1,nx+1): Psi2[i] /= PsiNorm     # normalized prob. density
   return PsiNorm

# main

a    = 5e0                                     # width of potential barrier
V0   = -50e0                                   # height of potential barrier
x0   = -20e0                                   # initial position of wave packet
sig  = 1e0                                      # half-width of packet
k0   = 10e0                                     # average wave number of packet
xmax = 100e0                                    # maximum x
hx   = 1e-1                                      # spatial step size
tmax = 5e0                                       # maximum propagation time
ht   = 5e-3                                      # time step
nout = 40                                        # output every nout steps

nx = 2*(int)(xmax/hx + 0.5) + 1                 # odd number of spatial nodes
nt = (int)(tmax/ht + 0.5)                        # number of time steps
nx2 = int(nx/2)

Psi  = [0]*(nx+1)                               # real part of wave function
Chi  = [0]*(nx+1)                               # imag part of wave function
Psi2 = [0]*(nx+1)                               # probability density
V = [0]*(nx+1)                                   # potential
x = [0]*(nx+1)                                   # spatial mesh

for i in range(1,nx+1):                         # tabulate spatial mesh and potential
   x[i] = (i-nx2-1)*hx
   V[i] = Pot(x[i],a,V0)

InitR(Psi,Chi,x,nx,x0,sig,k0)                   # initial wave packet

for it in range(1,nt+1):                         # time loop
   t = it*ht
   PropagQGS(Psi,Chi,V,nx,hx,ht)                # propagate by Gauss-Seidel solver
```

```
PsiNorm = ProbDensR(Psi,Chi,Psi2,nx,hx)                # probability density

if (it % nout == 0 or it == nt):                       # output every nout steps
    fname = "scatter_{0:4.2f}.txt".format(t)
    out = open(fname,"w")
    out.write("t = {0:4.2f}\n".format(t))
    out.write("         x              V          PsiR       PsiI       Psi2\n")
    for i in range(1,nx+1):
        out.write("{0:10.5f}{1:10.5f}{2:10.5f}{3:10.5f}{4:10.5f}\n".\
                  format(x[i],V[i],Psi[i],Chi[i],Psi2[i]))
    out.close
```

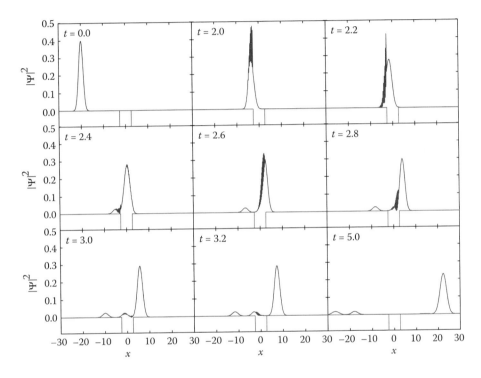

FIGURE 13.13 Scattering of a Gaussian wave packet on a square potential well, calculated with Program 13.15, using the Gauss–Seidel method implemented in routine `PropagQGS`. The average energy is equal to the absolute well depth ($k_0^2/2 = -V_0 = 50$).

13.5 Initial-Value Problems for Hyperbolic Differential Equations

The prototypical hyperbolic PDE encountered in science and engineering is the classical wave equation, having as 1D form:

$$\frac{\partial^2 u}{\partial t^2} = c^2 \frac{\partial^2 u}{\partial x^2}, \tag{13.122}$$

where c is the phase velocity of the wave. Distinctively, when grouped on the same side, the second-order time and space derivatives in the wave equation have opposite signs.

Although the above equation is spatially 1D, it can be actually assumed to model a 3D planar "slab" geometry, with explicit dependence on the Cartesian coordinate transverse to the slab, namely x, and no dependence on the other two coordinates, y and z, along which, the extent of the system may be considered virtually infinite. In such a geometry, a plane wave is a constant-frequency wave, whose wave fronts (constant-phase surfaces) are infinite parallel (y, z)-planes of constant amplitude, normal to the propagation direction, x.

To model a concrete physical situation, the wave equation needs to be complemented with *initial* and *boundary conditions*. The *initial conditions* specify the solution and its first-order time derivative over the whole spatial domain at some initial moment t_0:

$$u(x, t_0) = u^0(x), \quad x \in [-x_{\max}, x_{\max}], \tag{13.123}$$

$$\frac{\partial u}{\partial t}(x, t_0) = v^0(x). \tag{13.124}$$

For simplicity, we consider Dirichlet-type boundary conditions, implying constant solution values on the boundaries:

$$u(x_{\min}, t) = u^0_{x_{\min}}, \quad t > t_0, \tag{13.125}$$

$$u(x_{\max}, t) = u^0_{x_{\max}}. \tag{13.126}$$

In the following, we approximate problem (13.122–13.126) by finite-difference schemes based on regular space and time grids:

$$x_i = x_{\min} + (i - 1)h_x, \quad i = 1, 2, \dots, N_x, \tag{13.127}$$

$$t_n = nh_t, \quad n = 0, 1, 2, \dots, \tag{13.128}$$

where h_t stands for the time step size, N_x is the number of spatial mesh points, and the spatial step size is defined by

$$h_x = (x_{\max} - x_{\min})/(N_x - 1). \tag{13.129}$$

For the second spatial derivative at the space–time node (x_i, t_n), we use the second-order central-difference formula already employed in the preceding sections:

$$\left(\frac{\partial^2 u}{\partial x^2} \right)_{i,n} = \frac{u^n_{i+1} - 2u^n_i + u^n_{i-1}}{h_x^2} + O(h_x^2),$$

where $u^n_i \equiv u(x_i, t_n)$ and only information from time step t_n and from points located symmetrically about the reference point is required. The time derivative is approximated by a similar formula with respect to time:

$$\left(\frac{\partial^2 u}{\partial t^2} \right)_{i,n} = \frac{u^{n+1}_i - 2u^n_i + u^{n-1}_i}{h_t^2} + O(h_t^2).$$

Application of the above difference schemes yields the following $O(h_x^2 + h_t^2)$ approximation of the wave equation (13.122) at the space–time node (x_i, t_n):

$$\frac{u^{n+1}_i - 2u^n_i + u^{n-1}_i}{h_t^2} = c^2 \frac{u^n_{i+1} - 2u^n_i + u^n_{i-1}}{h_x^2}. \tag{13.130}$$

Here from, also making the notation

$$\lambda = \left(\frac{ch_t}{h_x} \right)^2, \tag{13.131}$$

one can express the solution propagated at time step t_{n+1} for all *interior* mesh points x_i in terms of values from the two preceding time steps, t_n and t_{n-1}:

$$u_i^{n+1} = \lambda u_{i-1}^n + 2(1 - \lambda) u_i^n + \lambda u_{i+1}^n - u_i^{n-1}, \tag{13.132}$$

$$i = 2, 3, \ldots, N_x - 1.$$

As for the boundary values, by virtue of the Dirichlet conditions, they remain constant throughout the propagation, and this is expressed simply by

$$u_1^{n+1} = u_1^n, \quad u_{N_x}^{n+1} = u_{N_x}^n. \tag{13.133}$$

In matrix notation, System 13.132–13.133 takes the form:

$$\mathbf{u}^{n+1} = \mathbf{B} \cdot \mathbf{u}^n - \mathbf{u}^{n-1}, \quad n = 0, 1, 2, \ldots, \tag{13.134}$$

where

$$\mathbf{B} = \begin{bmatrix} 1 & 0 & & & 0 \\ \lambda & 1 - 2\lambda & \lambda & & \\ & \ddots & \ddots & \ddots & \\ & & \lambda & 1 - 2\lambda & \lambda \\ 0 & & & 0 & 1 \end{bmatrix}, \quad \mathbf{u}^n = \begin{bmatrix} u_1^n \\ u_2^n \\ \vdots \\ u_{N_x-1}^n \\ u_{N_x}^n \end{bmatrix}, \tag{13.135}$$

and the propagation matrix $\mathbf{B}$ features tridiagonal structure.

Since the components of the propagated solution $\mathbf{u}^{n+1}$ can be expressed independently, the method is said to be *explicit* and it is essentially similar to the explicit FTCS method developed in Section 13.3.1 for the diffusion equation.

In spite of the formal similarity, there is, nevertheless, a nontrivial difference between the explicit methods for parabolic and hyperbolic PDEs, namely, the occurrence in the latter of the second-previous solution, $\mathbf{u}^{n-1}$, alongside with the previous one, $\mathbf{u}^n$. Once the propagation started, this aspect is easily handled by continuously storing the two most recent solutions.

A somewhat sensitive aspect concerns the *initiation* of the propagation. In fact, along with the initial solution, $\mathbf{u}^0$, one needs to provide the first propagated solution, $\mathbf{u}^1$, instead of the first-time derivative, $\partial \mathbf{u}^0 / \partial t$. An effective way to do this, maintaining the overall second-order precision of the method, starts with the Taylor series of the solution components with respect to time:

$$u_i^{n+1} = u_i^n + h_t \left(\frac{\partial u}{\partial t} \right)_{i,n} + \frac{1}{2} h_t^2 \left(\frac{\partial^2 u}{\partial t^2} \right)_{i,n} + O(h_t^3).$$

In particular, for the first propagation step, we have:

$$u_i^1 = u_i^0 + h_t \left(\frac{\partial u}{\partial t} \right)_{i,0} + \frac{1}{2} h_t^2 \left(\frac{\partial^2 u}{\partial t^2} \right)_{i,0} + O(h_t^3).$$

While the first-time derivative is readily available from the initial condition (13.124), the second derivative may be replaced with the second *spatial* derivative using the wave equation (13.122):

$$u_i^1 = u_i^0 + h_t v_i^0 + \frac{1}{2} h_t^2 c^2 \left(\frac{\partial^2 u}{\partial x^2} \right)_{i,0} + O(h_t^3),$$

where $v_i^0 = v^0(x_i)$. For the second spatial derivative, we can now employ the same second-order scheme (13.5) as above:

$$u_i^1 = u_i^0 + h_t v_i^0 + \frac{1}{2} h_t^2 c^2 \frac{u_{i+1}^0 - 2u_i^0 + u_{i-1}^0}{h_x^2} + O(h_t^3 + h_x^2).$$

Identifying the parameter λ, we finally find the expression for the components of the first propagated solution $\mathbf{u}^1$ in terms of the initial solution $\mathbf{u}^0$ and the initial time derivative $\mathbf{v}^0$:

$$u_i^1 = \frac{1}{2} \lambda u_{i-1}^0 + (1 - \lambda) u_i^0 + \frac{1}{2} \lambda u_{i+1}^0 + h_t v_i^0, \tag{13.136}$$

$$i = 2, 3, \ldots, N_x - 1.$$

According to the general methodology of the von Neumann stability analysis, we insert the generic eigenmode $\varepsilon_i^n = \xi(k)^n \exp(\iota k i h_x)$ as a solution into the difference equation (13.132) and find:

$$\xi = \left[\lambda \exp(-\iota k h_x) + 2(1 - \lambda) + \lambda \exp(\iota k h_x) \right] \xi - \xi^{-1}.$$

By combining the exponential terms into $2\lambda \cos(kh_x)$ and using the identity $1 - \cos x = 2 \sin^2(x/2)$, the amplification factor is found to satisfy:

$$\frac{1}{2} \left(\xi + \xi^{-1} \right) = 1 - 2\lambda \sin^2(kh_x/2). \tag{13.137}$$

Since the right-hand side is real, $\xi + \xi^{-1}$ must be real, as well. This requires that $|\xi| = |\xi|^{-1}$ and leads to

$$\frac{1}{2} \left(\xi + \xi^{-1} \right) = |\xi| \cos \varphi,$$

where φ is an arbitrary phase. Minding the stability condition $|\xi| < 1$, the above relation imposes on the right-hand side of Equation 13.137:

$$-1 \leq 1 - 2\lambda \sin^2(kh_x/2) \leq 1,$$

which can be satisfied only provided that

$$0 \leq \lambda \leq 1. \tag{13.138}$$

In view of definition (13.131) of λ, there results the *stability condition* for the propagation of the wave function $u(x, t)$ based on the explicit difference scheme (13.132)–(13.133):

$$h_t \leq \frac{h_x}{c}. \tag{13.139}$$

The routine `PropagWave` (Listing 13.16) employs Algorithm 13.131–13.133 to solve an elementary Cauchy problem for the wave equation 13.122, propagating two initial solutions, u0[] and u1[], over the interval ht. The constant phase velocity c, the spatial step size h_x, and the number of spatial nodes N_x are passed to the routine via arguments c, hx, and, respectively, nx. The propagated solution is returned

Listing 13.16 Explicit Propagator for the Classical Wave Equation (Python Coding)

```
#===========================================================================
def PropagWave(u0, u1, u, nx, c, hx, ht):
#---------------------------------------------------------------------------
#  Propagates the solutions u0[] and u1[] of the wave equation
#      d2u/dt2 = c^2 d2u/dx2,   c - phase velocity (constant)
#  over the time interval ht, using the explicit difference scheme on a
#  spatial grid with nx nodes and spacing hx. Returns the solution in u[].
#---------------------------------------------------------------------------
   lam = c*ht/hx; lam = lam*lam
   lam2 = 2e0*(1e0 - lam)

   u[1] = u0[1]; u[nx] = u0[nx]                 # Dirichlet boundary conditions
   for i in range(2,nx):                        # propagate solution at interior points
      u[i] = lam*u1[i-1] + lam2*u1[i] + lam*u1[i+1] - u0[i]
```

in the array u[] and the driving propagation loop from the main program is supposed to shift backward the content of the arrays u[], u1[], and u0[], aiming to free u[] in preparation for a new propagation step.

A typical main program for the solution of the 1D wave equation is presented in Listing 13.17. The program starts by retrieving the two initial solutions u0[] and u1[] by a call to the user function Init.

Listing 13.17 Finite-Difference Solution of the Classical Wave Equation (Python Coding)

```
# Solve 1D wave equation by the explicit finite-difference method
from math import *
from pde import *
from graphlib import *

#===========================================================================
def Init(u0, u1, x, nx, c, dk, hx, ht):
#---------------------------------------------------------------------------
#  Returns initial solutions u0 and u1, for the first two time steps
#      u0(x,0) = sin(x*dk) / (x*dk)                  initial solution
#      v0(x,0) = 0                                   initial time derivative
#  x - spatial mesh, nx - number of nodes
#  c - phase velocity of wave, dk - wave number interval
#  hx - x-spacing, ht - time step size
#---------------------------------------------------------------------------
   for i in range(1,nx+1):                                      # time step 0
      u0[i] = sin(dk*x[i])/(dk*x[i]) if dk*x[i] else 1e0   # initial solution

   lam = c*ht/hx; lam = lam*lam                                 # time step 1
   lam2 = 2e0*(1e0 - lam)
   u1[1] = u0[1]; u1[nx] = u0[nx]                     # constant boundary values
   for i in range(2,nx):
      v0 = 0e0                                        # initial time derivative
      u1[i] = 0.5e0*(lam*u0[i-1] + lam2*u0[i] + lam*u0[i+1]) - ht*v0

# main

c    = 10e0                                               # phase speed of wave
dk   = 1e0                                             # wave number interval
xmax = 100e0                                                        # maximum x
hx   = 5e-2                                                 # spatial step size
```

```
tmax = 40e0                                              # maximum propagation time
ht   = 5e-3                                                        # time step
nout = 500                                              # output every nout steps

nx = 2*(int)(xmax/hx + 0.5) + 1                     # odd number of spatial nodes
nt = (int)(tmax/ht + 0.5)                              # number of time steps
nx2 = int(nx/2)

u0 = [0]*(nx+1)                                            # initial solution
u1 = [0]*(nx+1)                                     # first propagated solution
u  = [0]*(nx+1)                                                   # solution
x  = [0]*(nx+1)                                              # spatial mesh

for i in range(1,nx+1): x[i] = (i-nx2-1)*hx                 # spatial mesh

Init(u0,u1,x,nx,c,dk,hx,ht)                            # initial wave packet

GraphInit(1200,600)

for it in range(1,nt+1):                                     # time loop
   t = it*ht
   PropagWave(u0,u1,u,nx,c,hx,ht)                     # propagate solution

   for i in range(1,nx): u0[i] = u1[i];  u1[i] = u[i]       # shift solutions

   if (it % nout == 0 or it == nt):                  # output every nout steps
      fname = "wave_{0:4.2f}.txt".format(t)
      out = open(fname,"w")
      out.write("t = {0:4.2f}\n".format(t))
      out.write("      x            u\n")
      for i in range(1,nx+1):
         out.write("{0:10.5f}{1:10.5f}\n".format(x[i],u[i]))
      out.close

      GraphClear()
      title = "Wave propagation  t = {0:4.2f}".format(t)
      Plot(x,u,nx,"blue",1,0.15,0.95,0.15,0.90,"x","u",title)
      GraphUpdate()

MainLoop()
```

It then controls the propagation of the solution within the temporal loop, by repeated calls to the routine PropagWave. Each time step is completed by the mentioned backward shift of the arrays u[], u1[], and u0[].

For illustration, we solve the classical wave equation:

$$\frac{\partial^2 u}{\partial t^2} = c^2 \frac{\partial^2 u}{\partial x^2}, \quad t > 0, \tag{13.140}$$

and follow the evolution of the initial wave packet centered at the origin:

$$u(x,0) = \frac{\sin x\Delta k}{x\Delta k}, \quad -x_{\max} \le x \le x_{\max}, \tag{13.141}$$

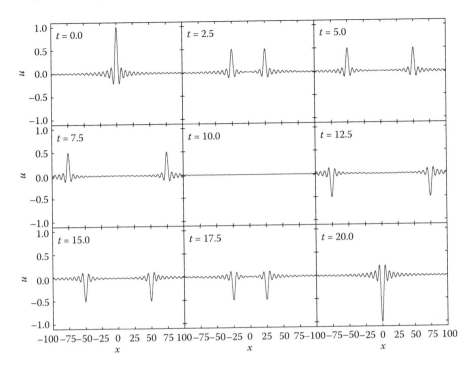

FIGURE 13.14 Numerical solution of problem (13.140–13.142) for the classical wave equation, yielded by Program 13.17.

with zero initial time derivative, $(\partial u/\partial t)(x, 0) = 0$, and subject to constant-value Dirichlet boundary conditions:

$$u(\pm x_{max}, t) = \frac{\sin x_{max} \Delta k}{x_{max} \Delta k}, \quad t > 0. \tag{13.142}$$

Δk denotes the width of the interval of wave numbers contributing to the wave packet, and the larger its value, the smaller the packet's half-width.

Specifically, we employ $c = 10$, $\Delta k = 1$, and $x_{max} = 100$, and propagate the solution with the step sizes $h_x = 5 \cdot 10^{-2}$ and $h_t = 5 \cdot 10^{-3}$ up to $t_{max} = 40$. The chosen parameters correspond to $\lambda = 1$, which guarantees the stable propagation of the solution.

Output is produced every $n_{out} = 500$ time steps to files having the current propagation time marked in their names. Snapshots are shown in Figure 13.14 and illustrate the reflection of the two emerging independent solutions on the domain boundaries, as well as the inverted reconstruction of the wave packet at the origin at $t = 20$.

Continuing the propagation up to $t_{max} = 40$, the original wave packet is reconstructed at the origin with an amplitude reduced to 0.99958 (instead of 1), which provides an estimate of $4 \cdot 10^{-3}$ for the relative error.

13.6 Implementations in C/C++

Listing 13.18 shows the content of the file pde.h, which contains equivalent C/C++ implementations of the Python functions developed in the main text and included in the module pde.py. The corresponding routines have identical names, parameters, and functionalities.

Listing 13.18   Solvers for Boundary-Value and Initial-Value Problems for PDEs (pde.h)

```
//--------------------------------- pde.h ---------------------------------
// Contains routines for solving partial differential equations.
// Part of the numxlib numerics library. Author: Titus Beu, 2013
//-------------------------------------------------------------------------
#ifndef _PDE_
#define _PDE_

#include <math.h>
#include "linsys.h"

//===========================================================================
int Poisson0(double **u, double x[], double y[], int nx, int ny,
             double eps, double Func(double,double))
//---------------------------------------------------------------------------
// Solves the 2D Poisson equation in Cartesian coordinates with Dirichlet
// boundary conditions on a regular grid with (nx x ny) nodes (x[],y[]) using
// the Gauss-Seidel method. The solution u[][] is converged with relative
// precision eps. An error index is returned: 0 - normal execution.
// Calls: Func(x,y) - RHS of Poisson equation
//---------------------------------------------------------------------------
{
   const int itmax = 10000;                         // max no. of iterations
   double **f;
   double eij, err, hx, hy, kx, ky, kxy, uij;
   int i, it, j;

   f = Matrix(1,nx,1,ny);

   hx = (x[nx]-x[1])/(nx-1); kx = 1e0/(hx*hx);             // mesh spacings
   hy = (y[ny]-y[1])/(ny-1); ky = 1e0/(hy*hy);
   kxy = 2e0*(kx + ky);

   for (j=2; j<=ny-1; j++)                    // RHS of PDE for interior points
      for (i=2; i<=nx-1; i++) f[i][j] = Func(x[i],y[j]);

   for (it=1; it<=itmax; it++) {              // Gauss-Seidel iteration loop
      err = 0e0;
      for (j=2; j<=ny-1; j++) {
         for (i=2; i<=nx-1; i++) {                    // interior mesh points
            uij = (kx*(u[i-1][j] + u[i+1][j]) +
                   ky*(u[i][j-1] + u[i][j+1]) - f[i][j]) / kxy;
            eij = uij ? 1e0 - u[i][j]/uij : uij - u[i][j];   // local error
            if (fabs(eij) > err) err = fabs(eij);          // maximum error
            u[i][j] = uij;
         }
      }
      if (err <= eps) break;                             // convergence test
   }

   FreeMatrix(f,1,1);

   if (it > itmax) {
      printf("Poisson0: max. number of iterations exceeded !\n"); return 1;
   }
   return 0;
}

//===========================================================================
```

```
int PoissonXY(double **u, double x[], double y[], int nx, int ny,
              double eps, double Func(double,double),
   void CondX(double, double&, double&, double&, double&, double&, double&),
   void CondY(double, double&, double&, double&, double&, double&, double&))
//---------------------------------------------------------------------------
// Solves the 2D Poisson equation in Cartesian coordinates on a regular grid
// with (nx x ny) nodes (x[],y[]) using the Gauss-Seidel method. The solution
// u[][] is converged with relative precision eps.
// An error index is returned: 0 - normal execution.
// Calls: Func(x,y) - RHS of Poisson equation; boundary conditions:
//        CondX(y,alf_min,bet_min,gam_min,alf_max,bet_max,gam_max)
//        CondY(x,alf_min,bet_min,gam_min,alf_max,bet_max,gam_max)
//---------------------------------------------------------------------------
{
#define Save eij = uij ? 1e0 - u[i][j]/uij : uij - u[i][j]; \
             if (fabs(eij) > err) err = fabs(eij); \
             u[i][j] = uij;              // error estimate and solution storage

   const int itmax = 10000;                        // max no. of iterations
   double **f;
   double *betXmin, *betXmax, *gamXmin, *gamXmax;
   double *betYmin, *betYmax, *gamYmin, *gamYmax;
   double alf_min, bet_min, gam_min, alf_max, bet_max, gam_max;
   double eij, err, hx, hy, kx, ky, kxy, uij;
   int i, it, j;

   f = Matrix(1,nx,1,ny);
   betXmin = Vector(1,ny); betXmax = Vector(1,ny);
   gamXmin = Vector(1,ny); gamXmax = Vector(1,ny);
   betYmin = Vector(1,nx); betYmax = Vector(1,nx);
   gamYmin = Vector(1,nx); gamYmax = Vector(1,nx);

   hx = (x[nx]-x[1])/(nx-1); kx = 1e0/(hx*hx);          // mesh spacings
   hy = (y[ny]-y[1])/(ny-1); ky = 1e0/(hy*hy);
   kxy = 2e0*(kx + ky);

   for (j=2; j<=ny-1; j++)                     // RHS of PDE for interior points
      for (i=2; i<=nx-1; i++) f[i][j] = Func(x[i],y[j]);
                                               // boundary conditions
   for (i=1; i<=nx; i++) {                     // lower and upper boundaries
      CondY(x[i],alf_min,bet_min,gam_min,alf_max,bet_max,gam_max);
      betYmin[i] = bet_min/(alf_min*hy + bet_min);
      gamYmin[i] = gam_min/(alf_min + bet_min/hy);
      betYmax[i] = bet_max/(alf_max*hy + bet_max);
      gamYmax[i] = gam_max/(alf_max + bet_max/hy);
   }
   for (j=2; j<=ny-1; j++) {                   // left and right boundaries
      CondX(y[j],alf_min,bet_min,gam_min,alf_max,bet_max,gam_max);
      betXmin[j] = bet_min/(alf_min*hx + bet_min);
      gamXmin[j] = gam_min/(alf_min + bet_min/hx);
      betXmax[j] = bet_max/(alf_max*hx + bet_max);
      gamXmax[j] = gam_max/(alf_max + bet_max/hx);
   }

   for (it=1; it<=itmax; it++) {               // Gauss-Seidel iteration loop
      err = 0e0;
      j = 1;                                   // lower boundary
      for (i=1; i<=nx; i++) {
         uij = betYmin[i]*u[i][2] + gamYmin[i]; Save
```

```
      }
      for (j=2; j<=ny-1; j++) {
         i = 1;                                              // left boundary
         uij = betXmin[j]*u[i+1][j] + gamXmin[j]; Save
         for (i=2; i<=nx-1; i++) {                    // interior mesh points
            uij = (kx*(u[i-1][j] + u[i+1][j]) +
                   ky*(u[i][j-1] + u[i][j+1]) - f[i][j]) / kxy; Save
         }
         i = nx;                                            // right boundary
         uij = betXmax[j]*u[i-1][j] + gamXmax[j]; Save
      }
      j = ny;                                               // upper boundary
      for (i=1; i<=nx; i++) {
         uij = betYmax[i]*u[i][ny-1] + gamYmax[i]; Save
      }
      if (err <= eps) break;                               // convergence test
   }

   FreeMatrix(f,1,1);
   FreeVector(betXmin,1); FreeVector(betXmax,1);
   FreeVector(gamXmin,1); FreeVector(gamXmax,1);
   FreeVector(betYmin,1); FreeVector(betYmax,1);
   FreeVector(gamYmin,1); FreeVector(gamYmax,1);

   if (it > itmax) {
      printf("PoissonXY: max. number of iterations exceeded !\n"); return 1;
   }
   return 0;
}

//===========================================================================
int Poisson2D(double **u, double x[], double y[],
              int imin[], int imax[], int nx, int ny, double eps,
              double Func(double,double))
//---------------------------------------------------------------------------
// Solves the 2D Poisson equation in Cartesian coordinates on a regular grid
// with (nx x ny) nodes (x[],y[]), with limiting indexes along x specified by
// imin[j] and imax[j] (j=1,ny), using the Gauss-Seidel method. The solution
// u[][] is converged with relative precision eps.
// An error index is returned: 0 - normal execution.
// Calls: Func(x,y) - RHS of Poisson equation
//---------------------------------------------------------------------------
{
   const int itmax = 10000;                         // max no. of iterations
   double **f;
   double eij, err, hx, hy, kx, ky, kxy, uij;
   int i, it, j;

   f = Matrix(1,nx,1,ny);

   hx = (x[nx]-x[1])/(nx-1); kx = 1e0/(hx*hx);             // mesh spacings
   hy = (y[ny]-y[1])/(ny-1); ky = 1e0/(hy*hy);
   kxy = 2e0*(kx + ky);

   for (j=2; j<=ny-1; j++)                      // RHS of PDE for interior points
      for (i=imin[j]+1; i<=imax[j]-1; i++) f[i][j] = Func(x[i],y[j]);

   for (it=1; it<=itmax; it++) {                 // Gauss-Seidel iteration loop
      err = 0e0;
```

```
      for (j=2; j<=ny-1; j++) {
         for (i=imin[j]+1; i<=imax[j]-1; i++) {          // interior mesh points
            uij = (kx*(u[i-1][j] + u[i+1][j]) - f[i][j]) / kxy;
            if (i >= imin[j-1] && i <= imax[j-1]) uij += ky*u[i][j-1] / kxy;
            if (i >= imin[j+1] && i <= imax[j+1]) uij += ky*u[i][j+1] / kxy;

            eij = uij ? 1e0 - u[i][j]/uij : uij - u[i][j]; // error estimate
            if (fabs(eij) > err) err = fabs(eij);      // max. component error
            u[i][j] = uij;
         }
      }
      if (err <= eps) break;                            // convergence test
   }

   FreeMatrix(f,1,1);

   if (it > itmax) {
      printf("Poisson2D: max. number of iterations exceeded !\n"); return 1;
   }
   return 0;
}

//===========================================================================
void PropagFTCS(double u0[], double u[], int nx, double D, double hx,
                double ht)
//---------------------------------------------------------------------------
// Propagates the solution u0[] of the diffusion equation
//    du/dt = D d2u/dx2,  D - diffusion coefficient (constant)
// over the time intervat ht, using an explicit FTCS difference scheme on
// a spatial grid with nx nodes and spacing hx. Returns the solution in u[].
//---------------------------------------------------------------------------
{
   double lam, lam1;
   int i;

   lam = D * ht/(hx*hx);
   lam1 = 1e0 - 2e0*lam;

   u[1] = u0[1]; u[nx] = u0[nx];
   for (i=2; i<=(nx-1); i++)
      u[i] = lam*u0[i-1] + lam1*u0[i] + lam*u0[i+1];
}

//===========================================================================
void PropagBTCS(double u0[], double u[], int nx, double D, double hx,
                double ht)
//---------------------------------------------------------------------------
// Propagates the solution u0[] of the diffusion equation
//    du/dt = D d2u/dx2,  D - diffusion coefficient (constant)
// over the time intervat ht, using an implicit BTCS difference scheme on
// a spatial grid with nx nodes and spacing hx. Returns the solution in u[].
//---------------------------------------------------------------------------
{
   static double *a, *b, *c;
   double lam, lam1;
   static int init = 1;                                 // initialization switch
   int i;

   if (init)
```

```
           { a = Vector(1,nx); b = Vector(1,nx); c = Vector(1,nx); init = 0; }

      lam = D * ht/(hx*hx);
      lam1 = 1e0 + 2e0*lam;

      b[1] = 1e0; c[1] = 0e0;          // build coefficients of discretized system
      u[1] = u0[1];
      for (i=2; i<=nx-1; i++) {
         a[i] = -lam; b[i] = lam1; c[i] = -lam;
         u[i] = u0[i];
      }
      a[nx] = 0e0; b[nx] = 1e0;
      u[nx] = u0[nx];

      TriDiagSys(a,b,c,u,nx);                   // solve tridiagonal discretized system
   }

   //==========================================================================
   void PropagCN(double u0[], double u[], int nx, double D, double hx, double ht
        )
   //--------------------------------------------------------------------------
   // Propagates the solution u0[] of the diffusion equation
   //     du/dt = D d2u/dx2,   D - diffusion coefficient (constant)
   // over the time intervat ht, using the Crank-Nicolson difference scheme on
   // a spatial grid with nx nodes and spacing hx. Returns the solution in u[].
   //--------------------------------------------------------------------------
   {
      static double *a, *b, *c;
      double lam, lam1, lam2;
      static int init = 1;                                    // initialization switch
      int i;

      if (init)
         { a = Vector(1,nx); b = Vector(1,nx); c = Vector(1,nx); init = 0; }

      lam = 0.5e0 * D * ht/(hx*hx);
      lam1 = 1e0 + 2e0*lam; lam2 = 1e0 - 2e0*lam;

      b[1] = 1e0; c[1] = 0e0;          // build coefficients of discretized system
      u[1] = u0[1];
      for (i=2; i<=(nx-1); i++) {
         a[i] = -lam; b[i] = lam1; c[i] = -lam;
         u[i] = lam*u0[i-1] + lam2*u0[i] + lam*u0[i+1];
      }
      a[nx] = 0e0; b[nx] = 1e0;
      u[nx] = u0[nx];

      TriDiagSys(a,b,c,u,nx);                   // solve tridiagonal discretized system
   }

   //==========================================================================
   void PropagDiff(double u0[], double u[], double D[], int nx, double hx,
             double ht, int iopBC1, int iopBC2, double Jdiff1, double Jdiff2)
   //--------------------------------------------------------------------------
   // Propagates the solution u0[] of the diffusion equation
   //     du/dt = D d2u/dx2,   D - diffusion coefficient (spatially variable)
   // over the time intervat ht, using the Crank-Nicolson difference scheme on
   // a spatial grid with nx nodes and spacing hx. Returns the solution in u[].
   // iopBC1, iopBC2 - left/right boundary condition: 0 - Dirichlet, 1 - Neumann
```

```
// Jdiff1, Jdiff2 - left/right boundary fluxes for Neumann conditions
//-------------------------------------------------------------------------
{
   static double *a, *b, *c;
   double f, lam, lam1, lam2, uint, uint0;
   static int init = 1;                              // initialization switch
   int i;

   if (init)
      { a = Vector(1,nx); b = Vector(1,nx); c = Vector(1,nx); init = 0; }

   f = 0.5e0 * ht/(hx*hx);

   b[1] = 1e0; c[1] = -iopBC1;     // build coefficients of discretized system
   u[1] = iopBC1 ? hx*Jdiff1/D[1] : u0[1];
   for (i=2; i<=(nx-1); i++) {
      lam = D[i] * f;
      lam1 = 0.5e0 * (D[i] + D[i-1]) * f;
      lam2 = 0.5e0 * (D[i] + D[i+1]) * f;
      a[i] = -lam1;
      b[i] = 1e0 + 2e0*lam;
      c[i] = -lam2;
      u[i] = lam1*u0[i-1] + (1e0 - 2e0*lam)*u0[i] + lam2*u0[i+1];
   }
   a[nx] = -iopBC2; b[nx] = 1e0;
   u[nx] = iopBC2 ? -hx*Jdiff2/D[nx] : u0[nx];

   TriDiagSys(a,b,c,u,nx);              // solve tridiagonal discretized system

   for (i=1; i<=nx; i++) if (u[i] < 0e0) u[i] = 0e0;    // keep solution >= 0

   uint0 = 0.5e0*(u0[1] + u0[nx]);                      // integral of old solution
   for (i=2; i<=nx-1; i++) uint0 += u0[i];              // trapezoidal rule
   uint0 *= hx;
   if (iopBC1 == 1) uint0 += Jdiff1*ht;                 // contribution of left flux
   if (iopBC2 == 1) uint0 -= Jdiff2*ht;                 // contribution of right flux

   uint = 0.5e0*(u[1] + u[nx]);                         // integral of new solution
   for (i=2; i<=nx-1; i++) uint += u[i];                // trapezoidal rule
   uint *= hx;

   f = uint0/uint;                                      // normalization factor
   if (f < 0e0) f = 0e0;
   for (i=1; i<=nx; i++) u[i] *= f;                     // normalize solution
}

//============================================================================
void PropagQTD(dcmplx Psi[], double V[], int nx, double hx, double ht)
//----------------------------------------------------------------------------
// Propagates the solution PSI = Psi + i Chi of the 1D Schrodinger equation
//    i d/dt PSI(x,t) = [-(1/2) d2/dx2 - V(x)] PSI(x,t),
// over the time interval ht. Uses the Crank-Nicolson scheme on a grid with
// nx nodes and spacing hx and solves the tridiagonal discretized system by
// LU factorization. Uses complex arithmetic.
//----------------------------------------------------------------------------
{
   static dcmplx *a, *b, *c;            // diagonals of discretized system
   dcmplx Psii, Psi1;
   double lam, W;
```

```
   static int init = 1;
   int i;

   if (init)
      { a = CVector(1,nx); b = CVector(1,nx); c = CVector(1,nx); init = 0; }

   lam = ht/(4e0*hx*hx);

   b[1] = 1e0; c[1] = 0e0;            // build coefficients of discretized system
   Psii = Psi[1];
   for (i=2; i<=(nx-1); i++) {
      Psi1 = Psii; Psii = Psi[i];             // save initial wave packet values
      W = 2e0*lam + 0.5e0*ht*V[i];
      a[i] = -lam;
      b[i] = dcmplx(W,-1e0);
      c[i] = -lam;
      Psi[i] = lam*Psi1 - dcmplx(W,1e0)*Psii + lam*Psi[i+1]; // constant term
   }
   a[nx] = 0e0; b[nx] = 1e0;
                                         // solve tridiagonal discretized system
   TriDiagSys(a,b,c,Psi,nx);            // solution Psi: propagated wave packet
}

//===========================================================================
int PropagQGS(double Psi[], double Chi[], double V[], int nx, double hx,
              double ht)
//---------------------------------------------------------------------------
// Propagates the solution PSI = Psi + i Chi of the 1D Schrodinger equation
//    i d/dt PSI(x,t) = [-(1/2) d2/dx2 - V(x)] PSI(x,t),
// over the time interval ht. Uses the Crank-Nicolson scheme on a grid with
// nx nodes and spacing hx and solves the discretized system by Gauss-Seidel
// iterations. Uses real arithmetic.
//---------------------------------------------------------------------------
{
   const double eps = 1e-6;                              // relative tolerance
   const int itmax = 100;                              // max no. of iterations
   static double *Psi0, *Chi0, *W;
   double err, erri, lam, Psii, Chii;
   static int init = 1;
   int i, it;

   if (init) {
      Psi0 = Vector(1,nx); Chi0 = Vector(1,nx); W = Vector(1,nx); init = 0;
   }

   lam = ht/(4e0*hx*hx);

   for (i=2; i<=nx-1; i++) {              // explicit 0th-order approximations
      W[i] = 0.5e0*ht*V[i] + 2e0*lam;
      Psi0[i] = Psi[i] - lam*(Chi[i-1] + Chi[i+1]) + W[i]*Chi[i];
      Chi0[i] = Chi[i] + lam*(Psi[i-1] + Psi[i+1]) - W[i]*Psi[i];
   }

   for (it=1; it<=itmax; it++) {                   // Gauss-Seidel iteration loop
      err = 0e0;
      for (i=2; i<=nx-1; i++) {
         Psii = Psi0[i] - lam*(Chi[i-1] + Chi[i+1]) + W[i]*Chi[i];
         Chii = Chi0[i] + lam*(Psi[i-1] + Psi[i+1])   W[i]*Psii;
                        // local error estimate based on probability density
```

```
          erri = fabs((Psii*Psii+Chii*Chii) - (Psi[i]*Psi[i]+Chi[i]*Chi[i]));
          if (erri > err) err = erri;                   // maximum error estimate
          Psi[i] = Psii;
          Chi[i] = Chii;
       }
       if (err <= eps) break;                           // convergence test
    }

    if (it > itmax) {
       printf("PropagQMGS: max. number of iterations exceeded !\n"); return 1;
    }
    return 0;
}


//============================================================================
void PropagWave(double u0[], double u1[], double u[], int nx, double c,
                double hx, double ht)
//----------------------------------------------------------------------------
// Propagates the solutions u0[] and u1[] of the wave equation
//    d2u/dt2 = c^2 d2u/dx2,   c - phase velocity (constant)
// over the time interval ht, using the explicit difference scheme on a
// spatial grid with nx nodes and spacing hx. Returns the solution in u[].
//----------------------------------------------------------------------------
{
   double lam, lam2;
   int i;

   lam = c*ht/hx; lam = lam*lam;
   lam2 = 2e0*(1e0 - lam);

   u[1] = u0[1]; u[nx] = u0[nx];            // Dirichlet boundary conditions
   for (i=2; i<=nx-1; i++)                  // propagate solution at interior points
      u[i] = lam*u1[i-1] + lam2*u1[i] + lam*u1[i+1] - u0[i];
}

#endif
```

13.7 Problems

The Python and C/C++ programs for the following problems may import the functions developed in this chapter from the modules pde.py and, respectively, pde.h, which are available as supplementary material. For creating runtime plots, the graphical routines contained in the libraries graphlib.py and graphlib.h may be employed.

PROBLEM 13.1

Use the routine Poisson0 described in Section 13.2 to solve the problem associated with the Poisson equation

$$\frac{\partial^2 u}{\partial x^2} + \frac{\partial^2 u}{\partial y^2} = \cos(x+y) - \cos(x-y) \tag{13.143}$$

in the domain $\mathcal{D} = [-\pi, \pi] \times [-\pi, \pi]$, subject to the boundary conditions

$$\begin{aligned}
u(\pm\pi, y) &= 0, \quad y \in [-\pi, \pi], \\
u(x, \pm\pi) &= 0, \quad x \in [-\pi, \pi].
\end{aligned} \tag{13.144}$$

Define a user function Func to return the right-hand side of the equation. Start with the trivial initial approximation $u^0(x, y) = 0$ for all $(x, y) \in \mathcal{D}$, and use an integration mesh with 51×51 nodes (50×50 intervals) and a relative tolerance $\varepsilon = 10^{-5}$. Compare the numerical solution with the exact one, $u(x, y) = \sin x \sin y$, and estimate the relative error from the maxima of the two solutions. Plot the numerical solution using the routine Contour from the libraries graphlib.py and, respectively, graphlib.h.

Double the number of intervals by taking 101×101 nodes and assess the reduction of the maximum error.

Solution

The Python implementation is presented in Listing 13.19 and the C/C++ version is available as supplementary material (P13-Poisson0.cpp). The contour plot of the solution is shown in Figure 13.15.

Listing 13.19 Solution of the Poisson Problem 13.143–13.144 (Python Coding)

```
# Solves the 2D Poisson equation in a rectangular domain
from math import *
from pde import *
from graphlib import *

def Func(x, y):                                    # RHS function for Poisson0
   return cos(x+y) - cos(x-y)

# main

xmin = -pi; xmax = pi; ymin = -pi; ymax = pi              # domain boundaries
nx = 51; ny = 51                                    # number of mesh points
eps = 1e-5                                      # relative solution tolerance

u = [[0]*(ny+1) for i in range(nx+1)]                          # solution
x = [0]*(nx+1); y = [0]*(ny+1)                      # mesh point coordinates

hx = (xmax-xmin)/(nx-1)
for i in range(1,nx+1): x[i] = xmin + (i-1)*hx              # x-mesh points
hy = (ymax-ymin)/(ny-1)
for j in range(1,ny+1): y[j] = ymin + (j-1)*hy              # y-mesh points

for j in range(1,ny+1):                    # initial approximation of the solution
   for i in range(1,nx+1): u[i][j] = 0e0           # and boundary conditions

Poisson0(u,x,y,nx,ny,eps,Func)

out = open("Poisson.txt","w")                              # open output file
out.write("       x             y              u\n")
for j in range(1,ny+1):
   for i in range(1,nx+1):
      out.write(("{0:10.5f}{1:10.5f}{2:14.5e}\n").format(x[i],y[j],u[i][j]))
out.close()

umin = umax = u[1][1]                       # minimum and maximum of the solution
for j in range(1,ny+1):
   for i in range(1,nx+1):
      if (u[i][j] < umin): umin = u[i][j]
      if (u[i][j] > umax): umax = u[i][j]
```

```
GraphInit(800,800)
Contour(u,nx,ny,xmin,xmax,ymin,ymax,umin,umax, \
        0.15,0.85,0.15,0.85,"x","y","Poisson")
MainLoop()
```

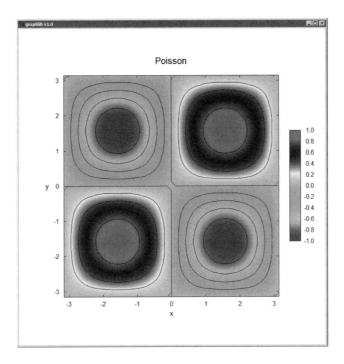

FIGURE 13.15   Finite-difference solution of the discretized Poisson problem 13.143–13.144 (for Problem 13.1).

PROBLEM 13.2

Consider the problem associated with the Poisson equation

$$\frac{\partial^2 u}{\partial x^2} + \frac{\partial^2 u}{\partial y^2} = -\left[\cos(x+y) + \cos(x-y)\right] \tag{13.145}$$

in the domain $\mathcal{D} = [-\pi,\pi] \times [-\pi/2,\pi/2]$, with the boundary conditions

$$\begin{aligned}
u(\pm\pi,y) &= \cos y, \quad y \in [-\pi/2,\pi/2], \\
u(x,\pm\pi/2) &= 0, \quad x \in [-\pi,\pi].
\end{aligned} \tag{13.146}$$

Solve this problem using the `PoissonXY` routine described in Section 13.2, defining the user functions `Func`, `CondX`, and `CondY` to return the right-hand side of the equation and, respectively, the boundary coefficients. Start with the trivial initial approximation $u^0(x,y) = 0$ for all $(x,y) \in \mathcal{D}$. Use an integration mesh with 51×26 nodes (50×25 intervals) and impose a relative tolerance $\varepsilon = 10^{-5}$. Compare the numerical solution with the exact one, $u(x,y) = \cos x \cos y$, and estimate the relative error from the maxima of the two solutions. Plot the numerical solution using the routine `Contour` from the libraries `graphlib.py` and, respectively, `graphlib.h`.

Double the number of mesh intervals in both directions (taking 101×51 nodes), and compare the solution in terms of accuracy with the previous estimate.

Solution

The Python implementation is presented in Listing 13.20 and the C/C++ version is available as supplementary material (P13-Poisson1.cpp). The contour plot of the solution is shown in Figure 13.16.

Listing 13.20   Solution of the Poisson Problem 13.145–13.146 (Python Coding)

```
# Solves the 2D Poisson equation in a rectangular domain
from math import *
from pde import *
from graphlib import *

def Func(x, y):                                        # RHS function for PoissonXY
   return -(cos(x+y) + cos(x-y))

def CondX(y):                      # Coefficients for left and right boundaries
   alf_min = 1e0; bet_min = 0e0; gam_min = -cos(y)
   alf_max = 1e0; bet_max = 0e0; gam_max = -cos(y)

   return (alf_min, bet_min, gam_min, alf_max, bet_max, gam_max)

def CondY(x):                      # Coefficients for lower and upper boundaries
   alf_min = 1e0; bet_min = 0e0; gam_min = 0e0
   alf_max = 1e0; bet_max = 0e0; gam_max = 0e0

   return (alf_min, bet_min, gam_min, alf_max, bet_max, gam_max)

# main

xmin = -pi; xmax = pi; ymin = -pi/2; ymax = pi/2          # domain boundaries
nx = 51; ny = 26                                       # number of mesh points
eps = 1e-5                                         # relative solution tolerance

u = [[0]*(ny+1) for i in range(nx+1)]                           # solution
x = [0]*(nx+1); y = [0]*(ny+1)                        # mesh point coordinates

hx = (xmax-xmin)/(nx-1)
for i in range(1,nx+1): x[i] = xmin + (i-1)*hx             # x-mesh points
hy = (ymax-ymin)/(ny-1)
for j in range(1,ny+1): y[j] = ymin + (j-1)*hy             # y-mesh points

for j in range(1,ny+1):                      # initial approximation of the solution
   for i in range(1,nx+1): u[i][j] = 0e0

PoissonXY(u,x,y,nx,ny,eps,Func,CondX,CondY)

out = open("Poisson.txt","w")                                # open output file
out.write("      x          y          u\n")
for j in range(1,ny+1):
   for i in range(1,nx+1):
      out.write(("{0:10.5f}{1:10.5f}{2:14.5e}\n").format(x[i],y[j],u[i][j]))
out.close()

umin = umax = u[1][1]                         # minimum and maximum of the solution
for j in range(1,ny+1):
   for i in range(1,nx+1):
      if (u[i][j] < umin): umin = u[i][j]
      if (u[i][j] > umax): umax = u[i][j]

GraphInit(1200,600)
```

```
Contour(u,nx,ny,xmin,xmax,ymin,ymax,umin,umax, \
        0.15,0.85,0.15,0.85,"x","y","Poisson")
MainLoop()
```

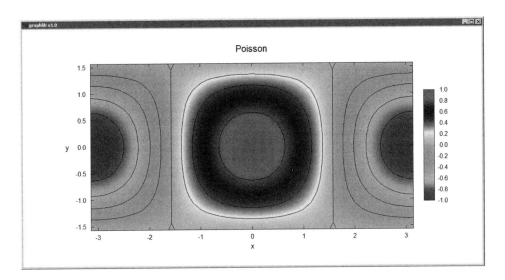

FIGURE 13.16   Finite-difference solution of the discretized Poisson problem 13.145–13.146 (for Problem 13.2).

PROBLEM 13.3

Consider a thin square metal plate with the side measuring $L = 10$ cm. The temperature varies along the left boundary according to $u(0, y) = 100 \sin(\pi y/L)$, showing a maximum in the middle and vanishing at the corners and along the other three sides. Determine the temperature profile in the plate by solving the stationary heat equation

$$\frac{\partial^2 u}{\partial x^2} + \frac{\partial^2 u}{\partial y^2} = 0 \tag{13.147}$$

in the domain $\mathcal{D} = [0, L] \times [0, L]$, with the boundary conditions

$$
\begin{aligned}
u(0, y) &= 100 \sin\left(\pi y/L\right), & u(L, y) &= 0, & y &\in [0, L], \\
u(x, 0) &= u(x, L) = 0, & & & x &\in [0, L].
\end{aligned}
\tag{13.148}
$$

Use the `PoissonXY` solver developed in Section 13.2. Define the user functions `CondX` and `CondY` to return the boundary coefficients and start with the initial approximation $u^0(x, y) = 0$ for all $(x, y) \in \mathcal{D}$. Use an integration mesh with 41×41 nodes (40×40 intervals) and impose a relative tolerance $\varepsilon = 10^{-5}$ on the solution. Create the contour plot of the solution.

Double the number of mesh intervals in both directions (taking 81×81 nodes) and estimate the relative error from the two solution values at the center of the plate.

Solution

The Python and C/C++ implementations are available as supplementary material (`P13-HeatSquare.py` and `P13-HeatSquare.cpp`). The contour plot of the solution on the 41×41-node grid is shown in Figure 13.17.

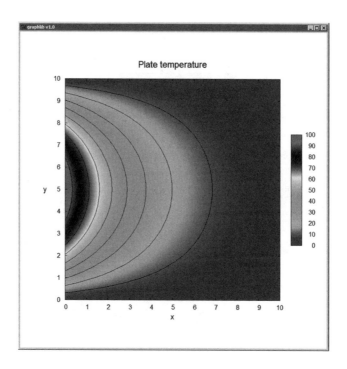

FIGURE 13.17   Finite-difference solution of the discretized heat problem 13.147–13.148 (for Problem 13.2).

PROBLEM 13.4

Consider a conducting disk of negligible thickness and radius $a = 5$ cm. The disk's temperature decreases linearly with x on the left semiperimeter, from $100°$C for $x = -a$ to $0°$C for $x = 0$, and remains fixed at $0°$C on the other semiperimeter. Determine the temperature distribution on the disk's surface by solving the stationary heat equation

$$\frac{\partial^2 u}{\partial x^2} + \frac{\partial^2 u}{\partial y^2} = 0 \tag{13.149}$$

in the square domain $\mathcal{D} = [-a, a] \times [-a, a]$, subject to the boundary conditions

$$u(x, y) = \begin{cases} 100(a + x)/a, & x^2 + y^2 = a^2, \quad x < 0, \\ 0, & x^2 + y^2 = a^2, \quad x \geq 0. \end{cases} \tag{13.150}$$

Use the solver Poisson2D (Listing 13.4) with an integration mesh with $N_x = N_y = 51$ nodes (50×50 intervals) and a relative tolerance $\varepsilon = 10^{-5}$. Define a function to determine the bounding line indexes $i_{\min}^j$ and $i_{\max}^j$ for $j = 1, \ldots, N_y$ and to initialize the solution with the boundary values. Create the contour plot of the solution.

Double the number of mesh intervals in both directions (taking 101×101 nodes) and estimate the relative error from the two solution values at the disk's center.

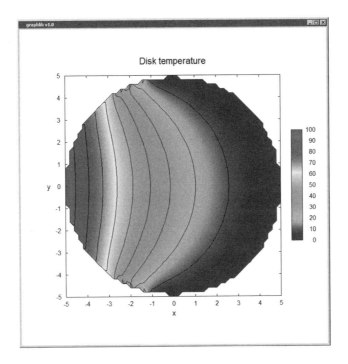

FIGURE 13.18   Finite-difference solution of the discretized heat problem 13.149–13.150 (for Problem 13.4).

Solution

The discretized boundary conditions read:

$$u_{i,1}^{(r)} = 0, \quad i = i_{\min}^1, \ldots, i_{\max}^1, \quad j = 1,$$

$$u_{i_{\min},j}^{(r)} = 100(N_x - i_{\min}^j)/(N_x - 1), \quad u_{i_{\max},j}^{(r)} = 0, \quad j = 2, \ldots, N_y - 1,$$

$$u_{i,N_y}^{(r)} = 0, \quad i = i_{\min}^{N_y}, \ldots, i_{\max}^{N_y}, \quad j = N_y. \tag{13.151}$$

The Python implementation is presented in Listing 13.21 and the C/C++ version is available as supplementary material (P13-HeatDisk.cpp). The contour plot of the solution is shown in Figure 13.18.

Listing 13.21   Solution of the Stationary Heat Problem 13.149–13.150 (Python Coding)

```
# Steady-state temperature distribution in a thin conducting disk
from math import *
from pde import *
from graphlib import *

def Func(x, y):                                    # RHS function for Poisson2D
    return 0e0

#===============================================================================
def BoundCondEllipse(u, imin, imax, nx, ny):
#-------------------------------------------------------------------------------
#  Determines the boundary indexes imin[j] and imax[j] (j=1,ny) for an
#  elliptical domain, determined by the number of mesh points nx and ny, and
#  sets values on the boundaries for the function u
```

```
#-----------------------------------------------------------------------
   x0 = 0.5e0*(nx + 1e0)                           # center of dimensionless ellipse
   y0 = 0.5e0*(ny + 1e0)
   a = x0 - 1e0                                                        # semi-axes
   b = y0 - 1e0

   for j in range(1,ny+1):                                    # boundary indexes
      yj = j - y0                                        # relative y-coordinate
      xi = a * sqrt(1e0 - yj*yj/(b*b))                 # x-coordinate on ellipse
      if (xi == 0e0): xi = 0.75e0                   # correction for 1-point boundary

      imin[j] = int(x0 - xi + 0.5e0)                          # left boundary index
      imax[j] = int(x0 + xi + 0.5e0)                         # right boundary index
                                                                # boundary values
   for i in range(imin[1],imax[1]+1): u[i][1] = 0e0                    # bottom
   for j in range(2,ny):
      u[imin[j]][j] = 100e0 * (nx - 2*imin[j])/(nx-1)                  # left
      u[imax[j]][j] = 0e0                                              # right
   for i in range(imin[ny],imax[ny]+1): u[i][ny] = 0e0                 # top

# main

a = 5                                                          # disk radius
xmin = -a; xmax = a; ymin = -a; ymax = a                  # domain boundaries
nx = 51; ny = 51                                       # number of mesh points
eps = 1e-5                                        # relative solution tolerance

u = [[0]*(ny+1) for i in range(nx+1)]                             # solution
x = [0]*(nx+1); y = [0]*(ny+1)                          # mesh point coordinates
imin = [0]*(ny+1)                                 # indexes of y-line boundaries
imax = [0]*(ny+1)

hx = (xmax-xmin)/(nx-1)
for i in range(1,nx+1): x[i] = xmin + (i-1)*hx                    # x-mesh points
hy = (ymax-ymin)/(ny-1)
for j in range(1,ny+1): y[j] = ymin + (j-1)*hy                   # y-mesh points

for j in range(1,ny+1):                                  # initialize solution
   for i in range(1,nx+1): u[i][j] = -1e99

BoundCondEllipse(u,imin,imax,nx,ny)                          # boundary values

for j in range(2,ny):                      # initial approximation of the solution
   for i in range(imin[j]+1,imax[j]): u[i][j] = 0e0             # interior nodes

Poisson2D(u,x,y,imin,imax,nx,ny,eps,Func)

out = open("Poisson.txt","w")                                 # open output file
out.write("        x           y            u\n")
for j in range(1,ny+1):
   for i in range(1,nx+1):
      out.write(("{0:10.5f}{1:10.5f}{2:14.5e}\n").format(x[i],y[j],u[i][j]))
out.close()

umin = umax = u[imin[1]][1]              # minimum and maximum of the solution
for j in range(1,ny+1):
   for i in range(imin[j],imax[j]+1):
      if (u[i][j] < umin): umin = u[i][j]
      if (u[i][j] > umax): umax = u[i][j]
```

```
GraphInit(800,800)
Contour(u,nx,ny,xmin,xmax,ymin,ymax,umin,umax, \
        0.15,0.85,0.15,0.85,"x","y","Disk temperature")
MainLoop()
```

PROBLEM 13.5

Consider a conducting isosceles triangle of negligible thickness, with the horizontal side $a = 10$ cm and the height $b = 8$ cm. The triangle's temperature is fixed at $100°$C on the bottom side and at $0°$C, on the lateral sides. Determine the temperature distribution on the triangle's surface by solving the stationary heat equation

$$\frac{\partial^2 u}{\partial x^2} + \frac{\partial^2 u}{\partial y^2} = 0 \tag{13.152}$$

in the square domain $\mathcal{D} = [0, a] \times [0, b]$, subject to the boundary conditions

$$u(x, y) = \begin{cases} 100, & y = 0, \\ 0, & y \geq 0. \end{cases} \tag{13.153}$$

Use the solver `Poisson2D` (Listing 13.4) with an integration mesh with 51×41 nodes (50×40 intervals) and impose a relative tolerance $\varepsilon = 10^{-5}$ on the solution. Devise a function to optimally determine the bounding line indexes i^j_{min} and i^j_{max} for $j = 1, \ldots, N_y$ and to initialize the solution with the boundary values. Create the contour plot of the solution using the routine `Contour`.

Double the number of mesh intervals in both directions (taking 101×81 nodes) and estimate the relative error from the two solution values at the triangle's center.

Solution
The discretized boundary conditions read:

$$u^{(r)}_{i,1} = 100, \quad i = i^1_{min}, \ldots, i^1_{max}, \quad j = 1, \tag{13.154}$$

$$u^{(r)}_{i_{min},j} = u^{(r)}_{i_{max},j} = 0, \quad j = 2, \ldots, N_y - 1, \tag{13.155}$$

$$u^{(r)}_{i,N_y} = 0, \quad i = i^{N_y}_{min}, \ldots, i^{N_y}_{max}, \quad j = N_y. \tag{13.156}$$

The Python implementation is presented in Listing 13.22 and the C/C++ version is available as supplementary material (`P13-HeatTriangle.cpp`). The contour plot of the solution is shown in Figure 13.19.

Listing 13.22   Solution of the Stationary Heat Problem 13.152–13.153 (Python Coding)

```
# Steady-state temperature distribution in a thin triangular conducting plate
from math import *
from pde import *
from graphlib import *

def Func(x, y):                              # RHS function for Poisson2D
    return 0e0

#============================================================================
def BoundCondTriangle(u, imin, imax, nx, ny, iapex):
#----------------------------------------------------------------------------
#  Determines the boundary indexes imin[j] and imax[j] (j=1,ny) for a
```

```
#    triangular domain, determined by the number of mesh points nx, ny, and the
#    apex i-index iapex, and sets values on the boundaries for the function u
#-----------------------------------------------------------------------------
   amin = (iapex-1e0)/(ny-1e0)                              # slope for left side
   amax = -(nx-iapex)/(ny-1e0)                              # slope for right side

   for j in range(1,ny+1):                                  # determine line-indexes
      imin[j] = int(amin*(j-1) + 1e0 + 0.5e0)                    # left boundary index
      imax[j] = int(amax*(j-1) + nx + 0.5e0)                     # right boundary index
                                                            # boundary values
   for i in range(imin[1],imax[1]+1): u[i][1] = 100e0            # bottom
   for j in range(2,ny): u[imin[j]][j] = u[imax[j]][j] = 0e0   # left & right
   for i in range(imin[ny],imax[ny]+1): u[i][ny] = 0e0             # top

# main

xmin = 0e0; xmax = 10e0; ymin = 0e0; ymax = 8e0             # domain boundaries
nx = 51; ny = 41                                     # number of mesh points
iapex = 26                                        # i-index of triangle apex
eps = 1e-5                                       # relative solution tolerance

u = [[0]*(ny+1) for i in range(nx+1)]                          # solution
x = [0]*(nx+1); y = [0]*(ny+1)                        # mesh point coordinates
imin = [0]*(ny+1)                                # indexes of y-line boundaries
imax = [0]*(ny+1)

hx = (xmax-xmin)/(nx-1)
for i in range(1,nx+1): x[i] = xmin + (i-1)*hx                # x-mesh points
hy = (ymax-ymin)/(ny-1)
for j in range(1,ny+1): y[j] = ymin + (j-1)*hy                # y-mesh points

for j in range(1,ny+1):                                  # initialize solution
   for i in range(1,nx+1): u[i][j] = -1e99

BoundCondTriangle(u,imin,imax,nx,ny,iapex)                   # boundary values

for j in range(2,ny):                      # initial approximation of the solution
   for i in range(imin[j]+1,imax[j]): u[i][j] = 0e0           # interior nodes

Poisson2D(u,x,y,imin,imax,nx,ny,eps,Func)

out = open("Poisson.txt","w")                              # open output file
out.write("        x            y            u\n")
for j in range(1,ny+1):
   for i in range(1,nx+1):
      out.write(("{0:10.5f}{1:10.5f}{2:14.5e}\n").format(x[i],y[j],u[i][j]))
out.close()

umin = umax = u[imin[1]][1]                   # minimum and maximum of the solution
for j in range(1,ny+1):
   for i in range(imin[j],imax[j]+1):
      if (u[i][j] < umin): umin = u[i][j]
      if (u[i][j] > umax): umax = u[i][j]

GraphInit(800,800)
Contour(u,nx,ny,xmin,xmax,ymin,ymax,umin,umax, \
        0.15,0.85,0.15,0.85,"x","y","Plate temperature")
MainLoop()
```

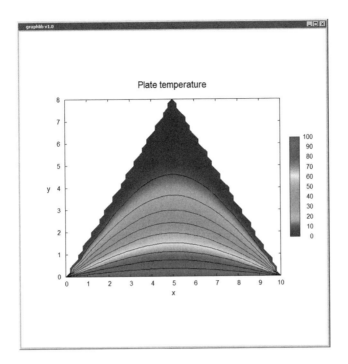

FIGURE 13.19   Finite-difference solution of the discretized heat problem 13.154–13.156 (for Problem 13.5).

PROBLEM 13.6

Use the routine `PropagDiff` (Listing 13.10), included in the files `pde.py` and, respectively, `pde.h`, to solve the diffusion equation

$$\frac{\partial u}{\partial t} = \frac{\partial}{\partial x}\left(D\frac{\partial u}{\partial x}\right), \quad x \in [0, L], \quad t > 0, \tag{13.157}$$

with $L = 1$, for a stack of three homogeneous layers occupying the intervals $[0, x_1)$, $[x_1, x_2)$, and $[x_2, L]$, with $x_1 = 0.4L$ and $x_2 = 0.6L$. Consider the nominal values of the diffusion coefficient $D_1 = D_3 = 0.1$ in the middle of the outer layers and $D_2 = 0.001$ in the middle of the inner layer, and use the continuous functional dependence

$$D(x) = D_1 + (D_2 - D_1)f_{sw}(x) \tag{13.158}$$

to switch smoothly between D_1 and D_2. The switch function $f_{sw}(x)$ is defined by

$$f_{sw}(x) = \left[1 + \exp\left(\frac{|x - c| - w}{d}\right)\right]^{-1},$$

where c is the center of layer 2, w is its width, and d is the distance over which the switch function drops from 1 (in layer 2) to 0 (in layers 1 and 3).

Take a step-like initial solution, with the entire diffusing substance concentrated in the first layer:

$$u(x, 0) = \begin{cases} 1, & 0 \le x < x_1, \\ 0, & x_1 \le x \le L. \end{cases} \tag{13.159}$$

Consider Neumann boundary conditions with equal in and out boundary fluxes,

$$J_{\text{diff}}^{(1)} = -\left[D\frac{\partial u}{\partial x}\right]_{x=0} = 1, \quad J_{\text{diff}}^{(2)} = -\left[D\frac{\partial u}{\partial x}\right]_{x=L} = 1, \qquad (13.160)$$

producing no net accumulation of substance in the system. Use a spatial mesh with $N_x = 101$ nodes (100 intervals with the spacing $h_x = 10^{-2}$) and run the program up to $t_{\max} = 2$ with a time step $h_t = 10^{-3}$. Print and plot the solution every $n_{\text{out}} = 100$ time steps. Explain the final concentration profile.

Increase alternatively the in and out fluxes, $J_{\text{diff}}^{(1)}$ and $J_{\text{diff}}^{(2)}$, by 20% and explain the qualitative change in the final profile.

Solution

The Python implementation is presented in Listing 13.23 and the C/C++ version is available as supplementary material (P13-Diffusion2.cpp). The graphical output is shown in Figure 13.20.

Listing 13.23 Solution of the Diffusion Problem 13.157–13.160 for a Layer Stack (Python Coding)

```
# Solves the diffusion equation for spatially variable diffusion coefficient
from math import *
from pde import *
from graphlib import *

def Init(u, x, D, nx):                                          # initialization
   global L                                         # extent of spatial domain
   x1 = 0.4e0*L                                          # limit of layer 1
   x2 = 0.6e0*L                                          # limit of layer 2
   c = 0.5e0*(x1+x2)                                       # center of layer 2
   w = 0.5e0*(x2-x1)                                    # half-width of layer 2
   d = 0.01e0 * L                             # drop distance of switch function
   D1 = 0.1e0; D2 = 0.001e0              # nominal D coef. in layers 1 (3) and 2

   for i in range(1,nx+1):           # switch function: 1/0 inside/outside layer 2
      fsw = 1e0 / (1e0 + exp((fabs(x[i]-c)-w)/d))
      D[i] = D1 + (D2 - D1) * fsw
      u[i] = 1e0 if x[i] <= x1 else 0e0

# main

L        = 1e0                                     # [0,L] spatial domain
iopBC1 = 1; iopBC2 = 1                     # left/right boundary condition type
Jdiff1 = 1e0; Jdiff2 = 1e0                      # left/right boundary fluxes
nx       = 101                              # number of spatial mesh points
tmax     = 2e0                                       # propagation time
ht       = 1e-3                                          # time step
nout     = 100                                # output every nout steps

nt = (int)(tmax/ht + 0.5)                              # number of time steps
u0 = [0]*(nx+1); u = [0]*(nx+1)                                  # solution
x = [0]*(nx+1)                                            # spatial mesh
D = [0]*(nx+1)                                       # difusion coefficient

hx = L/(nx-1)
for i in range(1,nx+1): x[i] = (i-1)*hx                        # spatial mesh

Init(u0,x,D,nx)            # initialization: diffusion coefficient, solution

GraphInit(1000,700)
```

```
out = open("diffusion.txt","w")
out.write("        x    ")
for i in range (1,nx+1): out.write("{0:10.5f}".format(x[i]))    # print x-mesh
out.write("\n")
out.write("        t              u\n")

for it in range(1,nt+1):                                        # time loop
    t = it*ht                                                   # propagate solution
    PropagDiff(u0,u,D,nx,hx,ht,iopBC1,iopBC2,Jdiff1,Jdiff2)

    for i in range(1,nx+1): u0[i] = u[i]                        # shift solutions

    if (it % nout == 0 or it == nt):                            # output every nout steps
        out.write("{0:10.5f}".format(t))
        for i in range (1,nx+1): out.write("{0:10.5f}".format(u[i]))
        out.write("\n")

        GraphClear()
        title = "Diffusion  t = {0:4.2f}".format(t)
        Plot(x,u,nx,"blue",1,0.15,0.95,0.50,0.90,"None","u",title)
        Plot(x,D,nx,"red",1,0.15,0.95,0.08,0.48,"x","D","")
        GraphUpdate()
out.close

MainLoop()
```

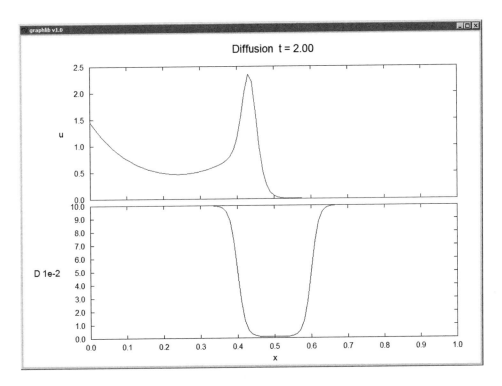

FIGURE 13.20 Solution of the discretized diffusion problem 13.157–13.160 and spatial distribution of the diffusion coefficient (lower panel) for a stack of three layers at time $t = 2.0$ (for Problem 13.6).

PROBLEM 13.7

Investigate the time evolution of the quantum Gaussian wave packet

$$\Psi(x,0) = \frac{1}{\sqrt{(2\pi)^{1/2}\sigma}} e^{-(x-x_0)^2/(4\sigma^2)} e^{ik_0 x}, \qquad (13.161)$$

scattered by the square potential

$$V(x) = \begin{cases} V_0, & |x| \le a/2, \\ 0, & |x| > a/2. \end{cases} \qquad (13.162)$$

Employ the Crank–Nicolson scheme for discretizing the Schrödinger equation and the LU factorization technique for tridiagonal systems described in the main text and implemented in routine `PropagQTD`. Build Python and C/C++ codes similar to program 13.14, adding graphic capabilities to visualize the probability density profiles.

Set $\hbar = 1$ and $m = 1$, and consider for the potential barrier the parameters $a = 5$ and $V_0 = 25$, and for the wave packet, the initial position $x_0 = -20$, the half-width $\sigma = 1$, and the average wave vector $k_0 = 10$ (the energy is $\hbar^2 k^2/2m = 2V_0$). Use a spatial grid in the range $(-x_{\max}, x_{\max})$, with $x_{\max} = 100$ and nodes spaced regularly by $h_x = 5 \cdot 10^{-2}$. Propagate the wave packet up to $t_{\max} = 5$, with a time step $h_t = 5 \cdot 10^{-3}$.

Print and plot the probability density every $n_{\text{out}} = 40$ time steps. Discuss the change in the final distribution of the probability density in relation to the case presented in the main text, in which the total energy was equal to the barrier height ($V_0 = 50$).

Solution

The Python implementation is presented in Listing 13.24 and the C/C++ version is available as supplementary material (`P13-ScatterQTD1.cpp`). Snapshots of the time evolution of the probability density of the wave packet are shown in Figure 13.21.

Listing 13.24   Scattering of a Quantum Wave Packet by a Potential Barrier (Python Coding)

```
# Reflexion/transmission of quantum wave packet using tridiagonal solver
from math import *
from cmath import *
from pde import *
from graphlib import *

def Pot(x, a, V0):                                          # Potential
    return V0 if fabs(x) <= 0.5e0*a else 0e0

#=====================================================================
def Init(Psi, x, nx, x0, sig, k0):
#---------------------------------------------------------------------
#   Initial Gaussian wave packet
#       Psi(x,0) = 1/sqrt(sqrt(2*pi)*sig) * exp[-(x-x0)^2/(4*sig^2)] * exp(ikx)
#   x0 - position of center, sig - half-width, k0 - average wave number
#---------------------------------------------------------------------
    a = 1e0/sqrt(sqrt(2e0*pi)*sig)
    b =-1e0/(4e0*sig*sig)
    for i in range(1,nx+1):
        dx = x[i] - x0
        f = a * exp(b*dx*dx)
        if (f < 1e-10): f = 0e0
        Psi[i] = f * complex(cos(k0*x[i]),sin(k0*x[i]))

#=====================================================================
```

```
def ProbDens(Psi, Psi2, nx, hx):
#-----------------------------------------------------------------------------
#  Calculates the probability density Psi2[] of the wave function Psi[]
#-----------------------------------------------------------------------------
   for i in range(1,nx+1):
      Psi2[i] = abs(Psi[i])*abs(Psi[i])      # unnormalized probability density
      if (Psi2[i] <= 1e-10): Psi2[i] = 0e0

   PsiNorm = 0.5e0*(Psi2[1] + Psi2[nx])          # integral by trapezoidal rule
   for i in range(2,nx): PsiNorm += Psi2[i]
   PsiNorm *= hx

   for i in range(1,nx+1): Psi2[i] /= PsiNorm      # normalized prob. density
   return PsiNorm

# main

a    = 5e0                                     # width of potential barrier
V0   = 25e0                                    # height of potential barrier
x0   = -20e0                               # initial position of wave packet
sig  = 1e0                                      # half-width of packet
k0   = 10e0                                 # average wave number of packet
xmax = 100e0                                               # maximum x
hx   = 5e-2                                        # spatial step size
tmax = 5e0                                    # maximum propagation time
ht   = 5e-3                                                # time step
nout = 40                                      # output every nout steps

nx = 2*(int)(xmax/hx + 0.5) + 1                # odd number of spatial nodes
nt = (int)(tmax/ht + 0.5)                        # number of time steps
nx2 = int(nx/2); nx4 = int(nx/4) + 1

Psi  = [complex(0,0)]*(nx+1)                               # wave function
Psi2 = [0]*(nx+1)                                   # probability density
V = [0]*(nx+1)                                              # potential
x = [0]*(nx+1)                                          # spatial mesh

for i in range(1,nx+1):                 # tabulate spatial mesh and potential
   x[i] = (i-nx2-1)*hx
   V[i] = Pot(x[i],a,V0)

Init(Psi,x,nx,x0,sig,k0)                                # initial wave packet

GraphInit(1000,700)

for it in range(1,nt+1):                                        # time loop
   t = it*ht
   PropagQTD(Psi,V,nx,hx,ht)          # propagate solution by tridiagonal solver

   PsiNorm = ProbDens(Psi,Psi2,nx,hx)             # probability density

   if (it % nout == 0 or it == nt):               # output every nout steps
      fname = "scatter_{0:4.2f}.txt".format(t)
      out = open(fname,"w")
      out.write("t = {0:4.2f}\n".format(t))
      out.write("      x         V        PsiR      PsiI      Psi2\n")
      for i in range(1,nx+1):
         out.write("{0:10.5f}{1:10.5f}{2:10.5f}{3:10.5f}{4:10.5f}\n".\
                   format(x[i],V[i],Psi[i].real,Psi[i].imag,Psi2[i]))
```

```
    out.close

    GraphClear()
    title = "Scattering of wave packet  t = {0:4.2f}".format(t)
    Plot(x[nx4:],Psi2[nx4:],nx2,"blue",1,0.15,0.95,0.50,0.90,
         "None","Psi2",title)
    Plot(x[nx4:],V[nx4:],nx2,"red",1,0.15,0.95,0.08,0.48,"x","V","")
    GraphUpdate()

MainLoop()
```

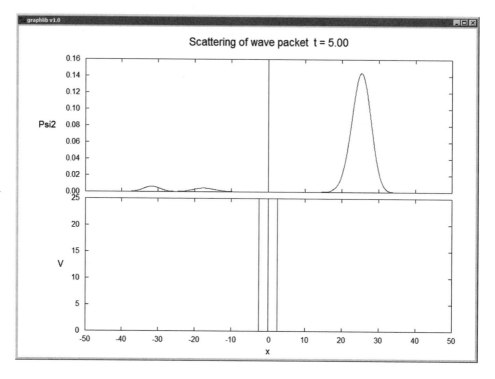

FIGURE 13.21 Scattering of a quantum Gaussian wave packet by a square potential, calculated by Program 13.14, using the LU decomposition technique implemented in routine PropagQTD. Upper panel: the energy is equal to twice the barrier height, $V_0 = 25$. Lower panel: the energy is equal to the absolute value of the well depth, $V_0 = -50$ (for Problem 13.7).

PROBLEM 13.8

Solve the classical wave equation

$$\frac{\partial^2 u}{\partial t^2} = c^2 \frac{\partial^2 u}{\partial x^2}, \quad t > 0, \tag{13.163}$$

for the initial wave packet

$$u(x, 0) = \frac{\sin x \Delta k}{x \Delta k}, \quad -x_{\max} \leq x \leq x_{\max}, \tag{13.164}$$

with the initial time derivative $(\partial u/\partial t)(x, 0) = 0$ and satisfying the Dirichlet boundary conditions

$$u(\pm x_{\max}, t) = \frac{\sin x_{\max} \Delta k}{x_{\max} \Delta k}, \quad t > 0. \tag{13.165}$$

Consider the phase velocity $c = 10$ and the domain extent $x_{max} = 100$, and propagate the solution with the step sizes $h_x = 5 \cdot 10^{-2}$ and $h_t = 5 \cdot 10^{-3}$ up to $t_{max} = 40$. Verify that the chosen parameters satisfy the von Neumann stability condition.

For shaping the initial wave packet, use comparatively the wave number interval widths $\Delta k = 0.5$, 1, and 5. Discuss the effect of changing Δk and use in each case the amplitude of the final wave packet, which is reconstructed at the origin, to obtain an error estimate. Correlate the error estimates with the half-widths of the initial wave packets.

Print and plot the output every $n_{out} = 500$ time steps.

Solution

The Python implementation is presented in Listing 13.25 and the C/C++ version is available as supplementary material (P13-Wave.cpp). Snapshots of the time evolution of the wave packets for $\Delta k = 0.5$ and $\Delta k = 5$ are shown in Figures 13.22 and 13.23.

Listing 13.25 Finite-Difference Solution of the Classical Wave Equation (Python Coding)

```python
# Solve 1D wave equation by the explicit finite-difference method
from math import *
from pde import *
from graphlib import *


#===========================================================================
def Init(u0, u1, x, nx, c, dk, hx, ht):
#---------------------------------------------------------------------------
#  Returns initial solutions u0 and u1, for the first two time steps
#      u0(x,0) = sin(x*dk) / (x*dk)                    initial solution
#      v0(x,0) = 0                                 initial time derivative
#  x - spatial mesh, nx - number of nodes
#  c - phase velocity of wave, dk - wave number interval
#  hx - x-spacing, ht - time step size
#---------------------------------------------------------------------------
   for i in range(1,nx+1):                                 # time step 0
      u0[i] = sin(dk*x[i])/(dk*x[i]) if dk*x[i] else 1e0   # initial solution

   lam = c*ht/hx; lam = lam*lam                            # time step 1
   lam2 = 2e0*(1e0 - lam)
   u1[1] = u0[1]; u1[nx] = u0[nx]                    # constant boundary values
   for i in range(2,nx):
      v0 = 0e0                                       # initial time derivative
      u1[i] = 0.5e0*(lam*u0[i-1] + lam2*u0[i] + lam*u0[i+1]) - ht*v0


# main

c     = 10e0                                        # phase speed of wave
dk    = 1e0                                        # wave number interval
xmax  = 100e0                                             # maximum x
hx    = 5e-2                                        # spatial step size
tmax  = 40e0                                     # maximum propagation time
ht    = 5e-3                                              # time step
nout  = 500                                     # output every nout steps

nx = 2*(int)(xmax/hx + 0.5) + 1              # odd number of spatial nodes
nt = (int)(tmax/ht + 0.5)                           # number of time steps
nx2 = int(nx/2)

u0 = [0]*(nx+1)                                     # initial solution
u1 = [0]*(nx+1)                               # first propagated solution
u  = [0]*(nx+1)                                          # solution
```

```
x   = [0]*(nx+1)                                           # spatial mesh

for i in range(1,nx+1): x[i] = (i-nx2-1)*hx              # spatial mesh

Init(u0,u1,x,nx,c,dk,hx,ht)                            # initial wave packet

GraphInit(1200,600)

for it in range(1,nt+1):                                    # time loop
    t = it*ht
    PropagWave(u0,u1,u,nx,c,hx,ht)                    # propagate solution

    for i in range(1,nx): u0[i] = u1[i];  u1[i] = u[i]     # shift solutions

    if (it % nout == 0 or it == nt):              # output every nout steps
        fname = "wave_{0:4.2f}.txt".format(t)
        out = open(fname,"w")
        out.write("t = {0:4.2f}\n".format(t))
        out.write("      x              u\n")
        for i in range(1,nx+1):
            out.write("{0:10.5f}{1:10.5f}\n".format(x[i],u[i]))
        out.close

        GraphClear()
        title = "Wave propagation  t = {0:4.2f}".format(t)
        Plot(x,u,nx,"blue",1,0.15,0.95,0.15,0.90,"x","u",title)
        GraphUpdate()

MainLoop()
```

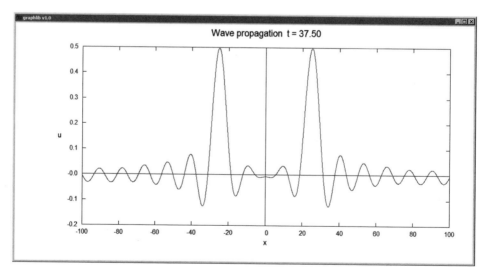

FIGURE 13.22 Numerical solution of problem 13.163–13.165 for the classical wave equation, yielded by Program 13.25 with $\Delta k = 0.5$ (for Problem 13.8).

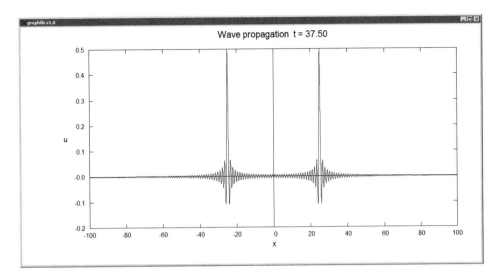

FIGURE 13.23   Numerical solution of problem 13.163–13.165 for the classical wave equation, yielded by Program 13.25 with $\Delta k = 5$ (for Problem 13.8).

References and Suggested Further Reading

Ames, W. F. 1992. *Numerical Methods for Partial Differential Equations* (3rd ed.). *Computer Science and Scientific Computing.* San Diego, New York: Academic Press.

Beu, T. A. 2004. *Numerical Calculus in C* (3rd ed., in Romanian). Cluj-Napoca: MicroInformatica.

Beu, T. A. 2007. Numerical solutions of the diffusion equation. In O.-G. Piringer and A. Baner (eds.), *Plastic Packaging Materials for Food: Barrier Function, Mass Transport, Quality Assurance, and Legislation.* Weinheim: Wiley-VCH.

Burden, R. and J. Faires. 2010. *Numerical Analysis* (9th ed.). Boston: Brooks/Cole, Cengage Learning.

Crank, J. 1975. *The Mathematics of Diffusion* (2nd ed.). Oxford: Clarendon Press.

Goldberg, A., H. M. Schey, and J. L. Schwartz. 1967. Computer-generated motion pictures of one-dimensional quantum-mechanical transmission and reflection phenomena. *American Journal of Physics* 35(3), 177–186.

Golub, G. H. and J. M. Ortega. 1992. *Scientific Computing and Differential Equations: An Introduction to Numerical Methods.* San Diego, New York, Boston, London, Sydney, Tokyo, Toronto: Academic Press.

Hoffman, J. D. 2001. *Numerical Methods for Engineers and Scientists* (2nd ed.). New York, Basel: Marcel Dekker.

Koonin, S. E. and D. C. Meredith. 1990. *Computational Physics: Fortran Version.* Redwood City, CA: Addison-Wesley.

Landau, R. H., M. J. Páez, and C. C. Bordeianu. 2007. *Computational Physics: Problem Solving with Computers* (2nd ed.). Weinheim: Wiley-VCH.

Press, W. H., S. A. Teukolsky, W. T. Vetterling, and B. P. Flannery. 2007. *Numerical Recipes: The Art of Scientific Computing* (3rd ed.). Cambridge: Cambridge University Press.

von Neumann, J. 1950. Numerical integration of the barotropic vorticity equation. *Tellus* 2(4), 237–254.

Appendix A: Dynamic Array Allocation in C/C++

Listing A.1 shows the content of the header file `memalloc.h`, which is included in all the C/C++ programs developed in this book for performing dynamic memory allocation for vectors and matrices of types `double`, `int`, and `complex`. The overall concepts are similar to those developed in Press et al. (2002) (for details, see Chapter 2).

Listing A.1   Functions for Dynamic Allocation of Vectors and Matrices (`memalloc.h`)

```
//------------------------------ memalloc.h ------------------------------
// Functions for dynamic memory allocation of vectors and matrices.
// Part of the numxlib numerics library. Author: Titus Beu, 2013
//------------------------------------------------------------------------
#ifndef _MEMALLOC_
#define _MEMALLOC_

#include <stdlib.h>
#include <stdio.h>
#include <complex>

using namespace std;
typedef complex<double> dcmplx;

//========================================================================
double *Vector(int imin, int imax)
//------------------------------------------------------------------------
// Allocates a double vector with indices in the range [imin,imax]
//------------------------------------------------------------------------
{
   double *p;
                                       // assign block start to array pointer
   p = (double*) malloc((size_t) ((imax-imin+1)*sizeof(double)));
   if (!p) {
      printf("Vector: allocation error !\n");
      exit(1);
   }
   return p - imin;                                       // adjust for offset
}

//========================================================================
void FreeVector(double *p, int imin)
//------------------------------------------------------------------------
// Deallocates a double vector allocated with Vector, with offset imin
//------------------------------------------------------------------------
{
   free((void*) (p+imin));                               // compensate for offset
}

//========================================================================
```

```
double **Matrix(int imin, int imax, int jmin, int jmax)
//--------------------------------------------------------------------------
// Allocates a double matrix, with row and column indices in the range
// [imin,imax] x [jmin,jmax]
//--------------------------------------------------------------------------
{
   int i, ni = imax-imin+1, nj = jmax-jmin+1;   // numbers of rows and columns
   double **p;
                                                // allocate array of row pointers
   p = (double**) malloc((size_t)(ni*sizeof(double*)));
   if (!p) {
      printf("Matrix: level 1 allocation error !\n");
      exit(1);
   }
   p -= imin;                                            // adjust for row offset
                                    // assign block start to 1st row pointer
   p[imin] = (double*) malloc((size_t)(ni*nj*sizeof(double)));
   if (!p[imin]) {
      printf("Matrix: level 2 allocation error !\n");
      exit(2);
   }
   p[imin] -= jmin;                     // adjust 1st row pointer for column offset
                                        // define row pointers spaced by row length
   for (i=imin+1; i<=imax; i++) p[i] = p[i-1] + nj;

   return p;
}


//==========================================================================
void FreeMatrix(double **p, int imin, int jmin)
//--------------------------------------------------------------------------
// Deallocates a double matrix allocated with Matrix, with row and column
// offsets imin and jmin
//--------------------------------------------------------------------------
{
   free((void*) (p[imin]+jmin));                           // deallocate block
   free((void*) (p+imin));                       // deallocate array of row pointers
}


//==========================================================================
int *IVector(int imin, int imax)
//--------------------------------------------------------------------------
// Allocates an int vector with indices in the range [imin,imax]
//--------------------------------------------------------------------------
{
   int *p;
                                        // assign block start to array pointer
   p = (int*) malloc((size_t) ((imax-imin+1)*sizeof(int)));
   if (!p) {
      printf("IVector: allocation error !\n");
      exit(1);
   }
   return p - imin;                                            // adjust for offset
}


//==========================================================================
void FreeIVector(int *p, int imin)
//--------------------------------------------------------------------------
// Deallocates an int vector allocated with IVector, with offset imin
```

```
//----------------------------------------------------------------------------
{
   free((void*) (p+imin));                                      // compensate for offset
}


//============================================================================
int **IMatrix(int imin, int imax, int jmin, int jmax)
//----------------------------------------------------------------------------
// Allocates an int matrix, with row and column indices in the range
// [imin,imax] x [jmin,jmax]
//----------------------------------------------------------------------------
{
   int i, ni = imax-imin+1, nj = jmax-jmin+1;  // numbers of rows and columns
   int **p;
                                                 // allocate array of row pointers
   p = (int**) malloc((size_t)(ni*sizeof(int*)));
   if (!p) {
      printf("Matrix: level 1 allocation error !\n");
      exit(1);
   }
   p -= imin;                                              // adjust for row offset
                                         // assign block start to 1st row pointer
   p[imin] = (int*) malloc((size_t)(ni*nj*sizeof(int)));
   if (!p[imin]) {
      printf("Matrix: level 2 allocation error !\n");
      exit(2);
   }
   p[imin] -= jmin;                   // adjust 1st row pointer for column offset
                                         // define row pointers spaced by row length
   for (i=imin+1; i<=imax; i++) p[i] = p[i-1] + nj;

   return p;
}


//============================================================================
void FreeIMatrix(int **p, int imin, int jmin)
//----------------------------------------------------------------------------
// Deallocates an int matrix allocated with IMatrix, with row and column
// offsets imin and jmin
//----------------------------------------------------------------------------
{
   free((void*) (p[imin]+jmin));                               // deallocate block
   free((void*) (p+imin));                        // deallocate array of row pointers
}


//============================================================================
dcmplx *CVector(int imin, int imax)
//----------------------------------------------------------------------------
// Allocates a complex vector with indices in the range [imin,imax]
//----------------------------------------------------------------------------
{
   dcmplx *p;
                                          // assign block start to array pointer
   p = (dcmplx*) malloc((size_t) ((imax-imin+1)*2*sizeof(double)));
   if (!p) {
      printf("CVector: allocation error !\n");
      exit(1);
   }
   return p - imin;                                              // adjust for offset
```

```
}

//================================================================================
void FreeCVector(dcmplx *p, int imin)
//--------------------------------------------------------------------------------
// Deallocates a complex vector allocated with CVector, with offset imin
//--------------------------------------------------------------------------------
{
   free((void*) (p+imin));                                    // compensate for offset
}


//================================================================================
dcmplx **CMatrix(int imin, int imax, int jmin, int jmax)
//--------------------------------------------------------------------------------
// Allocates a complex matrix, with row and column indices in the range
// [imin,imax] x [jmin,jmax]
//--------------------------------------------------------------------------------
{
   int i, ni = imax-imin+1, nj = jmax-jmin+1;  // numbers of rows and columns
   dcmplx **p;
                                              // allocate array of row pointers
   p = (dcmplx**) malloc((size_t)(ni*2*sizeof(double*)));
   if (!p) {
      printf("CMatrix: level 1 allocation error !\n");
      exit(1);
   }
   p -= imin;                                         // adjust for row offset
                                  // assign block start to 1st row pointer
   p[imin] = (dcmplx*) malloc((size_t)(ni*nj*2*sizeof(double)));
   if (!p[imin]) {
      printf("CMatrix: level 2 allocation error !\n");
      exit(2);
   }
   p[imin] -= jmin;                  // adjust 1st row pointer for column offset
                                  // define row pointers spaced by row length
   for (i = imin+1; i <= imax; i++) p[i] = p[i-1] + nj;

   return p;
}


//================================================================================
void FreeCMatrix(dcmplx **p, int imin, int jmin)
//--------------------------------------------------------------------------------
// Deallocates a complex matrix allocated with CMatrix, with row and column
// offsets imin and jmin
//--------------------------------------------------------------------------------
{
   free((void*) (p[imin]+jmin));                               // deallocate block
   free((void*) (p+imin));                          // deallocate array of row pointers
}

#endif
```

Appendix B: Basic Operations with Vectors and Matrices

Listings B.1 and B.2 show the modules matutil.py and matutil.h, which contain auxiliary Python and, respectively, C/C++ functions for performing basic operations with vectors and matrices.

Listing B.1 Utility Routines for Basic Operations with Vectors and Matrices (matutil.py)

```
#----------------------------------- matutil.py ----------------------------------
#  Contains utility routines for basic operations with vectors and matrices.
#  Author: Titus Beu, 2013
#---------------------------------------------------------------------------------
from math import *

#=================================================================================
def MatRead(a, n, m):
#---------------------------------------------------------------------------------
#  Reads from the keyboard the elements of matrix a[1:n][1:m]
#---------------------------------------------------------------------------------
   for i in range(1,n+1):
      for j in range (1,m+1):
         print("[",i,"][",j,"] = ",end=""); a[i][j] = input()

#=================================================================================
def MatPrint(a, n, m):
#---------------------------------------------------------------------------------
#  Prints the elements of matrix a[1:n][1:m] on the display
#---------------------------------------------------------------------------------
   for i in range(1,n+1):
      for j in range (1,m+1): print('{0:11.2e}'.format(a[i][j]),end="")
      print()

#=================================================================================
def MatPrintTrans(a, n, m):
#---------------------------------------------------------------------------------
#  Prints the elements of transposed matrix a[1:n][1:m] on the display
#---------------------------------------------------------------------------------
   for j in range(1,m+1):
      for i in range (1,n+1): print('{0:11.2e}'.format(a[i][j]),end="")
      print()

#=================================================================================
def MatZero(a, n, m):
#---------------------------------------------------------------------------------
#  Zeros the elements of matrix a[1:n][1:m]
#---------------------------------------------------------------------------------
   for j in range(1,m+1):
      for i in range (1,n+1): a[i][j] = 0e0

#=================================================================================
```

```
def MatCopy(a, b, n, m):
#-------------------------------------------------------------------------------
#  Copies matrix a[1:n][1:m] into matrix b[1:n][1:m]
#-------------------------------------------------------------------------------
   for i in range(1,n+1):
      for j in range (1,m+1): b[i][j] = a[i][j]


#===============================================================================
def MatTrans(a, n):
#-------------------------------------------------------------------------------
#  Replaces the square matrix a[1:n][1:n] by its transpose
#-------------------------------------------------------------------------------
   for i in range(2,n+1):
      for j in range(1,i):
         t = a[i][j]; a[i][j] = a[j][i]; a[j][i] = t


#===============================================================================
def MatDiff(a, b, c, n, m):
#-------------------------------------------------------------------------------
#  Returns the difference of matrices a and b in c[1:n][1:m]
#-------------------------------------------------------------------------------
   for i in range(1,n+1):
      for j in range (1,m+1): c[i][j] = a[i][j] - b[i][j]


#===============================================================================
def MatProd(a, b, c, n, l, m):
#-------------------------------------------------------------------------------
#  Returns the product of matrices a[1:n][1:l] and b[1:l][1:m] in c[1:n][1:m]
#-------------------------------------------------------------------------------
   for i in range(1,n+1):
      for j in range (1,m+1):
         t = 0e0
         for k in range(1,l+1): t += a[i][k] * b[k][j]
         c[i][j] = t


#===============================================================================
def MatPow(m, a, b, n):
#-------------------------------------------------------------------------------
#  Returns the m-th power of the square matrix a[1:n][1:n] in b[1:n][1:n]
#-------------------------------------------------------------------------------
   c = [[0]*(n+1) for i in range(n+1)]                    # work array

   MatCopy(a,b,n,n)                                        # case m = 1

   for k in range(2,m+1):                                 # repeated multiplication
      MatProd(a,b,c,n,n,n)
      MatCopy(c,b,n,n)


#===============================================================================
def MatNorm(a, n, m):
#-------------------------------------------------------------------------------
#  Returns the max norm of matrix a[1:n][1:m], i.e. max|a[i][j]|
#-------------------------------------------------------------------------------
   norm = 0e0
   for i in range(1,n+1):
      for j in range(1,m+1):
         if (norm < fabs(a[i][j])): norm = fabs(a[i][j])
   return norm
```

```python
#===========================================================================
def VecPrint(a, n):
#---------------------------------------------------------------------------
#  Prints the elements of vector a[1:n] on the display
#---------------------------------------------------------------------------
    for i in range(1,n+1): print('{0:11.2e}'.format(a[i]),end="")
    print()


#===========================================================================
def VecZero(a, n):
#---------------------------------------------------------------------------
#  Zeros the elements of vector a[1:n]
#---------------------------------------------------------------------------
    for i in range (1,n+1): a[i] = 0e0


#===========================================================================
def VecCopy(a, b, n):
#---------------------------------------------------------------------------
#  Copies vector a[1:n] into vector b[1:n]
#---------------------------------------------------------------------------
    for i in range(1,n+1): b[i] = a[i]


#===========================================================================
def VecDiff(a, b, c, n):
#---------------------------------------------------------------------------
#  Returns the difference of vectors a and b in c[1:n]
#---------------------------------------------------------------------------
    for i in range(1,n+1): c[i] = a[i] - b[i]


#===========================================================================
def VecNorm(a, n):
#---------------------------------------------------------------------------
#  Returns the 2-norm of a vector a[1:n]
#---------------------------------------------------------------------------
    norm = 0e0
    for i in range(1,n+1): norm += a[i] * a[i]
    norm = sqrt(norm)
    return norm


#===========================================================================
def MatVecProd(a, b, c, n):
#---------------------------------------------------------------------------
#  Returns the product of matrix a[1:n][1:n] and vector b[1:n] in c[1:n]
#---------------------------------------------------------------------------
    for i in range(1,n+1):
        t = 0e0
        for j in range (1,n+1): t += a[i][j] * b[j]
        c[i] = t
```

Listing B.2   Utility Routines for Basic Operations with Vectors and Matrices (`matutil.h`)

```c
//------------------------------ matutil.h ---------------------------------
// Contains utility routines for basic operations with vectors and matrices.
// Author: Titus Beu, 2013
//--------------------------------------------------------------------------
#ifndef _MATUTIL_
#define _MATUTIL_
```

```
#include <stdio.h>
#include <math.h>
#include "memalloc.h"

//=================================================================================
void MatRead(double **a, int n, int m)
//---------------------------------------------------------------------------------
// Reads from the keyboard the elements of matrix a[1:n][1:m]
//---------------------------------------------------------------------------------
{
   int i, j;

   for (i=1; i<=n; i++)
      for (j=1; j<=m; j++) { printf("[%i][%i]=",i,j); scanf("%f",&a[i][j]); }
}

//=================================================================================
void MatPrint(double **a, int n, int m)
//---------------------------------------------------------------------------------
// Prints the elements of matrix a[1:n][1:m] on the display
//---------------------------------------------------------------------------------
{
   int i, j;

   for (i=1; i<=n; i++) {
      for (j=1; j<=m; j++) printf("%12.2e",a[i][j]);
      printf("\n");
   }
}

//=================================================================================
void MatPrintTrans(double **a, int n, int m)
//---------------------------------------------------------------------------------
// Prints the elements of transposed matrix a[1:n][1:m] on the display
//---------------------------------------------------------------------------------
{
   int i, j;

   for (j=1; j<=m; j++) {
      for (i=1; i<=n; i++) printf("%12.2e",a[i][j]);
      printf("\n");
   }
}

//=================================================================================
void MatZero(double **a, int n, int m)
//---------------------------------------------------------------------------------
// Zeros the elements of matrix a[1:n][1:m]
//---------------------------------------------------------------------------------
{
   int i, j;

   for (i=1; i<=n; i++)
      for (j=1; j<=m; j++) a[i][j] = 0e0;
}

//=================================================================================
void MatCopy(double **a, double **b, int n, int m)
// ---------------------------------------------------------------------------------
```

```
// Copies matrix a[1:n][1:m] into matrix b[1:n][1:m]
//------------------------------------------------------------------------
{
   int i, j;

   for (i=1; i<=n; i++) {
      for (j=1; j<=m; j++) b[i][j] = a[i][j];
   }
}


//========================================================================
void MatTrans(double **a, int n)
//------------------------------------------------------------------------
// Replaces the square matrix a[1:n][1:n] by its transpose
//------------------------------------------------------------------------
{
   double t;
   int i, j;

   for (i=2; i<=n; i++)
      for (j=1; j<=(i-1); j++) {
         t = a[i][j]; a[i][j] = a[j][i]; a[j][i] = t;
      }
}


//========================================================================
void MatDiff(double **a, double **b, double **c, int n, int m)
//------------------------------------------------------------------------
// Returns the difference of matrices a and b in c[1:n][1:m]
//------------------------------------------------------------------------
{
   int i, j;

   for (i=1; i<=n; i++)
      for (j=1; j<=m; j++) c[i][j] = a[i][j] - b[i][j];
}


//========================================================================
void MatProd(double **a, double **b, double **c, int n, int l, int m)
//------------------------------------------------------------------------
// Returns the product of matrices a[1:n][1:l] and b[1:l][1:m] in c[1:n][1:m]
//------------------------------------------------------------------------
{
   double t;
   int i, j, k;

   for (i=1; i<=n; i++)
      for (j=1; j<=m; j++) {
         t = 0e0;
         for (k=1; k<=l; k++) t += a[i][k] * b[k][j];
         c[i][j] = t;
      }
}


//========================================================================
void MatPow(int m, double **a, double **b, int n)
//------------------------------------------------------------------------
// Returns the m-th power of the square matrix a[1:n][1:n] in b[1:n][1:n]
//------------------------------------------------------------------------
```

```
{
   double **c;
   int k;

   c = Matrix(1,n,1,n);                                    // work array

   MatCopy(a,b,n,n);                                       // case m = 1

   for (k=2; k<=m; k++) {                         // repeated multiplication
      MatProd(a,b,c,n,n,n);
      MatCopy(c,b,n,n);
   }

   FreeMatrix(c,1,1);
}

//=============================================================================
double MatNorm(double **a, int n, int m)
//-----------------------------------------------------------------------------
// Returns the max norm of matrix a[1:n][1:m], i.e. max|a[i][j]|
//-----------------------------------------------------------------------------
{
   double norm;
   int i, j;

   norm = 0e0;
   for (i=1; i<=n; i++)
      for (j=1; j<=m; j++)
         if (norm < fabs(a[i][j])) norm = fabs(a[i][j]);
   return norm;
}

//=============================================================================
void VecPrint(double a[], int n)
//-----------------------------------------------------------------------------
// Prints the elements of vector a[1:n] on the display
//-----------------------------------------------------------------------------
{
   int i;

   for (i=1; i<=n; i++) printf("%12.2e",a[i]);
   printf("\n");
}

//=============================================================================
void VecZero(double a[], int n)
//-----------------------------------------------------------------------------
// Zeros the elements of vector a[1:n]
//-----------------------------------------------------------------------------
{
   int i;

   for (i=1; i<=n; i++) a[i] = 0e0;
}

//=============================================================================
void VecCopy(double a[], double b[], int n)
//-----------------------------------------------------------------------------
// Copies vector a[1:n] into vector b[1:n]
```

```
//------------------------------------------------------------------------
{
   int i;

   for (i=1; i<=n; i++) b[i] = a[i];
}

//========================================================================
void VecDiff(double a[], double b[], double c[], int n)
//------------------------------------------------------------------------
// Returns the difference of vectors a and b in c[1:n]
//------------------------------------------------------------------------
{
   int i;

   for (i=1; i<=n; i++) c[i] = a[i] - b[i];
}

//========================================================================
double VecNorm(double a[], int n)
//------------------------------------------------------------------------
// Returns the 2-norm of a vector a[1:n]
//------------------------------------------------------------------------
{
   double norm;
   int i;

   norm = 0e0;
   for (i=1; i<=n; i++) norm += a[i] * a[i];
   norm = sqrt(norm);
   return norm;
}

//========================================================================
void MatVecProd(double **a, double b[], double c[], int n)
//------------------------------------------------------------------------
// Returns the product of matrix a[1:n][1:n] and vector b[1:n] in c[1:n]
//------------------------------------------------------------------------
{
   double t;
   int i, j;

   for (i=1; i<=n; i++) {
      t = 0e0;
      for (j=1; j<=n; j++) t += a[i][j] * b[j];
      c[i] = t;
   }
}

#endif
```

Appendix C: Embedding Python in C/C++

The topic of embedding Python code in C/C++ programs actually exceeds the scope of the book, yet it is briefly addressed below to offer an overview on the mechanisms lying at the core of creating Python graphics from C/C++ programs, as explained in Section 3.6.

From a more general perspective, *mixed-language programming* implies developing programs in two or more languages. It can certainly be worthwhile, since it enables combining complementary advantages of different languages (ease of use, portability, runtime performance, etc.), or calling existing code (libraries) in a language from code written in another language. It is, nevertheless, a complex undertaking and should be resorted to only under well-justified circumstances.

Embedding Python can extend the functionality of C/C++ programs by plug-ins which are generally more easy to develop, debug, and use (see, e.g., Du 2005; Fötsch 2012; Python 3.x 2013a,b). Specifically, embedding Python graphics libraries (such as the `graphlib.py` module developed in Chapter 3) is of particular interest, since it combines the best of both worlds: the portability of Python graphics and the efficiency of compiled C/C++ code.

In the following, we develop the basic ideas of embedding Python by a cookbook approach. We consider the Python module `operations.py` from Listing C.1, which consists of two functions: `Sum`—returning the sum of its arguments x and y, and `Average`—returning the average of the n values received by way of vector x. The first function is meant to exemplify the procedure for scalar arguments, while the second, for vector arguments.

The C++ program (Listing C.2) is supposed to simply calculate the sum of two values, which are to be passed to the Python function `Sum` as a *tuple*. To this end, the program includes the header file `Python.h` and defines several objects of type `PyObject`: pModule—the Python module object, pFunc—the Python function reference, pArgs—the tuple of arguments, pValue—the returned value.

The Python interpreter is invoked by a call to function `Py_Initialize`. The program then imports the module `operations.py` by applying the `PyImport_Import` method (the conversion of the module's name to the Python Unicode format is done by way of function `PyUnicode_FromString`). Once the module is imported, its reference pModule is used by the method `PyObject_GetAttrString` to obtain the reference pFunc to the Python function `Sum`.

Listing C.1   Module `operations.py` Containing Functions Called from C/C++ (Python Coding)

```python
def Sum(x, y):                               # returns the sum of x and y
    s = x + y
    return s

def Average(x, n):                # returns the average of x[1] through x[n]
    s = 0e0
    for i in range(1,n+1): s += x[i]
    s /= n
    return s
```

Listing C.2 Calling Python Functions with Scalar Arguments from C++ Code (C/C++ Coding)

```cpp
// Calling Python functions with scalar arguments from C++ code
#include <iostream>
#include <Python.h>

int main()
{
    PyObject *pModule, *pFunc, *pArgs, *pValue;  // pointers to Python objects
    double x, y;

    x = 2.0; y = 3.0;                                        // values to be added

    Py_Initialize();                                         // initialize Python
                                         // import module "operations.py"
    pModule = PyImport_Import(PyUnicode_FromString("operations"));
    if (!pModule) std::cout << "Failed to load <operations.py>";
                                         // get reference to function Sum
    pFunc = PyObject_GetAttrString(pModule, "Sum");

    pArgs = PyTuple_New(2);                           // define tuple of 2 arguments
    PyTuple_SetItem(pArgs, 0, PyFloat_FromDouble(x));       // set 1st argument
    PyTuple_SetItem(pArgs, 1, PyFloat_FromDouble(y));       // set 2nd argument

    pValue = PyObject_CallObject(pFunc, pArgs);       // call Python function

    std::cout << x << " + " << y << " = " << PyFloat_AsDouble(pValue);

    Py_DECREF(pArgs);                                        // clean up
    Py_DECREF(pFunc);
    Py_DECREF(pModule);

    Py_Finalize();                                           // finish Python
}
```

The tuple of arguments to be passed to the Python function is populated with the help of the PyTuple_SetItem method. The conversion from the C/C++ type double to the float type of Python is accomplished by the function PyFloat_FromDouble.

The actual call of the Python function Sum is carried out by PyObject_CallObject, having as arguments the function object pFunc and the argument list pArgs. Before any further use, the returned value, pValue, needs to be converted into the type double by a call to PyFloat_AsDouble.

In the spirit of object-oriented programming, it is essential to dispose all the Python objects in reversed order of their creation, and this is done by means of the Py_DECREF method. Finally, the Python interpreter is released by calling Py_Finalize().

Similar in concept, the program C.3 differs in that one of the arguments of the called function, Average, is a vector. To account for this difference, an additional pointer px to a Python list is created. The list is effectively allocated with PyList_New and its items are set by the method PyList_SetItem.

For a Python function, say Sum, to be conveniently used in a C/C++ program, it needs to be interfaced by a *wrapper function* meant to hide all the communication details (Listing C.3). The robust way to implement such a mechanism is actually by way of two C++ classes, namely a wrapper class for the

Listing C.3 Calling Python Functions with Vector Arguments from C++ Code (C/C++ Coding)

```cpp
// Calling Python functions with vector arguments from C++ code
#include <iostream>
#include <Python.h>

int main()
{
    PyObject *pModule, *pFunc, *pArgs, *pValue;  // pointers to Python objects
    double x[10];
    int i, n;
    PyObject *px;                                // pointer to Python list

    Py_Initialize();                                      // initialize Python
                                          // import module "operations.py"
    pModule = PyImport_Import(PyUnicode_FromString("operations"));
    if (!pModule) std::cout << "Failed to load <operations.py>";
                                     // get reference to function Average
    pFunc = PyObject_GetAttrString(pModule, "Average");

    n = 5;                                       // values to be averaged:
    for (i=1; i<=n; i++) { x[i] = 0.1 * i; }            // x[1] to x[n]
                                              // prepare Python list
    px = PyList_New(n+1);        // pointer to list (offset 0 => n+1 components)
    for (i=1; i<=n; i++) {
        PyList_SetItem(px, i, PyFloat_FromDouble(x[i]));      // populate list
    }

    pArgs = PyTuple_New(2);                      // define tuple of 2 arguments
    PyTuple_SetItem(pArgs, 0, px);                      // set 1st argument
    PyTuple_SetItem(pArgs, 1, PyLong_FromLong(n));       // set 2nd argument

    pValue = PyObject_CallObject(pFunc, pArgs);        // call Python function

    std::cout << " Average = " << PyFloat_AsDouble(pValue);

    Py_DECREF(pArgs);                                          // clean up
    Py_DECREF(pFunc);
    Py_DECREF(pModule);

    Py_Finalize();                                        // finish Python
}
```

Python interpreter itself and a wrapper for calling the functions of the Python module. The corresponding implementation can be seen in Listing C.4.

The constructor of the class `Python` is responsible for initializing the Python interpreter, while the constructor of the class `PyOperations` creates an instance of the Python interpreter and imports the `operations.py` module. The whole machinery for getting the reference to the Python function, for preparing and passing the tuple of arguments, and for retrieving the result has been moved to the C++ wrapper function bearing the same name, `Sum`.

The main program is just supposed to declare an object of type `PyOperations` and to call its method `Sum` to perform the summation.

Listing C.4 Calling Python Functions via C++ Wrapper Classes (C/C++ Coding)

```cpp
// Calling Python functions via C++ wrapper classes
#include <iostream>
#include <Python.h>
//=================================================================================
class Python                            // wrapper class for Python interpreter
{
   public:
   Python() { Py_Initialize(); }                // constructor: initialize Python
   ~Python() { Py_Finalize(); }                   // destructor: finish Python
};
//=================================================================================
class PyOperations                   // wrapper class for module operations.py
{
   Python* contxt;
   PyObject* pModule;

   public:
   PyOperations()                                                // constructor
   {
      contxt = new Python();                  // create new instance of Python
      pModule = PyImport_Import(PyUnicode_FromString("operations"));
      if (!pModule) std::cout << "Failed to load <operations.py>";
   }
   virtual ~PyOperations()                                         // destructor
   {
      if (pModule) Py_DECREF(pModule);
      if (contxt) delete contxt;
   }
//---------------------------------------------------------------------------------
   double Sum(double x, double y)           // wrapper for Python function Sum
   {
      PyObject *pFunc, *pArgs, *pValue;
                                                     // get reference to Sum
      pFunc = PyObject_GetAttrString(pModule, "Sum");

      pArgs = PyTuple_New(2);                   // define tuple of 2 arguments
      PyTuple_SetItem(pArgs, 0, PyFloat_FromDouble(x));   // set 1st argument
      PyTuple_SetItem(pArgs, 1, PyFloat_FromDouble(y));   // set 2nd argument

      pValue = PyObject_CallObject(pFunc, pArgs);     // call Python function

      Py_DECREF(pArgs);                                          // clean up
      Py_DECREF(pFunc);

      return PyFloat_AsDouble(pValue);                       // return result
   }
};

int main()
{
   PyOperations c;                            // object of type PyOperations
   double x, y;

   x = 2.0, y = 3.0;                                  // values to be added
   std::cout << x << " + " << y << " = " << c.Sum(x,y);   // call method Sum
}
```

References and Suggested Further Reading

Du, J. 2005. Embedding Python in C/C++: Part I. http://www.codeproject.com/Articles/11805/Embedding-Python-in-C-C-Part-I.

Fötsch, M. 2012. Embedding Python: Tutorial, Part 1. http://realmike.org/blog/tag/embedding/.

Python 3.x 2013a. Python Programming Language Official Website. http://www.python.org/.

Python 3.x 2013b. Python v3.3.2 Documentation: Embedding Python in C++. http://docs.python.org/3.3/extending/embedding.html.

Appendix D: The Numerical Libraries `numxlib.py` and `numxlib.h`

The numerical libraries `numxlib.py` and `numxlib.h` are located, respectively, in the folders `/INP/modules/` and `/INP/include/`. They have identical structure and include the partial Python and, respectively, C/C++ libraries developed in Chapters 4 through 13. The contained routines have identical names and functionalities. There is a single exception to the identical naming convention, regarding the files `random1.py` and `random.h`, which is due to the existence of a standard Python module `random.py`. In addition to the corresponding partial libraries, `numxlib.h` also includes `memalloc.h`, containing dynamic memory allocation routines for vectors and matrices (see Appendix A). The concrete content of the files `numxlib.py` and `numxlib.h` is provided in Listings D.1 and D.2, respectively.

Listing D.1 The `numxlib.py` Numerics Library for Python (Python Coding)

```python
#--------------------------------- numxlib.py ---------------------------------
#  Library of numerical methods for Python
#  Author: Titus Beu, 2013
#------------------------------------------------------------------------------

from sort import *                              # sorting and indexing

from specfunc import *                          # special functions

from roots import *                             # roots of equations

from linsys import *                            # systems of linear equations

from eigsys import *                            # eigenvalue problems

from modfunc import *                           # modeling of tabulated functions

from integral import *                          # integration of functions

from random1 import *                           # generation of random numbers

from ode import *                               # ordinary differential equations

from pde import *                               # partial differential equations

from utils import *                             # utility functions
```

Listing D.2   The numxlib.h Numerics Library for Python (C/C++ Coding)

```
//------------------------------ numxlib.h -----------------------------
// Library of numerical methods for C/C++
// Author: Titus Beu, 2013
//-----------------------------------------------------------------------
#ifndef _NUMXLIB_
#define _NUMXLIB_

#include "memalloc.h"                         // dynamic allocation of arrays

#include "sort.h"                                   // sorting and indexing

#include "specfunc.h"                                  // special functions

#include "roots.h"                                    // roots of equations

#include "linsys.h"                            // systems of linear equations

#include "eigsys.h"                                   // eigenvalue problems

#include "modfunc.h"                      // modeling of tabulated functions

#include "integral.h"                          // integration of functions

#include "random.h"                          // generation of random numbers

#include "ode.h"                         // ordinary differential equations

#include "pde.h"                         // partial differential equations

#include "utils.h"                                    // utility functions

#endif
```

The detailed structure of numxlib.py is described below, being similar to that of numxlib.h.

numxlib.py	Library of numerical methods for Python.
sort.py	Contains routines for sorting, indexing, and ranking numeric sequences.
BubbleSort	Sorts a numeric sequence using bubble sort.
InsertSort	Sorts a numeric sequence by direct insertion.
QuickSort	Sorts a numeric sequence using Quicksort.
Index	Indexes a numeric sequence by direct insertion.
Rank	Ranks a numeric sequence based on the array of indexes.
Index2	Performs the indexing of two correlated arrays by direct insertion.
specfunc.py	Contains routines for evaluating special functions.
Chebyshev	Evaluates Chebyshev polynomials of the first kind and their derivatives using the recurrence relation.
Legendre	Evaluates Legendre polynomials and their derivatives using the recurrence relation.
aLegendre	Evaluates associated Legendre functions.

Laguerre	Evaluates Laguerre polynomials and their derivatives using the recurrence relation.
aLaguerre	Evaluates associated Laguerre polynomials.
Hermite	Evaluates Hermite polynomials and their derivatives using the recurrence relation.
SpherY	Evaluates the real and imaginary parts of spherical harmonics.
SBessy	Evaluates spherical Neumann functions.
SBessj	Evaluates spherical Bessel functions.
roots.py	Contains routines for determining real roots of real functions.
Bisect	Determines a real root of a function by the bisection method.
FalsPos	Determines a real root of a function by the false position method.
Iter	Determines a real root of a function by the method of successive approximations.
Newton	Determines a real root of a function by the Newton–Raphson method using the analytical derivative.
NewtonNumDrv	Determines a real root of a function by the Newton–Raphson method using the numerical derivative.
Secant	Determines a real root of a function by the secant method.
BirgeVieta	Determines real roots of polynomials by the Birge–Vieta method.
Jacobian	Calculates the Jacobian of a system of multidimensional functions using central finite differences.
NewtonSys	Solves a system of nonlinear equations by Newton's method.
linsys.py	Contains routines for solving systems of linear equations.
Gauss	Solves a matrix equation by Gaussian elimination with partial pivoting.
GaussJordan0	Solves a matrix equation by Gauss–Jordan elimination with partial pivoting.
GaussJordan1	Solves a matrix equation by Gauss–Jordan elimination with partial pivoting and matrix inversion.
GaussJordan	Solves a matrix equation by Gauss–Jordan elimination with complete pivoting and matrix inversion.
LUFactor	Performs LU factorization of a matrix by Doolittle's method.
LUSystem	Solves a linear system using the LU decomposition of its matrix.
MatInv	Performs matrix inversion using LU factorization.
MatTriInv	Inverts a triangular matrix.
Cholesky	Performs the Cholesky factorization of a symmetric positive-definite matrix.
CholeskySys	Solves a positive-definite linear system using the Cholesky decomposition of its matrix.
MatSymInv	Inverts a positive-definite matrix using the Cholesky factorization.
TriDiagSys	Solves a linear system with tridiagonal matrix.
GaussSeidel	Solves a system of linear equations by the Gauss–Seidel method.
eigsys.py	Contains routines for solving eigenvalue problems.
Jacobi	Solves an eigenvalue problem for a real symmetric matrix using the Jacobi method.
EigSym	Solves a generalized eigenvalue problem for real symmetric (positive-definite) matrices.
EigSort	Sorts a set of eigenvalues and eigenvectors according to the eigenvalues.

modfunc.py	Contains routines for modeling tabular dependences by interpolation or regression.
Lagrange	Evaluates the Lagrange interpolating polynomial for a set of data points.
Neville	Evaluates the Lagrange interpolating polynomial by Neville's method.
Spline	Performs cubic spline interpolation.
LinFit	Performs linear regression on a set of data points.
MultiFit	Performs multilinear regression on a set of data points.
PolFit	Performs polynomial regression on a set of data points.
MarqFit	Performs nonlinear regression based on the Levenberg–Marquardt method.
integral.py	Contains 1D and 3D integrators for real functions with real variables.
qTrapz	Function integrator based on the trapezoidal rule.
qSimpson	Function integrator based on Simpson's rule.
qTrapzCtrl	Adaptive integrator based on the trapezoidal rule.
qSimpsonCtrl	Adaptive integrator based on Simpson's rule.
qRomberg	Adaptive integrator based on Romberg's method.
qImprop1	Adaptive integrator for improper integrals of the first kind (with infinite limits).
qImprop2	Adaptive integrator for improper integrals of the second kind (with singular integrable limits).
qMidPoint	Adaptive integrator based on the midpoint rule.
xGaussLeg	Generates abscissas and weights for Gauss–Legendre quadratures.
qGaussLeg	Gauss–Legendre integrator.
xGaussLag	Generates abscissas and weights for Gauss–Laguerre quadratures.
qGaussLag	Gauss–Laguerre integrator.
qTrapz3D	3D integrator based on the trapezoidal rule.
xSimpson	Generates integration points and weights for Simpson's rule.
qSimpson3D	3D integrator based on Simpson's rule.
qGaussLeg3D	3D integrator based on Gaussian quadratures.
qSimpsonAng	Integrator in spherical angular coordinates based on Simpson's rule.
qSimpsonSph	Integrator in spherical coordinates based on Simpson's rule.
qGaussSph	Integrator in spherical coordinates based on Gaussian quadratures.
qSimpsonCyl	Integrator in cylindrical coordinates based on Simpson's rule.
random1.py	Contains routines for generating pseudorandom numbers.
randLCG1	Linear congruential random number generator based on specifications from Press et al. (2002).
randLCG2	Linear congruential random number generator based on specifications from Rapaport (2004).
randMCG	Multiply-with-Carry random number generator based on specifications of George Marsaglia.
randNrm	Generates random numbers with normal distribution based on the central limit theorem.
randNrm2	Returns two random numbers with normal distribution.
randMet	Random number generator based on the Metropolis method.
ode.py	Contains routines for solving systems of ordinary differential equations.

Euler	Euler propagator for systems of first-order ODEs.
EulerPC	Euler predictor–corrector propagator for systems of first-order ODEs.
RungeKutta	Fourth-order Runge–Kutta solver for Cauchy problems for systems of first-order ODEs.
RKadapt	Adaptive ODE solver using step-halving and the Runge–Kutta method.
RKFehlberg	Adaptive ODE solver based on the Runge–Kutta–Fehlberg method.
Euler1	Euler propagator for a second-order ODE.
EulerCromer1	Euler–Cromer propagator for a second-order ODE.
Verlet1	Verlet propagator for a second-order ODE.
Euler2	Euler propagator for a system of second-order ODEs.
EulerCromer	Euler–Cromer propagator for a system of second-order ODEs.
Verlet2	2D velocity Verlet propagator for a single particle.
Verlet	Velocity Verlet propagator for a system of particles in interaction.
EulerCromerQM	Euler–Cromer integrator for the 1D Schrödinger equation.
Numerov	Numerov integrator for the 1D Schrödinger equation.
Propag	Propagates the solution of a second-order ODE on a regular mesh using the Euler–Cromer method.
Shoot	Solves a two-point boundary-value problem for a second-order ODE using the shooting method.
ShootQM	Solves the 1D Schrödinger equation using the shooting method.
Bilocal	Linear bilocal problem solver based on the finite-difference method.
pde.py	Contains routines for solving partial differential equations.
Poisson0	Solver for the 2D Poisson equation in Cartesian coordinates with Dirichlet boundary conditions.
PoissonXY	Solver for the 2D Poisson equation in Cartesian coordinates with arbitrary boundary conditions.
Poisson2D	Solver for the 2D Poisson equation in Cartesian coordinates on an irregular domain.
PropagFTCS	Explicit forward-time central-space propagator for the diffusion equation.
PropagBTCS	Implicit backward-time central-space propagator for the diffusion equation.
PropagCN	Crank–Nicolson propagator for the diffusion equation.
PropagDiff	Crank–Nicolson propagator for the diffusion equation with variable diffusion coefficient.
PropagQTD	Crank–Nicolson propagator for the Schrödinger equation based on tridiagonal solver.
PropagQGS	Crank–Nicolson propagator for the Schrödinger equation using the Gauss–Seidel method.
PropagWave	Explicit forward-time central-space propagator for the classical wave equation.
utils.py	Contains utility routines.
Nint	Returns the nearest integer to its real-type argument.
Sign	Transfers the sign of its second argument onto the absolute value of the first argument.
Magn	Returns the order of magnitude of its argument as a power of 10.
Index	Indexes a numeric sequence by direct insertion

Appendix E: The Graphics Library graphlib.py Based on Tkinter

Listing E.1 shows the Python library graphlib.py based on Tkinter, which is used for producing runtime graphics throughout the book. The module contains functions for plotting real functions of one variable (Plot, MultiPlot, and HistoBin), as well as of two variables (Contour). Functions are also provided for rendering systems of particles as connected spheres (PlotParticles) and 3D structures defined by triangular planes (PlotStruct).

The module graphlib.py uses a series of auxiliary functions, which are included in the module utils.py, shown in Listing E.2.

Listing E.1   The Python Graphics Library (graphlib.py)

```
#---------------------------------- graphlib.py ----------------------------------
#  Graphics library based on Tkinter.
#  Author: Titus Beu, 2013
#---------------------------------------------------------------------------------
from math import *
from utils import *
from tkinter import *

root = Tk()                                            # create Tk root widget
root.title("graphlib v1.0")


#=================================================================================
def MainLoop():                                        # creates Tk event loop
   root.mainloop()


#=================================================================================
def GraphUpdate():                                     # updates Tk root widget
   root.update()


#=================================================================================
def GraphInit(nxwin, nywin):        # creates Canvas widget and returns object
   global w, nxw, nyw                            # make canvas object available

   nxw = nxwin; nyw = nywin                                        # canvas size
   w = Canvas(root, width=nxw, height=nyw, bg = "white")      # create canvas
   w.pack()                                            # make canvas visible
   return w


#=================================================================================
def GraphClear():                         # deletes content of Canvas widget
   global w, nxw, nyw                               # canvas object and size
   w.delete(ALL)


#=================================================================================
def Limits(xmin, xmax, maxint):
#---------------------------------------------------------------------------------
```

```
#   Replaces the limits xmin and xmax of a real interval with the limits of
#   the smallest extended inteval which includes a number <= maxint of
#   subintervals of length d * 10**p, with d = 1, 2, 5 and p integer.
#
#   scale - scale factor (10**p)
#   nsigd - relevant number of significant digits
#   nintv - number of subintervals
#
#   Returns: xmin, xmax, scale, nsigd, nintv
#---------------------------------------------------------------------------
   eps = 1e-5                                       # relative precision criterion
   xfact = [0.5e0,0.5e0,0.4e0]                               # 0.5*0.5*0.4 = 0.1

   if (abs(xmax - xmin) < 10e0*eps*abs(xmax)):
      corrmin = 1e0 - 10e0 * Sign(eps,xmin)
      corrmax = 1e0 + 10e0 * Sign(eps,xmax)
      xmin *= corrmin
      xmax *= corrmax
                                                           # initial scale factor
   factor = 1e0/(eps * min(Magn(xmin),Magn(xmax))) if (xmin * xmax) else \
            1e0/(eps * max(Magn(xmin),Magn(xmax)))

   corrmin = 1e0 + Sign(eps,xmin)                                   # corrections
   corrmax = 1e0 - Sign(eps,xmax)
   for i in range(1,100):                                   # multiply iteratively
      xmins = floor(xmin * factor * corrmin)               # factor with xfact[]
      xmaxs = ceil (xmax * factor * corrmax)               # until the no. of
      xnint = abs(xmaxs - xmins)                           # subintervals becomes
      if (xnint <= maxint): break                          # xnint <= maxint
      modi = i % 3
      factor = factor * xfact[modi]

   factor = 1e0 / factor                                         # scale factor
   xmin = xmins * factor                                         # xmin and xmax
   xmax = xmaxs * factor
   scale = max(Magn(xmin),Magn(xmax))                            # scale factor
   factor = max(abs(xmins),abs(xmaxs))
   for i in range(1,modi+1): factor = factor / xfact[i]
   nsigd = int(log10(factor) + 1)                          # no. of significant digits
   nintv = Nint(xnint)                                      # no. of subintervals

   return (xmin, xmax, scale, nsigd, nintv)

#===========================================================================
def FormStr(x, scale, nsigd):
#---------------------------------------------------------------------------
#   Formats the number x (with factor scale) to nsigd significant digits
#   returning the mantissa in mant[] and the exponent of 10 in expn[].
#
#   Returns: mant, expn
#---------------------------------------------------------------------------
   ndigmax = 5                                              # maximum no. of digits
   mant = expn = ""

   n = Nint(log10(scale))                                            # exponent
   if ((n < -1) or (n > 3)):
      expn = repr(n)
      x = x / scale                                         # divide x by scale
      n = 0
```

```
   n += 1                                    # no. of digits before decimal point
   ndig = min(ndigmax,max(nsigd,n))                    # total no. of digits
   ndec = ndig - n                                       # no. of decimals
   x = round(x,ndec)
   mant = "{0:{1:}.{2:}f}".format(x,ndig,ndec)

   return (mant, expn)

#===============================================================================
def Plot(x, y, n, col, sty, fxmin, fxmax, fymin, fymax, xtext, ytext, title):
#-------------------------------------------------------------------------------
#  Plots a real function of one variable specified by a set of (x,y) points.
#  The x and y-domains are extended to fit at most 10 intervals expressible
#  as d * 10^p, with d = 1, 2, 5 and p integer.
#
#  x[]   - abscissas of tabulation points (x[1] through x[n])
#  y[]   - ordinates of tabulation points (y[1] through y[n])
#  n     - number of tabulation points
#  col   - plot color ("red", "green", "blue" etc.)
#  sty   - plot style: 0 - scatter plot, 1 - line plot, 2 - polar plot,
#                      3 - drop lines, 4 - histogram
#  fxmin - min fractional x-limit of viewport (0 < fxmin < fxmax < 1)
#  fxmax - max fractional x-limit of viewport
#  fymin - min fractional y-limit of viewport (0 < fymin < fymax < 1)
#  fymax - max fractional y-limit of viewport
#  xtext - x-axis title; for "None" - axis is not labeled
#  ytext - y-axis title; for "None" - axis is not labeled
#  title - plot title
#-------------------------------------------------------------------------------
   global w, nxw, nyw                              # canvas object and size
   maxintx = maxinty = 10            # max. number of labeling subintervals

   xmin = min(x[1:n+1]); xmax = max(x[1:n+1])              # domain limits
   ymin = min(y[1:n+1]); ymax = max(y[1:n+1])

   ixmin = Nint(fxmin*nxw); iymin = Nint((1e0-fymin)*nyw)  # viewport limits
   ixmax = Nint(fxmax*nxw); iymax = Nint((1e0-fymax)*nyw)

   if (sty == 2):                                          # polar plot
      xmin = min(xmin,ymin); xmax = max(xmax,ymax)      # make domain square
      xmax = max(abs(xmin),abs(xmax)); xmin = -xmax
      ymin = xmin; ymax = xmax
      if (ixmax-ixmin > iymin-iymax):            # adjust viewport limits
         c = 0.5*(ixmin+ixmax)
         d = 0.5*(iymin-iymax)
         ixmin = c - d; ixmax = c + d
      if (ixmax-ixmin < iymin-iymax):
         c = 0.5*(iymin+iymax)
         d = 0.5*(ixmax-ixmin)
         iymin = c + d; iymax = c - d

   w.create_rectangle(ixmin,iymax,ixmax,iymin)             # draw plot frame

   nfont = min(int((ixmax-ixmin)/60.) + 3,12)                 # font size
   font1 = ("Helvetica",nfont)                         # axis label font
   font2 = ("Helvetica",Nint(1.2*nfont))               # axis title font
   font3 = ("Helvetica",Nint(1.4*nfont))               # plot title font
   if ((ixmax-ixmin) < 3 * (iymin-iymax)):
```

```
      w.create_text((ixmin+ixmax)/2,iymax-3*nfont,text=title,font=font3)
   else:
      w.create_text(ixmax,iymax,text=title,font=font2,anchor="ne")
      maxinty = max(5,(iymin-iymax)/(ixmax-ixmin)*maxintx)
                                                              # X-AXIS
   (xmin,xmax,scale,nsigd,nintv) = Limits(xmin,xmax,maxintx)  # extend limits
   ax = (ixmax-ixmin)/(xmax-xmin)                       # x scaling coefficients
   bx = ixmin - ax*xmin

   tic =(ixmax-ixmin)/100.                                      # tic length
   h = (xmax-xmin)/nintv; htic = ax * h                     # labeling step
   iytext = iymin + 1.5*nfont
   for i in range(0,nintv+1):                                  # label axis
      ix = Nint(ixmin + i*htic)
      w.create_line(ix,iymin,ix,iymin-tic)                        # tics
      w.create_line(ix,iymax,ix,iymax+tic)
      if (xtext != "None"):
         (mant,expn) = FormStr(xmin+i*h,scale,nsigd)
         w.create_text(ix,iytext,text=mant,font=font1)           # labels

   if (xtext != "None"):
      if ((scale<0.1) or (scale>1000.0)): xtext = xtext + " 1e" + expn
      ixtext = (ixmin+ixmax)/2
      iytext = iytext + 2*nfont
      w.create_text(ixtext,iytext,text=xtext,font=font2)        # x title
                                                              # Y-AXIS
   if (ymin == 0.0 and ymax == 0.0): ymin = -1e0; ymax = 1e0     # horizontal
   if (abs(ymax-ymin) < 1e-5*abs(ymax)): ymin *= 0.9; ymax *= 1.1
   (ymin,ymax,scale,nsigd,nintv) = Limits(ymin,ymax,maxinty)  # extend limits
   ay = (iymax-iymin)/(ymax-ymin)                       # y scaling coefficients
   by = iymin - ay*ymin

   h = (ymax-ymin)/nintv; htic = ay * h                    # labeling step size
   ixtext = ixmin - nfont
   for i in range(0,nintv+1):                                  # label axis
      iy = Nint(iymin + i*htic)
      w.create_line(ixmin,iy,ixmin+tic,iy)                        # tics
      w.create_line(ixmax,iy,ixmax-tic,iy)
      if (ytext != "None"):
         (mant,expn) = FormStr(ymin+i*h,scale,nsigd)
         w.create_text(ixtext,iy,text=mant,font=font1,anchor="e")  # labels

   if (ytext != "None"):
      if ((scale<0.1) or (scale>1000.0)): ytext = ytext + " 1e" + expn
      ixtext = ixtext - (3*nfont/4) * (len(mant) + 2)       # skip labels + 2
      iytext = (iymin+iymax)/2                              # vertical middle
      w.create_text(ixtext,iytext,text=ytext,font=font2,anchor="e") # y title
                                                              # draw axes
   if (xmin*xmax < 0): w.create_line(Nint(bx),iymin,Nint(bx),iymax)  # y-axis
   if (ymin*ymax < 0): w.create_line(ixmin,Nint(by),ixmax,Nint(by))  # x-axis

   tic = 2*tic/3
   if (sty == 4): hx = ax * (x[2] - x[1])
   ix0 = Nint(ax*x[1]+bx); iy0 = Nint(ay*y[1]+by)               # 1st point
   for i in range(1,n+1):
      ix = Nint(ax*x[i]+bx); iy = Nint(ay*y[i]+by)             # new point
      if (sty == 0):                                         # scatter plot
         w.create_rectangle(ix-tic,iy-tic,ix+tic,iy+tic, \
                            fill="",outline=col)
```

```
        if (sty == 1 or sty == 2):                              # line or polar plot
           w.create_line(ix0,iy0,ix,iy,fill=col)
        if (sty == 3):                                          # drop lines
           w.create_line(ix,by,ix,iy,fill=col)
        if (sty == 4 and i < n):
           w.create_rectangle(ix+1,iy,ix+hx-1,by,fill="",outline=col)
        ix0 = ix; iy0 = iy                                      # save point

#===========================================================================
def MultiPlot(x, y, sig, n, col, sty, nplot, maxint, \
              xminp, xmaxp, ioptx, yminp, ymaxp, iopty, \
              fxmin, fxmax, fymin, fymax, xtext, ytext, title):
#---------------------------------------------------------------------------
#  Plots nplot real functions of one variable given by sets of (x,y) points.
#  The coordinate sets are stored contiguously in the arrays x[] and y[].
#  The x and y-domains are extended to fit at most maxint intervals
#  expressible as d * 10^p, with d = 1, 2, 5 and p integer.
#
#  x[]     - abscissas of tabulation points for all functions
#  y[]     - ordinates of tabulation points
#  sig[]   - error bars of the tabulation points (useful for sty == 4)
#  n[]     - ending index for the individual plots (nmax = n[nplot])
#  col[]   - plot color ("red", "green", "blue" etc.)
#  sty[]   - plot style 0 - scatter plot with squares
#                       1 - line plot;  -1 - dashed line
#                       2 - polar plot; -2 - dashed line
#                       3 - drop lines
#                       4 - error bars; -4 - including line plot
#  nplot   - number of plots
#  maxint  - max. number of labeling intervals
#  ioptx   - 0 - resize x-axis automatically
#            1 - resize x-axis based on user interval [xminp,xmaxp]
#  iopty   - 0 - resize y-axis automatically
#            1 - resize y-axis based on user interval [yminp,ymaxp]
#  fxmin   - min fractional x-limit of viewport (0 < fxmin < fxmax < 1)
#  fxmax   - max fractional x-limit of viewport
#  fymin   - min fractional y-limit of viewport (0 < fymin < fymax < 1)
#  fymax   - max fractional y-limit of viewport
#  xtext   - x-axis title; for "None" - axis is not labeled
#  ytext   - y-axis title; for "None" - axis is not labeled
#  title   - plot title
#---------------------------------------------------------------------------
   global w, nxw, nyw                                           # canvas object and size

   nmax = n[nplot]
   if (ioptx):                                                  # x-domain limits
      xmin = xminp; xmax = xmaxp                                   # input values
   else:
      xmin = min(x[1:nmax+1]); xmax = max(x[1:nmax+1])         # bounding values
   if (iopty):                                                  # y-domain limits
      ymin = yminp; ymax = ymaxp                                   # input values
   else:
      ymin = min(y[1:nmax+1]); ymax = max(y[1:nmax+1])         # bounding values

      for iplot in range(1,nplot+1):     # extend y-domain for error bar plots
         if (abs(sty[iplot]) == 4):
            i0 = 1 if iplot == 1 else n[iplot-1] + 1
            for i in range(i0,n[iplot]+1):
               ymin = min(ymin,y[i]-sig[i])
```

```
                    ymax = max(ymax,y[i]+sig[i])

   ixmin = Nint(fxmin*nxw); iymin = Nint((1e0-fymin)*nyw)      # viewport limits
   ixmax = Nint(fxmax*nxw); iymax = Nint((1e0-fymax)*nyw)

   if (2 in sty[1:]):                                            # polar plot
      xmin = min(xmin,ymin); xmax = max(xmax,ymax)          # make domain square
      xmax = max(abs(xmin),abs(xmax)); xmin = -xmax
      ymin = xmin; ymax = xmax
      if (ixmax-ixmin > iymin-iymax):                      # adjust viewport limits
         c = 0.5*(ixmin+ixmax)
         d = 0.5*(iymin-iymax)
         ixmin = c - d; ixmax = c + d
      if (ixmax-ixmin < iymin-iymax):
         c = 0.5*(iymin+iymax)
         d = 0.5*(ixmax-ixmin)
         iymin = c + d; iymax = c - d

   w.create_rectangle(ixmin,iymax,ixmax,iymin)                # draw plot frame

   nfont = min(int((ixmax-ixmin)/60.) + 3,12)                     # font size
   font1 = ("Helvetica",nfont)                                # axis label font
   font2 = ("Helvetica",Nint(1.2*nfont))                      # axis title font
   font3 = ("Helvetica",Nint(1.4*nfont))                      # plot title font
   if ((ixmax-ixmin) < 3 * (iymin-iymax)):
      w.create_text((ixmin+ixmax)/2,iymax-3*nfont,text=title,font=font3)
   else:
      w.create_text(ixmax,iymax,text=title,font=font2,anchor="ne")
                                                                  # X-AXIS
   (xmin,xmax,scale,nsigd,nintv) = Limits(xmin,xmax,maxint)   # extend limits
   ax = (ixmax-ixmin)/(xmax-xmin)                         # x scaling coefficients
   bx = ixmin - ax*xmin

   tic =(ixmax-ixmin)/100.                                      # tic length
   h = (xmax-xmin)/nintv; htic = ax * h                      # labeling step
   iytext = iymin + 1.5*nfont
   for i in range(0,nintv+1):                                  # label axis
      ix = Nint(ixmin + i*htic)
      w.create_line(ix,iymin,ix,iymin-tic)                        # tics
      w.create_line(ix,iymax,ix,iymax+tic)
      if (xtext != "None"):
         (mant,expn) = FormStr(xmin+i*h,scale,nsigd)
         w.create_text(ix,iytext,text=mant,font=font1)            # labels

   if (xtext != "None"):
      if ((scale<0.1) or (scale>1000.0)): xtext = xtext + " 1e" + expn
      ixtext = (ixmin+ixmax)/2
      iytext = iytext + 2*nfont
      w.create_text(ixtext,iytext,text=xtext,font=font2)        # x title
                                                                  # Y-AXIS
   if (ymin == 0.0 and ymax == 0.0): ymin = -1e0; ymax = 1e0    # horizontal
   if (abs(ymax-ymin) < 1e-5*abs(ymax)): ymin *= 0.9; ymax *= 1.1
   (ymin,ymax,scale,nsigd,nintv) = Limits(ymin,ymax,maxint)   # extend limits
   ay = (iymax-iymin)/(ymax-ymin)                         # y scaling coefficients
   by = iymin - ay*ymin

   h = (ymax-ymin)/nintv; htic = ay * h                      # labeling step size
   ixtext = ixmin - nfont
   for i in range(0,nintv+1):                                  # label axis
```

```
      iy = Nint(iymin + i*htic)
      w.create_line(ixmin,iy,ixmin+tic,iy)                              # tics
      w.create_line(ixmax,iy,ixmax-tic,iy)
      if (ytext != "None"):
          (mant,expn) = FormStr(ymin+i*h,scale,nsigd)
          w.create_text(ixtext,iy,text=mant,font=font1,anchor="e")     # labels

   if (ytext != "None"):
      if ((scale<0.1) or (scale>1000.0)): ytext = ytext + " 1e" + expn
      ixtext = ixtext - (3*nfont/4) * (len(mant) + 2)        # skip labels + 2
      iytext = (iymin+iymax)/2                               # vertical middle
      w.create_text(ixtext,iytext,text=ytext,font=font2,anchor="e") # y title
                                                                  # draw axes
   if (xmin*xmax < 0): w.create_line(Nint(bx),iymin,Nint(bx),iymax)  # y-axis
   if (ymin*ymax < 0): w.create_line(ixmin,Nint(by),ixmax,Nint(by))  # x-axis

   tic = 2*tic/3
   for iplot in range(1,nplot+1):
      icol = col[iplot]
      isty = sty[iplot]
      i0 = 1 if iplot == 1 else n[iplot-1] + 1
      ix0 = Nint(ax*x[i0]+bx); iy0 = Nint(ay*y[i0]+by)             # 1st point
      if (isty == 0):
          w.create_rectangle(ix0-tic,iy0-tic,ix0+tic,iy0+tic, \
                             fill="",outline=icol)
      for i in range(i0,n[iplot]+1):
          ix = Nint(ax*x[i]+bx); iy = Nint(ay*y[i]+by)            # new point
          if (isty == 0):                                       # scatter plot
             w.create_rectangle(ix-tic,iy-tic,ix+tic,iy+tic, \
                                fill="",outline=icol)
          if (abs(isty) == 1 or abs(isty) == 2):          # line or polar plot
             if (isty > 0): w.create_line(ix0,iy0,ix,iy,fill=icol)
             else:          w.create_line(ix0,iy0,ix,iy,fill=icol,dash=(4,4))
          if (isty == 3):                                       # drop lines
             w.create_line(ix,by,ix,iy,fill=icol)
          if (abs(isty) == 4):                                  # error bars
             isig = Nint(ay*sig[i])
             w.create_line(ix,iy-isig,ix,iy+isig,fill=icol)
             w.create_line(ix-tic,iy-isig,ix+tic,iy-isig,fill=icol)
             w.create_line(ix-tic,iy+isig,ix+tic,iy+isig,fill=icol)
             if (isty > 0):
                w.create_oval(ix-tic,iy-tic,ix+tic,iy+tic, \
                             fill="white",outline=icol)
             else:
                w.create_line(ix0,iy0,ix,iy,fill=icol)
          ix0 = ix; iy0 = iy                                      # save point

#===============================================================================
def RGBcolors(ncolstep):
#-------------------------------------------------------------------------------
#  Generates ncol = 1280/ncolsep RGB colors in icol[1] through icol[ncol]
#
#  Returns: icol, ncol
#-------------------------------------------------------------------------------
   ncol = int(1280/ncolstep)
   icol = [0]*(ncol+1)

   i = 0
   bc = 255; rc = 0
```

```
      for gc in range(0,256,ncolstep):
         i += 1; icol[i] = "#%02x%02x%02x" % (rc,gc,bc)

      gc = 255; rc = 0
      for bc in range(255,-1,-ncolstep):
         i += 1; icol[i] = "#%02x%02x%02x" % (rc,gc,bc)

      bc = 0; gc = 255
      for rc in range(0,256,ncolstep):
         i += 1; icol[i] = "#%02x%02x%02x" % (rc,gc,bc)

      bc = 0; rc = 255
      for gc in range(255,-1,-ncolstep):
         i += 1; icol[i] = "#%02x%02x%02x" % (rc,gc,bc)

      gc = 0; rc = 255
      for bc in range(0,256,ncolstep):
         i += 1; icol[i] = "#%02x%02x%02x" % (rc,gc,bc)

      return (icol, ncol)

#===============================================================================
def ColorLegend(fmin, fmax, ixmin, ixmax, iymin, iymax):
#-------------------------------------------------------------------------------
#  Draws and labels the color legend for the interval [fmin,fmax] in the
#  rectangle [ixmin,ixmax] x [iymin,iymax]. [fmin,fmax] is extended to fit at
#  most 10 intervals expressible as d * 10^p, with d = 1, 2, 5 and p integer.
#
#  Returns: fmin, fmax, icol, ncol, nintv
#-------------------------------------------------------------------------------
   global w, nxw, nyw                                   # canvas object and size

   nfont = min(int((iymin-iymax)/20.),12)
   font = ("Helvetica",nfont)                                       # label font

   ncolstep = 2
   (icol,ncol) = RGBcolors(ncolstep)

   hcol = float(ncol) / (iymax - iymin)
   for iy in range(iymax,iymin+1):
      ic = min(max(1,int((iy-iymin)*hcol)),ncol)
      w.create_line(ixmin,iy,ixmax,iy,fill=icol[ic])
   w.create_rectangle(ixmin,iymin,ixmax,iymax)

   (fmin,fmax,scale,nsigd,nintv) = Limits(fmin,fmax,10)         # extend limits
   ay = (iymax-iymin)/(fmax-fmin)                          # scaling coefficients
   by = iymin - ay*fmin
   h = (fmax-fmin)/nintv; htic = ay * h                       # labeling step size
   ixtext = ixmax + (nsigd+1)*nfont
   for i in range(0,nintv+1):                                     # label legend
      iy = Nint(iymin + i*htic)
      (mant,expn) = FormStr(fmin+i*h,scale,nsigd)
      w.create_text(ixtext,iy,text=mant,font=font,anchor="e")          # labels

   if ((scale<0.1) or (scale>1000.0)):
      ytext = " x 1e" + expn
      w.create_text(ixtext,iymin+2*nfont,text=ytext,font=font,anchor="e")

   return (fmin, fmax, icol, ncol, nintv)
```

```
#===========================================================================
def Contour(z, nx, ny, xmin, xmax, ymin, ymax, zmin, zmax, \
            fxmin, fxmax, fymin, fymax, xtext, ytext, title):
#---------------------------------------------------------------------------
#  Plots a function z(x,y) defined in [xmin,xmax] x [ymin,ymax] and tabulated
#  on a regular Cartesian grid with (nx-1)x(ny-1) mesh cells as contour plot.
#  The level curves result by inverse linear interpolation inside the cells.
#
#  z     - tabulated function values
#  nx    - number of x-mesh points
#  ny    - number of y-mesh points
#  zmin  - minimum level considered
#  zmin  - maximum level considered
#  fxmin - minimum relative viewport abscissa (0 < fxmin < fxmax < 1)
#  fxmax - maximum relative viewport abscissa
#  fymin - minimum relative viewport ordinate (0 < fymin < fymax < 1)
#  fymax - maximum relative viewport ordinate
#  xtext - x-axis title; xtext = "" - axis is not labeled
#  ytext - y-axis title; ytext = "" - axis is not labeled
#  title - plot title
#---------------------------------------------------------------------------
   global w, nxw, nyw                               # canvas object and size
   xg = [0]*8; yg = [0]*8

   ixmin = Nint(fxmin*nxw); iymin = Nint((1e0-fymin)*nyw)   # viewport coords
   ixmax = Nint(fxmax*nxw); iymax = Nint((1e0-fymax)*nyw)

   f = (xmax-xmin)/(ymax-ymin)                # scale viewport proportionally
   if (f < float(ixmax-ixmin)/(iymin-iymax)):              # shorter x-axis
      ixmax = ixmin + ixmax                         # correct ixmin and ixmax
      ixmin = int(0.5e0*(ixmax + (iymax-iymin)*f))
      ixmax = ixmax - ixmin
   else:                                                   # shorter y-axis
      iymax = iymin + iymax                         # correct iymin and iymax
      iymin = int(0.5e0*(iymax + (ixmax-ixmin)/f))
      iymax = iymax - iymin

   nfont = min(int((ixmax-ixmin)/60.) + 3,12)                   # font size
   font1 = ("Helvetica",nfont)                          # axis label font
   font2 = ("Helvetica",Nint(1.2*nfont))                # axis title font
   font3 = ("Helvetica",Nint(1.4*nfont))                # plot title font
   w.create_text((ixmin+ixmax)/2,iymax-3*nfont,text=title,font=font3)

   iyc = (iymax + iymin)/2.                                # color legend
   iyd = min((ixmax - ixmin + iymin - iymax)/8.,(iymin - iymax)/3.)
   ixd = iyd/5.
   ix1 = Nint(ixmax +   ixd); iy1 = Nint(iyc + iyd)
   ix2 = Nint(ixmax + 2*ixd); iy2 = Nint(iyc - iyd)
   (zmin,zmax,icol,ncol,nintv) = ColorLegend(zmin,zmax,ix1,ix2,iy1,iy2)
   if (nintv <= 6): nintv *= 2

   hx = (ixmax-ixmin)/(nx-1e0)                                  # draw grid
   hy = (iymax-iymin)/(ny-1e0)
   hz = (zmax-zmin)/(ncol-1e0)
   hv = (zmax-zmin)/nintv
   for iz in range(1,ncol+nintv+1):                         # loop over levels
      if (iz <= ncol):                                      # color shades
         z0 = zmin + (iz-1)*hz
```

```
                col = icol[iz]
          else:                                                      # lines
            z0 = zmin + (iz-ncol)*hv
          for i in range(1,nx):                           # loop over cells [i][j]
            for j in range(1,ny):                            # corner coordinates
              ix = Nint(ixmin + (i-1)*hx); ixh = Nint(ixmin + i*hx)
              iy = Nint(iymin + (j-1)*hy); iyh = Nint(iymin + j*hy)
              z1 = z[i][j]   - z0; z2 = z[i+1][j]   - z0   # rel. corner values
              z3 = z[i][j+1] - z0; z4 = z[i+1][j+1] - z0

              if (z1 < 0e0 or z2 < 0e0 or z3 < 0e0 or z4 < 0e0):
                ng = 0                          # level line passes through the cell
                if (z1*z2 <= 0):                      # intersection with lower edge
                  if (z1 != z2):
                    ng += 1; xg[ng] = Nint(ix + z1/(z1-z2)*hx); yg[ng] = iy
                  else:                                 # line coincides with edge
                    ng += 1; xg[ng] = ix ; yg[ng] = iy
                    ng += 1; xg[ng] = ixh; yg[ng] = iy

                if (z1*z3 <= 0):                        # intersection with left edge
                  if (z1 != z3):
                    ng += 1; xg[ng] = ix; yg[ng] = Nint(iy + z1/(z1-z3)*hy)
                  else:                                 # line coincides with edge
                    ng += 1; xg[ng] = ix; yg[ng] = iy
                    ng += 1; xg[ng] = ix; yg[ng] = iyh

                if (z2*z4 <= 0):                       # intersection with right edge
                  if (z2 != z4):
                    ng += 1; xg[ng] = ixh; yg[ng] = Nint(iy + z2/(z2-z4)*hy)
                  else:                                 # line coincides with edge
                    ng += 1; xg[ng] = ixh; yg[ng] = iy
                    ng += 1; xg[ng] = ixh; yg[ng] = iyh

                if (z3*z4 <= 0):                       # intersection with upper edge
                  if (z3 != z4):
                    ng += 1; xg[ng] = Nint(ix + z3/(z3-z4)*hx); yg[ng] = iyh

                if (iz <= ncol):
                  polygon0 = ()
                  if (z1 >= 0e0): polygon0 += ix , iy
                  if (z2 >= 0e0): polygon0 += ixh, iy
                  if (z3 >= 0e0): polygon0 += ix , iyh
                  if (z4 >= 0e0): polygon0 += ixh, iyh
                  for ig in range(1,ng+1):      # fill the polygon above level
                    for jg in range(1,ng):
                      polygon = polygon0 + (xg[ig],yg[ig],xg[jg],yg[jg])
                      w.create_polygon(polygon,fill=col,outline=col)
                else:
                  for ig in range(1,ng+1):                     # draw levele line
                    for jg in range(1,ng):
                      w.create_line(xg[ig],yg[ig],xg[jg],yg[jg])
              else:          # all corner values exceed level - fill entire cell
                if (iz <= ncol):
                  w.create_rectangle(ix,iy,ixh,iyh,fill=col,outline=col)

    w.create_rectangle(ixmin,iymax,ixmax,iymin)                    # draw border
                                                                   # X-AXIS
    maxint = 10
    (xmin1,xmax1,scale,nsigd,nintv) = Limits(xmin,xmax,maxint) # extend limits
```

```
   h = (xmax1-xmin1)/nintv
   if (xmin1 < xmin): xmin1 += h; nintv -= 1
   if (xmax1 > xmax): xmax1 -= h; nintv -= 1
   ax = (ixmax-ixmin)/(xmax-xmin)                           # scaling coefficients
   bx = ixmin - ax*xmin

   tic =(ixmax-ixmin)/100.                                          # tic length
   h = (xmax1-xmin1)/nintv                                        # labeling step
   iytext = iymin + 1.5*nfont
   for i in range(0,nintv+1):                                      # label axis
      xi = xmin1 + i*h
      ix = Nint(ax*xi+bx)
      w.create_line(ix,iymin,ix,iymin-tic)                              # tics
      w.create_line(ix,iymax,ix,iymax+tic)
      if (xtext):
          (mant,expn) = FormStr(xi,scale,nsigd)
          w.create_text(ix,iytext,text=mant,font=font1)              # labels

   if ((scale<0.1) or (scale>1000.0)): xtext = xtext + " x 1e" + expn
   ixtext = (ixmin+ixmax)/2
   iytext = iytext + 2*nfont
   w.create_text(ixtext,iytext,text=xtext,font=font2)            # axis title
                                                                      # Y-AXIS
   (ymin1,ymax1,scale,nsigd,nintv) = Limits(ymin,ymax,maxint) # extend limits
   h = (ymax1-ymin1)/nintv
   if (ymin1 < ymin): ymin1 += h; nintv -= 1
   if (ymax1 > ymax): ymax1 -= h; nintv -= 1
   ay = (iymax-iymin)/(ymax-ymin)                           # scaling coefficients
   by = iymin - ay*ymin

   h = (ymax1-ymin1)/nintv                                    # labeling step size
   ixtext = ixmin - nfont
   for i in range(0,nintv+1):                                      # label axis
      yi = ymin1 + i*h
      iy = Nint(ay*yi+by)
      w.create_line(ixmin,iy,ixmin+tic,iy)                              # tics
      w.create_line(ixmax,iy,ixmax-tic,iy)
      if (ytext):
          (mant,expn) = FormStr(yi,scale,nsigd)
          w.create_text(ixtext,iy,text=mant,font=font1,anchor="e")   # labels

   if ((scale<0.1) or (scale>1000.0)): ytext = ytext + " x 1e" + expn
   ixtext = ixtext - (3*nfont/4) * (len(mant) + 2)          # skip labels + 2
   iytext = (iymin+iymax)/2                                       # vertical middle
   w.create_text(ixtext,iytext,text=ytext,font=font2,anchor="e") # axis title

#===============================================================================
def PlotParticles(x, y, z, r, col, n, dmax, \
                  xminp, xmaxp, ioptx, yminp, ymaxp, iopty, \
                  fxmin, fxmax, fymin, fymax, title):
#-------------------------------------------------------------------------------
#  Plots a system of particles as connected colored spheres
#
#  x,y,z[] - coordinates of particles
#  r[]     - radii of particles
#  col[]   - colors of particles ("red", "green", "blue" etc.)
#  n       - number of particles
#  dmax    - max inter-distance for which particles are connected
#  ioptx   - 0 - resize x-axis automatically
```

```
#              1 - resize x-axis to provided user interval (xminp,xmaxp)
#   iopty   - 0 - resize y-axis automatically
#              1 - resize y-axis to provided user interval (yminp,ymaxp)
#   fxmin   - min fractional x-limit of viewport (0 < fxmin < fxmax < 1)
#   fxmax   - max fractional x-limit of viewport
#   fymin   - min fractional y-limit of viewport (0 < fymin < fymax < 1)
#   fymax   - max fractional y-limit of viewport
#   title   - plot title
#-----------------------------------------------------------------------
   global w, nxw, nyw                               # canvas object and size
   ind = [0]*(n+1)

   if (ioptx):                                             # x-domain limits
      xmin = xminp; xmax = xmaxp                              # input values
   else:
      xmin = min(x[1:]); xmax = max(x[1:])                 # bounding values
   if (iopty):                                             # y-domain limits
      ymin = yminp; ymax = ymaxp                              # input values
   else:
      ymin = min(y[1:]); ymax = max(y[1:])                 # bounding values

   ixmin = Nint(fxmin*nxw); iymin = Nint((1e0-fymin)*nyw)  # viewport limits
   ixmax = Nint(fxmax*nxw); iymax = Nint((1e0-fymax)*nyw)

   nfont = int((ixmax-ixmin)/20.)                             # font size
   font = ("Helvetica",nfont)                                # title font
   w.create_text((ixmin+ixmax)/2,iymax-3*nfont,text=title,font=font)  # title

   f = (xmax-xmin)/(ymax-ymin)              # scale viewport proportionally
   if (f < float(ixmax-ixmin)/(iymin-iymax)):              # shorter x-axis
      ixmax = ixmin + ixmax                      # adjust ixmin and ixmax
      ixmin = int(0.5*(ixmax + (iymax-iymin)*f))
      ixmax = ixmax - ixmin
   else:                                                   # shorter y-axis
      iymax = iymin + iymax                      # adjust iymin and iymax
      iymin = int(0.5*(iymax + (ixmax-ixmin)/f))
      iymax = iymax - iymin

   ax = (ixmax-ixmin)/(xmax-xmin); bx = ixmin - ax*xmin      # scaling coeffs
   ay = (iymax-iymin)/(ymax-ymin); by = iymin - ay*ymin

   for i1 in range(1,n+1):                                    # draw bonds
      for i2 in range(1,i1):
         dx = x[i1] - x[i2]; dy = y[i1] - y[i2]; dz = z[i1] - z[i2]
         d = sqrt(dx*dx + dy*dy + dz*dz)
         if ( d <= dmax):
            ix1 = Nint(ax*x[i1]+bx); iy1 = Nint(ay*y[i1]+by)
            ix2 = Nint(ax*x[i2]+bx); iy2 = Nint(ay*y[i2]+by)
            w.create_line(ix1,iy1,ix2,iy2,fill="slate gray",width=10)
            w.create_line(ix1,iy1,ix2,iy2,fill="gray",width=6)
            w.create_line(ix1,iy1,ix2,iy2,fill="white",width=2)

   Index(z,ind,n)                        # index particles in increasing order of z

   for i in range(1,n+1):             # draw particles in increasing order of z
      i1 = ind[i]
      ix = Nint(ax*x[i1]+bx); iy = Nint(ay*y[i1]+by)
      rgb = w.winfo_rgb(col[i1])            # rgb representation of dimmed color
      (rc, gc, bc) = (int(rgb[0]/512), int(rgb[1]/512), int(rgb[2]/512))
```

```
        r0 = int(ax * r[i1])                            # starting radius
        for j in range(r0,0,-1):                        # decrease radius
            rgb = '#%02x%02x%02x' % (rc, gc, bc)           # increase luminosity
            w.create_oval(ix-j,iy-j,ix+j,iy+j,fill=str(rgb),outline=str(rgb))
            (rc, gc, bc) = (min(255,rc+6), min(255,gc+6), min(255,bc+6))

#===============================================================================
def PlotStruct(x, y, z, n, ind1, ind2, ind3, n3, \
               xminp, xmaxp, ioptx, yminp, ymaxp, iopty, \
               fxmin, fxmax, fymin, fymax, title):
#-------------------------------------------------------------------------------
#  Renders a 3D structure defined by nodes and triangular surfaces
#
#  x,y,z[] - coordinates of nodes
#  n       - number of nodes
#  ind1[]  - index of 1st node of each triangle
#  ind2[]  - index of 2nd node of each triangle
#  ind3[]  - index of 3rd node of each triangle
#  n3      - number of triangles
#  ioptx   - 0 - resize x-axis automatically
#            1 - resize x-axis to provided user interval (xminp,xmaxp)
#  iopty   - 0 - resize y-axis automatically
#            1 - resize y-axis to provided user interval (yminp,ymaxp)
#  fxmin   - min fractional x-limit of viewport (0 < fxmin < fxmax < 1)
#  fxmax   - max fractional x-limit of viewport
#  fymin   - min fractional y-limit of viewport (0 < fymin < fymax < 1)
#  fymax   - max fractional y-limit of viewport
#  title   - plot title
#-------------------------------------------------------------------------------
   global w, nxw, nyw                                # canvas object and size
   zmax = [0]*(n3+1)
   cosn = [0]*(n3+1)
   ind  = [0]*(n3+1)

   if (ioptx):                                       # x-domain limits
      xmin = xminp; xmax = xmaxp                         # input values
   else:
      xmin = min(x[1:]); xmax = max(x[1:])              # bounding values
   if (iopty):                                       # y-domain limits
      ymin = yminp; ymax = ymaxp                         # input values
   else:
      ymin = min(y[1:]); ymax = max(y[1:])              # bounding values

   ixmin = Nint(fxmin*nxw); iymin = Nint((1e0-fymin)*nyw)  # viewport coords
   ixmax = Nint(fxmax*nxw); iymax = Nint((1e0-fymax)*nyw)

   nfont = int((ixmax-ixmin)/20.)                             # font size
   font = ("Helvetica",nfont)                                 # title font
   w.create_text((ixmin+ixmax)/2,iymax-3*nfont,text=title,font=font)  # title

   f = (xmax-xmin)/(ymax-ymin)                   # scale viewport proportionally
   if (f < float(ixmax-ixmin)/(iymin-iymax)):              # shorter x-axis
      ixmax = ixmin + ixmax                          # correct ixmin and ixmax
      ixmin = int(0.5*(ixmax + (iymax-iymin)*f))
      ixmax = ixmax - ixmin
   else:                                                   # shorter y-axis
      iymax = iymin + iymax                          # correct iymin and iymax
      iymin = int(0.5*(iymax + (ixmax-ixmin)/f))
      iymax = iymax - iymin
```

```
   ax = (ixmax-ixmin)/(xmax-xmin); bx = ixmin - ax*xmin      # scaling coeffs
   ay = (iymax-iymin)/(ymax-ymin); by = iymin - ay*ymin

   for i in range(1,n3+1):                        # max z and cos(n) of triangles
      i1 = ind1[i]; i2 = ind2[i]; i3 = ind3[i]

      zmax[i] = max(z[i1],z[i2],z[i3])                       # max z of triangle

      ux = x[i2] - x[i1]; vx = x[i3] - x[i1]            # defining vectors
      uy = y[i2] - y[i1]; vy = y[i3] - y[i1]
      uz = z[i2] - z[i1]; vz = z[i3] - z[i1]

      nx = uy * vz - uz * vy                              # normal to triangle
      ny = uz * vx - ux * vz
      nz = ux * vy - uy * vx
      if (nz < 0e0): nx = -nx; ny = -ny; nz= - nz # choose face toward viewer
      cosn[i] = (nx + ny + nz) / sqrt(3e0*(nx*nx + ny*ny + nz*nz))

   Index(zmax,ind,n3)                # index triangles in increasing order of z

   for i in range(1,n3+1):                     # draw triangles - the remotest first
      indi = ind[i]
      i1 = ind1[indi]; i2 = ind2[indi]; i3 = ind3[indi]
      ix1 = Nint(ax*x[i1]+bx); iy1 = Nint(ay*y[i1]+by)
      ix2 = Nint(ax*x[i2]+bx); iy2 = Nint(ay*y[i2]+by)
      ix3 = Nint(ax*x[i3]+bx); iy3 = Nint(ay*y[i3]+by)
      polygon = ix1, iy1, ix2, iy2, ix3, iy3, ix1, iy1
      d = int(128 * cosn[indi])
      rc = 64 + d; gc = 64 + d; bc = min(128 + d,255)
      col = str('#%02x%02x%02x' % (rc, gc, bc))
      w.create_polygon(polygon,fill=col,outline="slate gray",width=1)

#===============================================================================
def HistoBin(xnew, a, b, x, y, n, iopt):
#-------------------------------------------------------------------------------
#  Bins data for a histogram to be plotted by function Plot (with sty = 4)
#
#  xnew - new value to be binned
#  a, b - limits of total binning interval
#  x[]  - bin boundaries (x[1] = a, x[n] = b)
#  y[]  - frequency of values in the bins: y[i] in [x[i],x[i+1]), i = 1,n-1
#  n    - number of bin boundaries
#  iopt - option: 0 - zeros bins
#                 1 - bins new value xnew
#                 2 - normalizes histogram
#-------------------------------------------------------------------------------
   h = (b-a)/(n-1)                                          # bin size

   if (iopt == 0):                                     # initialize bins
      for i in range(1,n+1):
         x[i] = a + (i-1)*h                            # bin boundaries
         y[i] = 0e0

   elif (iopt == 1):                                   # bin new values
      i = (int) ((xnew-a)/h) + 1                       # bin index
      if ((i >= 1) and (i < n)): y[i] += 1         # increment bin value

   elif (iopt == 2):                                   # normalize histogram
```

```
       s = 0e0
       for i in range(1,n): s += y[i]              # sum of unnormalized bin values
       for i in range(1,n): y[i] /= s                   # normalize bin values
```

Listing E.2   Auxiliary Functions (`utils.py`)

```
#--------------------------------- utils.py ---------------------------------
#  Contains utility routines.
#  Part of the numxlib numerics library. Author: Titus Beu, 2013
#----------------------------------------------------------------------------
from math import *

# Nearest integer
def Nint(x): return int(floor(x + 0.5))

# Transfers the sign of y to the absolute value of x
def Sign(x,y): return (abs(x) if y > 0 else -abs(x))


#============================================================================
def Magn(x):
#----------------------------------------------------------------------------
# Returns the order of magnitude of x as 10^n
#----------------------------------------------------------------------------
   return 0e0 if x == 0e0 else \
          10e0**int(log10(abs(x))) if abs(x) >= 1e0 else \
          0.1e0**(int(abs(log10(abs(x))))+1e0)


#============================================================================
def Index(x, ind, n):
#----------------------------------------------------------------------------
#  Ascending indexing of array x[1..n] in ind[] by insertion sort
#----------------------------------------------------------------------------
   for i in range(1,n+1): ind[i] = i               # initialize index array

   for ipiv in range(2,n+1):                              # loop over pivots
      xpiv = x[ipiv]                          # save pivot to free its location
      i = ipiv - 1                                  # initialize sublist counter
      while ((i > 0) and (x[ind[i]] > xpiv)):      # scan to the left of pivot
         ind[i+1] = ind[i]                    # item > pivot: shift to the right
         i -= 1

      ind[i+1] = ipiv                         # insert pivot into last freed location
```

Appendix F: The C++ Interface to the Graphics Library `graphlib.py`

The Python graphics library `graphlib.py` can be seamlessly accessed from C++ programs by means of the interface file `graphlib.h`, presented in Listing F.1. `graphlib.h` contains two wrapper classes: one, called `Python`, for invoking the Python interpreter, and one, called `PyGraph`, for the actual calls to the functions of the module `graphlib.py`.

The methods of the class `PyGraph` are wrappers (interfaces) to the graphics functions contained in the module `graphlib.py`: MainLoop, GraphUpdate, GraphInit, GraphClear, Plot, MultiPlot, Contour, PlotParticles, and PlotStruct.

HistoBin is a stand-alone function with identical functionality with the function of the same name from `graphlib.py`.

Listing F.1   C++ Interface `graphlib.h` to the Graphics Library `graphlib.py`.

```
//------------------------------ graphlib.h ------------------------------
// Contains C++ interface classes for the graphics library graphlib.py.
// Author: Titus Beu, 2013
//------------------------------------------------------------------------
#ifndef _GRAPHLIB_
#define _GRAPHLIB_

#include <iostream>
#include <Python.h>

//========================================================================
class Python                        // wrapper class for the Python interpreter
{
   public:
   Python(int argc, wchar_t** argv)         // constructor: initialize Python
   {
      Py_Initialize();
      PySys_SetArgv(argc, argv);
   }

   ~Python() { Py_Finalize(); }                 // destructor: finish Python
};

//========================================================================
class PyGraph                       // wrapper class for calls to graphlib.py
{
   Python* contxt;
   PyObject* pModule;

   public:
   PyGraph(int argc, wchar_t** argv)        // constructor: import graphlib.py
```

```
      {
         contxt = new Python(argc, argv);
         pModule = PyImport_Import(PyUnicode_FromString("graphlib"));
         if (!pModule) std::cout << "Failed to load <graphlib.py>";
      }
   virtual ~PyGraph()                            // destructor: release graphlib.py
      {
         if (pModule) Py_DECREF(pModule);
         if (contxt) delete contxt;
      }

//=============================================================================
   void MainLoop()                                    // creates Tk event loop
      {
         PyObject *pFunc, *pArgs;

         pFunc = PyObject_GetAttrString(pModule,"MainLoop");
         pArgs = PyTuple_New(0);
         Py_XINCREF(pArgs);

         PyObject_CallObject(pFunc, pArgs);

         Py_XDECREF(pArgs);
         Py_DECREF(pFunc);
      }

//=============================================================================
   void GraphUpdate()                                 // updates Tk root widget
      {
         PyObject *pFunc, *pArgs;

         pFunc = PyObject_GetAttrString(pModule,"GraphUpdate");
         pArgs = PyTuple_New(0);
         Py_XINCREF(pArgs);

         PyObject_CallObject(pFunc, pArgs);

         Py_XDECREF(pArgs);
         Py_DECREF(pFunc);
      }

//=============================================================================
   void GraphInit(int nxwin, int nywin)               // creates Canvas widget
      {
         PyObject *pFunc, *pArgs;

         pFunc = PyObject_GetAttrString(pModule,"GraphInit");

         pArgs = PyTuple_New(2);
         PyTuple_SetItem(pArgs, 0, PyLong_FromLong(nxwin));
         PyTuple_SetItem(pArgs, 1, PyLong_FromLong(nywin));
         Py_XINCREF(pArgs);

         PyObject_CallObject(pFunc, pArgs);

         Py_XDECREF(pArgs);
         Py_DECREF(pFunc);
      }
```

```
//===========================================================================
   void GraphClear()                              // deletes content of Canvas widget
   {
      PyObject *pFunc, *pArgs;

      pFunc = PyObject_GetAttrString(pModule,"GraphClear");
      pArgs = PyTuple_New(0);
      Py_XINCREF(pArgs);

      PyObject_CallObject(pFunc, pArgs);

      Py_XDECREF(pArgs);
      Py_DECREF(pFunc);
   }


//===========================================================================
   void Plot(double x[], double y[], int n, const char *col, int sty,
             double fxmin, double fxmax, double fymin, double fymax,
             const char *xtext, const char *ytext, const char *title)
//---------------------------------------------------------------------------
// Plots a real function of one variable specified by a set of (x,y) points.
// The x and y-domains are extended to fit at most 10 intervals expressible
// as d * 10^p, with d = 1, 2, 5 and p integer.
//
// x[]   - abscissas of tabulation points (x[1] through x[n])
// y[]   - ordinates of tabulation points (y[1] through y[n])
// n     - number of tabulation points
// col   - plot color ("red", "green", "blue" etc.)
// sty   - plot style: 0 - scatter plot, 1 - line plot, 2 - polar plot,
//                     3 - drop lines, 4 - histogram
// fxmin - min fractional x-limit of viewport (0 < fxmin < fxmax < 1)
// fxmax - max fractional x-limit of viewport
// fymin - min fractional y-limit of viewport (0 < fymin < fymax < 1)
// fymax - max fractional y-limit of viewport
// xtext - x-axis title; for "None" - axis is not labeled
// ytext - y-axis title; for "None" - axis is not labeled
// title - plot title
//---------------------------------------------------------------------------
   {
      int i;
      PyObject *pFunc, *pArgs, *px, *py;

      pFunc = PyObject_GetAttrString(pModule,"Plot");

      px = PyList_New(n+1);                              // 0-offset arrays
      py = PyList_New(n+1);
      for (i=1; i<=n; i++) {
         PyList_SetItem(px, i, PyFloat_FromDouble(x[i]));
         PyList_SetItem(py, i, PyFloat_FromDouble(y[i]));
      }

      pArgs = PyTuple_New(12);
      PyTuple_SetItem(pArgs, 0, px);
      PyTuple_SetItem(pArgs, 1, py);
      PyTuple_SetItem(pArgs, 2, PyLong_FromLong(n));
      PyTuple_SetItem(pArgs, 3, PyUnicode_FromString(col));
      PyTuple_SetItem(pArgs, 4, PyLong_FromLong(sty));
      PyTuple_SetItem(pArgs, 5, PyFloat_FromDouble(fxmin));
      PyTuple_SetItem(pArgs, 6, PyFloat_FromDouble(fxmax));
```

```
        PyTuple_SetItem(pArgs, 7, PyFloat_FromDouble(fymin));
        PyTuple_SetItem(pArgs, 8, PyFloat_FromDouble(fymax));
        PyTuple_SetItem(pArgs, 9, PyUnicode_FromString(xtext));
        PyTuple_SetItem(pArgs,10, PyUnicode_FromString(ytext));
        PyTuple_SetItem(pArgs,11, PyUnicode_FromString(title));
        Py_XINCREF(pArgs);

        PyObject_CallObject(pFunc, pArgs);

        Py_XDECREF(pArgs);
        Py_DECREF(py);
        Py_DECREF(px);
        Py_DECREF(pFunc);
    }

//============================================================================
    void MultiPlot(double x[], double y[], double sig[], int n[],
                   const char *col[], int sty[], int nplot, int maxint,
                   double xminp, double xmaxp, int ioptx,
                   double yminp, double ymaxp, int iopty,
                   double fxmin, double fxmax, double fymin, double fymax,
                   const char *xtext, const char *ytext, const char *title)
//----------------------------------------------------------------------------
// Plots nplot real functions of one variable given by sets of (x,y) points.
// The coordinate sets are stored contiguously in the arrays x[] and y[].
// The x and y-domains are extended to fit at most maxint intervals
// expressible as d * 10^p, with d = 1, 2, 5 and p integer.
//
// x[]    - abscissas of tabulation points for all functions
// y[]    - ordinates of tabulation points
// sig[]  - error bars of the tabulation points (useful for sty == 4)
// n[]    - ending index for the individual plots (nmax = n[nplot])
// col[]  - plot color ("red", "green", "blue" etc.)
// sty[]  - plot style 0 - scatter plot with squares
//                     1 - line plot;  -1 - dashed line
//                     2 - polar plot;  -2 - dashed line
//                     3 - drop lines
//                     4 - error bars;  -4 - including line plot
// nplot  - number of plots
// maxint - max. number of labeling intervals
// ioptx  - 0 - resize x-axis automatically
//          1 - resize x-axis based on user interval [xminp,xmaxp]
// iopty  - 0 - resize y-axis automatically
//          1 - resize y-axis based on user interval [yminp,ymaxp]
// fxmin  - min fractional x-limit of viewport (0 < fxmin < fxmax < 1)
// fxmax  - max fractional x-limit of viewport
// fymin  - min fractional y-limit of viewport (0 < fymin < fymax < 1)
// fymax  - max fractional y-limit of viewport
// xtext  - x-axis title; for "None" - axis is not labeled
// ytext  - y-axis title; for "None" - axis is not labeled
// title  - plot title
//----------------------------------------------------------------------------
    {
        int i;
        PyObject *pFunc, *pArgs, *px, *py, *ps, *pn, *pcolor, *pstyle;

        pFunc = PyObject_GetAttrString(pModule,"MultiPlot");

        px = PyList_New(n[nplot]+1);                          // 0-offset arrays
```

```
      py = PyList_New(n[nplot]+1);
      ps = PyList_New(n[nplot]+1);
      for (i=1; i<=n[nplot]; i++) {
         PyList_SetItem(px, i, PyFloat_FromDouble(x[i]));
         PyList_SetItem(py, i, PyFloat_FromDouble(y[i]));
         PyList_SetItem(ps, i, PyFloat_FromDouble(sig[i]));
      }

      pn     = PyList_New(nplot+1);                            // 0-offset arrays
      pcolor = PyList_New(nplot+1);
      pstyle = PyList_New(nplot+1);
      for (i=1; i<=nplot; i++) {
         PyList_SetItem(pn    , i, PyLong_FromLong(n[i]));
         PyList_SetItem(pcolor, i, PyUnicode_FromString(col[i]));
         PyList_SetItem(pstyle, i, PyLong_FromLong(sty[i]));
      }

      pArgs = PyTuple_New(21);
      PyTuple_SetItem(pArgs, 0, px);
      PyTuple_SetItem(pArgs, 1, py);
      PyTuple_SetItem(pArgs, 2, ps);
      PyTuple_SetItem(pArgs, 3, pn);
      PyTuple_SetItem(pArgs, 4, pcolor);
      PyTuple_SetItem(pArgs, 5, pstyle);
      PyTuple_SetItem(pArgs, 6, PyLong_FromLong(nplot));
      PyTuple_SetItem(pArgs, 7, PyLong_FromLong(maxint));
      PyTuple_SetItem(pArgs, 8, PyFloat_FromDouble(xminp));
      PyTuple_SetItem(pArgs, 9, PyFloat_FromDouble(xmaxp));
      PyTuple_SetItem(pArgs,10, PyLong_FromLong(ioptx));
      PyTuple_SetItem(pArgs,11, PyFloat_FromDouble(yminp));
      PyTuple_SetItem(pArgs,12, PyFloat_FromDouble(ymaxp));
      PyTuple_SetItem(pArgs,13, PyLong_FromLong(iopty));
      PyTuple_SetItem(pArgs,14, PyFloat_FromDouble(fxmin));
      PyTuple_SetItem(pArgs,15, PyFloat_FromDouble(fxmax));
      PyTuple_SetItem(pArgs,16, PyFloat_FromDouble(fymin));
      PyTuple_SetItem(pArgs,17, PyFloat_FromDouble(fymax));
      PyTuple_SetItem(pArgs,18, PyUnicode_FromString(xtext));
      PyTuple_SetItem(pArgs,19, PyUnicode_FromString(ytext));
      PyTuple_SetItem(pArgs,20, PyUnicode_FromString(title));
      Py_XINCREF(pArgs);

      PyObject_CallObject(pFunc, pArgs);

      Py_XDECREF(pArgs);
      Py_DECREF(ps);
      Py_DECREF(py);
      Py_DECREF(px);
      Py_DECREF(pFunc);
   }

//==============================================================================
   void Contour(double **z, int nx, int ny,
                double xmin, double xmax, double ymin, double ymax,
                double zmin, double zmax,
                double fxmin, double fxmax, double fymin, double fymax,
                const char *xtext, const char *ytext, const char *title)
//------------------------------------------------------------------------------
// Plots a function z(x,y) defined in [xmin,xmax] x [ymin,ymax] and tabulated
// on a regular Cartesian grid with (nx-1)x(ny-1) mesh cells as contour plot.
```

```
// The level curves result by inverse linear interpolation inside the cells.
//
// z     - tabulated function values
// nx    - number of x-mesh points
// ny    - number of y-mesh points
// zmin  - minimum level considered
// zmin  - maximum level considered
// fxmin - minimum relative viewport abscissa (0 < fxmin < fxmax < 1)
// fxmax - maximum relative viewport abscissa
// fymin - minimum relative viewport ordinate (0 < fymin < fymax < 1)
// fymax - maximum relative viewport ordinate
// xtext - x-axis title; xtext = "" - axis is not labeled
// ytext - y-axis title; ytext = "" - axis is not labeled
// title - plot title
//----------------------------------------------------------------------------
   {
      int i, j;
      PyObject *pFunc, *pArgs, *pz, *pz1;

      pFunc = PyObject_GetAttrString(pModule,"Contour");

      pz = PyList_New(nx+1);                              // 0-offset arrays
      for (i=1; i<=nx; i++) {
         pz1 = PyList_New(ny+1);
         for (j=1; j<=ny; j++)
            PyList_SetItem(pz1, j, PyFloat_FromDouble(z[i][j]));
         PyList_SetItem(pz, i, pz1);
      }

      pArgs = PyTuple_New(16);
      PyTuple_SetItem(pArgs, 0, pz);
      PyTuple_SetItem(pArgs, 1, PyLong_FromLong(nx));
      PyTuple_SetItem(pArgs, 2, PyLong_FromLong(ny));
      PyTuple_SetItem(pArgs, 3, PyFloat_FromDouble(xmin));
      PyTuple_SetItem(pArgs, 4, PyFloat_FromDouble(xmax));
      PyTuple_SetItem(pArgs, 5, PyFloat_FromDouble(ymin));
      PyTuple_SetItem(pArgs, 6, PyFloat_FromDouble(ymax));
      PyTuple_SetItem(pArgs, 7, PyFloat_FromDouble(zmin)),
      PyTuple_SetItem(pArgs, 8, PyFloat_FromDouble(zmax));
      PyTuple_SetItem(pArgs, 9, PyFloat_FromDouble(fxmin));
      PyTuple_SetItem(pArgs,10, PyFloat_FromDouble(fxmax));
      PyTuple_SetItem(pArgs,11, PyFloat_FromDouble(fymin));
      PyTuple_SetItem(pArgs,12, PyFloat_FromDouble(fymax));
      PyTuple_SetItem(pArgs,13, PyUnicode_FromString(xtext));
      PyTuple_SetItem(pArgs,14, PyUnicode_FromString(ytext));
      PyTuple_SetItem(pArgs,15, PyUnicode_FromString(title));
      Py_XINCREF(pArgs);

      PyObject_CallObject(pFunc, pArgs);

      Py_XDECREF(pArgs);
      Py_XDECREF(pz1);
      Py_XDECREF(pz);
      Py_XDECREF(pFunc);
   }

//============================================================================
   void PlotParticles(double x[], double y[], double z[], double rad[],
                      char *col[], int n, double dmax,
```

```
                        double xminp, double xmaxp, int ioptx,
                        double yminp, double ymaxp, int iopty,
                        double fxmin, double fxmax, double fymin, double fymax,
                        const char *title)
//---------------------------------------------------------------------------
// Plots a system of particles as connected colored spheres
//
// x,y,z[] - coordinates of particles
// r[]      - radii of particles
// col[]    - colors of particles ("red", "green", "blue" etc.)
// n        - number of particles
// dmax     - max inter-distance for which particles are connected
// ioptx    - 0 - resize x-axis automatically
//            1 - resize x-axis to provided user interval (xminp,xmaxp)
// iopty    - 0 - resize y-axis automatically
//            1 - resize y-axis to provided user interval (yminp,ymaxp)
// fxmin    - min fractional x-limit of viewport (0 < fxmin < fxmax < 1)
// fxmax    - max fractional x-limit of viewport
// fymin    - min fractional y-limit of viewport (0 < fymin < fymax < 1)
// fymax    - max fractional y-limit of viewport
// title    - plot title
//---------------------------------------------------------------------------
   {
      int i;
      PyObject *pFunc, *pArgs, *px, *py, *pz, *pr, *pc;

      pFunc = PyObject_GetAttrString(pModule,"PlotParticles");

      px = PyList_New(n+1);                                      // 0-offset arrays
      py = PyList_New(n+1);
      pz = PyList_New(n+1);
      pr = PyList_New(n+1);
      pc = PyList_New(n+1);
      for (i=0; i<=n; i++) {
         PyList_SetItem(px, i, PyFloat_FromDouble(x[i]));
         PyList_SetItem(py, i, PyFloat_FromDouble(y[i]));
         PyList_SetItem(pz, i, PyFloat_FromDouble(z[i]));
         PyList_SetItem(pr, i, PyFloat_FromDouble(rad[i]));
         PyList_SetItem(pc, i, PyUnicode_FromString(col[i]));
      }

      pArgs = PyTuple_New(18);
      PyTuple_SetItem(pArgs, 0, px);
      PyTuple_SetItem(pArgs, 1, py);
      PyTuple_SetItem(pArgs, 2, pz);
      PyTuple_SetItem(pArgs, 3, pr);
      PyTuple_SetItem(pArgs, 4, pc);
      PyTuple_SetItem(pArgs, 5, PyLong_FromLong(n));
      PyTuple_SetItem(pArgs, 6, PyFloat_FromDouble(dmax));
      PyTuple_SetItem(pArgs, 7, PyFloat_FromDouble(xminp));
      PyTuple_SetItem(pArgs, 8, PyFloat_FromDouble(xmaxp));
      PyTuple_SetItem(pArgs, 9, PyLong_FromLong(ioptx));
      PyTuple_SetItem(pArgs,10, PyFloat_FromDouble(yminp));
      PyTuple_SetItem(pArgs,11, PyFloat_FromDouble(ymaxp));
      PyTuple_SetItem(pArgs,12, PyLong_FromLong(iopty));
      PyTuple_SetItem(pArgs,13, PyFloat_FromDouble(fxmin));
      PyTuple_SetItem(pArgs,14, PyFloat_FromDouble(fxmax));
      PyTuple_SetItem(pArgs,15, PyFloat_FromDouble(fymin));
      PyTuple_SetItem(pArgs,16, PyFloat_FromDouble(fymax));
```

```
      PyTuple_SetItem(pArgs,17, PyUnicode_FromString(title));
      Py_XINCREF(pArgs);

      PyObject_CallObject(pFunc, pArgs);

      Py_XDECREF(pArgs);
      Py_DECREF(pFunc);
   }

//==============================================================================
   void PlotStruct(double x[], double y[], double z[], int n,
                   int ind1[], int ind2[], int ind3[], int n3,
                   double xminp, double xmaxp, int ioptx,
                   double yminp, double ymaxp, int iopty,
                   double fxmin, double fxmax, double fymin, double fymax,
                   const char *title)
//------------------------------------------------------------------------------
// Renders a 3D structure defined by nodes and triangular surfaces
//
// x,y,z[] - coordinates of nodes
// n       - number of nodes
// ind1[]  - index of 1st node of each triangle
// ind2[]  - index of 2nd node of each triangle
// ind3[]  - index of 3rd node of each triangle
// n3      - number of triangles
// ioptx   - 0 - resize x-axis automatically
//           1 - resize x-axis to provided user interval (xminp,xmaxp)
// iopty   - 0 - resize y-axis automatically
//           1 - resize y-axis to provided user interval (yminp,ymaxp)
// fxmin   - min fractional x-limit of viewport (0 < fxmin < fxmax < 1)
// fxmax   - max fractional x-limit of viewport
// fymin   - min fractional y-limit of viewport (0 < fymin < fymax < 1)
// fymax   - max fractional y-limit of viewport
// title   - plot title
//------------------------------------------------------------------------------
   {
      int i;
      PyObject *pFunc, *pArgs, *px, *py, *pz, *p1, *p2, *p3;

      pFunc = PyObject_GetAttrString(pModule,"PlotStruct");

      px = PyList_New(n+1);                                 // 0-offset arrays
      py = PyList_New(n+1);
      pz = PyList_New(n+1);
      for (i=0; i<=n; i++) {
         PyList_SetItem(px, i, PyFloat_FromDouble(x[i]));
         PyList_SetItem(py, i, PyFloat_FromDouble(y[i]));
         PyList_SetItem(pz, i, PyFloat_FromDouble(z[i]));
      }
      p1 = PyList_New(n3+1);
      p2 = PyList_New(n3+1);
      p3 = PyList_New(n3+1);
      for (i=0; i<=n3; i++) {
         PyList_SetItem(p1, i, PyLong_FromLong(ind1[i]));
         PyList_SetItem(p2, i, PyLong_FromLong(ind2[i]));
         PyList_SetItem(p3, i, PyLong_FromLong(ind3[i]));
      }

      pArgs = PyTuple_New(19);
```

```
      PyTuple_SetItem(pArgs,  0, px);
      PyTuple_SetItem(pArgs,  1, py);
      PyTuple_SetItem(pArgs,  2, pz);
      PyTuple_SetItem(pArgs,  3, PyLong_FromLong(n));
      PyTuple_SetItem(pArgs,  4, p1);
      PyTuple_SetItem(pArgs,  5, p2);
      PyTuple_SetItem(pArgs,  6, p3);
      PyTuple_SetItem(pArgs,  7, PyLong_FromLong(n3));
      PyTuple_SetItem(pArgs,  8, PyFloat_FromDouble(xminp));
      PyTuple_SetItem(pArgs,  9, PyFloat_FromDouble(xmaxp));
      PyTuple_SetItem(pArgs,10, PyLong_FromLong(ioptx));
      PyTuple_SetItem(pArgs,11, PyFloat_FromDouble(yminp));
      PyTuple_SetItem(pArgs,12, PyFloat_FromDouble(ymaxp));
      PyTuple_SetItem(pArgs,13, PyLong_FromLong(iopty));
      PyTuple_SetItem(pArgs,14, PyFloat_FromDouble(fxmin));
      PyTuple_SetItem(pArgs,15, PyFloat_FromDouble(fxmax));
      PyTuple_SetItem(pArgs,16, PyFloat_FromDouble(fymin));
      PyTuple_SetItem(pArgs,17, PyFloat_FromDouble(fymax));
      PyTuple_SetItem(pArgs,18, PyUnicode_FromString(title));
      Py_XINCREF(pArgs);

      PyObject_CallObject(pFunc, pArgs);

      Py_XDECREF(pArgs);
      Py_DECREF(pFunc);
   }
};

//===========================================================================
void HistoBin(double xnew, double a, double b,
              double x[], double y[], int n, int iopt)
//---------------------------------------------------------------------------
// Bins data for a histogram to be plotted by function Plot (with sty = 4)
//
// xnew - new value to be binned
// a, b - limits of total binning interval
// x[]  - bin boundaries (x[1] = a, x[n] = b)
// y[]  - frequency of values in the bins: y[i] in [x[i],x[i+1]), i = 1,n-1
// n    - number of bin boundaries
// iopt - option: 0 - zeros bins
//                1 - bins new value xnew
//                2 - normalizes histogram
//---------------------------------------------------------------------------
{
   static double h, s;
   int i;

   if (iopt == 0) {                                        // initialize bins
      h = (b-a)/(n-1);                                           // bin size
      for (i=1; i<=n; i++) {
         x[i] = a + (i-1)*h;                               // bin boundaries
         y[i] = 0.e0;
      }
   } else if (iopt == 1) {                                // bin new values
      i = (int) ((xnew-a)/h) + 1;                               // bin index
      if ((i >= 1) && (i < n)) y[i] += 1;              // increment bin value
   } else if (iopt == 2) {                              // normalize histogram
      s = 0e0;
      for (i=1;i<=n-1;i++) s += y[i];
```

```
        for (i=1;i<=n-1;i++) y[i] /= s;
    }
}

#endif
```

Appendix G: List of Programs by Chapter

The file names of the application programs contain the complete information on their location within the folder structure descending from the root folder `/INP/`. The general format of the file names is `Pnn-name.py` or `Pnn-name.cpp`, whereby `nn` indicates the chapter folder `/INP/Chnn/` and the extension indicates the respective subfolder, that is, `/INP/Chnn/Python/` or `/INP/Chnn/C/`.

Listing	**/INP/Ch02/Python**	Listing	**/INP/Ch02/C**
2.1	`P02-Fact_global.py`	2.2	`P02-Fact_global.cpp`
2.3	`P02-Fact_comb.py`	2.4	`P02-Fact_comb.cpp`
2.7	`P02-Swap.py`	2.9	`P02-Swap.c`
2.8	`P02-Swap_list.py`	2.10	`P02-Swap.cpp`
2.18	`P02-MatOp.py`	2.19	`P02-MatOp.cpp`

Listing	**/INP/Ch03/Python**	Listing	**/INP/Ch03/C**
3.1	`P03-Tk.py`		
3.2	`P03-Tk_Label.py`		
3.3	`P03-Tk_Canvas0.py`		
3.4	`P03-Tk_Canvas.py`		
3.5	`P03-Tk_Clock0.py`		
3.6	`P03-Tk_Clock.py`		
3.8	`P03-Plot0.py`		
	`P03-Plots.py`		`P03-Plots.cpp`
3.10	`P03-MultiPlot.py`	3.17	`P03-MultiPlot.cpp`
3.11	`P03-Contour.py`	3.18	`P03-Contour.cpp`
3.12	`P03-PlotParticles.py`		
3.13	`P03-PlotStruct.py`		
3.15	`P03-Histogram.py`		
	`P03-Plot.py`	3.16	`P03-Plot.cpp`

Listing	**/INP/Ch04/Python**	Listing	**/INP/Ch04/C**
4.2	`P04-Sort0.py`		`P04-Sort0.cpp`
4.7	`P04-Sort1.py`		`P04-Sort1.cpp`
4.9	`P04-Sort2.py`		`P04-Sort2.cpp`
4.12	`P04-SortQuart.py`	4.13	`P04-SortQuart.cpp`
4.15	`P04-Sort3.py`	4.16	`P04-Sort3.cpp`

Listing	**/INP/Ch05/Python**	Listing	**/INP/Ch05/C**
5.2	`P05-Poly.py`		`P05-Poly.cpp`
5.12	`P05-PolyPlot.py`	5.13	`P05-PolyPlot.cpp`
5.16	`P05-SpherYPlot.py`	5.17	`P05-SpherYPlot.cpp`
5.18	`P05-SpherY.py`	5.19	`P05-SpherY.cpp`

Index